W9-BSF-419

CHEMICAL TECHNICIANS' READY REFERENCE HANDBOOK

OTHER McGRAW-HILL BOOKS OF INTEREST

AUSTIN • *Shreve's Chemical Process Industries, 5th Edition*
CHOPEY & HICKS • *Handbook of Chemical Engineering Calculations*
CONSIDINE • *Process Instruments and Controls Handbook, 3d Edition*
DEAN • *Handbook of Organic Chemistry*
DEAN • *Lange's Handbook of Chemistry, 13th Edition*
FREEMAN • *Hazardous Waste Minimization*
GRANT & GRANT • *Grant & Hackh's Chemical Dictionary, 5th Edition*
KISTER • *Distillation Operation*
MILLER • *Flow Measurement Engineering Handbook, 2d Edition*
NALCO • *The Nalco Water Handbook, 2d Edition*
PERRY & GREEN • *Perry's Chemical Engineers' Handbook, 6th Edition*
SCHWEITZER • *Handbook of Separation Techniques for Chemical Engineers, 2d Edition*
SEYMOUR • *Engineering Polymer Sourcebook*
SHINSKEY • *Process Control Systems, 3d Edition*
SHUGAR & DEAN • *The Chemist's Ready Reference Handbook*

CHEMICAL TECHNICIANS' READY REFERENCE HANDBOOK

Third Edition

GERSHON J. SHUGAR, B.S., M.A., Ph.D.
Professor, Essex County College
Newark, New Jersey

JACK T. BALLINGER, B.S., M.S.
Professor, St. Louis Community College at Florissant Valley
St. Louis, Missouri

Consulting Editors

RONALD A. SHUGAR, B.S., M.D.
Medical Associates
Edison, New Jersey

LAWRENCE BAUMAN, B.S., D.D.S.
Former Professor, NYU School of Dentistry
Fanwood, New Jersey

ROSE SHUGAR BAUMAN, B.S.
Science Writer
Watchung, New Jersey

McGraw-Hill, Inc.

New York St. Louis San Francisco Auckland Bogotá
Caracas Hamburg Lisbon London Madrid
Mexico Milan Montreal New Delhi Paris
San Juan São Paulo Singapore
Sydney Tokyo Toronto

Library of Congress Cataloging-in-Publication Data

Shugar, Gershon J., date
Chemical technicians' ready reference handbook/Gershon J. Shugar, Jack T. Ballinger; consulting editors, Ronald A. Shugar, Lawrence Bauman, Rose Shugar Bauman.—3rd ed.
p. cm.
Updated ed. of: Chemical technicians' ready reference handbook/ Gershon J. Shugar . . . [et al.]. 2nd ed. c1981.
ISBN 0-07-057183-X
1. Chemistry—Manipulation—Handbooks, manuals, etc.
I. Ballinger, Jack T. II. Chemical technicians' ready reference handbook. III. Title.
QD61.S58 1990
542—dc20 89-43670
CIP

234567890 DOH/DOH 9876543210

ISBN 0-07-057183-X

The sponsoring editor for this book was Gail F. Nalven, the editing supervisor was Susan Thomas, the designer was Elliot Epstein, and the production supervisor was Suzanne W. Babeuf. It was set in Caledonia by University Graphics, Inc.

Printed and bound by R. R. Donnelley & Sons/Harrisonburg.

CONTENTS

PREFACE

The first and second editions of the *Chemical Technicians' Ready Reference Handbook* elicited a flood of constructive comments and suggestions from the scientific community. Many of these letters suggested the inclusion of more instrumental analysis and calibration information especially aimed at the practicing chemist and/or chemical laboratory technician. We have added comprehensive chapters on gas chromatography (GC), high-performance liquid chromatography (HPLC), infrared spectroscopy (IR), atomic absorption spectroscopy (AA), and an introduction to nuclear magnetic resonance spectroscopy (NMR).

Many letters requested additional easy-to-understand information on laboratory safety: chemical-waste disposal, material safety data sheets (MSDS), National Fire Protection Association (NFPA) codes, Right-to-Know regulations, etc. Several sections have been included in this third edition in response to the scientific community's laboratory safety concerns and responsibilities.

As was the original purpose of the Handbook, this third edition is designed to be an "omnibook" for all chemical laboratory personnel and students ranging from high school to graduate school. The Handbook is designed to give "every single step" to be followed in most of the conventional laboratory procedures. It provides a refresher guide for the professionally trained chemical technician, chemist, chemical engineer, and laboratory supervisor. Special sections including basic laboratory mathematics, introduction to statistics, organic nomenclature, and a glossary of relevant terms have been expanded to meet today's laboratory needs.

Jack T. Ballinger
Gershon J. Shugar

ACKNOWLEDGMENTS

The authors have several individuals to acknowledge and thank for their substantial contributions toward the completion of this Handbook. We are indebted to Mr. Kenneth Chapman at the American Chemical Society for his suggestion and coordination of getting the writing team of Shugar and Ballinger together. Kenneth Chapman has been involved in chemical education and training for almost twenty-five years and is still recognized as "the authority" in chemical technology training.

The authors wish to give a special thanks to the original authors (Ronald A. Shugar, Lawrence Bauman, and Rose Shugar Bauman) involved in the first and second editions of this handbook; without their efforts this Handbook would not have been possible. We would also like to thank Gail Nalven and Susan Thomas at McGraw-Hill Book Company for their guidance and assistance in the production of this rather voluminous manuscript. We would also like to thank Harold Crawford, Editor in Chief—Handbooks and Technical Books, at McGraw-Hill Book Company for his recognition of the continued need for the third edition of this updated and expanded chemical technology resource.

The following companies, publishers, and organizations have contributed to this reference book:

Ace Scientific Supply Company
1420 East Linden Avenue
Linden, New Jersey 07036
Nate Schurer, Director of Marketing

Central Scientific Company
Chicago, Illinois 60623

Fisher Scientific Company
711 Forbes Avenue
Pittsburgh, Pennsylvania 15219
Michael J. Boyles, Director of Public Relations

Lab Safety Supply
P.O. Box 1368
Janeville, Wisconsin 53547-1368
Claire A. Gorayeb, Catalog & Advertising Manager

Matheson Gas Products
30 Seaview Drive
P.O. Box 1587
Secaucus, New Jersey 07094
Gene Brady—Manager, Advertising Services

National Fire Protection Association
Batterymarch Park
Quincy, Massachusetts 02269-9101
Dennis J. Berry, Associate General Counsel

Prentice Hall, Inc.
Harry G. Hajian and Robert L. Pecsok, authors:
Modern Chemical Technology, rev. ed. vol. 6, © 1973
Prentice Hall Building
Englewood Cliffs, New Jersey 07632
Maria Armand, Permissions Editor

SGA Scientific Company
735 Broad Street
Bloomfield, New Jersey 07003
William Geyer, President

Varian Associates, Inc.
220 Humboldt Court
Sunnyvale, California 94089
Judith Farrell, Product Promotion Manager

Finally, we would like to thank all the laboratory technicians involved in making our lives safer, healthier, more productive, and perhaps even easier. These dedicated individuals work daily to provide us with a safer environment and better quality of life. We invite comments, criticisms, and suggestions from these technicians and their employers as well. Good luck in the laboratory.

CHEMICAL TECHNICIANS' READY REFERENCE HANDBOOK

1
THE CHEMICAL TECHNICIAN

THE ROLE OF THE CHEMICAL TECHNICIAN

The occupation of "chemical technician" is difficult to define because the formal training, experience, and duties vary from company to company and region to region. For example, some chemical company employees can profess to be a "chemical technician" simply if they work with chemicals in any capacity, while other companies award this title to their employees only after much formalized education and years of on-the-job training. Various U.S. companies call their "chemical technicians" "laboratory technicians," "analysts," "laboratorians," "junior chemists," "research technicians," "quality control technicians," "quality assurance technicians," "pilot plant operators," and even "chemists." Many situations exist in which employees holding bachelor of science degrees in biology or chemistry have been classified as chemical technicians. This practice is especially prevalent in research laboratories that utilize large numbers of graduate degreed chemists and/or chemical engineers. The American Chemical Society (ACS) has tried for many years to clarify and standardize this "chemical technician" title by suggesting a minimum level of formalized training through an ACS-accredited program as is currently done with chemists (bachelor of science degrees). Until that process is completed, perhaps a better way to define a "chemical technician" would be through a job description.

1. Follow laboratory safety practices.

   (a) Know and properly use all personal-safety laboratory equipment (goggles, respirators, gloves, clothing, etc.).

   (b) Know and properly use all laboratory safety equipment (hoods, fire extinguishers, showers, eyewash fountains, etc.).

   (c) Know and follow Material Safety Data Sheet (MSDS) regulations.

   (d) Handle all chemical and radioactive substances properly.

   (e) Handle glassware, laboratory hardware, and equipment properly.

   (f) Be knowledgeable of and comply with all federal, state, local, and company chemical and hazardous-waste disposal procedures.

2. Follow proper procedures for housekeeping and maintenance of laboratory.

(**a**) Maintain a clean and safe laboratory working environment.

(**b**) Clean and store glassware properly.

(**c**) Inventory and order chemicals and supplies.

(**d**) Perform routine maintenance and calibration on laboratory equipment.

3. Apply and understand mathematical calculations.

(**a**) Perform routine calculations in basic mathematics and algebra using a scientific calculator.

(**b**) Know the technique of dimensional analysis: be able to convert metric to English units and English to metric.

(**c**) Calculate and express chemical concentrations (molarity, molality, normality, percentage by weight, percentage by volume, parts per million, etc.).

(**d**) Balance simple chemical equations and do stoichiometric calculations.

(**e**) Calculate routine statistical expressions.

(**f**) Plot and interpret graphical data.

4. Operate conventional laboratory equipment.

(**a**) Perform mass measurements using all types of balances.

(**b**) Perform volume measurements using volumetric equipment.

(**c**) Assemble laboratory glassware equipment (distillation, extraction, reflux, etc.) properly.

(**d**) Use compressed-gas cylinders, regulators, fittings, etc. correctly.

(**e**) Operate general laboratory equipment (balances, pH meters, burners, vacuum pumps, etc.) properly.

(**f**) Operate analytical laboratory instrumentation (spectrophotometers, gas chromatographs, etc.) properly.

5. Demonstrate cognitive skills.

(**a**) Follow and understand technical instructions and procedures.

(**b**) Know chemical terminology plus inorganic and organic nomenclature.

(**c**) Collect, store, and obtain representative samples.

(d) Prepare and standardize solutions.

(e) Do literature searches in chemically related references.

6. Demonstrate computer skills.

    (a) Understand a computer language, usually BASIC.

    (b) Operate a computer terminal.

    (c) Enter, store, and retrieve computer data.

    (d) Prepare written reports using a word processor.

7. Demonstrate communication skills.

    (a) Prepare and present oral reports.

    (b) Maintain instrument service and calibration records.

    (c) Inventory and catalog laboratory samples.

    (d) Maintain a laboratory notebook properly.

THE LABORATORY NOTEBOOK

The laboratory notebook (Fig. 1-1) is the one place where the information the technician acquires is recorded. That means that *all data* are recorded directly into the notebook as they are obtained. They are not taken down on stray sheets of paper.

The notebook is the diary and record of the technician. Because lawsuits may occur in which the notebook may play a key role in the decision of the court, it is imperative that the notebook record be above question. Many cases have been won or lost because of the credibility of the data contained in the notebook. Therefore, it is of paramount importance that the following rules be observed. (Of course, they are provided solely as guidelines. Each company uses its own procedure and sets forth its own format. If no formal procedure exists, however, these rules should be used.)

1. The laboratory notebook should be a numbered, bound book, preferably with automatic carbon or other duplication method. The pages should be consecutively numbered, and no original pages should ever be ripped out. The data should be recorded in ink, in detail, and all observations and ideas should be included. By utilizing a duplicate-copy notebook, a copy of what has been recorded may be kept at the technician's disposal for ready reference, while the original bound book can be placed in the proper place for security and ready reference by all other members of the staff. The cover of each record book should indicate the dates of the first and last entries made in that book.

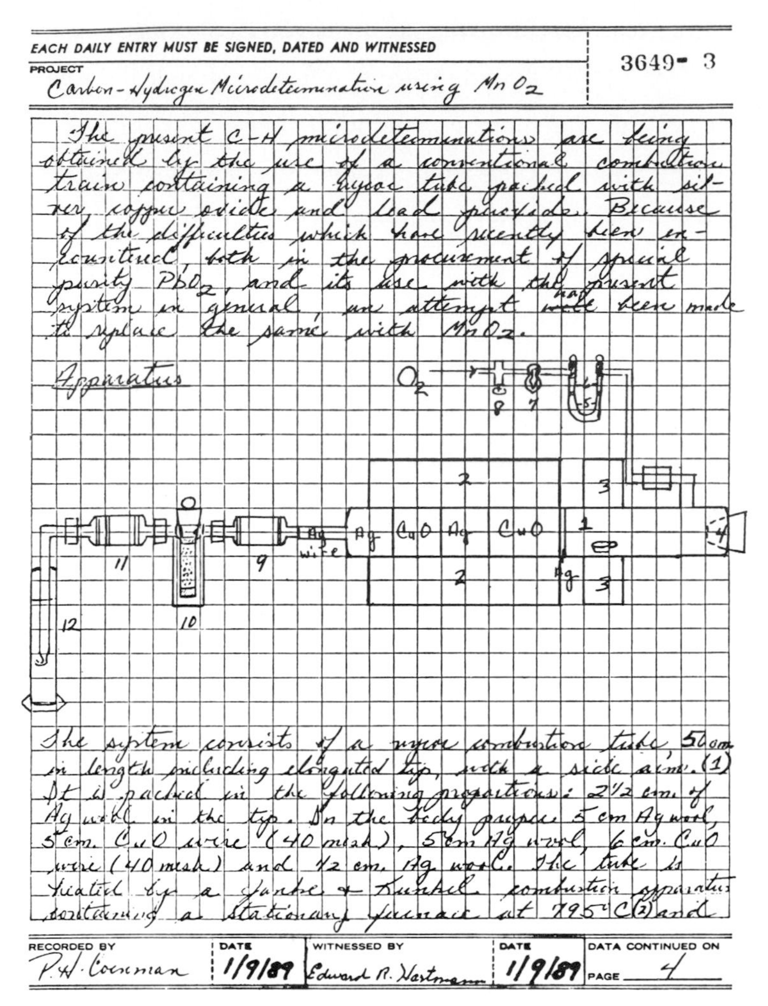

EACH DAILY ENTRY MUST BE SIGNED, DATED AND WITNESSED

3649- 3

PROJECT

Carbon-Hydrogen Microdetermination using MnO_2

The present C-H microdeterminations are being obtained by the use of a conventional combustion train containing a Vycor tube packed with silver, copper oxide and lead peroxide. Because of the difficulties which have recently been encountered, both in the procurement of special purity PbO_2, and its use with the present system in general, an attempt ~~will~~ has been made to replace the same with MnO_2.

Apparatus

The system consists of a Vycor combustion tube, 56 cm in length including elongated tip, with a side arm.(1) It is packed in the following proportions: 2½ cm. of Ag wool in the tip. In the body proper 5 cm Ag wool, 5 cm. CuO wire (40 mesh), 5 cm Ag wool, 6 cm. CuO wire (40 mesh) and ½ cm. Ag wool. The tube is heated by a Janke & Kunkel combustion apparatus containing a stationary furnace at 795°C(2) and

RECORDED BY	DATE	WITNESSED BY	DATE	DATA CONTINUED ON
P. H. Coenman	1/9/89	Edward R. Hartmann	1/9/89	PAGE 4

FIGURE 1-1 Sample page from a laboratory notebook.

2. All the information that is acquired should be entered, regardless of how trivial it may appear at the time.

3. Each page should be dated and signed by both the technician and his or her supervisor.

4. The objective of a procedure should be stated at the top of the first page of the sequence of pages used.

5. A diagram of the apparatus or equipment to be used should be sketched, followed by a short summation of the procedure to be followed.

6. All raw data should be recorded neatly and directly in the notebook. Mistakes may be crossed out but *never erased.* All data should be entered in the notebook immediately. Delay leads to forgotten entries. All data regarding the starting materials should be recorded without delay, and all changes in any data obtained from the samples should be entered.

7. Entries should always be specific. The technician should never generalize, so that no question can arise in the future. One of the purposes of a record book is to enable one to duplicate what has been done, and the omission of relevant data can cause needless delays and costly repetitive work.

8. Information obtained from automatic recording devices, such as charts, can be filed in an appropriate secure location, but all pertinent information should be noted from the charts and recorded in the notebook, and the appropriate reference should be indexed on the charts or recording paper so that the information can be retrieved easily when it is needed.

9. Calculations of the raw data may be carried out on other paper or on a calculator, and the results must then be recorded. *All* calculations should be checked by either the person who performs the work or a competent coworker.

PROFESSIONAL ORGANIZATIONS FOR CHEMICAL TECHNICIANS

Recognizing the professional role of chemical laboratory technicians in research and industry, the ACS Division of Industrial and Chemical Engineering established the Committee on Technician Activities (CTA) in 1966. The CTA was charged with the responsibility of improving the professional image and status of laboratory technicians and formed the first Technician Affiliate Group (TAG) in 1969 and the National Conference of Chemical Technician Affiliates (NCCTA) in 1971. Most progressive employers recognize the importance of these professional organizations to their technicians and encourage their participation. The objectives of the NCCTA are to coordinate technician activities (symposium, workshops, etc.) at national ACS meetings, support local TAG activities, provide continuing education programs, and generate technical publications specifically for laboratory technicians. For additional information on these professional organizations, contact the ACS in Washington, D.C., at (202) 872-8734.

PROFESSIONAL REFERENCES AND RESOURCES

Much of the information that a technician needs can be found in various reference handbooks, indexes, dictionaries, government publications, abstracting services, etc.:

1. *Lange's Handbook of Chemistry*, 13th ed., John A. Dean (ed.). McGraw-Hill Book Co. (1985). 1792 pages, ISBN 0-07-016192-5. *Partial*

Table of Contents—Mathematical and Statistical Tables: General Information and Conversion Tables: Inorganic Chemistry: Organic Chemistry: Atomic and Molecular Structure: Analytical Chemistry: Electrochemistry: Spectroscopy: Thermodynamic Properties: Physical Properties.

2. *CRC Handbook of Chemistry and Physics,* 70th ed. (1988—edited annually), CRC Press. 2488 pages, ISBN 0-8493-0469-0. The new edition retains all the important information of previous editions and includes information on rare earth metal properties: vapor pressure, boiling points, magnetic properties, thermal expansion and conductivity data, etc. In addition updates are included on semiconductors and their properties, and on other topics.

3. *The Chemist's Companion,* Arnold J. Gordon and Richard A. Ford, John Wiley & Sons (1973). 537 pages, ISBN 0-471-31590-7. *Partial Table of Contents*—Mathematical and Numerical Information: Experimental Techniques: Chromatography: Photochemistry: Spectroscopy: Kinetics and Energetics: Properties of Atoms and Bonds.

4. *Physical Chemistry Source Book,* McGraw-Hill Science Reference Series, McGraw-Hill Book Co. (1988). 500 pages and 260 illustrations, ISBN 0-07-045504-X. *Partial Table of Contents*—Chemical Thermodynamics: Chemical Reactions: Surface Chemistry: Transport Processes: Matter: Structure and Properties: Electrochemistry: Electroanalytical Chemistry: Cells and Batteries: Optical Phenomena: Specialized Fields of Study.

5. *Handbook of Applied Chemistry,* V. Hopp and I. Hennig, McGraw-Hill Book Co. (1983). 950 pages, ISBN 0-07-030320-7. *Partial Table of Contents*—Organic/Inorganic Raw Materials and Large-Scale Industrial Processes: Organic Chemistry Classification and Nomenclature.

6. *Grant & Hackh's Chemical Dictionary,* 5th ed., Roger and Claire Grant (eds.), McGraw-Hill Book Co. (1987). 656 pages, ISBN 0-07-024067-1. Contains over 55,000 entries covering all branches of chemistry and the related sciences: physics, medicine, biology, agriculture, pharmacy, and minerology, as well as IUPAC and CAS nomenclature. Contains many new entries and information on space science, biotechnology, and cytotoxic drugs.

7. *Improving Safety in the Chemical Laboratory—A Practical Guide,* Jay A. Young (ed.), John Wiley & Sons (1987). 368 pages, ISBN 0-471-84693-7. *Partial Table of Contents*—Organization for Safety in Laboratories: Precautionary Labels and Material Safety Data Sheets: Doing It Right: The 95% Solution: Safety Inspections: Safety Audits: Flammability: Combustibility: Chemical Reactivity: Instability and Incompatible Combinations; How Chemicals Harm Us: Handling and Management of Hazardous Research Materials: Other Hazards: Storage of Laboratory Chemicals: Federal Regulations: Air Sampling in the Chemical Laboratory: Safe Disposal of Hazardous Waste: Designing Safety into the Laboratory: Laboratory Hoods: Using Audio-Visual Materials in Safety Training: Laboratory Safety Library Holdings.

8. *Handbook of Ventilation for Contaminant Control*, 2nd ed., Henry J. McDermott (ed.), Butterworth Publishers (1985). 416 pages, ISBN 0-250-40641-1. *Partial Table of Contents*—OSHA Ventilation Standards: Indoor Air Pollution: Ventilation as Source and Solution: Ventilation for High-Toxicity or High-Nuisance Contaminants: Fans: Air Cleaner Selection: How Local Exhaust Systems Work: Hood Selection and Design: Hazard Assessment: Saving Ventilation Dollars: Testing: Symbols: Abbreviations: Glossary: Heating and Cooling.

9. *Handbook of Hazard Communication & OSHA Requirements*, George C. and Robert C. Lowry, Lewis Publishers (1985). Includes updates for 1986 and 1987. 150 pages, ISBN 0-87371-022-3. *Partial Table of Contents*—Legal Responsibilities and Penalties: Hazard Identification: Physical Hazard Characterization: Label Design: Written Hazard Communication Program: Employee Training: Consequences: Text of OSHA Hazard Communication Standard: Substances Regulated by Reference: Lethal Dose Equivalencies: MSDS Form: 1986 Update: Hazard Communication for Nonmanufacturers: Public "Right to Know": New Compliance Guidelines: Key Court Decisions: Carcinogenicity Determination: 1987 Update: Trade Secrets Revisions: New Developments in Labeling. OSHA MSDS, Expanded Coverage. Enforcement.

10. *Rapid Guide to Hazardous Chemicals in the Workplace*, N. Irving Sax and Richard J. Lewis, Van Nostrand Reinhold (1987). 256 pages, ISBN 0-442-28220-6. Gives rapid access to information on 700 chemicals frequently used in the laboratory. Covers OSHA permissible exposure limits, ACGIH threshold limit value, MAKS, and DOT hazard classification. 1-2-3 rating instantly shows degree of hazard. Includes alphabetical list of chemical entries, appendixes, cross-references.

11. *First Aid Manual for Chemical Accidents—For Use with Nonpharmaceutical Chemicals*, M. J. Lefevre, Van Nostrand Reinhold (1982). 218 pages, ISBN 0-879-3336-7. Color-coded sections give quick access for nearly 500 chemicals. Includes toxicology and symptoms of overexposure, first aid procedures, instructions to follow in in the event of poisoning by unknown chemicals, a list of U.S. government publications, and a glossary of commercial names.

12. *Dangerous Properties of Industrial Materials*, 7th ed., N. Irving Sax and Richard J. Lewis, Van Nostrand Reinhold (1988). 4000 pages, ISBN 0-442-28020-3. A three-volume set containing over 20,000 entries. Includes a toxic hazard review for each chemical, DOT numbers and classifications, CAS number index, and complete chemical name index with synonyms. Exposure standards include OSHA, PELS, ACGIH TLV, STEL, BEI, and MAKS.

13. *Merck Index,* 10th ed., Merck & Co. (1983). 2175 pages, ISBN 911910-27-1. Covers over 10,000 chemicals, drugs, and biological substances. A partial listing of contents includes abbreviations, monographs on individual compounds, organic reaction names, chemical abstract names and registry numbers, formula indexes, and atomic weights. Tables cover subjects including biochemical and immunological abbreviations, company register, radioactive isotopes, maximum allowable concentrations of air contaminants, isotonic solutions, indicators, standard buffers for pH calibration, pH values of standard solutions.

14. *Chemical Abstracts,* American Chemical Society, Office of Chemical Abstracts, Columbus, Ohio. Abstracted continuously since 1907. *Chemical Abstracts* have been "computerized" since 1966 and are listed by subject, author, and formula; there is also a patent index. The most complete reporting of all chemistry and chemically related publications in the United States and other countries. This journal is published weekly and the abstract consists of the following:

(**a**) The abstract number

(**b**) Title (English or literal translation)

(**c**) Author name(s)

(**d**) Address

(**e**) Journal or publication title

(**f**) Year published

(**g**) Volume number

(**h**) Page numbers

(**i**) Language used

PROFESSIONAL SOURCES OF INFORMATION

1. American Board of Industrial Hygienists (ABIH)
P.O. Box 16153
Lansing, MI 48901
(517) 321-2638

2. American Petroleum Institute (API)
2101 L Street, N.W.
Washington, DC 20037
(202) 682-8000

3. American Society for Testing and Materials (ASTM)
   1916 Race Street
   Philadelphia, PA 19103
   (215) 299-5400

4. American Chemical Society (ACS)
   Chemical Health and Safety Division
   1155 16th Street, N.W.
   Washington, DC 20036
   (202) 872-4401

5. Centers for Disease Control (CDC)
   1600 Clifton Road, N.E.
   Atlanta, GA 30333
   (404) 639-3535

6. Chemical Manufacturers Association (CMA)
   2501 M Street, N.W.
   Washington, DC 20037
   (202) 887-1100

7. Compressed Gas Association, Inc.
   1235 Jefferson Davis Highway
   Arlington, VA 22202
   (703) 979-0900

8. Government Printing Office
   Washington, DC 20036
   (202) 783-3238

9. National Fire Protection Assocation (NFPA)
   Batterymarch Park
   Quincy, MA 02269
   (617) 770-3000

10. National Institute for Occupational Safety & Health (NIOSH)
    4676 Columbia Parkway
    Cincinnati, OH 45226
    (513) 533-8236

11. National Safety Council (NSC)
    444 N. Michigan Avenue
    Chicago, IL 60611
    (312) 527-4800

12. Nuclear Regulatory Commission (NRC)
    1717 H Street, N.W.
    Washington, DC 20555
    (301) 492-7000

13. Occupational Safety & Health Administration (OSHA)
    200 Constitution Avenue
    Washington, DC 20210
    (202) 523-7075

14. U.S. Environmental Protection Agency (EPA)
    401 M Street, S.W.
    Washington, DC 20460
    (202) 382-4700

15. U.S. Food and Drug Administration (FDA)
    5600 Fishers Lane
    Rockville, MD 20857
    (301) 443-1544

OCCUPATIONAL SAFETY AND HEALTH ADMINISTRATION

The U.S. Department of Labor's Occupational Safety and Health Administration (OSHA) is responsible for enforcing various federal laws dealing with safety and hazards in the workplace. Specifically, in 1983, federal regulation (29 *CFR* 1910.1200) was promulgated requiring a "Hazard Communication Standard" to ensure that the hazards of all chemicals produced or imported are evaluated, and that information concerning their hazards is transmitted to employers and employees. This transmittal of information must be comprehensive and include container labeling, Material Safety Data Sheets (MSDS), and employee training. This law affects all employees, but especially laboratory technicians. ". . . (a) Employers shall ensure that labels on incoming containers of hazardous chemicals are not removed or defaced; (b) Employers shall maintain any material safety data sheets that are received with incoming shipments of hazardous chemicals, and ensure that they are readily accessible to laboratory employees; (c) Employers shall ensure that laboratory employees are appraised of the hazards of the chemicals in their workplaces . . ." This particular section of the OSHA regulation is sometimes referred to as the "Right to Know" law. The U.S. Environmental Protection Agency (EPA) has expanded this "Right to Know" one step further to include the entire community surrounding the manufacturers. Since 1986, the EPA's Superfund Amendments and Reauthorization Act (SARA) has required all employers to develop emergency plans and guidelines for informing community authorities of chemical hazards or crisis situations.

MATERIAL SAFETY DATA SHEETS

All chemical manufacturers and importers are required by law to provide a Material Safety Data Sheet (MSDS) for each hazardous chemical produced (see Fig. 1-2). Employers are required to: (1) Identify each chem-

FIGURE 1-2
Material Safety Data Sheets. (*Photo courtesy Lab Safety Supply Co.*)

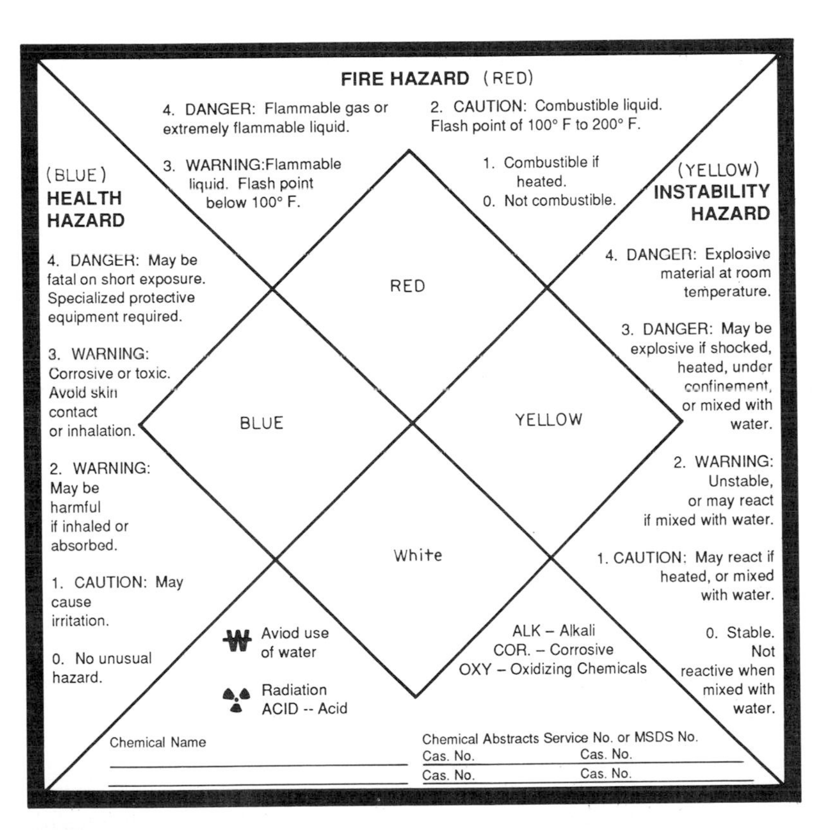

ACID	acids
ALK	alkali
COR	corrosive
OXY	oxidizing chemical
p	polymerization
-W-	avoid use of water
☢	radioactive

FIGURE 1-3
NFPA hazard labels. (*a*) Color-coded diamonds. (*b*) Symbols and abbreviations. (*Courtesy of National Fire Protection Association*)

ical's manufacturer, its address, emergency telephone numbers, and date prepared. (2) List all hazardous ingredients that make up at least 1% of the mixture and list the threshold limit value (TLV), the permissible exposure limit (PEL), and the Chemical Abstract Service (CAS) number (consult the Glossary of this reference for an explanation of these common MSDS terms). (3) Identify each substance by its physical properties: odor, appearance, boiling point, etc. (4) Provide information about fire and explosive hazards. (5) Specify the chemical or reactivity properties of

TABLE 1-1 Identification of NFPA Numerical Codes
(Courtesy of National Fire Protection Association)*

Identification of Health Hazard Color Code: BLUE		Identification of Flammability Color Code: RED		Identification of Reactivity (Stability) Color Code: YELLOW	
Type of Possible Injury		Susceptibility of Materials to Burning		Susceptibility to Release of Energy	
Signal		Signal		Signal	
4	Materials which on very short exposure could cause death or major residual injury even though prompt medical treatment were given.	4	Materials which will rapidly or completely vaporize at atmospheric pressure and normal ambient temperature, or which are readily dispersed in air and which will burn readily.	4	Materials which in themselves are readily capable of detonation or of explosive decomposition or reaction at normal temperatures and pressures.
3	Materials which on short exposure could cause serious temporary or residual injury even though prompt medical treatment were given.	3	Liquids and solids that can be ignited under almost all ambient temperature conditions.	3	Materials which in themselves are capable of detonation or of explosive reaction but require a strong initiating source or which must be heated under confinement before initiation or which react explosively with water.
2	Materials which on intense or continued exposure could cause temporary incapacitation or possible residual injury unless prompt medical treatment is given.	2	Materials that must be moderately heated or exposed to relatively high ambient temperatures before ignition can occur.	2	Materials which in themselves are normally unstable and readily undergo violent chemical change but do not detonate. Also materials which may react violently with water or which may form potentially explosive mixtures with water.
1	Materials which on exposure would cause irritation but only minor residual injury even if no treatment is given.	1	Materials that must be preheated before ignition can occur.	1	Materials which in themselves are normally stable, but which can become unstable at elevated temperatures and pressures or which may react with water with some release of energy but not violently.
0	Materials which on exposure under fire conditions would offer no hazard beyond that of ordinary combustible material.	0	Materials that will not burn.	0	Materials which in themselves are normally stable, even under fire exposure conditions, and which are not reactive with water.

*Reprinted with permission from NFPA 704-1985, *Identification of the Fire Hazards of Materials*, Copyright©, 1985, National Fire Protection Association, Quincy, MA 02269. This reprinted material is not the complete and official position of the NFPA on the referenced subject which is represented only by the standard in its entirety.

each material with regard to possible reactions with other materials. (6) Provide health-hazard information such as effects of overexposure and any necessary emergency or first aid treatment. (7) Include a section outlining the steps required to clean up a spill and methods of proper waste disposal. (8) List any special protective equipment necessary while working with this material. (9) If necessary, include a section dealing with long-term storage or handling of this material.

NATIONAL FIRE PROTECTION ASSOCIATION

The OSHA Hazard Communication Standard requires a substantial amount of information on each chemical. The National Fire Protection Association (NFPA) has developed a labeling system for quickly identifying these hazards and properties. These NFPA diamond-shaped labels are divided into four smaller, colored diamonds as is shown in Fig. 1-3*a*. Each color stands for a particular hazard from the MSDS requirements listed above: blue, health; red, fire; yellow, reactivity; and white, special problems. A numbering scale from 0 to 4 is used to represent the level of hazard within each category. Table 1-1 identifies the NFPA numerical codes for *Health Hazard*, *Flammability*, and *Reactivity*. In addition, in the *Special Hazard* category the symbols and abbreviations in Fig. 1-3*b* are used.

2
LABORATORY SAFETY

GENERAL LABORATORY SAFETY RULES

Every "chemical technician" should know and observe the following rules:

1. *Never attempt to carry out an experiment without knowing the safety rules and procedures that apply to the laboratory work.* Determine the potential hazards of all chemicals using Material Safety Data Sheets (MSDS) and any other appropriate information on chemicals, equipment, and procedures.

2. *Never attempt to perform an experiment without knowing the type of personal-safety equipment necessary:*

(**a**) *Wear appropriate eye protection at all times.* Ordinary prescription glasses are not adequate protection; glasses equipped with hardened-glass or plastic safety lenses with side shields (Fig. 2-1) are recommended. Even safety glasses equipped with side shields are considered inadequate if a splashing hazard exists; goggles or a face shield (Fig. 2-2) would be more appropriate. Specialized glasses are available for working with ultraviolet light, lasers, welding, glassblowing, etc. Contact lenses should never be worn in the work environment as laboratory vapors can concentrate under the lenses and cause eye damage.

(**b**) *Wear appropriate protective gloves.* Skin contact is a potential source of burn, toxic, or corrosive exposure; the use of protective apparel (gloves, face shields, specialized clothing, footwear, etc.) can minimize this hazard. Gloves should be inspected before using and washed, if necessary, appropriately before removal. Chemical-resistant gloves are commercially available in four materials (natural rubber, neoprene, nitrile, and vinyl), while insulated gloves (Fig 2-3) made of synthetic materials are available for temperature extremes. Insulated gloves made of asbestos are no longer recommended because of OSHA regulations on this carcinogenic material.

(**c**) *Wear appropriate protective apparel.* Safety shoes are suggested for all laboratory workers, while sandals, cloth sneakers, and open-toed shoes are not recommended. Protective clothing ranging from

FIGURE 2·1
Safety glasses with side shield. (*Photo courtesy Lab Safety Supply Co.*)

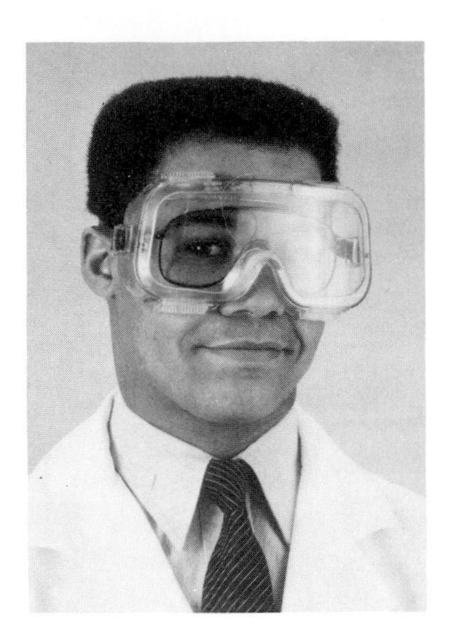

FIGURE 2·2
Face shield. (*Photo courtesy Lab Safety Supply Co.*)

rubber aprons and laboratory coats to entire disposable outer garments (Tyvek®) are commercially available (Fig. 2.4). Laboratory workers should know the appropriate technique for the removal and disposal of contaminated protective apparel.

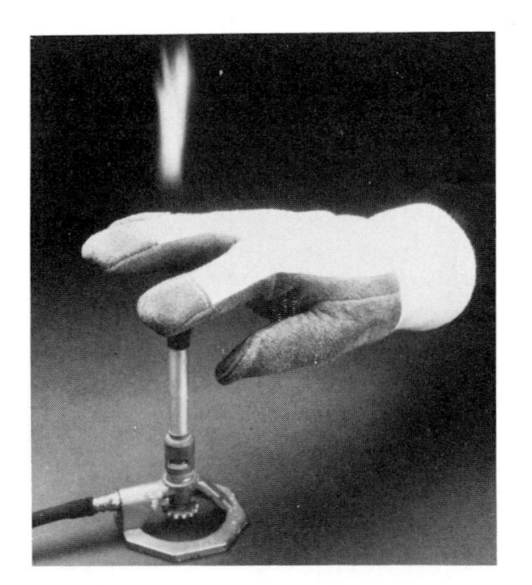

FIGURE 2·3
Insulated gloves. (*Photo courtesy Lab Safety Supply Co.*)

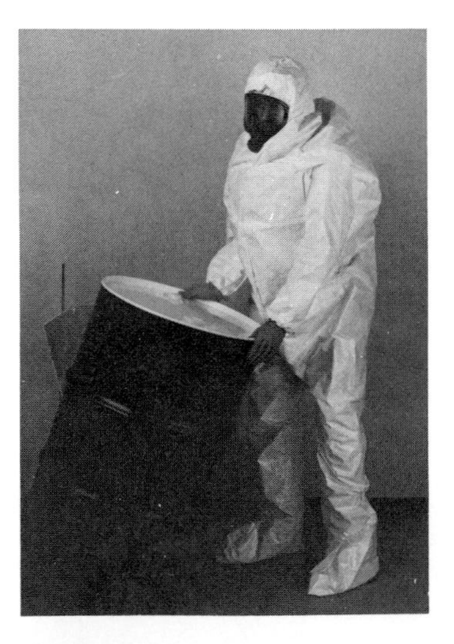

FIGURE 2·4
Disposable protective garment. (*Photo courtesy Lab Safety Supply Co.*)

3. *Never work without knowing the location and operation of all emergency laboratory equipment* (eyewash fountains, fire extinguishers, fire alarms, safety showers, safety shields, respirators, etc.). (See Figs. 2-5 and 2.6.) Each technician should know how to obtain emergency assistance and should know emergency procedures: evacuation procedures, alarm system, shutdown and personnel return procedures, etc. Safety drills should be a routine practice for all laboratories. Laboratories that do not maintain a regular medical staff should have personnel trained in first aid (See Appendix A, "Laboratory First Aid.")

4. *Never consume food, beverages, or smoke in areas where chemicals are being used or stored.* In addition, never use laboratory glassware or equipment (refrigerators, ovens, etc.) for food storage.

5. *Never pipet solutions by mouth.* Use a pipet bulb or an aspirator to provide the vacuum (Figs. 2-7 and 2-8).

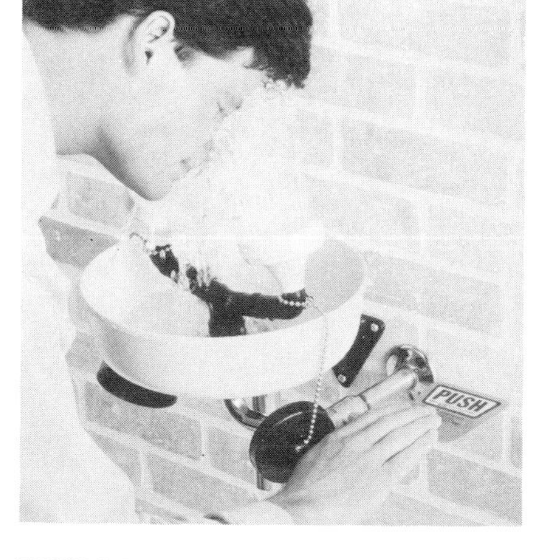

FIGURE 2-5
Eyewash fountain. (*Photo courtesy Lab Safety Supply Co.*)

FIGURE 2-6
Safety shower. (*Photo courtesy Lab Safety Supply Co.*)

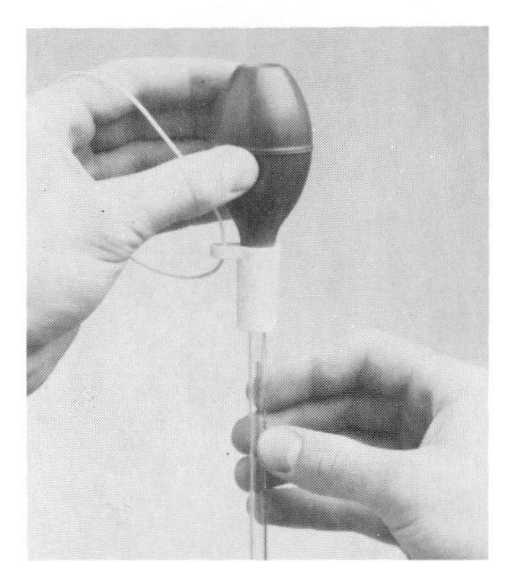

FIGURE 2-7
Pipet bulb. (*Photo courtesy Lab Safety Supply Co.*)

FIGURE 2-8
Safety sign: never pipet by mouth. (*Photo courtesy Lab Safety Supply Co.*)

6. *Never allow horseplay.* Never distract other laboratory workers or attempt practical jokes with compressed air, chemicals, etc.

7. *Never use any chemicals found in unlabeled containers.* Conversely, never store chemicals in a container without labeling it.

8. *Never wear loose-fitting clothing or unconfined hair in the laboratory.*

9. *Never leave the laboratory after using chemicals without washing thoroughly.*

10. *Never work alone in a laboratory building or leave chemical reactions unattended without arranging appropriate safeguards.*

11. *Never dispose of excess chemicals or reagents without consulting the MSDS and your supervisor.*

12. *Never use an open flame to heat flammable materials.* In general, open-flame heating devices are not as safe as nonsparking electrical equipment.

13. *Never point the open end of a test tube at yourself or any other person while the tube is being heated or during a reaction.*

14. *Never pour water into concentrated acid (especially sulfuric acid, Fig. 2-9).* The acid might spatter and the glass container may break because of the excessive heat generated.

15. *Never bring chemicals into contact with the skin.* Use a spatula (Fig. 2-10) or other sampling device.

16. *Never perform experiments or reactions that could produce objectionable or unknown gases without using a fume hood.*

FIGURE 2-9
Never pour water into concentrated acid.

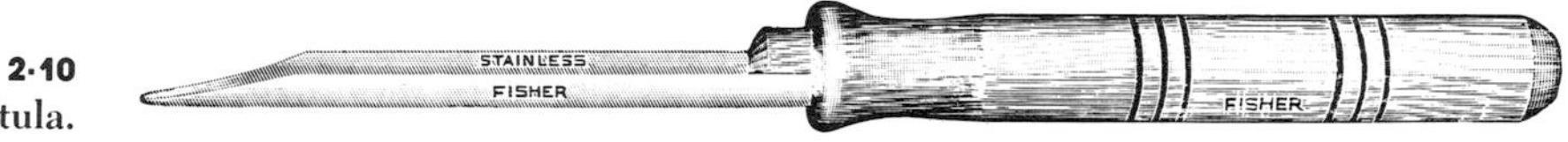

FIGURE 2-10
Spatula.

17. *Never return excess chemicals to their original container.* This problem of disposing of excess chemicals can be minimized by taking only the amount of material required for the reaction.

18. *Never sniff a chemical directly.* If necessary, gently fan the vapors (Fig. 2-11) toward your nose.

19. *Never heat "soft" glass containers (most bottles, funnels, graduated cylinders, thick-walled glassware, etc.) in an open flame.* Unlike Kimax® or Pyrex®, this type of glassware (Fig. 2-12) is not designed to withstand high temperatures or thermal shock.

20. *Never attempt to insert a glass tube or thermometer into a cork or rubber stopper without a lubricant (water or glycerine).* Glass tubing should be "fire-polished" and held in a cloth or special insertion device (Fig. 2-13) to minimize the hazard due to breakage.

21. *Never allow hazardous situations to occur without immediate and proper warning to fellow workers by signs (Fig. 2-14), barriers, announcements, etc.*

FIGURE 2-11
Never smell chemicals directly; fan vapors toward nose.

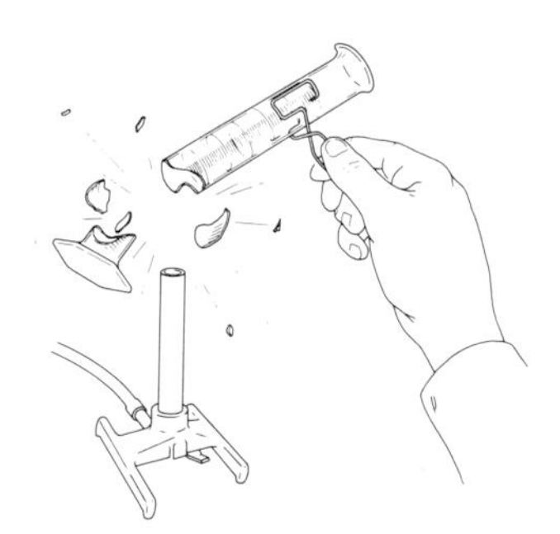

FIGURE 2-12
Never heat "soft" glass with an open flame.

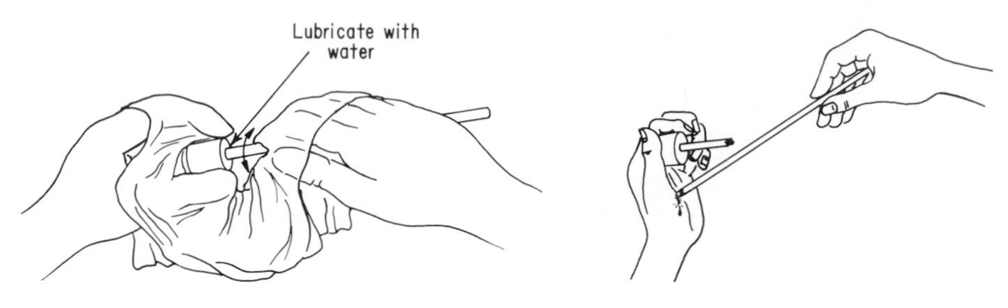

FIGURE 2·13
Always use a lubricant and cloth or insertion device for glass tubing and thermometers.

FIGURE 2·14
Hazard and safety signs.

Stockroom Safety Practices

Your stockroom should be adequate and efficiently planned for safe operation.

Checklist for a safe stockroom:

- ☐ Well-planned with wide aisles and adequate lighting
- ☐ No blind alleys
- ☐ All necessary personal-safety equipment available
- ☐ MSDS and other information readily available
- ☐ Well-marked exits
- ☐ Orderly and clean
- ☐ Properly ventilated (see section on ventilation)

- ☐ Equipped with adequate fire-safety equipment (refer to the section on fire safety)
- ☐ Heavy items stored near the floor
- ☐ Proper storage of glass apparatus and tubing (never projecting beyond shelf limits)
- ☐ Fragile and bulky equipment secured to shelving
- ☐ Shelving fitted with ledges to prevent glass bottles from falling off
- ☐ Preplanned grouping and separation of liquids and hazardous chemicals
- ☐ Drums securely mounted horizontally with container of sand or absorbent material beneath faucet
- ☐ Static-electricity ground on all drums (refer to Chap. 27, "Basic Electricity")
- ☐ Compressed-gas cylinders capped and securely supported (refer to Chap. 4, "Compressed Gases in the Laboratory")
- ☐ Safety ladders used
- ☐ No excessive heat, because of fire hazard
- ☐ No waste accumulation of any kind (refer to the section on safety practices for waste disposal)
- ☐ Good housekeeping with vigilant and constant maintenance of good storage facilities
- ☐ Commonsense behavior of all personnel using storage facilities

---

FIRE SAFETY

Introduction

Fire is the rapid combination of a combustible substance with oxygen, accompanied by the evolution of heat and light. When most substances burn, the actual combustion takes place only after the solid or liquid material has been vaporized or has been decomposed by the heat to produce a gas. The visible flame is the buring gas or vapor. Handle fire with care. All burners should be placed safely away from all flammable materials, and technicians should not expose their hair, clothing, or other flammable objects to them.

For combustion to take place you must have three things:

1. Fuel
2. Oxidizer (oxygen, in the case of air)

3. An ignition source (to bring the temperature up to the ignition point of the fuel)

Fires will not start if any one of these three components is missing. However, once the fire has started, the heat evolved by the fire itself becomes the ignition source.

Terminology

Flashback The rapid combustion of heavy vapors of organic compounds which collect in areas distant from their source, and when burning, lead the flame back to their source to cause a large fire or explosion.

Flash point The lowest temperature at which a compound in an open vessel will give off sufficient vapors to produce a momentary flash of fire when a flame,a spark, an incandescent wire, or another source of ignition is brought near the surface of the liquid.

Ignition temperature The lowest temperature at which the vapor over the surface of the liquid will ignite and continue to burn, if ignited.

Auto-ignition temperature The lowest temperature at which a vapor will ignite spontaneously when mixed with air.

Ignition Sources

Static Electricity

To eliminate static electricity ground all electrical equipment (refer to Chap. 27, "Basic Electricity").

Electrical Equipment

In electrical equipment, two basic sources of ignition are:

1. Arcing which may occur between component parts, such as contacts, switches, and circuit breakers. Other danger points include hot plates, electrical equipment such as mixers, heating mantles, refrigerators, centrifuges, etc., and sparks from motor starters, nails scraping cement, etc.

2. Surface temperature which results from the resistance of a conductor to the passage of electricity (refer to Chap. 27, "Basic Electricity").

Friction

Slippage of power-transmitting equipment such as belt drives, V belts, rubbing of metal-to-metal surfaces (unoiled or ungreased mechanical equipment) may cause friction.

Mechanical Sparks

To eliminate mechanical sparks always use suitable "nonsparking" tools and (if applicable) nonsparking materials for equipment such as fans, fan housings, scoops, or other moving parts where an impact might produce a spark.

Flames and Hot Surfaces

1. Never use solid fuel, gas, oil, or electrical heating equipment which involves a naked flame when you are working with volatile flammable liquids or combustible dusts. Even steam and hot-water piping should be properly installed to reduce the probability of heavy dust accumulations.

2. Maintenance, installation, and repair work must be carried out under strict supervision with safety guidelines.

Makeshift Safety Measures

You *cannot minimize* the inherent dangers of ignition of flammable substances by moving controls, lights, switches, relays, etc., to nearby areas with the mistaken idea that this renders them "no longer dangerous." This type of "jury rigging" gives you a false sense of security. Although containers of flammables are supposed to be tightly closed, many of the lids and covers are loose or become so. In some instances cotton balls or porous corks are substituted for proper stoppers and covers; sometimes, bottles are broken. Thus it is possible for explosive gas concentrations to come in contact with the arcing devices *"safely" moved to another location,* and the result could be fire, explosion, injury to personnel, and damage to the laboratory facilities.

CAUTION Do not be lulled into a sense of false security by makeshift safety measures.

Precautions

1. Do not use electrical equipment that is defective or has defective wiring.

2. Use explosionproof equipment, wiring, and controls where applicable.

3. Shut off *all nonexplosionproof* electrical equipment before working with volatile and flammable substances.

4. Never overload electric circuits, particularly multiple-socket installations.

5. Turn off all lights and lamps which are not equipped with explosionproof enclosures.

6. Never pour volatile flammable liquids down sinks. (Refer to section on waste-chemical disposal.)

Relative Flammability of Organic Compounds

Many organic compounds are flammable, but it is their vapors that burn, not the liquids themselves. Before flammable organic compounds can burn, they must be converted to the vapor state and mixed with oxygen or air in the proper proportions to support combustion. The concentration of the vapor in air has a lower limit, where there is not enough vapor to burn (lean mixture), and an upper limit, where the concentration of the vapor is too high (too rich a mixture) to burn.

TABLE 2-1
Relative Flammability Ratings of Solvents

Solvent	*Flammability rating*
Ethyl ether	100
Benzine	95–100
Benzene	95–100
Gasoline	95–100
Petroleum ether	95–100
Acetone	90
Toluene	75–80
Ethylene dichloride	60–70
Turpentine	40–50
Kerosene	40
Trichloroethylene	1–2
Tetrachloromethane	0
Tetrachloroethane	0
Tetrachloroethylene	0

The flammability ratings of the solvents in Table 2-1 are based upon a scale of arbitrary units, with ethyl ether having a value of 100 and tetrachloromethane having a value of 0.

National Fire Protection Association Classifications

Laboratories routinely use chemicals that the National Fire Protection Association (NFPA) describes as combustible Class I, Class II, and Class III liquids (see Table 2-2). The quantity and flammability of especially these Class I and II liquids makes the laboratory potentially hazardous. The total laboratory storage of these combustible materials is determined by company policy and local fire codes, but as a general rule, amounts of these combustible liquids should be limited to 1 L or less in unprotected glass or plastic containers and the total stored volume limited to 1 gal per hundred square feet of floor space. Using 1-L metal containers, larger volumes (5 gal maximum) of these ignitable liquids for each hundred square feet of surface can be stored with reasonable safety.

The NFPA has established four categories or "classes of fires" (Table 2-3). In addition to the letter code, Underwriters Laboratories (UL) tests

TABLE 2-2
NFPA Classification of Combustible Chemicals
(Courtesy of National Fire Protection Association)

Class of liquid	*Flash point*
I	At or below 100°F.
II	At or above 100°F but below 140°F
III	At or above 140°F.

TABLE 2-3
NFPA Classification of Fires
(Courtesy of National Fire Protection Association)

Class of fire	*Materials involved*
A	Routine combustibles such as wood, cloth, paper, rubber, and plastics
B	All flammable liquids and gases common to most laboratories (Class I, II, and III liquids, greases, solvents, paints, etc.)
C	Energized electrical equipment and apparatus (hot plates, ovens, instruments, etc.)
D	Combustible metals (magnesium, potassium, sodium, lithium aluminum hydride, etc.)

and assigns numerical value to each size and type of fire extinguisher, indicating the approximate square-foot area of fire-extinguishing capacity. For example, a 5-**B** extinguisher should be able to extinguish a 5-ft^2 area of class **B** combustibles.

Types of Fire Extinguishers

There are five basic types of fire extinguishers and they are classified by the type of fire for which they are suited:

1. *Water extinguishers* (Fig. 2-15) are effective against Class **A** fires, but should never be used for extinguishing organic liquid (Class **B**), electrical (Class **C**), or metal (Class **D**) fires. Obviously, this type of fire extinguisher would be of very limited use and possibly hazardous in a chemistry laboratory. This common soda–acid fire extinguisher usually contains sodium bicarbonate and sulfuric acid. When inverted, the acid and bicarbonate mix and react violently, forming carbon dioxide, which forces the solution out of the nozzle with hopefully great force.

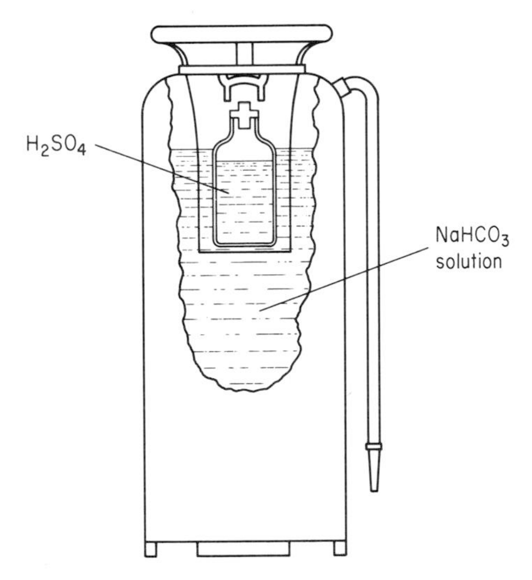

FIGURE 2-15
Water extinguisher.

2. *Carbon dioxide extinguishers* (Fig. 2-16) are effective against burning liquids (Class B) and electrical (Class C) fires. They are especially recommended around chemical instrumentation containing delicate electronic systems and/or optics. Caution should be used with this type of extinguisher because the tremendous force of the escaping gas can break expensive glassware. Operator asphyxiation can also be a problem in closed areas. This type of extinguisher is not very effective against paper or trash fires and should not be used on some combustible metal and metal hydride fires (sodium, potassium, lithium aluminum hydride, etc.) because it produces water from atmospheric condensation.

3. *Dry-powder extinguishers* (Fig. 2-17) are effective against liquids (Class B) and electrical (Class C) fires and are especially useful when large volumes of burning liquids are involved. These extinguishers are usually filled with an inorganic compound like sodium bicarbonate or monoammonium phosphate under nitrogen pressure; shaker-type canisters are also available for small fires. This type of extinguisher is not very effective against paper and trash fires or metal fires. Even though effective against electrical fires, they are not recommended for instrumentation fires because of the difficulty in cleaning delicate electronics and/or optical systems afterwards.

4. *Met-L-X extinguishers* (Fig. 2-18) are specialized for burning metal (Class D) type fires. They contain a granulated sodium chloride formulation which tends to be very effective against burning metals, metal

FIGURE 2-16
Carbon dioxide extinguisher. (*Photo courtesy Lab Safety Supply Co.*)

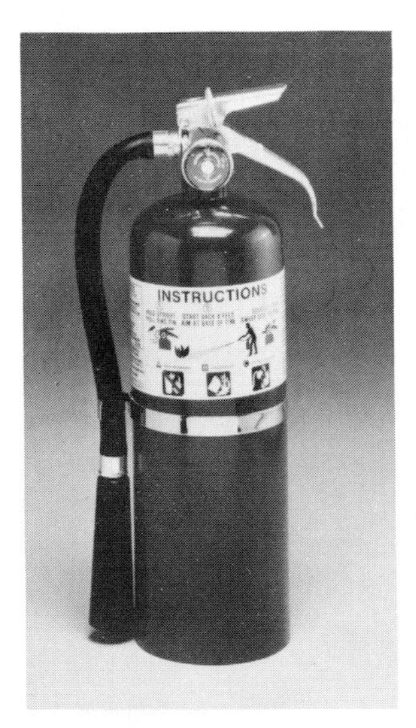

FIGURE 2-17
Dry-powder extinguisher. (*Photo courtesy Lab Safety Supply Co.*)

FIGURE 2-18
Met-L-X extinguisher. (*Photo courtesy Lab Safety Supply Co.*)

hydrides, and difficult organometallic compound fires. However, these extinguishers tend not to be very effective against the Class A, B, or C types of fire. Met-L-X extinguishers come in two varieties: a conventional gas-cartridge-operated device and a simple shaker-type canister. Obviously, the shaker-type is intended for small fires and the powder should be permitted to gently fall on the burning metal surface. Dry sand is also very effective for this type of fire.

5. *Halogenated hydrocarbon extinguishers* represent a new type of "clean" fire extinguisher which is effective against Class A, B, and C types of fire. Halon® (Allied trademark) is a relatively nontoxic, dense halocarbon mixture which is most effective in poorly ventilated spaces, since it prevents oxygen from getting to the fire source. Some caution is necessary because unpredictably hazardous decomposition products can be generated in a fire. These extinguishers are relatively expensive, but very effective and desirable around instrumentation fires.

6. *Automatic fire-extinguishing systems* are an excellent investment in areas where the potential for fire is high (gas manifold systems, solvent storage, chemical waste storage, etc.). As was discussed in the classification of fire extinguishers, the more common water-sprinkler type might not be advisable around organic liquids, combustible metals, etc. More effective Halon®, dry-powder, and carbon dioxide extinguishers are available in several automatic models (Fig. 2-19) intended to be suspended from the laboratory ceiling and equipped with a fusible (approximately

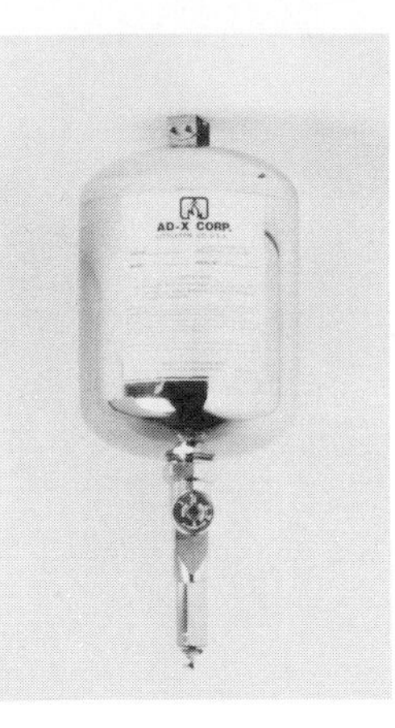

FIGURE 2-19
Automatic fire extinguisher system. (*Photo courtesy Lab Safety Supply Co.*)

60°C) metal trigger. Some precautions should be given to technicians about exposure to high concentrations of carbon dioxide or halogenated hydrocarbons after activation of these devices.

EXPLOSION SAFETY

Introduction

Explosions invariably accompany a fire, especially where combustibles are stored. An explosion differs from a fire because an ingredient of pressure is added. Explosions occur in closed volumes, requiring fuel, oxidizer, and ignition. They yield heat, light, and pressure.

The three most common explosion hazards are:

1. Exothermic reactions which get out of control (fire and explosion)
2. Explosions during evaporation of ethereal solutions because of peroxide residues
3. Explosions due to heating polynitro compounds, diazo compounds, diazonium compounds, peroxides, metallic acetylides, and perchlorates

CAUTION Use safety shields (Fig. 2-20) and safety glasses. Work with the smallest possible quantities of the dangerous substances.

Exothermic Reactions and Explosions

Exothermic reactions evolve heat.

1. As the rate of reaction increases, the rate of evolution of heat increases.

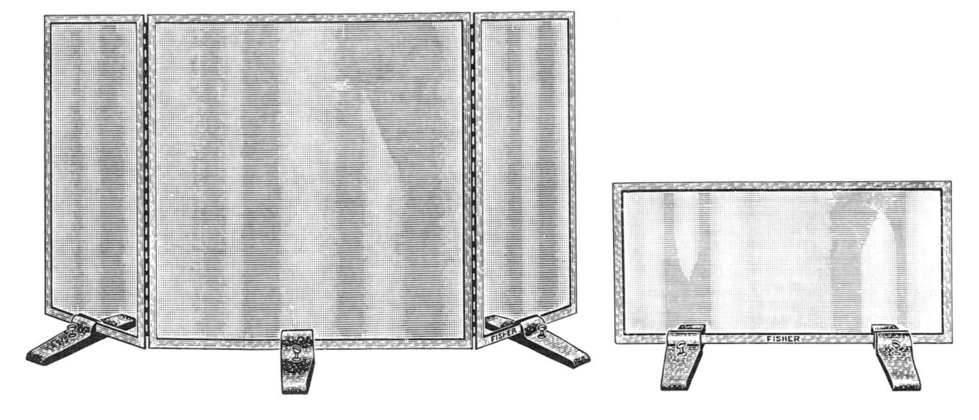

FIGURE 2-20
Safety shield made of shatterproof glass to protect laboratory personnel performing potentially explosive reactions.

2. If the rate of evolution of heat is greater than the rate at which the heat can be removed by cooling, the temperature rises and the mixture may boil over, vaporize, or explode.

To control exothermic reactions:

1. *Control the rate of addition of reactants.* Reactants are sometimes added too fast—faster than they react at the beginning. As a consequence reactants may accumulate, and then the reaction rate may accelerate and get out of control. The rate of reaction can be controlled by the rate of addition provided that the temperature rises when a reactant is added and drops when the addition is slowed or stopped. If this is the case, then wait until the temperature starts to drop after the addition of a small portion of the reactant, then add the next portion. Continue in this manner until all reactant has been added.

2. *Remove the heat by cooling.* Run the reaction in small batches in a large flask which provides a large surface area. The flask can be immersed in a cooling bath with a swirling motion to lower the temperature. Then combine the results of the small batches.

CAUTION Scaling up small runs to large batches creates the problem of heat-transfer control. It is much more difficult to cool large batches than small ones.

Explosive Mixtures

Almost any combustible solid, liquid, or gas can produce an explosive mixture under the right conditions. In the average laboratory there are more than 150 common flammable or explosive materials. There is also a combination of (1) percentage composition of the combustible-air mixture, and (2) a temperature which will spark a fire or an explosion (if confined).

EXAMPLES Five gallons of gasoline *or its equivalent* (such as volatile flammable aromatic and aliphatic hydrocarbons) can *explode* with as much force as *415 lb of dynamite*.

One ounce of ethyl alcohol or ethyl ether can render *12 ft³ of air explosive*.

CAUTION Very small quantities of flammable liquids can create a fire or explosion hazard.

Dust-Explosion Principles

Some materials which are normally considered noncombustible (such as aluminum, zinc, and similar metals) are very explosive in the form of extremely fine dust suspended in air. Any combustible solid will burn more readily as the ratio of surface area to volume is increased. Flammable dusts approach the ultimate in this respect: the higher ratio of surface area to volume causes an increase in burning rate. The rate of the release of heat is accelerated, causing explosions.

CAUTION When working with explosive substances:

1. Work with small quantities.

2. Use safety hood, safety shield or safety window, and safety mesh cloth wrapping.

RESPIRATORY SAFETY

There are several types of both emergency and nonemergency respirators.

1. *Chemical-cartridge respirators* (Fig. 2-21) are primarily effective only against gases and concentration levels specified by the manufacturer. The contaminant is usually adsorbed in a sorbent material, like activated charcoal; thus the cartridge must be replaced periodically. Saturation is usu-

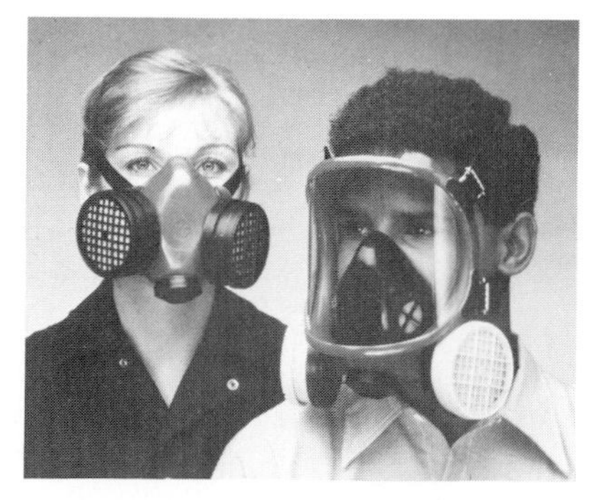

FIGURE 2-21 Chemical-cartridge respirator. (*Photo courtesy Lab Safety Supply Co.*)

ally indicated by difficulty in breathing or the odor of contaminant passing through the cartridge. Technicians should test their individual mask for proper fit before entering contaminated areas. This type of respirator is not recommended in laboratory atmospheres containing less than 19.5% oxygen.

2. *Dust and particulate respirators* (Fig. 2-22) use fiber filters to trap the contaminants as recommended by the manufacturer. This type of respirator is usually disposable and affords little protection against gaseous and other chemical contaminants. Toxic-dust, nuisance-dust, asbestos-removal, surgical masks of this type are not 100% effective in removing particulates.

3. *Supplied-air respirators* are effective protection against a wide range of gases, chemicals, and particulate contaminants and can be used in an oxygen-deficient atmosphere. A supply of fresh air is furnished to the face mask under slight positive pressure to prevent side entry of contaminants. OSHA regulation (29 *CFR* 1910.134) requires that a safety harness and an escape system in case of compressor failure be provided for the wearer of this type of respirator. Some hazard is involved in using this type of respirator with the compressor air quality, length of air hose, etc.

4. *Self-contained breathing apparatus* is the ultimate in protective respirators for emergency work. This equipment consists of a full-face mask (Fig. 2-23) connected to a cylinder of compressed air. In general, there are no restrictions as to contaminants, concentrations, or atmospheric oxygen concentration. However, the tank does have a limited capacity (usually 5 to 30 min), they are bulky and heavy, and additional safety gear may

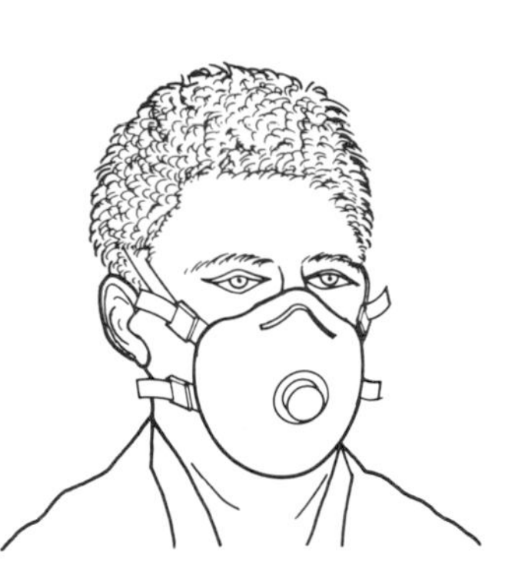

FIGURE 2-22
Dust and particulate respirator. (*Photo courtesy Lab Safety Supply Co.*)

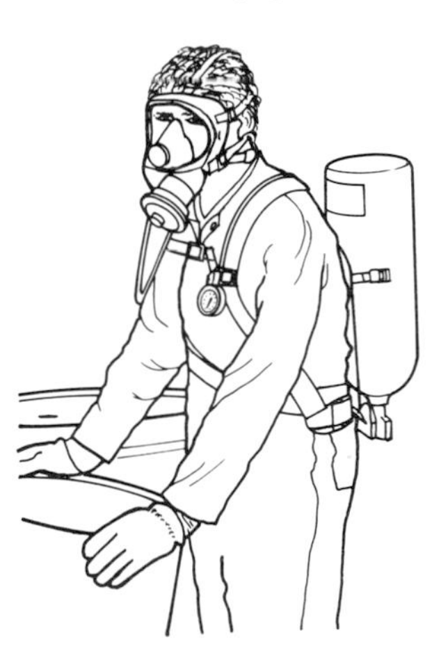

FIGURE 2-23
Self-contained breathing apparatus. (*Photo courtesy Lab Safety Supply Co.*)

be required if certain contaminants are present. It is strongly recommended that all chemically related organizations maintain this type of respirator and trained personnel.

MISCELLANEOUS LABORATORY SAFETY EQUIPMENT

Safety ladders can prevent many accidents caused in the laboratory by technicians using unsafe objects for climbing. The hazard in falling is compounded by the usual involvement of chemicals. A safety ladder should be available. A safety ladder provides a high degree of stability in climbing or reaching because it will not collapse or slip on slick floors.

Fire blankets are available in many laboratories. Fire blankets are rarely used to snuff out a worker's burning clothing; this should be a last resort since the blanket tends to hold heat in and perhaps increase the severity of the burn. The blanket is a first aid measure to help prevent shock in a traumatized victim. Clothing fires are extinguished faster by using a safety shower or dropping to the floor and rolling.

Safety showers should be available in all areas where chemicals are being handled and/or a fire hazard exists. Every laboratory worker should know the location and operation of this safety device, before the need ever arises. The shower should be of large enough capacity to handle more than one victim and should be tested on a routine basis.

Eyewash fountains are absolutely vital in a chemical environment. These fountains should provide a spray or soft stream of aerated water for an extended period of time. Ideally, they should be located in or very near the safety shower as both devices could be necessary at the same time. Again, each laboratory worker should be familiar with the location and operation of this device.

TOXIC CHEMICALS

Introduction

A *toxic chemical* is any substance that has the ability to damage, alter, or interfere with a human being's metabolic systems. For simplicity, the words "toxic" and "poison" are synonymous. Toxicologists have determined data on test animals as to lethal dosages per body mass and means of exposure. For example, the *lethal dose* (LD) of a toxin may be received by ingestion, injection, or through skin contact. A standardized expression of toxicity is LD_{50}, which represents the quantity of toxic material found necessary to cause death in 50% of the test animals. This term is normalized by including *the animal's body weight.* For example, the toxicity (LD_{50}) of mercury is expressed as 50 mg/kg. *Lethal concentration* (LC) of a toxin is similar to LD but refers to concentrations in air. The Federal

Hazardous Substance Act (FHSA) defines any substance as "highly toxic" if the LD_{50} is 50 mg/kg or less when administered orally or when the lethal concentration (LC_{50}) is 200 ppm or less when administered as a gas.

Guidelines to Minimize Toxic Exposure

A chemical technician can minimize exposure to toxic substances by following some general guidelines:

1. Know the chemical and toxic properties of all materials involved before starting, using the MSDS.

2. Substitute a less toxic substance whenever possible (see Table 2-4). For example, rinse glassware with hexane, instead of benzene.

3. Always use a fume hood and periodically test its efficiency (see section below on ventilation).

4. Know that toxic exposure can occur through respiration, ingestion, or through the skin; always use personal protective equipment (gloves, masks, spatulas, etc.).

TABLE 2-4
Relative Toxicity of Common Solvents (On Inhalation)

Degree of toxicity	*Solvent*
Relatively harmless: seldom cause injuries in everyday use	Ethyl acetate
	Ethyl alcohol
	Ethyl chloride
	Ethyl ether
	Heptanes
	Hexanes
	Mineral spirit
	Pentanes
	Petroleum benzine
	Petroleum ether
Mildly hazardous: can be endured for a short time within maximum permissible concentrations	Amyl acetate
	Amyl alcohol
	Butyl alcohol
	Cumene
	Cycloheptane
	Ethylene oxide
	Hydrogenated cyclic naphthas
Mildly hazardous (*continued*)	Nitroethane
	Perchloroethylene
	Tetrahydronaphthalene
	Toluene
	Xylene
Definitely hazardous: not to be inhaled even for a short time	Benzene
	Carbon bisulfide
	Tetrachloromethane (carbon tetrachloride)
	Dimethyl sulfate
	Formaldehyde
	Methyl alcohol
	Nitrobenzene
	Pentachloroethane
	Phenol
	Tetrachloroethane

5. Avoid excessive exposure to all chemicals, even those considered not to be high on any toxicity listing. These lists have been known to change.

6. Drinking alcoholic beverages on the job is never permissible for obvious reasons. Furthermore, ethyl alcohol has a synergistic effect with many other solvents and should be avoided.

7. Routinely monitor the laboratory atmosphere for specific contaminants and their concentration levels. This can be done quickly with a *gas analyzer* (Fig. 2-24) by manually drawing air through some specific colorimetric reagents contained in a calibrated glass tube or simply by wearing a *vapor-monitor badge.*

FIGURE 2-24
Gas analyzer (colorimetric, manual model).

NIOSH/OSHA Guide to Chemical Hazards

The American Conference of Governmental Industrial Hygienists (ACGIH) provides a comprehensive listing of potentially hazardous chemical substances and physical agents in the work environment. These *Threshold limit values* (TLVs) refer to airborne concentrations of substances and conditions under which nearly all workers may be repeatedly exposed without adverse effects. The TLV for a gaseous contaminant is reported in parts per million (ppm) and particulate matter is normally reported in milligrams per cubic meter (mg/m^3). For example, the TLV for mercury is 0.05 mg/m^3.

There are actually three categories of TLVs:

***Threshold limit value—time-weighted average* (TLV—TWA),** defined as "the time-weighted average concentration for a normal 8-hour workday or 40-hour workweek, to which nearly all workers may be repeatedly exposed, day after day, without adverse effects."

***Threshold limit value—short-term exposure* (TLV—STE),** defined as "the maximum concentration to which workers can be exposed for a period of up to 15 minutes continuously without suffering from (1) irritation, (2) chronic or irreversible tissue change, (3) narcosis of sufficient degree to increase accident proneness. . . ."

***Threshold limit value—ceiling* (TLV—C),** defined as "the concentration that should not be exceeded even instantaneously."

Other terms such as "permissible exposure limit" (PEL) and "immediately dangerous to life or health" (IDLH) are also used extensively. The listings of these various exposure limits are too extensive for this reference, but copies are available from the American Conference of Governmental Industrial Hygienists, Inc., P.O. Box 1937, Cincinnati, Ohio 45201.

VENTILATION

Introduction

The cardinal rule for working with toxic or hazardous substances is that these materials should be handled in such a way as to prevent their escape into the laboratory environment. Generally, a room ventilation system capable of changing the laboratory air 4 to 12 times per hour by exhausting to the outside is considered adequate. Airflow patterns and exchange rates should be periodically evaluated in the work environment. Aerosol generators and commercial smoke sources are available for these studies. Toxic and flammable substances should be stored in cabinets fitted with auxiliary ventilation, and laboratory equipment that might discharge toxic vapors should be externally vented.

Laboratory Fume Hoods

In addition to proper room ventilation, *laboratory fume hoods* should be readily available for the effective containment of flammable and toxic vapors. Federal, state, and local agencies have established air-movement standards for laboratory fume hoods. Generally, the recommended air-face velocity is between 60 and 100 linear feet per minute. Relatively inexpensive air-velocity meters (*velometers* and *anemometers*) are commercially available to determine the efficiency of fume hoods. Fume hoods tend to be expensive to operate in that heated or cooled laboratory air is being exhausted continuously. Recently, "supplementary-air hoods" have been developed which direct a blanket of outside air vertically to the hood face between the technician and the sash. Some hoods can exhaust as much as 70% supplemental air, but require an industrial hygienist or ventilation engineer to balance the total room ventilation. The hood sash (sliding glass door) has a tremendous effect on hood efficiency; the hood is least efficient when the sash is fully opened, and air-velocity measurements should be made at different sash heights to determine the optimum ventilation. Hoods are not intended for the storage of chemicals or equipment; these materials can adversely affect the airflow and performance of a hood.

Specialized Fume Hoods

Some fume hoods are equipped for very specialized purposes: *explosion-proof hoods* are manufactured with explosionproof electrical equipment; *perchloric acid fume hoods* have seamless, stainless steel linings and built-in water-spray systems for continuous washdown of exhausted contaminants; *radioisotope fume hoods* are similar to the perchloric acid hoods but have very efficient filtering assemblies on their exhaust.

WASTE-CHEMICAL DISPOSAL

Introduction

No other topic in laboratory safety has generated as much concern and confusion in recent years as chemical-waste disposal. The days of simply dumping chemicals down the drain with lots of water are over. *Always consult the MSDS for specific disposal guidelines on each waste chemical.* New regulations are continuously being introduced from federal, state, and local agencies covering all forms of chemical disposal ranging from air and water discharge to the incineration and burial of waste. Very strict federal regulations on hazardous waste are provided in the Resource Conservation and Recovery Act of 1976 (RCRA). Fines and/or imprisonment, plus loss of future access to disposal sites, can result if a company fails to provide the required information, provides erroneous information, fails to keep required records, etc. *Only general guidelines can be suggested in this reference, since each locality will have very specific governmental regulations and corporate policies.* Very hazardous wastes are usually converted (oxidized, reduced, neutralized, etc.) to a less dangerous substance before being placed in containers. The EPA has developed many chemical conversion procedures for highly toxic substances, especially known carcinogens.

Disposing of certain chemical wastes into the laboratory sewer systems may be permissible, but a knowledge of local regulations is mandatory. Usually, only water-soluble substances like acids and bases can be disposed of through the sewer system, but these must be diluted to a pH range of 3–11 and even then the rate of disposal is limited. Extreme caution should be observed with sewer disposal techniques because drain systems are usually interconnected and synergetic reactions can produce hazardous results.

Solid wastes are usually less voluminous and easier to dispose of than liquid waste, if identified and segregated accordingly. Waste solvents that are not corrosive or reactive and contain no solids can usually be collected in ordinary glass or metal containers. Most waste-disposal companies require the segregation and identification of all liquid wastes by type (chlorinated solvents, hydrocarbons, etc.). Many waste-disposal companies rely on incineration as their ultimate disposal method, and chlorinated solvents tend to produce hydrogen chloride gas which would violate local air-pollution codes. It should be noted that the transporting of this hazardous chemical waste requires that the user package this material in accordance (Fig. 2-25) with U.S. Department of Transportation (DOT) regulations (49 *CFR* Parts 100–199). As the complexity and costs of waste disposal have increased, many companies have implemented recycling processes to recover large quantities of waste chemicals.

FIGURE 2-25
DOT labeling.

Laboratory Practices for Waste Disposal

When you have completed procedures and reactions, you will have residues, slurries (watery mixtures of insoluble matter, precipitates, pigments, etc.), and waste solutions which must be disposed of.

Always consult the MSDS and your supervisor before attempting to dispose of waste chemicals.

CAUTION Do not pour concentrated acids, bases, or slurries into the sink without considering what you are doing. Think first!

Acid or basic solution wastes: Pour into your waste-disposal sink while *running water continuously from the faucet to dilute the acid or base.* When you have finished pouring the waste into the sink, *flush with large volumes of water to dilute any corrosive effect.*

Organic wastes, residues: *These are insoluble in water.* Discard the bulk of the residues in a waste-disposal safety container. Discard all volatile solvents in a waste-solvent receptacle which will contain the vapors and will not constitute a fire hazard. Volatile solvents are those solvents which vaporize readily at relatively low temperatures. The vapors which result can be toxic, nauseating, irritating, or flammable—or can have unpleasant side effects.

Destroy waste sodium and potassium properly by adding them slowly to absolute alcohol.

Avoid indiscriminate disposal of wastes. Always consider the possibility of spontaneous reactions, explosions, and fire. Waste receptacles are usually marked to indicate what should be put into them.

Segregate wastes in clearly marked safety containers to avoid any possible chemical reaction.

Protect maintenance personnel. When working with extremely dangerous materials (fluorides, cyanides, etc.) always follow your wastes step by step to their final disposition.

Refuse should not remain overnight and accumulate in the laboratory.

Safety Practices for Chemical Spills in the Laboratory

Spills of chemicals anywhere are dangerous, regardless of their nature, because they cause conditions which can result in fire, accident, and/or toxic fumes, depending upon the substance.

Solid, Dry Substances

These substances can be swept together, brushed onto a shovel, dustpan, or cardboard, and then deposited in the proper waste container.

Acid (Solutions)

Acid spills should be diluted with water and flushed to floor drains. Soda ash or sodium bicarbonate solid or solution can be used to neutralize any residual acid; this is followed by flushing with water.

CAUTION When water is poured on spills of concentrated sulfuric acid, there is the problem of heat generation and spattering. *Deluge* with water, to dilute the acid and minimize heat generation and spattering.

Alkali Solutions

These should be flushed with water to a floor drain. A mop and bucket can be used, but avoid any splattering when squeezing out the mop. Flush mop and bucket, replacing water frequently.

WARNING *Alkali solution makes the floor slippery.* Clean sand can also be used. Throw sand over the spill and sweep up. The wet sand is then discarded.

Volatile Solvents

Volatile-solvent spills evaporate very rapidly because of the extremely large surface area. This kind of spill can create a fire hazard, if the solvent is flammable, and it will invariably cause a high, (possibly) dangerous concentration of fumes in the laboratory. These fumes can have serious physiological effects when breathed. They can also form explosive mixtures with air. (Refer to the section on explosive mixtures.) Cleanup procedures are as follows:

1. If minor quantities of solvents are spilled, wipe up the liquid with rags or toweling and discard them in the proper waste receptacle. (Refer to the section on safety practices for waste disposal.)

2. If a large amount of solvent is involved in the spill, use a mop and pail. Squeeze out the mop in the pail and continue as needed. (Refer to the section on safety practices for waste disposal to discard the waste solvent and to the section on explosive mixtures.)

Oily Substances

These substances should be mopped up to remove the excess liquid, and the waste substance should be disposed of in a proper waste container. Select a nonflammable volatile solvent for the substance, pour some on an absorbent rag, and then wipe up the spilled substance. You may have to rinse your rag in a pail of the solvent to remove all the spilled material, because oily floors are slippery and dangerous. Finally, a thorough detergent-water scrub will clean up any oily remains.

Mercury

Mercury spills are one of the most common sources of mercury vapor in the laboratory air. As a result of a spill, mercury may be distributed over a wide area, exposing a large surface area of the metal. In any mercury spill *unseen* droplets are trapped in crevices. Unless the laboratory has adequate and reasonable ventilation, the combined mercury vapor concentrations may exceed the recommended limit.

CAUTION Surfaces which are apparently free of mercury will harbor microscopic droplets. Vibrations increase vaporization. Smoking in contaminated areas is very hazardous: tamping a cigarette causes mercury to adhere to the tobacco, and when the cigarette is inhaled, the mercury intake is increased.

Cleanup procedure is as follows:

1. Push droplets together to form pools (Fig. 2-26).

2. To pick up the mercury, use a suction device made from a filter flask, rubber stopper, and several pieces of rubber and glass tubing.

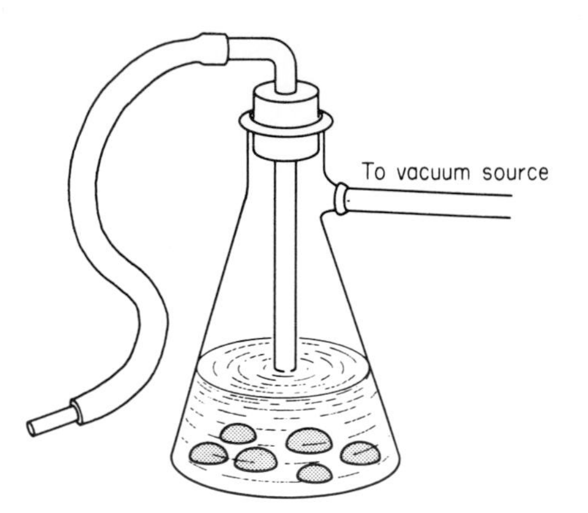

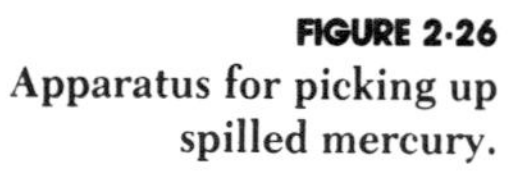

FIGURE 2-26
Apparatus for picking up spilled mercury.

3. If there are many crevices and cracks in the floor which trap small droplets of mercury that cannot be picked up by the suction device, seal over the cracks with a thick covering of floor wax or an aerosol hair spray. The covering will dramatically reduce vaporization. Sulfur dust can also be used to fix mercury. Several mercury-spill cleanup kits are commercially available.

3
HANDLING CHEMICALS AND SOLUTIONS

INTRODUCTION

For successful analytical work, the availability of reagents and solutions of established purity is of prime importance. A freshly opened bottle of some reagent-grade chemical can be used with confidence in most applications; whether the same confidence is justified when this bottle is half full depends entirely upon the care with which it was handled after being opened. The rules that are given here will be successful in preventing contamination of reagents only if they are conscientiously followed.

CAUTION *Always read the label twice before using any reagent.*

GENERAL GUIDELINES

1. Select the best available grade of chemical for analytical work. If there is a choice, pick the smallest bottle that will supply the desired quantity of substance.

2. Replace the top of every container immediately after removal of reagent; do not rely on having this done by someone else.

3. Stoppers should be held between the fingers and should never be set on the desk top.

4. Never return any excess reagent or solution to a bottle; the minor saving represented by a return of an excess is indeed a false economy compared with the risk of contaminating the entire bottle.

5. Do not insert pipets into a bottle containing a reagent chemical. Instead, shake the bottle vigorously with the cap in place to dislodge the contents; then pour out the desired quantity.

6. Keep the reagent shelf and the laboratory balances clean. Immediately clean up any spilled chemicals, even though others may be making the same transfer of reagent in the same area. (Refer to Chap. 2, "Laboratory Safety.")

Personal Safety

Always wear protective safety goggles, aprons, shoes, etc. when handling chemicals, in case of accident.

Always flush the outside of acid bottles with water and dry them well before using. (Wet bottles are slippery.)

REMOVAL OF SOLID MATERIALS FROM GLASS-STOPPERED BOTTLES

Method 1

This method can only be used with those bottles which have a hollow glass stopper capable of containing some of the material.

1. Rotate the bottle in an inclined position so that some of the material will enter the hollow glass stopper (Fig. 3-1). It may be necessary to tap the bottle gently to break up solidified surface material or to actually open the bottle and break up the solid with a clean spatula.

2. Position the bottle so that when the stopper is removed, some of the material will remain in the stopper (Fig. 3-2).

3. Place the bottle on the table. Gently tap the tilted stopper with a finger, pencil, or small spatula to dislodge enough of the desired material (Fig. 3-3).

4. Repeat to get the required amount of solid.

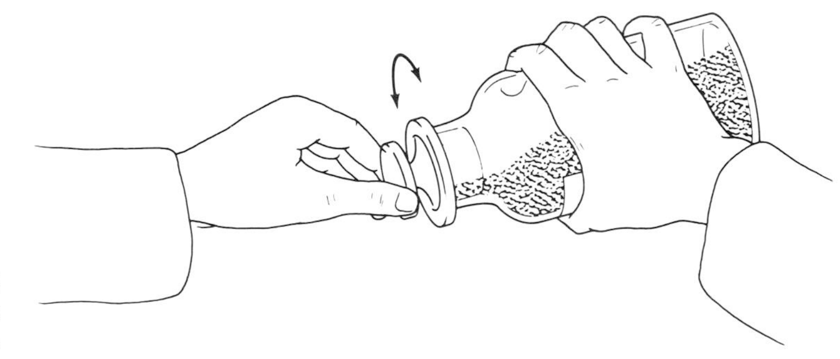

FIGURE 3-1 Rotate the bottle in an inclined position.

FIGURE 3-2 Remove stopper so that some of the material remains in it.

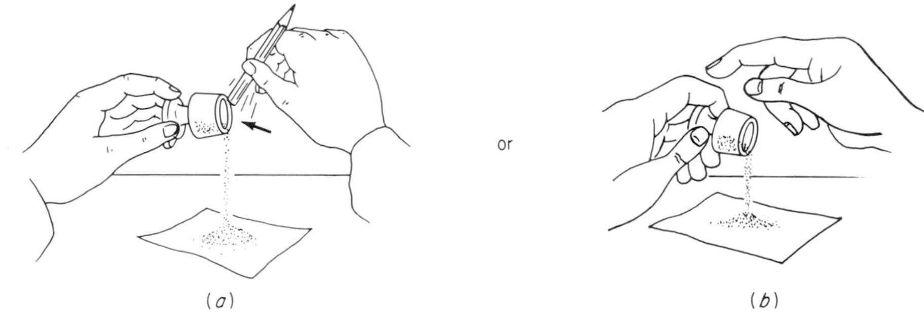

FIGURE 3-3
Gently tap tilted stopper to dislodge desired amount of material (*a*, *b*).

5. Return to the bottle *only that excess material which remains in the stopper*. Discard any excess which has come in contact with anything else.

6. Replace the stopper in the bottle.

Method 2

1. Remove the glass stopper by gently twisting, tapping the stopper gently to loosen, if necessary.

2. Use an *absolutely clean spatula* and dig out material, always laying the stopper upside down on the desk top. (See Fig. 3-4.)

3. Tap the spatula gently to cause the desired amount of material to fall off.

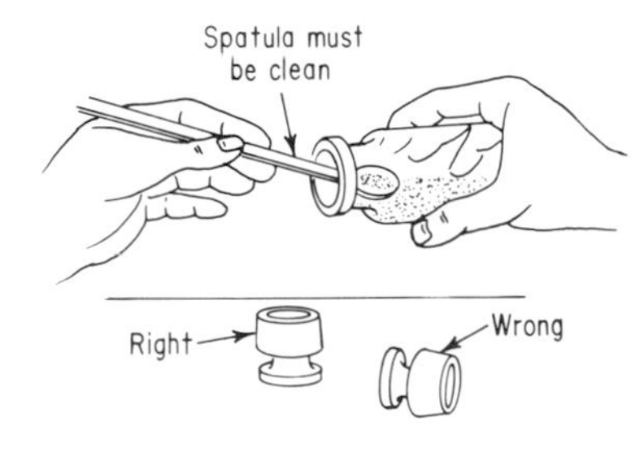

FIGURE 3-4
Proper procedure to dig out material.

Spatulas

Scoop-type spatulas are used to transfer larger quantities of solids, especially from narrow-necked bottles or containers. (See Fig. 3-5*a*.)

Spatulas are available in a variety of shapes, sizes, and designs to be used for special manipulations, from micro size to very large for production jobs. (See Fig. 3-5*b* to *d*.)

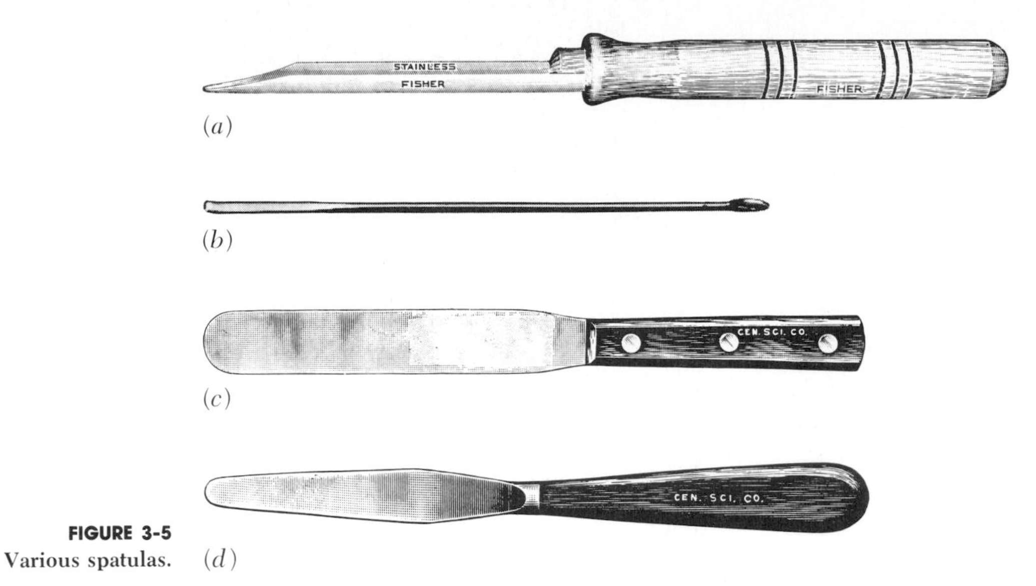

FIGURE 3-5
Various spatulas.

Method 3

1. Tap the jar lightly on the table top while gently rotating to loosen the material.

2. Remove the stopper and place upside down on a clean surface

3. Hold the jar over the container and roll and tilt the jar until enough material has fallen out.

POURING LIQUIDS FROM BOTTLES

Method 1

1. Loosen the stopper by gently twisting it.

2. Grasp the stopper either between the second and third fingers as in Fig. 3-6 or between the palm and fingers as in Fig. 3-7.

3. Pour the liquid as needed.

4. Replace the stopper immediately in the bottle. Never lay it on the desk.

Method 2

An alternative procedure for pouring from a bottle:

1. Tilt the bottle so that the contents wet the stopper. Rotate the stopper completely.

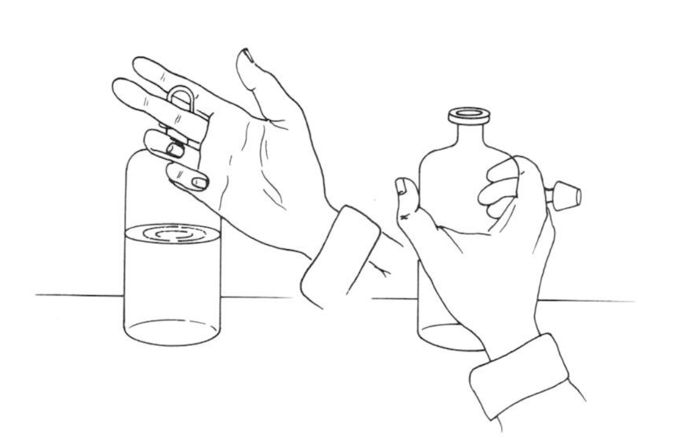

FIGURE 3-6
Handling a glass stopper.

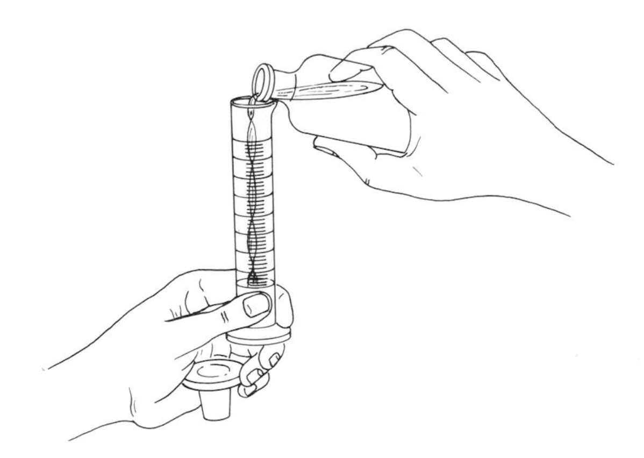

FIGURE 3-7
Alternative method of handling a glass stopper.

2. Withdraw the stopper, then moisten the inside of the neck of the bottle with the wetted stopper.

3. Reinsert the stopper and again rotate completely so that all contact surfaces are wetted. Remove the stopper using standard procedure, with the back of the hand.

4. Pour the liquid. The wetted inner surface of the neck and lip of the bottle enables the liquid to flow smoothly out of the bottle without gushing.

POURING LIQUIDS FROM BEAKERS OR OTHER CONTAINERS

1. Hold a glass stirring rod against the pouring lip of the beaker.

2. Tilt the container, allowing liquid to flow around the stirring rod which guides the liquid to the receiver. (See Fig. 3-8.)

FIGURE 3-8
Use of stirring rod to guide liquid.

3. When the desired amount of liquid has been poured, position the pouring beaker vertically, allowing the last liquid to drain off the lip and down the rod.

TRANSFERRING SOLUTIONS INTO CONTAINERS FROM PIPETS OR MEDICINE DROPPERS

1. Fill the pipet or medicine dropper. (Review the section on pipets in Chap. 25, "Volumetric Analysis.")

2. Hold it above the solution to which it is to be added. *Do not* immerse it in the solution because it will then become contaminated by the solution. (See Fig. 3-9.)

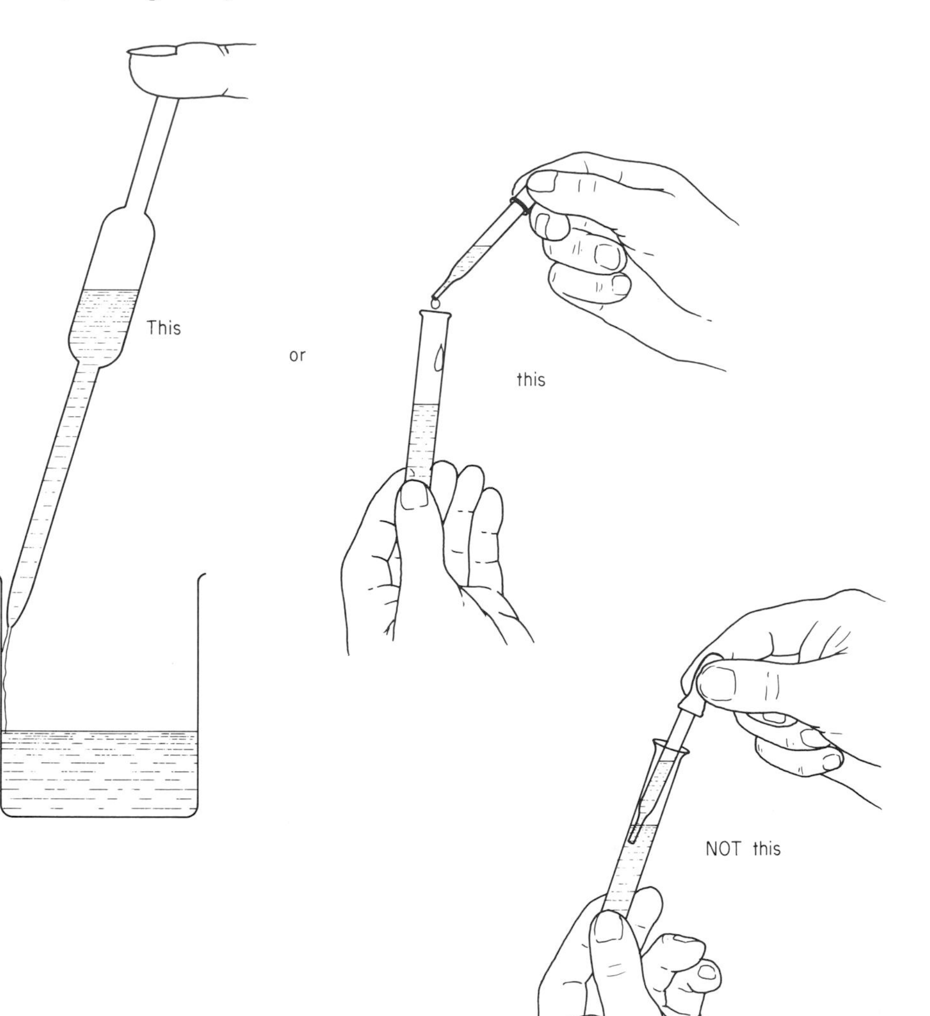

FIGURE 3-9
Proper use of a medicine dropper or pipet.

SHAKING A TEST TUBE

To mix substances in a test tube, always use a suitable, clean, cork or rubber stopper; never use your bare finger. The liquids may be corrosive and damage your skin, or your finger may be dirty and contaminate the solution. Shake with an up-and-down motion. (See Fig. 3-10.)

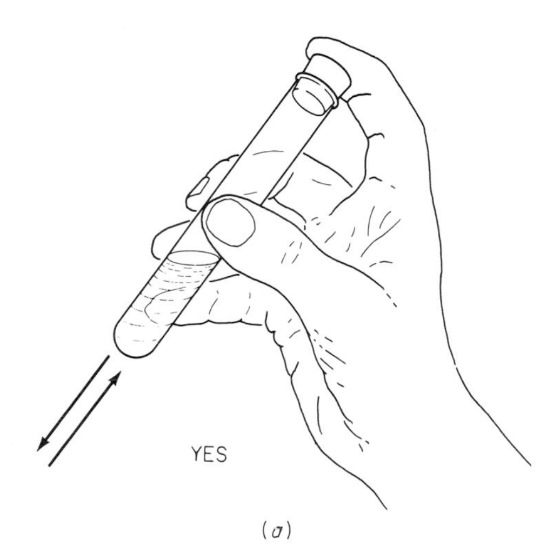

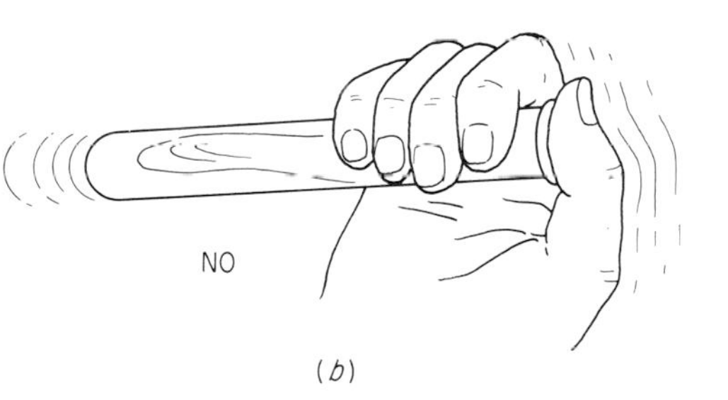

FIGURE 3-10 Proper way (*a*) and improper way (*b*) to shake a test tube.

CAUTION Use care when the stopper is removed. Pressure may have built up, and the liquid may foam out or spew from the test tube.

MIXING SOLUTIONS IN A TEST TUBE

1. Use a glass rod (Fig. 3-11), preferably with a small rubber policeman.
2. Agitate sideways, so that liquid is not spilled (Fig. 3-12).

FIGURE 3-11
Stirring a solution in a test tube. Don't poke out the bottom.

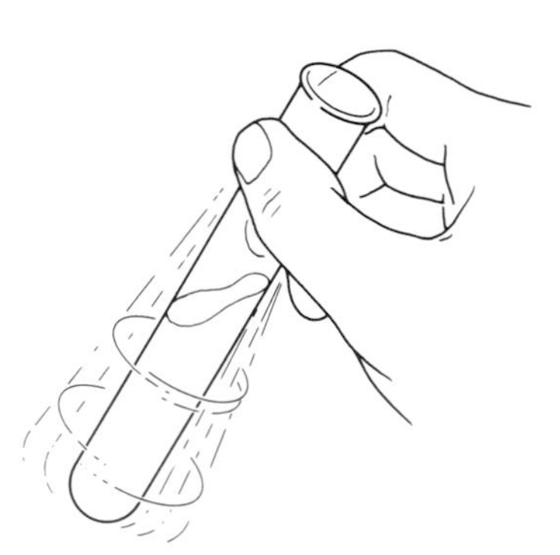

FIGURE 3-12
Agitate a solution sideways in a test tube.

SAFE HANDLING OF LIQUIDS

Carboys and Large Bottles

Handling

Carboys and large bottles of acids, caustics, flammable liquids, and corrosive materials are dangerous to handle and to move. They are extremely fragile, and a slight mechanical shock or jar can cause breakage. When this happens, the broken glass and the corrosive material may cause injury. It is also very difficult to pour from them because of spillage.

Safety carboy tilters (Fig. 3-13) enable you to pour faster, safer, and easier. You also minimize spills and leakage by using a safe pouring spout

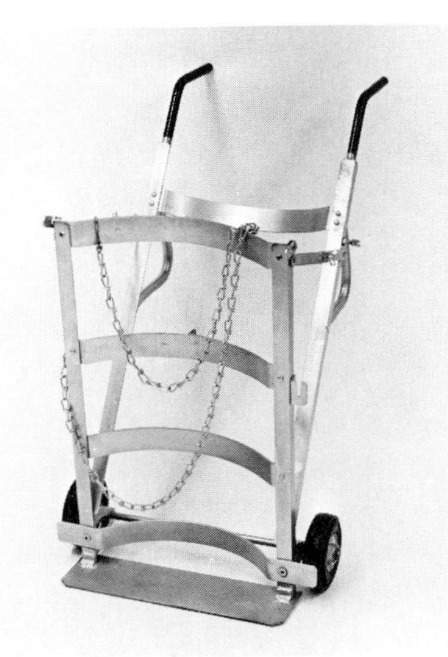

FIGURE 3-13
Carboy tilter.

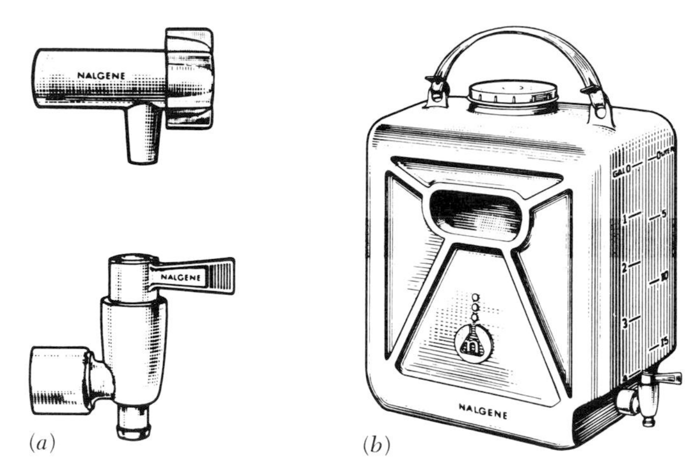

FIGURE 3-14
(*a*) Carboy spout.
(*b*) Plastic safety container with spout.

(Fig. 3-14) which permits the entry of air and prevents spurts and splashes.

Safety hand pumps (Figs. 3-15 and 3-16) enable you to pump liquids out of carboys and large bottles safely and easily. Furthermore you can pump out exactly the volume you wish.

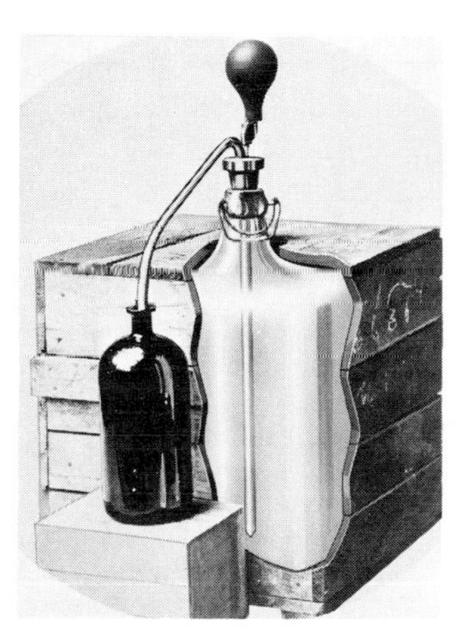

FIGURE 3-15
A positive-pressure acid pump. Squeezing the rubber bulb forces acid through the exit tube. The instantaneous relief valve stops the flow; this aids one to make precise measurements and is a good safey device. Be sure the stopper is secure.

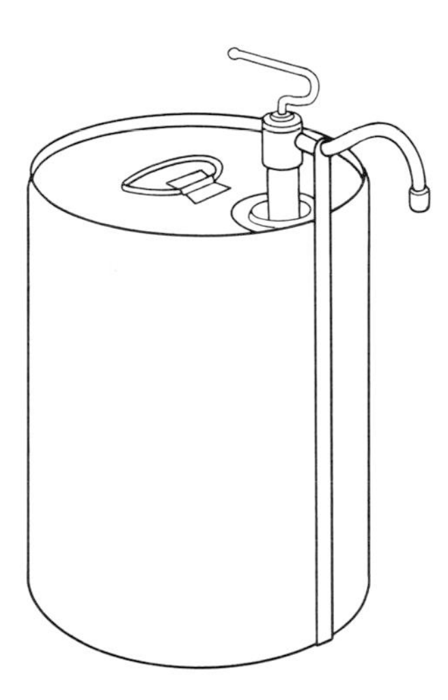

FIGURE 3-16
A positive-acting pump which delivers solvents quickly and safely from cans or larger containers. A vertical movement of the handle causes the pumping action.

Transporting

Moving large bottles and carboys is a dangerous operation because of the danger of bottle breakage and liquid spillage. Accidents can and do happen; therefore always use safety carts and safety bottle carriers (Fig. 3-17). The safety bottle carriers securely hold the bottle in place and cushion the bottle from shocks to prevent breakage. They should always be used when more than one large bottle is being moved on the same cart.

(*a*)

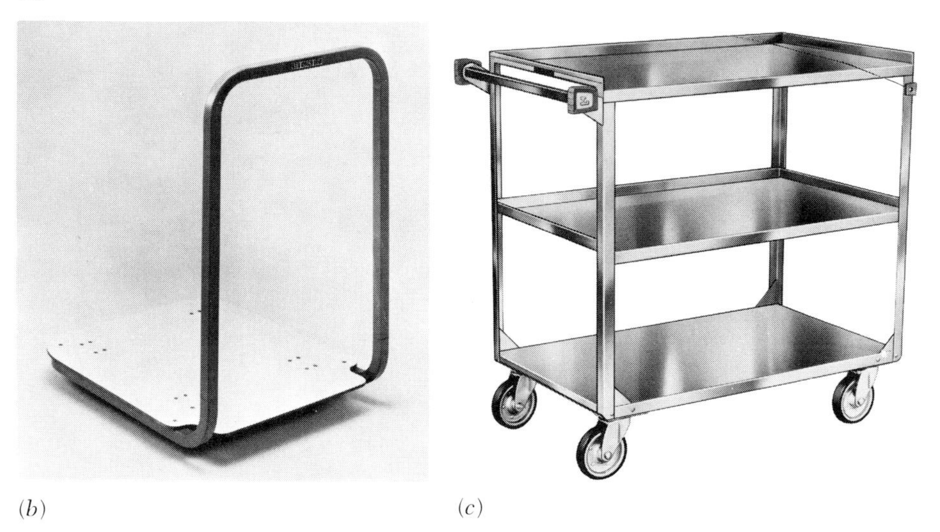

(*b*) (*c*)

FIGURE 3-17 Safe transportation for carboys and large bottles: (*a*) plastic protective shield for carboy; (*b*) carboy transporter; (*c*) movable table.

Drums

Handling

Many chemicals, such as solvents, plasticizers, raw materials, etc., are shipped in 55-gal drums, which weigh around 800 lb when full. You can easily be injured and cause a serious accident when you try to withdraw the material from the drum without the proper equipment (Fig. 3-18).

Drum cradles (Fig. 3-19) enable you to position the drums horizontally at a convenient height for safe liquid withdrawal.

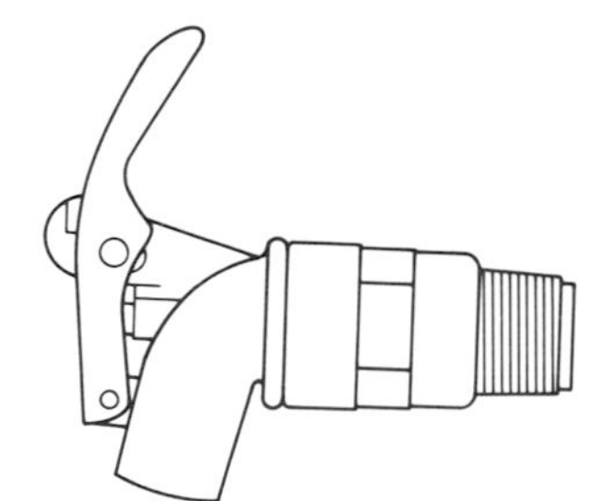

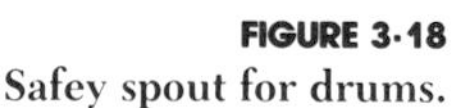

FIGURE 3-18
Safey spout for drums.

FIGURE 3-19
Safey cradle for drum which is fitted with flame arrestor and safety drip can.

Volatile flammable-solvent vapors escaping from openings in drums can be accidentally ignited, causing fire. This hazard can be eliminated by the use of flame arrestors installed in the drum openings (Fig. 3-19).

Leakage through faulty drum spigots or by careless and sloppy work can cause hazards. Always place a safety drip can beneath the spigot. (Fig. 3-19).

Transporting

To transport large drums and barrels you should use the proper handling truck (Fig. 3-20). You can safely handle the drum, because the locking device is designed to prevent accidents, and the load rests on the wheels, not on you.

Safety Cans

Safety cans (Fig. 3-21) prevent leakage or evaporation of solvents and automatically permit escape of excessive gas pressure. Use of these cans

FIGURE 3-20
Truck for moving drums and barrels.

FIGURE 3-21
Lightweight safety storage cans for storing flammable liquids. The positive spring-lid closure prevents both loss due to spillage and hazard from fire.

for storage and transportation of flammables reduces the hazards to a minimum.

Safety Pails

Always transport acids and caustics in safety pails, made of plastic or molded rubber. Never use glass because of the dangers involved in the event of accidental breakage.

USE OF PUMPS FOR TRANSFERRING LIQUIDS

In the laboratory two kinds of pumps are used: the centrifugal pump and the volume-displacement pump. Centrifugal and gear pumps must be fabricated out of materials which will not corrode and will not contaminate any solutions being transferred. Usually, stainless steel or plastic component pumps are used.

Centrifugal Pumps

In general, the centrifugal pump finds the widest use for recirculating liquids and general transfer work (Fig. 3-22). When pumps are used to transfer reaction solutions, it is important to avoid contamination, to clean the pump after use, and to get the centrifugal pump primed. Before centrifugal pumps will work, they must be primed and the inlet pipe and impeller must be completely filled with liquid. This can present a problem when working with corrosive solutions.

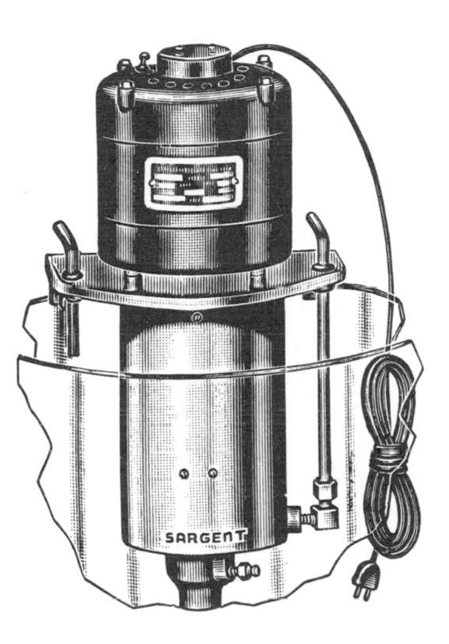

FIGURE 3-22
A centrifugal pump used to constantly circulate water in constant-temperature baths. It cannot be used for corrosives.

Volume-Displacement Pumps

There are several types of volume-displacement pumps: gear or modified-gear types, flexible-contact impeller type, housing type, or flexible-tubing squeeze type.

The gear-type volume-displacement pump does not have to be primed (Fig. 3-23). When started, the pump will prime itself, provided that it is in reasonably good condition and that the gear drive is not too far above the surface of the solution. The problem of using a clean pump, free of contaminants, always faces the technician.

FIGURE 3-23
All-purpose transfer pump for corrosive liquids. It is self-priming and has a flexible, replaceable line. It pumps a fixed volume per minute unless a variable-speed motor is used.

The volume-displacement pump offers the technician the best means of transferring corrosive solutions, because the mechanical components of the pump never come in contact with the solution. The squeezing action of these pumps moves the liquid to the desired container, the liquid coming in contact only with the tubing (Fig. 3-24). Various kinds of flexible plastic tubing are available to meet the need. (Refer to Chap. 17, "Plastic Labware.")

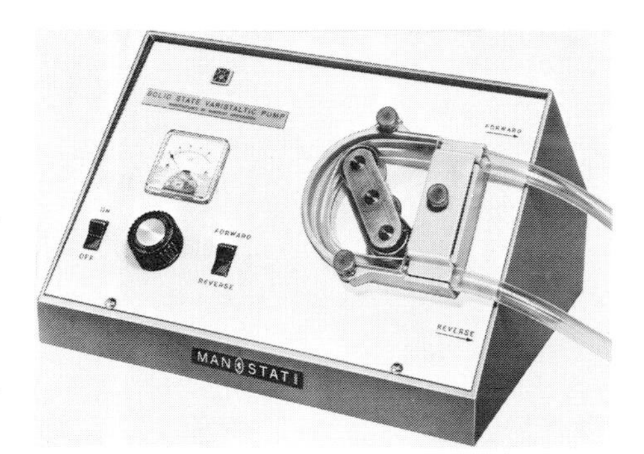

FIGURE 3-24
Transfer pump for positive transfer of corrosive, sterile, or acid fluids. The fluids have no contact of any kind with the pump. The pump acts to squeeze the tubing, causing the transfer of the liquid. The speed and direction of pumping can be varied.

Metering Pumps

Metering pumps are available in a wide range of designs and capacities, providing adjustable flow ranges from fractions of a milliliter per minute to fractions of liters per minute. They are self-priming, capable of lifting a solution to heights equivalent to about 20 ft of water. Their output stroke is adjustable, and they are made in a variety of materials (plastics, stainless steel, Monel, etc.). The selection is determined by the procedure.

Finger Pumps

When solutions and slurries must be pumped or transferred and must be protected from any possible contamination, the answer may be the flexible-tubing finger pump. Solutions never come in contact with any portion

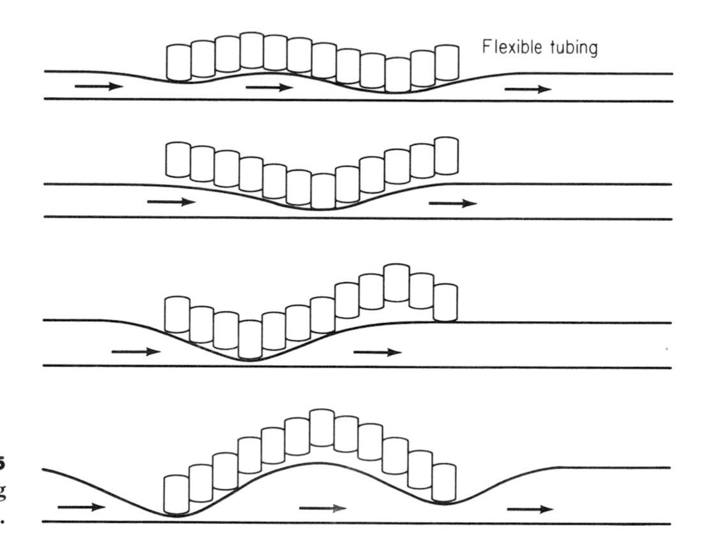

FIGURE 3-25
Sequence of squeezing action in finger pumps.

of the pump; therefore no contamination of the solution results, and the pump does not have to be cleaned. A series of metal fingers actuated by an electric motor gently forces the solution through the tubing. (See Fig. 3-25.) These pumps can be used to move corrosive liquids, to pump several solutions at the same time, and to feed and mix solutions. They also find use in metering operations for column chromatography.

4
COMPRESSED GASES IN THE LABORATORY*

INTRODUCTION

Most laboratories use compressed liquids and gases which are contained under very high pressures (typically 100 to 2500 psi) in metal cylinders. A *compressed gas* is defined by the U.S. Department of Transportation (DOT) as "any material or mixture having in the *container an absolute pressure* exceeding 40 psi at 70°F (21°C) or regardless of the pressure at 70°F, having an absolute pressure exceeding 104 psi at 130°F (54°C); or any liquid flammable material having a vapor pressure exceeding 40 psi absolute at 100°F (38°C) as determined by ASTM Test D-323." These compressed substances are potentially dangerous because they are pressurized, flammable, corrosive, toxic, and/or extremely cold. When using compressed gases, wear appropriate protective equipment (review Chap. 2, "Laboratory Safety"). Gas masks should be kept available for immediate use when working with potentially toxic gases. These masks should be placed in a convenient location, not likely to become contaminated, and should be approved by the U.S. Bureau of Mines for the service intended. Those involved in the handling of compressed gases should become familiar with the proper application and limitations of the various types of respirators. Many industrial accidents have occurred from the mishandling of these cylinders and their contents. Table 4-1 lists the most common Occupational Health and Safety Administration (OSHA) violations involving compressed gases.

Figure 4-1 gives a pictorial overview of general handling procedures for gas cylinders, and the most common general precautions are listed below.

GENERAL CYLINDER-HANDLING PRECAUTIONS

1. Before using any compressed gases, read all cylinder label information and Material Safety Data Sheets (MSDS) (review Chap. 1, "The Chemical Technician," under Material Safety Data Sheets) associated with the compressed gas. Do not accept cylinders that do not identify the contents by name; do not rely on color code for content identification.

* Courtesy Matheson Gas Products, Inc., *Guide to Safe Handling of Compressed Gases,* East Rutherford, N.J. 07073, 1983.

TABLE 4-1
Common OSHA Violations Involving Compressed Gas

1. Unsecured cylinders
2. Cylinders stored without protective caps
3. Noncompatible gases (such as hydrogen and oxygen) stored together
4. Cylinder valves open when cylinder is not in use (an attached regulator with a closed discharge valve is not sufficient)
5. Fire extinguishers not present during welding, burning, or brazing operations
6. No safety showers and eyewash fountains where corrosive gases are used
7. No gas masks and/or self-contained breathing apparatus conveniently located near areas where toxic gases are used or stored

2. The cylinder cap should be left on each cylinder until it has been secured against a wall or bench, or placed in a cylinder stand, and is ready to be used. Figure 4-2*a* and *b* demonstrates two cylinder-securing devices.

3. Cylinders may be stored in the open, but should be protected from the ground beneath to prevent rusting. Cylinders may be stored in the sun, except in localities where extreme temperatures prevail; in the case of certain gases, the supplier's recommendation for shading should be observed. If ice or snow accumulates on a cylinder, thaw at room temperature or with water at a temperature not exceeding 125°F (52°C).

4. Avoid dragging, rolling, or sliding cylinders, even for short distances. They should be moved by a suitable hand truck.

5. Never tamper with safety devices in valves or cylinders.

6. Do not store full and empty cylinders together. In addition, serious suckback can occur when an empty cylinder is attached to a pressurized system.

7. No part of a cylinder should be subjected to a temperature higher than 125°F (52°C). A flame should never be permitted to come in contact with any part of a compressed-gas cylinder.

8. Cylinders should not be subjected to artificially created low temperatures (−30°C or lower), since many types of steel will lose their ductility and impact strength at lower temperatures. Special stainless steel cylinders are available for low-temperature use.

9. Do not place cylinders where they may become part of an electric circuit.

10. Bond and ground all cylinders, lines, and equipment used with flammable compressed gases.

11. Use compressed gases only in a well-ventilated area. Table 4-2 contains a partial listing of compressed-gas hazards. The smallest cylinder

FIGURE 4-1
General procedures and cautions for handling compressed gases. (*Courtesy Matheson Gas Products, Inc.*)

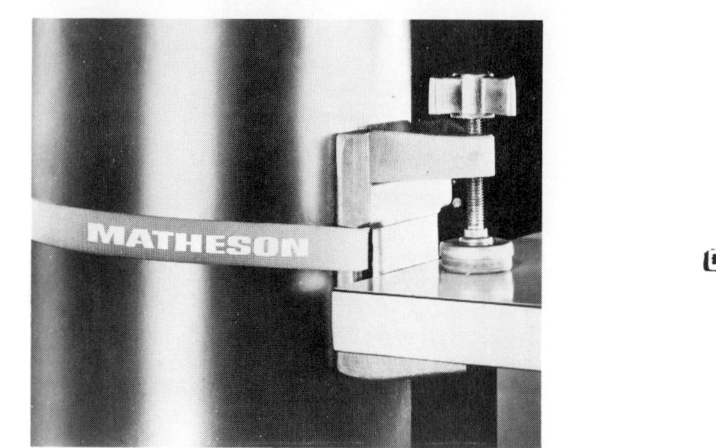

FIGURE 4-2
(*a*) Bench clamp for gas cylinder. (*b*) Safety wall chain for cylinders.

practical (for example, a lecture-bottle size) of potentially toxic gases should be selected for use.

12. Cylinders should be used in rotation as received from the supplier. Storage areas should be set up to permit proper inventory rotation.

13. When discharging gas into a liquid, a trap or suitable check valve should be used to prevent liquid from getting back into the cylinder or regulator.

14. When returning empty cylinders, close the valve before shipment, leaving some positive pressure in the cylinder. Replace any valve outlet and protective caps originally shipped with the cylinder. Mark or label the cylinder "EMPTY" or "MT" (or utilize standard DOT "empty" labels) and store in a designated area for return to the supplier.

15. Always leak-check installed regulators, manifolds, connections, etc. with appropriate leak-detection solutions or devices. Never use an open flame for leak detection.

16. Never drop cylinders or permit them to strike each other violently.

17. Eyewash fountains, safety showers, gas masks, respirators, and/or resuscitators should be located nearby but out of the immediate area that is likely to become contaminated in the event of a large release of gas.

18. Fire extinguishers, preferably of the dry-chemical type, should be kept close at hand and should be checked periodically to ensure their proper operation.

19. Low-boiling-point liquids or cryogenic materials (dry ice, fluorocarbons, liquid oxygen, liquid nitrogen, etc.) can cause frostbite on contact with living tissue and should be handled with care.

TABLE 4-2
Hazardous Properties of Compressed Gases*

Gas	*Hazard*			*Gas*	*Hazard*		
	Toxic	*Flammable*	*Corrosive*		*Toxic*	*Flammable*	*Corrosive*
Acetylene		X		Isobutane		X	
Air				Isobutylene		X	
Allene		X		Krypton			
Ammonia	X	X	X	Methane	X	X	
Argon				Methylacetylene		X	
Arsine	X	X		Methyl bromide	X	X	
Boron trichloride	X		X	Methyl chloride	X	X	
Boron trifluoride	X		X	Methyl mercaptan	X	X	
1,3-Butadiene		X		Monoethylamine	X	X	
Butane		X		Monomethylamine	X	X	
Butenes		X		Neon			
Carbon dioxide				Nickel carbonyl	X	X	
Carbon monoxide	X	X		Nitric oxide	X		X
Carbonyl sulfide	X	X		Nitrogen			
Chlorine	X		X	Nitrogen chloride	X		X
Cyanogen	X	X		Nitrogen trioxide	X		X
Cyclopropane	X	X		Nitrosyl chloride	X		X
Deuterium		X		Nitrous oxide			
Diborane	X	X		Oxygen			
Dimethylamine	X	X	X	Ozone	X		
Dimethyl ether		X		Phosgene	X		X
Ethane		X		Phosphine	X	X	
Ethyl acetylene	X	X		Propane		X	
Ethyl chloride	X	X		Propylene		X	
Ethylene		X		Silane	X	X	
Ethylene oxide	X	X		Silicon tetrafluoride	X		X
Fluorine	X		X	Sulfur dioxide	X		X
Germane	X	X		Sulfur hexafluoride			
Helium				Sulfur tetrafluoride	X		X
Hexafluoropropene	X			Sulfuryl fluoride	X		X
Hydrogen		X		Tetrafluoroethylene		X	
Hydrogen bromide	X		X	Trimethylamine	X	X	
Hydrogen chloride	X		X	Vinyl bromide	X	X	
Hydrogen fluoride	X		X	Vinyl chloride	X	X	
Hydrogen selenide	X	X		Vinyl fluoride	X	X	
Hydrogen sulfide	X	X		Xenon			

* From "Safe Handling of Compressed Gases in Laboratory and Plant," copyright © 1974 by Matheson Gas Products, East Rutherford, N.J.

FIGURE 4-3. Wrenches for compressed-gas cylinders and regulators. Always use the correct tool for the connection (*no pipe wrenches*).

20. Use the proper wrench or tool (Fig. 4-3) for attaching a cylinder regulator or valve. A pipe wrench should *never* be used, and excessive force to tighten a regulator should be avoided.

CYLINDER MARKINGS

The DOT has developed regulations and markings to assure compressed-gas-cylinder safety. Figure 4-4 shows a typical gas cylinder's parts and markings. The cylinder cap (1) protects the main cylinder valve. The valve handwheel (2) is used to open and close the cylinder valve. Certain valves are not equipped with handwheels (for example, those on acetylene tanks)

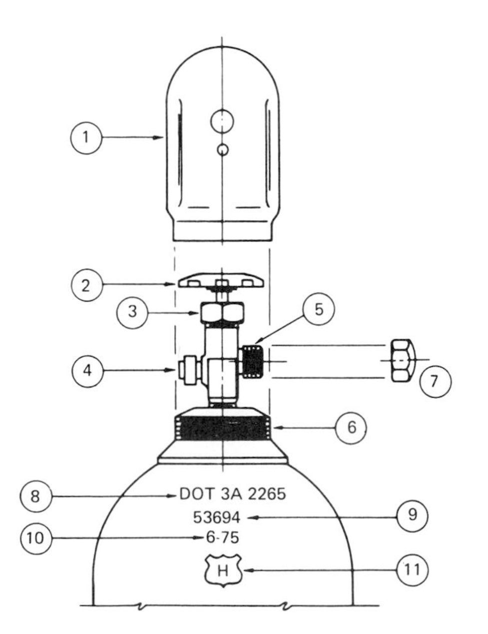

FIGURE 4-4 Compressed-gas cylinder parts and DOT markings. (*Courtesy Matheson Gas Products, Inc.*)

and require special wrenches for operation. The valve packing nut (3) contains a packing gland and packing around the stem. It is adjusted only occasionally and is usually tightened if leakage is observed around the valve stem. It should not be tampered with when used in conjunction with diaphragm-type valves. A safety device (4) permits gas to escape if the temperature gets high enough to endanger the cylinder by increased unsafe pressures. The valve outlet connection (5) connects to pressure- and/or flow-regulating equipment. Various types of connections are provided to prevent interchange of equipment for incompatible gases, usually identified by *CGA (Compressed Gas Association) number:* for example, CGA No.350 is used for hydrogen service. A cylinder collar (6) holds the cylinder cap at all times, except when regulating equipment is attached to the cylinder valve. The valve outlet cap (7) protects valve threads from damage and keeps the outlet clean: it is not used universally. Specification number (8) signifies that the cylinder conforms to DOT specification DOT-3A, governing material of construction, capacities, and test procedures, and that the service pressure for which the cylinder is designated is 2265 psig at 70°F. The cylinder serial number is indicated by (9), and (10) indicates the date (month and year; in this case, June 1975) of initial hydrostatic testing. For most gases, hydrostatic pressure tests are performed on cylinders every 5 years to determine their fitness for further use. The original inspector's insignia for conducting hydrostatic and other required tests to approve the cylinder under DOT specifications is shown by (11).

CYLINDER SIZES

Normally, compressed-gas cylinders range in capacity from approximately 2 to 100 L as shown in Fig. 4-5. The lecture bottle (see Fig. 4-6) is the

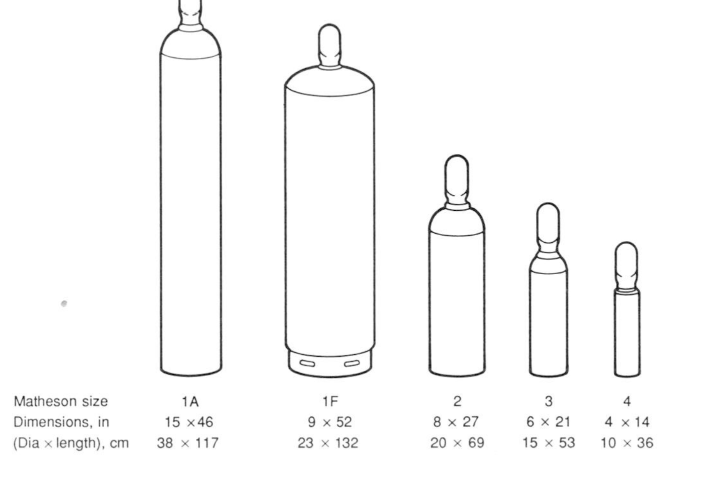

FIGURE 4-5 Compressed-gas cylinder sizes. A variety of cylinders are available from different vendors. Typically, the smallest is shown by No. 4 (0.08 ft^3, 2.35 L) and the largest is shown by example size No. 1F (3.87 ft^3, 109.6 L).

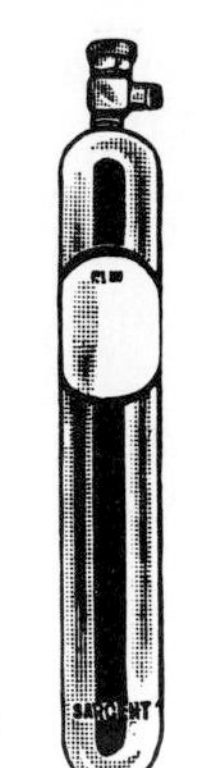

FIGURE 4-6
The lecture bottle.

safest, best package for gases in small quantities. For noncorrosive gases, it is equipped with a leak-free packless valve with handwheel control which provides good metering characteristics. No wrenches are required for cylinder valve operation, and hookup is simple. Lecture bottles for noncorrosive gases have both internal and external threads for compatibility with both new and old needle valves and other accessories. The external thread permits the orientation of the regulator to any position.

CAUTION Special care must be used in changing lecture-bottle equipment from one gas service to another since outlets are the same, regardless of gas.

WHAT TO DO WITH LEAKING CYLINDERS

Inert Gases

Leaking cylinders of inert gases, such as argon, helium, nitrogen, etc., do not represent a hazard unless they are situated in confined places with no ventilation.

Acid Gases

Acid gases are corrosive and toxic. Therefore, put on appropriate protective equipment (face shield, rubber gloves, breathing equipment) before transporting the leaking cylinder to a safe out-of-doors area or a hood with forced ventilation.

Alkaline Gases

The alkaline gases are corrosive, flammable, and toxic. Put on appropriate protective equipment (face mask, rubber gloves, breathing equipment) before transporting the cylinder to a hood with forced ventilation or to a safe out-of-doors area.

CHARACTERISTICS OF COMMON GASES

Oxygen

Oxygen is colorless, odorless, and tasteless. It is slightly soluble in water, and it is a poor conductor of heat and electricity. *Use with extreme caution.* It supports combustion and will combine chemically with practically all of the known elements, except the rare gases.

Nitrogen

Nitrogen is a colorless, odorless gas. For most practical purposes it is considered chemically inert. It does not react readily with other elements, does not burn, and will not support combustion or respiration. It will chemically combine with the more active metals, lithium and magnesium, to form nitrides; and with hydrogen and oxygen and other elements at high temperatures. It is slightly soluble in water and is a poor conductor of heat and electricity.

Helium

Helium is chemically inert. It is a colorless, odorless, tasteless gas.

Hydrogen

Gaseous hydrogen is colorless, odorless, and tasteless. It will diffuse rapidly through many plastic materials. It is extremely flammable, burning in air with a pale, bluish flame that is nearly invisible. Although it is nontoxic, it can cause asphyxiation in confined spaces.

COMPRESSED AIR IN THE LABORATORY

Compressed air is a valuable utility to have in the chemical laboratory, especially good, clean, oil-free air. If compressed air is available at outlets, it is reasonable to assume that the air has been mechanically compressed in an air compressor, such as is found in the garage service station. This means that the air has been compressed in a piston, which must be oiled, and probably contains water vapor, oil, and possibly some debris from the intake air.

Compressed air can be filtered through glass-wool plugs in tubing and dried by passing it through gas-drying towers (refer to Chap. 21, "Distillation.")

Clean, filtered compressed air is a very useful substance to have; it can be used for drying glassware, for accelerating evaporation of solvents, and to introduce as needed into reaction mixtures.

CAUTION Avoid excessive pressure in glass labware, which can explode if outlets become accidentally blocked.

Clean, dry compressed air can be purchased from all compressed-gas manufacturers and distributors, and its cleanliness and purity can be specified as needed. When compressed-air cylinders are used, follow all safety precautions and rules for using other compressed gases.

PROPER DISCHARGE OF CYLINDER CONTENTS*

Liquefied Gases†

For controlled removal of the liquid phases of a liquefied gas, a manual control valve is used. Special liquid-flow regulators are also available.

Rapid removal of the gas phase from a liquefied gas may cause the liquid to cool too rapidly, causing the pressure and flow to drop below the required level. In such cases, cylinders may be heated in a water bath with the temperature controlled to go no higher than 125°F (52°C). Safety relief devices should be installed in all liquid-transfer lines to relieve sudden, dangerous hydrostatic or vapor-pressure buildups.

Nonliquefied Gases

The most common device used to reduce pressure to a safe value for gas removal is a pressure regulator. . . . Delivery pressure will exactly balance the delivery-pressure spring to give a relatively constant delivery pressure.

EQUIPMENT FOR CONTROL AND REGULATION OF GASES

Gas Regulators: Use and Operation*

Purpose of a Regulator

A gas-pressure regulator is a precision instrument designed to reduce high source pressures (cylinders or compression systems) to a safe value, one consistent with a system's design. Each regulator will control a chosen delivery pressure within the bounds of the regulator's delivery-pressure range.

This constant delivery pressure prevents the overpressurization of any apparatus downstream of the regulator and permits stable flow rates to be established according to requirements.

---

* Courtesy Matheson Gas Products, Inc., *Guide to Safe Handling of Compressed Gases*, East Rutherford, N.J., 1983.

† Liquefied gases, such as butane, are liquid in the cylinder under pressure at room temperature. Nonliquefied gases are in the gaseous state even under pressure at room temperature.

CAUTION Pressure regulators are not flow regulators. They do maintain a constant pressure, which in turn will maintain a constant flow *provided the pressure downstream does not vary.* Use flow controllers to control flow, if the downstream pressures are subject to variations.

Three criteria are used to measure the performance of a regulator:

1. The regulator's ability to maintain a constant delivery pressure, regardless of the rate of gas discharge. All regulators will show a drop in delivery pressure with increased flow. The smaller the drop, the better the regulator performance.

2. The regulator's ability to maintain a constant delivery pressure as source pressure varies. This is very important.

3. The *lockup* of the regulator. This is defined as the final pressure attained by a system when all flow is stopped. It is usually slightly above the delivery pressure when set at flowing conditions. All regulators are chosen to give the best possible lockup performance, with but slight deviation from delivery pressure.

How the Regulator Works

A regulator reduces gas pressure by the counteraction of gas pressure on a diaphragm against the compression of a spring which can be adjusted externally with the pressure-adjusting screw (Figs. 4-7 and 4-8).

In operation, the pressure-adjusting screw is turned to exert force on the spring and diaphragm. This force is transmitted to the valve assembly, pushing the valve away from the seat. The high-pressure gas will flow past the valve into the low-pressure chamber. When the force of gas pressure on the diaphragm equals the force of the spring, the valve and seat assem-

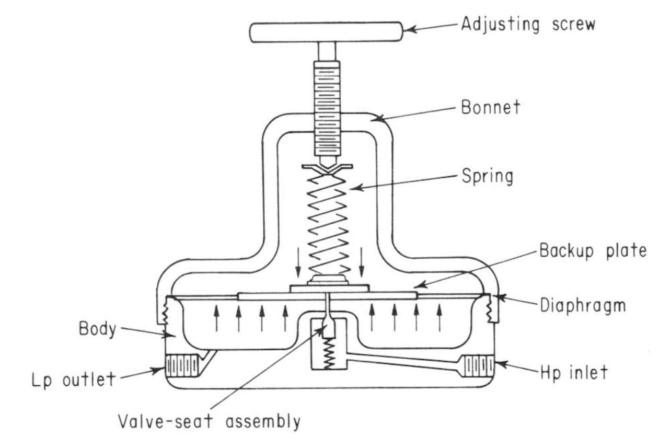

FIGURE 4-7 Schematic of a single-stage regulator (typical construction).

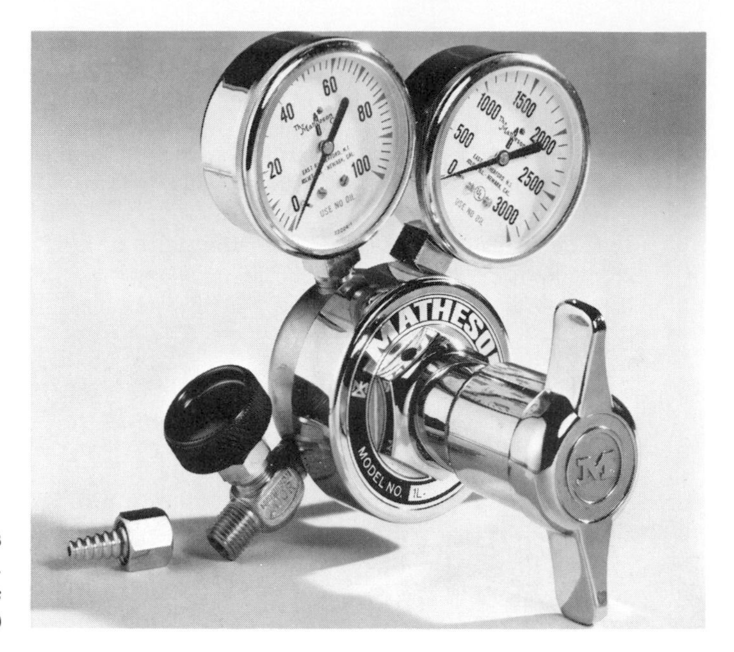

FIGURE 4-8
Two-stage gas regulator. (*Courtesy Matheson Gas Products, Inc.*)

blies close, preventing the flow of additional gas into the low-pressure chamber.

Removal of gas from the low-pressure chamber will permit downward deflection of the diaphragm, opening the valve assembly and thereby permitting a pressure increase in the low-pressure chamber. This constant throttling action permits a pressure balance in the regulator's low-pressure chamber, thus yielding a steady delivery pressure relatively independent of normal flow fluctuations and falling cylinder pressure.

NOTE Every gas cylinder has a matching connector on the regulator. If the inlet of the regulator does not fit the cylinder outlet, *do not force* the fitting. It may not be the correct one for the situation.

Pressure-Reduction Stages

Controlled pressure reduction requires the use of two-stage pressure reduction. Two stages of reduction constitute the same action in series, with the delivery pressure from one stage becoming the source pressure for the second stage.

Most gas regulators employed for use on high-pressure cylinders are of either the single- or two-stage variety.

Generally, the reduction of pressure in two stages permits a closer control of the delivery pressure over a wider range of inlet pressures.

Installation

1. Before connecting the regulator to the cylinder-valve outlet, be sure the regulator has the proper connection to fit the cylinder valve. If there is some doubt about the connection's being correct, check the maker's catalog for valve-outlet designation and description. Inspect the regulator inlet and cylinder-valve outlet for foreign matter. Remove foreign matter with a clean cloth *except in the case of oxygen*. In the case of oxygen, open the cylinder valve slightly to blow any dirt out of the outlet. A dirty oxygen-regulator inlet can be rinsed clean in fresh tetrachloromethane and blown dry with oil-free nitrogen.

2. With a flat-faced wrench tighten the regulator-inlet connection nut to the cylinder-valve outlet. (Depending on gas service, the regulator inlet may be a right-hand thread or a left-hand thread. Make sure that proper identification of the mating connections has been made.) *Do not force the threads*. Some regulator connections require the use of a flat gasket to provide a leaktight seal between the regulator and valve outlet. In this instance, gaskets are supplied with the regulator and should be replaced when they become worn. When utilizing Teflon® gaskets, do not exert excessive force in tightening the connection or the gasket may force its way into the valve opening and impede the discharge of gas.

3. Close the regulator by releasing the pressure-adjusting screw. Turn counterclockwise until the screw turns freely without tension.

4. Check to see that the needle valve on the regulator outlet is closed.

5. Attach tubing or piping to the regulator-valve outlet. Except for high-pressure regulators a hose end is provided with the regulator. Regulators supplied with Tylok connections accept standard ¼-in outside diameter (OD) copper or stainless steel tubing.

CAUTION Regulators and valves used with oxygen must not come into contact with oil and grease. In case of such contamination do not connect the regulator. This problem must be referred to personnel trained in handling this situation.

Operation

1. Slowly open the cylinder valve until full cylinder pressure is registered on the tank gauge. (In the case of liquefied gases a tank gauge is not usually provided.) It is recommended that the cylinder valve be fully opened to prevent limiting of flow to the regulator, which would result in the failure of the regulator to maintain required delivery pressure.

2. Adjust the delivery pressure to the desired pressure setting by turning the pressure-adjusting screw clockwise and noting the delivery pressure as registered on the delivery-pressure gauge.

3. The flow may now be regulated by proper adjustment of the needle valve.

Shutdown

1. Close the cylinder valve.

2. Relieve all the pressure from the regulator through the needle valve until both gauges register zero.

3. Turn the adjusting screw counterclockwise until the screw turns freely without tension.

4. Close the regulator-outlet needle valve.

Dismantling

1. If the regulator will not be used for a while, store in a clean, dry location, free of corrosive fumes.

2. If the regulator has been used with corrosive or flammable gases, flush with dry nitrogen. This can be done by screwing in the pressure-adjusting screw (clockwise), opening the outlet valve, and directing a stream of dry nitrogen into the regulator inlet by means of a flexible tube or rubber hose. After flushing, turn out the adjusting screw and close the outlet valve.

3. Cap or seal the regulator inlet or simply store it in its original plastic bag; this will prevent dirt from clogging the regulator inlet and will extend the life of the regulator.

Troubleshooting

Regulators should be checked periodically to ensure proper and safe operation. This periodic check will vary, depending on gas service and usage.

Regulators in noncorrosive gas service such as nitrogen, hydrogen, and helium require relatively little maintenance, and a quick check on a monthly basis is usually adequate. Regulators in corrosive-gas service such as hydrogen chloride, chlorine, and hydrogen sulfide require considerably more checking. Once a week is recommended.

The procedure for checking out any regulator is as follows:

1. Be sure gauges read zero when all pressure is drained from the system.

2. Open the cylinder valve and turn the adjusting screw counterclockwise; the high-pressure gauge should read the cylinder pressure.

3. Close the regulator-outlet needle valve and wait 5 to 10 min; the delivery-pressure gauge should not indicate a pressure increase. A pressure increase would indicate leakage across the internal valve system.

4. Next, turn the adjusting screw clockwise until a nominal delivery pressure is indicated. Inability to attain a proper delivery-pressure setting or abnormal adjustment of the screw indicates improper operation, which may be attributed to blockage of the gas passage or a leak in the low-pressure side of the regulator. Continued wear on a regulator valve-and-seat assembly will cause a rise above a set delivery pressure, termed *crawl.* A regulator exhibiting crawl should not be used.

5. Close cylinder valve and observe pressure on both inlet and delivery side of the regulator after 5 or 10 min. A drop in the pressure reading after this period of time may indicate a leak in the system, possibly at the inlet or through the needle valve, safety devices, or diaphragm.

6. An excessive fall in delivery pressure under operating conditions and normal flow indicates an internal blockage.

Any deviation from the normal in the preceding checkout will require servicing by reputable repair personnel.

WARNING A regulator, valve, or other equipment that has been used with another gas should never be used with oxygen. A regulator or control should never be used on more than one gas unless the user is fully familiar with the properties of the gases involved or has obtained assurance from the gas supplier that the interchange is permissible and there is no safety hazard.

Repair

When a regulator shows signs of wear it should be serviced only by reputable repair personnel. Detailed drawings of all regulators and recommended parts lists are available for those equipped to do their own repairs.

A complete overhaul for regulators in noncorrosive-gas service is recommended once a year, and for regulators in corrosive-gas service every 3 to 6 months.

NOTE Regulators in corrosive-gas service (hydrogen chloride, chlorine, etc.) which are used only intermittently should be adequately flushed with dry nitrogen and stored in a dry area at room temperature when not in use to prevent excessive corrosion of the metal parts.

Gauges

Since the performance of pumps will be measured by gauges, a brief comment on a few of the most popular types of gauges is needed. The gauge must be appropriately placed. If it is placed near the pump, it measures the vacuum at this location. If the system is large or has restrictions, the gauge reading may not be an accurate measure of the test area. If the gauge is placed in the test area (as it should be) and there are restrictions in the system, the pump performance may be unjustly judged as too slow. (Refer to Chap. 5, "Pressure and Vacuum.")

All the gauges except the McLeod measure the total pressure exerted by both gases and vapors, and different gases give different readings at the same pressure. Since all mechanical vacuum pumps are tested with the McLeod gauge, because it is a primary standard, any other type of gauge will indicate a higher pressure because it registers both gas and vapor. The exact reading of the gauge will depend, therefore, upon the gases and vapors present and the calibration of the gauge.

Manual Flow Controls

Where intermittent flow control is needed and an operator will be present at all times, a manual type of flow control may be used. This type of control is simply a valve which is operated manually to deliver the proper amount of gas. Fine flow control can be obtained, but it must be remembered that dangerous pressures can build up in a closed system or in one that becomes plugged, since there is no provision for automatic prevention of excessive pressures.

Safety Devices

It is necessary to provide further supplementary safety devices to prevent overpressurizing of lines and to prevent suckback of materials into the cylinder controls—possibly into the cylinder itself. Aside from the possibility of causing rapid corrosion, the reaction of a gas with material that has been sucked back may be violent enough to cause extensive equipment and cylinder damage.

Traps and Check Valves

The danger of suckback can be eliminated by providing a trap (Fig. 4-9) which will hold all material that can possibly be sucked back, or by using a check valve or suitable vacuum break. Check valves prevent the return

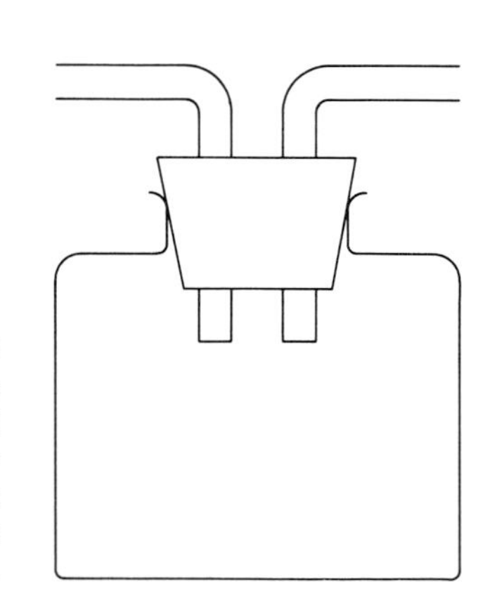

FIGURE 4-9
Safety bottle used to trap materials that are sucked back. Absolutely necessary in protecting vacuum systems and filtrates from contamination.

flow of gas and thus keep foreign matter out of gas lines, regulators, and cylinders ahead of the valve. The valves are springloaded.

Quick Couplers

Quick couplers permit regulators, needle valves, and other components of a gas system to be connected and interconnected to cylinder outlets quickly and safely without the use of wrenches.

Safety Relief Valves

Pressure increases due to uncontrolled reactions or unexpected surges of pressure can be relieved by means of a safety relief device installed in the gas line.

For experiments conducted in glassware, such a pressure-relief device can be improvised by using a U tube filled with mercury (or other inert liquid), with one end attached by means of a T tube to the gas line, and with the other end free to exhaust into an open flask which will contain the mercury in case of overpressure (Fig. 4-10).

Flowmeters

Flowmeters (Fig. 4-11) are used in fluid systems to indicate the rate of flow of the fluid by registering the scale graduation at the center of the spherical float. Flowmeters do not control the rate of flow of the fluid unless they are specifically equipped with control valves or flow controllers. See the section on flowmeters in Chap. 37, "Gas Chromatography," for calibration procedure.

Flowmeters are designed for specific gases and for varying amounts of flow. The center of the float is the register of the fluid flow. The higher

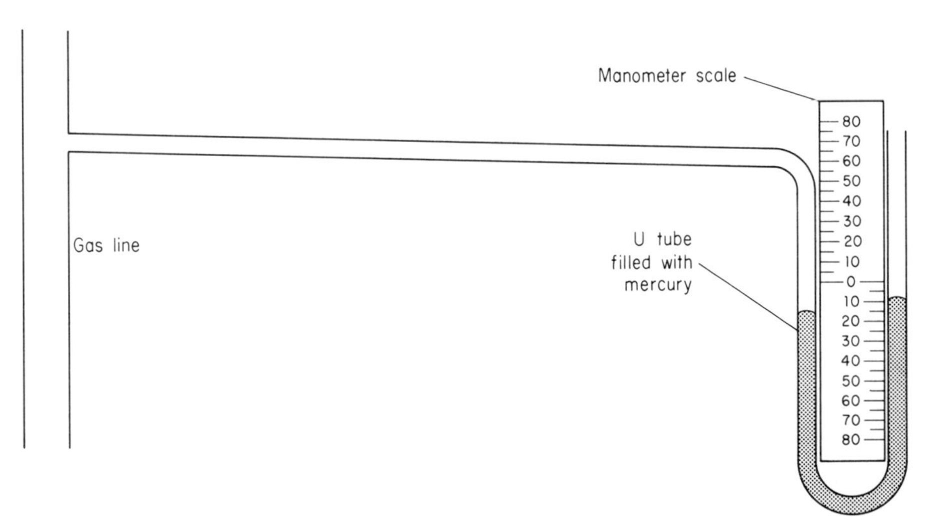

FIGURE 4-10
Pressure-relief U tube.

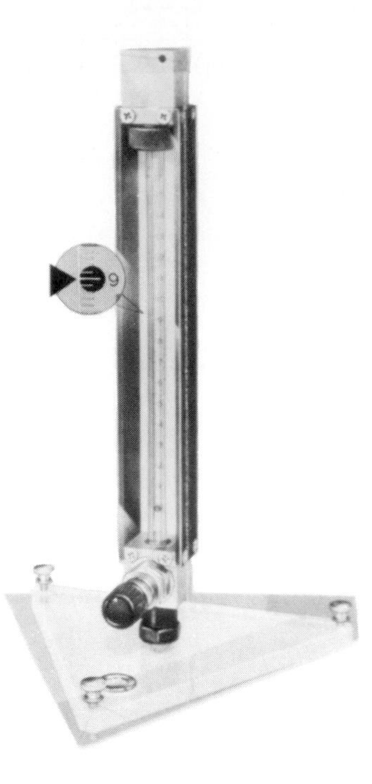

FIGURE 4-11
Gas or liquid flowmeter.

the float rises, the greater the flow rate. Parallax can affect the accuracy and reproducibility of measurements.

Gas-Leak Detectors

Gas leaks in systems can cause problems in flow control and can totally invalidate analytical results, especially in gas chromatography. The gas-leak detector enables you to pinpoint gas leaks easily and quickly. Some types feature meter readouts; others emit an audio signal when the leak is located. The gas-leak detector can detect helium which is leaking at the rate of 1×10^{-6} mL/s, a leak too small to bubble, yet one which could be really critical.

Gas-Washing Bottles

The flow of gases can be monitored by means of a gas-washing bottle which contains a liquid through which the gas is passed (Fig. 4-12). The liquid must be inert to the gas, and the gas must not be soluble in the liquid. Solubility of common laboratory gases in water is shown in Table 4-3. The rate of flow of the gas can be measured and adjusted by counting the bubbles per minute.

Procedure

Connect the gas-washing bottle to be used as a gas-flow monitor as shown in Fig. 4-13.

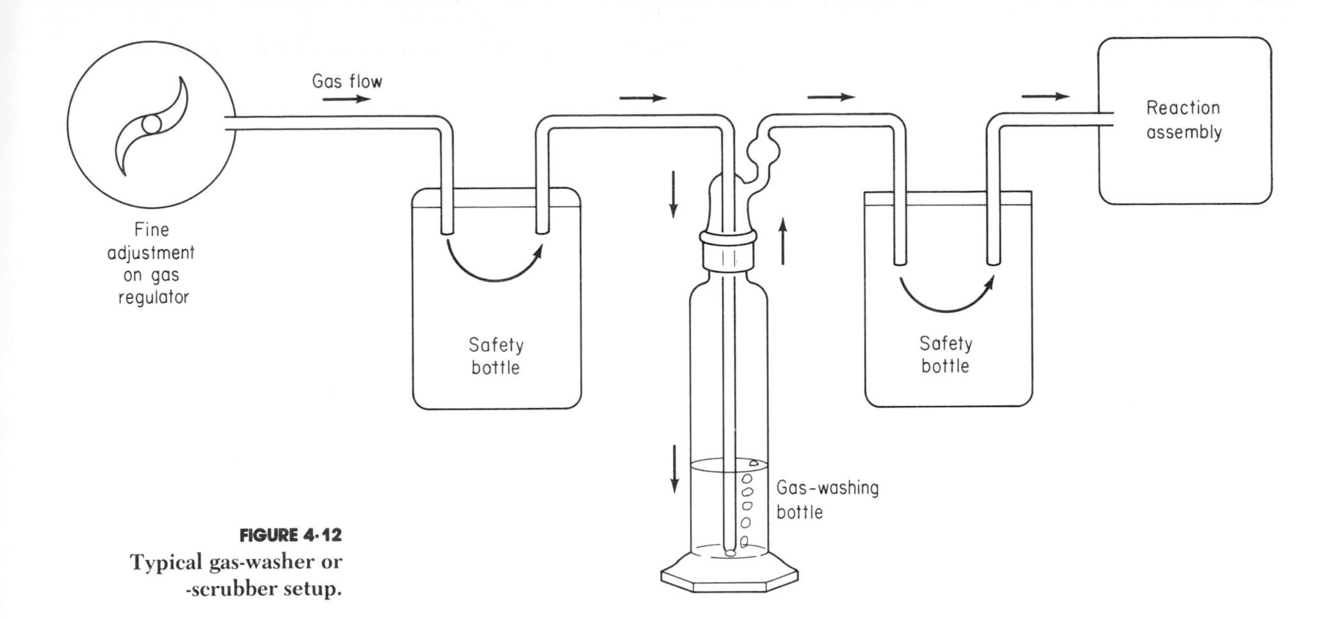

FIGURE 4·12
Typical gas-washer or -scrubber setup.

TABLE 4-3
Solubility of Common Laboratory Gases in Water

Class of solubility at 25°C	*Formula of gas*	*Approximate solubility, L/100 L H_2O*
Extremely soluble	HCl	50,000
	NH_3	75,000
	SO_2	4,000
Moderately soluble	Cl_2	250
	CO_2	100
	H_2S	250
Very slightly soluble	H_2	2
	O_2	3
	N_2	1+
	CO	2

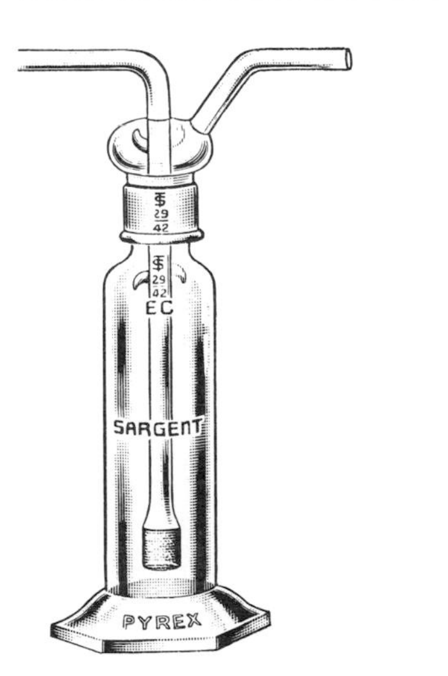

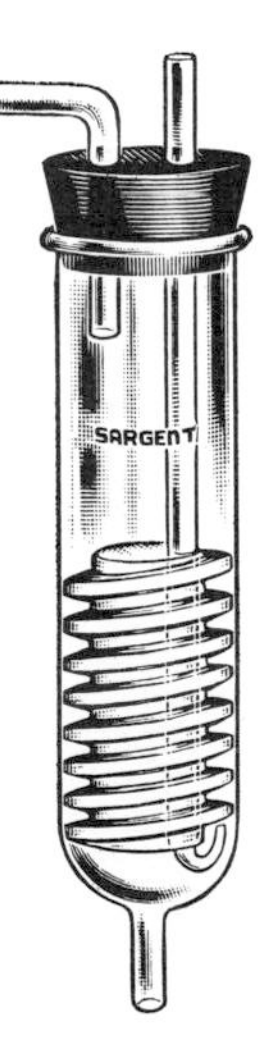

FIGURE 4·13
Two types of gas-washing bottle.

CAUTION Always insert a safety trap between the gas-washing bottle and the pressure regulator. Without the safety bottle or trap the reaction liquid could back up into the wash bottle and contaminate the regulator.

COLLECTING AND MEASURING GASES

Apparatus

Gases evolved in a reaction can be collected in an inverted buret, previously filled with an inert liquid (see Fig. 4-14). The evolved gas cannot be soluble or react with the liquid. As the gas displaces the liquid, the rate of evolution can be calculated from the volume of gas and the time.

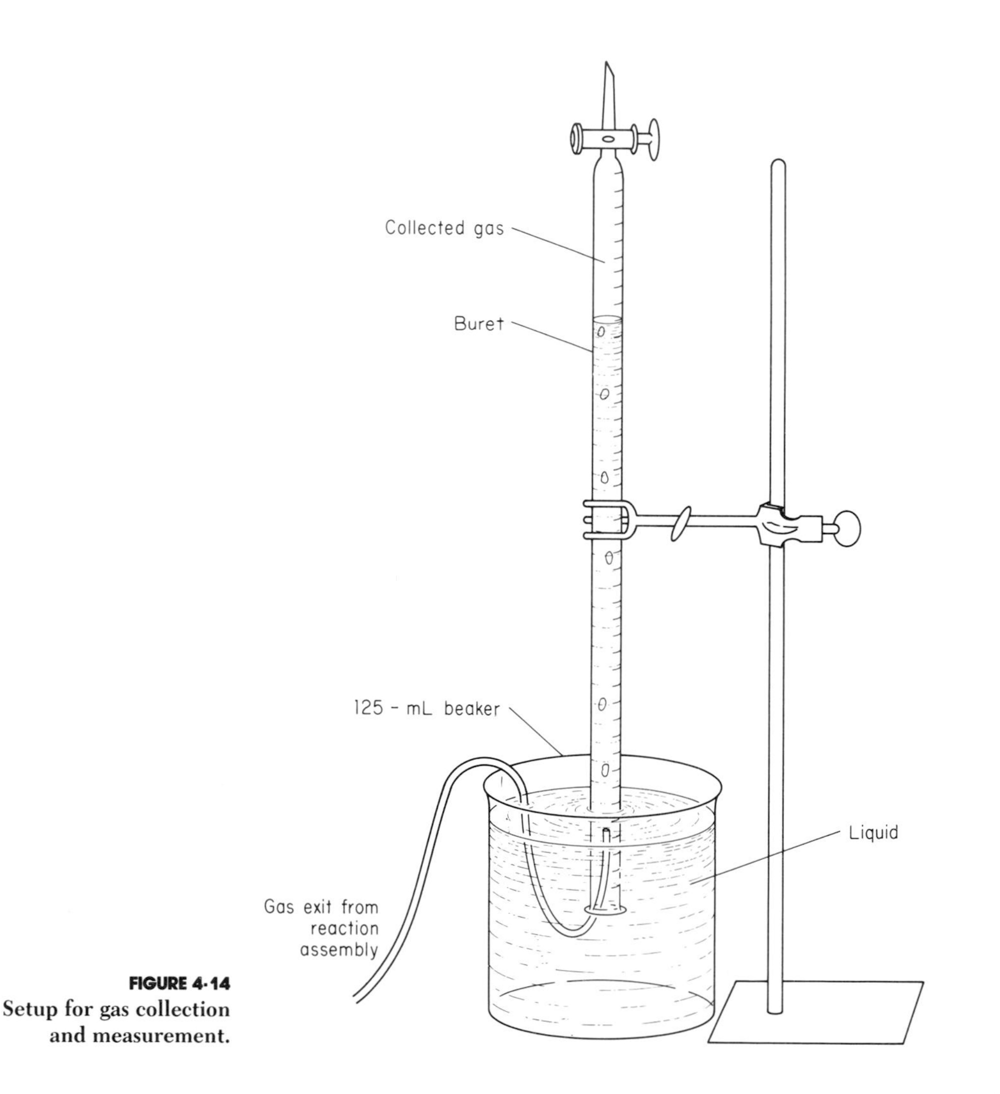

FIGURE 4-14
Setup for gas collection and measurement.

Gases have greater compressibility and greater thermal expansion than liquids and solids. Gas volumes depend sensitively on the pressure and temperature.

DRYING AND HUMIDIFYING GASES

If a dry gas is needed for the reaction assembly, concentrated sulfuric acid can be used as the liquid in the gas bubbler. The sulfuric acid will dry the gas being bubbled through it (refer to sections on drying, Chap. 15).

Gases can be dried by passing them through a drying tower or tube packed with an absorbent, such as Drierite®, $CaCl_2$, and so on. (See Fig. 4-15.)

If the gas is to be humidified (moistened), water is used in the gas bubbler. The amount of moisture absorbed by the gas depends on the rate of flow, temperature of the gas and liquid, and size of the bubbles.

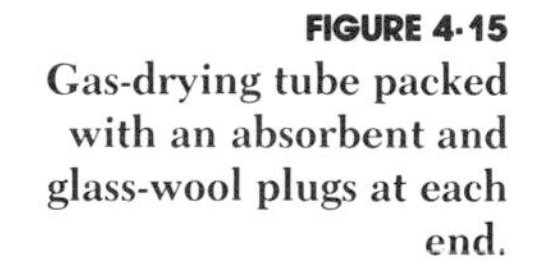

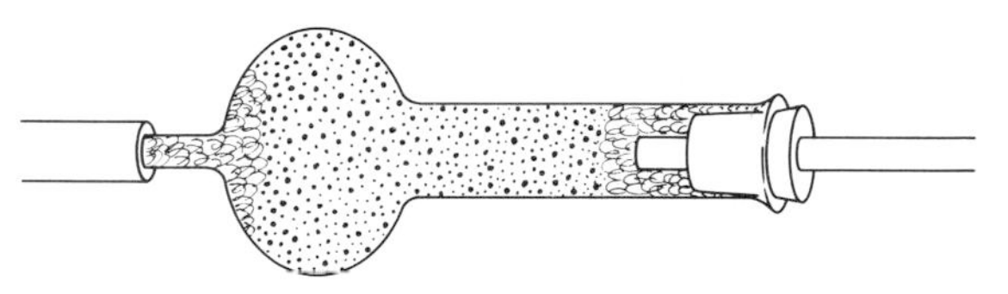

FIGURE 4-15
Gas-drying tube packed with an absorbent and glass-wool plugs at each end.

Types of Gas Dispersers

Gas dispersers range from a simple piece of glass tubing immersed beneath the liquid to more sophisticated and efficient types. The choice of gas disperser depends upon the reaction conditions, the reactants, the gas, the rate of flow of the gas, and the rate of reaction desired. The smaller the size of the gas bubbles, the greater the area of contact between the gas and the liquid.

A *straight piece of glass tubing* (Fig. 4-16*a*) provides the largest bubbles and shows the least tendency to clog.

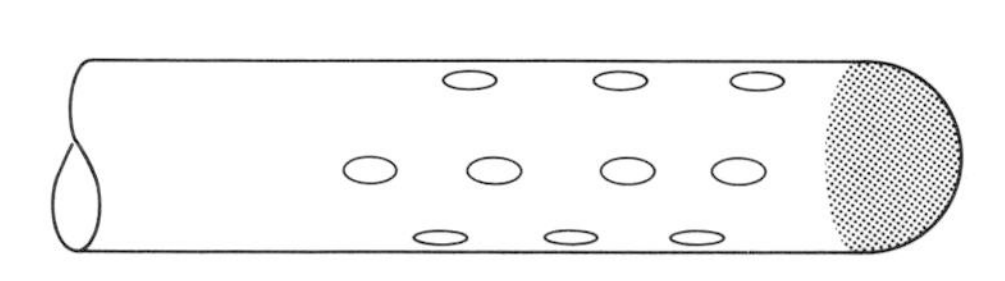

FIGURE 4-16
(*a*) Straight-glass-tube type of gas disperser. (*b*) Gas disperser consisting of closed-end tube with holes in side.

A piece of *tubing sealed at one end and with small holes on the sides* (Fig. 4-16*b*) provides smaller bubbles and greater dispersion than straight tubing.

A *fritted disk* (Fig. 4-17) gives very small gas bubbles, depending on its porosity. It affords the greatest gas contact with the liquid but also has the greatest tendency to clog.

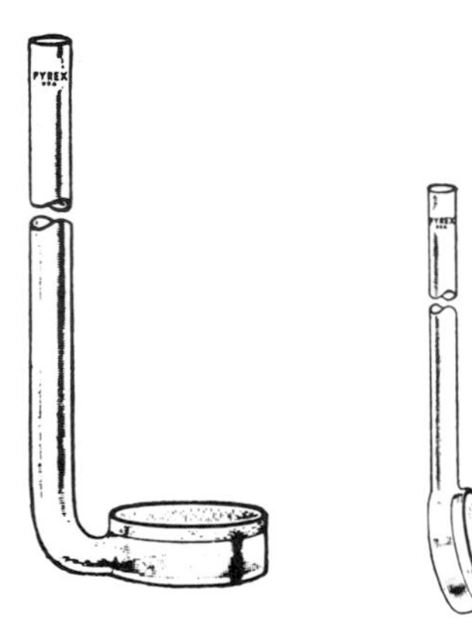

FIGURE 4-17
Fritted-glass gas dispersers.

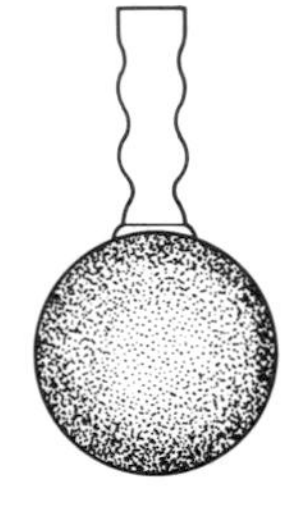

FIGURE 4-18
Gas-diffusing stone.

Gas-Diffusing Stones

Porous stones (Fig. 4-18) are made of inert crystalline fused aluminum oxide, and they come in cylindrical or spherical shapes fitted with a tubing fitting.

Filter Candle

A filter candle (Fig. 4-19) is a porous ceramic device that permits gas passage. It disperses gases into extremely fine bubbles for maximum contact

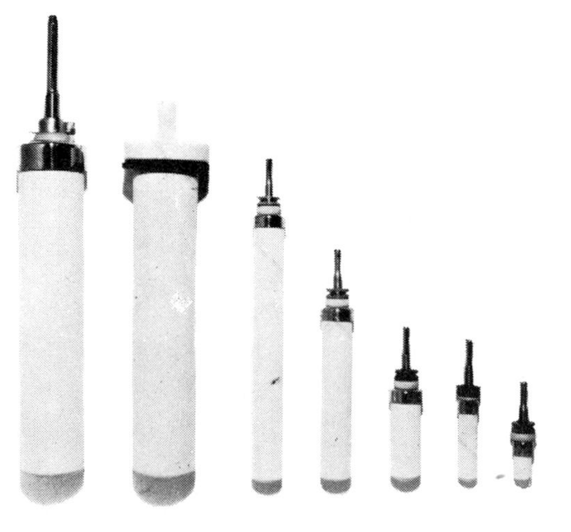

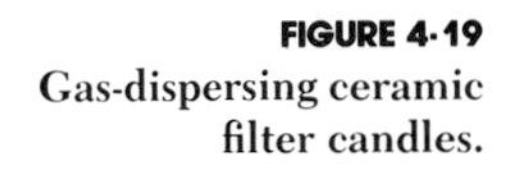

FIGURE 4-19
Gas-dispersing ceramic filter candles.

with liquids, and is also used to filter gases and remove entrained solid particles.

CONTROL OF GASES EVOLVED IN A REACTION

In many reactions, noxious or corrosive gases are evolved in a chemical reaction. They should not be allowed to escape in the laboratory because of health reasons and because the gases will corrode the equipment. Such reactions should be performed in a well-ventilated hood.

Precautions

1. Always use a hood when working with toxic or irritating chemicals.

2. The major source of accidents is spillage of corrosive chemicals on the clothing and skin. Immediately flood with excessive amounts of water and then consult the medical service.

3. Anything or any operation that must be forced should be examined very carefully. The application of excessive force to make something work can lead to accidents and broken equipment. *Always think! Always be on guard!*

Alternative Methods

Method 1

When a hood is not available and noxious or corrosive fumes are emitted from a reaction flask or from a concentration-solution evaporation, an *emergency hood* composed of a glass funnel, rubber tubing, an aspirator, and a water pump should be assembled as shown in Fig. 4-20.

Method 2

Pass the gases through a gas-washing bottle which contains an absorbent for the gases. Use a basic absorbent (NaOH or NH_4OH solution) for acid gases. Use an acid absorbent (dilute HCl or H_2SO_4) for basic gases.

Method 3

Aspirate the opening from which the gases are exiting the reaction assembly with a T or Y tube which is attached to a water aspirator (Fig. 4-21). One end should be open to the atmosphere to maintain atmospheric pressure in the assembly.

Method 4

Pass the gas over an absorbing solution (Fig. 4-22). Gaseous HCl is very soluble in water. Suspend a funnel just over the surface of a container of water. When absorbing gases that are very soluble in water, do not immerse the funnel because there may be so much gas absorbed or dissolved that the water may be drawn back into the reaction assembly.

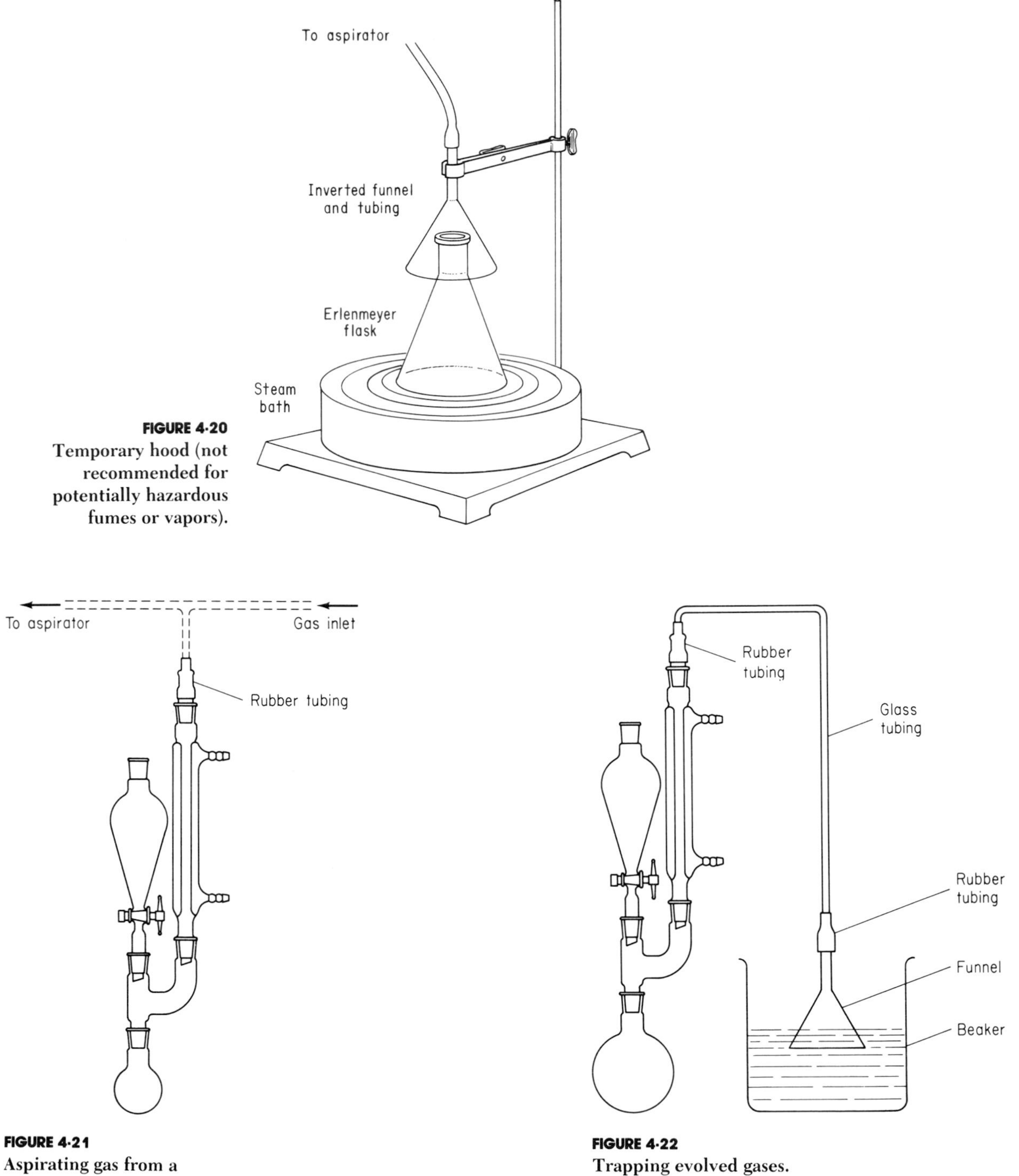

FIGURE 4-20
Temporary hood (not recommended for potentially hazardous fumes or vapors).

FIGURE 4-21
Aspirating gas from a reaction assembly.

FIGURE 4-22
Trapping evolved gases.

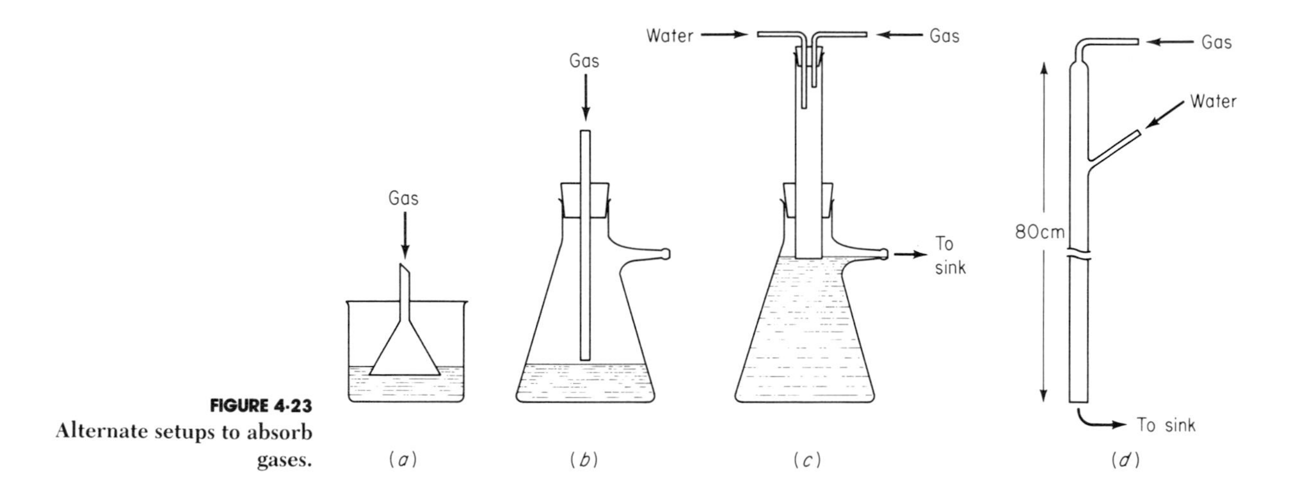

FIGURE 4-23
Alternate setups to absorb gases.

Alternative setups for absorbing gases are shown in Fig. 4-23.

To absorb a gas which is not vigorously soluble in water, the funnel may be immersed beneath the liquid surface (Fig. 4-23*a*).

To absorb a moderately soluble gas, the setup shown in Fig. 4-23*b* may be used, with the tube ending *above* the liquid.

To absorb very large volumes of gas, or rapidly evolved gas, the setups shown in Fig. 4-23*c* and *d* may be used. In the former, the constant input of water overflows at a constant level from the outlet of the suction flask. Water level is above lower end of a large-bore tube to prevent escape of the gas into the atmosphere. In the latter, the flowing water absorbs the gas, exiting to the drain. A tube about 80 cm long and about 25 mm in diameter is generally a convenient size to use in this setup.

5
PRESSURE AND VACUUM

INTRODUCTION

Pressure is defined as force per unit area. Thus the appropriate units would be pounds per square inch (lb/in^2 or psi) and, in the SI system, the *pascal (Pa)* or *newton per square meter* ($N \cdot m^{-2}$). The SI unit of force equals the SI unit of mass times the SI unit of acceleration ($kg \cdot m/s^2$) and is called the *newton*. Pressure therefore would be expressed in the SI system as newtons per unit of area (m^2).

A relative expression of pressure is the *atmosphere (atm)* which at sea level will cause a column of mercury to rise 750 mm in an evacuated tube (see the section on the mercury barometer) and is equivalent to 14.696 psi and 101,325 Pa. These and other atmosphere equivalents and conversions are given in Tables 5-1A and B.

TABLE 5-1A
Atmosphere Equivalents

1 atm = 14.696 lb/in^2 (psi)*
= 29.921 in Hg†
= 76 cm Hg
= 760 mm Hg (exact definition)
= 760 torr
= 33.899 ft H_2O
= 10.33 m
= 1.01325×10^5 Pa
= 101.325 kPa (exact definition)
= 1.0133 bars
= 1.01325×10^6 dyn/cm^2

Other pressure equivalents are:

1 torr = 1 mm Hg
= 0.03937 in Hg
= 0.53524 in H_2O
= 0.0013 bar
= 1000 μm Hg
= 1.000×10^3 μm Hg

* Here, lb really stands for pound-force, a force or weight unit rather than a mass unit, and the measurement is sometimes written as lbf/in^2.
† All measurements assume mercury (Hg) to have a density of 13.5951 g/cm^3.

TABLE 5-1B
Conversion Factors for Pressure Units*

Units	*Pa (SI)*	*bar*	*atm*	*cm Hg*	*dyn/cm³ (cgs)*	*kg/cm²*	*psi*	*torr*
1 Pa (SI)	1	10^{-6}	9.86923×10^{-6}	7.50064×10^{-4}	10	1.01972×10^{-5}	1.45038×10^{-4}	7.50064×10^{-3}
1 bar	10^{5}	1	0.986923	75.0064	10^{6}	1.01972	14.5038	7.50064×10^{2}
1 atm†	1.01325×10^{5}	1.01325	1	76.0002	1.01325×10^{6}	1.03323	14.6960	7.60002×10^{2}
1 cm Hg (0%C)	1.33322×10^{3}	1.33322×10^{-2}	1.31579×10^{-2}	1	1.33322×10^{4}	1.35951×10^{-2}	0.193367	10
1 dyn/cm² (cgs)	10^{-1}	10^{-6}	9.86923×10^{-7}	7.50064×10^{-5}	1	1.01972×10^{-6}	1.45038×10^{-5}	7.50064×10^{-4}
1 kg/cm²	9.80665×10^{4}	0.980665	0.967841	73.5561	9.80665×10^{5}	1	14.2233	7.35561×10^{2}
1 psi	6.89476×10^{-3}	6.89476×10^{-2}	6.80460×10^{-2}	5.17151	6.89476×10^{4}	7.03070×10^{-2}	1	51.7151
1 torr (mm Hg at °C)‡	1.33322×10^{2}	1.33322×10^{-3}	1.31579×10^{-3}	10^{-1}	1.33322×10	1.35951×10^{-3}	1.93367×10^{-2}	1

* Conversion to SI units taken from ASTM Metric Practice Guide, U.S. Department of Commerce, *Natl. Bur. Standards Handbook* **102,** 39 (Mar. 10, 1967).
† 1 atm (standard atmosphere) = 101.325 kPa.
‡ 1 torr = 101.325/760 kPa.

MEASURING ATMOSPHERIC PRESSURE

Mercury Barometer

Atmospheric pressure is measured by a barometer (Figs. 5-1 and 5-2) which actually measures the weight of the air above the instrument. In a mercury barometer, the mercury will fall, creating a vacuum above the mercury, until the weight of the column of mercury is exactly counterbalanced by the pressure of the surrounding atmosphere. A commercial mercury barometer (Fig. 5-2) is accurate for general laboratory work. Generally a thermometer is attached as well as a vernier reading to 0.1 mm. (Refer to vernier procedures in the section on Chainomatic® balances in Chap. 14, "The Balance.")

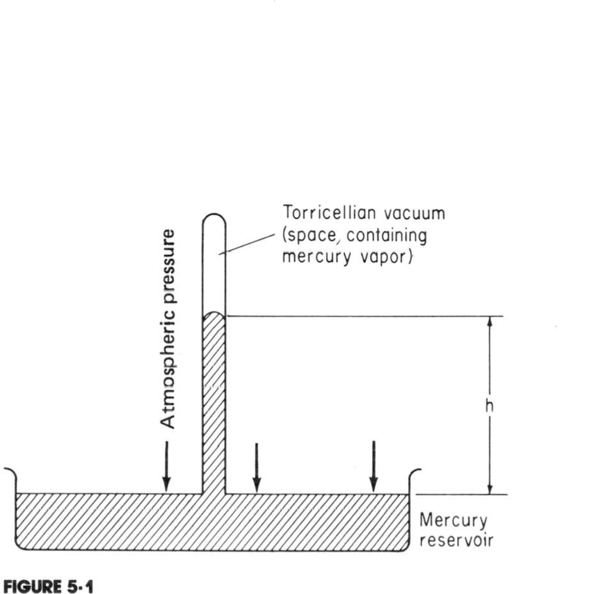

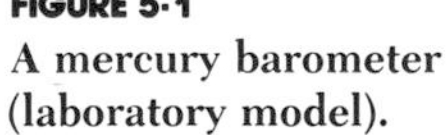

FIGURE 5-1
A mercury barometer (laboratory model).

FIGURE 5-2
Commercial mercury barometer.

Corrections for the Mercury Barometer

The barometer should be corrected for temperature when you are doing precision work because there are differences in the expansion of the metallic mercury and the brass scale indicator on the barometer. To make these corrections for barometer readings, subtract the appropriate correction from the barometric reading at the required temperature. (See Table 5-2.)

TABLE 5-2
Temperature Corrections for Barometer Readings

Temperature, °C	Barometer reading (0°C), mm						
	640	660	680	700	720	740	760
16	1.7	1.7	1.8	1.8	1.9	1.9	2.0
18	1.9	1.9	2.0	2.1	2.1	2.2	2.2
20	2.1	2.2	2.2	2.3	2.3	2.4	2.5
22	2.3	2.4	2.4	2.5	2.6	2.7	2.7
24	2.5	2.6	2.7	2.7	2.8	2.9	3.0
26	2.7	2.8	2.9	3.0	3.0	3.1	3.2
28	2.9	3.0	3.1	3.2	3.3	3.4	3.5
30	3.1	3.2	3.3	3.4	3.5	3.6	3.7

Aneroid Barometer

Atmospheric pressure can also be measured by a more convenient device called the *aneroid barometer* (Fig. 5-3), which consists of a partially evacuated cylinder made of very thin spring metal, corrugated in concentric circles, and designed to cave in at the center. Inside the bellows is a spring which prevents the complete collapse of the bellows. As atmospheric pressure increases, pushing in the bellows, lever and linkage mechanisms amplify the movement to the indicating point on a calibrated scale. When the pressure decreases on the spring, the bellows expands, and again the movement is transmitted to the pointer. This type of barometer is not as accurate as the mercury barometer, which should be used for precision work.

FIGURE 5-3
The aneroid barometer indicates atmospheric pressures in inches of mercury and is used only in weather forecasting.

ABSOLUTE PRESSURE AND GAUGE PRESSURE

Pressure intensities are related to absolute zero pressure or local atmospheric pressure:

1. Gauge pressure: psig
2. Absolute pressure: psia
3. Inches of water, inches of mercury, millimeters of mercury, torr, or pascals for vacuum scales below atmospheric pressure

Zero absolute pressure means the complete absence of any and all molecules, and therefore corresponds to a complete absence of pressure. It is at present a purely theoretical concept.

Gauge pressure, which is normally used to designate pressures *above* atmospheric pressure, does not include atmospheric pressure. At atmospheric pressure, the gauge reads zero. Thus, gauge pressures and absolute pressures differ from each other by the zero-point location. (See Fig. 5-4 and Table 5-3.) To convert a gauge pressure to the equivalent absolute pressure, add the atmospheric pressure.

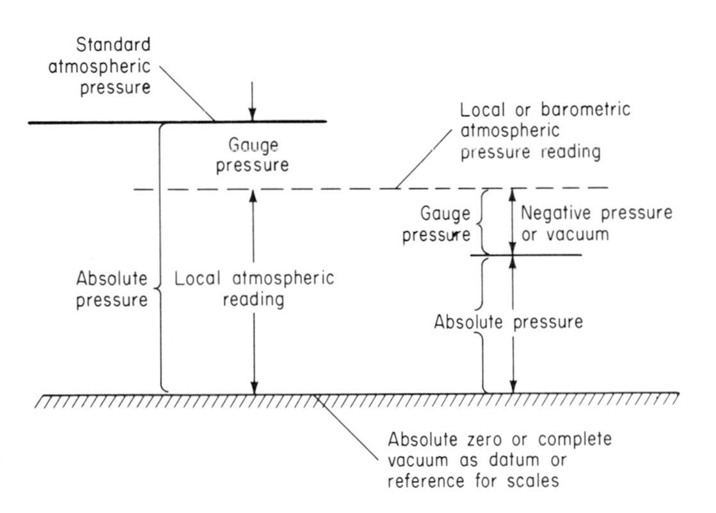

FIGURE 5-4 Scales for pressure measurements.

TABLE 5-3 Relation between Gauge Pressure, Absolute Pressure, and Pressure Expressed in Atmospheres

Gauge pressure, psig	*Absolute pressure, psia*	*Atmospheres, atm*
−14.70	0	0
0	14.70	1
10.00	24.70	1.67
14.70	29.40	2

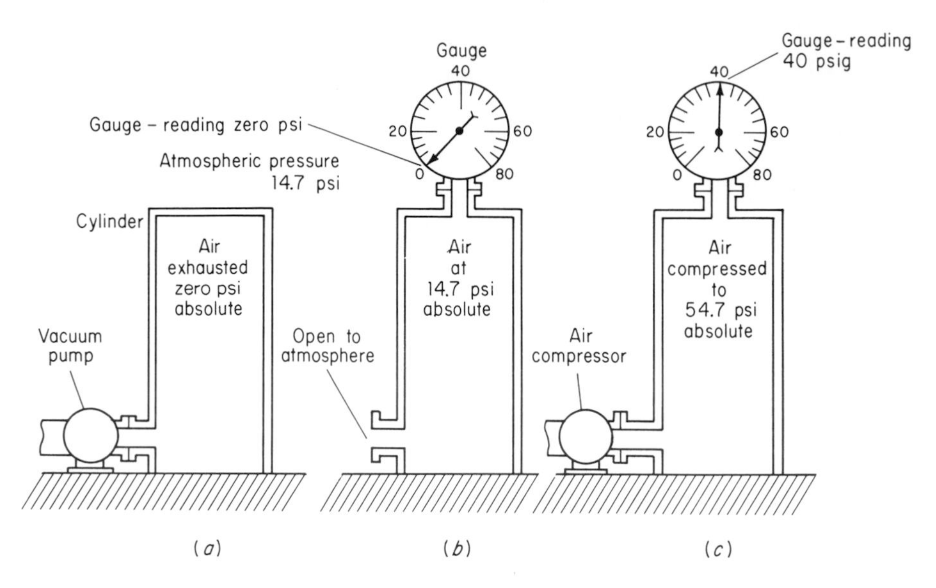

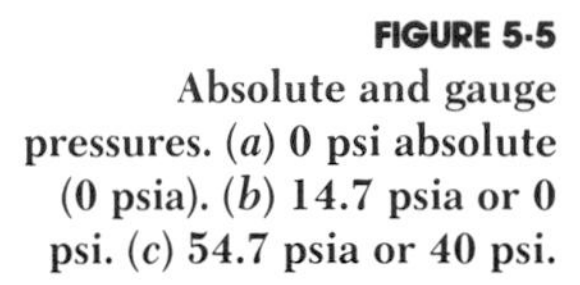

FIGURE 5-5
Absolute and gauge pressures. (*a*) 0 psi absolute (0 psia). (*b*) 14.7 psia or 0 psi. (*c*) 54.7 psia or 40 psi.

Absolute pressure = gauge pressure + local atmospheric pressure

and vice versa:

Gauge pressure = absolute pressure − local atmospheric pressure

Relationship of absolute and gauge pressure is shown in Fig. 5-5.

GAS LAWS

Boyle's law states that the pressure (P) of any ideal gas times its volume (V) will give a constant value, if the temperature (T) is held constant.

$PV = \text{constant} \qquad (T \text{ constant})$

Charles' law states that the volume (V) of a given mass of gas will be proportional to its temperature (T) expressed in Kelvin.

$V \propto T \qquad (T \text{ in Kelvin})$

Gay-Lussac's law states that the pressure (P) of any gas will change directly as the temperature (T) changes, if the volume is held constant.

$P \propto T \qquad (V \text{ constant})$

Avogadro's law states that equal volumes (V) of gases at the same temperature (T) and pressure (P) contain equal numbers of molecules. If *standard temperature and pressure conditions (STP)* of 0°C (273 K) and 1 atm are used, then Avogadro's law can be restated as: *one mole of any gas at STP occupies 22.4 liters.*

Dalton's law states that the total pressure of a gas mixture is the sum of the pressures from the individual gases under the same conditions.

$$P_{total} = P_1 + P_2 + P_3 + \cdots$$

Graham's law states that the rate (r) of diffusion (or effusion) of gases is inversely proportional to the square root of their molecular weights (m), or their densities (D). *Diffusion* is defined as the mixing of one gas into another by the random motion of the gas molecules. *Effusion* is the process of a gas escaping from a container through a very small opening.

$$\frac{r_1}{r_2} = \sqrt{\frac{m_2}{m_1}} \quad \text{or} \quad \frac{r_1}{r_2} = \sqrt{\frac{D_2}{D_1}}$$

The combined gas law combines Boyle's, Charles', and Gay-Lussac's laws into a useful mathematical relationship for calculating changes in a gaseous substance at some starting conditions (1) to different conditions (2).

$$\frac{P_1V_1}{T_1} = \frac{P_2V_2}{T_2}$$

The ideal gas law is a mathematical application of Avogadro's law which provides a means for calculating the pressure, temperature, volume, or moles of any ideal gas.

$$PV = nRT$$

where P = pressure of the gas in atmosphere units
V = volume of the gas expressed in liters
n = number of moles of the gas
T = temperature of the gas in Kelvin
R = *ideal gas constant:* 0.08206 L·atm/mol·K*

PASCAL'S LAW

Pascal's law states that in a *confined fluid*, any externally applied pressure is transmitted equally in all directions. This principle is utilized in all hydraulic systems to move pistons to do work as well as to measure the pressure applied (see Fig. 5-6). Since the applied pressure is transmitted equally in all directions throughout the confined volume, the location of the pressure-indicating device (or vacuum-measuring device) is a matter of convenience and choice.

---

* Other ideal gas constants are available but require different units for pressure and volume in the calculation. For example: 8.3145 kPa·dm^3/mol·K, 8.3145 J/mol·K, and 1.987 cal/mol·K.

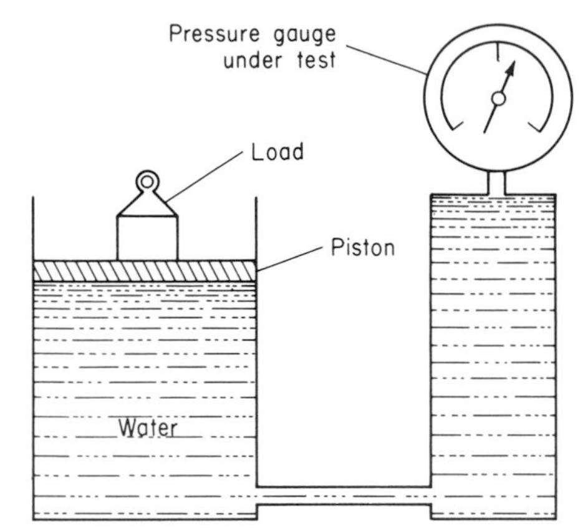

FIGURE 5-6
Pascal's law: transmission of pressure in a closed system.

VACUUM

Vacuum can be considered as space in which there are relatively few molecules. We say "relatively few" because there is no such thing as an absolute vacuum; every substance does exert a definite vapor pressure. Generally speaking, a vacuum is a state of reduced atmospheric pressure, i.e., some point below normal atmospheric pressure.

Graphic Symbols Used in Vacuum Technology

A shorthand method which is used to indicate the interconnections and components of a vacuum system in a flow diagram uses graphic symbols (Fig. 5-7). These symbols can provide a schematic flow diagram of the system, but do not specify the size, shape, or actual physical spatial location of the components. By using these standard uniform symbols, one can communicate accurate information regarding vacuum systems.

SOURCES OF VACUUM IN THE LABORATORY

There are three general kinds of equipment used to produce a vacuum in the laboratory:

1. The water aspirator
2. The mechanical vacuum pump, a rotary pump
3. The vapor-diffusion pump, filled with mercury or special silicone oils

THE WATER ASPIRATOR

The water aspirator is the most commonly used source of moderate vacuum in the laboratory because it is inexpensive, almost infallible, and does not require complicated traps to protect it. In fact, it does provide an easy

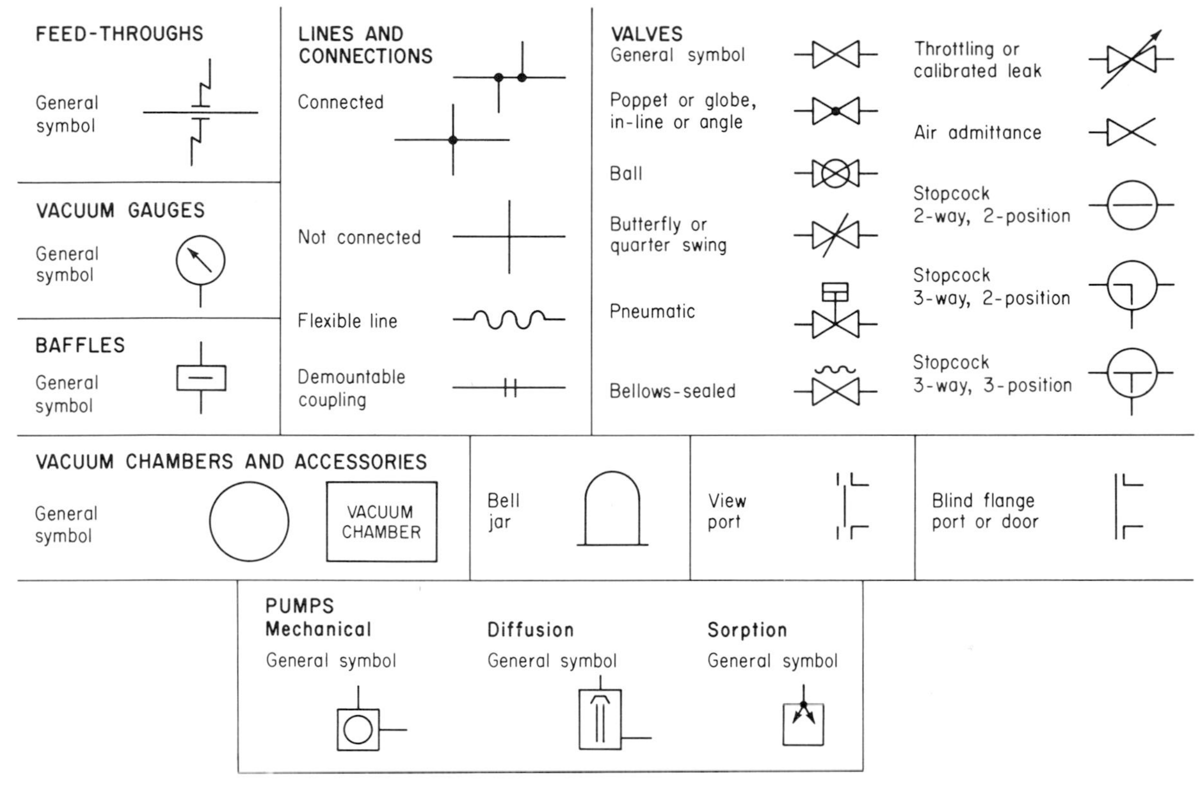

FIGURE 5-7
Symbols used in vacuum technology.

method of disposing of acidic gases and vapors, corrosive fumes, and irritating or nauseating gases, because they are dissolved in the water flow and are eliminated in the drain.

The aspirator works on *Bernoulli's principle,* which states that as the velocity of the fluid is increased, the lateral pressure to the flow is decreased, resulting in the formation of a partial vacuum by the rapidly moving water (Fig. 5-8*a*). The aspirator, however, cannot provide a pressure lower than the vapor pressure of the flowing water, and that pressure is dependent on the temperature of the water (Table 5-4). On a cold day (H_2O @ 10–15°C), the lowest pressure attainable is about 10 torr.

Rules for Using the Aspirator

1. Always insert a bottle trap between the aspirator and the apparatus under vacuum (Fig. 5-8*b*). The water pressure may drop suddenly, and when it does the pressure in the apparatus may become less than that of

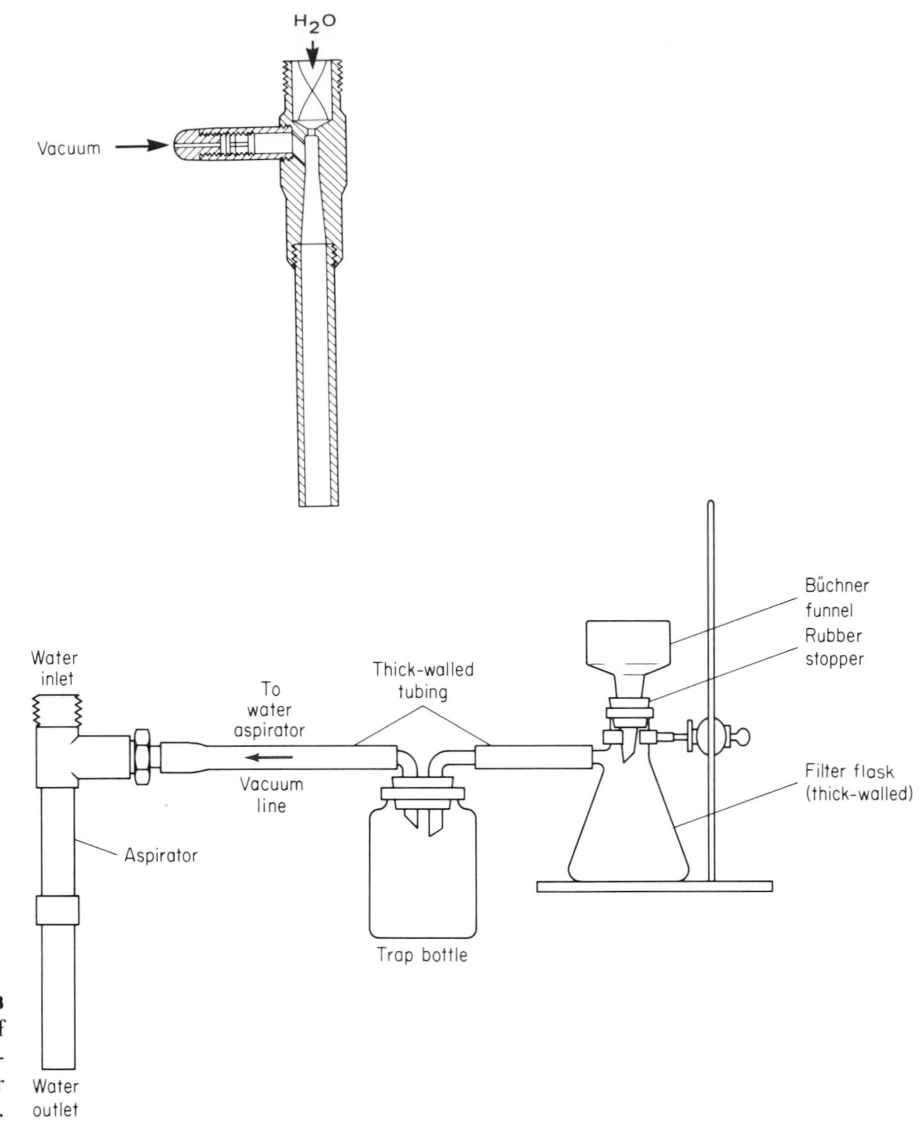

FIGURE 5·8
(*a*) Cutaway section of water aspirator. (*b*) Water-aspirator hookup for vacuum filtration.

the aspirator. Water would be drawn back from the aspirator into the apparatus, causing contamination.

2. Always disconnect the aspirator from the apparatus *before* turning off the water, otherwise the water will be drawn back into the apparatus as described above.

3. If water does back up for any reason, immediately disconnect the tubing from the aspirator as quickly as possible.

TABLE 5-4
The Vapor Pressure of Water of Various Temperatures

Temperature, °C	*Vapor pressure, torr*	*Temperature, °C*	*Vapor pressure, torr*	*Temperature, °C*	*Vapor pressure, torr*
−10 (ice)	2.0	22	19.8	45	71.9
− 5 (ice)	3.0	23	21.1	50	92.5
0	4.6	24	22.4	60	149.4
5	6.5	25	23.8	70	233.7
10	9.2	26	25.2	80	355.1
15	12.8	27	26.7	90	525.8
16	13.6	28	28.3	100	760.0
17	14.5	29	30.0	110	1,074.6
18	15.5	30	31.8	150	3,570.5
19	16.5	35	42.2	200	11,659.2
20	17.5	40	55.3	300	64,432.8
21	18.6				

THE MECHANICAL VACUUM PUMP*

A mechanical pump uses a rotary vane to produce a rough vacuum (approximately 10 torr, which is slightly better than the water aspirator). This device relies on the sweeping action of multiple vanes turning within a cylindrical housing. An electric motor usually provides the driving force. Pumps of this type may include oiling devices for lubricating or may use vanes of self-lubricating material. Large pumps of this type, together with ballast tanks, are used as the basis for central vacuum systems. A cross-sectional diagram of a typical rotary-vane, oil-sealed mechanical pump is shown in Figure 5-9*a*. Rotary-vane pumps are composed of either one or two stages, the number of stages being determined by the degree of vacuum desired. The stages of a two-stage pump are connected in series and mounted on a common shaft in such a manner that one stage backs up the other to obtain the best possible vacuum. Each stage consists of a rotor mounted concentrically on a shaft and located eccentrically in a generally cylindrical stator or ring, together with two movable vanes located diametrically opposite each other in slots of the rotor.

THE DIFFUSION PUMP*

A vapor-diffusion pump is similar in principle to a water aspirator; the steam which entrains the undesired gases consists of a heavy vapor gen-

* Courtesy of Sargent Welch Co., Skokie, Ill.

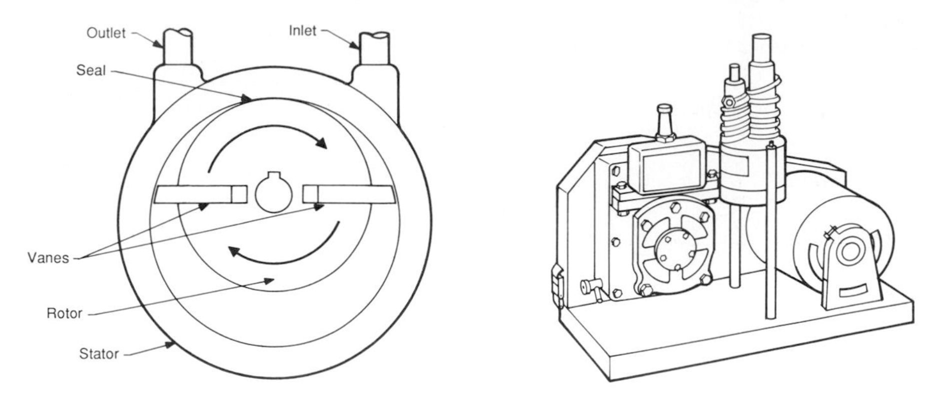

FIGURE 5-9
(*a*) Cross-sectional diagram of a typical mechanical (oil-sealed rotary-vane) vacuum pump.
(*b*) Combined mechanical–oil-diffusion vacuum pump.

erated by evaporation of a pump oil. The pump-oil vapor is condensed after serving this purpose and is returned to the boiling pot, thus comprising a closed cycle. A combined mechanical–oil-diffusion pump is shown in Fig. 5-9*b*. This two-stage mechanical pump is coupled to an all-metal, water-cooled oil-diffusion pump. This type of pump is capable of producing a vacuum in the 1×10^{-6} torr range without using a liquid nitrogen or other cold trap. Originally, most diffusion pumps used mercury, but this process is seldom used now because of possible toxicity hazards.

Diffusion-Pump Oils

Organic oils for diffusion pumps are superior to mercury, which has a vapor pressure of 10^{-4} torr at 0°C, 10^{-2} torr at 50°C, and 5 torr at 150°C. These diffusion-pump oils are esters (E), hydrocarbons (H), polyphenyl ethers (P), or silicones (S), having excellent stability toward heat and extremely low vapor pressures, as shown in Table 5-5. The molecules of these oils, while passing through the space where the gas molecules of the

TABLE 5-5
Characteristics of Diffusion-Pump Oils

Type	*Chemical structure*	*Vapor pressure (20°C), torr*
Apiezon A	H	10^{-6}
Octoil	E	2×10^{-7}
Apiezon B	H	5×10^{-8}
DC 702	S	5×10^{-8}
DC 704	S	1×10^{-8}
Apiezon C	H	5×10^{-9}
Convalex	P	Very low
Santovac 5	P	Very low

system are being evacuated, strike those gas molecules and push them toward the outlet. The gas molecules are then replaced by the vapor molecules from the pump until practically all have been exhausted. Diffusion pumps yield vacuums of 10^{-3} to 10^{-6} torr.

VACUUM PUMPS: USE, CARE, AND MAINTENANCE*

General

Modern high-vacuum pumps are exceptionally high-quality products. As with all precision-made apparatus, certain basic rules concerning their operation and application must be observed if full value and design performance are to be realized. If only wear of the moving parts were considered, most vacuum pumps would last 20 years or more.

A great deal of work and energy are expended each time a high-vacuum pump is turned on. The correct and full voltage must be present at the motor to secure rated performance. Some consideration must be given to the possibility of voltage drops resulting from overtaxed circuits or from using wires of improper size or length.

Proper care must be taken to prevent foreign materials from entering the pump. A small particle of glass, a metal filing, a globule of mercury, or even certain condensable gas vapors will impair the normal life expectancy of any pump. All gases being evacuated will pass through the pump and its oil unless adequate traps are employed. If gases are allowed to condense in the pump oil, not only will the rated ultimate vacuum not be reached, but corrosion of the finely machined parts is likely to occur.

Pump Oil

In most cases, depending on model, pumps are shipped already filled with oil. The extra oil supplied with these models is for "topping" purposes if necessary. All pumps are shipped with protective plastic caps over the intake and exhaust. Do not attempt to operate a pump with the exhaust cap in place, as high internal pressures will be built up. These caps may be either discarded or saved to seal the pump again if it is taken out of service and put into storage.

Vacuum-pump oil serves three vital functions: (1) it seals the internal portion of the pump from the atmosphere; (2) it lubricates the pump; and (3) it cools the pump by conducting heat to the outer housing. Use only approved high-vacuum oil for guaranteed results; this is a pure mineral oil with an extremely low vapor pressure, with the proper flow characteristics and lubricating properties to ensure the very best pump service.

---

* Used with permission of the Precision Scientific Group, Chicago, Ill.

A properly filled pump will have sufficient oil above the internal exhaust valve to seal the pump from the atmosphere and also to damp the sound of the gas as it passes through the valve. Since an overfilled pump may "spit" oil during operation and back the oil up beyond the intake during off cycles, it is better to start with less and add oil gradually to the desired level. A sight gauge is provided to indicate the amount of oil present.

Filling with Oil

When filling the pump for the first time or refilling it with fresh oil, it is important to clear out any oil that may be in the stators. For draining information see the section, When and How to Change Oil. Turn the pump pulley at least two revolutions clockwise (facing the pulley), or turn the pump on and allow it to operate under motor power for a few seconds. There need be no fear of damage since there is always sufficient oil present for lubricating purposes, and this action will clear any oil from the stators into the housing.

CAUTION If oil is added too fast at this point, spitting may occur. When the pump is quiet, alternately open and close the intake to the atmosphere. Each time it is opened, popping may occur. This noise is caused by the larger inrush and exhaust of air. When the pump has been closed for a few seconds, the noise should disappear. If it doesn't, add a little more oil and repeat the above procedure.

Do not overfill the pump. If any pump continually spits oil, it is overfilled. When the pump is shut off, that amount of oil above the exhaust valve will drain into the pump and open the pump to the atmosphere. This feature prevents oil from backing up into a system that is not separated from the pump by a valve. A pump cannot be expected to act as a valve and "hold" vacuum. Pumps are designed with sufficient volume in the stators and trap to accommodate the correct amount of oil above the exhaust valve. A properly filled pump will not have oil rising above the intake connection when it is shut off.

Since oil is continually being "consumed," either by entrainment in the gases leaving the pump or by backstreaming, it will be necessary to add oil from time to time. The best gauge for this is the sound of the pump with regard to the pumping noise discussed above.

Vacuum Connection

The simplest system consists of a pump connected directly to the vessel to be evacuated. The speed of evacuation will be a function of (1) the pump-down curve of the pump, (2) the vessel size and type, and (3) the length, diameter, and bends in the connecting tubing. The different types and methods of assembly are limited only by the imagination.

Basic rules that apply whenever practical are: (1) Choose the correct size of pump for the time cycle desired. (2) The vessel should be nonporous and have a smooth surface. (3) Eliminate condensable vapors with a trap. (4) Keep lines as short, as large in diameter, and as direct as possible. A vacuum plate (Fig. 5-10) is often used.

If the inlet of the vessel is smaller than the inlet of the pump, the line between them should be the same size as the pump inlet, with any reduction being made at the vessel, not at the pump. The biggest deterrent to full pumping speed is the use of small-diameter vacuum lines. Large-diameter vacuum tubing is available that is designed for quick installation and requires only finger tightening. It gives an excellent seal suitable for 10^{-6} torr or better.

NOTE All hoses should be clamped securely to prevent leakage (see Fig. 5-11). When insufficient torque is applied with a medium-sized screwdriver, irregularities in the pipe, hose, or fitting may permit leakage. Excessive torque may damage the hose or break the connection or fitting. Use average torque with a medium-sized screwdriver (about 15 to 30 in-lb).

Pump oil must be protected from contamination. The ultimate vacuum that a pump will achieve is partially related to the vapor pressure of the pump oil (see Fig. 5-12). Vacuum oil has been specifically selected for its

FIGURE 5-10
Vacuum plate for use with bell jars in performing vacuum and pressure manipulations; it is connected to a mechanical pump or water aspirator with tubing attached to the serrated nipple.

FIGURE 5-11
Tubing clamp used for positive attachment of pressure or vacuum tubing.

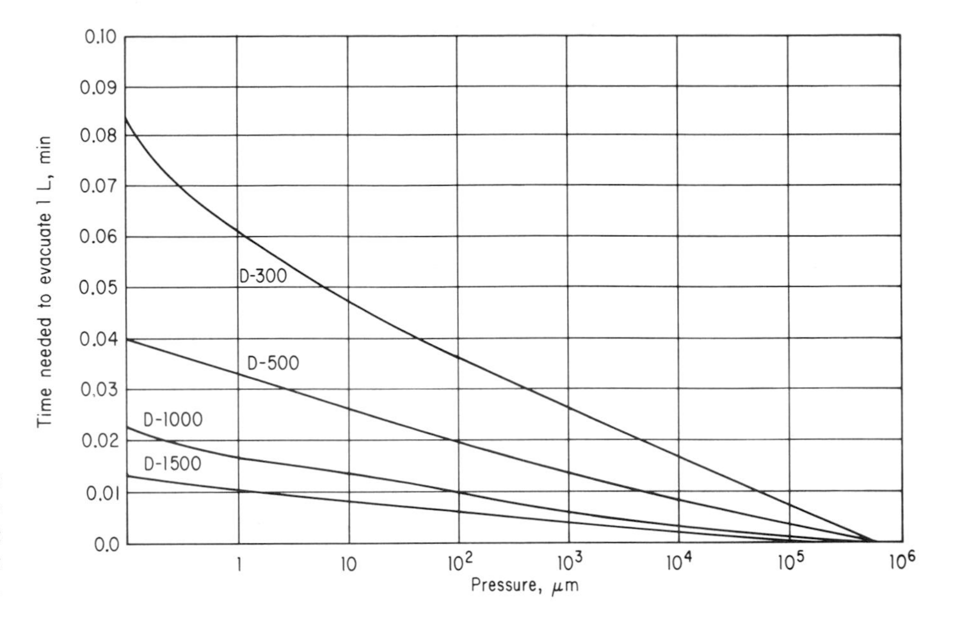

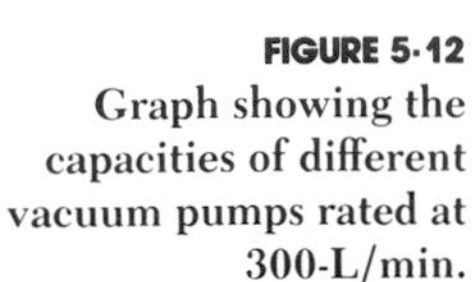

FIGURE 5-12
Graph showing the capacities of different vacuum pumps rated at 300-L/min.

low vapor pressure and its lubricating qualities. Do not substitute other oils. A trap should be installed between the pump and the vessel being evacuated. The simplest traps are cold types, using dry ice or liquid nitrogen. The necessary glassware for a cold trap may be obtained from any number of glass manufacturers. Since traps usually restrict the size of the line, they should be installed immediately in front of the pump. Traps will also prevent *backstreaming* of oil vapors from the pump to the vessel.

High-vacuum pumps should not be used to control the level of vacuum in a system. A single-stage pump should not operate continually above 100 μm (micrometer), and a double-stage pump should not operate continually above 50 μm.

Locating Leaks and Cracks in Vacuum Systems

Leaks and cracks in vacuum systems destroy their effectiveness and can be easily detected and located by the use of a high-frequency Tesla coil (Fig. 5-13). When the electrode is passed over the surface of evacuated vacuum-glassware assemblies, any leakage which is caused by a poor seal, a flaw in the glass, or an imperfect joint is pinpointed by a yellow glow at the point where the electrical discharge enters the system.

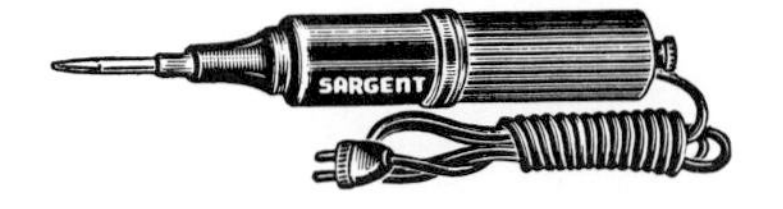

FIGURE 5-13
High-frequency Tesla coil.

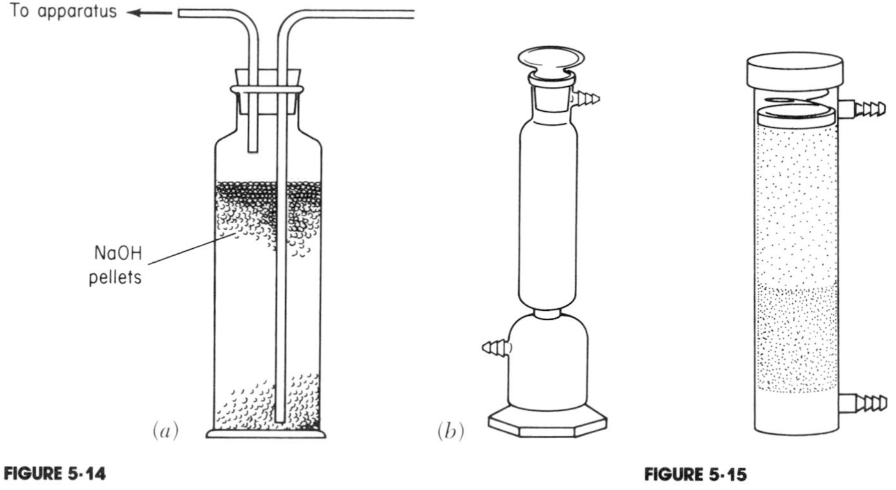

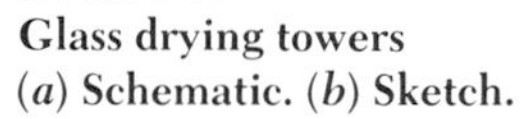

FIGURE 5-14
Glass drying towers
(*a*) Schematic. (*b*) Sketch.

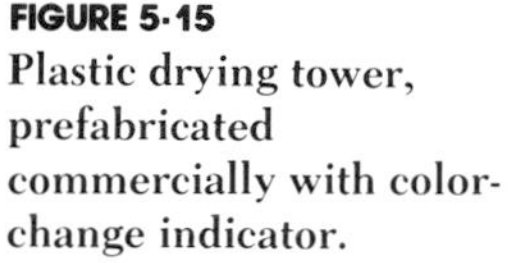

FIGURE 5-15
Plastic drying tower, prefabricated commercially with color-change indicator.

Drying Towers

Drying towers, made of heavy glass and filled with solid drying agents, are used to dry gases, thus preventing water vapor from entering the high-vacuum pump (see Fig. 5-14). The most common drying agents are solid potassium or sodium hydroxide, commercial Drierite® ($CaSO_4$), and Anhydrone [$Mg(ClO_4)_2 \cdot 2H_2O$].

Commercial drying towers, prefabricated and already filled with Drierite®, are also available. They are made of cylindrical acrylic plastic tubing fitted with plastic inlet and outlet nipples for rubber-hose connection (Fig. 5-15). Drierite's color changes from blue to pink when its moisture-absorbing capacity is exhausted. Drierite® can then be regenerated by removing it from the tower and heating it at 225°C for 2 h.

Electric Connection

The power requirements for each pump motor are listed on the motor nameplate. Average electrical specifications are given in Table 5-6. The smaller pumps (those with motors rated at less than 1 hp) are supplied with a switch, cord, and plug according to the electrical characteristics ordered. Motors rated at 1 hp and up, while sometimes capable of operation on 120V, should not be operated on less than 220V when used with a vacuum pump. The starting torque needed by any vacuum pump under load is quite high, and operation can be seriously impaired by low operating voltages or low ambient temperatures. When possible, it is suggested that a vacuum pump be connected to a separately fused circuit, free of other devices that may occasionally start simultaneously and take

TABLE 5-6 Average Electrical Specifications

Motor hp	*Operating voltage*	*Full load, A*	*Locked-rotor current*	*Minimum voltage*	*Recommended fuse*	*Recommended wire gauge*
⅓	120	6.8	48.8	108	25	Line cord
	240	3.4	24.4	216	15	supplied
¾	120	11.4	52.4	108	35	Line cord
	240	5.7	26.2	216	20	supplied
1	120	12.6	70	108	40	12
	240	6.3	35	216	20	14
	240 } 3-phase	3.2	21	216	15	14
	480 } 3-phase	1.6	10.5	432	15	14
1½	120	15.8	100	108	50	10
	240	7.9	50	216	25	14
	240 } 3-phase	5.6	34.4	216	15	14
	480 } 3-phase	2.8	17.2	432	15	14
2	120	22	106	108	70	10
	240	11	53	216	35	14
	240 } 3-phase	6.2	42	216	20	14
	480 } 3-phase	3.1	21	432	15	14

away needed power. The voltage specified is that which is needed at the motor, and care must be taken that the correct wire size is used and that the wire length will not cause a voltage drop. When pumps slow down or fail to start under vacuum, many times the malfunction can be traced directly to low voltage at the motor.

Motor Substitution

Motors of similar horsepower rating will not necessarily have the same starting torque or operating efficiency.

Since motors are available from many manufacturers and have various starting and running torques in every horsepower category, it is important that the correct selection be made for efficient operation.

Shutdown Procedure

A pump that is allowed to run continuously will last longer and remain cleaner. Never cause a pump to cycle on and off. If the desired vacuum has been reached in a system, block off the pump with a valve and allow it to continue operating at low pressure. This will actually bring about a better "wear-in" of the moving parts and reduce the possibility of corrosion, which is more likely to occur while the pump is standing idle. When a task is complete and the pump is to be removed from the system, bleed air into the system and allow the pump to come to atmospheric pressure. If the pump is to be stored before it is used again, drain, flush, and refill with fresh Precision high-vacuum oil. Stopper the intake and exhaust openings.

Gas Ballast

This is a device used to help prevent the condensation of contaminant vapors within the pump, thereby protecting the pump from the corrosive action of these condensed vapors. Air is bled into the pump via the adjustable valve on top of the pump just before the gas is exhausted through the oil. Because of the pressure increase, the exhaust valve is forced open without subjecting the contaminant vapor molecules to as great a compression. Compression of the vapor would cause condensation.

With the gas ballast open, the pump will not reach its rated vacuum because gas ballast is actually a controllable leak. The pump will also run warmer with the gas-ballast valve open because of the greater amount of gas (air) it is handling.

When all traces of the contaminant vapors have disappeared from the system and oil, the gas-ballast valve may be closed to permit the pump to attain its ultimate vacuum.

The amount of moisture that can successfully be handled by a gas ballast will vary with the size of the pump.

CAUTION If the contaminant vapor has reacted with the oil, it cannot be separated from the oil by the gas ballast. When this happens, the oil must be changed immediately.

Smoke Eliminators

At intermediate pressures, all high-vacuum pumps produce an intermittent pumping sound and an oil "smoke" can be observed coming from the exhaust port. While this is much less noticeable with Precision high-vacuum pumps, it may be objectionable in a given location. The "fog," or oil "mist," can be eliminated by adding the proper smoke eliminator to the exhaust side of the pump. It can also be discharged elsewhere by removing the external baffle over the exhaust port and piping the exhaust to another location. Smoke eliminators are highly efficient and will cause very little back pressure. An easily replaceable element, contained within the housing, should be inspected on a routine basis. Saturation with oil indicates replacement is needed, and the element cannot be allowed to become "nonporous" with an accumulation of any thick, tarlike foreign material. As the element becomes saturated, the oil will merely drain into the bottom portion of the housing. The drain plug can be unscrewed to remove any excess oil. If this excess is allowed to collect, the housing may fill and oil may be spit out as air passes through the filter.

The eliminator element should be replaced if (1) it becomes clogged by oil sludge or varnish-type materials, or (2) back pressure has increased through the element, resulting in lengthy pump-down time or difficulty in maintaining the required vacuum in the system.

Belt Guards

All vacuum pumps have facilities to permit belt-guard installation in minutes. All belt guards should be easily opened for access to the belt and pulleys. Belt guards are required by OSHA regulations, and removal can result in fines.

Maintenance

Risk of pump damage can be reduced and money will be saved in the long run if routine maintenance is performed on the pump. While operating, the pump should be protected from destructive vapors by a cold trap or a sorption trap before the pump intake for heavy vapor loads and by use of the gas ballast for lighter vapor loads. If these vapors contaminate or react with the oil, it should be changed promptly. During normal operation some oil will be lost from the pump. Periodically check and refill to the proper level when necessary.

Check the V belt for tightness, cleanliness, and wear. A loose belt will slip, and a tight belt will cause excess bearing wear. Belt tightness is adjusted by moving the motor toward or away from the pump. Check, by observation, for oil leaks at the drain valve, housing gasket, and shaft seal. If any leakage is observed, replace that component. If the pump is equipped with a smoke eliminator, check to see that the element is not clogged.

When and How to Change Oil

The most common cause of pump failure and unsatisfactory performance is contaminated oil. If water or acid vapors have been passed through the pump and the oil is allowed to stand for any length of time, severe corrosion and extensive damage to any pump may occur. The simplest way to determine if an oil change is needed is to connect a gauge directly to the pump and ascertain if the rated ultimate vacuum can be attained. Refer to the following discussion of gauges for an understanding of the various levels of vacuum and the limitations of different types of gauges.

An odor indicating the presence of a solvent indicates the need for an oil change. Light-brown, cloudy oil may indicate the presence of water. *When in doubt, change oil.* A blackish color does not necessarily indicate bad oil in smaller models. The longer the vanes are allowed to wear in, the more perfect the internal seat of the moving parts. The carbon that is thus added to the oil will discolor it but will not reduce its vapor pressure and will even add to its lubricity.

If the pump is used in only occasional service or is loaned to another operator, it is a good policy to drain and refill it with fresh oil before it is set aside for temporary storage. Many users find it convenient to attach a tag upon which the dates of oil changes may be recorded.

Draining a pump is quite simple. The pump should be operated so the oil will be warmed and become as thin as possible. Open the vacuum intake to the atmosphere, shut off the power, and open the drain valve. If possible, tip up the end opposite the valve to assure complete drainage. When the flow has stopped, turn the power on for a few seconds or turn the pulley by hand to clear any oil that may still be in the stator cavities. Sufficient oil film is present on the moving parts to prevent damage if the pump is operated under power for a few moments while draining. Remember, any contaminated oil left in the pump will only serve to contaminate the fresh oil when it is added. The use of *detergent* oil is *not recommended.*

Shaft-Seal Replacement

Careful seal installation pays dividends in excellent service. Less care means short service life, even failure after just minutes of operation. A new seal should be used any time the pump is disassembled. Do not misinterpret condensed oil vapor on the pump as an oil leak.

General Procedures

CAUTION Refer to the operating manual of your particular pump for specific directions.

Troubleshooting

Pump Runs Hot

Operating temperatures will be related to surrounding ambient conditions. Under normal conditions when operating at low pressures, depending on the model, pump-oil temperatures may be expected to be approximately 65 ± 10°C. While small models may run slightly cooler, the larger single-stage pumps may run slightly warmer. Operation is satisfactory if the oil temperature does not exceed 80°C. The most frequent cause of overheating is handling too large a volume of air for prolonged periods or operation with contaminated oil. A pump should never be used as a control to regulate a specific vacuum. If it is allowed to cycle on and off frequently, it will probably overheat and fail to start, since the motor's Thermoprotector will prevent the motor from operating. Check the following:

1. Oil level is low. Add oil.
2. Oil is gummy. Drain, flush, and refill with fresh oil.

3. Gas ballast is open. Close ballast valve.

4. The V belt is too tight. Loosen.

5. Abrasive particles have entered the pump. Disassemble and clean.

6. Pump is binding mechanically because of misalignment of parts damaged during shipping. Return for replacement.

Pump Is Noisy

Noise, of course, is a relative thing. When the noise level of a new pump is evaluated, it must be compared with another pump of comparable performance with regard to its free-air displacement and it must not be on a platform that will amplify the normal operating sound. The sound should be analyzed as to its probable origin with respect to the following points:

1. Oil level is low. Refer to paragraph on proper filling procedure, and add oil.

2. System is too large for the pump, causing prolonged operation at intermediate pressures with the resultant normal pumping sound. Add a smoke eliminator or select a larger pump.

3. System has pronounced leaks, causing prolonged operation at intermediate pressures with the resultant normal pumping sound. Locate and seal the leaks.

4. Exhaust valve is damaged or corroded. Return to the factory.

5. Vane springs are malfunctioning as a result of damage in shipment or the introduction of some foreign material. Return to the factory.

Pump Does Not Produce Expected Vacuum

At the factory, all pumps are tested on a McLeod gauge. All other types of gauge will give a higher reading. For an explanation, see the discussion of gauges. When a pump is connected to a system, the rated ultimate vacuum in most cases will not be achieved because of the configuration of the system. Always check to see that the oil in both the mechanical and diffusion pumps is at the proper level. There is always the possibility of leaks, and a complete check, with a leak detector if possible, should be made. Quite frequently material in the system (volatiles) will be releasing vapor (outgassing) at such a rate that the vacuum obtained will seem higher than expected. Water vapor in the air is a prime example of such foreign material. If the pump's performance must be substantiated, separate it from the system and gauge it directly. The following points may also be investigated:

1. Gas ballast is open. Close completely.

2. Plain grease instead of high-vacuum silicone grease was used at slip joints or seals.

3. Remove any excess vacuum grease at joints. Only a very thin film should be necessary.

4. Check seal around threaded hose connection at the pump.

5. Oil is contaminated or improper oil was used. Drain, flush, and refill.

6. Check gauge calibration.

Pump Will Not Start

Occasionally some users may find a mechanical vacuum pump that is difficult to start. Several considerations may enter into such a picture; it is rarely indicative of a defective pump. Most often it reflects the particular application and will be common to pumps of any manufacture. Proper sizing of the pump to the task to ensure a fairly rapid pump-down time will prevent motor overloading. Trapping out strongly reactive agents that would turn the high-vacuum oil into a gummy substance will avoid seizure. Proper oil level will prevent hydraulic bind if a pump is shut down. If condensable vapors are continually drawn through any pump, their accumulation will result in an improper oil level. It is suggested that the following points be checked:

1. Check fuse, line cord, and switch.

2. Check voltage at the motor to be sure there has not been excessive line loss.

3. Remove V belt and determine that the motor will operate. In cases where a motor's Thermoprotectors are defective, it may operate separately, but not under load. Repair or replace motor.

4. If the pump has been subjected to the abuse of frequent cycling on and off, ascertain that the motor has not overheated and that the malfunction is not a result of normal Thermoprotector cutout protection.

5. Drain a small amount of oil and see if it has been contaminated to a point where its viscosity has been increased.

6. Overfilling and abusive cycling have caused a hydraulic bind. With the switch off, slowly turn the pump clockwise (facing the pulley) by hand. If excessive oil has been drawn into the stator of the pump, it will be very difficult to turn. A hydraulic seizure, or "lock," has occurred. Drain off excess oil. If the pump can be turned two revolutions, it will start normally when the motor switch is turned on. If it cannot be turned, the pump is "frozen" and must be disassembled and repaired.

Volatiles

Under vacuum most liquids turn to vapor. There are many tables showing the vapor pressure of a liquid relative to temperature. For vacuum work, this means that as soon as that pressure is reached, the equilibrium shifts to the vapor phase. Water, the most common liquid, has a vapor pressure of about 17 torr at room temperature. This means that as long as there is any liquid water present, no vacuum pump can achieve a vacuum greater than 17 torr. The same phenomenon occurs with all other liquids.

If the pump is permitted to continue to run with the gas ballast open, the water will be separated from the oil and pass from the pump as a vapor. Until the water (or any other volatile) is out of the oil, the pump cannot reach its ultimate pressure. A suitable trap would prevent these vapors from reaching the pump.

Pumping Speed Is Too Slow

Since it is impossible to anticipate all types of systems wherein a pump may be employed, speed curves for all vacuum pumps are plotted according to American Vacuum Society standards. Every bend or restriction in connecting tubing, each valve or trap, will reduce the pumping speed. These factors as well as the length of connecting tubing should be considered when calculating speed of evacuation and selecting the proper size of pump. Additional considerations are:

1. Gas load is too great for the pump in use. Use larger pump.
2. There are leaks in the system. Locate and seal them off.
3. Oil is contaminated. Drain, flush, and refill the pump.
4. Material in system is outgassing. If possible, heat the material.

Outgassing

Gases adhere to the surface and are occluded in most solids. They are also present in liquids. In a vacuum these gases, commonly air, leave the surfaces and depths of these solids and liquids (including vacuum-pump oil) during normal operation. This increases the amount of gas a pump must handle to an amount greater than that of the system to be evacuated. Hence, a pump may evacuate a system at a slower rate than anticipated. Heating (or baking) the components will speed up the release of such gas but will increase the pump load during this accelerated release.

Pump Smokes

During normal operation some oil is continually being exhausted with the air from any vacuum pump. At intermediate pressures this oil mist is commonly referred to as "smoke." If operation is to be continued at this pressure level, add a smoke eliminator. This condition may also indicate a low oil level. Oil can be slowly added to the pump while it is running. Add oil through the exhaust port in small quantities until the sight gauge is two-thirds to three-fourths full. If oil is added too fast, it may cause spitting.

Pump Spits Oil from Exhaust Port

Oil level is too high. Drain oil until the proper level is reached. Oil expands with heat after the pump is started. If the pump was overfilled slightly while cold, it may spit some oil at operating temperature. Also, other liquids or condensed vapor may have entered the pump from other components in the system to raise the oil level.

Backstreaming

Most vacuum pumps are designed so that, when they are not overfilled, the oil will not back up beyond the intake connection. Backstreaming, however, occurs with all high-vacuum pumps. As the system is being evacuated, fewer and fewer gas molecules are passing through the lines, and oil vapor will pass in the opposite direction and may condense on the walls of the vessel being evacuated. To prevent this, a trap should be placed in the line just before the pump.

Pump Leaks Oil

Do not confuse an accumulation of condensed oil vapors with a leak. When in doubt, wipe off the pump and isolate the source:

1. Is it spitting or overfilled? See above; drain partially.
2. Check the drain valve. Tighten or replace it if necessary.
3. Check the housing gasket. Tighten the screws or replace it.
4. Check the cover gasket. Tighten the screws or replace it.
5. Check the shaft seal. Replace it if necessary.

Parts and Service

Should problems arise, contact your laboratory-supply dealer.

DEVICES TO MEASURE PRESSURE OR VACUUM

U-Tube Manometers

Open-End Manometers

An open-end or U-tube manometer (Fig. 5-16) is a pressure-measuring device consisting of a U tube filled with mercury or some other liquid, such as water, alcohol, or oil, which has one end open to the atmosphere and the other end attached to the system to be measured. The atmosphere

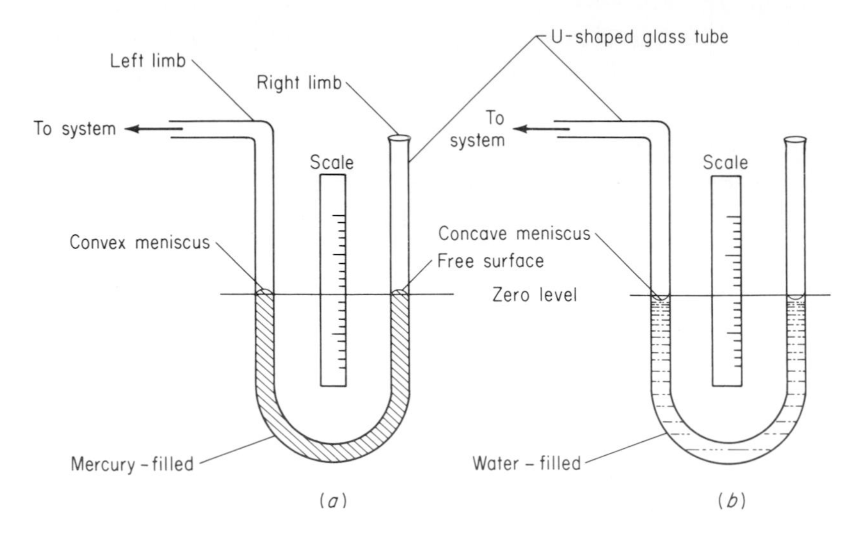

FIGURE 5-16
The U-tube manometer. When no pressure is applied to either limb, the manometer liquid level remains at zero reading on the scale; the level of the meniscus in the left limb is the same as that in the right limb.

exerts pressure on the open end, the system on the other end. When the pressure in the system is equal to the atmospheric pressure, the heights of the liquids are the same; there is no difference in the levels.

When there is a difference between the pressure of the atmosphere and that of the system, there is a difference in the levels of the liquid in the manometer, and the difference in the levels is a measure of the difference in pressure. When the pressure of the system is higher than atmospheric pressure, the end attached to the system has the lower level because of the higher pressure (Fig. 5-17).

CAUTION The open-end manometer is limited to the measurement of moderate higher-than-atmospheric pressures. Too high a system pressure exceeds the capacity of the manometer and will blow the liquid out of the open end.

When an open-end manometer is used for measurements of vacuum (low-pressure), the length of the columns must be able to show (theoretically) zero pressure, the complete absence of molecules. Since 1 atm = 76 cm Hg, a U-tube mercury-filled manometer should be (for practical purposes) about 100 cm tall (Fig. 5-18).

The difference in the heights of the mercury columns is the pressure on the system, stated in millimeters of mercury, centimeters of mercury, torr, or pascals.

Accuracy of the Manometer

The accuracy of the manometer is not affected by the size or shape of the tube, only by the difference in the height of the liquid (Fig. 5-19). This

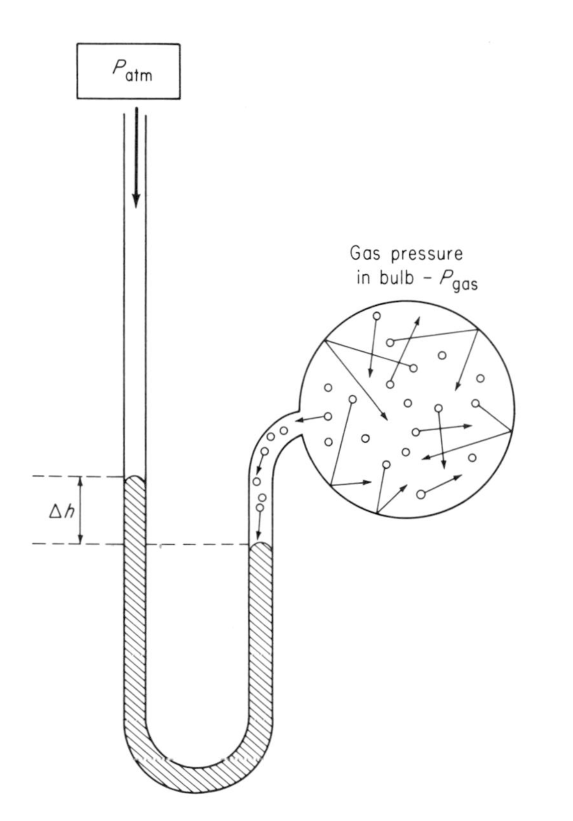

FIGURE 5-17
The open-end manometer is used for measurements of pressure greater than atmospheric pressure.

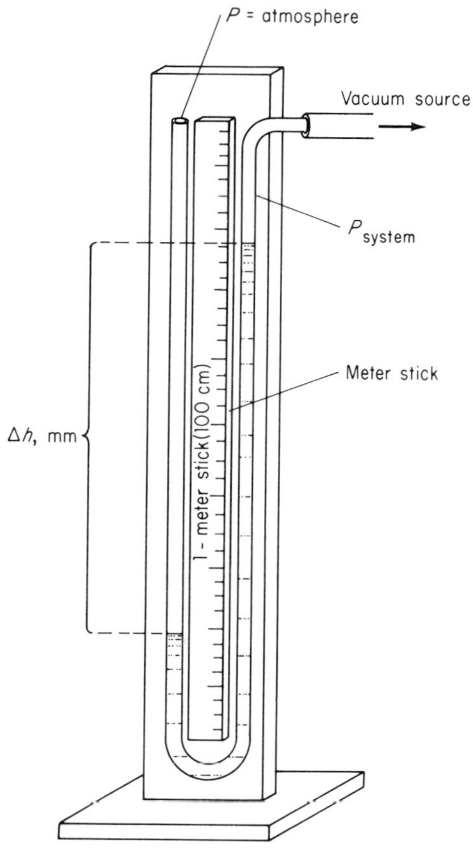

FIGURE 5-18
Commercially available open-end manometer.

FIGURE 5-19
The shape of the manometer tube has no effect upon the liquid levels. Identical levels are obtained for identical pressures, regardless of the shape or the diameter of the tube.

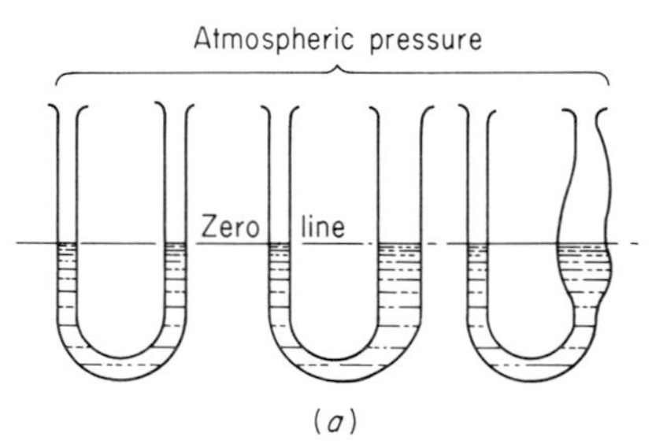

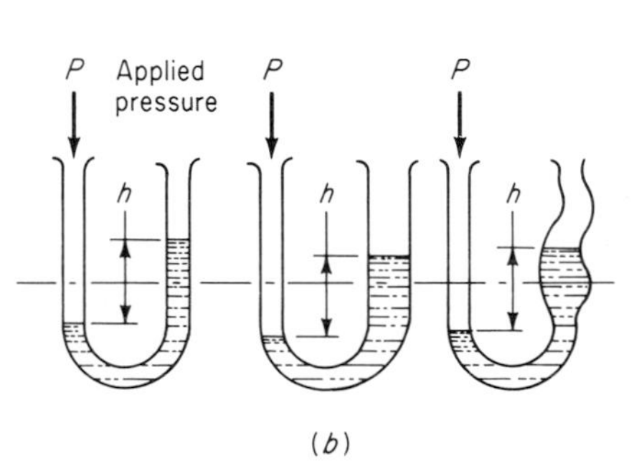

difference depends on the density or vapor pressure of the liquid used as well as on ambient temperature and viscosity.

Density of the Liquid Used

Under the same pressure or vacuum, U-tube manometers filled with low-density liquids, such as water or oil, will show greater difference in height than manometers filled with a high-density liquid, such as mercury; therefore use low-density liquids to measure small differences in pressure. The density of the liquid must be accurately known if it is used to measure pressures with an open-end U-tube manometer.

Vapor Pressure of the Measuring Liquid

The vapor pressure of the liquid used in the manometer must also be taken into consideration, because it exerts pressure.

Temperature of the Liquid Used

The ambient temperature and hence the temperature of the liquid in the manometer must be carefully controlled, because changes in temperature change the density of the liquid.

Viscosity of the Liquid Used

In a vertical manometer, the viscosity of the liquid used is relatively unimportant, but it is very significant when used in an inclined manometer. Furthermore, when very viscous liquids are used, sufficient time must elapse to allow the true level to be attained before taking the reading.

The Closed-End Manometer

The closed-end manometer (Fig. 5-20) is normally filled with mercury and is normally used to measure pressures less than atmospheric, such as those encountered in vacuum distillation (see Chap. 21, "Distillation and Evaporation"). When a closed-end manometer is used to measure these low pressures, it is usually connected into the system with a Y connector.

Reading the Reduced Pressure

As the pressure is reduced, the mercury will rise in the tube which is connected to the vacuum system. When the mercury column stops rising, the vacuum in the system is determined by reading the *difference* in the heights of the mercury columns in the two tubes. A U-tube manometer can be more accurately read than a Bourdon tube in the range of 0 to 10 torr. By modifications, such as using an auxiliary pump and measuring the differential pressure or by inclining the U tube, the accuracy of closed-end tables in this range can be improved. They are usually used only to 0.5 torr.

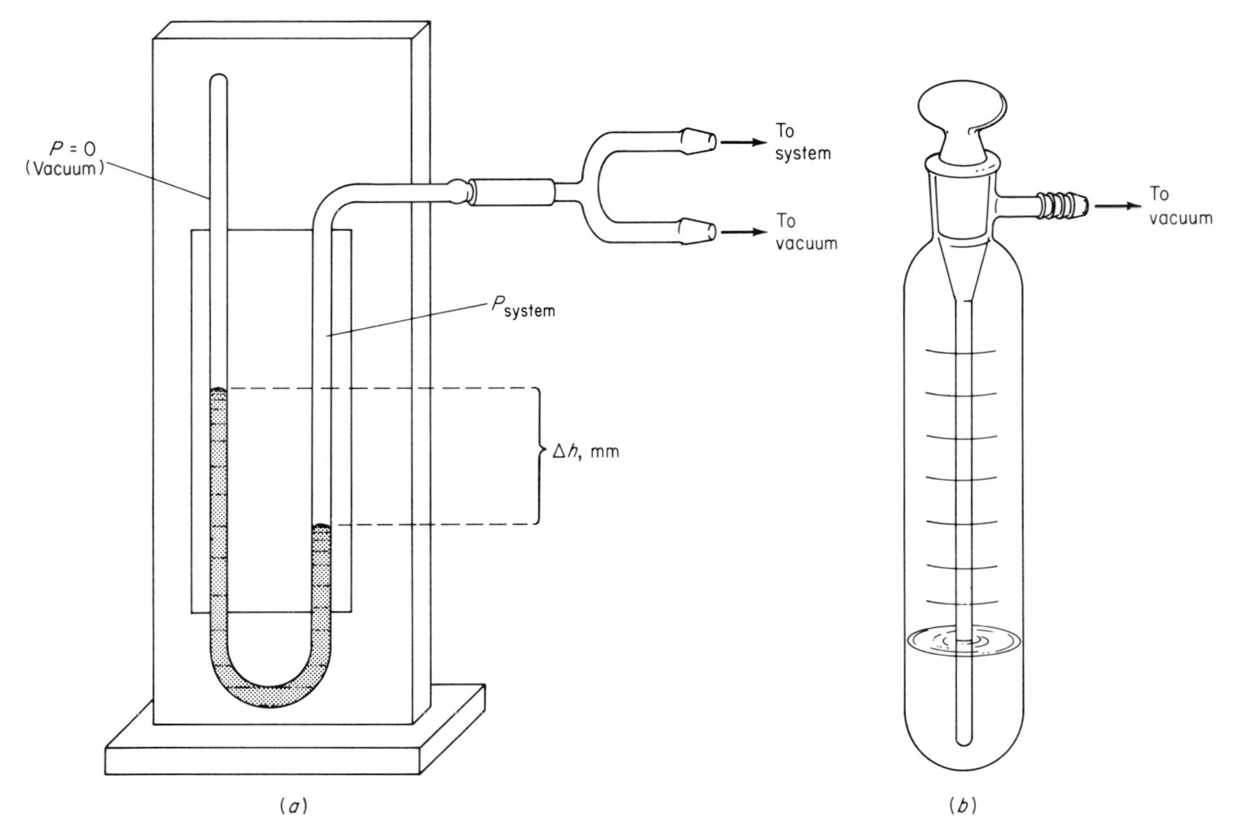

FIGURE 5-20
Closed-end manometer. (*a*) U-tube type. (*b*) Commercial type.

Filling the Manometer

The manometer can be filled easily by means of the following procedure (Fig 5-21):

1. Evacuate the system with a high-vacuum pump.
2. Tilt the mercury resevoir.
3. Gradually allow the pressure to rise by bleeding in air.
4. Mercury is forced into the evacuated tube by atmospheric pressure. The constriction (if present) slows down the flow of the mercury and so protects the closed end of the manometer from breakage caused by mercury shock.
5. The closed end of the tube must be completely filled. If an air bubble is present, repeat the procedure, again tilting the U tube.

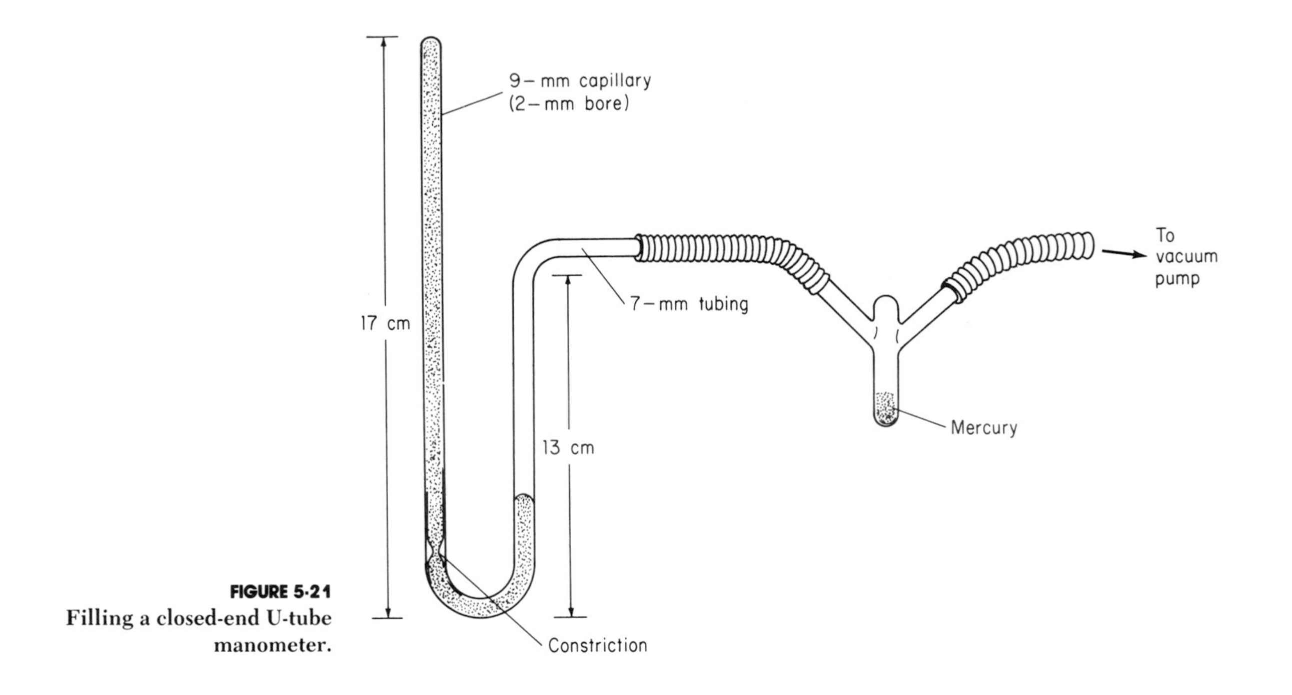

FIGURE 5-21
Filling a closed-end U-tube manometer.

Protecting the Manometer

When the vacuum is interrupted and the system is brought back to atmospheric pressure, care must be exercised. A sudden change in pressure may cause the mercury to break the closed end of the tube by striking it sharply as it rushes back (mercury shock). The constriction in the tube (if present) helps, but a bleeder arrangement is better. The bleeder may consist of a screw clamp on a piece of rubber tubing inserted in the trap (Fig. 5-22) or the needle-valve assembly of the bottom part of a Tirrill-type gas burner attached to the trap (Fig. 5-23). Either will work; the needle valve, however, provides better control of the rate of air intake.

Gauges

Bourdon-Type Gauges

Bourdon-type gauges are the simple mechanical types generally used to measure pressure. A Bourdon gauge consists of a length of thin-walled metal tubing, flattened into an elliptical cross section, and then rolled into a C shape. When pressure is applied, the tube tends to straighten out, and that motion is transmitted by levers and gears to a pointer, indicating the pressure (Fig. 5-24). As a vacuum gauge, it is usually used only to indicate the condition of the system. Bourdon gauges (Fig. 5-25) are not suitable for accurate high-vacuum measurements.

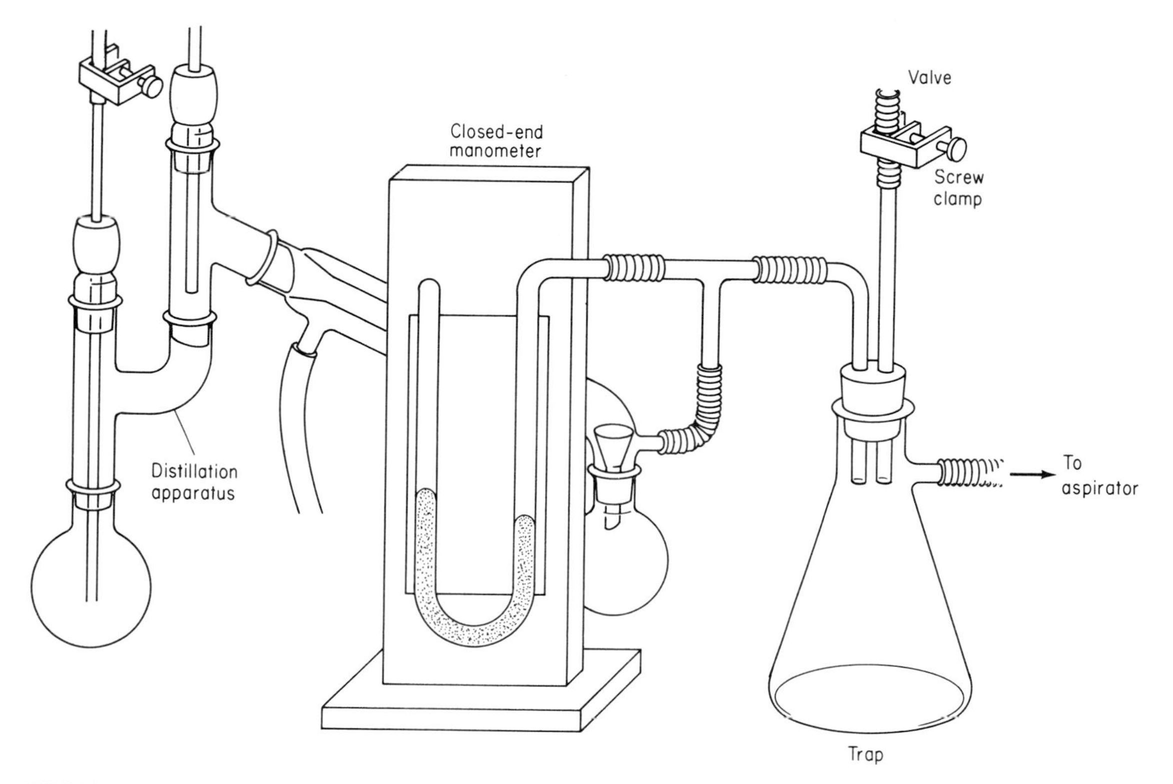

FIGURE 5-22
A valve with a screw clamp is used as protection for a manometer. It can be used to maintain a fixed desired pressure by bleeding air into the system as desired.

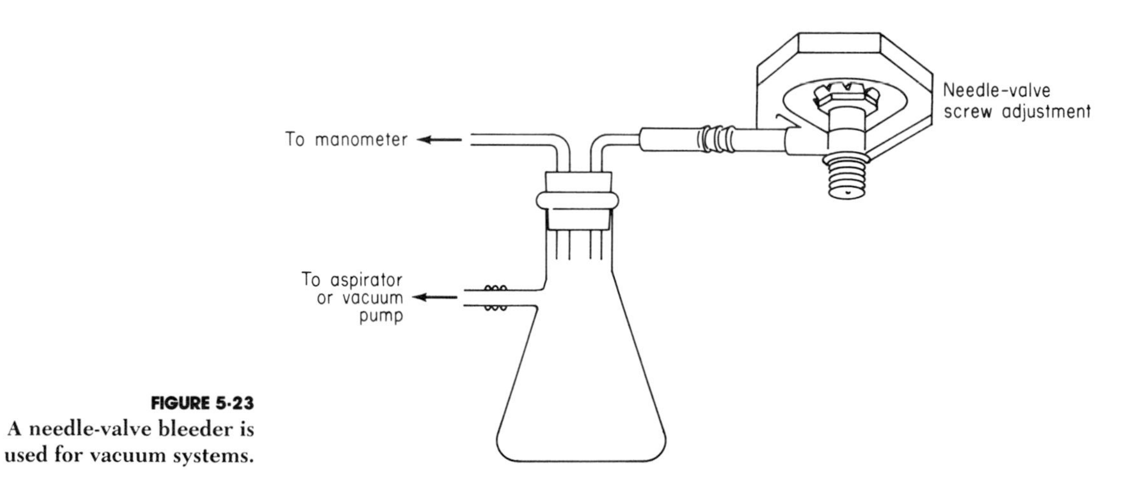

FIGURE 5-23
A needle-valve bleeder is used for vacuum systems.

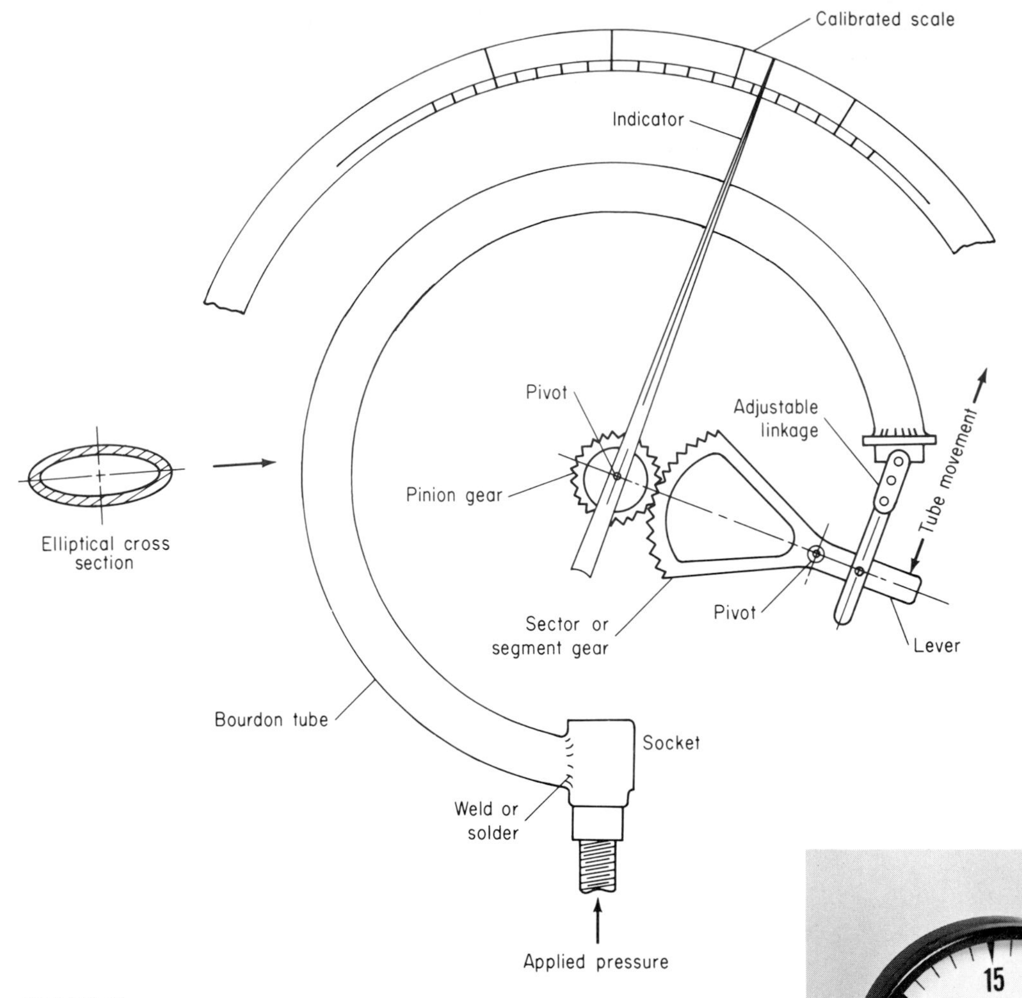

FIGURE 5·24
A Bourdon tube or pressure gauge.

FIGURE 5·25
A Bourdon-type vacuum gauge is an easy-to-read dial gauge, accurate to 2% for quick determination of pressure conditions.

The McLeod Gauge

The *McLeod gauge* (Fig. 5-26) is the primary standard for the absolute measurement of pressure. A chamber (part of the gauge) of known volume is evacuated and then filled with mercury. This chamber terminates in a sealed capillary which is calibrated in micrometers of mercury, absolute pressure. When the mercury fills the chamber, the gas therein is compressed to approximately atmospheric pressure, and any trapped vapors are liquefied and have no significant volume. In other words, this gauge does not measure the pressure caused by any vapors present. Its reading represents only the total pressure of gases.

A cold trap is sometimes employed in conjunction with this gauge to prevent the transfer of vapors from the pump to the gauge or from the gauge to the pump. These gauges, while accurate to approximately 10^{-5} torr, are

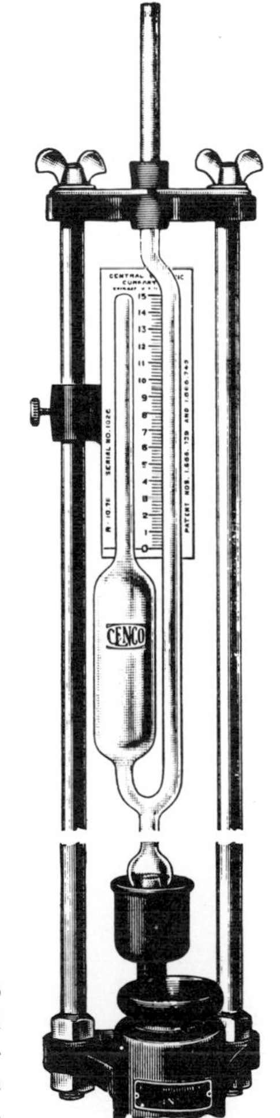

FIGURE 5-26
The McLeod commercial gauge is used for high-vacuum precision determinations of pressure.

not considered suitable for ordinary use because they do not read continuously and are usually fragile.

There are two fundamental methods used for reading McLeod gauges, one (Fig. 5-27) based on the uniform scale (linear) and the other (Fig. 5-28) on the nonuniform scale (square-law). Both methods involve the measurement of the difference in mercury levels after raising the mercury to a fixed reference point.

The Tilting McLeod Gauge

The *tilting McLeod gauge* (Fig. 5-29) is very compact and portable. When the gauge is tilted (Fig. 5-29*a*), mercury traps the sample of gas in the measuring tube (Fig. 5-29*b*).

The Pirani Gauge

A Pirani gauge measures the pressure of the gas by indicating the ability of the gas to conduct heat away from a hot filament. The greater the density (or pressure) of a gas, the greater the conduction of heat from the filament. As the temperature of the filament varies, so does its ability to

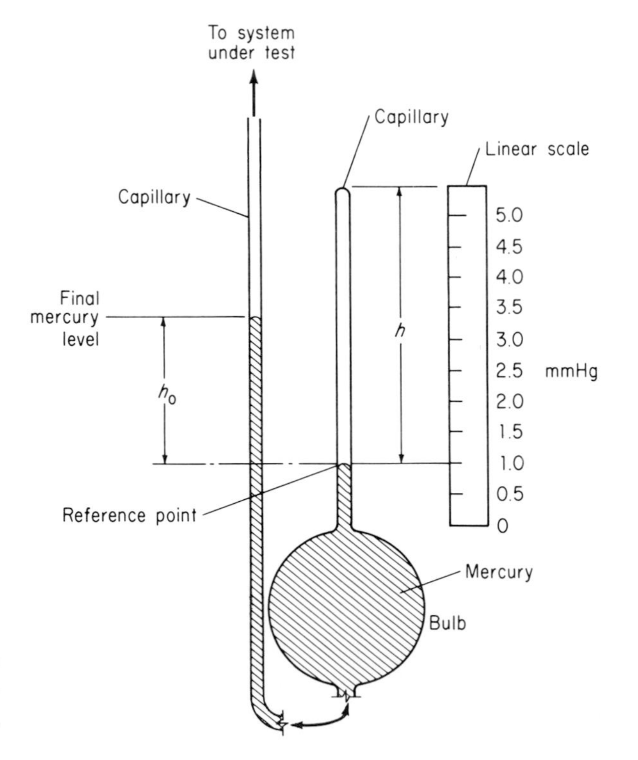

FIGURE 5-27
Uniform-scale method for reading McLeod gauges.

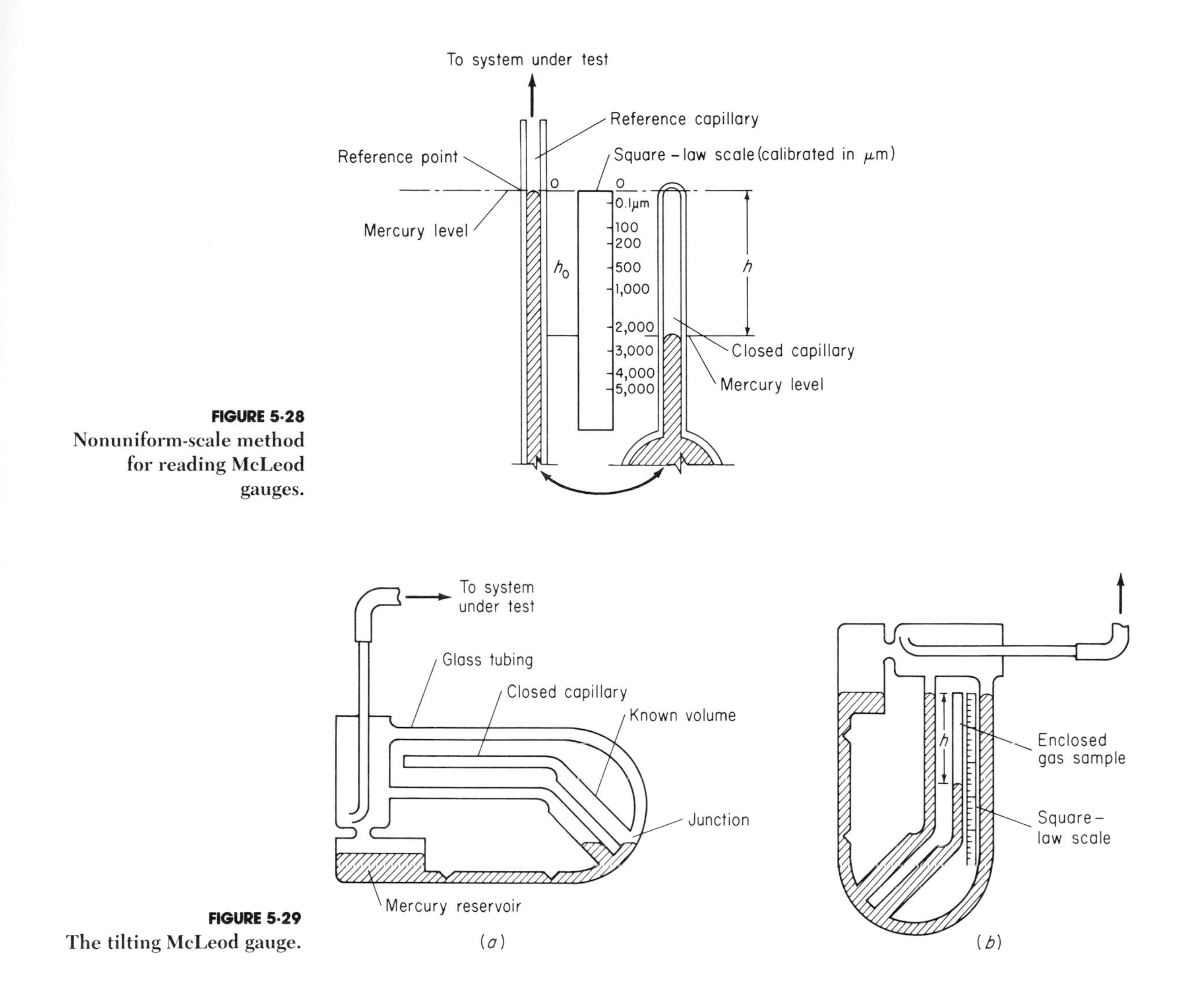

FIGURE 5-28
Nonuniform-scale method for reading McLeod gauges.

FIGURE 5-29
The tilting McLeod gauge.

carry current. Usually a Wheatstone-bridge circuit is employed with a microammeter calibrated to read in micrometers of mercury (Fig. 5-30).

Thus pressure is read on the basis of the current through the filament, which is a function of filament temperatures. Since heat is conducted away from the filament by vapors as well as gases, these gauges measure the presence of the vapors. Also, different gases have different thermal conductivities. Hence, in the same system, they will yield a higher reading than a McLeod gauge, which measures only the pressure caused by the gases. They are usually used in the pressure range of 1 μm Hg to 1 mm Hg.

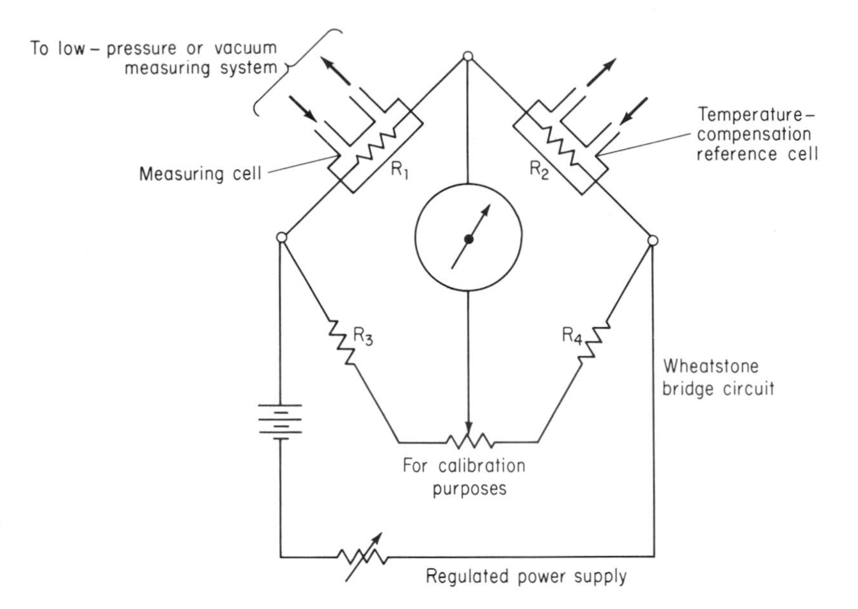

FIGURE 5-30
Simple circuit for a Pirani gauge.

The Thermocouple Gauge

A *thermocouple gauge* (Fig. 5-31) is similar to a Pirani gauge in many ways. The difference, basically, is that a thermocouple measures the filament temperature and the thermocouple output is shown on a meter calibrated in micrometers of mercury. Just as a Pirani gauge will give a different reading for the same pressure of different gases and vapors, so will a thermocouple gauge. The thermocouple gauge is more rugged, smaller, and

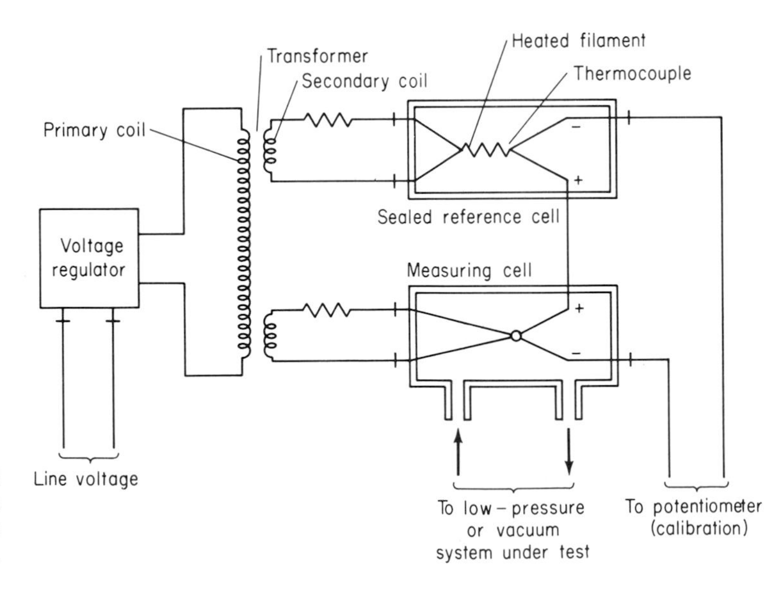

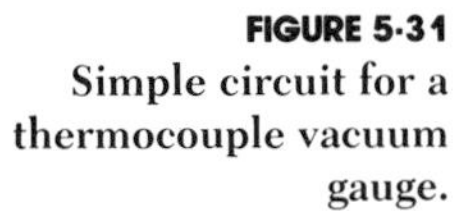
FIGURE 5-31
Simple circuit for a thermocouple vacuum gauge.

slightly less sensitive than the Pirani. Its range is approximately the same, 1 μm Hg to 1 mm Hg.

Ionization Gauges

Ionization gauges comprise another group of indicating devices. In general these gauges are more sophisticated, sensitive, and fragile than the Pirani and thermocouple types. They are usually used to measure beyond the range of the simpler gauges—up to 10^{-14} torr, depending on the exact type and design. These gauges (see Fig. 5-32) form ions of the gas molecules present, and the amount of current carried by this ionized gas depends upon the amount of gas present (in reality, the density or pressure of the gas). There are two basic types, the thermionic, which forms the (gas) ions by electrons emitted from a hot filament, and the cold-cathode type. The Bayard-Alpert gauge is a common example of the thermionic design. The Penning or Philips gauge is the most common example of the cold-cathode design. The gauges also give different readings for different gases or vapors at the same pressure.

There are also gauges utilizing radioactivity, viscosity, discharge tubes, and radiometer principles to measure pressure.

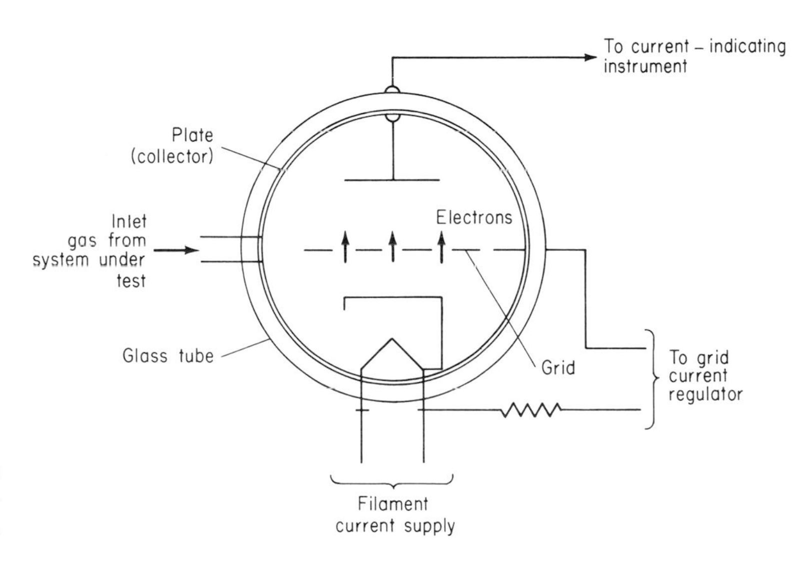

FIGURE 5-32 Ionization gauge for vacuum measurement.

MANOSTATS

The Cartesian Manostat

Manostats are devices which are used in vacuum distillation and in other types of vacuum systems to maintain a particular pressure. These devices are inserted in a system, and they open or close an orifice or capillary as necessary to compensate for leaks into or from the system. The cartesian

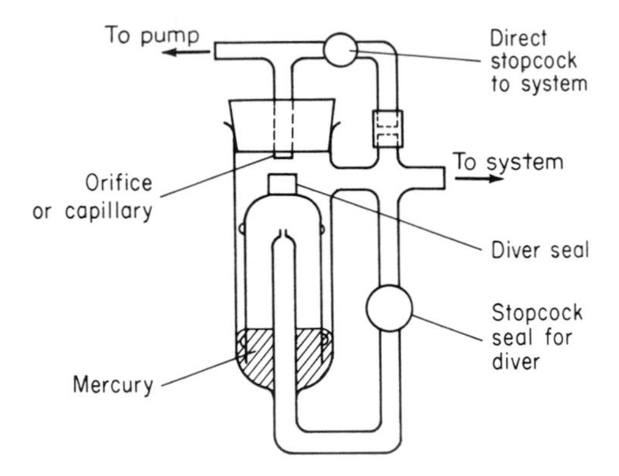

FIGURE 5-33 Schematic of a cartesian diver manostat.

type (diver) is a glass vessel which floats on a pool of mercury in a chamber connected to the controlled system (Fig 5-33). A sample of gas is trapped in the chamber at the desired pressure. Changes in the surrounding pressure affect the buoyancy of the float, causing it to rise or sink vertically, thereby sealing or opening a capillary or an orifice.

Procedure

1. Fill the reservoir with clean mercury until the float almost touches the orifice or capillary.

2. Adjust the stopcocks (there may be one or two) so that all are open.

3. Evacuate the system almost to the desired pressure by observing the manometer.

4. Close the stopcock which connects the vacuum pump directly to the system.

5. Close the other stopcock to seal off the diver assembly from the system when the pressure in the system is 1 to 2 torr above the desired pressure. This system will now maintain the desired pressure.

(**a**) The diver rises and closes the capillary when the pressure in the system is less than that in the diver assembly.

(**b**) The diver sinks and opens the capillary when the pressure in the system is greater than that in the diver assembly.

CAUTIONS Clean mercury must be used. Observe safety procedures in handling mercury.

The Noncartesian Manostat

How to Use for Vacuum Control (Fig. 5–34)

1. Open valve.

2. Bring system to desired vacuum.

3. Close valve.

4. Make fine adjustment for pressure by using orifice-height-positioning control.

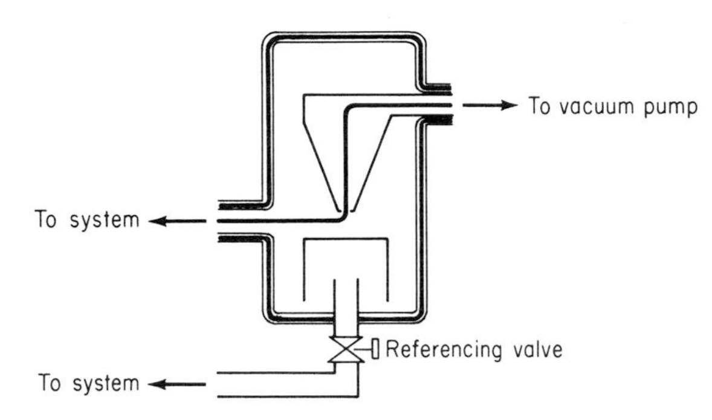

FIGURE 5-34
Setup for vacuum control.

How to Use for Pressure Control

1. Add auxiliary on-off valve and auxiliary needle valve to system as shown in the diagram (Fig. 5-35).

2. Closc cxhaust valvc.

3. Open reference valve and bring system to desired pressure.

4. Closc rcfcrcnccs valvc.

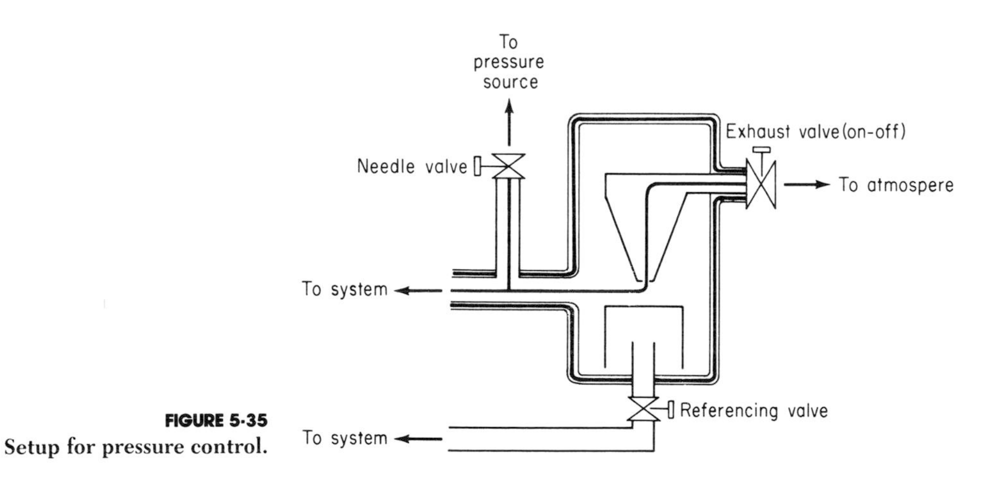

FIGURE 5-35
Setup for pressure control.

5. Open exhaust valve and adjust auxiliary needle valve to allow smallest detectable quantity of gas to pass to the pressure source.

6. Make fine adjustments by positioning the height of the orifice.

The Lewis Manostat

The Lewis manostat uses mercury itself to open and seal the system by the rise of mercury against a fritted glass insert (Fig. 5-36). If clean mercury is used, the operation is easier than the one using the cartesian diver.

CAUTIONS Use only clean mercury. Use adequate traps both before and after the manostat to prevent contamination of the mercury.

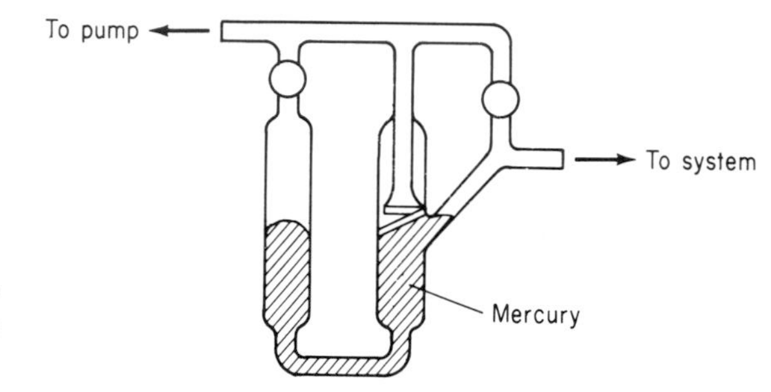

FIGURE 5-36 A modified Lewis manostat.

CRYOGENIC LIQUIDS

All gases can be reduced eventually to liquids and the liquids to solids by an appropriate decrease in temperature and/or an increase in the applied pressure. However, there is a temperature for all gases above which pressure alone cannot condense them into liquids. This temperature is known as the *critical temperature.* For example, carbon dioxide easily liquefies if sufficient pressure is applied while the gas is below 31.04°C; however, no amount of applied pressure can cause liquefaction of the gas above this temperature. The pressure required to liquefy a gas at its critical temperature is called the *critical pressure.* The critical pressure for carbon dioxide is 72.85 atm ($7.358 \cdot 10^6$ Pa). The density of a substance at its critical temperature and critical pressure is called its *critical density.* The critical density of carbon dioxide is 0.468 g/cm^3. The critical temperature, critical pressure, and critical density for several gases are given in Table 5-7.

TABLE 5-7 Critical Properties of Some Substances

Substance	*Critical temperature, °C*	*Critical pressure, atm*	*Critical density, g/cm^3*
Ammonia	132.4	111.3	0.235
Nitrogen-14	−146.89	33.54	0.311
Helium	−267.96	2.261	0.0693
Diethyl ether	193.6	35.90	0.265

Gases that are liquefied by applying pressure and stored in insulated containers are referred to as *cryogenic liquids.* These very cold liquids require specially designed vessels for storage. A common device is the *Dewar flask* which is a double-walled glass vessel with a silvered interior, similar to the lining found in a Thermos® jug. The extremely cold temperatures and potentially high pressures of these cryogenic liquids (gases) present many hazards.

CRYOGENIC HANDLING PRECAUTIONS

Always wear protective gloves, clothing, and eye shields when working with cryogenic fluids. Following are some of the hazards connected with cryogenic fluids.

- *Damage to living tissue.* Contact with cryogenic fluids can cause a localized solidification of tissue and produce a "burn" as painful as that received from a heat source. The extreme coldness causes a local arrest in the circulation of blood, and any exposed skin tissue should be restored to normal temperature as quickly as possible. This is most easily accomplished by simply immersing the damaged skin in water at approximately 45°C.
- *A high expansion rate upon vaporization.* Cryogenic liquids can expand to great volumes upon evaporation; for example, liquid methane expands to approximately 630 times its initial liquid volume when converted to its gaseous state. Therefore, if the mechanism of cooling this liquefied methane fails or is inadequate, the internal pressure within the storage container can increase rapidly. If proper ventilation is not provided, the container can explode.
- *An ability to liquefy other gases.* Cryogenic fluids are capable of condensing and even solidifying other gases. Gaseous air, for example, solidifies upon exposure to a number of cryogenic liquids. This solidification of air could present a hazard by blocking vent tubes on the cryogenic storage containers and preventing the release of pressure buildup.

6
BASIC LABORATORY MATHEMATICS

INTERNATIONAL SYSTEM OF UNITS (SI)

All laboratory measurements consist of two parts: a "number" and a "unit" of measure. By international agreement in 1960, scientists throughout the world have established a system of standardized units called the *International System* (called *le Système International* in French, and abbreviated *SI*). By agreement there are seven base units in the SI as indicated in Table 6-1.

TABLE 6-1
SI Base Units

Quantity	*Name*	*Symbol*
Length	meter	m
Mass	kilogram	kg
Time	second	s
Electric current	ampere	A
Temperature	kelvin	K
Amount of substance	mole	mol
Luminous intensity	candela	cd

These base units are not always a convenient size; thus prefixes are employed to change the magnitude of these units by multiples of 10. The most common SI prefixes used for these base units and their meanings are given in Table 6-2. The SI is actually based on an earlier convention proposed in France in 1791, and generally used by the scientific community, called the *metric system*. The metric system simply expresses all units as factors of 10, very much like the U.S. monetary system:

10 mills (*milli*dollars) = 1 cent
10 cents (*centi*dollars) = 1 dime
10 dimes (*deci*dollars) = 1 dollar

The SI prefixes given above and their respective multiplier meanings are given in Table 6-2. In Germany and some other countries, a comma indicates a decimal point; when this is the case, by SI agreement, spaces are used as separators instead of commas when numbers have four or more digits. This spacing technique is demonstrated in Table 6-2.

TABLE 6-2
SI Prefixes

Prefix	*Symbol*	*Meaning*	*Multiplier (numerical)*	*Multiplier (exponential)*
		Greater than 1		
tera	T	trillion	1 000 000 000 000	10^{12}
giga	G	billion	1 000 000 000	10^{9}
mega	M	million	1 000 000	10^{6}
kilo	k	thousand	1 000	10^{3}
hecto	h	hundred	100	10^{2}
deka	da	ten	10	10^{1}
		Less than 1		
deci	d	tenth	0.1	10^{-1}
centi	c	hundredth	0.01	10^{-2}
milli	m	thousandth	0.001	10^{-3}
micro	μ	millionth	0.000 001	10^{-6}
nano	n	billionth	0.000 000 001	10^{-9}
pico	p	trillionth	0.000 000 000 001	10^{-12}

DIMENSIONAL ANALYSIS

It is often necessary to convert one system of units to another, for example English to SI or even one SI expression to another SI form. This mathematical technique is often called *dimensional analysis,* the *factor-label method,* or the *unit-factor method.* This technique will be demonstrated with many example calculations showing how multiplication and/or division will cause all units to cancel except the desired conversion unit when appropriate conversion factors (Table 6-3) are used.

TABLE 6-3
Conversion Factors between SI and English Units

Length
1 meter = 1.0936 yards
1 centimeter = 0.39370 inch
1 inch = 2.54 centimeters (exactly)
1 kilometer = 1000 meters
1 kilometer = 0.62137 mile
1 mile = 5280 feet
1 mile = 1.6093 kilometers
1 angstrom = 1×10^{-10} meter
Mass
1 kilogram = 1000 grams
1 kilogram = 2.2046 pounds
1 pound = 453.59 grams
1 pound = 0.45359 kilogram

TABLE 6-3 *(Continued)*

1 pound =	16 ounces
1 ton =	2000 pounds
1 ton =	907.185 kilograms
1 metric ton =	1000 kilograms
1 metric ton =	2204.6 pounds
1 atomic mass unit =	1.66056×10^{-27} kilogram
*Volume**	
1 cubic meter =	1×10^{6} cubic centimeters
1 liter =	1000 cubic centimeters
1 liter =	1×10^{-3} cubic meters
1 liter =	1 cubic decimeter
1 liter =	1.0567 quarts
1 gallon =	4 quarts
1 gallon =	8 pints
1 gallon =	3.7854 liters
1 quart =	32 fluid ounces
1 quart =	0.94633 liter
Energy	
1 joule =	1 kilogram-meter^{2} second^{-2}
1 joule =	0.23901 calorie
1 joule =	9.4781×10^{-4} British thermal units (Btu)
1 joule =	1×10^{7} ergs
1 kilocalorie =	1000 calories
1 kilocalorie =	1 large calorie (nutritional)
1 calorie =	4.184 joules
1 calorie =	3.965 British thermal units
1 British thermal unit =	1055.06 joules
1 British thermal unit =	252.2 calories
Pressure	
1 pascal =	1 newton meter^{-2}
1 pascal =	1 kilogram meter^{-1} second^{-2}
1 atmosphere =	101.325 kilopascals
1 atmosphere =	760 torr
1 atmosphere =	760 millimeters of mercury
1 atmosphere =	33.8 feet water
1 atmosphere =	14.70 pounds inch^{-2}
1 bar =	1×10^{5} pascals
Temperature	
0 kelvins =	−273.15 degrees Celsuis
0 kelvins =	−459.67 degrees Fahrenheit
Kelvin =	Celsius + 273.15
Celsius =	(Fahrenheit − 32) $\times \frac{5}{9}$
Fahrenheit =	($\frac{9}{5} \times$ Celsius) + 32

* It shoud be noted that the unit of "volume" was not included in the seven SI base units. Volume is not considered to be a base unit, since it is simply a three-dimensional "length" expression. The volume expression *liter* is actually an old metric term which is equivalent to one cubic decimeter (1 dm^{3}).

EXAMPLE Express the mass of a 0.340-lb sample in grams.

Solution $0.340 \text{ lb} \times \frac{454 \text{ g}}{\text{lb}} = 154.36 \text{ g}$

Notice how the pound units will cancel each other when these two terms are multiplied together. It should also be mentioned that this answer is technically incorrect because it violates the significant-figure rules given on p. 130 (answer limited to three significant figures: 154 g).

Examples—Multiplication

The correct answer with proper number of significant figures is given in parentheses. (See the sections titled Significant Figures and Rounding in this chapter.)

EXAMPLE Change 15 in to centimeters.

Solution

1. 1 in = 2.54 cm
2. 15 in = 15 × 2.54 cm
3. 15 in = 38.10 cm (38 cm)

EXAMPLE Change 64 kg to pounds.

Solution

1. 1 kg = 2.2046 lb
2. 64 kg = 64 × 2.2046 lb
3. 64 kg = 141.0944 lb (140 lb)

EXAMPLE Change 8 km to miles.

Solution

1. 1 km = 0.6214 mi
2. 8 km = 8 × 0.2614 mi
3. 8 km = 4.9712 mi (5 mi)

EXAMPLE Change 12 gal to liters.

Solution

1. 1 gal = 3.785 L
2. 12 gal = 12 × 3.785 L
3. 12 gal = 45.420 L (45 L)

EXAMPLE Change 14.8 qt to milliliters.

Solution
1. 1 qt = 946 mL
2. 14.8 qt = 14.8 × 946 mL
3. 14.8 qt = 14000.8 mL (14,000 mL)

SUMMARY For the examples above, we have converted measurements by multiplication. In general, to convert a given measure with a larger unit to an equivalent measure with a smaller unit, determine the number of times the smaller unit is contained in the larger unit, and then multiply the given measure by this value. For instance, in the example showing how to change 12 gal to liters, the gallon is a larger unit than the liter. We have determined that 3.785 L is contained in 1 gal. Then we *multiply* the given measure 12 by the number 3.785 L.

Examples—Division

EXAMPLE Change 20 cm to inches.

Solution
1. 2.54 cm = 1 in
2. 20 cm = 20 ÷ 2.54
3. 20 cm = 7.87 in (7.9 in)

EXAMPLE Change 180 lb to kilograms.

Solution
1. 2.2046 lb = 1 kg
2. 180 lb = 180 ÷ 2.2046
3. 180 lb = 81.6475 kg (81.6 kg)

EXAMPLE Change 12 mi to kilometers.

Solution
1. 0.6214 mi = 1 km
2. 12 mi = 12 ÷ 0.6214
3. 12 mi = 19.3112 km (19 km)

EXAMPLE Change 9.0 L to gallons.

Solution
1. 3.785 L = 1 gal
2. 9.0 L = 9 ÷ 3.785
3. 9.0 L = 2.378 gal (2.4 gal)

Alternative solution

1. 1 L = 0.2642 gal
2. 9.0 L = 9 × 0.2642
3. 9.0 L = 2.378 gal (2.4 gal)

EXAMPLE Change 500 mL to quarts.

Solution

1. 946 mL = 1 qt
2. 500 mL = 500 ÷ 946
3. 500 mL = 0.529 qt (0.529 qt)

SUMMARY For the examples above, we have converted measurements by division. In general, to convert a given measure with a smaller unit to an equivalent measure with a larger unit, determine the number of times the smaller unit is contained in the larger unit, and then *divide* the given measure by this value. For instance, in the example showing how to change 500 mL to quarts, the milliliter is a smaller unit than the quart. We have determined that 946 mL is contained in 1 qt. Then we *divide* the given measure 500 by the number 946 mL.

CONVERSION FACTORS

Table 6-3 contains a listing of common conversion factors for the SI, the metric system, and the English system. A more comprehenisve listing of conversion factors is given in Appendix C, "Abbreviations, Symbols, and Tables."

SIGNIFICANT FIGURES

The proper handling of laboratory data is a prerequisite for any laboratory statistical analysis (see Chap. 7) and ultimately for the reporting of results. Several chapters in this text describe various laboratory measuring devices. Obviously, the different devices produce different degrees of accuracy. For example, microbalances are capable of measurements to eight figures (for example: 13.023456 g), while the same sample would be reported to only four figures (13.02 g) on a less accurate triple-beam balance. The number of *significant figures* in a laboratory measurement is the number of digits that are known accurately, plus one that is uncertain or doubtful. In the microbalance result above, the digits in the first seven places are known accurately but the last number (the 6 in the millionths place) is doubtful. The number of significant figures in a measurement is counted beginning with the first nonzero digit and stopping after including the first doubtful digit. The microbalance measurement has eight significant figures while the triple-beam balance has only four significant figures; the zero is a significant figure in both cases.

Calculations should never "improve" the precision, or number of significant figures, in an answer. Thus, a laboratory report should not contain any more significant figures in the final answer than found in the least precise data used to obtain that answer.

General rules for dealing with significant figures:

1. The number of significant figures in a measured quantity is the number of digits that are known accurately, plus one that is in doubt.
2. The concept of significant figures only applies to measured quantities and to results calculated from measured quantities. The concept is never applied to exact numbers or definitions.
3. All significant figures are counted from the first nonzero digit.
4. All significant figure counting ends with but includes the first doubtful digit (including zeros) encountered.
5. The decimal point location or units of measure used have no effect on the number of significant figures.
6. The answer in a multiplication or division problem must be rounded off to contain the same number of significant figures as the smallest number of significant figures in any factor.
7. The answer in an addition or subtraction problem must be rounded off to the first column that has a doubtful digit.

Rounding

If a calculation yields a result that would suggest more precision than the measurements from which it originated, "rounding off" to the proper number of significant figures is required. Examples and more details of appropriate rounding procedures will be given in the following example problems. Based on statistical probability, the following rounding rules should be applied:

1. If the digit following the last significant figure is greater than 5, the number is rounded up to the next higher digit.
2. If the digit following the last significant figure is less than 5, the number is rounded off to the present value of the last significant figure.
3. If the digit following the last significant figure is exactly 5, the number is rounded off to the nearest *even* digit.

ARITHMETIC IN THE U.S. CUSTOMARY SYSTEM

Adding Measurements

1. Arrange the measures in vertical columns.
2. Use like units of measure in the same column.
3. Add each column separately.
4. Simplify the result by using conversion formulas so that the smaller units will be changed to larger ones.

EXAMPLE Add:

5 yd	2 ft	9 in
6 yd	2 ft	8 in

Solution

1. 11 yd 4 ft 17 in
2. 11 yd 5 ft 5 in
   since 17 in = 1 ft + 5 in
3. 12 yd 2 ft 5 in
   since 5 ft = 1 yd + 2 ft

EXAMPLE Add:

6 gal	3 qt	1 pt
2 gal	1 qt	3 pt

Solution

1. 8 gal 4 qt 4 pt
2. 8 gal 6 qt
   since 4 pt = 2 qt
3. 9 gal 2 qt
   since 4 qt = 1 gal

Subtracting Measurements

1. Place the larger measure on top and the smaller measure on the bottom.
2. Place the like units of measure in the same column.
3. Begin subtracting with the column that has the smallest unit of measure.
4. Simplify the result by using conversion formulas so that the smaller units will be changed to larger ones.

EXAMPLE Subtract:

6 yd	2 ft	4 in
3 yd	1 ft	8 in

Solution

1. 3 yd 0 ft 8 in

2. Since we cannot subtract 8 from 4, we change 2 ft into 1 ft and 12 in; then we can subtract.

ARITHMETIC OF DECIMALS

The most important idea in the arithmetic of decimals is the positional principle. The place of each digit represents a particular power of 10. For example, the number 1969.38 means (1 × 1000) + (9 × 100) + (69 × 1) + (3 × 1)/10 + (8 × 1)/100. In general, we use the scale of notation shown in Fig. 6-1.

	1	9	6	9	and	3	8		
Ten - thousands	Thousands	Hundreds	Tens	Ones or units		Tenths	Hundredths	Thousandths	Ten - thousandths

FIGURE 6-1 Positional principle of decimal notation.

We read the number 1969.38 as one thousand, nine hundred sixty-nine and* thirty-eight hundredths.

Adding Decimals

1. Write the numbers in a vertical column so that the decimal points are directly in a vertical line.

2. Add the numbers in the same way as for whole numbers.

3. Place the decimal point in the sum directly beneath the vertical line of decimal points.

4. Round off your answer.

EXAMPLE

```
  5.807
 48.91
  3.5261
  0.081
118.9206
--------
177.2447
```

---

* The word "and" signifies the decimal point. It is used only to locate verbally the position of the decimal.

Solution

1. Bring down the decimal point.
2. Begin at the extreme right, obtaining 7 as the sum of 6 and 1.
3. Record your answer as 177.24 by the following rounding-off rule.

Round off your answer so that there are no more digits to the right of the decimal point than the smallest number of such digits found among the numbers being added.

In the previous example, the smallest number of digits to the right of the decimal point is two, which is found in 48.91. Therefore we round off our answer to two places as 177.24 by dropping all digits after the 0.24. If the first digit to be dropped (after the 4) is less than 5 (from 1 to 4) leave the 4 as is. If it is larger than 5 (from 5 to 9) change the last digit by adding 1.

Subtracting Decimals

(When both numbers are positive.)

1. Write the numbers so that the larger number is on top and the smaller number is on the bottom.

2. Be sure that the decimal points are lined up directly beneath each other.

3. Perform the subtraction by standard methods of arithmetic.

EXAMPLE

$$\begin{array}{r} 4625.8091 \\ \underline{398.7607} \\ 4227.0484 \end{array}$$

Solution

1. Place the larger number on top.
2. Be sure the decimals are lined up.
3. Perform the subtraction.

Multiplying Decimals

1. Multiply the numbers in the standard form of ordinary multiplication.

2. The number of decimal places in your answer (called the product) is equal to the total number of decimal places in your multiplier and multiplicand.

3. Round off your answer so that it contains *only* as many significant digits as are contained in the number having the smallest number of significant digits.

EXAMPLE Multiply: 3.845 by 16.7

Multiplicand	3.845	(three decimal places)
Multiplier	16.7	(one decimal place)

Solution

```
              26915
             23070
             3845
Product      64.2115   (four decimal places)
```

There are four significant digits in the multiplicand (top number) and three significant digits in the multiplier (bottom number). Thus, according to statement 3 of the procedure, our answer is rounded off so that it contains only three significant digits. Ans. = 64.2 (note that the digit that follows the 2 is less than 4; therefore the 2 is unchanged).

EXAMPLE Multiply:

```
   36.814
     2.05
   184070
  736280
 75.46870
```

Rounded off = 75.5 (note that the digit that follows the 4 is more than 5; it is 6; therefore the value of the 4 is increased by 1 to 5).

Solution

1. Multiply in standard form.

2. Note that there are three decimal places in the top number and two decimal places in the bottom number, so that the total number of decimal places in the answer is *five*.

3. There are five significant digits in the top number and three significant digits in the bottom number. Our answer is rounded off to three significant digits. In rounding off the product, 75.46870, we had to drop the 6 and change the 4 to a 5. Remember that when the first digit dropped is less than 5, the last digit retained remains unchanged, but when the first digit dropped is more than 5, the last digit retained is increased by 1. When the first digit dropped is exactly 5, then the last digit retained is increased by 1.

Multiplying Decimals by Powers of 10

1. To multiply a decimal by 10, 100, 1000, and so on, move the decimal point as many places to the right as there are zeroes in the multiplier. (These numbers can be conveniently written as 1×10, 1×10^2, 1×10^3, and so forth.)

2. To multiply a decimal by $\frac{1}{10}$ or 0.1, $\frac{1}{100}$ or 0.01, $\frac{1}{1000}$ or 0.001, and so on, move the decimal point as many places to the left as there are zeroes in the multiplier.

EXAMPLE Multiply: 36.821 by 100
3682.1 (Ans.)
Decimal is moved two places to the right.

EXAMPLE Multiply: 6.54 by 1/1000 (0.001)
0.00654 (Ans.)
Decimal is moved three places to the left.

EXAMPLE Multiply: 0.00186 by 0.00001
0.0000000186 (Ans.)
Decimal is moved five places to the left.

Dividing Decimals

1. Make the divisor a whole number by multiplying by whatever power of 10 moves the position of the decimal point to the extreme right of the divisor.

2. Multiply the dividend by the same number (power of 10) that you used in step 1; that is, move the decimal point in the dividend the same number of places to the right. Annex zeroes if necessary.

3. Place a decimal point in the quotient (answer) directly above the new position of the decimal point in the dividend.

4. Divide, using standard methods of division.

5. Round off your answer.

$$\text{Divisor}\overline{)\text{Dividend}}^{\text{Quotient} + \text{Remainder}}$$

Your check on answers to division problems is multiplication:

(Divisor × Quotient) + Remainder = Dividend

After counting the number of significant digits in each number, determine the smallest number of significant digits found in either the divisor or the dividend. Now round off the answer so that it contains only as many significant digits as are found in the number having the smallest number of significant digits.

EXAMPLE Divide: 3.5 into 19.856

$$3.5\overline{)19.856}$$

Move decimal point one place to the right in the divisor, and then one place to the right in the dividend.

The example is now: $35\overline{)198.56}$.

Since we multiplied both divisor and dividend by 10, note that a decimal point is placed in the quotient over the decimal point in the dividend.

$$\begin{array}{r} 5.673 \\ 35\overline{)198.56} \end{array}$$

There are two significant digits in the divisor and five significant digits in the dividend. Thus the answer is rounded off to two significant digits as 5.7.

EXAMPLE Divide: 0.061 by 8.2574

$8.2574\overline{)0.061}$

Move decimal point four places to the right in the divisor. To move four places to the right in the dividend, annex a zero to it.

The example is now: $82574\overline{)0610}$.

Apply standard methods of division to obtain the quotient of 0.00738. Now the smallest number of significant digits is two for both the divisor and the dividend. Therefore, our answer is rounded off to two significant digits, or 0.0074.

ARITHMETIC OF FRACTIONS

Introductory Concepts

1. A fraction is a symbol A/B, where A is the numerator and B is the denominator. The fraction A/B is read "A over B."

2. A proper fraction is a fraction where the numerator is smaller than the denominator: for example, $\frac{3}{5}$ and $\frac{19}{21}$.

3. An improper fraction is a fraction where the numerator is larger than or equal to the denominator: for example, $\frac{5}{4}$, $\frac{1}{1}$, and $\frac{18}{10}$.

4. A common fraction is either a proper fraction or an improper fraction.

Reducing Fractions to Simplest Terms

To reduce a fraction to lowest terms, divide both the numerator and the denominator by the largest possible number which divides both evenly.

EXAMPLES

$$\frac{6}{21} = \frac{2}{7}$$

$$\frac{45}{60} = \frac{3}{4}$$

NOTE In the first example, we divided both the numerator and the denominator by 3, and in the second example, we divided both the numerator and the denominator by 15.

Changing a Mixed Number into an Improper Fraction

A mixed number is a number that represents the sum of a whole number and a proper fraction: for example, $3\frac{1}{2}$ and $6\frac{3}{4}$.

Procedure

To change a mixed number into an improper fraction:

1. Multiply the whole number by the denominator of the fraction.
2. Add the numerator to this product.
3. Write this sum over the denominator of the fraction.

EXAMPLE Change $16\frac{5}{8}$ to an improper fraction.

$$16\,\frac{5}{8} = \frac{(16 \times 8) + 5}{8} = \frac{133}{8}$$

EXAMPLE Change $32\frac{11}{15}$ to an improper fraction.

$$32\,\frac{11}{15} = \frac{(32 \times 15) + 11}{15} = \frac{491}{15}$$

Changing an Improper Fraction into a Mixed Number

1. Divide the numerator by the denominator.
2. If there is a remainder, after this division, write it as the numerator of a fraction having the same denominator as the improper fraction.
3. Reduce the answer to lowest terms.

EXAMPLES

$$\frac{59}{9} = 6\,\frac{5}{9}$$

$$\frac{182}{8} = 22\,\frac{6}{8} = 22\,\frac{3}{4}$$

Equality of Fractions

Two fractions A/B and C/D are equal if and only if AD = BC. For example:

$$\frac{(A)\ 2}{(B)\ 4} = \frac{(C)\ 5}{(D)\ 10} \quad \text{because}$$

$$\frac{2}{4} = \frac{5}{10} \text{ yields}$$

$$\underset{2 \times 10}{(A) \times (D)} = \underset{5 \times 4}{(B) \times (C)} \qquad \text{or } 20 = 20$$

From A/B = C/D we can always conclude that $A \cdot D = B \cdot C$ (this is called cross-multiplication).

Multiplying Fractions

If A/B and C/D are two fractions, then their product

$$\frac{A}{B} \times \frac{C}{D} = \frac{A \times C}{B \times C}$$

EXAMPLE $\frac{3}{4} \times \frac{8}{10} = \frac{3 \times 8}{4 \times 10} = \frac{24}{40} = \frac{3}{5}$

NOTE The fraction 24/40 can be reduced to 3/5 by dividing both the numerator and the denominator by 8.

Dividing Fractions

If A/B and C/D are fractions, then:

$$\frac{A}{B} \div \frac{C}{D} = \frac{A \times D}{B \times C}$$

Procedure

Simply invert the second fraction and multiply. Reduce your answer.

EXAMPLE $\frac{3}{8} \div \frac{9}{20} = \frac{3}{8} \times \frac{20}{9} = \frac{3 \times 20}{8 \times 9} = \frac{60}{72} = \frac{5}{6}$

EXAMPLE $\frac{2}{3} \div \frac{4}{9} = \frac{2}{3} \times \frac{9}{4} = \frac{2 \times 9}{3 \times 4} = \frac{18}{12} = \frac{3}{2}$

Finding the Least Common Denominator for Fractions

The least common denominator (LCD) for several fractions is the smallest number which is divisible by each denominator of the fractions.

EXAMPLE Find the LCD for $\frac{2}{6}, \frac{3}{8}, \frac{4}{3}$.

The LCD is 24 because 6, 8, and 3 divide 24 evenly, and 24 is the *smallest* number which has this property. Note that 48 and 96 are not acceptable for the LCD although they are both common denominators (neither is the least common denominator).

EXAMPLE Find the LCD for $\frac{1}{5}$, $\frac{4}{3}$, $\frac{3}{2}$.

The LCD is 30 because 5, 3, and 2 divide 30 evenly, and 30 is the *smallest* number which can be evenly divided by 5, 3, and 2.

Adding Fractions with the Same Denominators

1. Add the numerators.
2. Divide the sum by the same denominator.
3. Reduce to lowest terms.

EXAMPLE $\frac{A}{B} + \frac{C}{B} = \frac{A + C}{B}$

EXAMPLE $\frac{2}{3} + \frac{7}{3} = \frac{2 + 7}{3} = \frac{9}{3} = \frac{3}{1} = 3$

EXAMPLE $15\frac{1}{8} + 12\frac{3}{8} = \frac{121}{8} + \frac{99}{8} = \frac{121 + 99}{8} = \frac{220}{8} = \frac{55}{2}$ or $27\frac{1}{2}$

Adding Fractions with Different Denominators

1. Find the LCD.
2. Change each fraction so that they all have the same denominator, which will be the LCD.
3. Use the procedure for adding fractions with like denominators.

EXAMPLE $\frac{3}{4} + \frac{2}{3} + \frac{5}{8}$

The LCD = 24.

$\frac{18}{24} + \frac{16}{24} + \frac{15}{24} = \frac{18 + 16 + 15}{24} = \frac{49}{24}$ or $2\frac{1}{24}$

NOTE $\frac{3}{4} = \frac{18}{24}$ $\frac{2}{3} = \frac{16}{24}$ $\frac{5}{8} = \frac{15}{24}$

EXAMPLE $12\frac{1}{2} + 5\frac{1}{3} + 2\frac{1}{16}$

$\frac{25}{2} + \frac{16}{3} + \frac{33}{16}$

The LCD is 48.

NOTE $\frac{25}{2} = \frac{600}{48}$ $\frac{16}{3} = \frac{256}{48}$ $\frac{33}{16} = \frac{99}{48}$

$$\frac{600}{48} + \frac{256}{48} + \frac{99}{48} = \frac{955}{48} \text{ or } 19\frac{43}{48}$$

Subtracting Fractions with Same Denominators

1. Subtract the numerators.
2. Write the difference over the same denominator.
3. Simplify the answer.

EXAMPLE $\frac{8}{12} - \frac{5}{12} = \frac{8-5}{12} = \frac{3}{12} = \frac{1}{4}$

EXAMPLE $18\frac{3}{5} - 5\frac{1}{5} = \frac{93}{5} - \frac{26}{5} = \frac{93-26}{5} = \frac{67}{5}$

Subtracting Fractions with Different Denominators

1. Find the LCD.
2. Change each fraction so that they all have the same denominator, which will be the LCD.
3. Apply the procedure for subtraction of fractions with the same denominators.

EXAMPLE $\frac{5}{8} - \frac{1}{2} = \frac{5}{8} - \frac{4}{8} = \frac{5-4}{8} = \frac{1}{8}$

EXAMPLE $\frac{17}{6} - \frac{3}{2} + \frac{8}{5}$ (LCD = 30)

$$\frac{85}{30} - \frac{45}{30} + \frac{48}{30} = \frac{85 - 45 + 48}{30} = \frac{88}{30} = \frac{44}{15}$$

ARITHMETIC OF PERCENTAGE

X% means a fraction whose numerator is X and whose denominator is 100. For example: 30% = $^{30}/_{100}$, 14% = $^{14}/_{100}$, and 100% = $^{100}/_{100}$ = 1.

Changing Percentages to Decimals

1. Move the decimal point two places to the left.
2. Omit the percent sign.

EXAMPLES 27% = 0.27

13.5% = 0.135

$3\% = 0.03$

$\frac{3\%}{8} = \frac{0.003}{8} = 0.00375$

$0.6\% = 0.006$

$3.8\% = 0.038$

Important Equivalents

$12\frac{1}{2}\% = 0.125 = \frac{1}{8}$

$25\% = 0.25 = \frac{1}{4}$

$33\frac{1}{3}\% = \frac{1}{3}$

$37\frac{1}{2}\% = 0.375 = \frac{3}{8}$

$87\frac{1}{2}\% = 0.875 = \frac{7}{8}$

$50\% = 0.50 = \frac{1}{2}$

$66\frac{2}{3}\% = \frac{2}{3}$

$62\frac{1}{2}\% = 0.625 = \frac{5}{8}$

$75\% = 0.75 = \frac{3}{4}$

$100\% = 1.00 = 1$

Changing Decimals to Percentages

1. Move the decimal point two places to the right.
2. Write in the percent sign.

EXAMPLES

$0.45 = 45\%$

$0.035 = 3.5\%$ or $3\frac{1}{2}\%$

$0.07 = 7\%$

$0.003 = 0.3\%$

$0.0875 = 8.75\%$ or $8\frac{3}{4}\%$

$\frac{0.001}{4} = \frac{1}{4}\%$

Changing Percentages to Fractions

1. The numerator is the given number without the percent sign.
2. Denominator = 100.
3. Reduce fraction to simplest form.

EXAMPLE $30\% = \frac{30}{100} = \frac{3}{10}$

EXAMPLE $6\frac{1}{4}\% = \frac{6\frac{1}{4}}{100} = \frac{25}{4} \div 100 = \frac{25}{4} \times \frac{1}{100} = \frac{1}{16}$

EXAMPLE $2.5\% = \frac{2.5}{100} = \frac{5}{2} \div 100 = \frac{5}{2} \times \frac{1}{100} = \frac{1}{40}$

Changing Fractions to Percentages

1. Change the fraction so that its denominator = 100.

2. The numerator of the resulting equivalent fraction represents the desired percentage.

EXAMPLES $\frac{6}{25}$: $\frac{6}{25} \times \frac{4}{4} = \frac{24}{100} = 24\%$

$\frac{1}{4}$: $\frac{1}{4} \times \frac{25}{25} = \frac{25}{100} = 25\%$

$\frac{7}{10}$: $\frac{7}{10} \times \frac{10}{10} = \frac{70}{100} = 70\%$

Alternate solution

1. Change fraction to a decimal by dividing the numerator by the denominator.

2. Move the decimal point in the answer (step 1) two places to the right and add the % symbol.

$\frac{4}{15}$: $\frac{4}{15} = 4 \div 15 = 0.266 = 0.27 = 27\%$

$\frac{3}{35}$: $\frac{3}{35} = 3 \div 35 = 0.0857 = 0.086 = 8.6\%$

Finding Percentage

1. Change the percent to a decimal.

2. Multiply the number by the decimal.

EXAMPLE Find 20% of 40.

Solution $0.20 \times 40 = 8$

EXAMPLE Find $33\frac{1}{3}\%$ of 96.

Solution $\frac{1}{3} \times 96 = 32$ or $0.33 \times 96 = 32$

EXAMPLE Find $\frac{1}{2}\%$ of 50.

Solution $0.005 \times 50 = 0.25$

Percentage Error

1. Find the experimental value E by conducting an experiment.
2. Find the standard value S from an established sourcebook.
3. Assuming E is greater than S ($E > S$), find $E - S$.
4. Divide $\frac{E - S}{S}$
5. Multiply by 100: $\left(\frac{E - S}{S}\right) \times 100 =$ percentage error.

If the experimental value is numerically greater than the standard value, the percentage error is *positive.*

If the experimental value is numerically less than the true value, the percentage error is *negative.*

EXAMPLE Find the percentage error if $E = 14.8$ cm and $S = 14.2$ cm.

Solution

1. $E - S$: $14.8 - 14.2 = 0.6$
2. $\frac{E - S}{S} = 0.042 = 0.04$
3. 4%

EXAMPLE Find the percentage error if $E = 2.87$ g and $S = 2.98$ g.

Solution

1. $E - S$: $2.87 - 2.98 = -0.11$
2. $\frac{E - S}{S} = \frac{-0.11}{2.98} = -0.0369 = -0.037$
3. -3.7%

LOGARITHMS

There are two kinds of logarithms: the *common logarithm* of a number is the power or exponent to which 10 must be raised to be equal to the number ($n = 10^x$). The second type of logarithm, called a *natural logarithm*, has a natural base of 2.718, sometimes called *base* e ($n = e^x$). To distinguish between the two bases, the common logarithm of x is written $\log_{10} x$ or simply $\log x$ and the natural logarithm is written $\log_e x$ or $\ln x$. A simple mathematical relationship exists between the two logarithmic forms:

$$\ln x = 2.303 \log x$$

The relationship between ordinary numbers and their common logarithmic expression is shown in Table 6-4. It should be noted from the table

TABLE 6-4
Numbers and Their Logarithms

Number	*Fraction*	*Exponential form*	*Logarithm*
1000	1000/1	10^3	3
100	100/1	10^2	2
10	10/1	10^1	1
1	1/1	10^0	0
0.1	1/10	10^{-1}	−1
0.01	1/100	10^{-2}	−2
0.001	1/1000	10^{-3}	−3

that the logarithm of 1 is zero, the logarithm of any number greater than 1 is a positive number, and the logarithm of a number less than 1 is a negative number. The table indicates that whole-number multiples of 10 have logarithms that are whole numbers. However most numbers are not whole-number multiples of 10 and require two sets of numbers to define the logarithmic expression.

EXAMPLE Express the number 200 in logarithmic form.

Solution **1.** First write the number in scientific form:
$\log 200 = \log (2 \times 10^2)$

2. Second, determine the logarithm of the numbers individually from a common logarithm table (Appendix B) and add them together.

$$\log 200 = \log 2 + \log 10^2 = 0.3010 + 2 - 2.3010$$

The number on the left of the decimal point in a logarithmic expression is called the *characteristic*, and the digits on the right of the decimal point are called the *mantissa*. The characteristic determines the order of magnitude (power) of the number, and the mantissa represents the number itself expressed as a value between 1 and 10. Fortunately, most scientific calculators have both common and natural logarithmic functions available. Refer to the section in this chapter on scientific calculators for more details.

THE SCIENTIFIC CALCULATOR

The hand-held calculator (Fig. 6-2) offers unprecedented computational power. The original models were capable only of performing the basic arithmetic functions: addition, subtraction, multiplication, and division. Now they can perform highly specialized operations such as computing the tangent of an angle; computing square roots, reciprocals, and scientific notation; extracting roots; determining factorial powers, angular functions, sines, arcsines, hyperbolic sines, cosines, and tangents; calculating arithmetic mean; and many, many others, depending upon the model.

FIGURE 6-2
A hand-held calculator *(Hewlett-Packard Company).*

Some calculators are preprogrammed, while more sophisticated ones can be programmed to meet almost any need.

The following material provides basic information about advanced calculators and a general procedure regarding their use.*

* This section used by permission of Hewlett-Packard Company.

Whatever the machine's language, a calculator's usefulness is very much determined by such things as its keyboard functions, memory capacity and display features. In this section you'll find information about each of these areas that will help you decide just what you need on an advanced calculator to meet your specific needs.

Keyboard Functions

All advanced calculators have keys for entering the digts 0–9 and executing an assortment of special-purpose mathematical functions and operations. They also have a cluster of instruction keys for such purposes as clearing the calculator and controlling both the display and memory.

The various instruction and special-purpose function keys can be classified into these broad categories:

1. General Keys—those which execute instructions and mathematical functions common to both scientific and business calculators.

2. Specialized Keys—those which execute special-purpose scientific, financial or statistical functions.

Figure 6-3 divides many keys commonly found on advanced calculators into these areas.

Many advanced machines have so many keyboard operations that individual keys are assigned one or more secondary functions. These secondary functions are identified by colored markings printed on or adjacent to the key and are selected by first pressing a prefix having the same color.

Although an advanced calculator may have 24 or more keys (as compared to 15 or 16 on a simple calculator), it is important to pay close attention to just what keys are included. Be sure to select a calculator which has operations, instructions and functions you most need.

Memory Registers

A memory register is an electronic storage circuit capable of retaining any number which can be keyed into the calculator. All advanced scientific and business calculators have at least one memory, and as you will recall from the discussion of calculator languages, RPN machines include a set of special number-processing memory registers called the operational stack.

You can probably handle most of your problem-solving requirments with an advanced machine having only one memory register, especially if it is complemented by a three- or four-level RPN operational stack. However, the computational ability of any advanced machine, particularly programmable models, is enhanced by multiple memory registers.

Even more important than the *number* of memory registers is the *ease* with which a number in the display can be arithmetically combined with the contents of the memory. This ability is called register arithmetic, and the best calculators have at least one fully addressable memory register which allows display-memory arith-

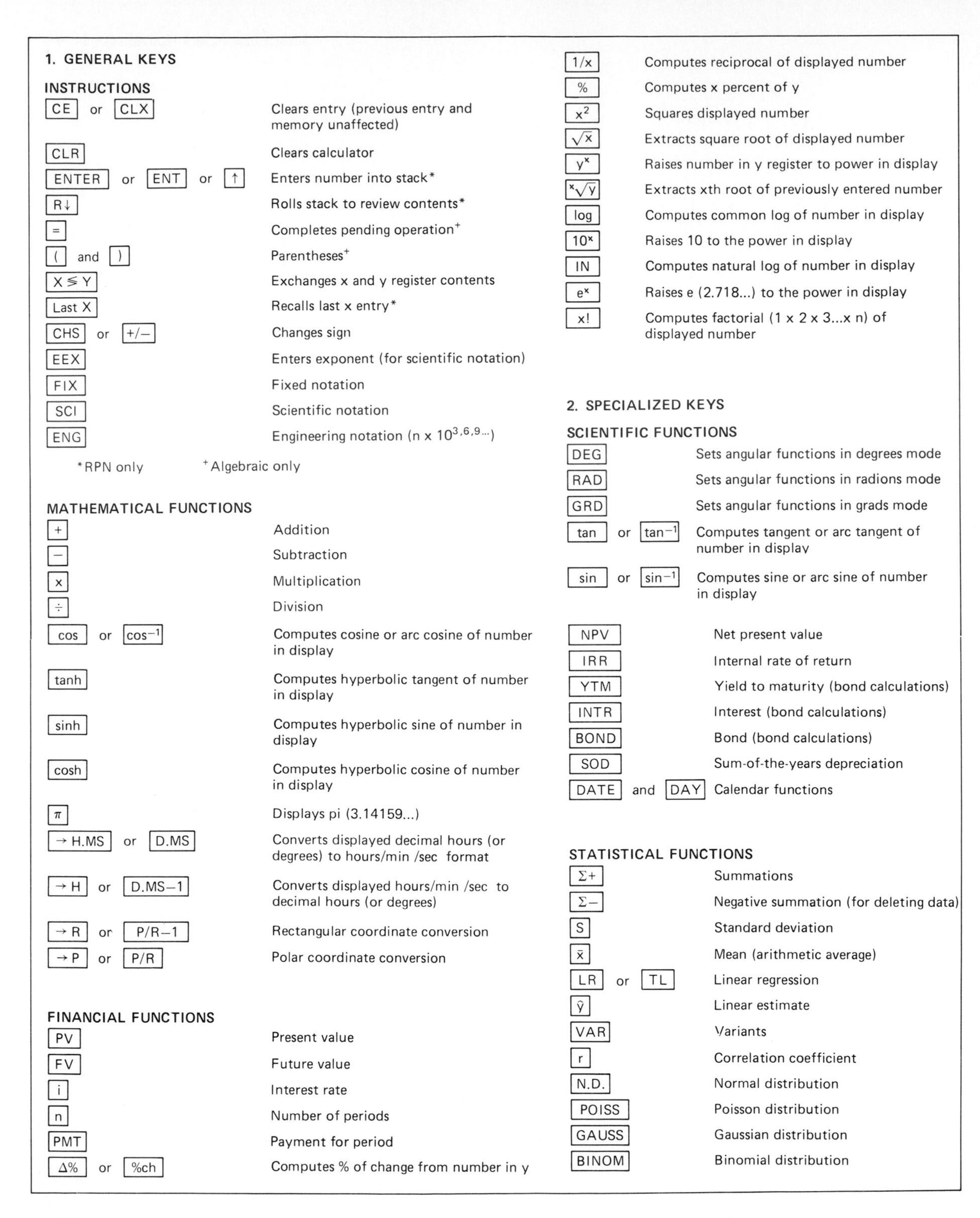

1. GENERAL KEYS

INSTRUCTIONS

CE or CLX	Clears entry (previous entry and memory unaffected)
CLR	Clears calculator
ENTER or ENT or ↑	Enters number into stack*
R↓	Rolls stack to review contents*
=	Completes pending operation+
(and)	Parentheses+
X ≶ Y	Exchanges x and y register contents
Last X	Recalls last x entry*
CHS or +/−	Changes sign
EEX	Enters exponent (for scientific notation)
FIX	Fixed notation
SCI	Scientific notation
ENG	Engineering notation (n x $10^{3,6,9...}$)

*RPN only +Algebraic only

MATHEMATICAL FUNCTIONS

+	Addition
−	Subtraction
x	Multiplication
÷	Division
cos or $\cos^{-1}$	Computes cosine or arc cosine of number in display
tanh	Computes hyperbolic tangent of number in display
sinh	Computes hyperbolic sine of number in display
cosh	Computes hyperbolic cosine of number in display
π	Displays pi (3.14159...)
→ H.MS or D.MS	Converts displayed decimal hours (or degrees) to hours/min /sec format
→ H or D.MS−1	Converts displayed hours/min /sec to decimal hours (or degrees)
→ R or P/R−1	Rectangular coordinate conversion
→ P or P/R	Polar coordinate conversion

FINANCIAL FUNCTIONS

PV	Present value
FV	Future value
i	Interest rate
n	Number of periods
PMT	Payment for period
Δ% or %ch	Computes % of change from number in y
1/x	Computes reciprocal of displayed number
%	Computes x percent of y
x^2	Squares displayed number
$\sqrt{x}$	Extracts square root of displayed number
y^x	Raises number in y register to power in display
$\sqrt[x]{y}$	Extracts xth root of previously entered number
log	Computes common log of number in display
10^x	Raises 10 to the power in display
IN	Computes natural log of number in display
e^x	Raises e (2.718...) to the power in display
x!	Computes factorial (1 x 2 x 3...x n) of displayed number

2. SPECIALIZED KEYS

SCIENTIFIC FUNCTIONS

DEG	Sets angular functions in degrees mode
RAD	Sets angular functions in radions mode
GRD	Sets angular functions in grads mode
tan or $\tan^{-1}$	Computes tangent or arc tangent of number in display
sin or $\sin^{-1}$	Computes sine or arc sine of number in display
NPV	Net present value
IRR	Internal rate of return
YTM	Yield to maturity (bond calculations)
INTR	Interest (bond calculations)
BOND	Bond (bond calculations)
SOD	Sum-of-the-years depreciation
DATE and DAY	Calendar functions

STATISTICAL FUNCTIONS

Σ+	Summations
Σ−	Negative summation (for deleting data)
S	Standard deviation
$\bar{x}$	Mean (arithmetic average)
LR or TL	Linear regression
$\hat{y}$	Linear estimate
VAR	Variants
r	Correlation coefficient
N.D.	Normal distribution
POISS	Poisson distribution
GAUSS	Gaussian distribution
BINOM	Binomial distribution

FIGURE 6-3

Common advanced calculator keyboard functions *(Hewlett-Packard Company)*.

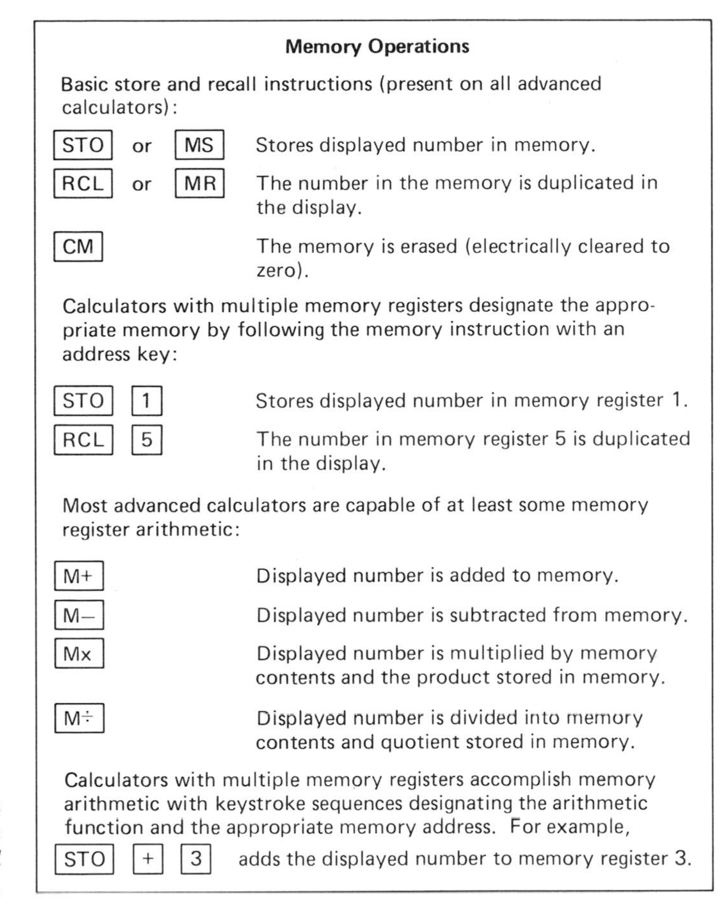

Memory Operations

Basic store and recall instructions (present on all advanced calculators):

STO or MS	Stores displayed number in memory.
RCL or MR	The number in the memory is duplicated in the display.
CM	The memory is erased (electrically cleared to zero).

Calculators with multiple memory registers designate the appropriate memory by following the memory instruction with an address key:

STO 1	Stores displayed number in memory register 1.
RCL 5	The number in memory register 5 is duplicated in the display.

Most advanced calculators are capable of at least some memory register arithmetic:

M+	Displayed number is added to memory.
M–	Displayed number is subtracted from memory.
Mx	Displayed number is multiplied by memory contents and the product stored in memory.
M÷	Displayed number is divided into memory contents and quotient stored in memory.

Calculators with multiple memory registers accomplish memory arithmetic with keystroke sequences designating the arithmetic function and the appropriate memory address. For example,

STO + 3 adds the displayed number to memory register 3.

FIGURE 6-4
Memory operations *(Hewlett-Packard Company).*

metic. The keystrokes which are fairly typical of conventional and addressable memory operations are shown in Fig. 6-4.

Since memory operations will involve perhaps 30 percent of the keystrokes you make on an advanced calculator, be sure to evaluate carefully the memory capabilities of the machines you are considering. A single memory register is adequate, particularly if you can perform memory arithmetic on it. But the optimum choice is a bank of six or more fully addressable memory registers.

Incidentally, calculators with multiple memory registers may have special operating restrictions you should know about. For example, some advanced calculators use one or more of their memory registers for such operations as tallying summations or deriving trigonometric functions. This may erase data you have previously stored in the affected registers.

Display Features

The display is the only means a calculator has for communicating with its user. The display announces when the machine is turned on and off, shows data entered and the results of calculations, flashes error signals, and indicates signs and exponents.

Most advanced calculators use a display format similar or identical to the one shown below (Fig. 6-5):

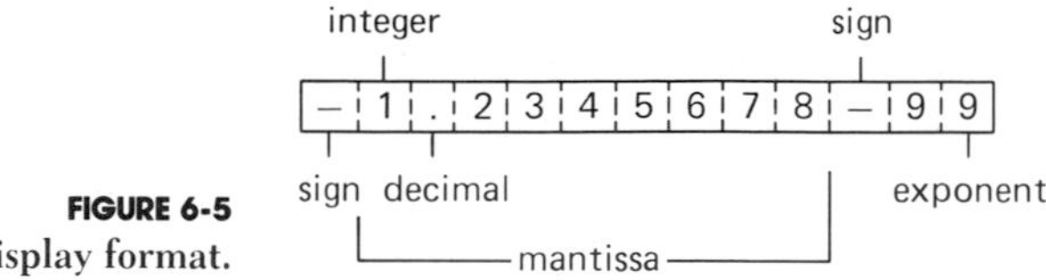

FIGURE 6-5
Display format.

Depending upon the calculator and the display format you select, the number 12,345 can be displayed in each of the following manners (see Fig. 6-6):

Fixed Point:

| 1 2 3 4 5 . 0 0 0 |

Scientific Notation:

| 1 . 2 3 4 5 0 0 0   0 4 |

(1.2345×10^4)

Engineering Notation:

| 1 2 . 3 4 5 0 0 0   0 3 |

(12.345×10^3)

FIGURE 6-6
Display variations.

Each format is correct, but depending on your use, one may be far more convenient than the others.

Fixed point display The fixed point display mode, which is common to most advanced calculators, permits you to select the number of digits to be displayed *after* the decimal point. If a numerical entry or result exceeds the number of digits selected, the calculator automatically rounds the least significant digit shown in the display while continuing to perform all calculations at the machine's maximum accuracy. *Both* features are useful, as you can see in this example.

Pencils cost .027 cents when bought in large lots. How much will a gross of pencils (144) cost, to the nearest cent?

$144 \times .027 = 3.888$

To solve this problem you would fix the number of decimal points at two, then multiply 144 and .027. Although the answer is 3.888, the calculator will automatically round it to give you the convenient answer of $3.89.

Scientific notation display Virtually all serious applications for advanced pocket calculators involve very small or very large numbers. Since a calculator display has a limited number of digits, most advanced models express numbers which exceed the capacity of the display in scientific notation.

Scientific notation is a simple shorthand method of expressing a number as a multiple of a power of 10. For example, 10,000 in scientific notation is 1×10^4. Numbers smaller than one require a negative exponent: e.g., 1×10^{-4} is the same as 0.0001. In each case, the exponent defines the number of digits which separate the number from the decimal point.

Here's an example of how scientific notation simplifies problem solving:

Light travels at the incredible velocity of approximately 30,000,000,000 centimeters in one second. How much time is required for a ray of light to traverse a space defined by an atom measuring 0.00001 centimeters in diameter?

To get the traverse time you would divide the distance by the velocity. The calculator will give you the answer: $3.3333333 \times 10^{-16}$. Without scientific notation, a 17-digit display would be required just to show the first 3 in fixed notation.

In addition to permitting the processing of numbers far smaller or larger than the digit capacity of its display, a calculator with scientific notation simplifies many types of computations.

Engineering notation display Engineers, astronomers, physicists and others who frequently encounter numbers expressed in increments of three decades (e.g., milli-, mega-, tera-, etc.) will find engineering notation a handy feature. This capability is found on a few of the newer, more advanced scientific calculators and, when selected, automatically converts entries and results to a modified form of scientific notation wherein the exponent of 10 is a multiple of three (e.g., 10^3, 10^9, 10^{12}, etc.). For example:

The fastest shutter setting on most quality cameras is $\frac{1}{1000}$ second or one *millisecond* (1×10^{-3} second). Most inexpensive cameras have a single shutter speed of $\frac{1}{60}$ second. How many milliseconds is the latter time?

The answer would be displayed as shown in Fig. 6-7.

Fixed: 0.016666667

Scientific: 1.6666666 02

Engineering: 16.666666 03

FIGURE 6-7
Display variations.

While all three forms of the answer are correct, the engineering notation version is a very obvious 167.7 milliseconds.

7
LABORATORY STATISTICS

INTRODUCTION

Statistics are an absolute necessity in the chemistry laboratory for determining the accuracy and precision of quantitative analytical data. Many new and specific mathematical terms will be encountered in the study of statistics. For example in the laboratory, *accuracy* means the "correctness" of a given analysis while *precision* indicates the "reproducibility" of an analytical procedure. The National Bureau of Standards described the difference between accuracy and precision: "Accuracy" has to do with closeness [of data] to the truth, "precision" only with closeness of readings to one another. An archery target provides a good analogy for these two often confused terms. If the arrows are all located directly in the bull's-eye, then both accuracy and precision are demonstrated. However, if all the arrows are clustered in the lower-left quadrant and not near the bull's-eye, one would have good precision but poor accuracy.

A major potential source of error in laboratory results can be the *sampling* procedure itself. The question to be asked is: Does the laboratory sample truly represent the material being tested? Many professional societies and governmental agencies (Association of Official Analytical Chemists, American Society for Testing and Materials (ASTM), U.S. Environmental Protection Agency (EPA), etc.) provide very specific instructions on obtaining, preparing, and storing representative samples. Consult Chap. 15, "Gravimetric Analysis," for more details on proper sampling techniques.

Chapter 6, "Basic Laboratory Mathematics," discusses some of the more elementary mathematical concepts needed in laboratory calculations. A major source of error with the advent of the electronic calculator involved significant-figure misuse. (See Significant Figures and Rounding in Chap. 6.)

STATISTICAL TERMINOLOGY

Some additional common terms in laboratory statistics are *absolute error, relative error, indeterminate errors, determinate errors, mean, median, mode, deviation, relative deviation, average deviation, standard deviation, relative standard deviation,* and *confidence limits,* or *confidence intervals.*

These terms are defined as follows:

Absolute error The difference between the true value and the measured value with the algebraic sign indicating whether the measured value is above (+) or below (−) the true value.

Relative error The absolute error (difference between the true and measured value) divided by the true value; usually expressed as a percentage.

Indeterminate errors Random errors that result from uncontrolled variables in an experiment and cannot normally be determined because they do not have a single source.

Determinate errors Those errors that can be ascribed to a particular cause and thus can usually be determined as being personal, instrumental, or method uncertainties.

Mean **(m)** The technique of "taking an average" by adding together the numerical values (x, y, z, etc.) of an analysis and dividing this sum by the number n of measurements used yields the mean.

$$m = \frac{x + y + z}{n}$$

Median The same data used for calculating the mean can be displayed in an increasing or decreasing series and the "middle" value simply selected as the median. The advantage over the mean is that the median will always be one of the actual measurements. Of course, if the total number of measurements is an even number, there will not be a single middle value; thus the median will be the average of the two middle values.

Mode The measurement value that appears most frequently in the series is called the *mode.*

Deviation As the name might imply, how much each measured value differs from the mean is referred to as the *deviation.* Mathematically, the deviation is calculated using the following equation where d_x is the deviation, m represents the mean, and x stands for the measured value. The vertical bars (| |) signify "absolute value"; thus the algebraic sign will always be positive.

$$d_x = |m - x|$$

Relative deviation **(d_R)** This value relates the deviation to the mean to indicate the magnitude of the variance. If the mean is a rather large number, then the deviation is not as critical as it would be in the case of a smaller mean. Mathematically, the relative deviation d_R can be calculated by dividing the deviation d by the mean m according to this equation:

$$d_R = \frac{d}{m}$$

The d_R can be multiplied by 100 to express the percent relative deviation, or multiplying by 1000 yields parts per thousand (ppt), which are used frequently in quantitative analysis to express precision of measurement.

$$\%d_R = \frac{d_R}{m} \times 100$$

$$\text{ppt } d_R = \frac{d_R}{m} \times 1000$$

***Average deviation* (d_A)** This value indicates the precision of all the measurements and is calculated by dividing the sum of all the individual deviations (d_x, d_y, d_z etc.) by the number n of deviations calculated.

$$d_A = \frac{d_x + d_y + d_z}{n} \ldots$$

***Standard deviation* (d_S)** Standard deviation is the most used of the deviation averaging techniques because it indicates *confidence limits*, or the *confidence interval*, for analyzing all data. The standard deviation can be calculated in five steps:

1. Determine the mean m.
2. Subtract the mean from each measured data item.
3. Square each difference.
4. Find the average of the squared terms in step 3.
5. Calculate the square root of the average found in step 4 by dividing by one less than the actual number of measurements.

In other words, the standard deviation is calculated by taking the square root of the quotient from the sum of all the squared individual deviations divided by one less than the number of measurements ($n - 1$) used in the analysis. Statistically it has been determined that as the number of measurements n exceeds 30, the $n - 1$ term can be simplified to n.

$$d_S = \sqrt{\frac{d_x^2 + d_y^2 + d_z^2 \ldots}{n - 1}}$$

EXAMPLE Calculate the standard deviation in these four weighings: 36.78 mg, 36.80 mg, 36.87 mg, and 36.94 mg.

$$\text{Mean } m = \frac{36.78 + 36.80 + 36.87 + 36.94}{4} = 36.85$$

Measurement	*Deviation* (m − x)	*Deviation squared* $(m - x)^2$
36.78	0.07	0.0049
36.80	0.05	0.0025
36.87	0.02	0.0004
36.94	0.09	0.0081
		0.0159

$$\text{Standard deviation } d_S = \sqrt{\frac{0.0159}{4-1}} = 0.07$$

Thus the answer should be reported as 36.85 ±0.07 mg

Relative standard deviation Relative standard deviation can be calculated by dividing the standard deviation by the mean. Like relative deviations, this ratio can be multiplied by 100 or 1000 to obtain the relative percent or parts per thousand deviation.

Confidence limits The interval around an experimental mean within which the true result can be expected to lie with a stated probability. Given a large enough number of measurements, a symmetrically shaped distribution curve (Gaussian) can be drawn by plotting the number of times a specific numerical measurement is found versus the numerical value of the measurements. Most measurements should be clustered around the mean as shown in Fig. 7-1 below. In this particular distribution, the frequency of observations is plotted on the ordinate and the standardized variable Z which is calculated by subtracting the mean value from the value of each result and dividing this difference by the standard

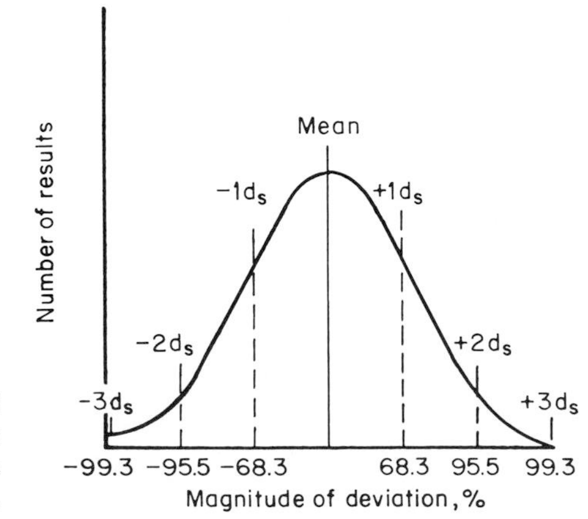

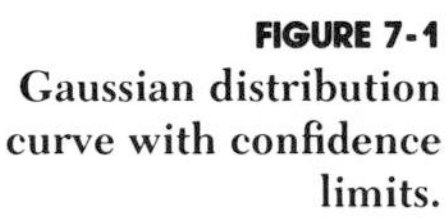
FIGURE 7-1
Gaussian distribution curve with confidence limits.

deviation [$Z = (x - m)/d_S$]. It can be shown that 68.3% of the area beneath this curve lies within one standard deviation d_S either way from the mean. Approximately 95.5% of all values will be within $\pm 2\ d_S$ and 99.3% within $\pm 3\ d_S$. These intervals are known as *confidence intervals.* Typically laboratory data is "rejected" or "retained" in the final laboratory report based on these predetermined confidence limits.

REJECTION OF LABORATORY DATA

A wide variety of statistical tests have been developed to determine whether data should be retained or rejected. The Q *Test* is one of the more popular methods and is useful even with a limited number of data. The ratio Q is calculated by arranging all data in increasing order and determining the difference between the questionable number and its nearest neighbor. This difference is then divided by the range (highest number − lowest number) of all the experimental data:

$$Q = \frac{\text{difference}}{\text{range}}$$

This ratio Q is compared to standard values of Q at a 90% confidence level for varying numbers of observations as is shown in Table 7-1. If the calculated Q ratio is equal to or exceeds the tabulated value for a given number of observations, the measurement is discarded.

For example, the following results were obtained from an analysis conducted to determine the amount of lead in a soil sample: 203, 204, 205, 206, 207, and 214 ppm. Should the last result, 214 ppm, be discarded? The questionable observation differs from its nearest neighbor, 207 ppm, by 7 ppm. The range in this experiment is from 214 to 203 ppm, or 11 ppm. Q would calculate to be $7/11$, or 0.64. The tabulated value for six observations is 0.56. Since the calculated Q is greater than the tabulated value for six observations, the questionable number should be rejected.

TABLE 7-1
Rejection Quotient Q at the 90% Confidence Limit

Number of observations	*Q*
3	0.94
4	0.76
5	0.64
6	0.56
7	0.51
8	0.47
9	0.44
10	0.41

METHOD OF LEAST SQUARES

A technician is frequently required to plot data on a graph, but can encounter data which does not fall on a straight line. One solution to this dilemma is simply to "eyeball" in the best straight line using a ruler and as many data points as possible. A better approach involves using statistics to define the best straight-line fit to the data. A straight-line relationship should be: $y = mx + b$, where y represents the dependent variables, x represents the independent variable, m is the *slope* of the curve, and b represents the ordinate (y axis) intercept. Mathematically, it has been determined that the best straight line through a series of data points is that line for which the sum of the squares of the deviations of the data points from the line is a minimum. Thus, this is known as the *method of least squares.* The actual calculations involve the use of differential calculus which is beyond the scope of this reference; however, most scientific calculators are capable of solving this type of calculation with minimum input from a technician. Normally, the calculator has a statistics function called *linear regression* (L.R.). Since the actual procedure varies for different models, consult the manual for your scientific calculator for the specific details.

QUALITY CONTROL STATISTICS

Statistics not only are useful for plotting graphs and rejecting unreliable laboratory data; they are also routinely used in laboratory quality control programs. Control samples are routinely analyzed along with all other laboratory samples. These standards, or *control samples,* have usually been analyzed many times or have been purchased as a standard from a commercial source. The laboratory management establishes a confidence level (for example: three standard deviations) that these control samples must be within, and a daily plot is maintained of the analytical results. With the central line normally representing the known concentration of the control, these "quality control charts" will indicate any sudden or even gradual trend for the analytical results to deviate. These control charts are also

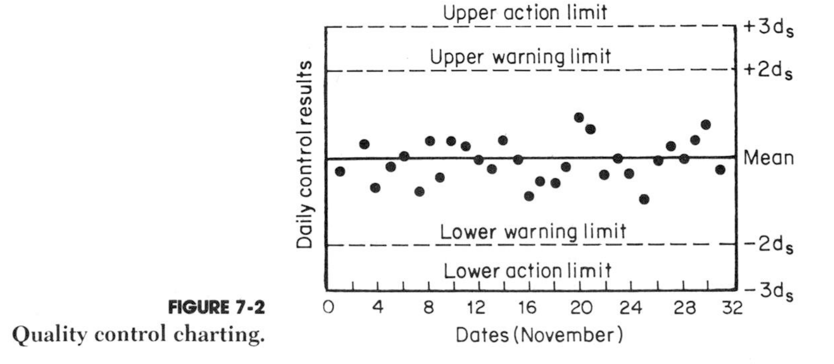

FIGURE 7-2
Quality control charting.

used in interlaboratory comparisons and auditing. A typical quality control chart is shown in Fig. 7-2.

REFERENCES

A more detailed treatment of laboratory statistics can be found in the following references:

Youdan, W. J., and Steiner, E. H., *Statistical Manual of the AOAC,* American Association of Official Analytical Chemists, Washington, D.C., 1975.

Industrial Hygiene Service Laboratory Manual, Technical Report No. 78, Department of Health, Education and Welfare, PHS, CDC, Cincinnati and Washington, D.C.

Laitinen, H. A., and Harris, W. E., "Statistics in Quantitative Analysis" and "Sampling," in *Chemical Analysis*, 2d ed., New York, McGraw-Hill Book Co., 1975, Chaps. 26 and 27.

8 COMPUTERS IN THE ANALYTICAL LABORATORY

INTRODUCTION

Computers have become an integral part of any modern chemical laboratory for tasks like routine calcuations, data acquisition, data manipulation, data storage, data reporting, instrument control, and robotics. The computer is especially useful in reducing the time and labor required of laboratory technicians to process laboratory information and data. Computers can also reduce the number of errors normally associated with the manual methods of computation. Computers are being used to control the parameters and in establishing the optimum operating conditions for laboratory instrumentation. Computers are also finding extensive use in averaging and integrating the tremendous output of data from sophisticated laboratory instruments into simplified formats. For a review of hand-held electronic calculators see Chap. 6's section entitled The Scientific Calculator.

There are two basic types of computers in use in the laboratory.

1. *Analog computers* solve problems by simulating the magnitude of a physical measurement in terms of analogous (proportional) voltages.

2. *Digital computers* solve problems based on an ability to handle num bers or "digits" and tend to be much more accurate than analog computers. Their operation is primarily arithmetically based on the binary system, which uses 2 as the base, instead of base 10 as is used with the decimal or SI/metric system. Any number used in the binary system can be represented by a series of 1s and 0s.

COMPUTER COMPONENTS AND TERMINOLOGY

Most computers consist of a *central processor unit (CPU)* and various *peripheral units.* The CPU normally contains the internal memory, control unit, and arithmetic unit. The peripheral units are various input-output devices, auxiliary memory, and long-term storage.

Definitions

Alphanumeric Set of computer characters which contain both letters and numbers.

ASCII Abbreviation for *American Standard Code for Information Inter-*

change. Industrywide code used to represent alphanumeric characters and to interface all computers and peripheral devices.

ROM *Read-only memory* program. Information located in the CPU that has been fixed by the manufacturer and cannot be changed by the user.

RAM *Random-access memory.* Data located in the CPU which can be changed by the user is located in random-access memory.

PROM *Programmable read-only memory.* Allows the user to modify the ROM established by the manufacturer.

Off-line computer Computer not dedicated to a particular instrument or application.

On-line computer Computer dedicated to a specific laboratory instrument and/or process. These computers allow technicians to operate much more sophisticated instrumentation such as gas chromatograph–mass spectrometers and Fourier transform infrared spectrophotometers.

Bit *Binary digit;* a single digit (1 or 0) in the binary system.

Byte A sequence of bits constituting a discrete piece of information. Bytes are usually no more than eight bits in length.

Interface A junction between the laboratory instrument and/or process and the computer; at the interface, signal matching or adjusting is accomplished to make the two systems compatible.

A–D converter An analog-to-digital coverter takes the typical "analog" signal from a laboratory instrument (for example, gas chromatograph millivolt signal) and converts it to equivalent "digital values" that the computer can process.

Operating system software Software designed to control tasks like the sequencing of jobs and the control of input-output laboratory devices and robotics.

Batch processing A system primarily used by mainframe (large) processors. Data is collected and stored on an input medium, for instance, a magnetic tape. Then the information is scheduled and run as a job where the output is stored on a peripheral device. This batch processing tends to optimize use of the CPU.

Real-time processing The computer is directly connected and usually dedicated to one or more laboratory instruments and/or processes. Data are introduced directly into the computer system, processed, and made available to the operator or to modify instrument conditions immediately.

Robotics An automated system used to perform repetitious, manual laboratory procedures. The system is usually a microprocessor-controlled robotic arm specifically designed for sample preparation or handling.

COMPUTER LANGUAGES

BASIC *(Beginner's All-Purpose Symbolic Instruction Code)* is the most popular language used with chemistry laboratory computers. The commands are in "English" and common algebraic terms, rather than in a symbolic code; thus the BASIC language is somewhat easier for a beginning user since the commands correspond to familiar "words." FORTRAN is another general language which is used in batch scientific programming. FORTRAN *(Formula Translation)* is based on algebraic notation and is easier to learn after BASIC since the structure and some of the commands are similar in both languages. COBOL is yet another general language which is used primarily in business and accounting applications. There are many other high-level (machine-independent) languages available: for example, PASCAL is finding increasing use in chemistry laboratories because it can be adapted to structured programming techniques in robotics.

COMPUTER SEARCHES

In 1980, the American Chemical Society (ACS) introduced a substance search service for access to the Chemical Abstracts Service's (CAS's) Chemical Registry System. This abstracting service currently contains more than 7 million chemical substances and more than 10 million chemical names. As new substances are reported, they are assigned a unique CAS registry number based on the substance's computer-readable molecular structure. Currently the CAS is adding approximately 8000 new substances each week. Once these substances have been entered into the computer registry system, they are available to the general public through the CAS ONLINE system. The substance database contains three files: (1) The "Registry File" contains the records of substances reported in the chemical literature and cited in *Chemical Abstracts* since 1965. (2) The "CA File" contains the *Chemical Abstracts* index, bibliographic references, and abstracts since 1967. (3) The "CAOLD File" contains references for substances cited in the chemical literature before 1967. For additional information about CAS ONLINE contact:

Chemical Abstracts Service
2540 Olentangy River Road
P.O. Box 3012
Columbus, OH 43210
(614) 421-3600

REFERENCES

Hepple, P., *The Application of Computer Techniques in Chemical Research*, Institute of Petroleum, London, 1972.

Klopfenstein, C. F., and Wilkins, C. L. (eds.), *Computers in Chemical and Biochemical Research,* Vols. 1 and 2, Academic Press, New York, 1972 and 1974.

Malmstadt, H. V., and Enke, C. G., *Digital Electronics for Scientists,* W. A. Benjamin, Menlo Park, Calif., 1969.

Perone, S. P., and Jones, D. O., *Digital Computers in Scientific Instrumentation: Application to Chemistry,* McGraw-Hill Book Co., New York, 1973.

Shames, M. H., *Automation in Clinical Chemistry,* an audio short course on tape, American Chemical Society, Washington, D.C., 1973.

Wipke, W. T., et al. (eds.), *Computer Representation and Manipulation of Chemical Information,* John Wiley & Sons, New York, 1974.

9

MEASURING TEMPERATURE

LIQUID THERMOMETERS

Temperatures are measured by instruments which indicate the intensity of the heat in a body. Heat will flow from a higher-temperature body to a lower-temperature body. The *liquid thermometer,* usually filled with mercury, is calibrated at the freezing point of water and the boiling point of water when taken under 1 atm of pressure.

There are four scales of calibration: Fahrenheit, Celsius (centigrade), SI (Kelvin), and Rankine. They are related as shown in Fig. 9-1. The lowest temperature theoretically obtainable is called the *absolute zero* of temperature.

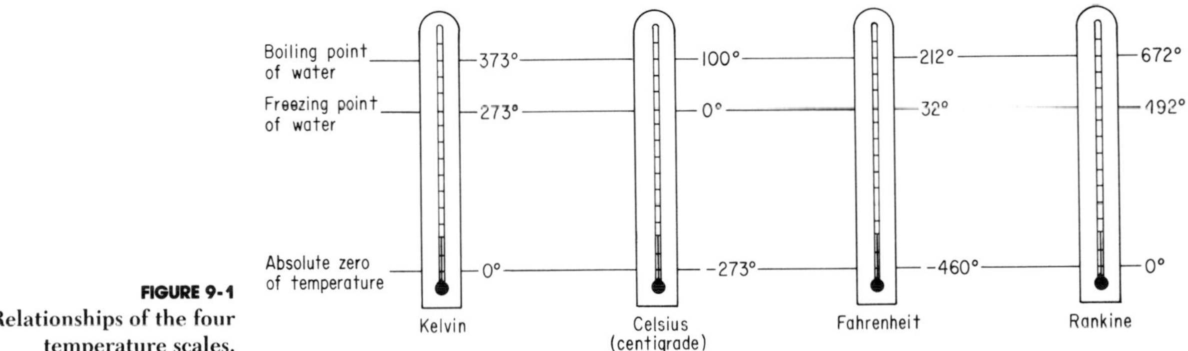

FIGURE 9-1
Relationships of the four temperature scales.

Conversion between Scales

Temperatures can be easily converted from one scale to another. Refer to Table 9-1 for temperature conversions from Celsius to Fahrenheit and from Fahrenheit to Celsius.

Fahrenheit to Celsius

1. Subtract 32° from Fahrenheit reading.
2. Multiply result by 5/9.

EXAMPLE $212°F = ?°C$

TABLE 9-1
Temperature-Conversion Table*

°C	*-140 to 20*	°F	°C	*21 to 55*	°F	°C	*56 to 90*	°F	°C	*91 to 340*	°F	°C	*350 to 690*	°F	°C	*700 to 1040*	°F
-95.5	**-140**	-220	-6.1	**21**	69.8	13.3	**56**	132.8	32.8	**91**	195.8	177	**350**	662	371	**700**	1292
-90.0	**-130**	-202	-5.6	**22**	71.6	13.9	**57**	134.6	33.3	**92**	197.6	182	**360**	680	377	**710**	1310
-84.4	**-120**	-184	-5.0	**23**	73.4	14.4	**58**	136.4	33.9	**93**	199.4	188	**370**	698	382	**720**	1328
-78.9	**-110**	-166	-4.4	**24**	75.2	15.0	**59**	138.2	34.4	**94**	201.2	193	**380**	716	388	**730**	1346
-73.3	**-100**	-148	-3.9	**25**	77.0	15.6	**60**	140.0	35.0	**95**	203.0	199	**390**	734	393	**740**	1364
-67.6	**-90**	-130	-3.3	**26**	78.8	16.1	**61**	141.8	35.6	**96**	204.8	204	**400**	752	399	**750**	1382
-62.2	**-80**	-112	-2.8	**27**	80.6	16.7	**62**	143.6	36.1	**97**	206.6	210	**410**	770	404	**760**	1400
-56.6	**-70**	-94	-2.2	**28**	82.4	17.2	**63**	145.4	36.7	**98**	208.4	216	**420**	788	410	**770**	1418
-51.1	**-60**	-76	-1.7	**29**	84.2	17.8	**64**	147.2	37.2	**99**	210.2	221	**430**	806	416	**780**	1436
-45.5	**-50**	-58	-1.1	**30**	86.0	18.3	**65**	149.0	37.8	**100**	212.0	227	**440**	824	421	**790**	1454
-40.0	**-40**	-40	-0.6	**31**	87.8	18.9	**66**	150.8				232	**450**	842	427	**800**	1472
-34.4	**-30**	-22	0	**32**	89.6	19.4	**67**	152.6	43	**110**	230	238	**460**	860	432	**810**	1490
-28.9	**-20**	-4				20.0	**68**	154.4	49	**120**	248	243	**470**	878	438	**820**	1508
-23.3	**-10**	14	0.6	**33**	91.4	20.6	**69**	156.2	54	**130**	266	249	**480**	896	443	**830**	1526
-17.8	**0**	32	1.1	**34**	93.2	21.1	**70**	158.0	60	**140**	284	254	**490**	914	449	**840**	1544
			1.7	**35**	95.0	21.7	**71**	159.8	66	**150**	302	260	**500**	932	454	**850**	1562
-17.2	**1**	33.8	2.2	**36**	96.8	22.2	**72**	161.6	71	**160**	320	266	**510**	950	460	**860**	1580
-16.7	**2**	35.6	2.8	**37**	98.6	22.8	**73**	163.4	77	**170**	338	271	**520**	968	466	**870**	1598
-16.1	**3**	37.4	3.3	**38**	100.4	23.3	**74**	165.2	82	**180**	356	277	**530**	986	471	**880**	1616
-15.6	**4**	39.2	3.9	**39**	102.2	23.9	**75**	167.0	88	**190**	374	282	**540**	1004	477	**890**	1634
-15.0	**5**	41.0	4.4	**40**	104.0	24.4	**76**	168.8	93	**200**	392	288	**550**	1022	482	**900**	1652
-14.4	**6**	42.8	5.0	**41**	105.8	25.0	**77**	170.6	99	**210**	410	293	**560**	1040	488	**910**	1670
-13.9	**7**	44.6	5.6	**42**	107.6	25.6	**78**	172.4	100	**212**	414	299	**570**	1058	493	**920**	1688
-13.3	**8**	46.4	6.1	**43**	109.4	26.1	**79**	174.2				304	**580**	1076	499	**930**	1706
-12.8	**9**	48.2	6.7	**44**	111.2	26.7	**80**	176.0	104	**220**	428	310	**590**	1094	504	**940**	1724
-12.2	**10**	50.0	7.2	**45**	113.0	27.2	**81**	177.8	110	**230**	446	316	**600**	1112	510	**950**	1742
-11.7	**11**	51.8	7.8	**46**	114.8	27.8	**82**	179.6	116	**240**	464	321	**610**	1130	516	**960**	1760
-11.1	**12**	53.6	8.3	**47**	116.6	28.3	**83**	181.4	121	**250**	482	327	**620**	1148	521	**970**	1778
-10.6	**13**	55.4	8.9	**48**	118.4	28.9	**84**	183.2	127	**260**	500	332	**630**	1166	527	**980**	1796
-10.0	**14**	57.2	9.4	**49**	120.2	29.4	**85**	185.0	132	**270**	518	338	**640**	1184	532	**990**	1814
-9.4	**15**	59.0	10.0	**50**	122.0	30.0	**86**	186.8	138	**280**	536	343	**650**	1202	538	**1000**	1832
-8.9	**16**	60.8	10.6	**51**	123.8	30.6	**87**	188.6	143	**290**	554	349	**660**	1220	543	**1010**	1850
-8.3	**17**	62.6	11.1	**52**	125.6	31.1	**88**	190.4	149	**300**	572	354	**670**	1238	549	**1020**	1868
-7.8	**18**	64.4	11.7	**53**	127.4	31.7	**89**	192.2	154	**310**	590	360	**680**	1256	554	**1030**	1886
-7.2	**19**	66.2	12.2	**54**	129.2	32.2	**90**	194.0	160	**320**	608	366	**690**	1274	560	**1040**	1904
-6.7	**20**	68.0	12.8	**55**	131.0				166	**330**	626						
									171	**340**	644						

* The conversion table may be used for converting degrees Fahrenheit to degrees Celsius or vice versa. Boldface numbers in the center refer to the known temperature in either Celsius or Fahrenheit. Equivalent temperature is found in the appropriate left or right column.
SOURCE: SGA Scientific, Inc., Bloomfield, N.J.

Solution

$$\begin{array}{r} 212°F \\ -\ 32° \\ \hline 180°F \end{array}$$

$\frac{5}{9} \times 180 = 100°C$ (Ans.)

Celsius to Fahrenheit

1. Multiply Celsius reading by $\frac{9}{5}$.
2. Add 32°.

EXAMPLE $100°C = ?°F$

Solution

$$\begin{array}{r} \frac{9}{5} \times 100°C = 180°C \\ +\ 32° \\ \hline 212°F \end{array}$$ (Ans.)

Celsius to SI (Kelvin)

1. Add the number 273 to the Celsius reading.
2. The value is the temperature reading in kelvins.

EXAMPLE $20°C = ?\ K$

Solution $273 + 20°C = 293\ K$ (Ans.)

Fahrenheit to SI

1. Convert Fahrenheit to Celsius.
2. Convert Celsius to SI.

Fahrenheit to Rankine

Add 460° to Fahrenheit temperature.

Celsius to Rankine

1. Convert Celsius to Fahrenheit.
2. Add 460° to Fahrenheit temperature.

Calibration of Thermometers

Calibrate laboratory thermometers by testing at the standard points.

0°C: Immerse thermometer in a well-stirred mixture of crushed ice and distilled water. The thermometer should read 0°C.

100°C: Fix thermometer above the surface of a beaker or flask of boiling water so that the mercury column is exposed to the vapor. The thermometer should read 100°C at a barometric pressure of 760 mmHg (torr).

TABLE 9-2
Corrections for Exposed Thread of Mercury for Total-Immersion Thermometers (Celsius)*

	$(T_o - T_m)$ *degrees*																			
l *degrees*	*10*	*20*	*30*	*40*	*50*	*60*	*70*	*80*	*90*	*100*	*110*	*120*	*130*	*140*	*150*	*160*	*170*	*180*	*190*	*200*
100	0.15	0.31	0.47	0.62	0.78	0.94	1.1	1.2	1.4	1.6	1.7	1.9	2.0	2.2	2.3	2.5	2.6	2.8	3.0	3.1
110	0.17	0.34	0.51	0.69	0.86	1.0	1.2	1.4	1.5	1.7	1.9	2.0	2.2	2.4	2.6	2.7	2.9	3.1	3.3	3.4
120	0.19	0.37	0.56	0.75	0.94	1.1	1.3	1.5	1.7	1.9	2.1	2.2	2.4	2.6	2.8	3.0	3.2	3.4	3.6	3.7
130	0.20	0.41	0.61	0.81	1.0	1.2	1.4	1.6	1.8	2.0	2.2	2.4	2.6	2.8	3.0	3.2	3.4	3.7	3.9	4.1
140	0.22	0.44	0.66	0.87	1.1	1.3	1.5	1.7	2.0	2.2	2.4	2.6	2.8	3.1	3.3	3.5	3.7	3.9	4.1	4.4
150	0.23	0.47	0.70	0.94	1.2	1.4	1.6	1.9	2.1	2.3	2.6	2.8	3.0	3.3	3.5	3.7	4.0	4.2	4.4	4.7
160	0.25	0.50	0.75	1.0	1.2	1.5	1.7	2.0	2.2	2.5	2.7	3.0	3.2	3.5	3.7	4.0	4.2	4.5	4.7	5.0
170	0.27	0.53	0.80	1.1	1.3	1.6	1.9	2.1	2.4	2.7	2.9	3.2	3.4	3.7	4.0	4.2	4.5	4.8	5.0	5.3
180	0.28	0.56	0.84	1.1	1.4	1.7	2.0	2.2	2.5	2.8	3.1	3.4	3.7	3.9	4.2	4.5	4.8	5.1	5.3	5.6
190	0.30	0.59	0.89	1.2	1.5	1.8	2.1	2.4	2.7	3.0	3.3	3.6	3.9	4.1	4.4	4.7	5.0	5.3	5.6	5.9
200	0.31	0.62	0.94	1.2	1.6	1.9	2.2	2.5	2.8	3.1	3.4	3.7	4.1	4.4	4.7	5.0	5.3	5.6	5.9	6.2
210	0.33	0.66	0.98	1.3	1.6	2.0	2.3	2.6	2.9	3.3	3.6	3.9	4.3	4.6	4.9	5.2	5.6	5.9	6.2	6.6
220	0.34	0.69	1.0	1.4	1.7	2.1	2.4	2.7	3.1	3.4	3.8	4.1	4.5	4.8	5.1	5.5	5.8	6.2	6.5	6.9
230	0.36	0.72	1.1	1.4	1.8	2.2	2.5	2.9	3.2	3.6	3.9	4.3	4.7	5.0	5.4	5.7	6.1	6.5	6.8	7.2
240	0.37	0.75	1.1	1.5	1.9	2.2	2.6	3.0	3.4	3.7	4.1	4.5	4.9	5.2	5.6	6.0	6.4	6.7	7.1	7.5
250	0.39	0.78	1.2	1.6	1.9	2.3	2.7	3.1	3.5	3.9	4.3	4.7	5.1	5.5	5.8	6.2	6.6	7.0	7.4	7.8

* Emergent-stem corrections: When a thermometer calibrated for total immersion is used with only a portion of the stem immersed, the following formula may be used to calculate the true temperature: $T_c = T_o + 0.000156l(T_o - T_m)$ where T_c is the corrected temperature, T_o the observed temperature, l the length in degrees of the mercury column not immersed, and T_m the temperature of the middle point of the emergent thread. For convenience, the table shows the corrections $[0.000156l(T_o - T_m)]$ to be applied for various values of l and $(T_o - T_m)$.
SOURCE: CENCO.

Laboratory thermometers of the long-scale type are calibrated for *complete immersion of the mercury column* in the liquid or vapor.

Stem Correction of Thermometers

With glass-mercury thermometers, the depth of immersion is very important, because the amount of mercury in the stem is significant when compared to the amount in the bulb. For that reason, thermometers are marked "full immersion" or "partial immersion," with definite markings to indicate the required depth of immersion for correct readings. If they are not immersed to the specified mark, errors in the readings will occur, and they must be compensated for with a stem correction.

NOTE In melting-point or boiling-point determinations, the entire mercury column *is not* completely immersed in the vapor or liquid. Therefore, corrections must be made. See Table 9-2.

At temperatures 0 to 100°C, error is negligible; around 200°C, error may be 3 to 5°C; and around 300°C, error may be 10°C.

Correct these values by using the following formula (see Fig. 9-2):

Stem correction (in degrees) $= KN(T_o - T_m)$

where N = length of the exposed thermometer in degrees (the length not being exposed to the vapor of liquid); $T - T_1 = N$
T_o = observed temperature on the thermometer

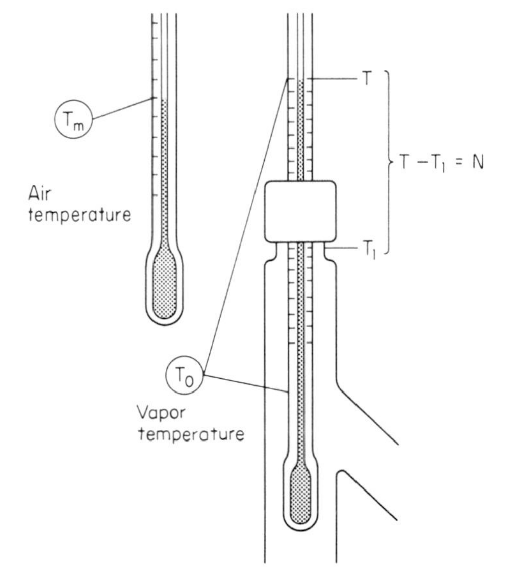

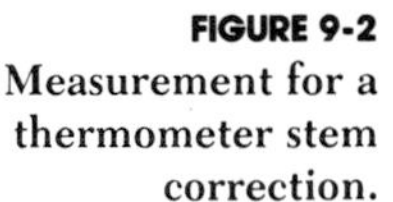
FIGURE 9-2 Measurement for a thermometer stem correction.

T_m = mean temperature of the exposed column (obtained by placing auxiliary thermometer alongside with its bulb midpont)

K = constant, characteristic of the particular kind of glass and the temperature (see Table 9-3)

TABLE 9-3
Values of K*

Temperature, °C	*Soft glass*	*Pyrex glass*
0–150°C	0.000158	0.000165
200	0.000159	0.000167
250	0.000161	0.000170
300	0.000164	0.000174
350		0.000178
400		0.000183
450		0.000188

* For Fahrenheit thermometers the average value for K is 0.00009.

EXAMPLE The temperature reads 250°C on a thermometer. What is the correct reading?

Solution The exposed column of the thermometer reads from 100 to 360°C, a difference of 250°. The temperature of the auxiliary thermometer reads 50°C. Substituting:

$$K = 0.0016$$
$$T_o = 250°\text{C}$$
$$T_m = 50°\text{C}$$
$$N = 360° - 110° = 250° \text{ (exposed column)}$$
$$\text{Correction} = KN(T_o - T_m)$$
$$= 0.0016 \times 250(250 - 50)$$
$$= 8.0°\text{C}$$

The corrected temperature should be 250 + 8.0°C = 258.0°C.

Types of Liquid Thermometers

There is a large variety of thermometers available. The common liquid thermometer is mercury-filled or alcohol-filled. (See Fig. 9-3.) In the lat-

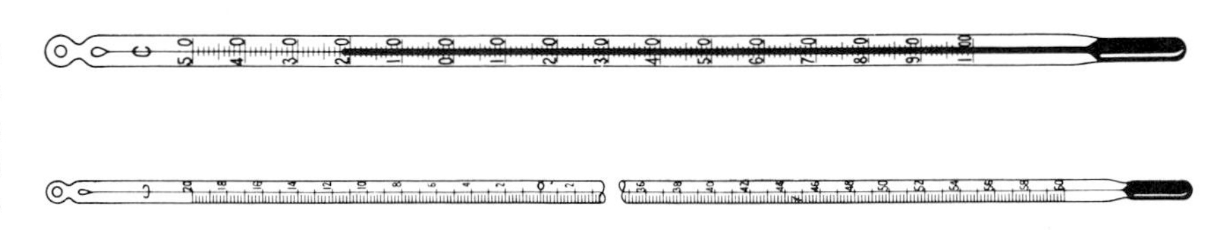

FIGURE 9-3 (*a*) General-purpose thermometer. (*b*) Example of a special-requirement thermometer.

TABLE 9-4
Representative Thermometers*

ASTM No.	*Test*	*Temperature range*	*Scale division, degree*	*Stem immersion*	*Length, ±5 mm*
1C	General use	−20 to +150°C	1	76 mm	322
1F	General use	0 to +302°F	2	76 mm	322
2C	General use	−5 to +300°C	1	76 mm	390
2F	General use	+20 to +580°F	2	76 mm	390
3C	General use	−5 to +400°C	1	76 mm	413
3F	General use	+20 to +760°F	2	76 mm	413
5C	Cloud and pour	−38 to +50°C	1	108 mm	231
5F	Could and pour	−36 to +120°F	2	108 mm	231
6C	Low cloud and pour	−80 to +20°C	1	76 mm	232
6F	Low cloud and pour	−112 to +70°F	2	76 mm	232
7C	Low distillation	−2 to +300°C	1	Total	386
7F	Low distillation	+30 to +580°F	2	Total	386
8C	High distillation	−2 to +400°C	1	Total	386
8F	High distillation	+30 to +760°F	2	Total	386
9C	Pensky-Martens, low range tag closed tester	−5 to +110°C	0.5	57 mm	287
9F	Pensky-Martens, low range tag closed tester	+20 to +230°F	1	57 mm	287
10C	Pensky-Martens, high range	+90 to +370°C	2	57 mm	287
10F	Pensky-Martens, high range	+200 to +700°F	5	57 mm	287
11C	Open flash	−6 to +400°C	2	25 mm	308
11F	Open flash	+20 to +760°F	5	25 mm	308
12C	Gravity	−20 to +102°C	0.2	Total	420

* This is not a complete list, but is given to indicate the large variety of thermometers available, with the specifications.

ter type the alcohol is generally dyed red so that the thermometer can be read more easily. Selection depends upon the particular requirements regarding scale, divisions, range, stem length and design of thermometer, precision desired, and immersion requirements. See Table 9-4 for a partial listing of representative thermometers.

The Beckmann differential thermometer (Fig. 9-4) does not read the temperature of the material. It reads *only the temperature difference* for a 5°C range. The thermometer is set by heating to the approximate temperature of the material; then the *change in temperature* to 0.01°C can be measured.

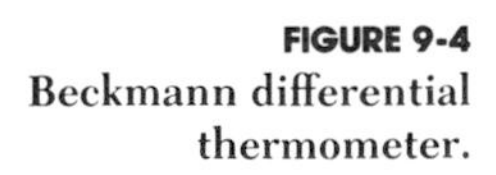

FIGURE 9-4
Beckmann differential thermometer.

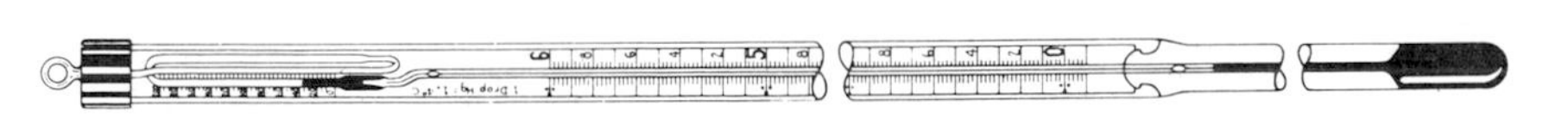

Reuniting the Mercury Column

Do not reject thermometers because the mercury column has become separated. Separation results from mechanical and thermal shock. Separated mercury columns cause *error* in thermometer readings; however, in some cases, the mercury can be reunited and the thermometer salvaged. Always inspect thermometers before using to detect separation in the capillary, bulb, and reservoir and to detect dispersed droplets on the bore.

CAUTION Remember thermometers are made of glass; they are fragile and will break easily. The bulb is especially thin and fragile. *Do not subject them to severe mechanical shock.*

Thermometer mercury columns can be reunited by the following procedures. Use any or all of these procedures or any combination of them which will work.

1. Immerse the bulb of the thermometer in a suitable freezing mixture of dry ice and acetone. All the mercury should be drawn into the reservoir by contraction. Remove and warm gently. The column should reunite. If unsuccessful, repeat this procedure several times.

2. Repeat procedure 1, but this time tap *gently* to dislodge gas bubbles.

3. Tap the thermometer at room temperature by holding the thermometer in your right hand, making a fist around the bulb (gently), then hitting your clenched fist into the palm of your other hand.

4. Turn the thermometer upside down, tap gently while heating the bulb intermittently, forcing the mercury downward. Turn right side up and repeat this procedure. Sometimes the mercury column will join together if there is a sufficiently large capillary.

5. Swing the thermometer rapidly in a circle. Centrifugal force may reunite the column. Do not "snap" your arm because the thermometer may break. The circular motion forces the upper section of the column to fall to the bottom.

6. Warm *gently* over a Bunsen flame with the bulb at a reasonable height above the flame. Heat expands the mercury into the upper expansion chamber. *Do not overfill the expansion reservoir* because the thermometer will break.

BIMETALLIC EXPANSION THERMOMETERS

When metals are heated, they expand; the amount of the expansion depends upon the temperature and the coefficients of expansion of the metals. When two metals having different coefficients of expansion are bonded together and heated, the strip of metal will distort and bend. This is the principle of thermostats, which can act as electrical regulators for

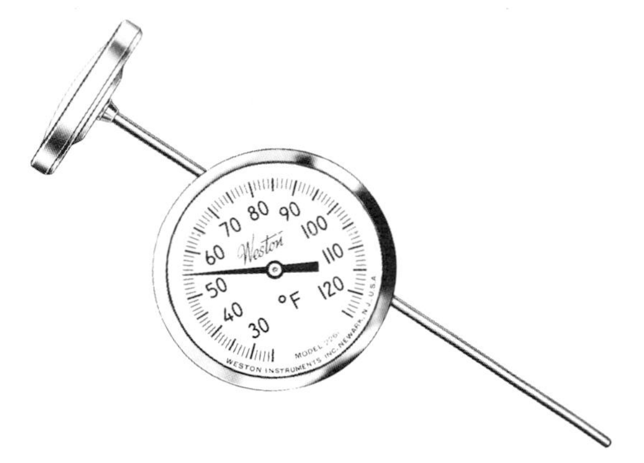

FIGURE 9-5
Dial-reading thermometer.

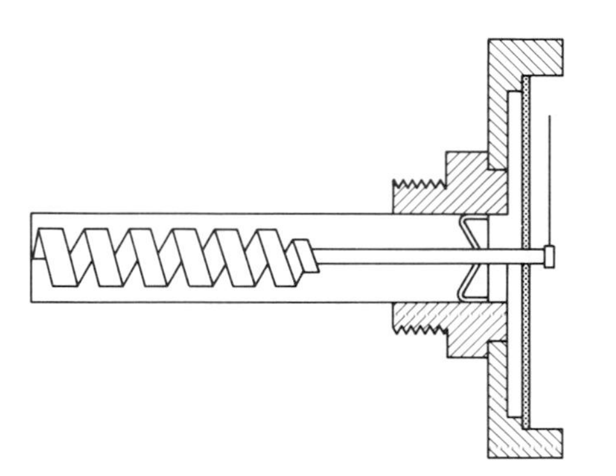

FIGURE 9-6
Industrial temperature indicator with a helical bimetal element.

temperature control, and bimetallic thermometers (Fig. 9-5). In these a rotary motion is developed by expansion of the bimetallic coil (Fig. 9-6).

CAUTION Do not use a dial-reading thermometer if the stem has been bent or deformed. This causes binding of the helical coil on expansion, retarding its motion and giving inaccurate readings on the dial. Attempting to straighten a bent stem is risky, because true alignment is extremely difficult and creases of the metal tube on the inside may hinder coil motion. Always check the accuracy of a dial-reading thermometer.

LIQUID-FILLED REMOTE INDICATING THERMOMETERS

Filled thermometers (excluding mercury-glass thermometers) are composed of a bulb, a capillary tube, a pressure-activated and sensing mechanism, and a mechanical-lever amplifying pointer. The sensor may be a Bourdon tube or a bellows device, and the system operates on the principle of expansion of liquids by heat. As the liquid expands, it actuates the

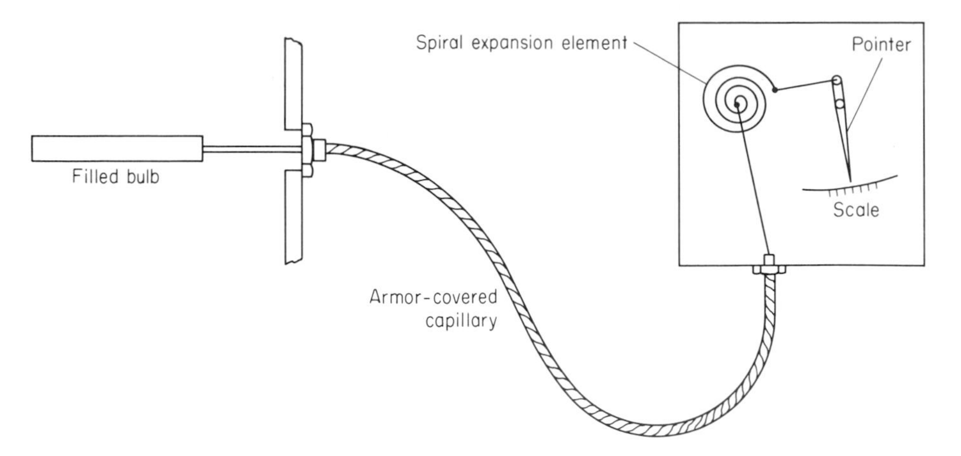

FIGURE 9-7
Simple thermal system for industrial temperature measurement.

sensor, which in turn moves the calibrated pointer or recording monitor, or the electric relay part of the instrumentation. (See Fig. 9-7.)

THERMISTORS

Thermistors (Fig. 9-8) are semiconductors of electricity which act to regulate the flow of electric current through them because their resistance to the flow of electricity changes significantly as the temperature changes; they have a high temperature coefficient of resistance. They are extremely sensitive, and when heat is applied to them externally, they convert the changes in heat to corresponding changes of voltage or current. They do not require reference or cold-junction compensation, which is required for thermocouples, and are available for a wide range of temperatures. They require a source of direct current incorporated in a simple bridge circuit with an indicating galvanometer.

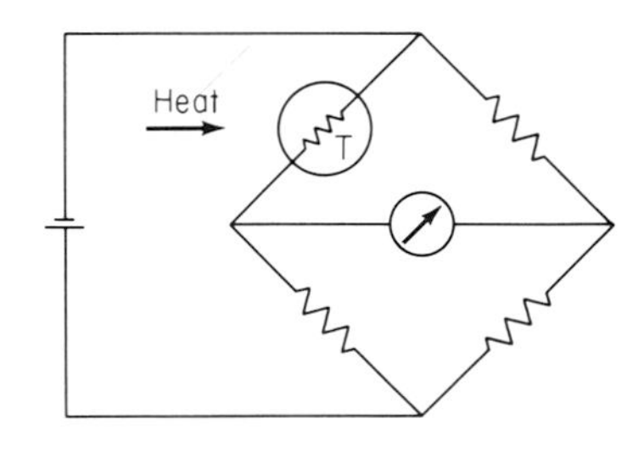

FIGURE 9-8
Typical thermistor temperature indicator circuit (simplified). Such units can indicate to a precision of 0.001°F.

Temperature Telemetry with Thermistors

The standard thermistor bridge or telemetry circuit (Fig. 9-9) can provide sufficient voltage (without amplification) to actuate and operate signal-scanning or -transmitting equipment.

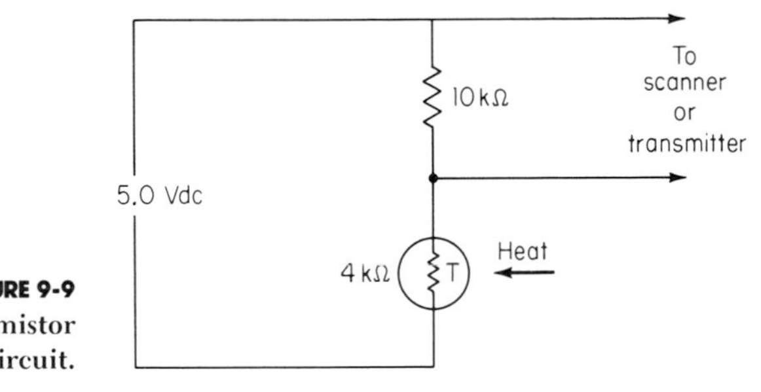

FIGURE 9-9
Typical thermistor telemetry circuit.

THERMOCOUPLES

Thermocouples are two dissimilar wires electrically fused together at one end (the temperature-indicating end) which generate an electric current when that end is heated (Fig. 9-10). The current is measured by an appropriate meter, called a *pyrometer*. Various combinations of specially selected metals and alloys provide broad temperature ranges of thermocouples (see Table 9-5), and they are available in various insulators and

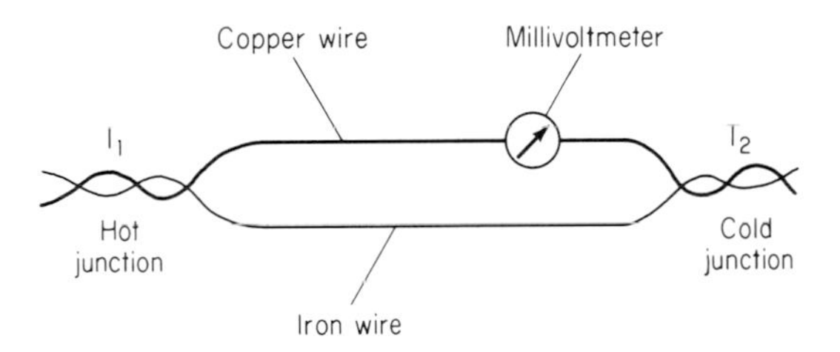

FIGURE 9-10
Simple thermocouple circuit.

TABLE 9-5
Useful Ranges of Thermocouples

Base metals	*°F*	*EMF (mV)*
Copper/Constantan	−300 to 750	−5.284 to 20.805
Iron/Constantan	−300 to 1600	−7.52 to 50.05
Chromel/Alumel	−300 to 2300	−5.51 to 51.05
Chromel/Constantan	32 to 1800	0 to 75.12
Platinum 10% rhodium/platinum	32 to 2800	0 to 15.979
Platinum 13% rhodium/platinum	32 to 2900	0 to 18.636
Platinum 30% rhodium/platinum 6% rhodium	100 to 3270	0.007 to 13.499
Platinel 1813/Platinel 1503	32 to 2372	0 to 51.1
Iridium/iridium 60% rhodium 40%	2552 to 3326	7.30 to 9.55
Tungsten 3% rhenium/tungsten 25% rhenium	50 to 4000	0.064 to 29.47
Tungsten/tungsten 26% rhenium	60 to 5072	0.042 to 43.25
Tungsten 5% rhenium/tungsten 26% rhenium	32 to 5000	0 to 38.45

sheaths. They are not as accurate as the thermistors, and the thermocouple must be matched to the pyrometer because of loop resistance of the thermocouple.

Pyrometers

The pyrometer is the indicator of the temperature signal of the thermocouple (Fig. 9-11). Some pyrometers incorporate amplifier circuits to eliminate thermocouple resistance problems and to allow narrow-span indication. Electrical cold junctions can compensate for and eliminate bimetallic cold junctions. The pyrometers can be obtained with various temperature ranges to attain maximum accuracy in the range desired. (Refer to Chap. 10, "Heating and Cooling.")

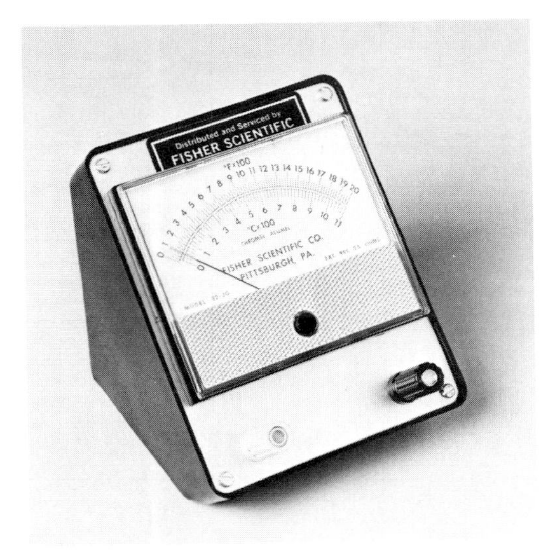

FIGURE 9-11
Pyrometer.

Digital Readouts for Temperatures

Thermocouple thermometers can be connected to digital readout units to allow direct reading of the temperature within 0.1°C or °F for temperatures ranging from −208°F to 2552°F. The readout can be made to indicate degrees Celsius or degrees Fahrenheit by the turn of a switch.

OPTICAL PYROMETERS

Radiant energy in the optical part of the electromagnetic spectrum can be measured with a special pyrometer. Optical pyrometers utilize a telescope system in which a small filament lamp, heated by a battery, is visually compared with the color of a very hot body. As the variable resistance is adjusted, the filament brightens or darkens (Fig. 9-12), and when a uni-

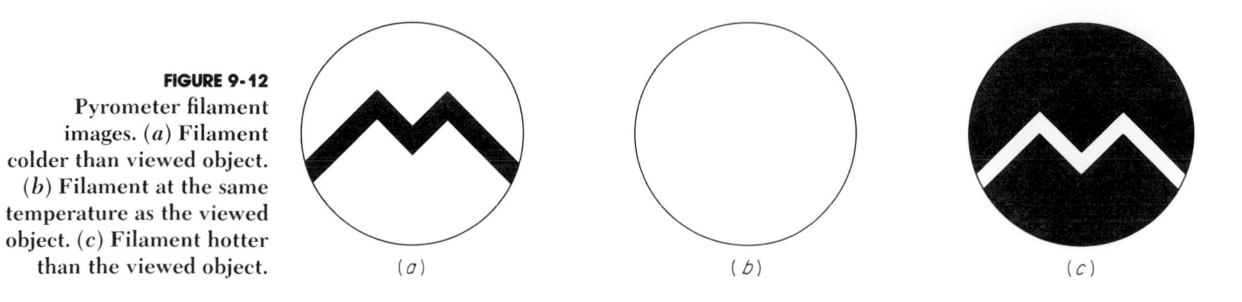

FIGURE 9-12 Pyrometer filament images. (*a*) Filament colder than viewed object. (*b*) Filament at the same temperature as the viewed object. (*c*) Filament hotter than the viewed object.

form picture of the superimposed filament and heated object appears, the temperature is read off a calibrated milliammeter. Pyrometers of this kind are very useful in ranges from about 800°C to about 1200°C, with an accuracy of about 5 percent.

OTHER TEMPERATURE INDICATORS: LABELS, LACQUERS, CRAYONS, PELLETS, AND LIQUID CRYSTALS

Labels

These temperature indicators are heat-sensitive and are available for a broad range of temperatures. When the temperature to which they are exposed reaches the rating of the indicator, they turn irreversibly black. The labels are available with either a single temperature rating or four temperature ratings; the crayons, lacquers, and pellets are single-temperature indicators.

Lacquers

The lacquer is mixed thoroughly in the bottle, and a thin smear is applied to the surface of the item before heating begins. It dries to a dull mark which liquefies sharply when the stated temperature is reached. When the heat is removed, the melted coating resolidifies, but now remains glossy and transparent. Only the appearance of the liquid melt is the temperature-indicating signal.

Crayons

When the item is marked with the crayon (below the crayon's temperature range) a dry, chalky mark results. When the stated temperature is reached, a liquid smear results. Color changes are disregarded. The melting of the mark is the signal that the temperature has been reached.

Pellets

Temperature-indicating pellets are individual pellets which are placed upon the piece to be checked before any heating is applied. When the stated temperature of the pellet is reached, the pellet begins to melt.

Liquid Crystals

The newest technique for detecting and measuring small changes in temperature (usually ambient) involves the use of special molecules called liquid crystals. These are usually secured to a plastic film which is placed on the object to be monitored. Temperature changes are indicated by changes in the color of the crystals. A color guide is supplied for use with the indicator. The color changes can be seen at a distance, and this makes it convenient to monitor several objects at the same time. A decided disadvantage is the limited range over which the crystals can be used. However, they are finding an increasing use in clinical work for just this reason.

10
HEATING AND COOLING

GENERAL GUIDELINES

How to Handle Hot Labware

Hot glassware can cause severe burns. These burns are often compounded when one attempts to hold hot glassware without adequate protection, because the item generally falls and breaks and may thus cause hot and corrosive liquids to splash and splatter. Use asbestos gloves or appropriate tongs to remove all glassware from heat. Crucible tongs are always used with hot crucibles going to and from a muffle furnace (Fig. 10-1*a*); utility tongs are used to handle small apparatus and crucibles (Fig. 10-1*b*); and beaker tongs are used to hold and carry small beakers (Fig. 10-1*c*).

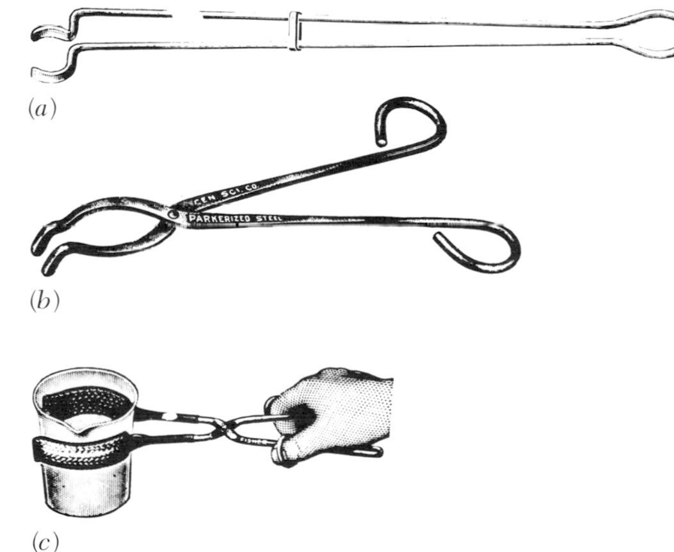

FIGURE 10-1 (*a*) Crucible tongs are flexible and long. (*b*) Utility tongs. (*c*) Beaker tongs.

Tips on Heating and Cooling

1. Always watch evaporation work closely. A vessel that is heated after evaporation has already completely occurred may crack or explode.

2. Do not place hot glassware on a cold or wet surface; it may break because of the sudden temperature change. Always use caution.

3. Do not heat badly scratched or etched glassware; it is likely to break.

4. Cool glassware slowly to prevent breakage. There is certain glassware, however, e.g., Vycor®, that can go from red heat into ice water without breakage.

5. Burns are caused by heat and also by ultraviolet or infrared radiation. Limit your exposure time when working with extravisual light.

6. When working with volatile materials, always remember that expansion and confinement of expansion results in an increase of pressure with explosion possibilities. That danger always exists in a closed system, even though external heat is not applied.

GAS BURNERS

The laboratory burner is the piece of equipment used more often than any other. Learn its parts (Fig. 10-2) and memorize the simple rules for using it:

1. Turn the gas cock on the gas-inlet pipe until it is wide open.

2. Open the thumbscrew gas adjustment on the burner as far as it will go.

3. Close the air holes with the movable sleeve.

4. Light the burner with a match or striker (Fig. 10-2*d*).

5. Regulate the thumbscrew and the movable sleeve to produce a nonluminous flame with a sharp-pointed inner blue cone (Fig. 10-3).

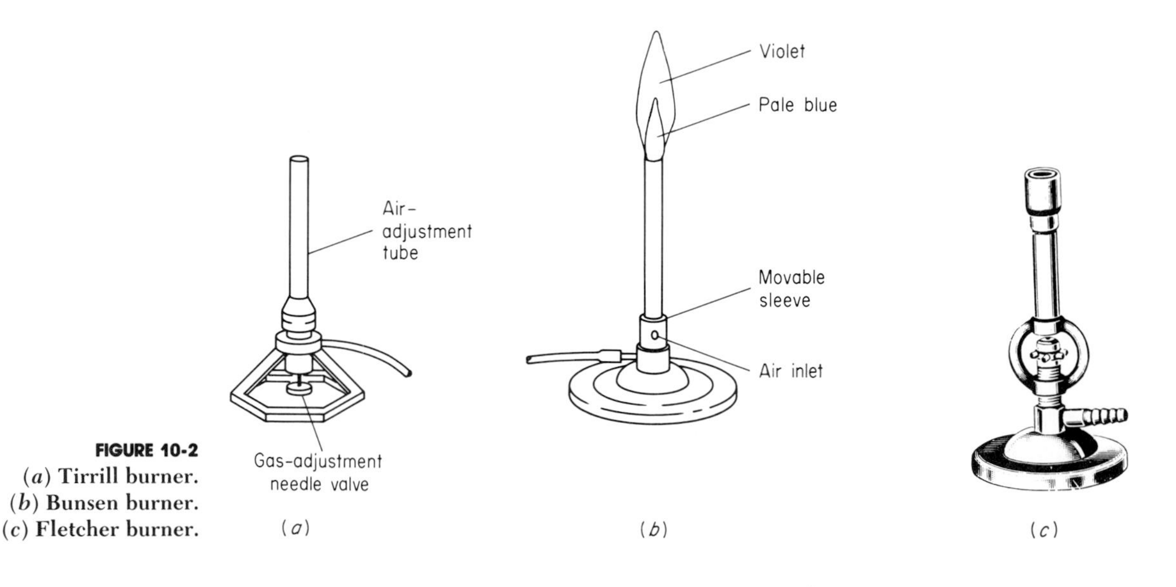

FIGURE 10-2
(*a*) Tirrill burner.
(*b*) Bunsen burner.
(*c*) Fletcher burner.

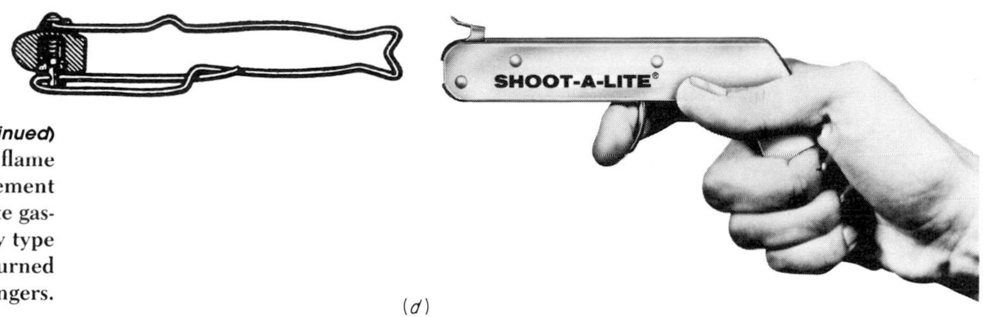

FIGURE 10-2 ***(continued)***
(*d*) Positive-acting flame strikers. Rapid movement of the flint will ignite gas-flame burners of any type and prevent burned fingers.

There are many kinds of burners used in the laboratory. The Bunsen, Tirrill, and Fletcher (Fig. 10-2) are the most common. When a very hot flame is needed, a Fisher burner (Fig. 10-4*a*) is very useful; an even hotter flame can be obtained with a blast burner (Fig. 10-4*b* and *d*).The halo support (Fig. 10-4*c*) is used for direct heating with any Fisher burner.

To help you select the proper burner you should know the type of gas in your laboratory; artificial, mixed, natural, or cylinder gas. Table 10-1 will then tell you at a glance the Btu range of each burner. For efficient com-

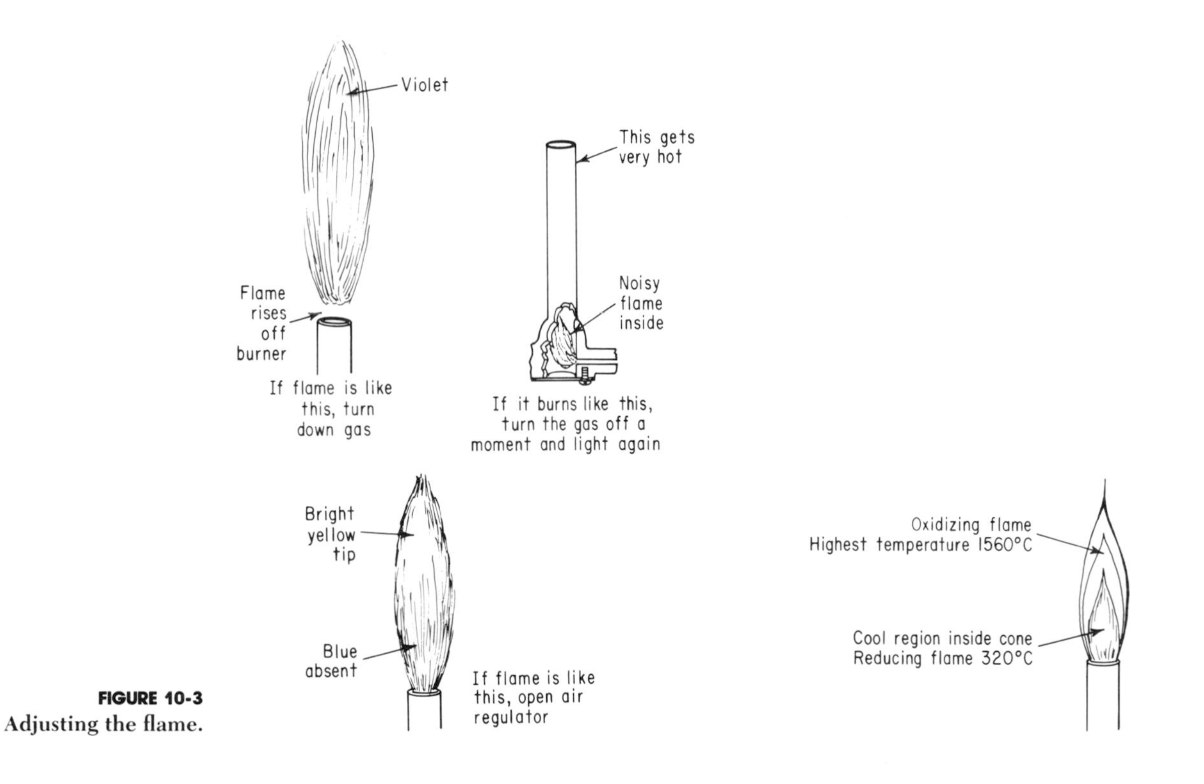

FIGURE 10-3
Adjusting the flame.

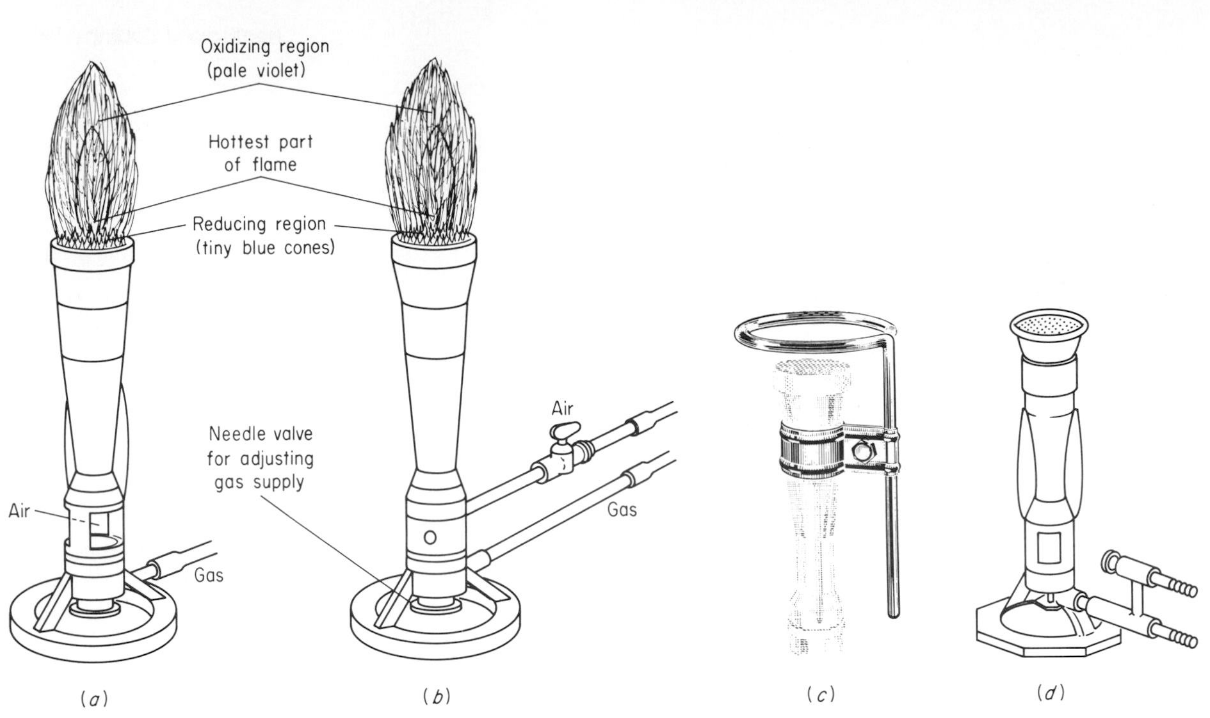

FIGURE 10-4
Burners that yield a very hot flame. (*a*) Fisher burner. (*b*) Fisher blast burner. (*c*) Halo support. (*d*) Meker blast burner.

TABLE 10-1
Selecting the Proper Burner

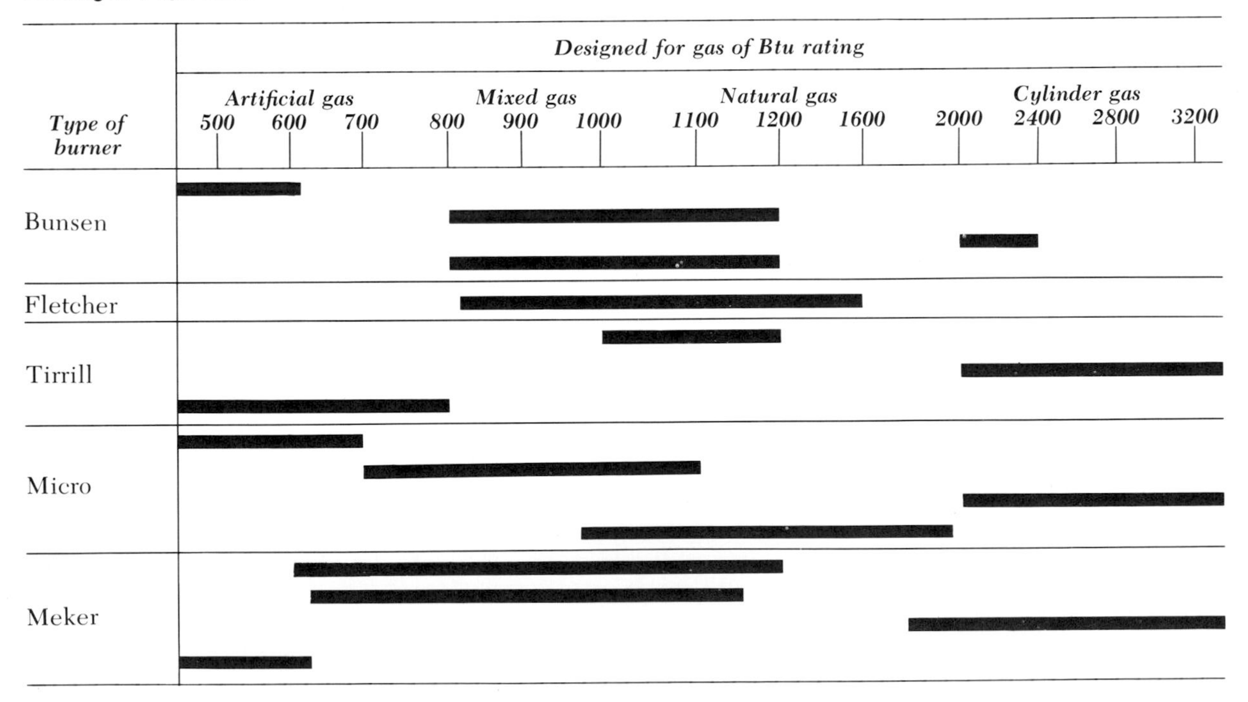

SOURCE: Corning Glass Works.

bustion of gas above 800 Btu, use burners equipped with stabilizer tops or with special heat-intensifier tops.

Heating with Open-Flame Gas Burners

1. Adjust burner to get a soft flame. It will heat slowly, but, more importantly, it will heat evenly and uniformly. Uniform heat can be a critical factor in some chemical reactions.

2. Adjust the ring stand or clamp holding the glassware so that the flame makes contact with the glass below the liquid level. Heating above the liquid level does nothing to promote even heating of the solution. It can cause thermal shock and breakage of the glassware.

3. Always use a wire gauze (Fig. 10-5) when heating laboratory glassware with a Bunsen, Fisher, Meker, or other gas-flame burner. The gauze distributes the heat for uniform heating and spreads the flame to prevent the glassware from being heated in one spot only.

4. Never heat liquids too fast. Fast heating may cause boilover, splattering, and injury in addition to loss of the solution.

CAUTION Be aware of the high temperatures that can be reached when working with electric and combustion devices. (See Table 10-2.)

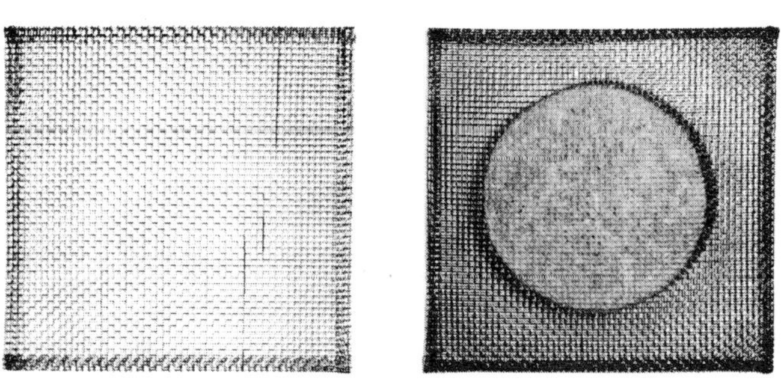

FIGURE 10-5
Plain and asbestos-centered wire gauzes.

HEATING NONFLAMMABLE LIQUIDS

In Test Tubes

1. Fill the test tube no more than one-half full.
2. Hold the test tube with a test-tube clamp (Fig. 10-6).
3. Point the mouth of the test tube *away* from you and anyone near you.

TABLE 10-2
Some High Temperatures Available in the Laboratory

Devices used	*Approximate temperature obtained, °C*
Electric devices:	
Nichrome wire resistance heating	1150
Silicon carbide resistance bars	1425
High-frequency induction	1650
DC electric arc (carbon electrodes)	3000
Combustion devices:	
Alcohol lamp	1350
Bunsen burner (natural gas)	1225
Meker burner (natural gas)	1300
Blast lamp (natural gas and compressed air)	1400
Torches	
Natural gas + O_2	1600
Acetylene + compressed air	2300
Acetylene + O_2	3100
Hydrogen + O_2	2600
Atomic hydrogen (Langmuir)	4000

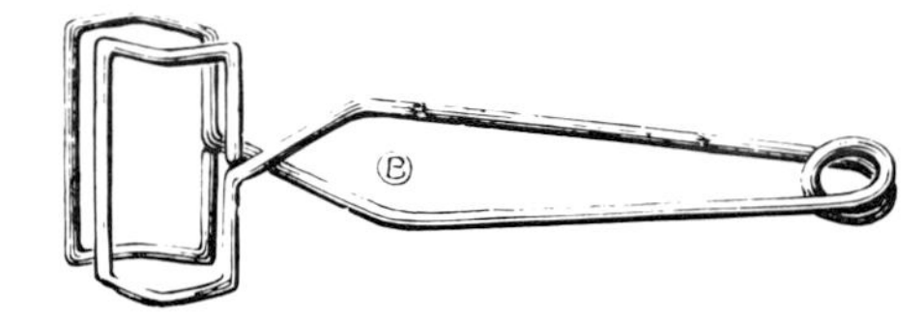

FIGURE 10-6
Test-tube clamp.

4. Place the test tube in the flame, and *move the test tube constantly.* If the test tube is not kept in motion, the liquid will get very hot and vapor will form and eject liquid violently (Fig. 10-7).

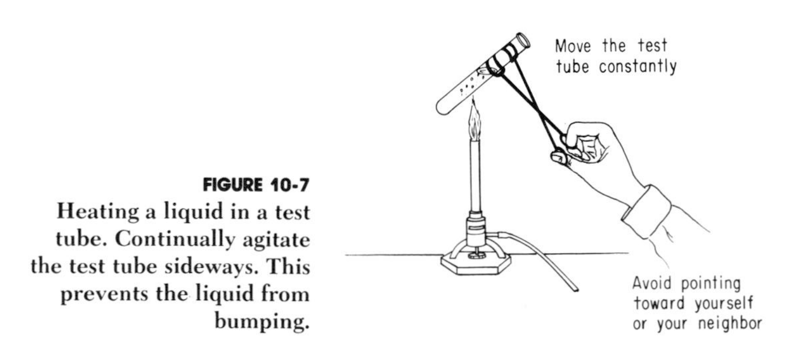

FIGURE 10-7
Heating a liquid in a test tube. Continually agitate the test tube sideways. This prevents the liquid from bumping.

In Beakers or Flasks

Method 1

1. Place the beaker on asbestos-centered wire gauze resting on a tripod or ring stand (Fig. 10-8).
2. Heat with a gas burner.

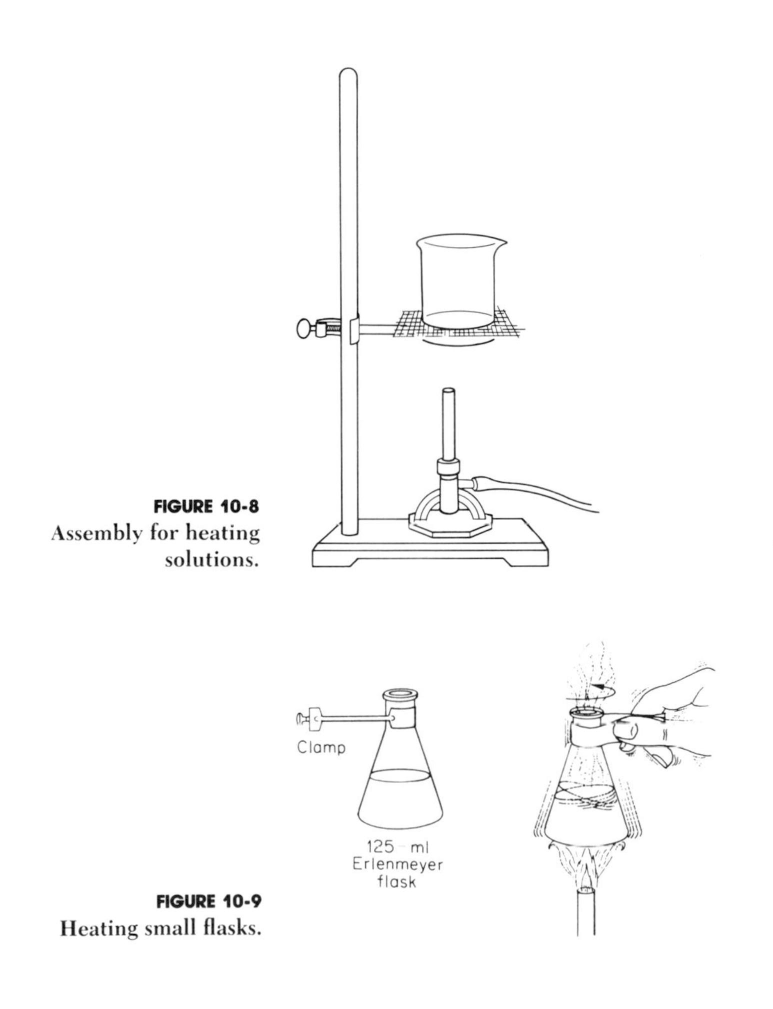

FIGURE 10-8
Assembly for heating solutions.

FIGURE 10-9
Heating small flasks.

Method 2

Grasp the neck of the flask as shown in Fig. 10-9 and heat.

BOILING

When a liquid boils, the temperature must be slightly above its boiling point before bubble initiation can begin. When bubbles form, they may either collapse (because they are too small) or grow (because they are large enough) and rise to the surface. When some liquids are heated, the temperature of the liquid may rise much higher than that of the boiling point without any bubbles forming. This is called *superheating.* Should a bubble start to form in this superheated state, it can grow almost instantaneously with a violent explosive force, which can jar glassware containers so vigorously that they will break and shatter.

Use of Boiling Stones and Boiling Chips

To prevent the liquid from overheating, boiling stones or boiling chips, which act as built-in bubbles, may be used. Not many boiling stones are needed, but enough are required so that there are always some on the bottom, even with the lifting action.

Boiling stones are essentially 99.6% pure silica, fused and bonded to form stones which have innumerable sharp projections for the release of vapor bubbles. They are chemically inert, and when they are used, they stop bumping, reduce the danger of breakage, and speed up distillation. They make possible sharper separations of fractions in distillations and are especially efficient for Kjeldahl digestions. *Boiling chips* are microporous, chemically inert chips, made of carbon or other materials, which are used for the same purpose as boiling stones. Clean chips of glass tubing can serve as boiling chips.

CAUTION Never add any boiling stones or boiling chips to hot liquids. Large amounts of vapor will be *formed immediately,* causing *frothing, spraying,* and *ejection* of the heated liquid from the flask. *Always cool your solutions* before you add the boiling stones.

HEATING ORGANIC LIQUIDS

Practically all organic liquids are flammable. The lower the boiling point, the more flammable they are. If the identity of the liquid is known, check the reference handbook for the boiling point. Assume *all organic liquids are flammable* until you have determined otherwise.

Choose the heating system depending upon (1) the flammability of the liquid, (2) the vessel used, and (3) the presence of fire hazard in the work area.

CAUTION *Vapors of flammable liquids must be kept away from open flames.* Never heat in *open* beakers or flasks with a gas burner. The heavier-than-air vapors will descend and catch fire.

Method 1

1. Heat a sand bath or a very high-boiling mineral oil with gas burner.
2. Turn off the gas burner.
3. Immerse the beaker or flask in the heated bath to heat the material.

Method 2

If temperatures up to 100°C are adequate, use a steam bath (Fig. 10-10).

1. Seat the flask on the steam bath after removing the proper number of supporting rings to give the maximum heating surface.
2. Pass steam into the *top* inlet.
3. Connect the bottom outlet to the drain to discharge steam condensate.

NOTE Water condenses in steam lines when they are not in use. Purge the water from the steam line first before connecting the steam line to the bath. Once the bath is heated, a slow steam flow will maintain the temperature of the liquid in the container. Avoid excessive steam flow; condensation can occur elsewhere in the laboratory and cause problems.

Method 3

Test tubes containing flammable liquids may be heated in a hot-water bath. The hot-water bath (Fig. 10-11) can be easily constructed from soft aluminum. It can be formed by hand, and the holes may be cut with a circular drill or knife.

FIGURE 10-10 The steam-heated or electrically heated water bath is used to heat solutions requiring temperatures not exceeding 100°C.

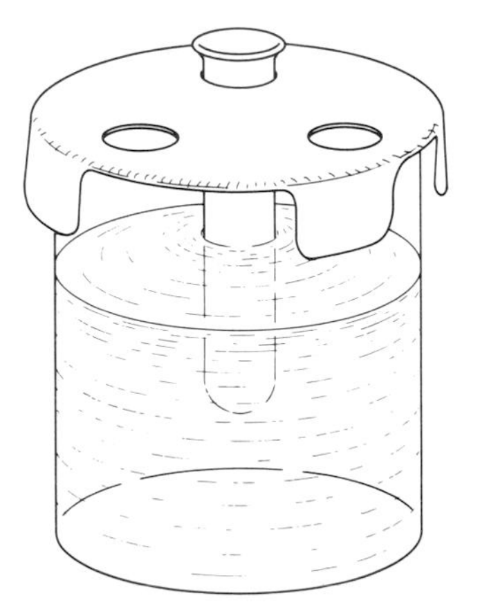

FIGURE 10-11 Test-tube hot-water bath.

FLAMELESS HEATING DEVICES

Besides the sand bath and the steam bath (see above) there are other methods for heating flammable liquids.

Heating mantles (Fig. 10-12) provide safe, intimate heat transfer, to heat most containers and are available in a variety of shapes and sizes. Heat is adjusted with a variable transformer.

Heating Mantles

The mantle consists of a simple, insulated electric-resistance heater and requires only an electric power supply. However, use of a variable transformer or an automatic controller is recommended to prevent overheating and to effect accurate temperature control. In some instances, it may be desirable to use a variable transformer and an automatic device together for very accurate control. Instructions for connection and operation of this arrangement are available.

For safety reasons, the three-prong plug of the cord on the mantles should always be inserted into a ground outlet to electrically ground the equipment. Since the exterior of some mantles is an insulator, these mantles are furnished with two wire cords.

With proper care and operation, the heating mantle will give long and efficient service. Chemical spillage, overheating, and general misuse will greatly reduce the life of the mantle. Maintenance of the mantle on a regular schedule is not required. Of course, any damage which occurs to the mantle should be repaired immediately. Mantles should be protected from chemical spillage and corrosive atmospheres so far as is practical.

The only limitation in the operation of the mantle is the upper temperature limit of 450°C for the glass fabric in some mantles and 650°C for other mantles. Most mantles are equipped with iron-Constantan thermocouples to measure the temperature of the fabric. Under normal conditions 650°C will not be exceeded.

In spherical mantles of 12-L capacity and larger, there are three heating circuits—two in the lower half and one in the upper half. The heat input in each circuit should be controlled with a suitable variable transformer. The two lower circuits furnish heat for boiling the liquid contents, while the upper circuit prevents condensation of the vapors. It will rarely be necessary to operate the upper circuit on more than 60 or 70 V. For low-boiling liquids the upper circuit need not be used. When the flask is full of liquid, the two bottom circuits may be operated at the rated voltage. When the liquid level falls below the halfway mark in the flask, the voltage in circuit No. 2 should be reduced to 70 or less. This will prevent superheating of the vapors and overheating of the glass fabric in circuit No. 2.

FIGURE 10-12
Heating mantles. (*a*) Flask mantle is available in varied sizes and designs. (*b*) Funnel type. (*c-f*) Asbestos mantles of varied designs, sizes, and supports.

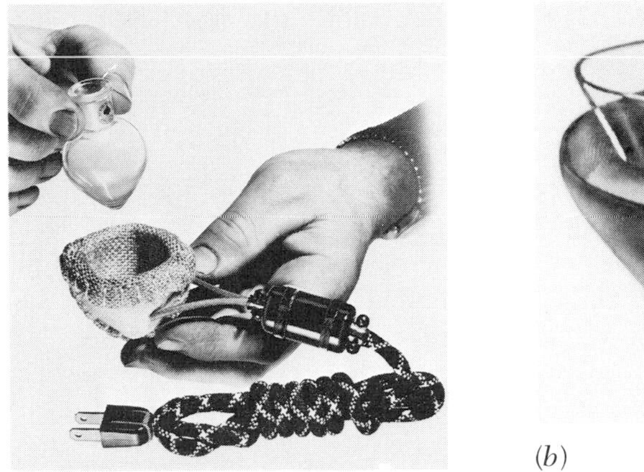

(*a*)

(*b*)

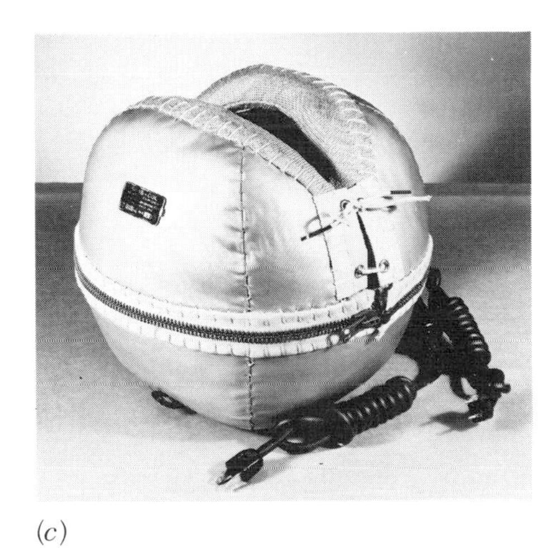

(*c*)

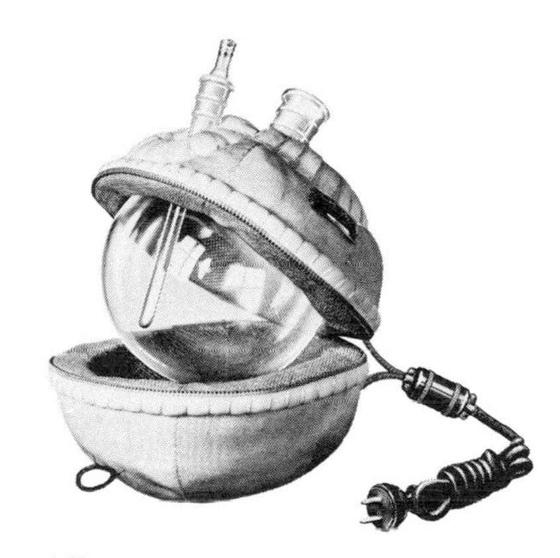

(*d*)

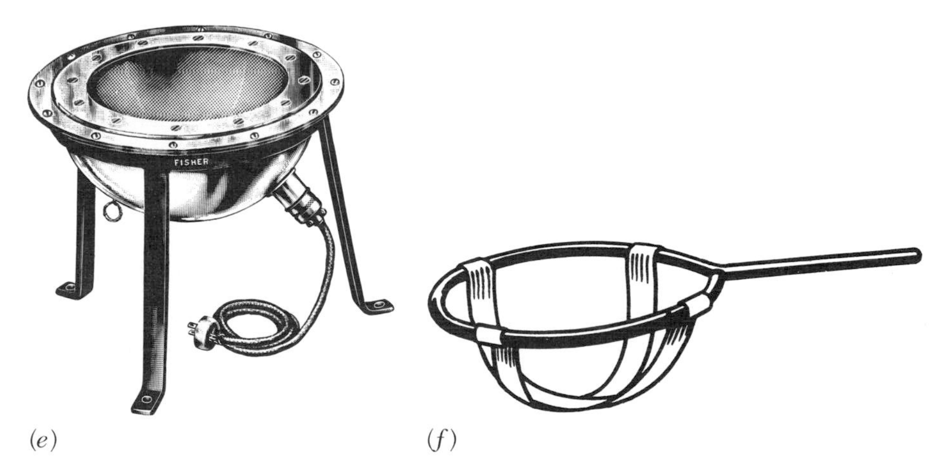

(*e*) (*f*)

Procedure

1. Select the correct mantle that fits the container snugly.

2. Connect the temperature-indicating meter and the transformer.

3. Adjust the transformer as needed to maintain the proper temperature of the reaction-liquid flask.

Precautions in Using Heating Mantles

1. Do not allow liquids in the flask to fall below the top of the mantle. The glass between the top of the liquid and the top of the mantle will become much hotter than the rest of the flask. That material which splashes on the extra-hot glass will become superheated and possibly decompose or otherwise cause problems.

2. Always allow the mantle to come to equilibrium heat. Mantles heat slowly. Do not increase the heat on the mantle too quickly; you may apply too much heat.

3. Always suspend flasks heated by mantles above the desk top. You may need to stop the heating immediately, and merely turning off the electricity does not stop the mantle from heating the flask. Mantles have a high heat capacity. When the flasks are suspended, the mantles can be removed quickly, removing the source of heat.

Heating Tapes

Flexible heating tapes (Fig. 10-13) are made of finely stranded insulated resistance wire fitted with electrical leads. They are rated for 110 or 220 V, and may be insulated with glass, rubber, silicone, or other polymers. Temperature control is imperative, and may be effected by variable transformers or automatic thermostatic controllers.

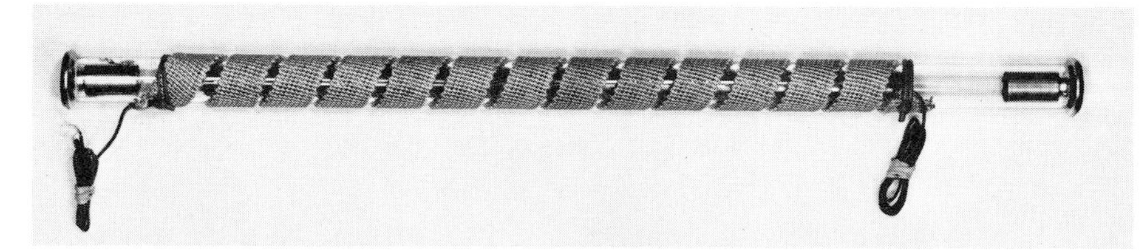

FIGURE 10-13
Heating tape.

Electrically Heated Oil Baths

Electrically heated oil baths using ordinary mineral oil can be used safely up to 200°C. Any of the liquids listed in Table 10-3 can be substituted for mineral oil, each having its particular advantages and disadvantages.

The bath is controlled with a variable-voltage transformer (Fig. 10-14), and the heating element is an immersion coil, which minimizes danger of

TABLE 10-3
Substances Which Can Be Used for Heating Baths

Medium	*Melting point, °C*	*Boiling point, °C*	*Useful range, °C*	*Flash point*	*Comments*
Water	0	100	0–100	None	Ideal
Silicone oil	−50	—	30–250	300	Some viscous at low temperature
Triethylene glycol	−5	287	0–250	310	Noncorrosive
Glycerin	18	290	−20 to 260	160	Water-soluble, nontoxic
Paraffin	50	—	60–300	199	Flammable
Dibutyl phthalate	—	340	150–320	—	Generally used

fire. Because it requires a fairly long time for oil baths to reach the desired temperature, it is advisable to preheat the unit partially before it is actually needed. Baths also reduce the possibility of charring, which can occur when reaction vessels are heated by means of a flame.

CAUTION Mineral oil may "flash" and burst into flame above 200°C, and the flame is not easily extinguished. Mineral oil eventually oxidizes and darkens, and water should be kept away from it. Water in hot oil will cause it to splatter, and hot oil burns are painful and dangerous. These baths may be very heavy, and a jack (Fig. 10-15*a*) can be used to adjust the height of the bath.

Hot Plates

There is a variety of hot plates, each designed to do a particular job or act as a general heating device (Fig. 10-15*b* to *d*). Those with exposed heating elements should never be used to heat flammable organic solvents, and

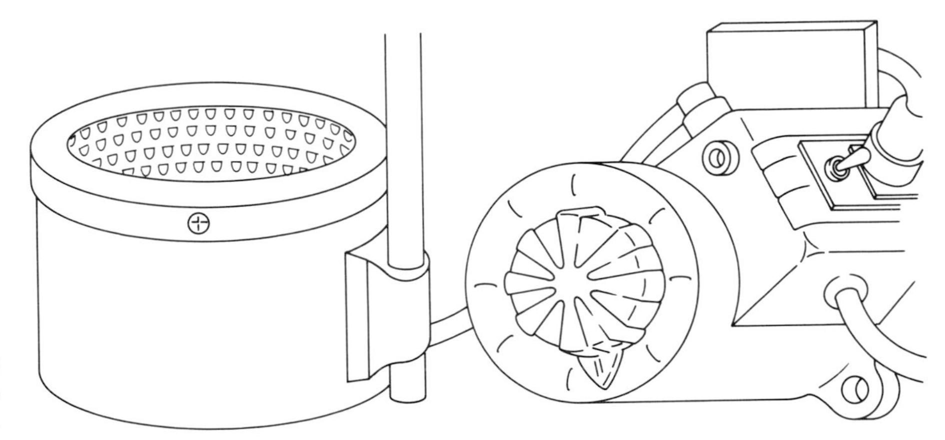

FIGURE 10-14
Oil bath with a variable transformer.

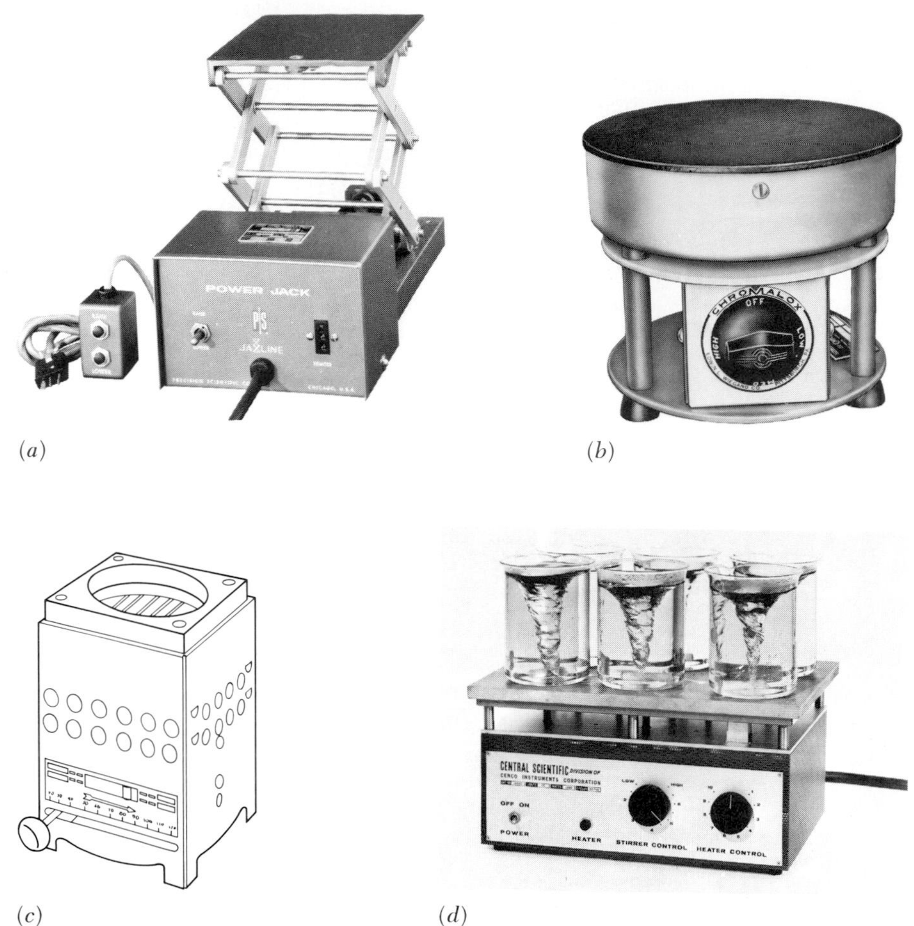

(a) (b) (c) (d)

FIGURE 10-15
(*a*) An electrically powered jack with remote control for use in hazardous areas. It lifts up to 100 lb about 9 in. (*b*) Adjustable-temperature heavy-duty hot plate for heating nonflammable substances. (*c*) Open-coil adjustable-temperature hot plate for heating nonflammable substances. (*d*) Hot plate with multiple head, magnetic mixer, and adjustable heat and mixing rate, for use with magnetic stirring bars.

care should always be exercised when heating those solvents with hot plates having enclosed elements. Some hot plates have a control to vary the heat in small increments, while others merely adjust from low to medium to high heat.

Some hot plates have a built-in magnetic stirrer whose rotation speed can be adjusted. The rotating magnet underneath the glass container rotates the plastic-coated (to prevent contamination) magnetic stirrer inside (Fig. 10-16). The rotation of the inner bar magnet mixes the solution, so that both heating and mixing operations can be carried on simultaneously. The rotating magnet also creates a turbulence in the liquid, which breaks up the large bubbles in boiling solutions, and inhibits bumping.

Precautions in Using Hot Plates

1. The entire top surface heats and remains hot for some time after the hot plate has been turned off. *(Remember this!)*

FIGURE 10-16
Teflon-coated magnetic stirring bars.

2. Be sure that the electric cord, plug, and connector are in good condition. Do not use a hot plate with defective electric wiring or components. You can get an electrical shock or fire can result from shorted wires. (Refer to Chap. 27, "Basic Electricity.")

Fluidized Baths

In this type of bath there is a fluidized bed which consists of a loosely packed mass of solid particles through which a flow of gas, usually air, is caused to pass vertically upward. (See Fig. 10-17.) When fluidization takes place, individual particles circulate freely and the mass appears to acquire the properties of a liquid. It displays buoyancy and viscosity and becomes an excellent heat-transfer medium. The temperature of the fluidized bed can be accurately controlled over a wide range. The aluminum oxide particles have no melting point or boiling point and provide an ideal constant-temperature bath.

Procedures

1. Turn on air and adjust velocity until the bed is fluidized.*
2. Immerse labware in the fluidized medium as desired.
3. Adjust heater control as desired. The bath is now operable.

Immersion Heaters

Various types of immersion heaters are used in the laboratory. Immersion heaters for general laboratory use (Fig. 10-18*a*) operate on 110 to 220 V,

---

* A pressure or fail-safe switch will automatically turn off the heaters if the air supply drops and fluidization is lost. Periodically check the fail-safe switch to determine that it is operating properly.

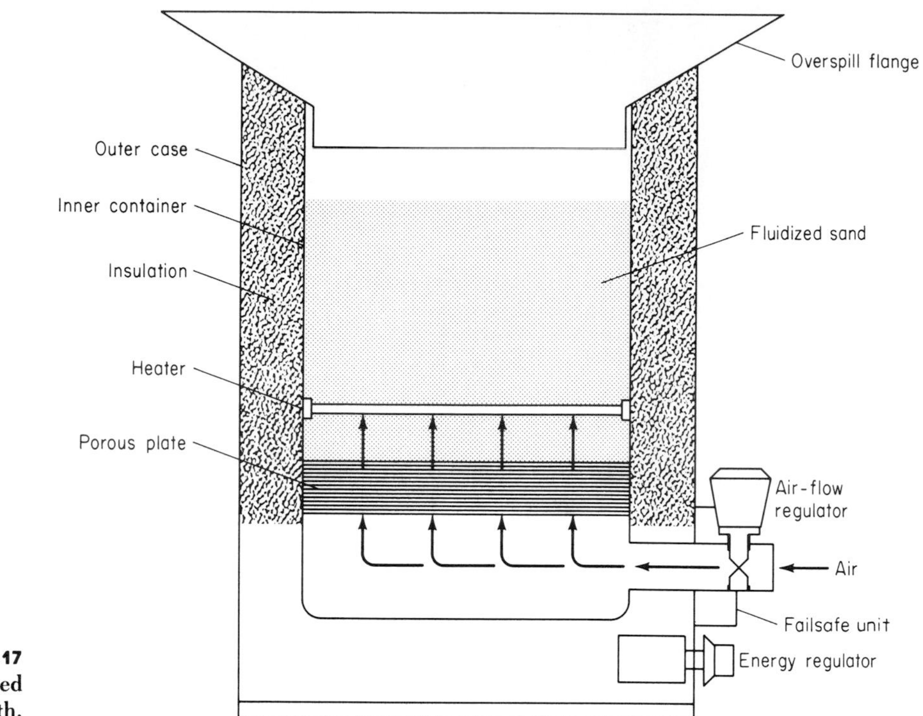

FIGURE 10-17
A schematic of a fluidized bath.

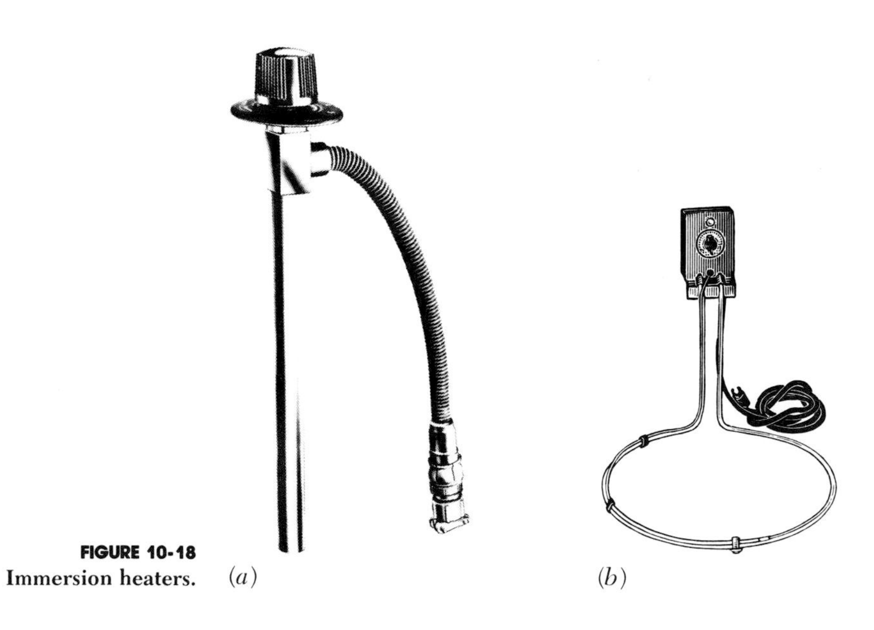

FIGURE 10-18
Immersion heaters. (*a*) (*b*)

from 100 to 2000 W capacity, and are usually actuated by an immersion thermostat. This type is a heater only.

A liquid immersion heater with variable temperature adjustment such as that shown in Fig. 10-18*b* is used for heating constant-temperature baths.

OTHER EQUIPMENT USED FOR HEATING AND DRYING

Muffle Furnaces

Muffle furnaces (Fig. 10-19) are electrically heated furnaces designed to operate continuously at temperatures ranging to 1200°C (2200°F). Temperatures, adjusted by rotating the control knob, are read from an indicating pyrometer which is usually calibrated in both scales. They are used for small melts of metals, for heat treatment, and for chemical analysis.

Hot-Air Blowers

Hot-air blowers (Fig. 10-20) provide flameless heat, the temperature available depending upon the construction and power of the blower. They are useful for drying wet labware, drying samples, heating plastic tubing for forming, and for various other chores. They attain operating temperatures almost immediately, and some have nozzle controls for directing and distributing the heated air.

FIGURE 10-19
Muffle furnace with a temperature-indicating dial, adjustable rheostat heat control, and automatic door closure.

FIGURE 10-20
Hot-air blower.

FIGURE 10-21
Infrared lamp.

Infrared Lamps and Ovens

Infrared lamps (250 W) (Fig. 10-21) can produce a directional heat beam with radiation in the near-infrared region. They are used to dry samples, providing controllable heat (by varying distance), and can be used in a gooseneck portable lamp stand.

Infrared ovens use 125-, 275-, or 350-W infrared lamps to dry samples and evaporate liquids. Shelves are adjustable to accommodate containers of different sizes and to vary the distance of the container from the lamp.

Electrically Heated Ovens

Ovens are used for all scientific laboratory and production procedures which involve drying, baking, preheating, aging, and curing and for all tests which include the application of controlled heat for certain periods of time. There are two general types: gravity convection and mechanical convection.

Gravity-Convection Ovens

These ovens basically comprise an insulated container which has heating elements located at the bottom (Fig. 10-22). The heated air rises through gravity convection.

FIGURE 10-22
Electrically heated oven for drying samples, baking, and sterilizing. Its temperature is automatically controlled to about a 2°C range, and it is fitted with an on-off switch and a pilot light.

Mechanical-Convection Ovens

These ovens include a mechanical blower system, which enables the unit to have certain advantages over the gravity type. Air is more rapidly circulated; closer uniformity of temperatures throughout the unit are possible; and there are finer control tolerances with smaller temperature drift. Mechanical-convection ovens can perform all the simple functions of gravity-convection units; however, they cost more initially, they take up more space, and they require more maintenance because of their more complicated structure.

CONSTANT-TEMPERATURE BATHS

Investigations that require the maintenance of constant temperature for a procedure use constant-temperature baths (Fig. 10-23).

Apparatus and Purpose

1. A sufficiently large glass or metal container is the bath.
2. Copper tubing through which cold water circulates, as required, is the cooling component.
3. Electric stirrers or circulating pumps give equalized heat distribution.
4. A thermoregulator permits temperature control to about 0.02°C (Fig. 10-24).
5. The electric heater, activated by the thermoregulator, heats the solution as required.
6. A thermometer reads the temperature.
7. The temperature range is −15 to 70°C.
8. Oil is substituted for water for temperatures from 100 to 250°C.

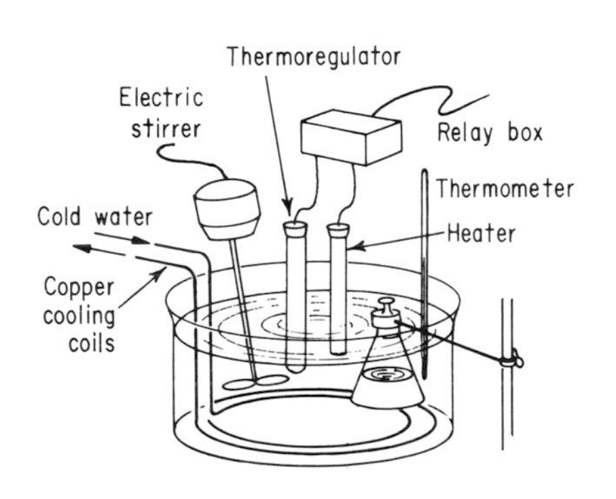

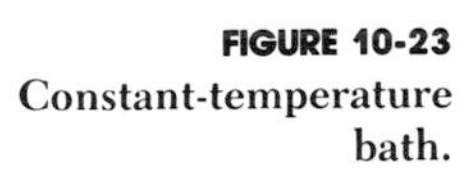

FIGURE 10-23 Constant-temperature bath.

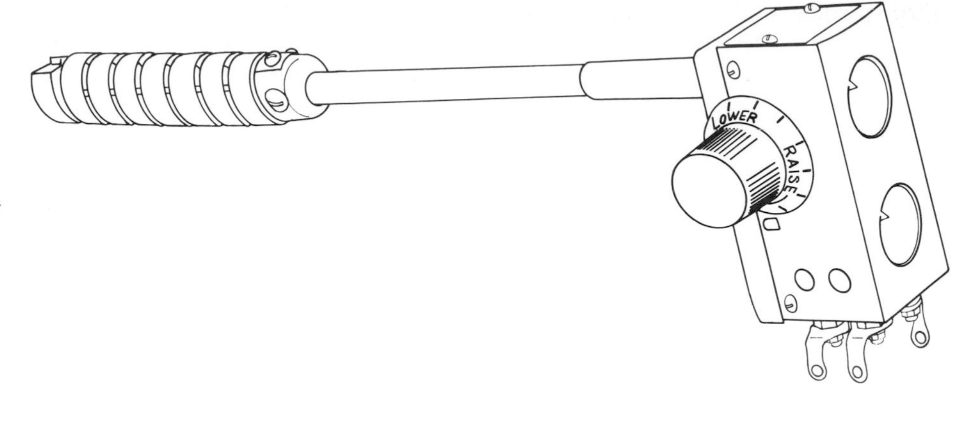

FIGURE 10-24
Electric-contact-type thermoregulator with open contacts or explosionproof construction. Immersion in liquid opens and closes contacts as temperature varies. Rheostat adjustment provides varied temperature settings.

COOLING BATHS

Salt-Ice Mixtures

When low temperatures are required in a procedure to collect a gas, carry out a reaction, or condense a volatile gas of distillate, the salt-ice mixtures listed in Table 10-4 can be used to attain the desired temperatures. Factors which contribute to achieving the stated final temperatures include the

1. Rate of mixing
2. Heat transfer of the container
3. Fineness of the crushed ice

Ice

Ice baths are actually ice-water baths. Large pieces of ice make poor contact with the walls of the vessel; therefore liquid water is needed to make the cooling medium efficient.

Thermos flasks or foam-insulated containers minimize heat transfer for maximum efficiency of the cooling bath.

TABLE 10-4
Cooling-Bath Mixtures

Substance	*Initial temp., °C*	*g/100 g H_2O*	*Final temp., °C*
KCl	0 (ice)	30	−10
NaCl	0 (ice)	33	−21
$MgCl_2$	0 (ice)	85	−34
$CaCl_2 \cdot 6H_2O$	0 (ice)	143	−55

Dry Ice

Dry ice (solid carbon dioxide) provides an easy way to obtain very low temperatures. The dry ice is crushed and mixed with ethanol, acetone, or xylene. When mixed with ethanol or acetone it can produce a temperature as low as −72°C.

CAUTION Dry ice is *dangerous. Handle with care! Do not handle* with bare hands or fingers. It will cause severe cold "burns."

Liquid Nitrogen

Liquid nitrogen, at a temperature of −195.8°C, can be used if the procedure requires such temperatures. However, the temperatures that can be reached with dry ice normally meet the need of the laboratory.

REFRIGERATED COOLERS

Units which contain refrigeration components can provide very low temperatures as needed without the use of dry ice. The design and construction of the unit determines the temperature range available, each unit having a rated cooling capacity. Coils can be immersed in properly selected liquids, such as ethylene glycol or silicone oils, to provide the desired cooling temperatures to lab equipment by means of circulating pumps.

11
MECHANICAL AGITATION

INTRODUCTION

Mechanical agitation is necessary when you must add materials a portion at a time so that they have immediate intimate contact with the bulk of the solution. Efficient mechanical agitation (1) reduces the time for completion of reaction, (2) can be used to control rate of reaction, and (3) improves yields of products.

MOTORS FOR LABORATORY MIXERS

There are many types of motor-powered laboratory stirrers (Fig. 11-1), and each motor is designed to perform under specified conditions with rated horsepower. Representative kinds are:

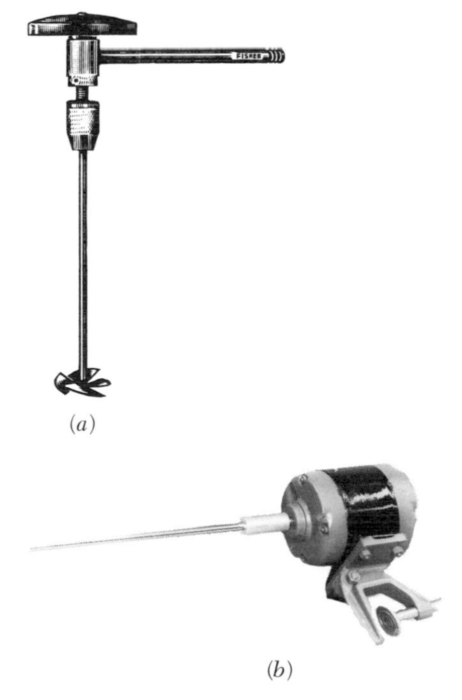

(*a*) (*b*) (*c*)

FIGURE 11-1 Stirrers. (*a*) A variable-speed stirrer, driven by compressed air, which is safe to use in explosive atmospheres. (*b*) This single-speed stirrer for general agitation in small-sized production has a direct drive, and a speed of about 1600 rpm. An adjustable clamp permits positioning for maximum agitation. (*c*) A variable-speed continuous-duty stirrer with a speed of 500 to 12,000 rpm. It has high torque for mixing thick emulsions and suspensions.

Light-duty, rheostat-controlled

High-torque, low-speed, rheostat-controlled

High-speed, rheostat-controlled

Variable-speed, friction-drive

High-torque, variable-speed, solid-state–controlled

High-torque, variable-speed, heavy-duty

Constant-speed, light-duty

Low-speed, light-duty

Constant-speed, heavy-duty

High-speed, light-duty

and many variations supplied by manufacturers to meet the need. Explosionproof motors and special-duty motors are also available.

CHOICE OF MOTOR DRIVE

Air-Operated

This type is used where explosive vapors are present. Vary the speed by adjusting the air pressure. Air stirrers eliminate the use of electric motors in hazardous locations.

Electric

The standard 1800-rpm fixed-speed type is used for mixing thin, nonviscous solutions.

Fixed reduced-speed types are used to mix viscous, heavy-bodied solutions and mixtures. Speeds depend on gear ratios.

Variable-speed, rheostat-controlled types usually vary from very slow to a full speed of 1800 rpm.

Inexpensive units do not have much power at low speeds; power increases as speed increases.

CAUTION Do not overload electric-motor mixers by using underpowered units for your operation; they will burn out. Laboratory mixers range from 1/30 to 1/8 hp.

CHOICE OF STIRRER

Selecting the *right* stirrer is important. The technician must determine if the stirrer is to be used for continuous or intermittent duty, because laboratory stirrers are rated either way.

Continuous duty: Those running for periods of 8 h or more

Intermittent-duty: Those running for a maximum time of about 1 h, with about the same time "off," or up to 8 h at 50 percent of the full load*

The physical design, shape, and size of the stirrer together with the speed of rotation determine the characteristics and efficiency of the mixing operation in a specific liquid solution. The choice of a particular mixer depends upon the:

1. Specific gravity and viscosity of the solution
2. Speed of the motor
3. Size, shape, and volume of the container
4. Shaft length of the stirrer
5. Material of shaft construction
6. The type of operation desired:
   (a) Blending
   (b) Shearing
   (c) Homogenizing
   (d) Mixing
   (e) Dispersion
   (f) Reaction
7. Characteristics of the reactants and products

Stirrers are generally made out of glass, Monel, stainless steel, or Teflon®.

Design depends upon the limitations of the equipment. Metal-propeller types are generally used for beakers, open containers, and open flasks. Other designs are used because the blades are narrow or collapse and the stirrer can be inserted through the narrow neck of a flask. (See Fig. 11-2.)

In connecting the mixer to the stirrer, a mechanical chuck will hold the driving shaft by friction for direct connection. A short piece of heavy-

---

* Mixers are rated by their horsepower. At full load they consume the rated wattage,

$$\text{Watts} = \text{volts} \times \text{amperes}$$

See Chap. 27, "Basic Electricity," for determining the horsepower, which can be measured in watts.

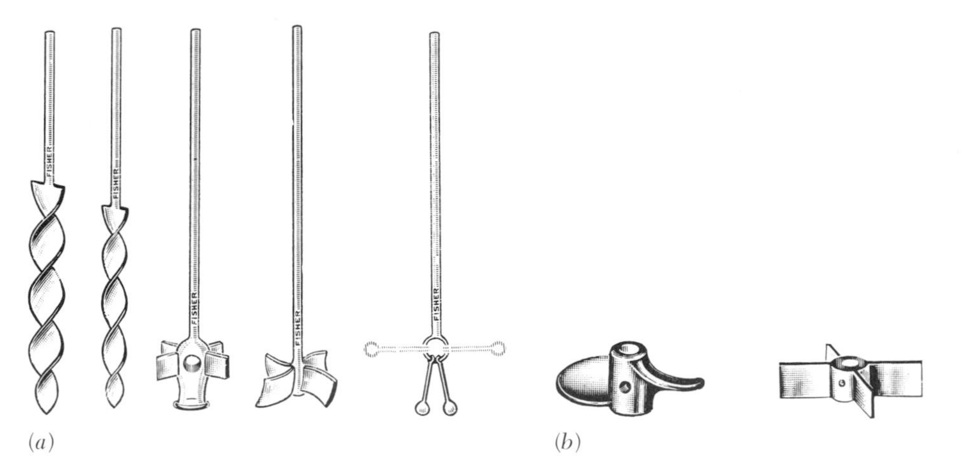

FIGURE 11-2
Stirrer accessories. (*a*) Glass stirrers in a variety of designs are used to mix thin to viscous solutions or to fit into narrow openings of containers. (*b*) Propellers of stainless steel or Monel are available in different shaft diameters and propeller designs, including two-, three-, and four-blade or turbine type.

walled rubber tubing connecting the mixer shaft to the stirrer shaft isolates the vibration of the mixer from the glass apparatus (Fig. 11-4*b*).

NOTE If the heavy-walled pressure-tubing connection is used to isolate the vibration of the mixer from the assembly, a stabilizing bearing must be used on the stirrer shaft (refer to section on bearing assemblies).

MAGNETIC MECHANICAL AGITATION

When operations prohibit use of mixer-stirrer apparatus, agitation can be accomplished by a rotating-magnet unit on which the glass beaker or flask rests (Fig. 11-3). A small steel bar (coated completely with inert Teflon®) is placed in the flask or beaker. When the magnetic rotator is turned on, the steel bar attracted to the magnet rotates in concert with the magnet, mixing the solution. (Refer to Chap. 10, "Heating and Cooling," p. 192.)

FIGURE 11-3
A magnetic stirrer.

BEARING ASSEMBLIES

The centrifugal force of an out-of-balance shaft or propeller assembly can cause undue vibrations which, if severe enough, can rupture the shaft or the glassware. A defective assembly should therefore be repaired as soon as possible; until repairs are made, it should be stabilized as shown in Fig. 11-4.

Assembled laboratory glass apparatus can also be damaged or broken by motor and shaft vibration. That vibration can be reduced by stabilizing the shaft rotation and isolating the motor from the glassware. A short piece of glass tubing which just slides over the stirrer shaft acts as a bearing; it is lubricated with glycerin. The tubing bearing can be anchored in a rubber stopper which is held by a clamp (Fig. 11-4*a*) or it can be part of an inter-joint standard-taper glass connection to the reaction flask (Fig. 11-4*b*).

Reaction mixtures in closed systems are isolated by means of mercury-sealed stirrer assemblies and compression sleeves pressing against the shaft (Figs. 11-5, 11-6, and 11-7).

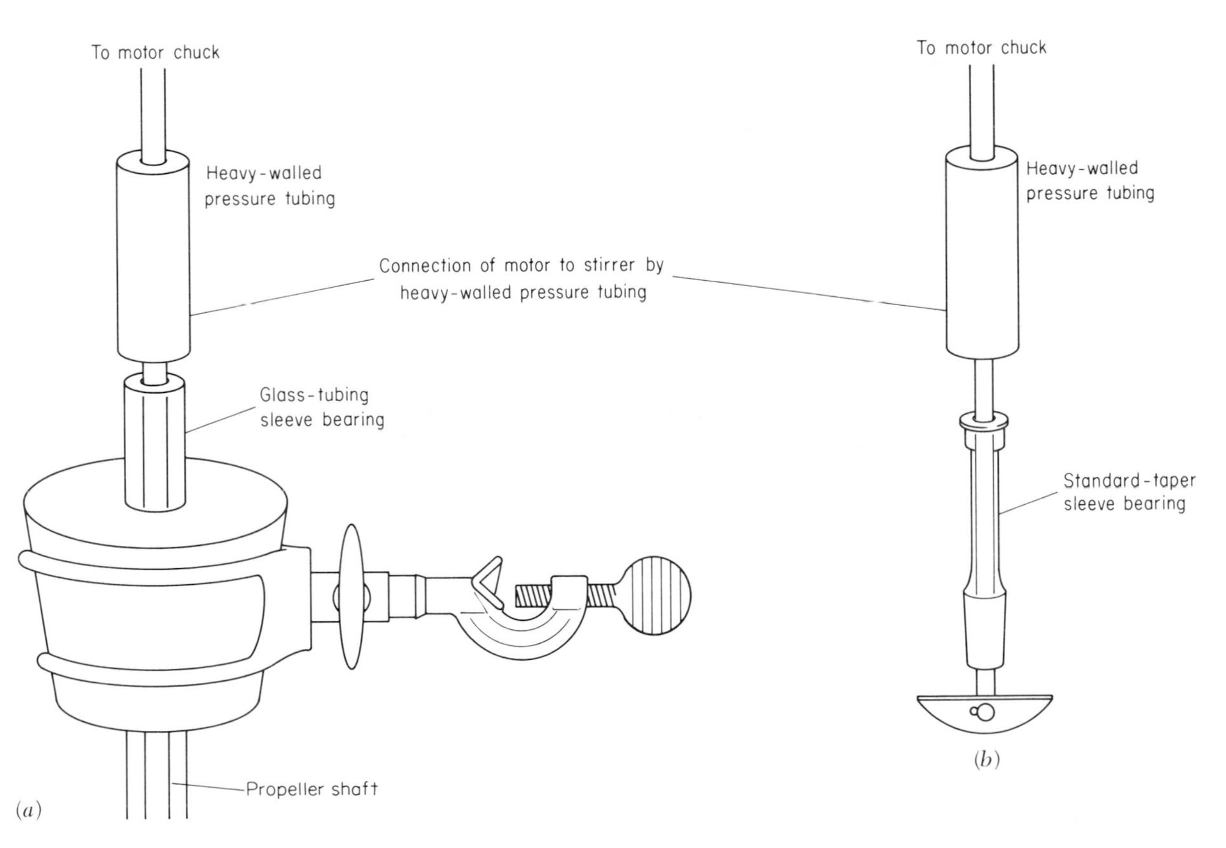

FIGURE 11-4
(*a*) Glass-tubing sleeve bearing. (*b*) Standard-taper sleeve bearings.

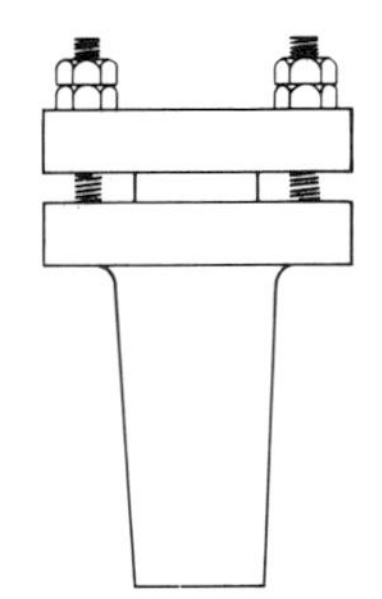

FIGURE 11-5
A stainless steel bearing assembly. This seals against the atmosphere by compression against the shaft.

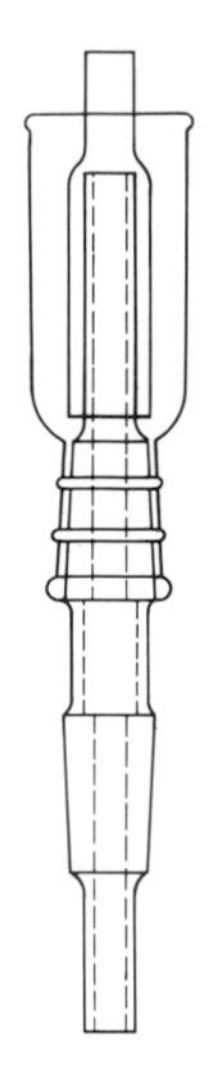

FIGURE 11-6
A mercury-sealed bearing assembly. This seals the system against the atmosphere by means of mercury in a cup. The rotating tube is affixed to the shaft with flexible rubber tubing (not shown), and the mercury fills the well to isolate the system and prevent any loss of volatile solvent. This kind of seal cannot be used for systems under pressure or vacuum.

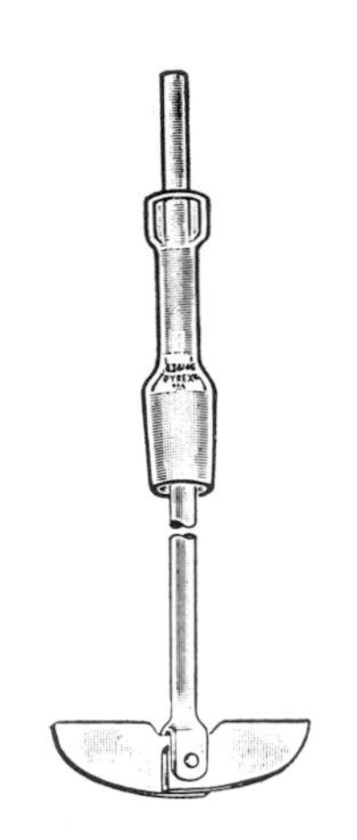

FIGURE 11-7
This glass seal, closely fitting the shaft to its housing (standard–taper joint), is lubricated with glycerin or silicone. It provides isolation of the system from the atmosphere and contains very moderate pressure and vacuum conditions while it is rotating.

Bearing assemblies are also used to isolate the system from the atmosphere: (1) to prevent escape of vapor, (2) to maintain inert atmospheres and exclude air, and (3) to stir and mix in closed systems.

MINISTIRRERS

A self-contained ministirrer can provide instant mixing and stirring of both thin and viscous liquids directly in the reaction flask (Fig. 11-8). It is self-powered and does not require a motor drive, couplings, bearings, or clamps. Such a stirrer may be obtained with interjoint standard-taper ground-glass fittings which enable it to be seated directly in the corresponding glass fitting of the flask. Various sizes of adapter sleeves are available as well as interchangeable stainless steel stirring paddles. A ministirrer can operate from the 12-V dc output of any transformer/rectifier actuated up to 240 V ac.

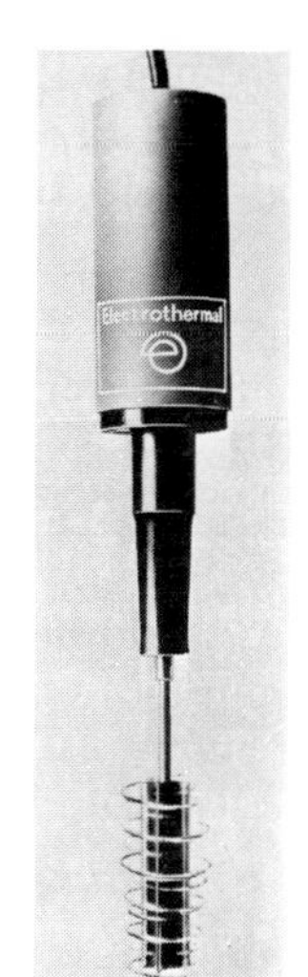

FIGURE 11-8
A ministirrer.

BLENDERS

Blenders can be used for rapid pulping or mixing of a very wide variety of substances. Extremely fine emulsions, suspensions, and solutions are formed quickly. Blenders can emulsify and disintegrate materials and are therefore used in procedures for the formulation of cosmetics, vitamins, paints and varnishes, and various other products. The blades are sharp and rotate at speeds over 15,000 rpm, cutting suitable substances and converting them into a homogeneous mass much more quickly than other equipment can. (See Fig. 11-9.)

FIGURE 11-9
A blender with speed varying to a maximum of 12,000 rpm. It is used for emulsifying, pulping, and blending foodstuffs, plant materials, and other substances into extremely fine suspensions, emulsions, and solutions.

PILOT-PLANT AGITATION

Agitation in the laboratory in beakers and flasks can easily be accomplished, and the selection of the agitator is not that critical, except under unusual conditions. However, in pilot-plant operations, mixers must be chosen according to (1) purpose, (2) degree of agitation desired, (3) properties of the fluid, and (4) size of the container and the volume of fluid.

The general classes of mixing operations are:

1. The blending of two or more liquids
2. The dispersion of immiscible components
3. The dissolving of soluble solids in a liquid
4. The suspension of nonsoluble solids in a fluid
5. The transfer of heat, either into the fluid batch or out of the fluid batch to heat exchangers
6. The extraction of substances from one immiscible fluid into another
7. The maintenance of desired reaction conditions to induce and complete crystallization of solids
8. The use of mixers to increase the solubility of gases in liquids

Mixing Efficiently in Tanks

Number of Propellers

The dimensions of the tank or container affect the efficiency of the mixing operation.

Containers of standard dimensions are effectively agitated by a single propeller (Fig. 11-10).

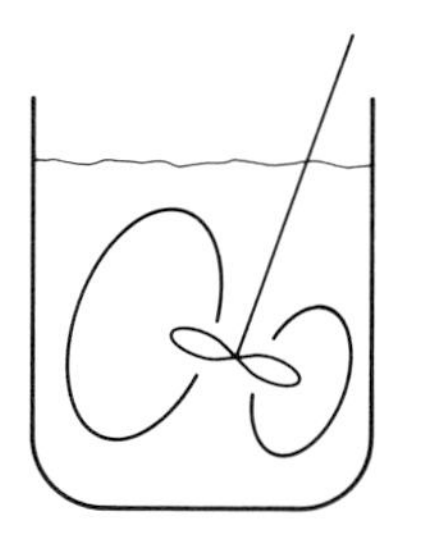

FIGURE 11-10
Standard tank.

Tall, cylindrical containers are not efficiently mixed by single-propeller agitation. Pockets are formed, and only part of the liquid mass is mixed (Fig. 11-11).

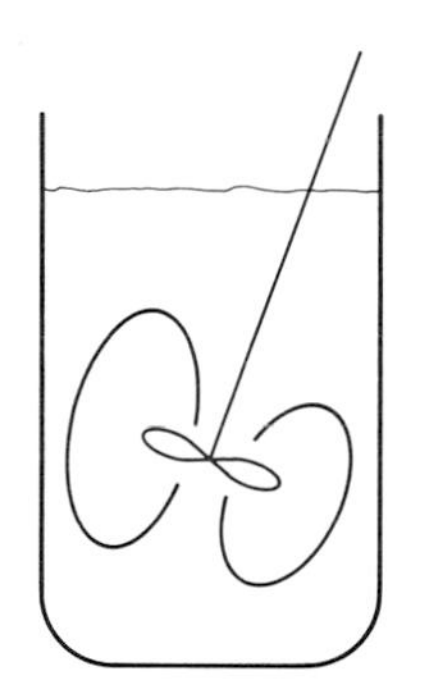

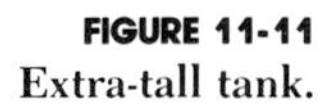

FIGURE 11-11
Extra-tall tank.

When taller tanks must be agitated efficiently, use dual (two) propellers on the same shaft (Fig. 11-12). The use of two propellers overcomes the pocketing action.

FIGURE 11-12
Extra-tall tanks require dual propellers.

Positioning the Agitator

Vertical-center agitation merely moves the material around, creating a vortex, but not effectively mixing the fluid (Fig. 11-13). Vortexing in a

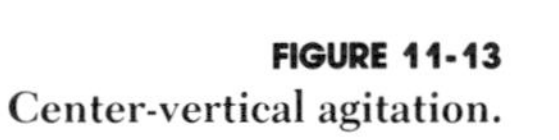

FIGURE 11-13
Center-vertical agitation.

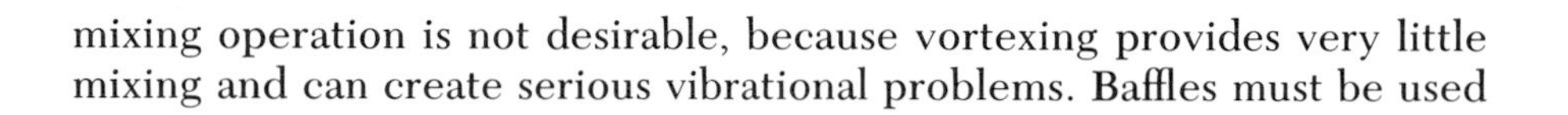

mixing operation is not desirable, because vortexing provides very little mixing and can create serious vibrational problems. Baffles must be used in vertical-center agitation; the baffles disturb the fluid flow symmetry, and thus contribute to good mixing.

Center-angled agitation is more effective; vortexing can be eliminated or minimized by adjusting the angle, and the fluid does turn over effectively (Fig. 11-14).

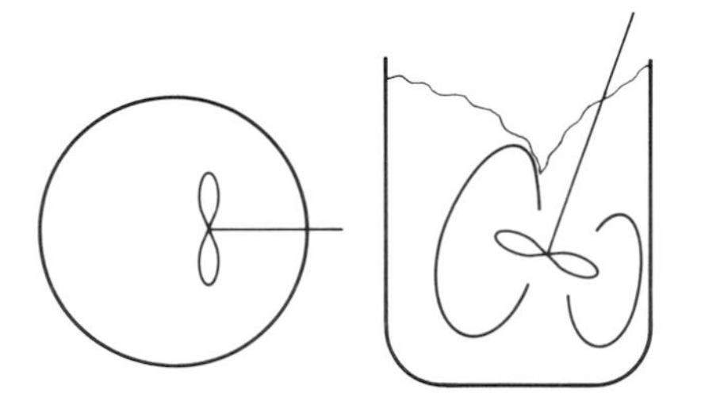

FIGURE 11-14
Center-angle agitation.

Side-entering angled agitation is preferred and most effective. The fluid not only rotates, but is turned over effectively, bottom to top, with good sweeping action on the bottom (Fig. 11-15).

FIGURE 11-15
Side-entering angled agitation.

Selection of Motor Speeds

Usually direct-drive agitators operate at 1800 rpm under normal loads. By the use of gear-reduction boxes, agitators can operate at any desired speed. In general, the selection of a high-speed or low-speed agitator depends upon the conditions of the reaction and the physical sizes of the components.

High-speed agitators use small propellers with motors designed to efficiently mix aqueous solutions and thin (light) liquids. Agitators that work at lower, intermediate speeds are suggested for viscous oils, paints, varnishes, syrups, etc.

Very-low-speed agitation with very high torque is used for extremely thick, viscous reaction fluids. It is also used when high-speed agitation would create undesirable foam, or when high-speed agitation would initiate undesired side effects, as in the mixing of milk and cream.

Turbine Mixers

These are flat-bladed turbine pumps which pump the liquid outward by centrifugal force; the liquid which is displaced by the turbine action is replaced by radial flow from the top and bottom of the fluid.

Turbines are always mounted exactly vertical. In an unbaffled tank, the turbine mixer is mounted off center (Fig. 11-16).

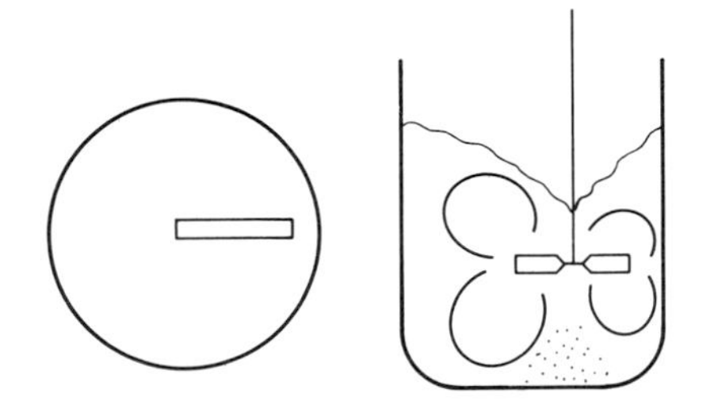

FIGURE 11-16 Off-center vertical mounting in unbaffled tank.

In a baffled tank (a tank with anchored blades to disturb fluid rotation), the turbine may be mounted vertical center (Fig. 11-17).

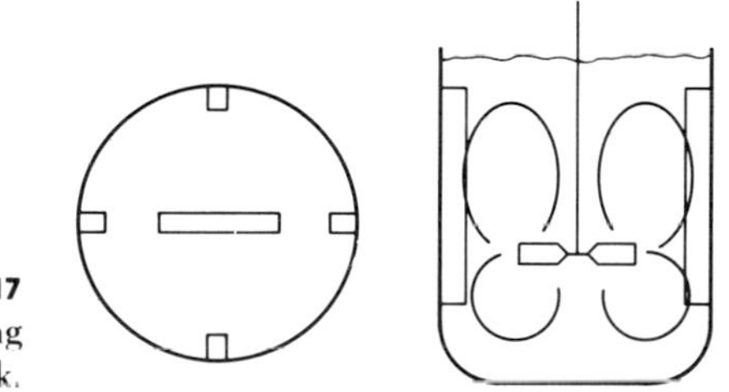

FIGURE 11-17 Vertical-center mounting of turbine in baffled tank.

12 LABORATORY FILTRATION

INTRODUCTION

Filtration is the process of removing material, often but not always a solid, from a substrate (liquid or gas) in which it is suspended. This process is a physical one; any chemical reaction is inadvertent and normally unwanted. Filtration is accomplished by passing the mixture to be processed through one of the many available sieves called *filter media*. These are of two kinds: surface filters and depth filters.

With the surface filter, filtration is essentially an exclusion process: particles larger than the filter's pore or mesh dimensions are retained on the surface of the filter; all other matter passes through. Examples are filter papers, membranes, mesh sieves, and the like. These are frequently used when the solid is to be collected and the filtrate is to be discarded. Depth filters, however, retain particles both on their surface and throughout their thickness; they are more likely to be used in industrial processes to clarify liquids for purification. A mat of Celite is an example. Filtration most commonly is used in one of four ways:

Solid-liquid filtration: The separation of solid particulate matter from a carrier liquid

Solid-gas filtration: The separation of solid particulate matter from a carrier gas

Liquid-liquid separation: A special class of filtration resulting in the separation of two immiscible liquids, one of them water, by means of a hydrophobic medium

Gas-liquid filtration: The separation of gaseous matter from a liquid in which it is usually, but not always, dissolved

In the laboratory, filtration is generally used to separate solid impurities from a liquid or a solution or to collect a solid substance from the liquid or solution from which it was precipitated or recrystallized. This process can be accomplished with the help of gravity alone (Fig. 12-1) or it can be speeded up by using vacuum techniques. (Refer to the section on vacuum filtration.) Vacuum filtration provides the force of atmospheric pres-

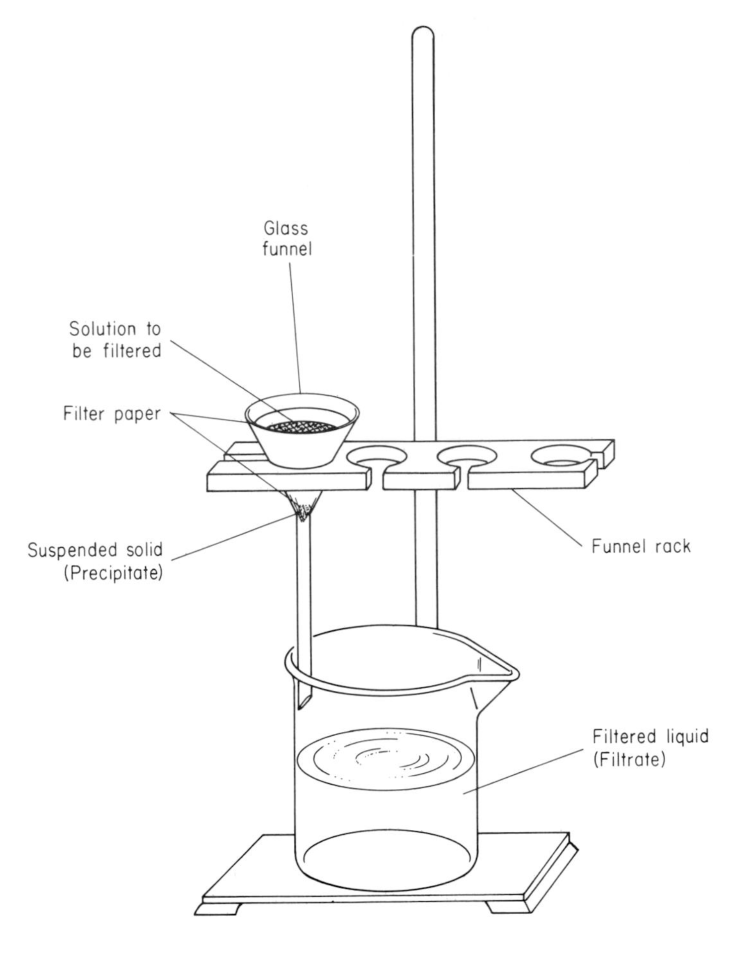

FIGURE 12-1
A gravity-filtration setup.

sure on the solution in addition to that of gravity, and thus increases the rate of filtration.

The efficiency of filtration depends on the correct selection of the method to be used, the various pieces of apparatus available, the utilization of the filter medium most appropriate for the particular process, and the use of correct laboratory technique in performing the manipulations involved. During filtration (as previously mentioned) a liquid is usually separated from a solid by pouring the liquid through a sieve, usually filter paper. The liquid passes through the paper, but the solid is retained. Although the carrier liquid is usually relatively nonreactive, it is sometimes necessary to filter materials from highly alkaline or acidic carrier liquids or to perform filtration under other highly reactive conditions. A variety of filter media exists from which it is possible to select whichever one best fits the particular objectives and conditions of a given process. The most common filter media are:

Paper

Fiberglass "papers" or mats

Gooch crucibles

Sintered-glass (or fritted-glass) crucibles and funnels

Porous porcelain crucibles

Monroe crucibles

Millipore® membranes

All of those listed are available in various porosities, and their use will be discussed later in this section.

FILTRATION METHODS

There are two general methods of filtration: gravity and vacuum (or suction). During gravity filtration the filtrate passes through the filter medium under the combined forces of gravity and capillary attraction between the liquid and the funnel stem. In vacuum filtration a pressure differential is maintained across the filter medium by evacuating the space below the filter medium. Vacuum filtration adds the force of atmospheric pressure on the solution to that of gravity, with a resultant increase in the rate of filtration. The choice of method to be used depends upon the following factors.

1. The nature of the precipitate
2. The time to be spent on the filtration
3. The degree to which it is necessary to retain all the precipitate
4. The extent to which one can tolerate the contamination of the precipitate with the filtrate

We will discuss each type later in the chapter. Information applicable to both methods follows.

FILTER MEDIA

Paper

There are several varieties or grades of filter paper (Fig. 12-2) for special purposes; there are qualitative grades, low-ash or ashless quantitative

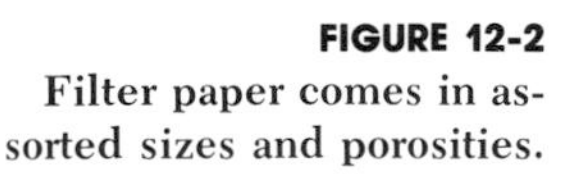

FIGURE 12-2
Filter paper comes in assorted sizes and porosities.

grades, hardened grades, and even glass-fiber "papers." For a given filtration, you must select the proper filter paper with regard to porosity and residue. Some of this information is given in Table 12-1.

Qualitative-grade papers will leave an appreciable amount of ash upon ignition (of the order of 0.7 to 1 mg from 9-cm circle) and are therefore unsuitable for applications in quantitative analysis where precipitates are to be ignited on the paper and weighed. They are widely used for clarifying solutions, filtration of precipitates which will later be dissolved, and general nonquantitative separations of precipitates from solution.

Low-ash or ashless quantitative-grade papers can be ignited without leaving an ash. The residue left by an 11-cm circle of a low-ash paper may be as low as 0.06 mg; an ashless-grade paper typically leaves 0.05 mg or less from an 11-cm circle. In most analytical procedures, this small mass can be considered negligible.

TABLE 12-1
Commonly Used Filter Papers*

W	*S&S*	*RA*	*Porosity*	*Speed*	*Use for*
Qualitative- or regular-grade papers					
4	604	202	Coarse	Very rapid	Gelatinous precipitates
1	595	271	Medium	Medium	Ordinary crystalline precipitates
3	602	201	Medium	Slow	Fine precipitates; used with Büchner funnels
Quantitative-grade papers (less than 0.1 mg ash)					
41	589 blue ribbon	. . .	Coarse	Very rapid	Gelatinous precipitates
40	589 white ribbon	. . .	Medium	Rapid	Ordinary crystalline precipitates
42	589 black ribbon	. . .	Fine	Slow	Finest crystalline precipitates

* Code W: Whatman; S & S: Schleicher and Shüll; RA: Reeve Angel.

Hardened-grade papers are designed for use in vacuum filtrations and are processed to have great wet strength and hard lintless surfaces. They are available in low-ash and ashless as well as regular grades.

Fiberglass papers are produced from very fine borosilicate glass and are used in Gooch, Büchner, or similar filtering apparatus to give a combination of very fine retention, very rapid filtration, and inertness to the action of most reagents to an extent not found in any cellulose paper.

All grades of filter paper are manufactured in a variety of sizes and in several degrees of porosity. Select the proper porosity for a given precipitate. If too coarse a paper is used, very small crystals may pass through, while use of too fine a paper will make filtration unduly slow. The main objective is to carry out the filtration as rapidly as possible, retaining the precipitate on the paper with a minimum loss.

Membrane Filters

Membrane filters are thin polymeric (plastic) structures with extraordinarily fine pores. These sheets of highly porous material are composed of pure, biologically inert cellulose esters or other polymeric materials. Such filters are distinctive in that they remove from a gas or liquid passing through them all particulate matter or microorganisms larger than the filter pores. With proper filter selection, they yield a filtrate that is ultraclean and/or sterile.

Membrane filters are available in a wide variety of pore sizes in a number of different materials. The range of pore sizes and the uniformity of pore size in a typical filter are shown in Table 12-2 and Fig. 12-3.

When liquids pass thorugh a Millipore® membrane filter, all contaminants larger than the filter-pore size are retained on the surface of the filter,

TABLE 12-2
Uniformity of Millipore® Filters

Filter pore size, μm	*Maximum "rigid" particle to penetrate, μm*	*Filter pore size, μm*	*Maximum "rigid" particle to penetrate, μm*
14	17	0.65	0.68
10	12	0.60	0.65
8	9.4	0.45	0.47
7	9.0	0.30	0.32
5	6.2	0.22	0.24
3	3.9	0.20	0.25
2	2.5	0.10	0.108
1.2	1.5	0.05	0.053
1.0	1.1	0.025	0.028
0.8	0.85		

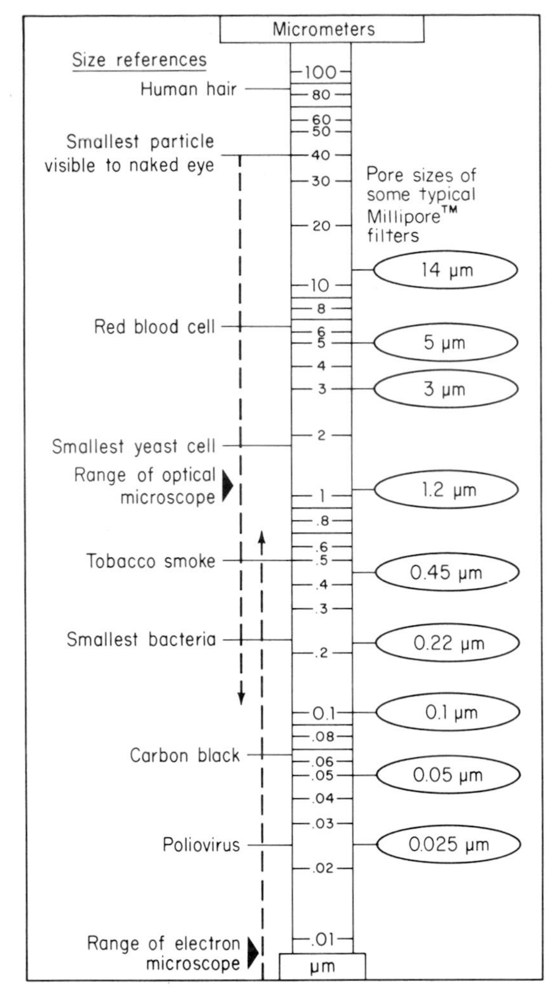

FIGURE 12-3
Scale comparing Millipore® filter-pore sizes with sizes of microbes and microparticles.

where they can be readily analyzed or counted. This is in sharp contrast with the action of a "depth" filter, which retains contaminants not only on its surface but also inside the filter matrix.

Fritted Glassware

The filtration of solids can be performed with funnels fitted with *fritted-glass* (also called *sintered-glass*) plates. Fritted glass is available in different porosities, and some of the problems encountered in using filter paper are minimized by using fritted-glass equipment. The grades of fritted glassware are listed in Table 12-3.

For optimum service, fritted ware must be maintained carefully. It is best to follow any manufacturer's instructions that may be enclosed with the

TABLE 12-3
Grades of Fritted Ware

Designation	*Nominal maximum pore size, μm*
Extra-coarse	170–220
Coarse	40–60
Medium	10–15
Fine	4–5.5
Very fine	2–2.5
Ultrafine	0.9–1.4

equipment. The following paragraph and table comprise an example of such instructions.

Cleaning—A new fritted filter should be washed by suction with hot hydrochloric acid and then rinsed with water before it is used. Clean all fritted filters immediately after use. Many precipitates can be removed from the filter surface simply by rinsing from the reverse side with water under pressure not exceeding 15 lb/in^2. Some precipitates tend to clog the fritted filter pores, and chemical means of cleaning are required. Suggested solutions are listed below.

Material	*Cleaning solution*
Fatty materials	Tetrachloromethane
Organic matter	Hot, concentrated cleaning solution
Mercury residue	Hot nitric acid
Silver chloride	Ammonia or sodium hyposulfite

See the section on vacuum filtration for a discussion of other types of crucibles and funnels.

FILTERING ACCESSORIES

Filter Supports

Some solutions which are to be filtered tend to weaken the filter paper, and at times, the pressure on the cone of the filter will break the filter paper, ruining the results of the filtration. The thin woven textile disks shown in Fig. 12-4 are used to support the tip of the filter-paper cone.

They are approximately of the same thickness as the filter paper, and therefore ensure close contact of the reinforced paper with the funnel

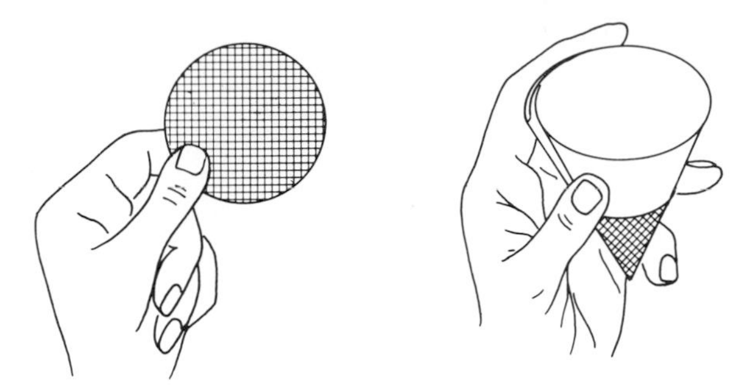

FIGURE 12-4
Filter-paper support.

walls. They are folded along with the filter paper when it is formed into the normal conical shape, and they can easily be removed from the wet filter paper after the filtration has been completed, if desired. This is done when the filter paper and the collected precipitate are to be ashed.

Filter Aids

During filtration, certain gummy, gelatinous, flocculent, semicolloidal, or extremely fine particulate precipitates often quickly clog the pores of a filter paper, and then the filtering action stops. Normal analytical determinations require positive clarification at maximum filtration rates. Therefore, when a necessary filtration is impeded or halted by the presence of recalcitrant particles, filter aids are employed to speed up the process. Filter aids consist of diatomaceous earth and are sold under the trade names of Celite® or FilterAid; they are extremely pure and inert powder-like materials which form a porous film or cake on the filter medium. In use, they are slurried or mixed with the solvent to form a thin paste and are then filtered through a Büchner funnel (with the paper already in place on the funnel) to form a film or cake about 3 to 4 mm thick. The troublesome filtration slurry is then filtered through the cake, and the gummy, gelatinous, or finely divided particulate precipitate is caught in the cake, whereupon the filtration proceeds almost normally. An alternative procedure involves the direct addition of the filter aid directly to the problem slurry with thorough mixing. The filter aid then speeds up the filtration by forming a porous film or cake on the filter medium and yields filtrates of brilliant crystalline clarity. All the suspended matter has been retained on the funnel, and the filtration normally proceeds rapidly.

Limitations of Filter Aids

1. They cannot be used when the object of the filtration is to collect a solid product, because the precipitate collected also contains the filter aid. Filter aids can be used only when the filtrate is the desired product.

2. Because they are relatively inert, they can be used in normally acidic

and basic solutions; however, they cannot be used in strongly alkaline solutions or solutions containing hydrofluoric acid.

3. Filter aids cannot be used when the desired substance is likely to precipitate from the solution.

Wash Bottles

Two types of wash bottles are (1) the Florence flask (now largely a laboratory curiosity), which operates on breath pressure (the stream of water is directed by moving the tip with the fingers) and (2) the plastic wash bottle operated by squeezing, the "squeeze bottle" (Fig. 12-5).

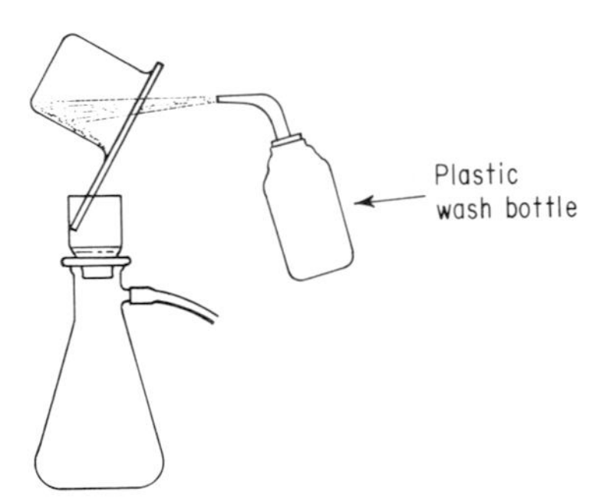

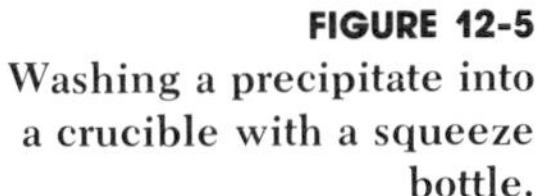

FIGURE 12-5
Washing a precipitate into a crucible with a squeeze bottle.

The polyethylene plastic wash bottle has all but displaced the traditional glass Florence flask wash bottle. It is available in a variety of colors for content identification, eliminating the hazard of choosing the wrong bottle. The flow of the bottle can be increased by cutting off part of the tip.

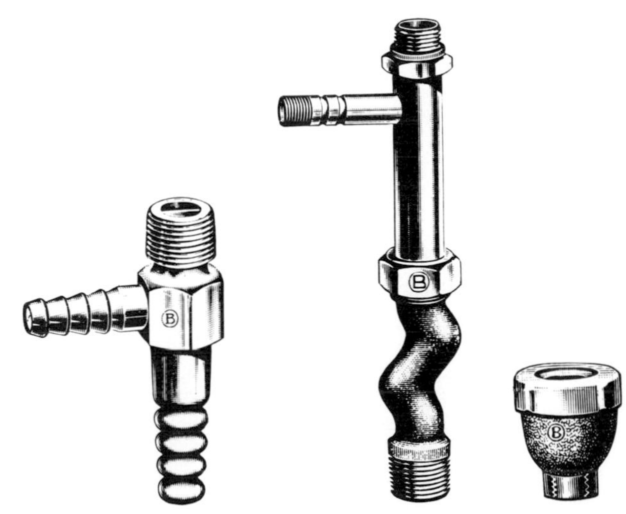

FIGURE 12-6
Filter pump and accessories.

Filter Pump and Accessories

A water pump (Fig. 12–6) is used for suction filtrations, general vacuum manipulations, and pipet cleaning. Accessories are used to couple to water lines and prevent splashing to existing water. (See The Water Aspirator in Chap. 5, "Pressure and Vacuum," p. 90.)

MANIPULATIONS ASSOCIATED WITH THE FILTRATION PROCESS

Whether one uses gravity or vacuum filtration, three operations must be performed: *decantation, washing,* and *transfer* (Fig. 12-7).

Decantation

When a solid readily settles to the bottom of a liquid and shows little or no tendency to remain suspended, it can be separated easily from the liquid by carefully pouring off the liquid so that no solid is carried along. This process is called *decantation.* To decant a liquid from a solid:

1. Hold the container (beaker, test tube, etc.) which has the mixture in it in one hand and have a glass stirring rod in the other (Fig. 12-7*a*).

2. Incline the beaker until the liquid has almost reached the lip (Fig. 12-7*b*).

3. Touch the center of the glass rod to the lip of the beaker and the end of the rod to the side of the container into which you wish to pour the liquid.

4. Continue the inclination of the beaker until the liquid touches the glass rod and flows along it into the second container. The glass rod enables you to pour the liquid from the beaker slowly enough that the solid is not carried along and also prevents the liquid from running back along the outside of the beaker form which it is being poured (Fig. 12-7*c*).

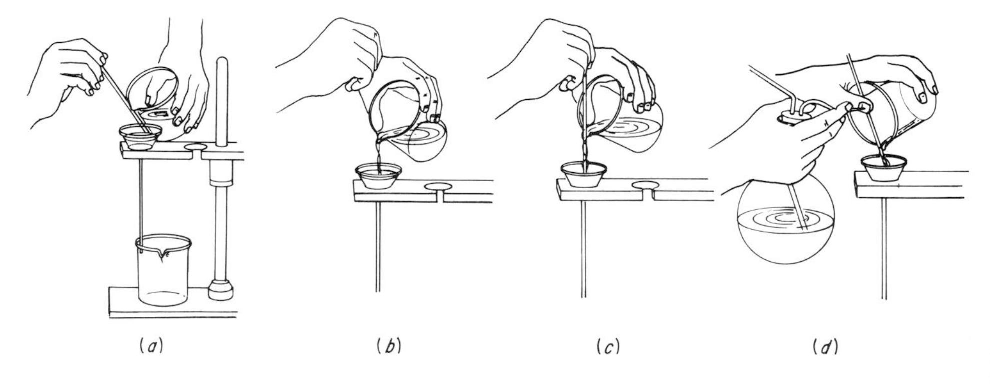

FIGURE 12-7 Gravity-filtering operation showing techniques for decantation and transfer of precipitates.

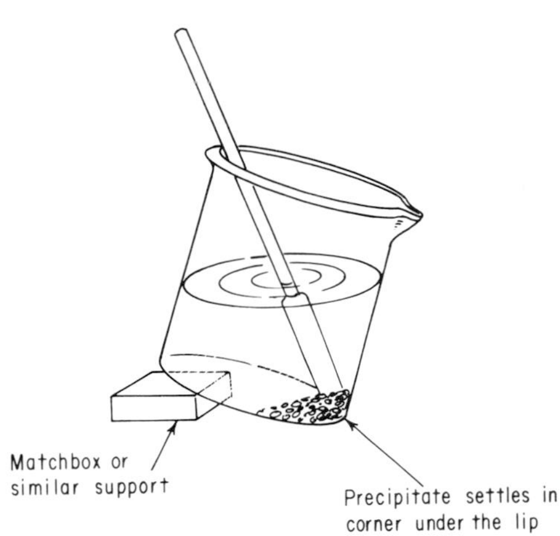

FIGURE 12-8
Supporting a beaker in a tilted position to allow the precipitate to settle prior to decantation.

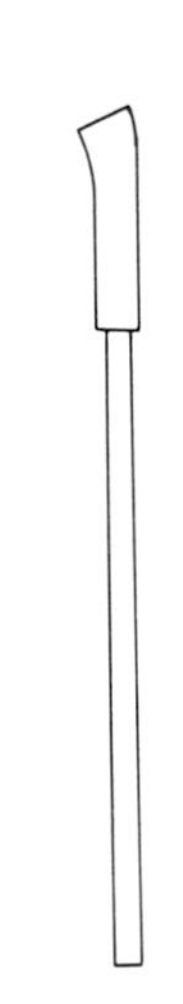

FIGURE 12-9
Stirring rod with rubber policeman (a piece of rubber tubing with a flattened end) is used to remove traces of solids from containers and to speed up solutions of solids in liquids when used as a stirrer. Also used to prevent scratching the inside of a vessel.

Washing

The objective of washing is to remove the excess liquid phase and any soluble impurities which may be present in the precipitate. Use a solvent which is miscible with the liquid phase but does not dissolve an apprecia-ble amount of the precipitate.

Solids can be washed in the beaker after decantation of the supernatant liquid phase. Add a small amount of the wash liquid and thoroughly mix it with the precipitate. Allow the solid to settle. Decant the wash liquid through the filter. Allow the precipitate to settle, with the beaker tilted slightly so that the solid accumulates in the corner of the beaker under the spout (Fig. 12-8). Repeat this procedure several times.

Several washings with small volumes of liquid are more effective in remov-ing soluble contaminants than a single washing using the total volume.

Transfer of the Precipitate

Remove the bulk of the precipitate from the beaker to the filter by using a stream of wash liquid from a wash bottle (Fig. 12-7*d*). Use the stirring rod (Fig. 12-9) to direct the flow of liquid into the filtering medium. The last traces of precipitate are removed from the walls of the beaker by scrubbing the surfaces with a rubber policeman attached to the stirring rod. All solids collected are added to the main portion in the filter paper (Fig. 12-10). If the precipitate is to be ignited, use small fragments of ash-less paper to scrub the sides of the beaker; then add these fragments to the bulk of the precipitate in the filter with the collected solid.

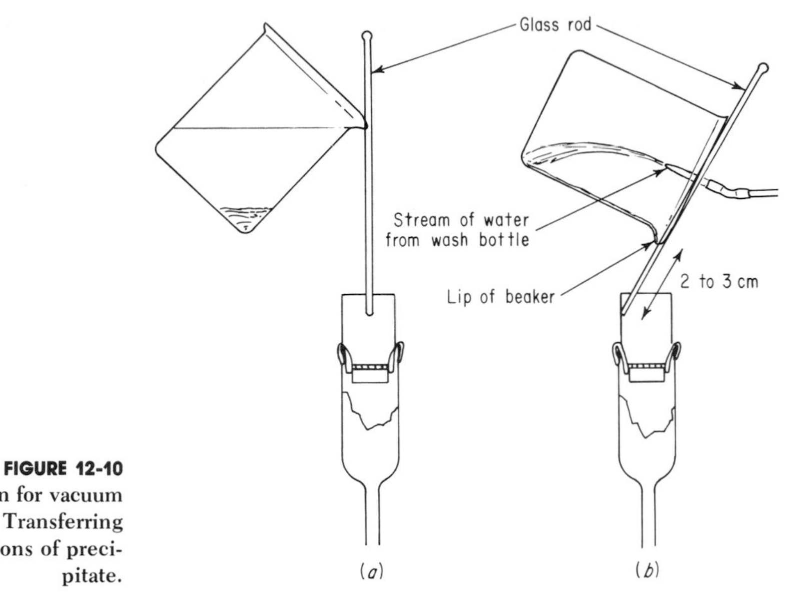

FIGURE 12-10
(*a*) Decantation for vacuum filtration. (*b*) Transferring the last portions of precipitate.

GRAVITY FILTRATION

During gravity filtration the filtrate passes through the filter medium under the forces of gravity and capillary attraction between the liquid and the funnel stem (Fig. 12-1). The most common procedure involves the use of filter paper and a conical funnel (see Fig. 12-11). The procedure is slow, but it is highly favored for gravimetric analysis over the more rapid vacuum filtration because there is better retention of fine particles of precipitate and less rupturing or tearing of the paper. Moreover, gravity filtration is generally the fastest and most preferred method for filtering gelatinous precipitates because these precipitates tend to clog and pack the pores of the filter medium much more readily under the additional force supplied during a vacuum filtration.

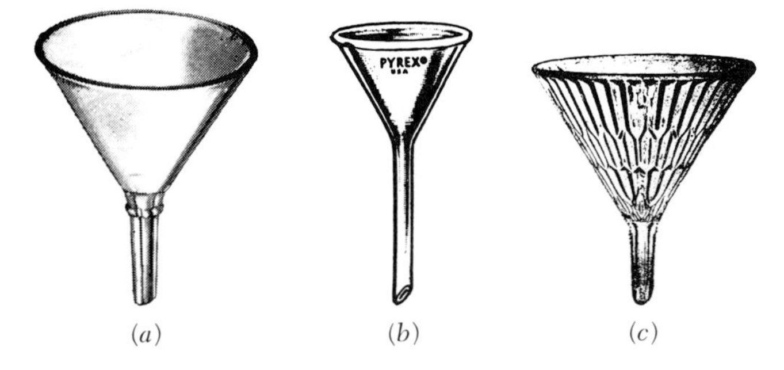

FIGURE 12-11
Glass funnels. (*a,b*) Varied size and stem length. (*c*) Funnel with heavy ribbed construction; raised ribs on inner surface facilitate rapid filtration.

Avoid accumulating precipitate on the filter paper during the early stages of the filtration process. This is necessary for rapid filtering, since the precipitate will be drawn into the pores of the paper, where it will impede the passage of solution and retard the rate of filtration. It is advisable to carry out precipitations in a beaker whenever possible because it has a pour spout to facilitate pouring of liquids without loss. The precipitate should be allowed to settle to the bottom of the beaker before filtration begins. The supernatant liquid phase, free of most of the suspended precipitate, is then poured onto the filter, leaving the precipitated solid essentially undisturbed. The bulk of the precipitate is not added until the last stages of the filtration as part of the *washing* process.

Procedure

Optimum filtering speed is achieved in gravity filtration by proper folding and positioning of the paper in the funnel (Figs. 12-12 and 12-13). If maximum speed is to be maintained, follow the suggestions given below.

1. Take maximum advantage of capillary attraction to assist in drawing the liquid phase through the paper. Use a long-stemmed funnel (Fig. 12-11*b*) and maintain a continuous column of water from the tip of the funnel stem to the undersurface of the paper. The tip of the funnel should touch

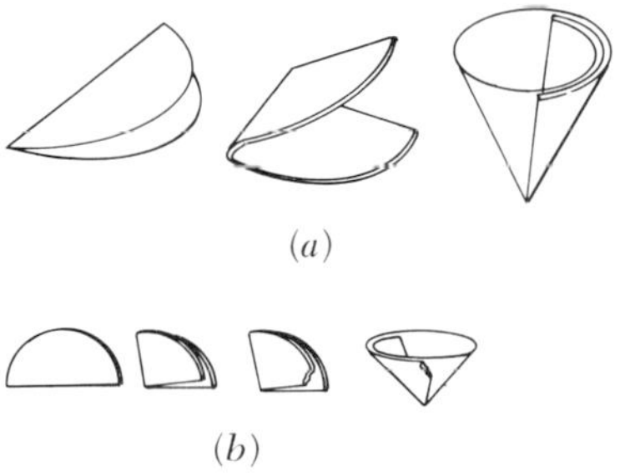

(*a*)

(*b*)

FIGURE 12-12
(*a*) Folding a filter paper. (*b*) Alternative method of folding filter paper. Steps in folding paper for use in filtering with a regular funnel. The second fold is not exactly at a right angle. Note the tear, which makes the paper stick better to the funnel.

FIGURE 12-13
Seating of a filter paper.

the side of the vessel which receives the filtrate; this procedure aids the filtration and minimizes any loss of filtrate that might be caused by splashing. Accurate fit of the folded filter helps maintain an airtight seal between the funnel and the top edge of the wet filter paper.

2. Expose as much of the paper as possible to provide free flow of liquid through the paper. If you fold the paper as shown in Fig. 12-12*b*, the paper will not coincide exactly with the walls of the funnel and the liquid will be able to flow between the paper and the glass. This type of fold will also help to maintain an airtight seal between the top edge of the filter paper and the funnel. Another way to expose most of the surface of the paper is to use a fluted filter paper. Fluting the filter paper increases the filtration in two ways. It permits free circulation of air in the receiving vessel and maintains pressure equalization. The solvent vapors present during the filtration of certain hot solutions can cause pressure to build up in the receiver and consequently decrease the speed of filtration. There are two ways of folding a flat circular piece of filter paper into a fluted filter. The first is shown in Fig. 12-14, the second in Fig. 12-15. The fluted filter in a funnel is shown in Fig. 12-16. Prepared fluted filters are also available commercially.

3. Many precipitates will spread over a wetted surface against the force of gravity; this behavior is known as *creeping*, and it can cause loss of precipitate. For this reason, be sure to pour the solution into the filter until the paper cone is no more than three-quarters filled. Never fill the cone completely. This precaution prevents loss of precipitate from both creep and overflow. It also provides an area near the top of the paper which is free of precipitate. By grasping this "clean" portion, you can remove the cone from the funnel and fold it for ignition (p. 236) without loss or contamination. See Fig. 12-17 for the best way to pour the supernatant liquid into the filter.

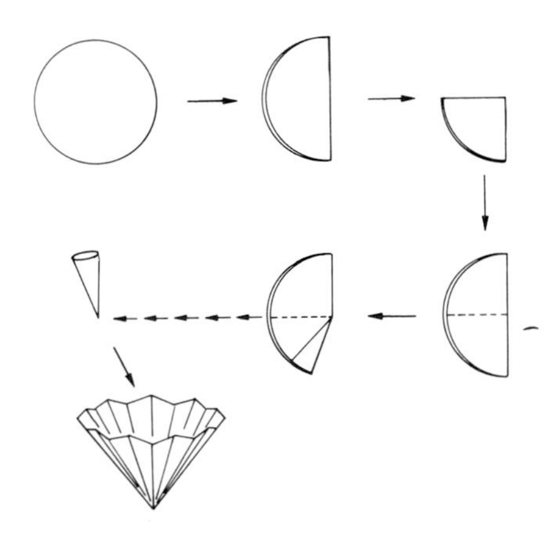

FIGURE 12-14
Folding a fluted filter. Fold the filter paper in half, then fold this half into eight equal sections, like an accordion. The fluted filter paper is then opened and placed in a funnel.

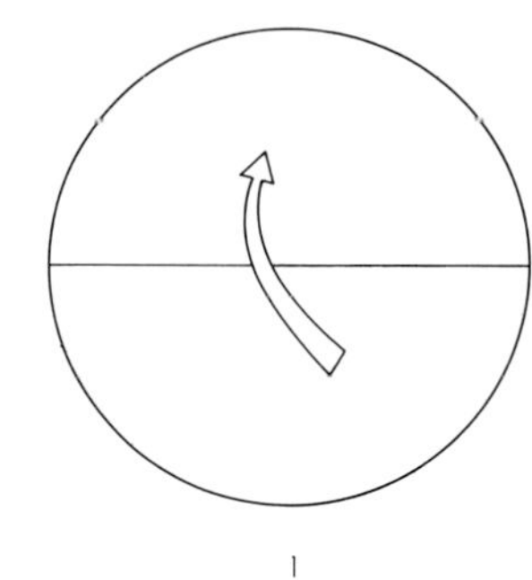

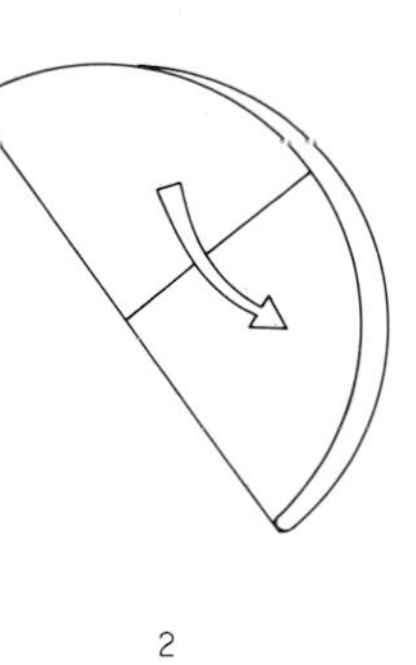

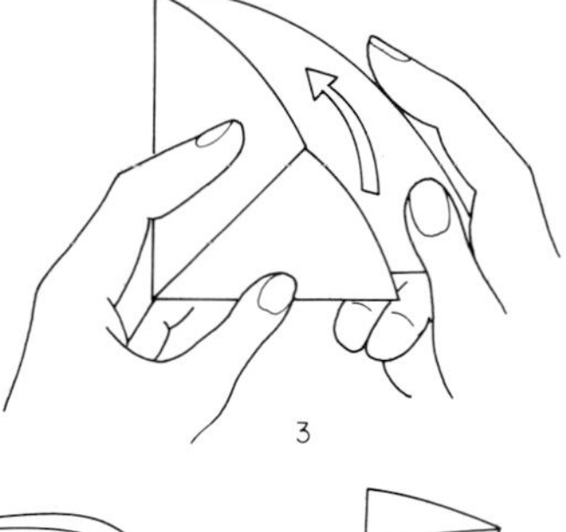

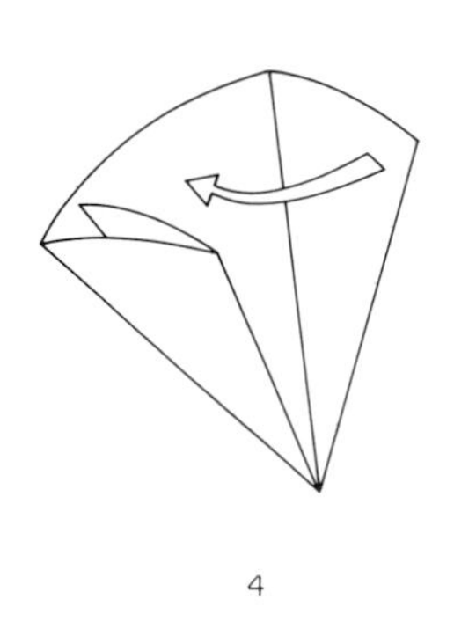

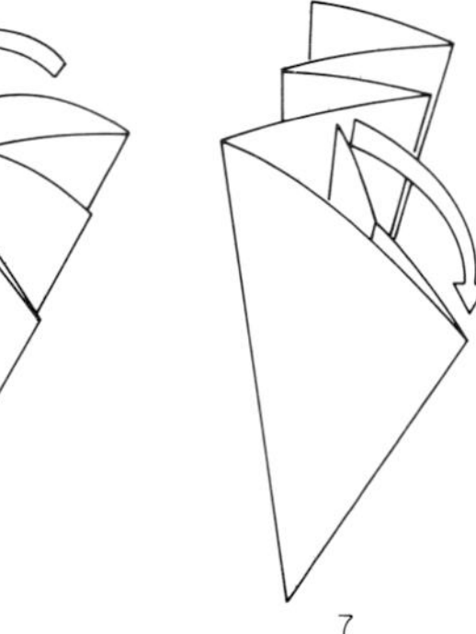

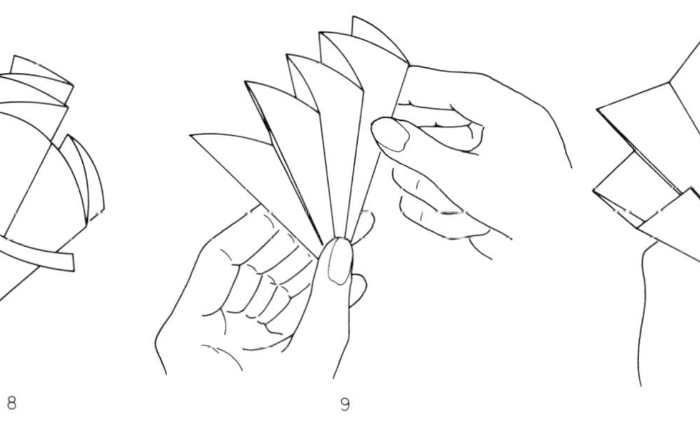

FIGURE 12-15
Alternate method of folding a fluted filter.

FIGURE 12-16
A fluted filter paper in a funnel.

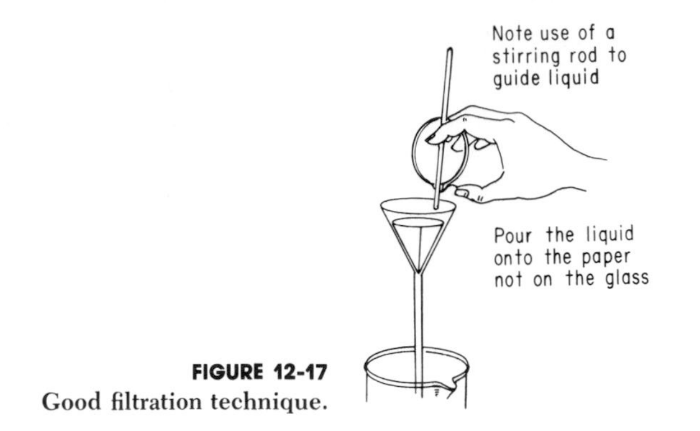

FIGURE 12-17
Good filtration technique.

VACUUM FILTRATION

Vacuum filtration is a very convenient way to speed up the filtration process, but the filter medium must retain the very fine particles without clogging up. The vacuum is normally provided by a water aspirator, although a vacuum pump, protected by suitable traps, can be used. Because of the inherent dangers of flask collapse from the reduced pressure, thick-walled filter flasks should be used, and the technician should be always on the alert for the possibility of an implosion.

Procedure

A typical setup for carrying out a vacuum filtration is shown in Fig. 12-18. This illustration shows the use of a Büchner funnel in which the wetted filter paper must be seated before the suction is applied. The funnel or crucible is fitted to a suction flask. The sidearm of the flask is connected to a source of vacuum such as a water aspirator. A water-trap bottle is inserted between the flask and the source of vacuum. When the vacuum

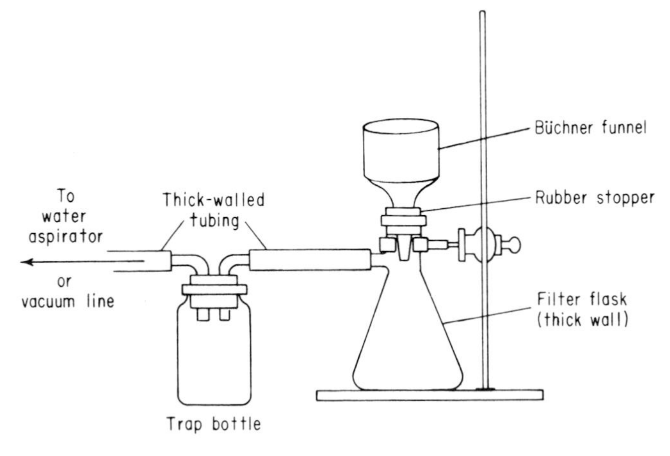

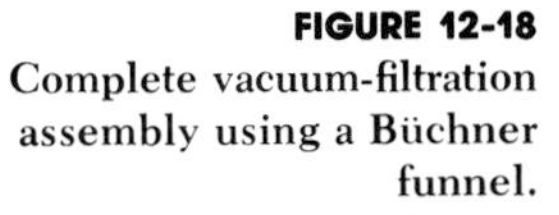

FIGURE 12-18
Complete vacuum-filtration assembly using a Büchner funnel.

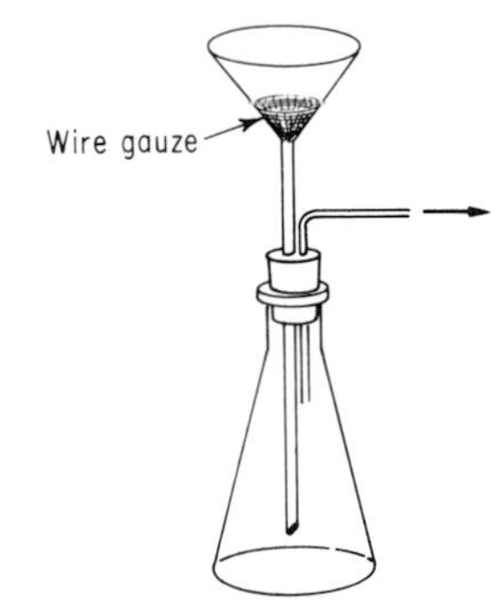

FIGURE 12-19
Funnel with a wire-gauze cone to be used as a support for the filter paper.

is turned on, the pressure difference between the filter medium and the atmosphere helps to speed up the filtration process.

CAUTIONS

1. Wear protective glasses when the assembly is under reduced pressure.

2. Be careful that the liquid level in the "safety trap" bottle is never as high as the inlet tubes.

Vacuum filtration is advantageous when the precipitate is crystalline. It should not be employed for gelatinous precipitates because the added pressure forces the particles into the pores of the filter medium, clogging them so much that no liquid can pass through.

Vacuum filtration can be performed with filter paper as well as with the various crucibles. The common conical funnel paper is easily ruptured at the apex of the cone when under the added stress of the vacuum. To strengthen the cone at the apex, a small metal liner is often inserted (see Fig. 12-19).

Equipment and Filter Media

Büchner Funnels

Büchner funnels (Fig. 12-20) are often used for vacuum filtration. They are not conical in shape but have a flat, perforated bottom. A filter-paper circle of a diameter sufficient to cover the perforations is placed on the flat bottom, moistened, and tightly sealed against the bottom by applying a slight vacuum.

When a Büchner funnel is used, the precipitate is allowed to settle, and the liquid phase is first decanted by pouring it down a stirring rod aimed at the center of the filter paper, applying only a light vacuum until sufficient solid has built up over the paper to protect it from breaking. The vacuum is then increased, and the remainder of the precipitate is added.

The precipitate can be washed by adding small amounts of wash liquid over the surface of the precipitate, allowing the liquid to be drawn

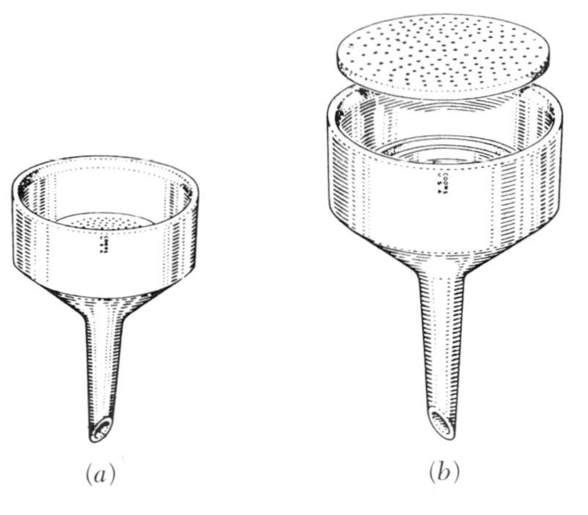

FIGURE 12-20
Büchner funnels. (*a*) Büchner suction funnel. (*b*) Büchner plain funnel with a removable plate, available in various sizes (14.5 to 308 mm).

through the solid slowly with the vacuum. Precipitates cannot be dried or ignited and weighed in Büchner funnels.

Büchner funnels are not applicable for gravimetric analysis because they do permit rapid filtration of large amounts of crystalline precipitates. They are extremely useful in synthetic work, however. Precipitates can be air-dried by allowing them to stand in the funnel and drawing a current of air from the room through the precipitate with the vacuum pump or water aspirator. The last traces of water can be washed from the precipitate with a suitable water-miscible, volatile solvent.

NOTE Solutions of very volatile liquids, ether, and hot solutions are not filtered very conveniently with suction. The suction may cause excessive evaporation of the solvent, which cools the solution enough to cause precipitation of the solute.

Wire-Gauze Conical Funnel Liner

When a Büchner funnel and filter flask are not available, a serviceable apparatus can be constructed and used for vacuum filtration. A small piece of fine wire gauze is bent into the form of a cone that fits into the funnel (Fig. 12-19). The filter paper is then wetted and pressed against the side of the funnel to make a good seal with the glass. Vacuum should be applied gently so that a hole will not be torn in the paper. It may be necessary to use a double thickness of filter paper to prevent tearing. Wrap the flask with a towel before applying vacuum.

Crucibles

Sintered-Glass Crucibles

Glass crucibles with fritted-glass disks sealed permanently into the bottom end are available in a variety of porosities (Fig. 12-21). With care they

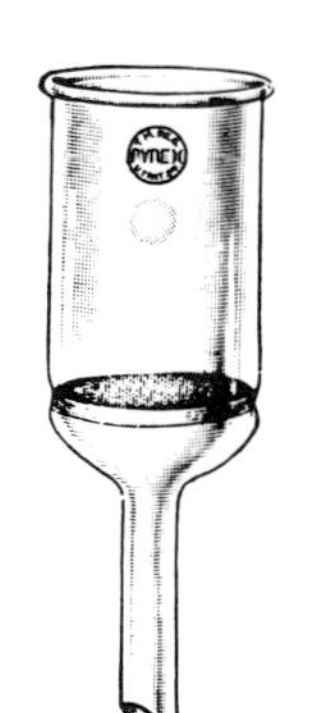

FIGURE 12-21
A funnel with a fritted-glass disk is used in vacuum filtration where the paper filter in a Büchner funnel would be attacked.

can be used for quantitative analyses requiring ignition to a temperature as high as 500°C.

Porous Porcelain and Monroe Crucibles

Porcelain crucibles (Fig. 12-22) with porous ceramic disks permanently sealed in the bottom are used in the same way as sintered-glass crucibles, except they may be ignited at extremely high temperatures. The Monroe crucible is made of platinum, with a burnished platinum mat serving as the filter medium. Advantages of these crucibles are

1. Their high degree of chemical inertness
2. Their ability to withstand extremely high ignition temperatures

These and other crucibles are seated in an adapter when filtering is performed (Fig. 12-23).

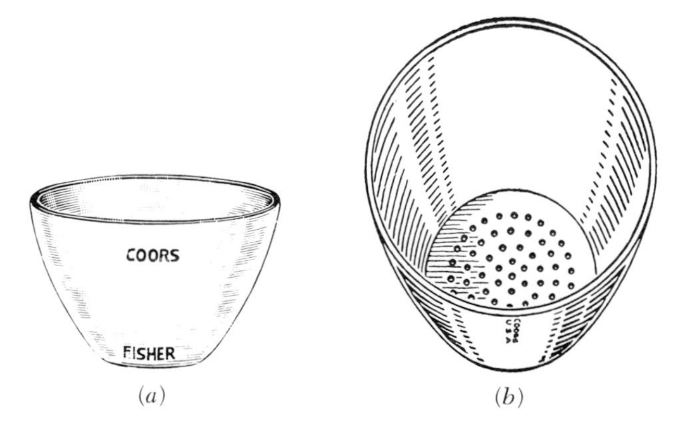

FIGURE 12-22
(*a*) Porcelain crucible used for ignition of samples in analysis. (*b*) Gooch crucible with perforated bottom.

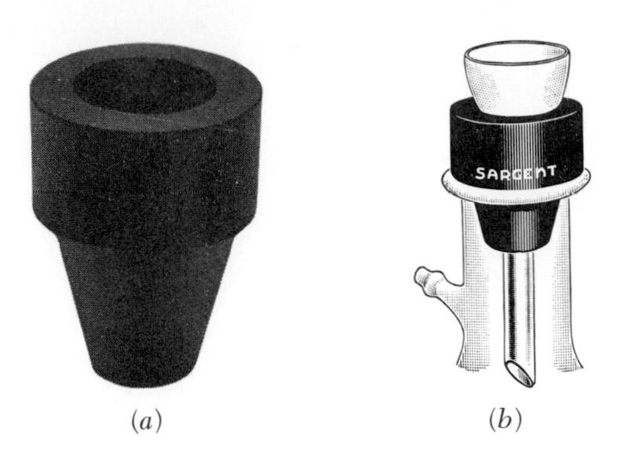

FIGURE 12-23
(*a*) Rubber adapter for suction filtration. (*b*) Adapter in use with a crucible in a suction flask.

(*a*) (*b*)

The Gooch Crucible

The Gooch crucible is a porcelain thimble with a perforated base. (See Fig. 12-22*b*.) The filter medium is either a mat of asbestos or a fiberglass paper disk. The mat is prepared by pouring a slurry of asbestos fiber suspended in water into the crucible and applying light suction.

Asbestos mats permit precipitates to be quantitatively ignited to extremely high temperatures without danger of reduction by carbon. With fiberglass mats, ignition temperatures above 500°C are not possible. Both filter media are resistant to attack by most chemicals.

The filter media used in these crucibles are quite fragile. Exercise extreme care when adding the liquid, so that the asbestos or glass paper will not be disturbed or broken, allowing the precipitate to pass through. Use a small, perforated porcelain disk over the asbestos mat or glass paper to deflect any stream of liquid poured into the crucible.

Platinum Crucibles

Platinum is useful in crucibles for specialized purposes. The chemically valuable properties of this soft, dense metal include its resistance to attack by most mineral acids, including hydrofluoric acid; its inertness with respect to many molten salts; its resistance to oxidation, even at elevated temperatures; and its very high melting point.

With respect to limitations, platinum is readily dissolved on contact with aqua regia and with mixtures of chlorides and oxidizing agents generally. At elevated temperatures, it is also dissolved by fused alkali oxides, peroxides, and to some extent hydroxides. When heated strongly, it readily alloys with such metals as gold, silver, copper, bismuth, lead, and zinc. Because of this predilection toward alloy formation, contact between heated platinum and other metals or their readily reduced oxides must be avoided. Slow solution of platinum accompanies contact with fused nitrates, cyanides, alkali, and alkaline-earth chlorides at temperatures above 1000°C; bisulfates attack the metal slightly at temperatures above 700°C. Surface changes result from contact with ammonia, chlorine, volatile chlorides, sulfur dioxide, and gases possessing a high percentage of carbon. At red heat, platinum is readily attacked by arsenic, antimony, and phosphorus, the metal being embrittled as a consequence. A similar effect occurs upon high-temperature contact with selenium, tellurium, and to a lesser extent, sulfur and carbon. Finally, when heated in air for prolonged periods at temperatures greater than 1500°C, a significant loss in weight due to volatilization of the metal must be expected.

RULES GOVERNING THE USE OF PLATINUMWARE

1. Use platinum equipment only in those applications which will not affect the metal. Where the nature of the system is in doubt, demonstrate the absence of potentially damaging components before committing platinumware to use.

2. Avoid violent changes in temperature; deformation of a platinum container can result if its contents expand upon cooling.

3. Supports made of clean, unglazed ceramic materials, fused silica, or platinum itself may be safely used in contact with incandescent platinum; tongs of Nichrome or stainless steel may be employed only after the platinum has cooled below the point of incandescence.

4. Clean platinumware with an appropriate chemical agent immediately following use; recommended cleaning agents are hot chromic acid solution for removal of organic materials, boiling hydrochloric acid for removal of carbonates and basic oxides, and fused potassium bisulfate for the removal of silica, metals, and their oxides. A bright surface should be maintained by burnishing with sea sand.

5. Avoid heating platinum under reducing conditions, particularly in the presence of carbon. Specifically, (*a*) do not allow the reducing portion of the burner flame to contact a platinum surface, and (*b*) char filter papers under the mildest possible heating conditions and with free access of air.

The Hirsch Funnel and Other Funnels

Hirsch funnels (Fig. 12-24), used for collecting small amounts of solids are usually made of porcelain. The inside bottom of the funnel is a flat plate with holes in it which supports the filter paper. Büchner and Hirsch fun-

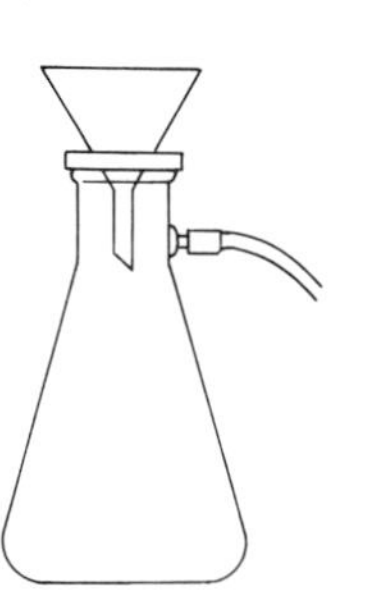

FIGURE 12-24 Suction filtration with a Hirsch funnel.

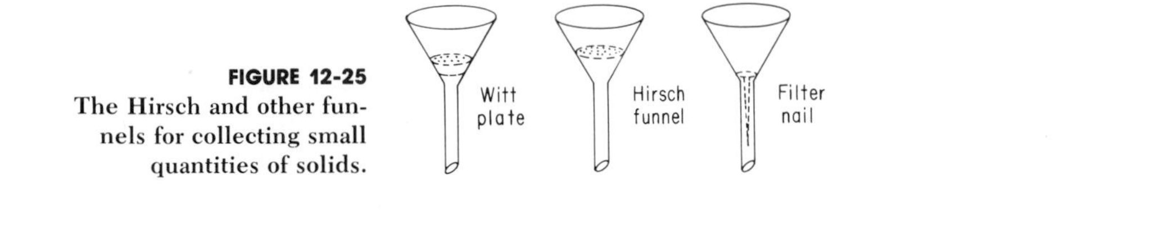

FIGURE 12-25
The Hirsch and other funnels for collecting small quantities of solids.

nels can also be obtained in glass with sintered-glass disks. Other funnels used like the Hirsch are the Witt and the filter nail (see Fig. 12-25).

When using these funnels, a rubber ring forms the seal between the funnel and the filter flask, which is connected to the vacuum line or to the aspirator. (A rubber stopper or cork can also be used instead of the rubber ring to fit the funnel to the filter flask.)

Large Scale Vacuum Filtration

Large amounts of material can be easily vacuum-filtered by using a table-top Büchner funnel. The filtrate is collected in a vacuum flask placed alongside the filter on the table top (see Fig. 12-26).

Alternatively, the vacuum flask can be located below the level of the funnel (Fig. 12-27).

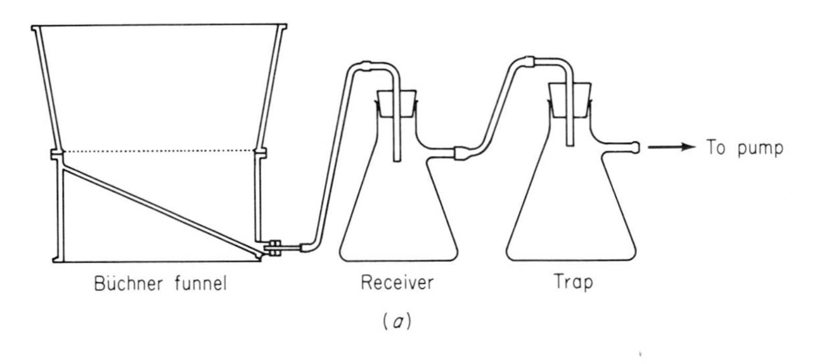

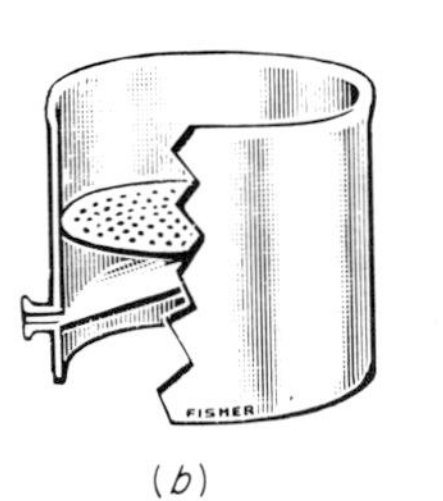

FIGURE 12-26
Vacuum filtration on a large scale. (*a*) Setup with table-top Büchner funnel. (*b*) Cutaway view of table-top Büchner funnel (ID 56–308 mm).

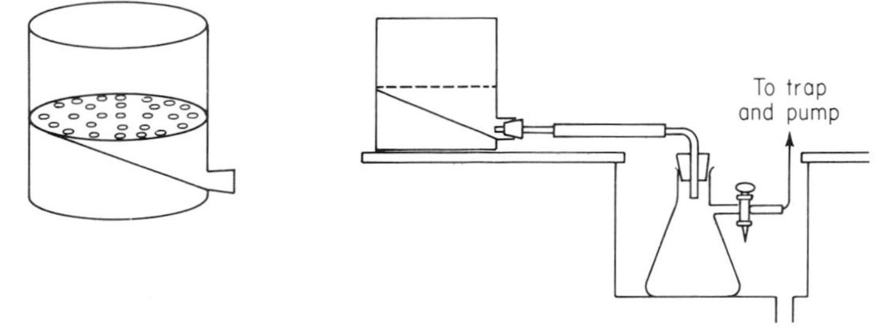

FIGURE 12-27 Alternative setup for vacuum filtration on a large scale.

FILTRATION FOR GRAVIMETRIC ANALYSIS

One of the most important applications of filtration is in gravimetric analysis, which pertains to the measurement of mass; the analysis therefore is completed by "massing" procedures. The substances which are massed are obtained (1) by forming an insoluble precipitate from the desired component; from its weight, the percentage calculations can be made, given the mass of the original sample, or (2) by distilling off a volatile component; the nonvolatile residue is then massed. Both the volatile and nonvolatile portions can be massed, and calculations can be made from the data so obtained.

A precipitate suitable for analytical procedures should have the following characteristics:

1. It should be relatively insoluble, to the extent that any loss due to its solubility would not significantly affect the result.
2. It should be readily filterable. Particle size should be large enough to be retained by a filter.
3. Its crystals should be reasonably pure, with easily removed solid contaminants.
4. It should have a known chemical composition or be easily converted to a substance which does have a known composition.
5. It should not be hygroscopic.
6. It should be stable.

Aging and Digestion of Precipitates

Aging and digestion of precipitates frequently help to make them suitable for analytical procedures.

Freshly formed precipitates are aged by leaving them in contact with the supernatant liquid at room temperature for a period of time. There are frequently changes in the surface: a decrease in the total surface area or removal of strained and imperfect regions. Both effects are due to recrystallization because small particles tend to be more soluble than large ones, and ions located in imperfect and strained regions are less tightly held than is normal and therefore tend to return to the solution. On *aging* they are deposited again in more perfect form. These changes cause a beneficial decrease in adsorbed foreign ions and yield a more filterable as well as purer precipitate.

Heating during the aging process is called *digestion*. Increasing the temperature greatly enhances digestion. The precipitate is kept in contact with the supernatant at a temperature near boiling for a period of time. Flocculated colloids usually undergo rapid aging, particularly on digestion, and a major portion of the adsorbed contaminants may often be removed.

Filtration and Ignition

Eventually, the substances precipitated and prepared for filtration as the early steps in a gravimetric determination must be filtered, dried, and massed; i.e., their mass must be accurately determined so that analytical calculations may be performed. Some precipitates are collected in tared crucibles and oven-dried to constant mass. The following procedures cover the handling of both types.

Preparation of Crucibles

All crucibles employed in converting a precipitate into a form suitable for weighing must maintain a substantially constant mass throughout the drying or ignition process; you must demonstrate that this condition applies before you start.

1. Inspect each crucible for defects, especially where the crucible has previously been subjected to high temperatures.

2. Place a porcelain crucible upright on a hard surface and gently tap with a pencil. You should hear a clear ringing tone indicating an intact crucible. A dull sound is characteristic of one that is cracked and should be discarded.

3. Clean the crucible thoroughly. Filtering crucibles are conveniently cleaned by backwashing with suction.

4. Bring the crucible to constant weight, using the same heating cycle as will be required for the precipitate. Agreement within 0.2 mg between consecutive measurements is considered constant mass.

5. Store it in a desiccator until it is needed.

Preparation of a Filter Paper

Fold the paper exactly in half; then make the second fold so that the corners fail to coincide for about 3 mm in each dimension. Tear off a small triangular section from the short corner to permit a better seating of the filter in the funnel. Open the paper so that a cone is formed and then seat it gently in the funnel with the aid of water from a wash bottle. There should be no leakage of air between paper and funnel, and the stem of the funnel should be filled with an unbroken column of liquid, a condition that markedly increases the rate of filtration.

Transfer of Paper and Precipitate to Crucible

When filtration and washing are completed, transfer the filter paper and its contents from the funnel to a tared crucible (Fig. 12-28). Use considerable care in performing this operation. The danger of tearing can be reduced considerably if partial drying occurs prior to transfer from the funnel.

First, flatten the cone along its upper edge; then fold the corners inward. Next, fold the top over. Finally, ease the paper and contents into the crucible so that the bulk of the precipitate is near the bottom.

Ashing of a Filter Paper

If a heat lamp is available, place the crucible on a clean, nonreactive surface; an asbestos pad covered with a layer of aluminum foil is satisfactory. Then position the lamp about 6 mm from the top of the crucible and turn it on. Charring of the paper will take place without further intervention; the process is considerably accelerated if the paper can be moistened with no more than one drop of strong ammonium nitrate solution. Removal of the remaining carbon is accomplished with a burner.

Considerably more attention must be paid to the process when a burner is employed to ash a filter paper. Since the burner can produce much higher temperatures, the danger exists of expelling moisture so rapidly in the initial stages of heating that mechanical loss of the precipitate occurs. A similar possibility arises if the paper is allowed to flame. Finally, as long

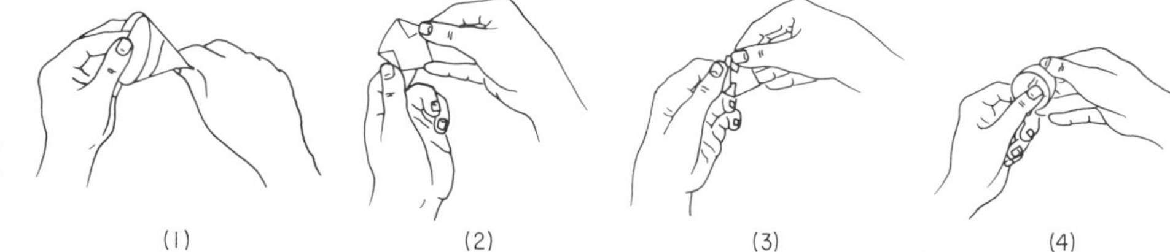

FIGURE 12-28
This is the way to transfer a filter paper and precipitate to a crucible.

as carbon is present, there is also the possibility of chemical reduction of the precipitate; this is a serious problem where reoxidation following ashing of the paper is not convenient.

In order to minimize these difficulties, the crucible is placed as illustrated in Fig. 12-29. The tilted position of the crucible allows for the ready access of air. A clean crucible cover should be located nearby, ready for use if necessary.

NOTE Always place the hot cover or crucible on a wire gauze—never directly on the desk top. The cold surface may crack the crucibles, and dirt, paint, etc., are easily fused into the porcelain, thus changing its mass.

Heating is then commenced with a small burner flame. This is gradually increased as moisture is evolved and the paper begins to char. The smoke that is given off serves as a guide to the intensity of heating that can be safely tolerated. Normally it will appear to come off in thin wisps. If the volume of smoke emitted increases rapidly, the burner should be removed temporarily; this condition indicates that the paper is about to flash. If, despite precautions, a flame does appear, it should be snuffed out immediately with the crucible cover. (The cover may become discolored owing to the condensation of carbonaceous products; these must ultimately be removed by ignition so that the absence of entrained particles of precipitate can be confirmed.) Finally, when no further smoking can be detected, the residual carbon is removed by gradually lowering the crucible into the full flame of the burner. Strong heating, as necessary, can then be undertaken. Care must be exercised to avoid heating the crucible in the reducing portion of the flame.

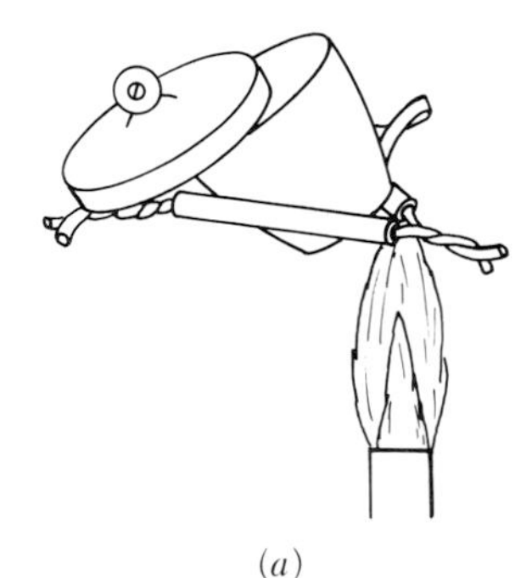

(*a*)

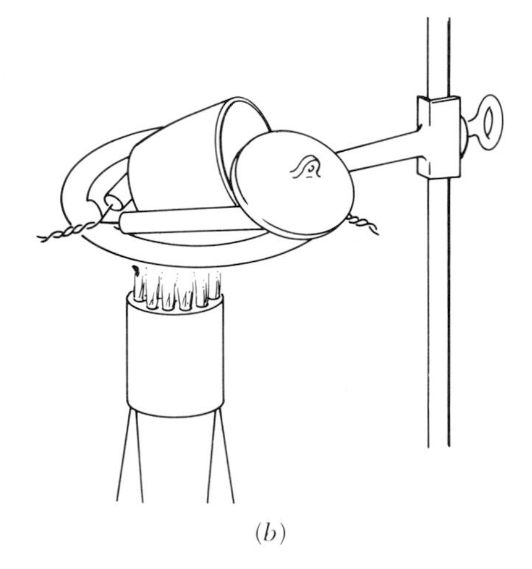

(*b*)

FIGURE 12-29 Ignition of a precipitate with access to air. (*a*) Start heating slowly from the side. (*b*) Do not let the flame enter the crucible.

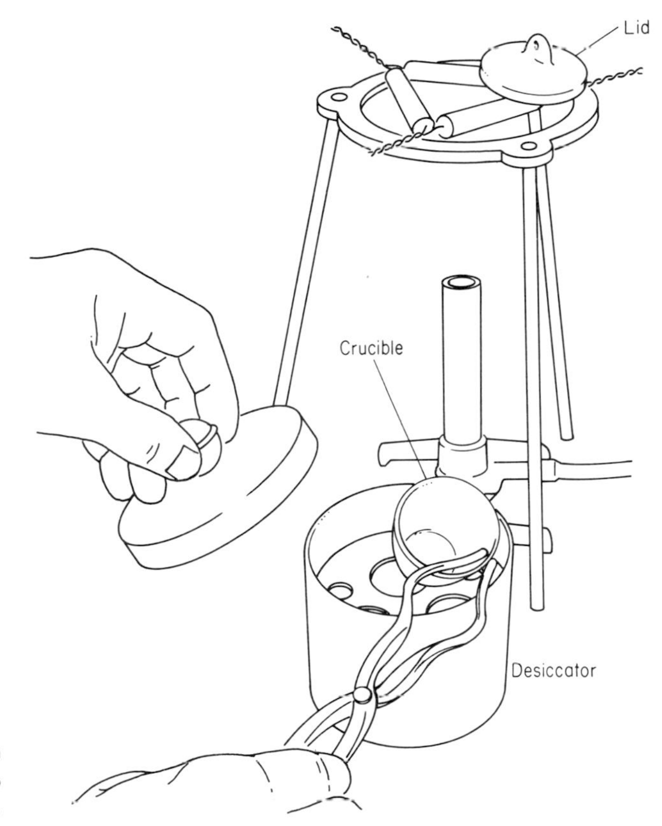

FIGURE 12-30
Cooling the crucible to constant mass.

Cooling the Crucible to Constant Mass

Place the warm crucible, contents, and lid in a desiccator,* containing an effective desiccant, to cool before determining its mass (Fig. 12-30). This procedure enables you to get balance readings to constant mass, especially in a humid atmosphere.

Summary

1. Oxidize the paper completely to CO_2 and H_2O.
2. Record the mass of the previously prepared crucible.
3. Arrange the crucible according to Fig. 12-29*a*.
4. Increase the temperature *slowly* until all the black carbon residue is burned away.

---

* For a complete discussion of desiccators, refer to Chap. 15, "Gravimetric Analysis," p. 283.

5. Position the burner so that the reducing gases of the flame are *not* deflected into the crucible (Fig. 12-29*b*).

6. Reposition the crucible to expose fresh portions to the highest temperature of the burner.

7. Final ignition converts the precipitate to the anhydrous oxide: (*a*) Remove the crucible cover. (*b*) Ignite at red heat for 30 min with a Fisher, Meker, or other high-temperature burner; or you may use a muffle furnace.

8. Cool the crucible in a desiccator (Fig. 12-30).

9. Determine the mass of the cool crucible.

10. Repeat steps 7 to 9 until a constant mass is reached.

13
RECRYSTALLIZATION

INTRODUCTION

Recrystallization is a procedure whereby organic compounds which are solid at room temperature are purified by being dissolved in a hot solvent and reprecipitated by allowing the solvent to cool. The solvent may be a pure compound or a mixture, and the selection of the solvent depends upon a number of important factors. If the growth of the crystals is very fast and not selective, the precipitation process does not aid in purification. When the crystals grow very slowly and consist of pure compounds, the precipitation process is a purifying one; this second type of precipitation is usually defined as *crystallization*.

The crystallization process is very slow and requires relatively long periods of time to ensure that no impurities will be trapped in the crystal lattice as the crystal grows. Ordinary precipitation is a relatively fast process and occurs in minutes or hours. In this case, any impurities in the solution are actually trapped as the precipitate forms, resulting in an impure crystal.

In the laboratory a solid is purified by recrystallization by dissolving it in a hot solvent, filtering the solution, and then allowing the desired crystals to form in the filtrate, while the impurities remain in solution.

REQUIREMENTS OF THE SOLVENT

In general solvents should:

1. Not react with the compound
2. Form desirable, well-formed crystals
3. Be easily removed from the purified crystals
4. Have high solvency for the desired substance at high temperatures and low solvency for that substance at low temperatures
5. Have high solvency for impurities

Solvency

The substance to be purified should be sparingly soluble in the solvent at room temperature, yet should be very soluble in the solvent at its boiling point. The solubility of a solute in a solvent is a function not only of the chemical structures of the solute and the solvent, but also of the temperature. In the majority of cases, the solubility of the solute in a solvent increases as the temperature increases, and in some cases the increase in solubility is very dramatic. This is the basis for the recrystallization method of purification. If the compound has been reported in the literature, its solubility in common solvents can be found in the reference. Normally, polar organic compounds (those which contain one or more $-OH$, $-COOH$, $-CONH_2$, $-NH_2$, or $-SH$ functional groups) tend to dissolve in polar solvents such as water, the lower-molecular-weight alcohols, or combinations of them. Nonpolar compounds tend to dissolve in nonpolar organic solvents, such as benzene, the petroleum ethers, hexanes, chlorohydrocarbons, etc. (See Table 13-1.) The general rule regarding solubility is that *like substances tend to dissolve in like substances*, but the molecule as a whole must be considered before making the decision. For example, a high-molecular-weight fatty acid, stearic acid, behaves more like a nonpolar substance than a polar one, because the $-COOH$ group is not the major part of the molecule.

In general, the following points should be considered:

1. A useful solvent is one that will dissolve a great deal of the solute at high temperatures and very little at low temperatures.

2. If a solvent dissolves too much solute at low temperatures, it is unsuitable. You will be working with such a small volume of solvent that you

TABLE 13-1
Solvent Polarity Chart

Relative polarity	*Compound formula*	*Group*	*Representative solvent compounds*
Nonpolar	R—H	Alkanes	Petroleum ethers, ligroin, hexanes
Increasing polarity ↓	Ar—H	Aromatics	Toluene, benzene
	R—O—R	Ethers	Diethyl ether
	R—X	Alkyl halides	Tetrachloromethane, chloroform
	R—COOR	Esters	Ethyl acetate
	R—CO—R	Aldehydes and ketones	Acetone, methyl ethyl ketone
	$R—NH_2$	Amines	Pyridine, triethylamine
	R—OH	Alcohols	Methanol, ethanol, isopropanol, butanol
	$R—COHN_2$	Amides	Dimethylformamide
	R—COOH	Carboxylic acids	Ethanoic acid
Polar	H—OH	Water	Water

will have a slush rather than a solution to filter. Furthermore, too much of the solute will not crystallize out at the low temperature and therefore much will be lost.

3. If too much solvent is required to dissolve the solute even at its boiling point, it may be possible to recrystallize several grams, but extremely large volumes of solvent would be required to recrystallize several hundred grams.

4. Quick tests of solubility are unreliable and are misleading: Some solutes dissolve very slowly in boiling solvents. A quick observation may be misleading and cause you to reject the solvent as being unsatisfactory. Give the solute sufficient time to dissolve; otherwise you may use too much solvent because you will add additional quantities unnecessarily.

5. The suitability of a solvent depends upon the establishment of equilibrium. Maximum solute will dissolve when equilibrium has been attained between the dissolved and solid solute.

Volatility

The volatility of a solvent determines the ease or difficulty of removing any residual solvent from the crystals which have formed. Volatile solvents may be removed easily by drying the crystals under vacuum or in an oven.

CAUTION The temperature of the oven must be carefully watched and controlled so that the temperature is well below the melting point of the recrystallized compound or the flash point of the solvent.

Solvents with a high boiling point should be avoided, if possible. They are difficult to remove and the crystals usually must be heated mildly under high vacuum to remove such solvents.

Some common solvents and their boiling points are listed in Tables 13-2 and 13-3.

Solvent Pairs

Miscible solvents of different solvent power yield a mixture which gives a usable solvent system (Table 13-4).

TABLE 13-2 Common Water-Miscible Solvents

Solvent	*Boiling point, °C*
Acetone	56.5
Methanol	64.7
Ethanol, 95%	78.1
Water	100
Dioxane	101
Acetic acid	118

TABLE 13-3
Common Water-Immiscible Solvents

Solvent	*Boiling point, °C*
Diethyl ether	34.6
Petroleum ether	40–60
Chloroform	61.2
Ligroin	65–75
Tetrachloromethane	76.7
Benzene	80.1
Ligroin	60–90

Principle

The solute is soluble in one solvent but relatively insoluble in the second solvent.

CAUTION The solvents must be miscible in all proportions.

Procedure

1. Dissolve the solute in the minimum amount of the hot (or boiling) solvent in which it has maximum solubility.

2. Add the second solvent (in which the solute is relatively insoluble) dropwise to the boiling solution of the solute obtained in step 1, until the boiling solution just begins to become cloudy.

3. Add more of the solute-dissolving solvent (step 1) dropwise to the boiling solution until the solution clears up.

4. Allow the clear solution to slowly cool. Crystals of the solute should form.

TABLE 13-4
Solvent Pairs

Solvent pairs
Benzene–ligroin
Ether–acetone
Acetone–water
Ethanol–water
Methanol–water
Ether–petroleum ether
Acetic acid–water
Methanol–ether
Ethanol–ether
Methanol–methylene chloride
Dioxane–water

NOTE If crystals do not form and an oil separates from the solvent mixture, refer to the section on inducing crystallization.

RECRYSTALLIZATION OF A SOLID

Selecting the Funnel

Gravity Filtration

Use either a short-stemmed or a stemless funnel. Long-stemmed funnels tend to cool the filtering solution, and crystallization then takes place in the stem, decreasing the flow rate and even clogging up the funnel.

Vacuum Filtration

Use a Büchner funnel, either porcelain or plastic. Jacketed Büchner funnels may be desirable to minimize any crystallization of the solute caused by evaporation under reduced pressure.

Heating the Funnel

Principle

If a hot recrystallization solution is poured through a cold funnel, the solvent cools and crystallization may sometimes take place in the funnel and its stem, clogging the funnel. Funnels (and thus solvents) can be heated or kept hot by the following procedures.

Method 1

Place a stemless or short-stemmed funnel in a beaker containing the pure solvent which is heated on a steam bath. Hot solvent can be poured

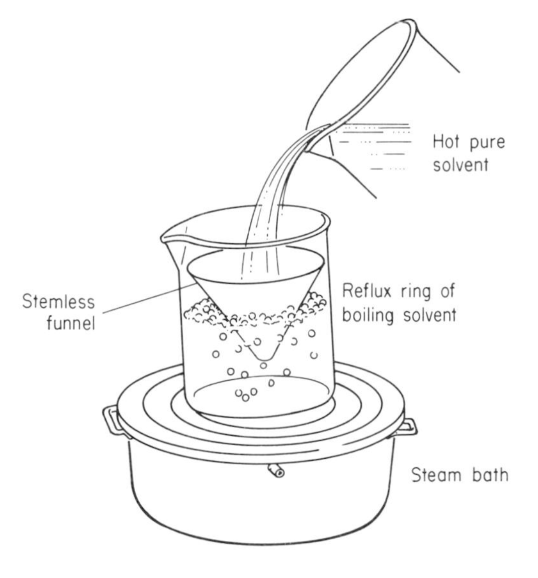

FIGURE 13-1
Heating and maintaining the temperature of a hot funnel.

through, and the reflux ring of the boiling solvent will heat the funnel (Fig. 13-1).

Method 2

Place a funnel and fluted filter paper in the neck of an Erlenmeyer flask which is heated on a steam bath to reflux the pure solvent in it (Fig. 13-2). The reflux ring will heat the funnel. When the recrystallization procedure is to be started, the heated funnel and filter paper are transferred to a funnel support.

Method 3

Pass hot water or steam through a jacketed Büchner funnel during vacuum filtration (Fig. 13-3).

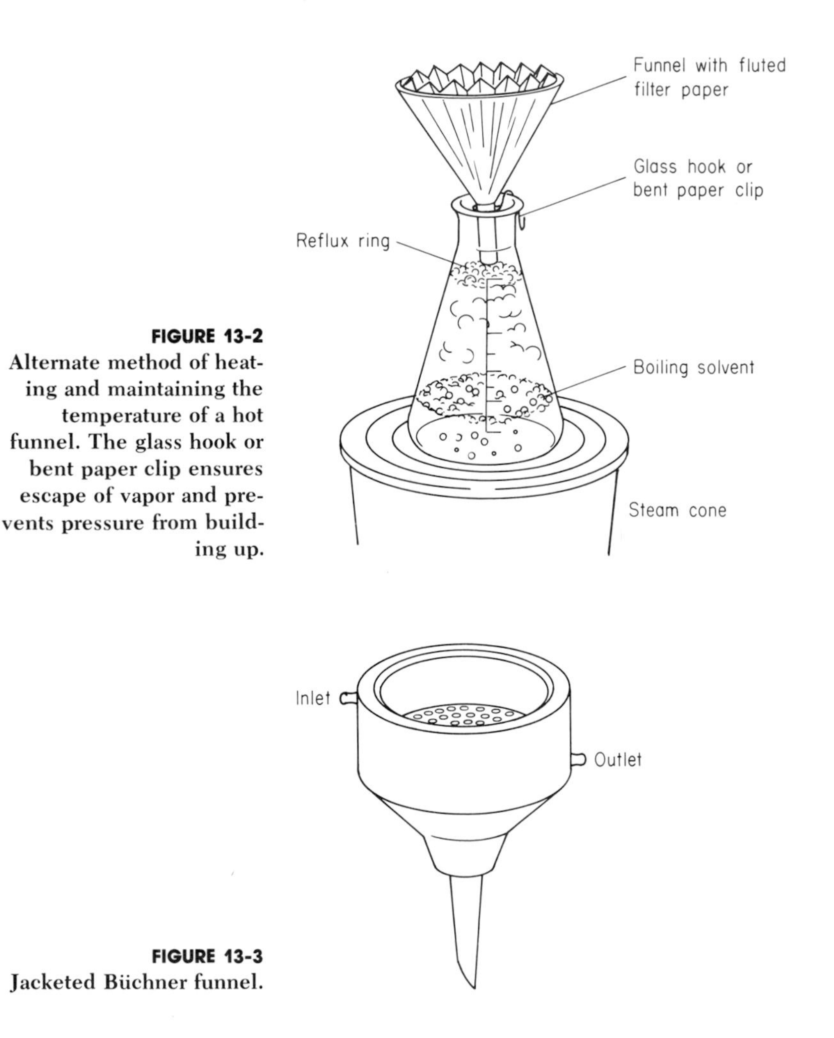

FIGURE 13-2 Alternate method of heating and maintaining the temperature of a hot funnel. The glass hook or bent paper clip ensures escape of vapor and prevents pressure from building up.

FIGURE 13-3 Jacketed Büchner funnel.

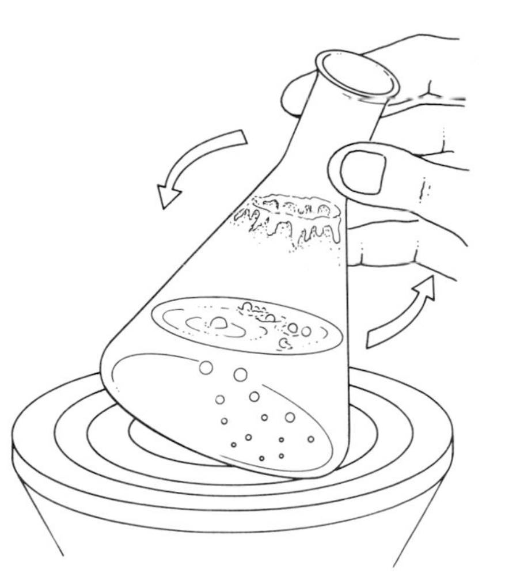

FIGURE 13-4
Heating the solvent for recrystallization.

Receiver Selection

It is desirable to use an Erlenmeyer flask instead of a beaker to collect the filtered crystallizing solution because:

1. The large opening of the beaker is conducive to catching dust and contaminating the product.

2. The Erlenmeyer flask receiver can be easily stoppered, and the content can thus be stored without loss of the solvent by evaporation. If a beaker is used and all the solvent mother liquor evaporates, the process of recrystallization will be ruined. All the impurities dissolved in the mother liquor will crystallize out and coat the crystals that are to be purified.

General Procedure

Recrystallizing

1. Select the most desirable solvent; refer to solubility tables.

2. Add the determined volume of solvent to the flask (no more than two-thirds the volume of the flask) and heat (Fig. 13-4). Add a few boiling stones if desired.

3. Add the minimum amount of hot solvent to the solute slowly to dissolve it. Boil, if necessary, to dissolve all the solute.

CAUTION Do not add too much solvent. Stop adding solvent when only a small quantity of solute remains and further small additions of solvent do not dissolve that remaining solid. Usually, the insoluble material is an impurity. Always allow enough time after adding portions of solvent for the solute to dissolve, because some materials dissolve slowly.

FIGURE 13-5
Preheating the funnel with hot solvent.

4. Preheat the filter funnel to prevent crystallization of the solid in the funnel (Fig. 13-5).

CAUTION Observe *all fire-hazard cautions* because of volatile fumes.

5. Filter the boiling solution through the preheated funnel. Add the solution in small increments, and keep the filtrate hot and in a state of reflux to prevent premature crystallization. Steam baths are suitable for those solvents that have a boiling point lower than 100°C.

6. Collect the filtrate in a flask; allow it to stand and cool.

(a) Cool or chill rapidly in a cooling bath for small crystals.

(b) Cool slowly to get large crystals.

7. Filter the crystals from the mother liquor by gravity or suction. Further crystals can be obtained by evaporation and concentration of the mother liquor.

CAUTION The later crystals may be impure as compared with those from the first crystallization.

8. Dry the crystals in a warm oven.

CAUTION Be careful that the oven is not hot enough to melt the crystals.

Completing Crystallization

If the crystals are collected by filtration immediately after the solution has come to room temperature, or to a lower one by chilling, some of the crystals may not have been collected. Some materials require only a few minutes to crystallize out, while others may require days. The degree of completion of crystallization under the conditions of the experiment can be determined only by practice. It is always a good practice, when working with new substances, to bottle, label, and save the filtrate for a reasonable period of time, and observe if any more crystals come out of solution.

Preventing Crystallization of a Solute during Extraction

Should crystals of an organic substance begin to crystallize out of an organic solvent during extraction, add additional solvent in small portions until the crystals redissolve. This may happen because of the solubility of water in the organic solvent or because a water-soluble solvent such as alcohol, which was a part of the organic solvent mixture, was itself extracted by the water.

Washing Crystals

After all the crystals and solution have been transferred to the filter, some cold, fresh solvent (the same used in the recrystallization) should be poured over the crystals to wash them. If this is not done, any soluble impurities in the solvent that remained on the wet crystals will be deposited on those crystals when the solvent evaporates. Usually one washing (or possibly two) with cold solvent will free the crystals of any possible contamination from this source.

If vacuum filtration has been used and the crystals have been pulled down into a tight cake, the crystals can be washed on the filter, or, better yet, resuspended in a minimum amount of fresh solvent and refiltered. If, however, washing is done in the suction filter, first disconnect the vacuum, then carefully break up the cake gently with a rubber policeman and add fresh solvent to form a wash slurry. Take care not to tear the filter paper. Finally, reapply suction and pull the wash liquid through the cake.

Inducing Crystallization

When the solute fails to crystallize and remains as an oil in the mother liquor, one of the following techniques or some combination of them may be helpful in inducing crystals to form.

1. Scratch the oil against the side of the beaker with a glass stirring rod. Use a fresly cut piece of glass rod (not fire-polished) with a vertical (up-and-down) motion in and out of the solution. Seed crystals or nuclei may develop which will cause crystallization to take place.

2. "Seed" the oil with some of the original material, finely powdered, by dropping some into the cooled flask.

3. Cool the solution in a freezing mixture. (To select combinations that yield extremely low temperatures, refer to Chap. 10, "Heating and Cooling," pp. 198–199.) Refrigerate for a long time.

4. Add crumbs of dry ice.

5. Let stand for a long period.

6. If a solvent pair has been used, oiling may be prevented and crystallization induced by adding a little more of the better solvent, or changing the solvent system.

DECOLORIZATION

Principle

Colored contaminants may sometimes be removed by adding finely powdered decolorizing charcoal, such as Norit, which adsorbs the contaminants. Soluble contaminants, not adsorbed, remain in solution in the mother-liquor filtrate.

General Procedure

1. Select the most desirable solvent; refer to solubility tables.

2. Place the substance to be purified in a suitably sized flask.

3. Add a determined volume of solvent (maximum two-thirds the volume of the flask) and a few boiling stones.

4. Add decolorizing carbon, 1% by weight of solute, if needed.

5. Boil until all crystals have dissolved.

6. Filter as quickly as possible through a fluted filter in the funnel (Fig. 13-6). Stemless funnels are best to use, because there is no stem in which crystallization can take place and clog the system. If necessary, warm the filter funnel to prevent crystallization of hot filtrate in the funnel.

7. Continue with steps 6–8, p. 248.

CAUTION Observe all fire-hazard cautions because of volatile fumes.

Additional Techniques and Hints

1. The decolorizing charcoal may be added after the solute has been dissolved in the minimum amount of solute. This is done to ensure complete solubilization of the solute.

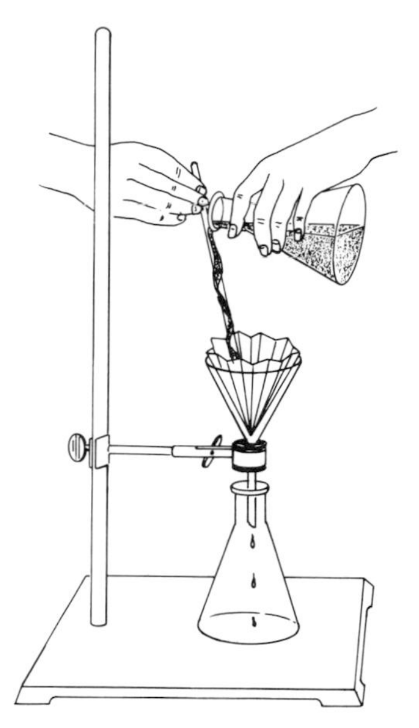

FIGURE 13-6
Experimental setup for recrystallization (decolorization filtration).

CAUTION Add the charcoal in small increments so that the solution does not froth and boil over.

2. If you are filtering with suction, use a heated or a jacketed Büchner funnel to prevent clogging caused by crystallization due to cooling. Use a layer of diatomaceous earth or Celite® to form a base on the paper in the Büchner funnel to trap and catch the finely divided particles of charcoal.

NOTE Observe all safety precautions relating to filtration under reduced pressure. (See Chap. 12, "Laboratory Filtration.")

Batch Decolorization

Slightly colored solutions may be decolorized by adsorption on alumina (Fig. 13-7).

Procedure

1. Prepare a short column packed with activated alumina.
2. Pour the colored solution in the top of the packed column.
3. Collect the decolorized solution at the bottom. If this doesn't work, carbon decolorization will be necessary (or you are dealing with a colored substance).

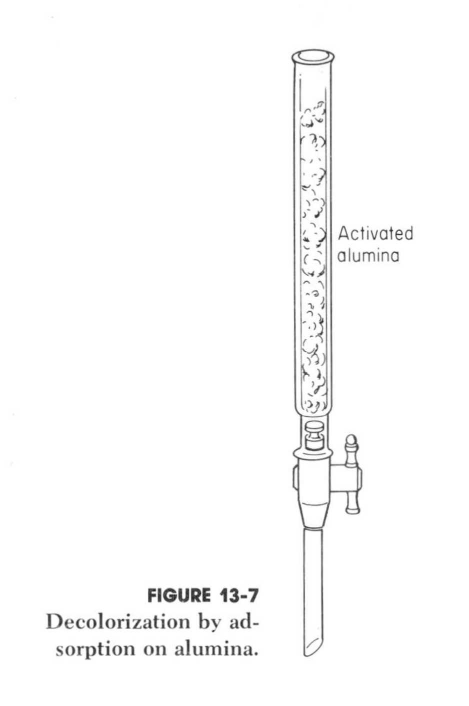

FIGURE 13-7
Decolorization by adsorption on alumina.

FRACTIONAL CRYSTALLIZATION

Principle

You will often encounter mixtures of two compounds which have similar solubilities, and both will crystallize out of solution. They can be separated by fractional crystallization, which is a multistep crystallization repeated as many times as necessary.

Procedure

1. Isolate the mixture of compounds which are to be separated by fractional crystallization.

2. Select a solvent in which one compound is more soluble than the other compound.

3. Recrystallize the mixture using normal procedures. The result will also be a mixture, but there will be enrichment of the less soluble compound.

4. Repeat the crystallization as often as necessary until one pure product is obtained.

NOTE "Seeding" a dilute solution with a pure crystal of the desired compound and then cooling the solution very slowly (without any mixture or agitation) may result in the crystallization of the desired compound, and leave the other compound in a supersaturated state.

CAUTIONS Carry fractional crystallizations out in very dilute solution; otherwise, no purification takes place. Crystallization from a concentrated solution leads to total recovery of both compounds.

LABORATORY USE OF PURIFICATION BY FRACTIONAL CRYSTALLIZATION

Different substances have different solubilities in the same solvent at different temperatures. The changes in the solubilities with temperature are not the same. Advantage can be taken of such solubility relationships to effect the separation and purification of organic compounds and inorganic salts.

EXAMPLE To separate KCl from NaCl, technicians take advantage of the marked change in the solubility of KCl with temperature as compared with NaCl (Fig. 13-8).

Procedure

1. Assume a mixture of 30 g NaCl and 50 g KCl in 100 g water at 100°C.

2. Cool the solution; precipitation of KCl starts to occur at about 70°C.

3. At 0°C, about 20 g KCl will have crystallized out of solution; most of the NaCl will remain in solution because it has a greater solubility in water than KCl.

4. Filter the solution to obtain almost pure KCl.

5. Recrystallize the KCl in the minimum amount of water, cooling again to 0°C and filtering.

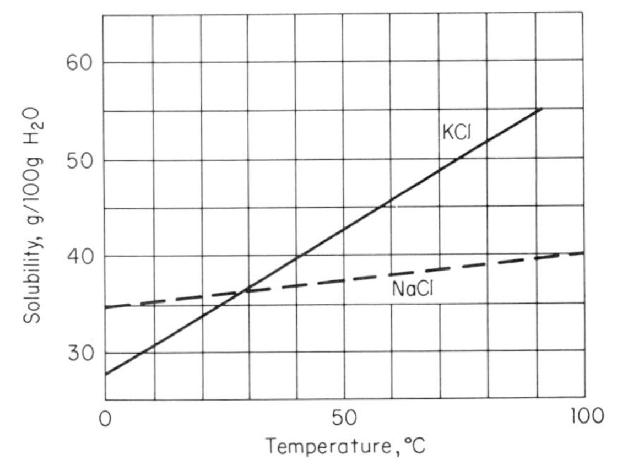

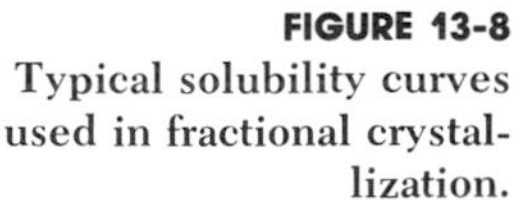
FIGURE 13-8 Typical solubility curves used in fractional crystallization.

Procedure for Other Compounds

1. Locate in the literature the solubility-temperature data of the compounds, if they are available. Or determine the solubility of the compounds involved by experiment.

2. Graph the solubilities of the compounds in 100 g of solvent versus the temperature.

3. Use the KC1–NaCl separation example to develop a working procedure.

NONCRYSTALLIZATION: "OILING" OF COMPOUNDS

Sometimes crystallization does not take place as expected, and the solid may "oil out"; that is, the substance does not crystallize, but becomes a supercooled, amorphous liquid mass.

The substance may melt in the solvent instead of dissolving in it. Oiling-out occurs when the boiling point of the solvent is too high, and the melted solute is insoluble in the solvent.

Some substances are very difficult to crystallize, regardless of the boiling point of the solvent, and again they become oils instead of crystals, as the solvent cools. Some techniques that may be helpful were discussed in the section on inducing crystallization, earlier in this chapter.

Substances which have very low melting points are extremely difficult to crystallize, and crystal formation above the melting point is impossible. Obviously such substances must be purified by some other method.

14
THE BALANCE

INTRODUCTION

Chemistry is a science of precision, a quantitative science. The most important single piece of apparatus available to the chemist is the balance. It is as important to the chemist as the microscope is to the biologist. Balances are mechanical devices used to determine the mass of objects. Because the mass to be determined ranges from kilograms to micrograms, the choice of the balance to be used for any determination is governed by the total mass of the object and the sensitivity desired. Therefore, the technician is always faced first with the decision of which balance to use. The precision required is the second decision.

All balances are expensive precision instruments, and you should use extreme care when handling and using them. Many kinds of balances are found in the chemical laboratory, ranging from rough measuring devices (the trip balances, the triple-beam balance) which are sensitive to 0.1 g to the analytical balances sensitive to fractions of a microgram.

Because balances are delicate instruments, the following comprehensive rules should be observed in caring for and using them. (These are general rules for all balances. Prudence will dictate which are not applicable in work with rougher measuring devices.)

1. Level the balance.

2. Inspect the balance to be certain that it is working properly. Use calibrated, undamaged masses.

3. Check the balance zero.

4. Be certain the beam is locked before removing or changing masses or objects to be massed.

5. Keep the balance scrupulously clean.

6. Work in front of the balance to avoid parallax errors.

7. Handle all masses and objects with forceps, never with fingers. Place the masses as close as possible to the center of the pans.

8. Avoid massing hot objects.

9. Release the locking mechanism slowly, avoiding jars.

10. Do not overload the balance.
11. Never place moist objects or chemicals directly on the balance pans.
12. Close the balance case (if part of the balance).
13. Triple count all masses to avoid error. Separate masses.
14. Record masses in notebook for addition. Never add mentally.

MEASUREMENT

To *measure*, by definition, is to determine the dimensions, capacity, or quantity of anything. A *measurement* is then the extent, capacity, or amount of something as determined by measuring. Any of our life experiences are studied by a system of measures and communicated to others by transferring a stimulus to an instrument which measures its intensity. For example, we transfer our feeling of temperature to others by comparing our body temperature with that of the surroundings and saying we are hot, cold, or comfortable. The stimulus is temperature and the instrument is our body's sense of feeling.

Laboratory instruments designed to measure are mere refinements of our body senses. For example, a balance permits us to determine the mass of an object more accurately than we could determine it by lifting.

In a quantitative determination we are accurately measuring some part of the whole, some constituent of the product. The amount of constituent can be measured by a volumetric, instrumental, or gravimetric technique, and the percent of a constituent must be a ratio of the amount of constituent to the amount of product. The amount of product is usually determined by mass, and the mass is measured by a balance. To distinguish the measurement of mass from the measurement of weight, technicians use the term "mass" as a verb and say that an object is *massed* rather than weighed.

DEFINITIONS OF TERMS

Mass An invariant measure of the quantity of matter in a object. The SI unit of mass is the kilogram, but in the laboratory gram quantities are more usual. Technicians properly use the term *mass* in discussing measurements made with a balance.

Weight The force of attraction exerted between an object and the earth. Weight equals mass times the gravitational attraction. Mass is proportional to weight, so we ordinarily interchange the terms, but the unit of weight is the newton.

Capacity The largest load on one pan for which the balance can be brought to equilibrium.

Precision (standard deviation) Degree of agreement of repeated measurements of the same quantity. It is a statistical value and is calculated:

$$S = \sqrt{\frac{\Sigma d^2}{f}}$$

where S = standard deviation
d = deviation between individual massing and average
$f = n - 1$

Precision and reproducibility are synonymous.

Readability The smallest fraction of a division at which the index scale can be read with ease.

Accuracy The agreement between the result of a measurement and the true value of the quantity measured.

The National Bureau of Standards described the difference between precision and accuracy: "Accuracy" has to do with closeness [of data] to the truth, "precision" only with closeness of readings to one another.

Factors Influencing Accuracy

1. Magnitude of the lever-arm error
2. Magnitude of error in scale indication due to variable load
3. Adjustment error of masses
4. Uniform value of divisions throughout the optical scale
5. Precision
6. Environmental factors

Factors 1 to 3 have no influence on the Mettler balance because of substitution.

Sensitivity The change in load required to produce a perceptible change in indication. It is therefore a ratio and is not to be used to discuss the quality of a measurement.

ERRORS IN DETERMINING MASS

1. *Changes in moisture or CO_2 content.* Some materials take up H_2O or CO_2 from the air during the massing process. Such materials must be massed in a closed system.

2. *Volatility of sample.* Materials which are volatile at room temperature will lose mass while on the balance. Such materials must be massed in a closed system.

3. *Electrification.* An object carrying a charge of static electricity is attracted to various parts of the balance, and an error in mass may occur. An antistatic brush might help in such cases.

4. *Temperature.* If an object is warm relative to the balance, convection currents cause the pan to be buoyed up, and the apparent mass is less than the true mass. Determine mass at room temperature, if possible.

5. *Buoyancy.* This error is due to the weight of air displaced by the object on the pan and is generally quite small.

TYPES OF BALANCES

Equal-Arm Balances

Principle of Operation

The equal-arm analytical balance (Fig. 14-1) acts like a first-class lever. The addition of mass to one side of such a lever at rest (in equilibrium) will cause it to become unbalanced. The force at the point of load is a product of the mass involved and the horizontal distance from the fulcrum through which it is acting. The lever again achieves its position of equilibrium when the force at the load site is exactly balanced on the opposite side of the fulcrum.

$$F_1 = F_2$$

where F_1 and F_2 are opposing forces.

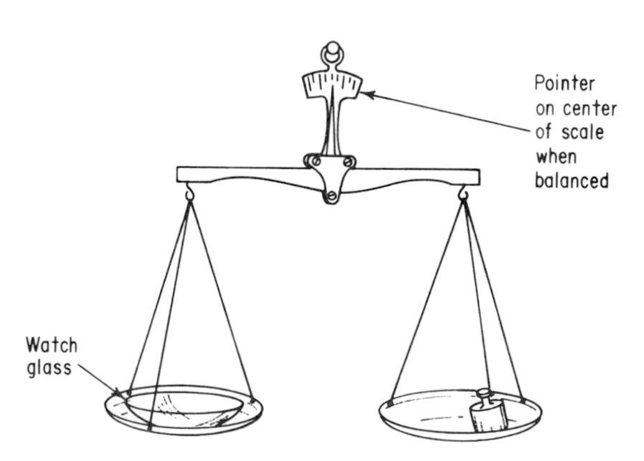

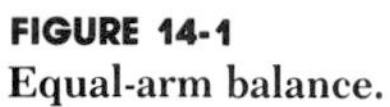

FIGURE 14-1
Equal-arm balance.

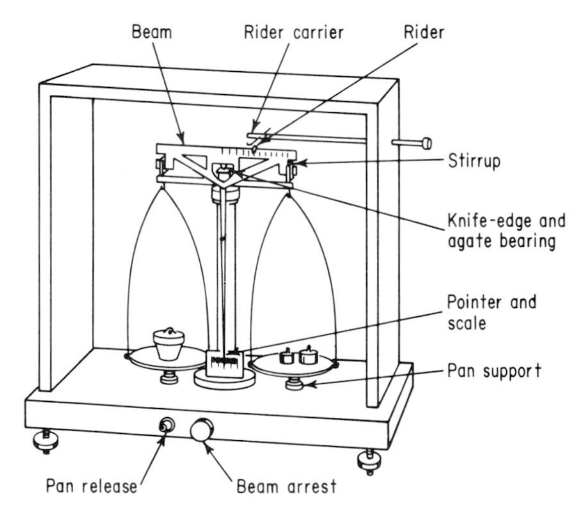

FIGURE 14-2
Controls and components of a laboratory analytical balance.

Since the force is dependent on the distance from the fulcrum, it is essential that the pans of the balance be exactly equidistant from the fulcrum.

The massing operation on an equal-arm balance then consists of duplicating under load the equilibrium position of the unloaded balance.

Rider balances (Fig. 14-2),* Chainomatic® balances, and keyboard balances are examples of equal-arm balances.

General Procedure

1. Find the rest point of the balance when there is *no load* on either pan.

(**a**) Raise the balance beam by turning the operating knob into the free-swinging position.

(**b**) Start the balance swinging 10 to 20 divisions by air current or by fanning gently with a piece of paper.

(**c**) Record 3 to 5 consecutive swing points of the pointer.

(**d**) Return the balance to the supported position by reversing the position of the operating knob.

2. Place the object to be massed (empty crucible) on the left pan of the balance. Handle with forceps.

3. Transfer the appropriate masses with ivory-tipped forceps to the center of the right pan. Adjust the masses on the right pan to 10 mg light.

4. Move the rider to bring the swinging pointer to rest at the *original no-load rest point.*

5. Record the total mass needed to achieve step 4.

6. You have obtained the mass of the crucible.

7. Place the material to be massed in the crucible.

8. Repeat operations 2 through 5.

9. You have obtained the mass of the substance and the mass of the crucible combined.

10. Subtract the mass in step 5 from that in step 9 to get the mass of the material.

---

* Although they are being replaced by more modern balances, these are still found in some laboratories.

The Triple-Beam Balance

The capacity of a triple-beam balance is 2610 g with attachment masses. Its sensitivity is 0.1 g. See Fig. 14-3.

Procedure

1. Observe all general massing procedures.
2. Slide all poises or riders to zero.
3. Zero the balance, if necessary, with balance-adjustment nuts.
4. Place the specimen on the pan of the balance.
5. Move the heaviest poise or rider to the first notch that causes the indicating pointer to drop; then move the poise back one notch, causing the pointer to rise.
6. Repeat procedure 4 with the next highest poise.
7. Repeat this procedure with the lightest poise, adjusting the poise position so that the indicator points to zero.
8. The mass of the specimen is equal to the sum of the values of all the poise positions, which are read directly from the position of the poises on the marked beams.

The Dial-O-Gram Balance

The capacity of a Dial-O-Gram® balance (Fig. 14-4) is 310 g. Its sensitivity is 0.01 g.

Procedure

1. Observe all general massing procedures, sliding poises to zero.
2. Rotate the dial to 10.0 g.

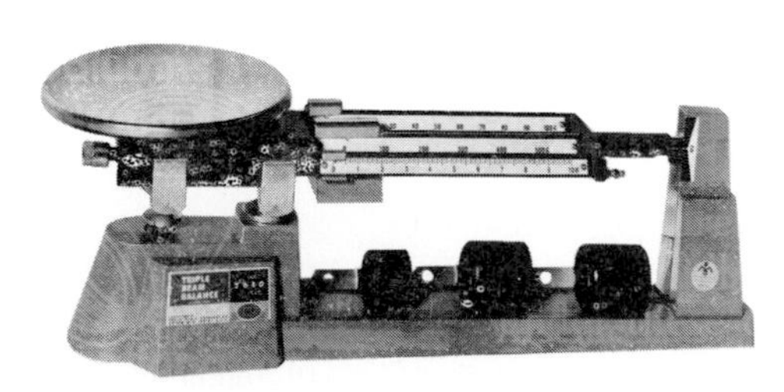

FIGURE 14-3
Two types of triple-beam balance.

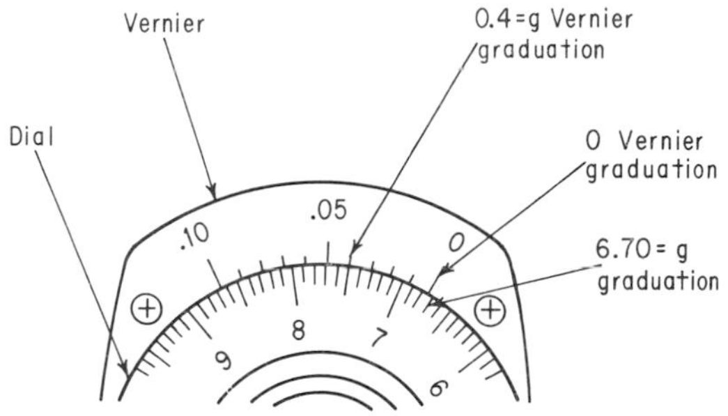

FIGURE 14-4
Dial-O-Gram control knob.®

FIGURE 14-5
Ainsworth balance, Chainomatic® type.

3. Move the 200-g poise on the rear beam to the first notch which causes the pointer to drop; then move it back one notch.

4. Move the 100-g poise to the first notch which causes the pointer to drop.

5. Rotate the dial knob until the pointer is centered.

6. Add the values of the 200-g poise, the 100-g poise, and the dial reading. Each graduation of the dial reads 0.1 g, with a vernier breaking the value down to 0.01 g. (See discussion of verniers below.)

Two-Pan Equal-Arm Chainomatic Balances

Adjustment of the height of the chain in this type of balance causes changes in the mass applied to the right-hand pan. It eliminates the use of masses less than 0.1 g. It will apply masses from 0 to 100 mg (0.1 g) to the right-hand pan. See Fig. 14-5.

Procedure

1. Use the procedure given under Two-Pan Equal-Arm Balances; except when adding masses of 100 mg or less, adjust the height of the chain indicator to get the same rest point as that of the original with *no load.*

2. The calibrated vernier gives the mass portion from 0 to 100 mg.

Using the Vernier on the Chainomatic Balance and Other Equipment

Principle

A vernier is used to measure accurately a fraction of the finest division on the main scale of a measuring instrument.

Estimation without Vernier

On the main scale, without a vernier, the sliding index indicates the portion on the scale corresponding to the measurement (Fig. 14-6).

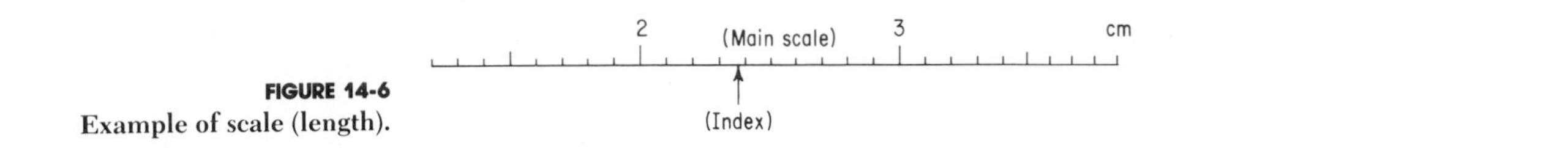

FIGURE 14-6
Example of scale (length).

1. The index points to a reading between 2.3 and 2.4 cm.
2. *Estimate* the index to be 8/10 of a division (8/10 of 0.1 cm, or 0.08 cm).
3. The *estimated* reading of the index pointer = 2.3 + 0.08 = 2.38 cm.

Exact Reading with Vernier

EXAMPLE 1

1. The index in Fig. 14-7 points to a reading between 2.3 and 2.4 cm (the same as was obtained in the estimated procedure). The zero mark on the vernier equals the simple index pointer, indicating that the reading is between 2.3 and 2.4 cm.

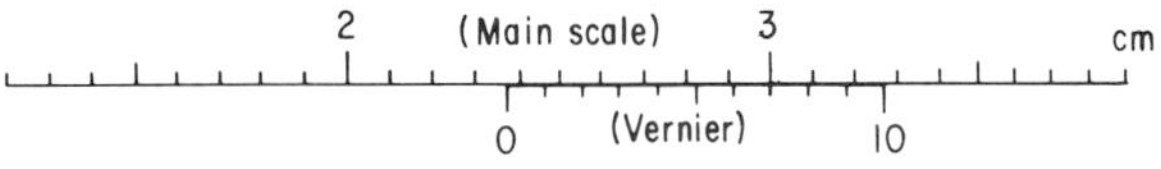

FIGURE 14-7
Example of vernier and scale (length).

2. The division of the vernier scale which coincides with a division of the main scale indicates the *exact* reading.

   (a) The vernier division which coincides is 7, which is exactly 0.07 cm.

   (b) The *accurate* reading is the sum of 2.3 cm and the exact 0.07 cm to give 2.37.

EXAMPLE 2

1. The vernier zero index (Fig. 14-8) is 2.7.

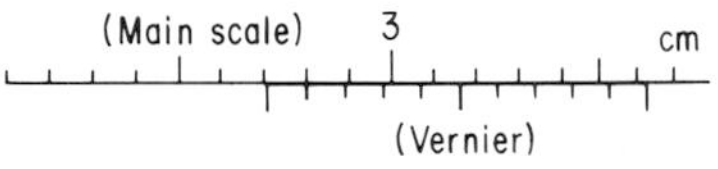

FIGURE 14-8
Additional example of vernier and scale (length).

2. The vernier reading is 0.01 cm (division 1 on the vernier coinciding with division 2.8 on the main scale).
3. The exact reading is 2.7 + 0.01 = 2.71.

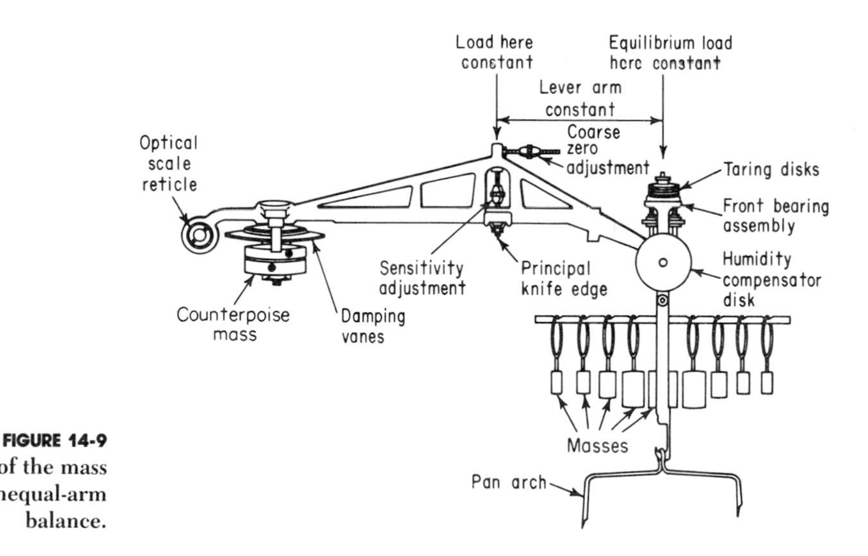

FIGURE 14-9
Diagram of the mass system of an unequal-arm balance.

Unequal-Arm Balance—Substitution

The principle of operation of the unequal-arm balance is substitution. The balance consists of an asymmetric beam. The maximum load is placed on both sides. On the shorter end are a pan and a full complement of masses. A counterpoise mass is used on the longer end to impart equilibrium to the system. When a load is placed on the pan, the analyst must remove an equivalent mass from the load side, within the range of the optical scale, to bring the balance into equilibrium. The total masses removed plus the optical-scale reading equal the mass on the pan. The Mettler single-pan balance operates on this principle. A diagram of its operating system appears in Fig. 14-9.

Single-Pan Analytical Balance

Basic Controls

1. Pan-arrest control

   (a) Assures constant position of the beam between and during massing.

   (b) Protects the bearing surface from excessive wear and injury due to shocks.

2. Arrest control—three positions

   (a) Arrest position is used when removing or placing objects on the pan, when the balance is being moved, and when the balance is not in use.

   (b) Partial arrest position is used to obtain preliminary balance.

   (c) Release position is used when the final massings are being made.

3. Zero-adjust knob

(a) Positions the optical scale to read zero when the pan is empty, because of minute changes in beam position.

4. Mass-setting knobs—two knobs which remove and replace masses from the beam

(a) One knob removes masses in 1- to 9-g increments.

(b) The second knob removes masses in 10-g increments, load limit 100 g.

5. Optical-scale adjustment

(a) Turn knob positions the optical scale relative to a reference line so that the final mass can be obtained to 0.1 mg.

General Procedure

1. Check to see that the balance is level.

2. Zero the balance in arrest position, with pans clear and all mass readings at zero.

3. Mass the object.

(a) Put the pan in arrest position.

(b) Place the object on the pan.

(c) Set to semiarrest position.

(d) Adjust 1- and 10-g control knobs until the mass is within 1 g of the object's mass.

(e) Return to arrest position, then to full release position with the arrest control.

(f) Obtain the final mass by adjustment of the optical-system adjustment knob.

Types of Single-Pan Analytical Balances

Following is a list of the single-pan analytical balances most frequently encountered by technicians.

Mettler Model H-5 (See Fig. 14-10.)

Sartorius Series 2400 (See Fig. 14-11.)

Stanton Unimatic CL2 (See Fig. 14-12.)

Ainsworth Magni Grad Type 21 (See Fig. 14-13.)

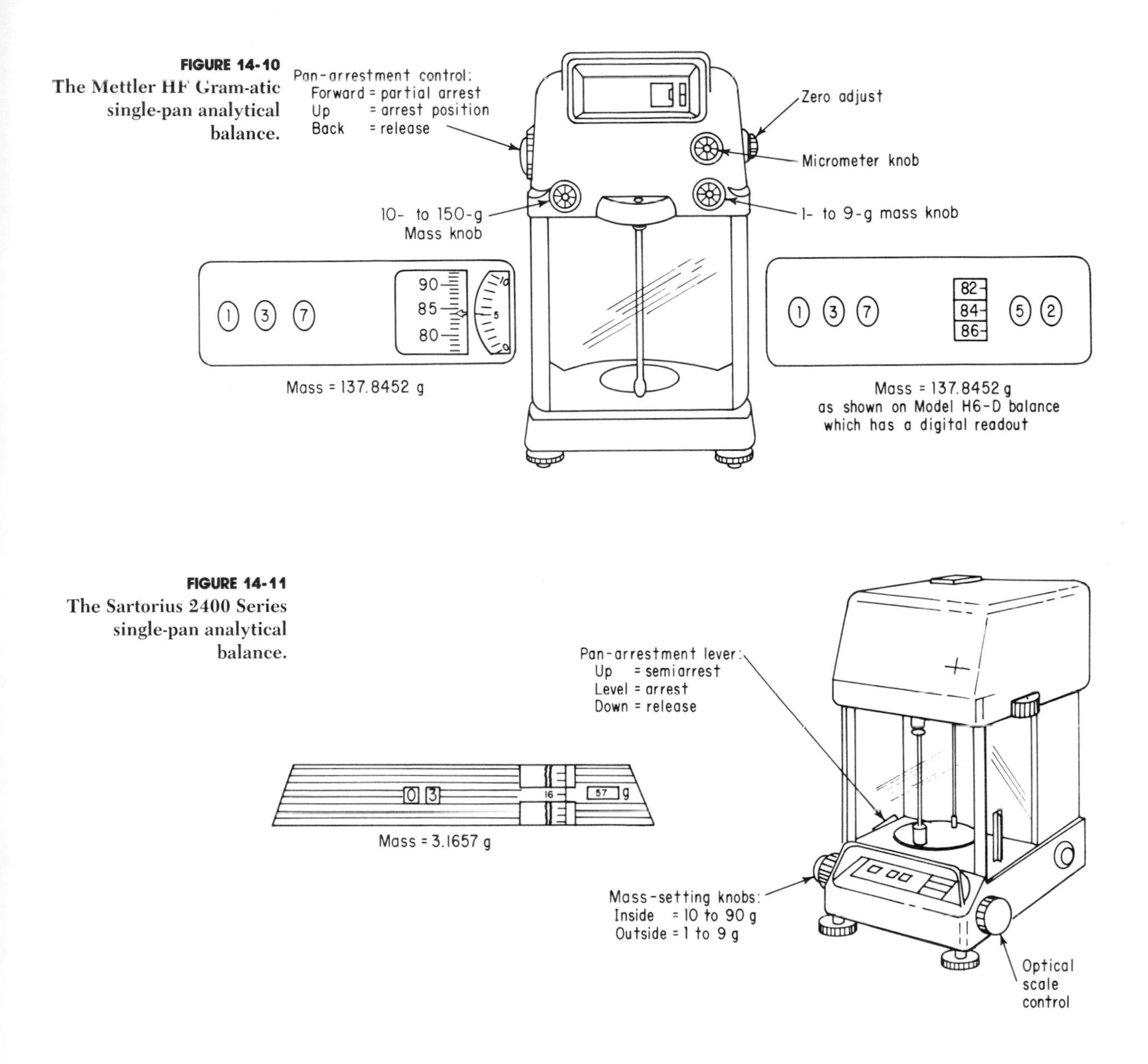

FIGURE 14-10 The Mettler HF Gram-atic single-pan analytical balance.

FIGURE 14-11 The Sartorius 2400 Series single-pan analytical balance.

Electronic Balances

The newest balances available are the electronic balances. These come in capacities from a few milligrams to kilograms. A single control bar turns the balance on to provide a digital readout of the mass (Fig. 14-14).

Procedure

1. Place sample on pan, with balance turned on.
2. Press zero-set button.
3. Read digital readout for mass.

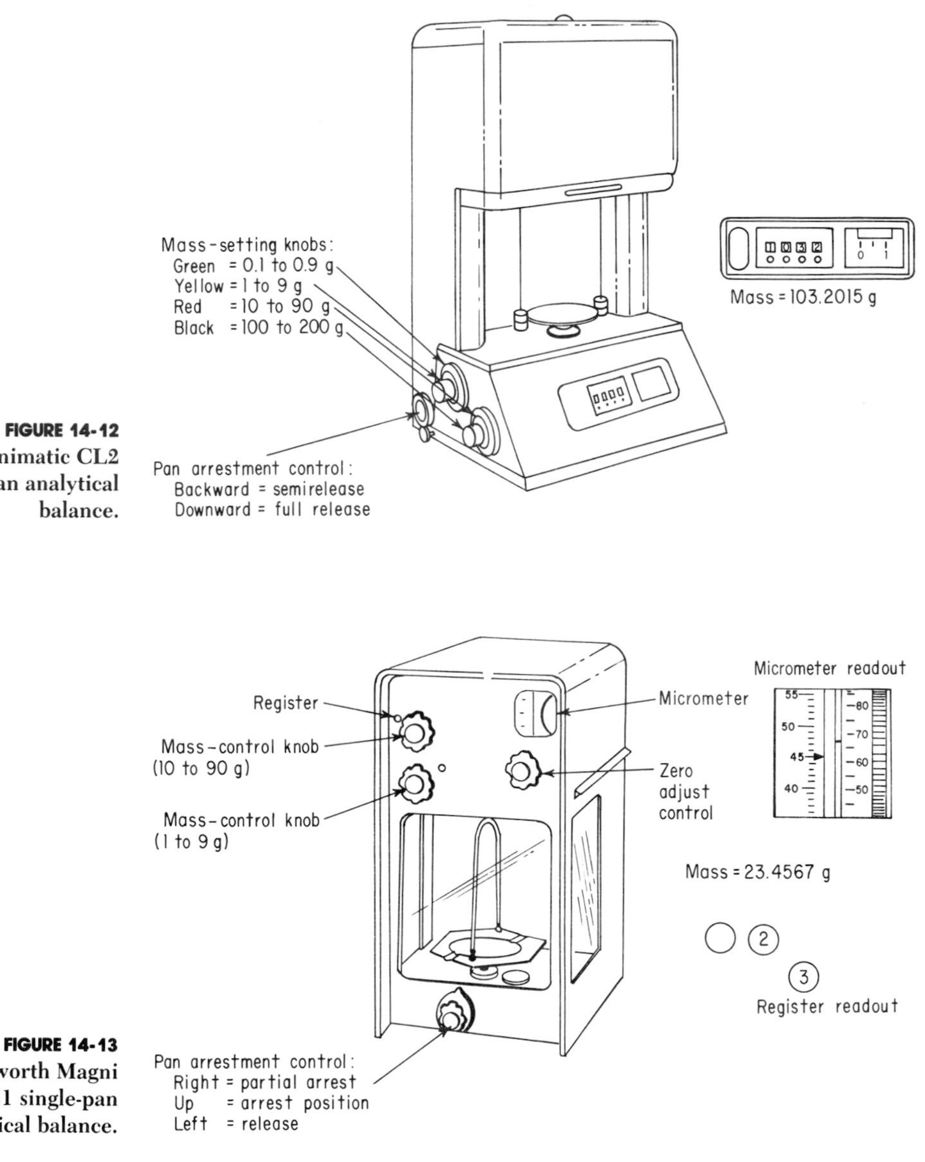

FIGURE 14-12
The Stanton Unimatic CL2 single-pan analytical balance.

FIGURE 14-13
The Ainsworth Magni Grad Type 21 single-pan analytical balance.

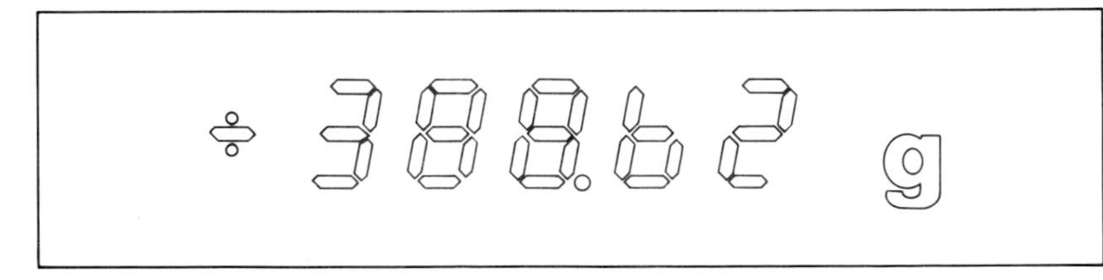

FIGURE 14-14
Digital readout from a new electronic balance.

Electronic balances can be coupled directly to computers or recording devices if necessary.

Other Types of Balances

Various types of balances encountered in the laboratory are shown in Figs. 14-15 to 14-18.

FIGURE 14-15
Single-pan, top-loading, substitution-type, rapid-reading balance with varied sensitivities and optical micrometer scale.

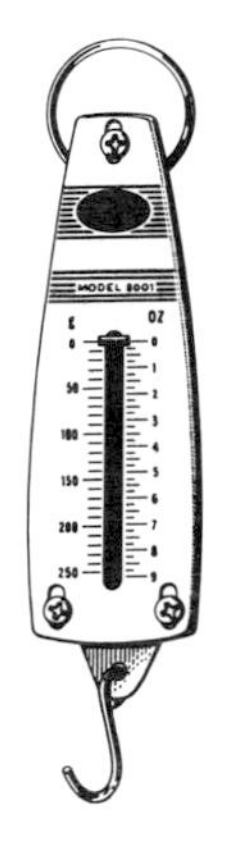

FIGURE 14-16
Portable spring balance for coarse massing of objects.

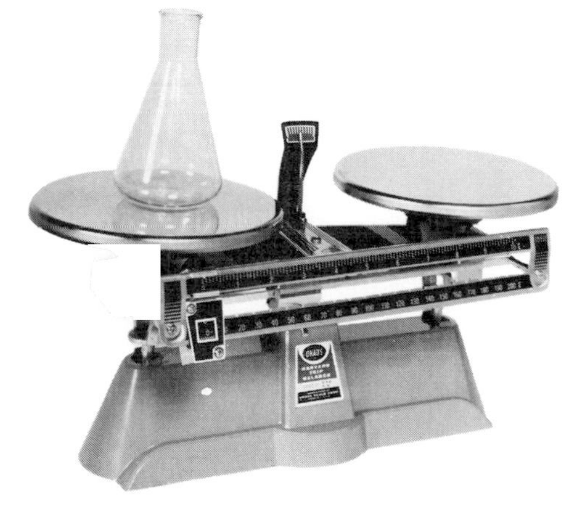

FIGURE 14-17
Double-pan balance for general laboratory massings, sensitivity 0.1 g; used with counterbalance containers.

FIGURE 14-18
Dial-reading torsion balance; allows rapid adjustment of the 10-g mass increment by rotation of the dial.

CHOOSING THE CORRECT BALANCE

Criteria for Choosing a Balance

1. Is the balance suitable for making the desired measurement?

EXAMPLE Is a Mettler Model B-5 (four-place balance) suitable for massing 10 mg of material?

The accuracy in the optical scale is ± 0.05 mg.
The error on a 10-mg sample is $\pm 0.5\%$

Conclusion: Analytical tests cannot tolerate 0.5 percent errors; therefore, a different balance must be chosen.

2. What balance is suitable to mass 10 mg of material?

Model B-5 accuracy in the optical scale is ± 0.05 mg.
The error on a 10-mg sample is $\pm 0.5\%$ (not tolerable).

Model B-6 accuracy in the optical scale is ± 0.02 mg.
The error on a 10-mg sample is $\pm 0.2\%$ (tolerable)

Conclusion: Model B-5 is not suitable. Model B-6 is suitable.

Model Calculation

Determination of chloride:

$$\%\ \text{Cl} = \frac{V \times N \times \text{milliequivalent mass} \times 100}{\text{mass of sample (g)}}$$

where V = volume of titrant (2 mL)
N = normality of titrant (1 N)

Milliequivalent mass of chloride = 0.035

Mass of sample, g = 0.010000 (no error assumed)

Assumption: The V and N of the titrant have no error.

$$\frac{2 \times 0.1 \times 0.035 \times 100}{0.010000} = 70.0\%$$

Four-place balance: $\dfrac{2 \times 0.1 \times 0.035 \times 100}{0.01015^{*}} = 69.64\%$

possible error = ± 0.37

* Inherent mass error in balance.

$$\text{Five-place balance: } \frac{2 \times 0.1 \times 0.035 \times 100}{0.01002^*} = 69.76\%$$

$$\text{possible error} = \pm 0.24$$

$$\text{Microbalance: } \frac{2 \times 0.1 \times 0.035 \times 100}{0.01002^*} = 69.98\%$$

$$\text{possible error} = \pm 0.02$$

ACCESSORIES

Accessories used in determining mass are shown in Figs. 14-19 to 14.21.

* Inherent mass error in balance.

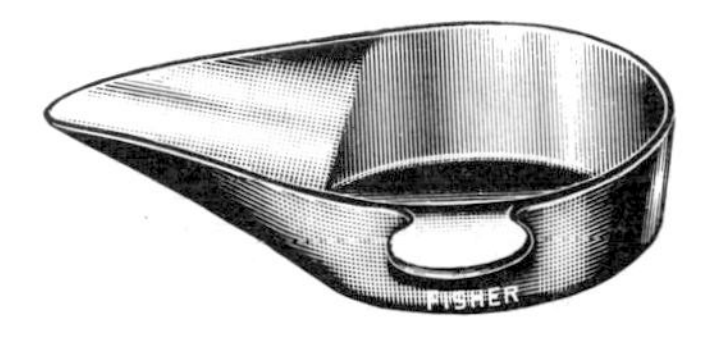

FIGURE 14-19
A balance scoop used to mass sample powders which are easily poured from the spout.

FIGURE 14-20
A disposable pressed-aluminum dish to be used for massing small quantities of liquids or solids.

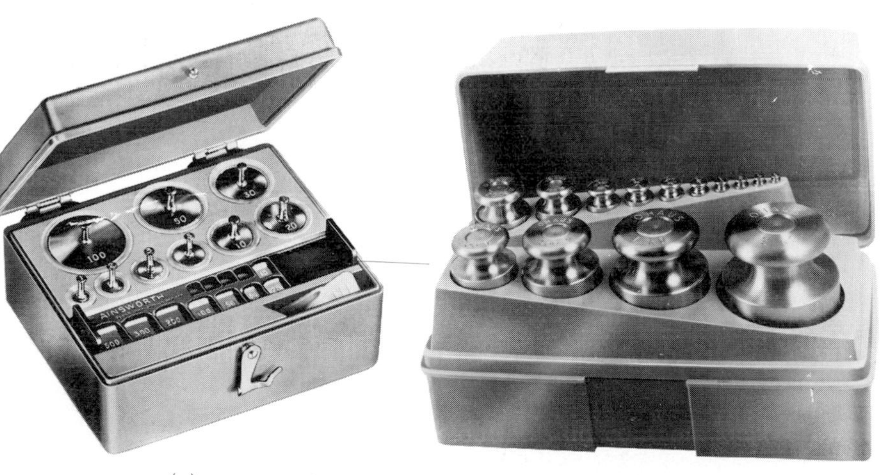

(*a*) (*b*)

FIGURE 14-21
Balance weights or masses. (*a*) Analytical balance masses from 1 mg to 100 g. Set comes with forceps. Handle masses *only* with forceps. Handle with care because they are precision masses. (*b*) One-piece brass masses to be used where accuracy permits (1 to 2000 g).

15
GRAVIMETRIC ANALYSIS

INTRODUCTION

Gravimetric analysis refers to the isolation of a specific substance from a sample and ultimately weighing the substance in a pure or known form. This substance is usually isolated by precipitating it in some insoluble form, by depositing it as a pure metal in electroplating, or by converting it to a gas which is then quantitatively absorbed. Review Filtration for Gravimetric Analysis in Chap. 12, "Laboratory Filtration," for details of this isolation technique. It is necessary that (1) the sought substance be completely removed from the sample, (2) the moisture content and/or other volatile components be determined, and (3) the sample be representative of the material being analyzed.

THE TECHNIQUES OF REPRESENTATIVE SAMPLING

Raw materials and products must be sampled in order to conduct analyses for components or to determine their purity. The size of the lot to be sampled can range from a few grams to thousands of pounds, and yet the sam ple used in the analysis must represent as closely as possible the average composition of the total quantity being analyzed.

The Gross Sample

The gross sample of the lot being analyzed is supposed to be a miniature replica in composition and in particle-size distribution. If it does not truly represent the entire lot, all further work to reduce it to a suitable laboratory size and all analytical procedures are a waste of time. The technique of sampling varies according to the substance being analyzed and its physical characteristics.

Basic Sampling Rules

1. Size of sample must be adequate, depending upon what is being measured, the type of measurement being made, and the level of contaminants.

2. The sample must be representative and reproducible; in static systems multilevel sampling must be made.

Sampling Gases

The size of the gross sample required for gases can be relatively small because any nonhomogeneity occurs at the molecular level. Relatively

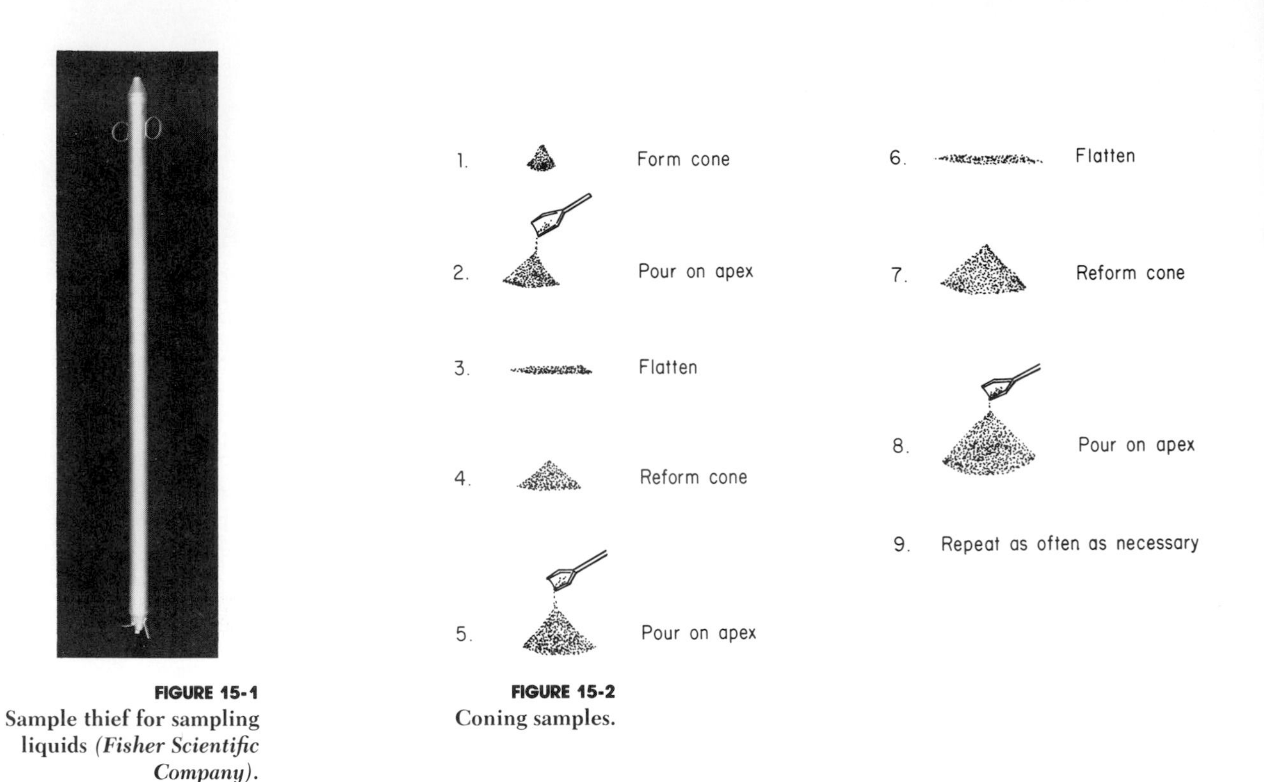

FIGURE 15-1
Sample thief for sampling liquids *(Fisher Scientific Company)*.

FIGURE 15-2
Coning samples.

small samples contain tremendous quantities of molecules. The major problem is that the sample must be representative of the entire lot, and this requires the taking of samples with a "sample thief" (much like that shown in Fig. 15-1) at various locations of the lot, and then combining the various samples into one gross sample.

Sampling Liquids

Liquids Being Pumped through Pipes

When liquids are pumped through pipes, a number of samples can be collected at various times and combined to provide the gross sample for analysis. Care should be taken that the samples represent a constant fraction of the total amount pumped and that all portions of the pumped liquid are sampled. (See Fig. 15-1 for sampling equipment.)

Homogeneous Liquids and Solutions in Containers

Homogeneous liquid solutions can be sampled relatively easily provided the material can be mixed thoroughly by means of agitators or mixing paddles. After adequate mixing, samples can be taken from the top and bottom and combined into one sample which is thoroughly mixed again; from this the final sample is taken for the analysis.

Sampling Nonhomogeneous Solids

The task of obtaining a representative sample from a lot of nonhomogeneous solids requires that:

1. A gross sample be taken
2. The gross sample be reduced to a representative laboratory-size sample
3. The sample be prepared for analysis

Two methods of doing this are described below.

Coning and Quartering

When very large lots are to be sampled, a representative sample can be obtained by coning and quartering (Fig. 15-2). The first sample is formed into a cone, and the next sample is poured onto the apex of the cone. The result is then adequately mixed, and a new cone is formed. As each successive sample is added to the re-formed cone, the total is mixed thoroughly and a new cone is formed prior to the addition of another sample.

After all the samples have been mixed by coning, the mass is flattened and a circular layer of material is formed. This circular layer is then quartered, and the alternate quarters are discarded. This is shown in Fig. 15-3. This process can be repeated as often as desired until a sample size suitable for analysis is obtained.

Rolling and Quartering

A representative cone of the sample is obtained by the coning procedure, and this sample is then placed on a flexible sheet of suitable size. The cone is flattened (Fig. 15-4), and then the entire mass of the sample is repeatedly rolled by pulling first one corner of the sheet over to the opposite corner, and then another corner over to its opposite corner (Fig. 15-5). The number of rollings required depends upon the size of the sample, the size of the particles, and the physical condition of the sample.

To collect the sample, raise all four corners of the sheet simultaneously and collect the sample in the middle of the sheet (Fig. 15-6). The resulting

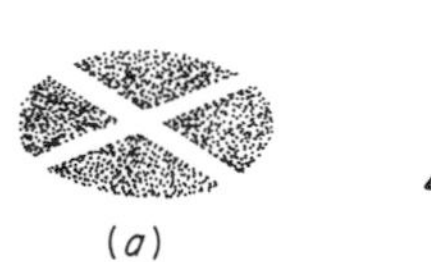
(a)

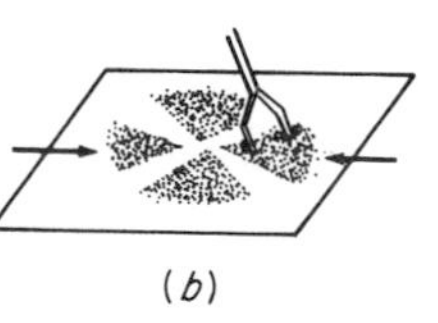
(b)

FIGURE 15-3
Quartering samples: select opposite quarters, discard other two quarters.

FIGURE 15-4
The cone is flattened after being formed.

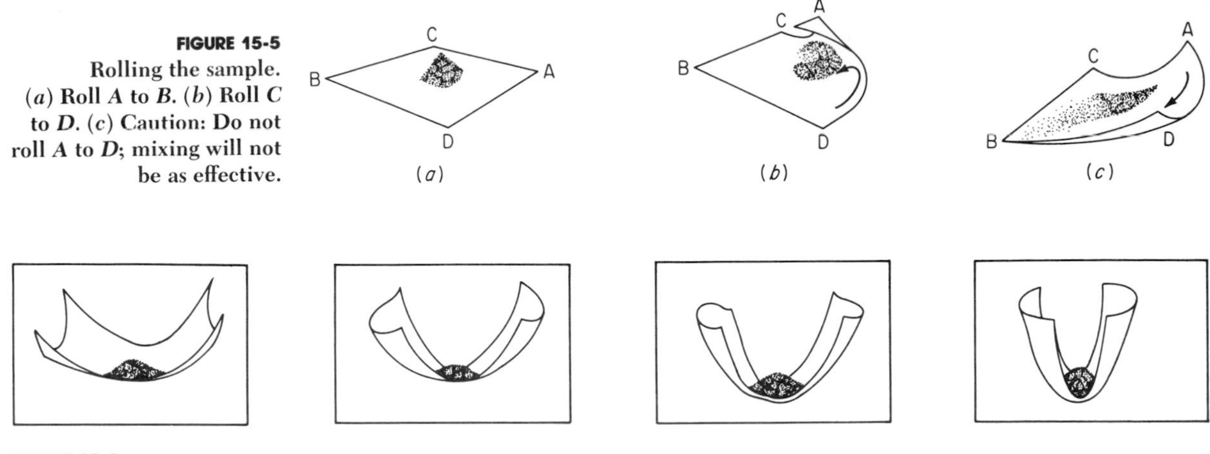

FIGURE 15-5 Rolling the sample. (*a*) Roll *A* to *B*. (*b*) Roll *C* to *D*. (*c*) Caution: Do not roll *A* to *D*; mixing will not be as effective.

FIGURE 15-6 Collecting the rolled sample.

rolled sample is flattened into a circular layer and quartered until a suitably sized sample is obtained.

Sampling Metals

Metals can be sampled by drilling the piece to be sampled at regular intervals from all sides, being certain that each drill hole extends beyond the halfway point. Additional samples can be obtained by sawing through the metal and collecting the "sawdust." Other equipment designed for sampling metal is shown in Figs. 15-7 and 15-8.

All particles collected are then mixed thoroughly and quartered; or the metal can be melted in a graphite crucible to provide the sample for analysis.

CAUTION Chips obtained solely from the surface may not represent the true composition of the object.

HANDLING THE SAMPLE IN THE LABORATORY

The technician may analyze an already prepared sample or prepare a new sample which is to be tested, analyzed, or evaluated. Each sample should be completely identified, tagged, or labeled so that no question as to its origin or source can arise.

Some of the information which may be on the sample is:

1. The number of the sample
2. The notebook experiment-identification number

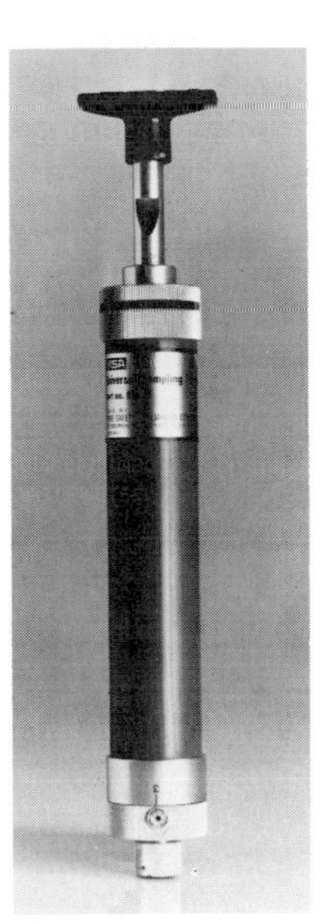

FIGURE 15-7
Sample gun for molten metals. The sample is withdrawn into a glass holder. When the sample cools, the glass is broken to obtain the sample (*Fisher Scientific Company*).

FIGURE 15-8
Solid sampler for small particles. The sampler is made of two concentric, slotted brass tubes. The sampler is inserted into a molten or powdered mass, and the tube is rotated to secure a solid core representative of the lot (*Seedboro Equipment Co.*).

3. The date
4. The origin, e.g., the technician's name, and cross-reference number
5. Weight or volume
6. Identifying code of the container
7. What is to be done with the sample, what determination is to be made, or what analysis is desired

Suitable containers are shown in Fig. 15-9.

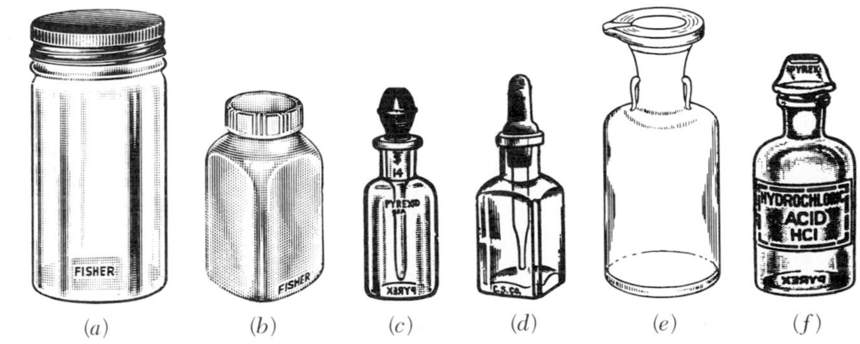

FIGURE 15-9
(*a*) and (*b*) Liquid- and solid-storage bottles for chemicals and samples; varied design, openings, and sizes. (*c-e*) Dropping bottles to dispense small volumes of liquids. (*f*) Liquid-storage bottle for laboratory acids, bases, reagents, and salts.

Pretreatment of Samples

As it arrives at the laboratory, the sample often requires treatment before it is analyzed, particularly if it is in the form of a solid. One of the objectives of this pretreatment is to produce a material so homogeneous that any small portion removed for the analysis will be identical to any other portion. This usually involves reduction of the size of particles to a few tenths of a millimeter and thorough mechanical mixing. Another objective of the pretreatment is to convert the substance to a form in which it is readily attacked by the reagents employed in the analysis; with refractory materials particularly, this involves grinding to a very fine powder. Finally, the sample may have to be dried or its moisture content may have to be determined, because this is a variable factor that is dependent upon atmospheric conditions as well as the physical state of the sample.

Crushing and Grinding

In dealing with solid samples, a certain amount of crushing or grinding is sometimes required to reduce the particle size. Unfortunately, these operations tend to alter the composition of the sample, and for this reason the particle size should be reduced no more than is required for homogeneity and ready attack by reagents.

Ball or jar mills are jars or containers, usually made of porcelain, fitted with a cover and gasket which can be securely fastened to the jar. The jar is half filled with flint pebbles or porcelain or metal balls, and then enough of the material which is to be ground is added to cover the pebbles or balls and the voids between them. The cover is fastened securely to seal the mill hermetically, and the jar is revolved on a rotating assembly. The length of time for which the material is ground depends upon the fineness desired and the hardness of the material. The jar is then emptied into a coarse-mesh screen to separate the pebbles or balls from the ground material.

Changes in Sample

Several factors may cause appreciable alteration in the composition of the sample as a result of grinding. Among these is the heat that is inevitably generated. This can cause losses of volatile components in the sample. In addition, grinding increases the surface area of the solid and thus increases its susceptibility to reactions with the atmosphere. For example, it has been observed that the iron(II) content of a rock may be altered by as much as 40% during grinding—apparently a direct result of atmospheric oxidation of the iron to iron(III).

The effect of grinding on the gain or loss of water from solids is considered in a later section.

Another potential source of error in the crushing and grinding of mixtures arises from the difference in hardness of the components of a sample. The softer materials are converted to smaller particles more rapidly than the hard ones; any loss of the sample in the form of dust will thus cause an alteration in composition. Furthermore, loss of sample in the form of flying fragments must be avoided, since these will tend to be made up of the harder components.

Screening

Intermittent screening of the material is often employed to increase the efficiency of grinding. In this operation, the ground sample is placed upon a wire or cloth sieve (Fig. 15-10) that will pass particles of the desired size. The residual particles are then returned for further grinding; the operation is repeated until the entire sample passes through the screen. This process will certainly result in segregation of the components on the basis of hardness, the toughest materials being last through the screen; it is obvious that grinding must be continued until the last particle has been passed. The need for further mixing after screening is also apparent. Sieves are made to conform with specifications of the American Society for Testing and Materials (ASTM). Made of stainless steel, bronze, nickel, or other materials, they are available in different sieve openings.

Grinding Surfaces

A serious error can arise during grinding and crushing as a consequence of mechanical wear and abrasion of the grinding surfaces. For this reason only the hardest materials such as hardened steel, agate, or boron carbide are employed for the grinding surfaces. Even with these, contamination of the sample is sometimes encountered.

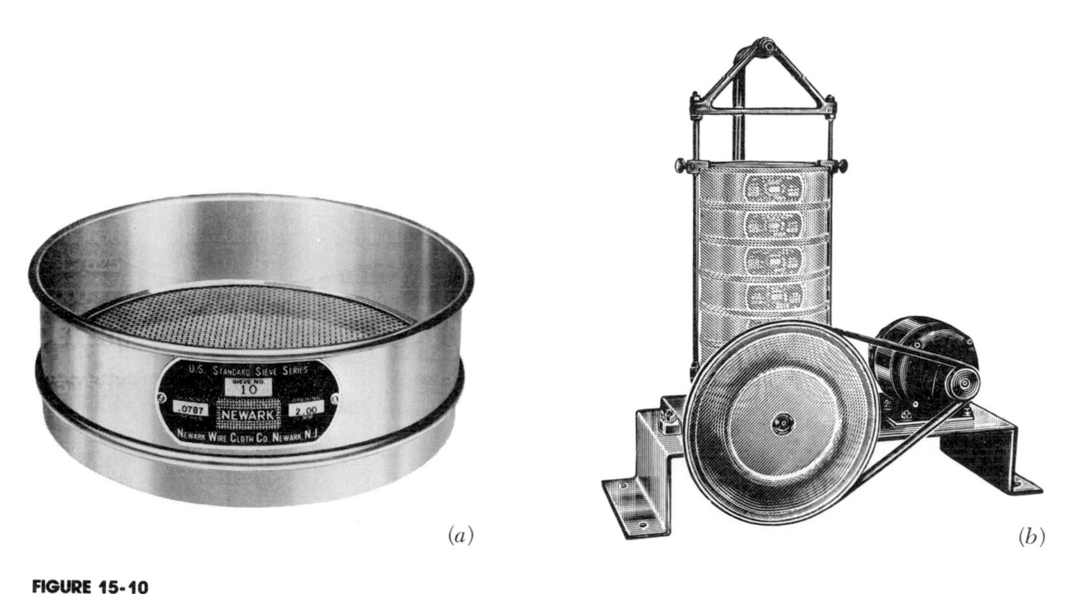

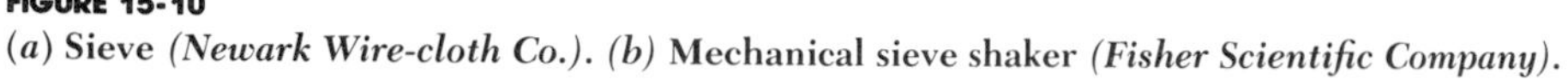

FIGURE 15-10
(*a*) Sieve *(Newark Wire-cloth Co.)*. (*b*) Mechanical sieve shaker *(Fisher Scientific Company)*.

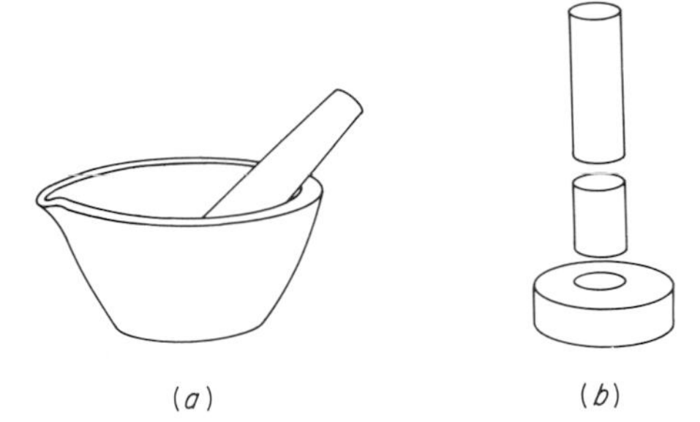

FIGURE 15-11
Mortars and pestles.
(*a*) Porcelain mortar.
(*b*) Agate mortar.

The mortar and pestle (Fig. 15-11), the most ancient of grinding tools, still find wide use in the analytical laboratory. These now come in a variety of sizes and shapes and are commonly constructed of glass, porcelain, agate, mullite, and other hard materials.

CAUTION

1. *Always clean the mortar and pestle thoroughly both before and after grinding each sample.*

2. *Never grind two materials together unless specifically told to do so.*

Minor and some major explosions have occurred and fingers and eyes have been lost or burned because technicians have failed to observe these two simple rules.

Flame Tests for Identification of Elements

Technicians often make a preliminary examination of a sample to determine the presence of certain common elements. This examination may include a flame test, which is easily performed as follows.

1. Obtain a loop of platinum or Nichrome wire mounted in a glass rod. Clean it carefully by dipping it into a small amount of concentrated hydrochloric acid (in a test tube) and then heating it in the blue flame of a Bunsen-type burner until it is cherry red. If the wire loop is clean, the flame will not change color; if the flame shows any color at all, repeat the process until you are satisfied that the loop is free of any contaminant. Replace the acid frequently.

2. Pour a small amount of the powdered sample into a clean watch glass.

3. Heat the *clean* wire loop to cherry redness and dip it into the sample. Some of the powder will cling to the loop. Tap the rod lightly to dislodge any excess sample.

4. Place the wire loop plus the sample back in the flame and reheat it to

TABLE 15-1
Flame Tests for Elements

Color of flame	*Element indicated*
Blue	
Azure	Lead, selenium, $CuCl_2$ (and other copper compounds when moistened with HCl); $CuBr_2$ appears azure blue, then is followed by green
Light blue	Arsenic and some of its compounds; selenium
Greenish-blue	$CuBr_2$; arsenic; lead; antimony
Green	
Emerald green	Copper compounds other than halides (when not moistened with HCl); thallium compounds
Blue-green	Phosphates moistened with sulfuric acid; B_2O_3
Pure green	Thallium and tellurium compounds
Yellow-green	Barium; possibly molybdenum; borates (with H_2SO_4)
Faint green	Antimony and ammonium compounds
Whitish green	Zinc
Red	
Carmine	Lithium compounds (masked by barium or sodium), are invisible when viewed through green glass, appear violet through cobalt glass
Scarlet	Calcium compounds (masked by barium), appear greenish when viewed through cobalt glass and green through green glass
Crimson	Strontium compounds (masked by barium), appear violet through cobalt glass, yellowish through green glass
Violet	Potassium compounds other than silicates, phosphates, and borates; rubidium and cesium are similar. Color is masked by lithium and/or sodium, appears purple-red through cobalt glass and bluish-green glass
Yellow	Sodium, even the most minute amounts; is invisible when viewed through cobalt glass

cherry redness. As shown in Table 15-1, certain elements will lend characteristic colors to the flame.

5. Be sure to clean the wire loop thoroughly between samples.

MOISTURE IN SAMPLES

The presence of water in a sample represents a common problem that frequently faces the analyst. This compound may exist as a contaminant from the atmosphere or from the solution in which the substance was formed, or it may be bonded as a chemical compound, a hydrate. Regardless of its origin, however, water plays a part in determining the composition of the sample. Unfortunately, particularly in the case of solids, the water content is a variable quantity that depends upon such things as humidity, temperature, and the state of subdivision. Thus, the constitution of a sample may change significantly with environment and method of handling.

In order to cope with the variability in composition caused by the presence of moisture, the analyst may attempt to remove the water by drying prior to weighing samples for analysis. Alternatively the water content may be determined at the time the samples are weighed out for analysis; in this way results can be corrected to a dry basis. In any event, most analyses are preceded by some sort of preliminary treatment designed to take into account the presence of water. There are many established tests used for this purpose.

FORMS OF WATER IN SOLIDS

It is convenient to distinguish among the several ways in which water can be held by a solid. Although it was developed primarily with respect to minerals, the classification of Hillebrand and his collaborators may be applied to other solids as well, and it forms the basis for the discussion that follows.

The *essential water* in a substance is that water which is an integral part of the molecular or crystal structure of one of the components of the solid. It is present in that component in stoichiometric quantities. Thus, the water of crystallization in stable solid hydrates (for example, $CaC_2O_4 \cdot 2H_2O$, $BaCl_2 \cdot 2H_2O$) qualifies as a type of essential water.

A second form is called *water of constitution*. Here the water is not present as such in the solid but rather is formed as a product when the solid undergoes decomposition, usually as a result of heating. This is typified by the processes

$$2KHSO_4 \rightarrow K_2S_2O_7 + H_2O$$

$$Ca(OH)_2 \rightarrow CaO + H_2O$$

Nonessential water is not necessary for the characterization of the chemical constitution of the sample and therefore does not occur in any sort of stoichiometric proportions. It is retained by the solid as a consequence of physical forces.

Adsorbed water is retained on the surface of solids in contact with a moist environment. The quantity is dependent upon humidity, temperature, and the specific surface area of the solid. Adsorption is a general phenomenon that is encountered in some degree with all finely divided solids. The amount of moisture adsorbed on the surface of a solid also increases with the amount of water in its environment. Quite generally, the amount of adsorbed water decreases as temperature increases, and in most cases it approaches zero if the solid is dried at temperatures above 100°C.

Equilibrium, in the case of adsorbed moisture, is achieved rather rapidly, ordinarily requiring only 5 or 10 min. This often becomes apparent to the chemist who weighs finely divided solids that have been rendered anhydrous by drying; a continuous increase in weight is observed unless the solid is contained in a tightly stoppered vessel.

A second type of nonessential water is called *sorbed water*. This is encountered with many colloidal substances such as starch, protein, charcoal, zeolite minerals, and silica gel. The amounts of sorbed water are often large compared with adsorbed moisture, amounting in some instances to as much as 20 percent or more of the solid. Interestingly enough, solids containing even this much water may appear to be perfectly dry powders. Sorbed water is held as a condensed phase in the interstices or capillaries of the colloidal solids. The quantity is greatly dependent upon temperature and humidity.

A third type of nonessential moisture is *occluded water*. Here, liquid water is entrapped in microscopic pockets spaced irregularly throughout the solid crystals. Such cavities often occur naturally in minerals and rocks.

Water may also be dispersed in a solid in the form of a solid solution. Here the water molecules are distributed homogeneously throughout the solid. Natural glasses may contain several percent of moisture in this form.

EFFECTS OF GRINDING ON MOISTURE CONTENT

Often the moisture content and thus the chemical composition of a solid is altered to a considerable extent during grinding and crushing. This will result in decreases in some instances and increases in others.

Decreases in water content are sometimes observed when one is grinding solids containing essential water in the form of hydrates; thus the water content of gypsum, $CaSO_4 \cdot 2H_2O$, is reduced from 20 to 5% by this treatment. Undoubtedly the change is a result of localized heating during the grinding and crushing of the particles.

Losses also occur when samples containing occluded water are reduced in particle size. Here, the grinding process ruptures some of the cavities and exposes the water so that it may evaporate.

More commonly perhaps, the grinding process is accompanied by an increase in moisture content, primarily because of the increase in surface area exposed to the atmosphere. A corresponding increase in adsorbed water results. The magnitude of the effect is sufficient to alter appreciably the composition of a solid. For example, the water content of a piece of porcelain in the form of coarse particles was zero, but after it had been ground for some time it was found to be 0.62%. Grinding a basaltic greenstone for 120 min changed its water content from 0.22 to 1.70%.

DRYING SAMPLES

Principle

Samples may be dried by heating them at 100 to 105°C or higher, if the melting point of the material is higher and the material will not decompose at that temperature. This procedure will remove the moisture bound to the surface of the particles.

Procedure

1. Label the beaker and the weighing bottle. (Remove cover from weighing bottle.)

2. Place the weighing bottle in the beaker, which is covered by a watch glass supported on glass hooks (Fig. 15-12).

3. Place in the oven for the required time at the temperature suggested.

DRYING COLLECTED CRYSTALS

Gravity-Filtered Crystals

Gravity-filtered crystals collected on a filter paper may be dried by the following methods.

CAUTION Be sure to label everything properly.

1. Remove the filter paper from the funnel. Open up the filter paper and flatten it on a watch glass of suitable size or a shallow evaporating dish. Cover the watch glass or dish with a large piece of clean, dry filter paper (secured to prevent wind currents from blowing it off) and allow the crystals to air-dry.

CAUTION Hygroscopic substances cannot be air-dried in this way.

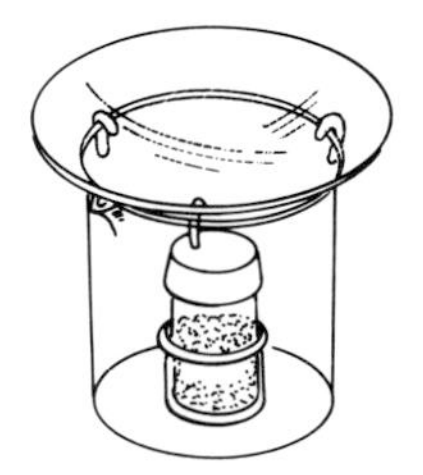

FIGURE 15-12 Arrangement for the drying of samples.

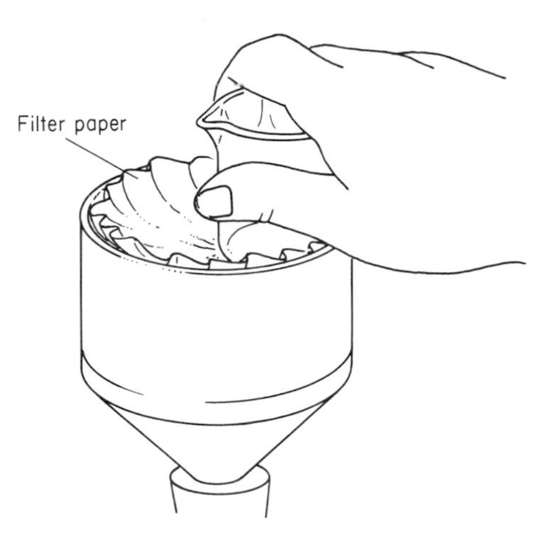

FIGURE 15-13
Pressing out excess moisture from the wet crystals is a method applicable to both gravity-filtered crystals and vacuum-filtered crystals as shown here.

2. Press out excess moisture from the crystals by laying filter paper on top of the moist crystals and applying pressure with a suitable object. (See Fig. 15-13).

3. Use a spatula to work the pasty mass on a porous plate (Fig. 15-14); then allow it to dry.

4. Use a portable infrared lamp to warm the sample and increase the rate of drying. Be sure the temperature does not exceed the melting point of the sample.

5. Use a desiccator filled with a desiccant (Fig. 15-15).

The Desiccator

A *desiccator* is a glass container filled with a substance which absorbs water (a *desiccant*); it is used to provide a dry atmosphere for objects and substances. Desiccators are employed to achieve and maintain an atmosphere of low humidity for the storage of samples, precipitates, crucibles, weighing bottles, and other equipment. Use a desiccator as follows:

1. Remove the cover by sliding sideways as in Fig. 15-16.

FIGURE 15-14
The porous plate absorbs the excess water.

FIGURE 15-15
Desiccator.

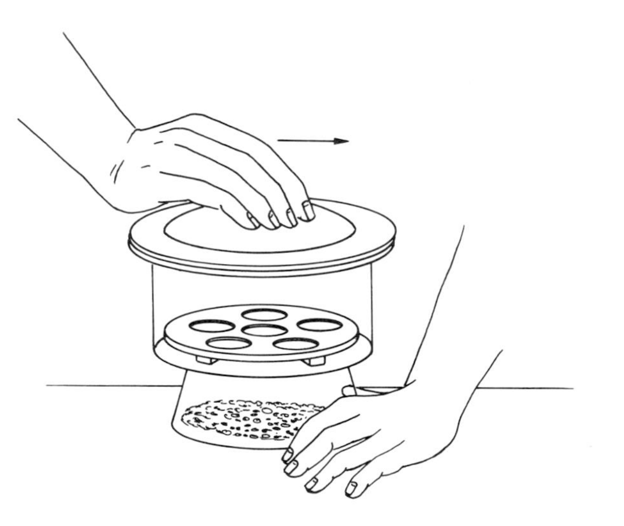

FIGURE 15-16
Removing the desiccator cover.

2. Place the object to be dried on the porcelain platform plate.

3. Regrease the ground-glass rim with petroleum jelly or silicone grease if necessary.

4. Slide the lid back in position.

CAUTION *Hot crucibles should never be inserted immediately in the desiccator. Allow to cool in air for 1 min prior to insertion.* If this caution is not observed, the air will be heated in the desiccator when it is closed. On cooling, a partial vacuum will result. When the desiccator is opened, a sudden rush of air may spill the sample.

NOTE *Vacuum desiccators* are equipped with side arms, so that they may be connected to a vacuum and the contents will be subject to a vacuum rather than to dried air. Vacuum-type desiccators should be used to dry crystals which are wet with organic solvents. Vacuum desiccators should not be used for substances which sublime readily.

A desiccator must be kept clean and its charge of desiccant must be frequently renewed to keep it effective. This is done as follows:

1. Remove the cover and the porcelain support plate.
2. Dump the waste desiccant in an appropriate waste receptacle.
3. Wash and dry the desiccator.
4. Refill with fresh desiccant (Fig. 15-17).
5. Regrease the ground-glass lid.
6. Replace the porcelain support.
7. Slide the lid into position on the desiccator.

Vacuum-Filtered Crystals

Vacuum-filtered crystals can be dried in one of two ways:

1. Remove the filter cake and use the procedures listed under drying gravity-collected crystals.

2. After all the crystals have been collected on the Büchner funnel, the funnel is covered loosely with an evaporating dish or a larger piece of

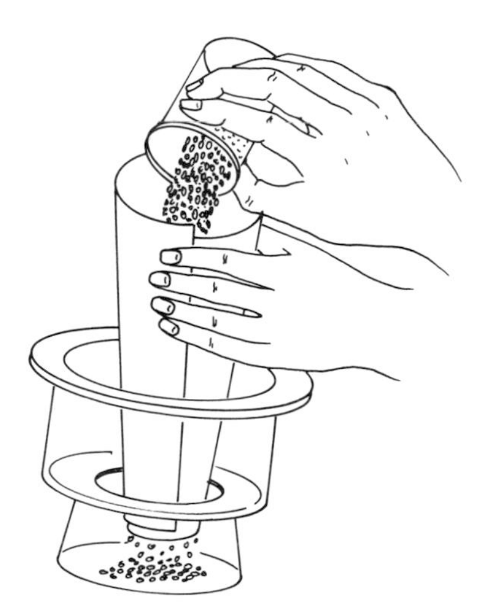

FIGURE 15-17
Filling the desiccator with fresh desiccant.

filter paper (secured to prevent its blowing off). The vacuum system is then maintained to pull air through the moist crystals, which will dry after a short period of time.

NOTE Less time will be required to dry the crystals if, while continuing the vacuum system, the covering is removed periodically and the cake is mixed (careful!) and evened out with a spatula before the cover is resecured.

Drying Crystals in a Centrifuge Tube

When very small quantities of crystals are collected by centrifugation, they can be dried by subjecting them to vacuum in the centrifuge tube while gently warming the tube (Fig. 15-18). This procedure prevents any loss of the small quantity of crystals collected, as would occur if you tried to transfer the crystals out of the tube with a rubber policeman.

Abderhalden Drying Pistol

Some substances retain water and other solvents so persistently that drying them in ordinary desiccators at room temperature will not remove the solvents. In these cases, the *drying pistol* (Fig. 15-19) works very well. The substance to be dried is placed in a boat which is inserted in the pistol, which is then connected to a source of vacuum. Also in the contained volume is a pocket for an effective adsorbing agent, such as P_4O_{10} for water, solid KOH or NaOH pellets for acid gases, or thin layers of paraffin wax

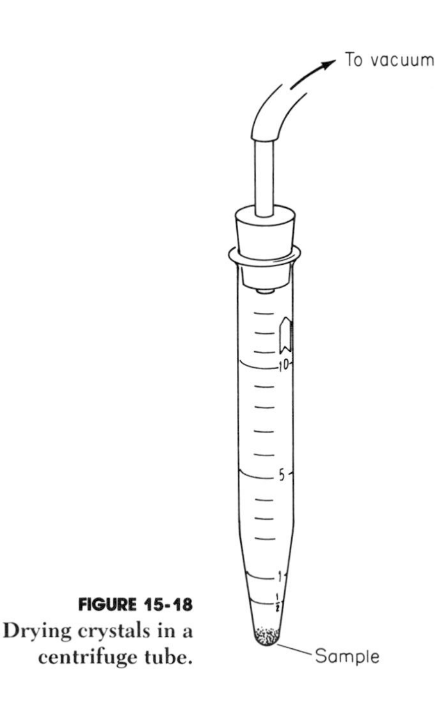

FIGURE 15-18 Drying crystals in a centrifuge tube.

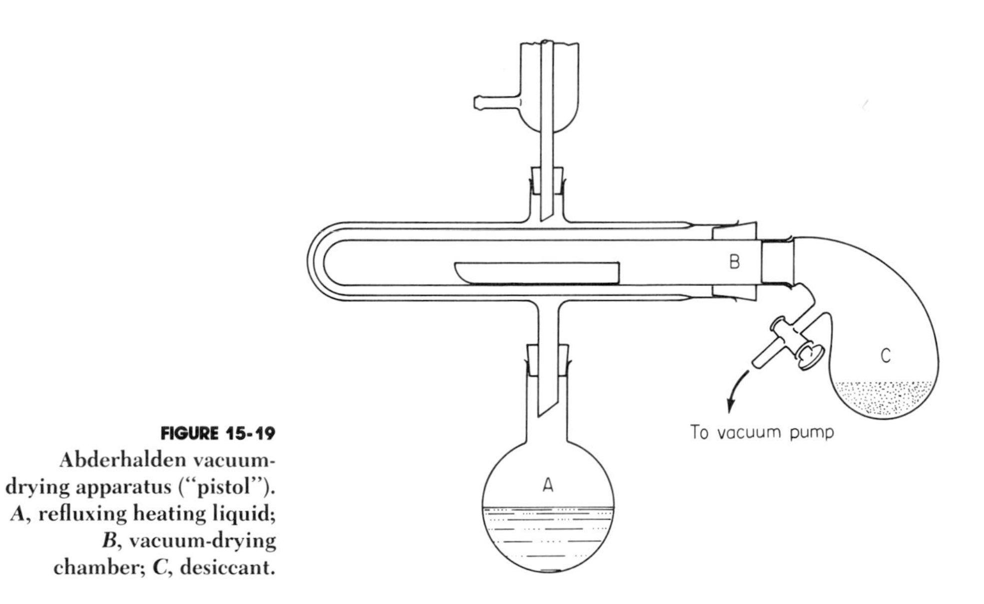

FIGURE 15-19 Abderhalden vacuum-drying apparatus ("pistol"). *A*, refluxing heating liquid; *B*, vacuum-drying chamber; *C*, desiccant.

for the removal of organic solvents. The temperature at which the substance is to be dried is determined by the boiling point of the refluxing liquid.

DRYING ORGANIC SOLVENTS

Water can be removed from organic liquids and solutions by treating the liquids with a suitable drying agent to remove the water. Water is soluble to some extent in all organic liquids, and any organic solvent which has been used in a water-immiscible–organic-solvent extraction will contain water. Each organic solvent will dissolve its own characteristic percentage of water. For example, ethyl ether will contain about 1.5% water. To "dry" or dehydrate organic solvents, we use *drying agents*.

NOTE The selection of drying agents must be carefully made. The drying agent selected should not react with the compound or cause the compound to undergo any reaction, but will *only* remove the water. The best drying agents are those which will react rapidly and irreversibly with water and will not react with or affect in any way the solvent or the solute dissolved in it.

Principle

Solid drying agents are added to wet organic solvents. They remove the water, and then the hydrated solid is separated from the organic solvent by decantation and filtration.

Procedure

1. Pour the organic liquid into a flask which can be stoppered. Add small portions of the drying agent, shaking the flask thoroughly after each addition. Add as much drying agent as is required.
2. Allow to stand overnight or for a predetermined time.
3. Filter the solid hydrate from the liquid with a funnel and filter paper.

NOTE Several operations may be required. Repeat if necessary.

Efficiency of Drying Operations

The efficiency of a drying operation is improved if the organic solvent is repeatedly exposed to fresh portions of the drying agent, just as the efficiency of an extraction operation is improved by repeated exposure of the solute to fresh extraction solvent. The efficiency of the drying operation is lowest when the wet solution is exposed to all of the drying agent at one time.

Some dehydrating agents are very powerful and dangerous, especially if the water content of the organic solvent is high. These should be used only after the wet organic solvent has been grossly predried with a weaker agent. If you are in doubt as to the advisability of using a particular dehydrating agent, always consult with your supervisor or with specialists in the field. (See Tables 15-2 and 15-3.)

Classification of Drying Agents

1. Those which form compounds with water of hydration (the hydrates can be returned to the anhydrous form by suitable heating to remove the water): Na_2CO_3, Na_2SO_4, anhydrous $CaCl_2$, $ZnCl_2$, NaOH, $CaSO_4$, H_2SO_4 (95%), silica gel, CaO.

NOTE Traces of acid remaining in wet organic-liquid reaction products are removed simultaneously with the water when *basic* drying agents are used: Na_2CO_3, NaOH, $Ca(OH)_2$.

TABLE 15-2
Intensity of Drying Agents

High intensity	*Moderate intensity*	*Low intensity*
$Mg(ClO_4)_2$*	KOH	Na_2CO_3
Molecular sieves	NaOH	Na_2SO_4
Metallic sodium (Na)	K_2CO_3	$MgSO_4$
P_4O_{10}	$CaSO_4$	
H_2SO_4 (conc)	CaO	
	$CaCl_2$	

* $Mg(ClO_4)_2$ is the most efficient drying agent available, but is an explosion hazard with easily oxidized or acidic organic compounds.

TABLE 15-3 Characteristics of Drying Agents

Compound	*Acidity*	*Comments*
Calcium sulfate	Neutral	General use, commercially available as Drierite®, very fast
Calcium chloride	Neutral	Reacts with N and O compounds, rapid; use for hydrocarbons and R-X
Magnesium sulfate	Neutral	General use; rapid (avoid acid-sensitive compounds)
Sodium sulfate	Neutral	General use; mild; high-capacity; gross dryer for cold solutions
Potassium carbonate	Basic	Use with esters, nitriles, ketones, alcohols (not for use with acidic compounds)
Sodium carbonate	Basic	Use with esters, nitriles, ketones, alcohols (not for use with acidic compounds)
Sodium hydroxide	Basic	Use only with inert compounds; very fast; powerful; good for amines
Potassium hydroxide	Basic	Use only with inert compounds; very fast; powerful; good for amines
Calcium oxide	Basic	Use for alcohols and amines; slow; efficient; not for use with acidic compounds
Tetraphosphorus decoxide	Acidic	Use only with inert compounds (ethers, hydrocarbons, halides); fast; efficient
Molecular sieves 3Å,4Å	Neutral	General use; high-intensity; predry with common agent
Sulfuric acid	Acidic	Very efficient; use for saturated hydrocarbons, aromatic hydrocarbons, and halides; reacts with olefins and basic compounds

2. Those which form new compounds by chemical reaction with water: metallic sodium, CaC_2, P_4O_{10}.

CAUTIONS

These dehydrating agents are extremely reactive and most efficient. *Handle with care.* They react with water to give NaOH, CaOH, and H_3PO_4, respectively.

1. *Do not use* them where either the drying agent itself or the product

that it forms will react with the compound or cause the compound itself to undergo reaction or rearrangement.

2. *Use to dry* saturated hydrocarbons, aromatic hydrocarbons, ethers.

3. The compounds to be dried should not have functional groups, such as $-OH^-$ and $-COOH^-$, which will react with the drying agent.

4. *Do not dry* alcohols with metallic sodium. *Do not dry* acids with NaOH or other basic drying agents. *Do not dry* amines (or basic compounds) with acidic drying agents. *Do not use* $CaCl_2$ to dry alcohols, phenols, amines, amino acids, amides, ketones, or certain aldehydes and esters.

Determining If the Organic Solvent Is "Dry"

Drying agents will clump together, sticking to the bottom of the flask when a solution is "wet." They will even dissolve in very wet solutions if an insufficient amount of them has been added. Wet solvent solutions appear to be cloudy; dry solutions are clear. If the solution is "dry," the solid drying agent will move about and shift easily on the bottom of the flask.

FREEZE-DRYING: LYOPHILIZATION

Some substances cannot be dried at atmospheric conditions because they are extremely heat-sensitive materials, but they can be freeze-dried. Freeze drying is a process whereby substances are subjected to high vacuum after they have been frozen, and under those conditions ice (water) will sublime. This leaves the nonsublimable material (everything but the water) behind in a dried state.

Commercial freeze-driers are available in some laboratories. They consist of a self-contained freeze-drying unit which will effectively remove volatile solvents. They may be simple ones, consisting merely of a vacuum pump, adequate vapor traps, and a receptacle for the material in solution. Others include refrigeration units to chill the solution plus more sophisticated instruments to designate temperature and pressure, plus heat and cold controls and vacuum-release valves.

Freeze-driers, as the name indicates, are usually used to remove all the volatile solvents or water, but they can be used to remove smaller amounts as required.

Freeze-drying procedures are excellent for drying or concentrating heat-sensitive substances. They differ from ordinary vacuum distillation in that the solution or substance to be dried must be frozen to a solid mass first. It is under these conditions that the water is selectively removed by sublimation, the ice going directly to the water-vapor state.

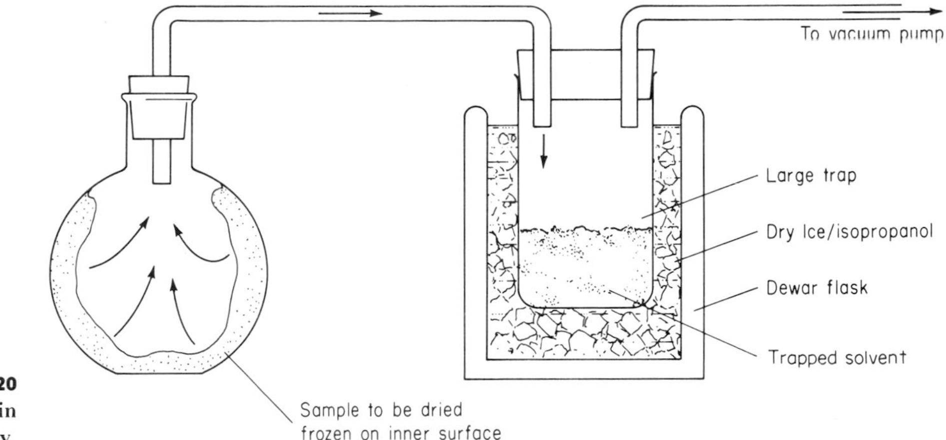

FIGURE 15-20
Setup for freeze-drying in the laboratory.

Procedure

1. Freeze the solution, spreading it out on the inner surface of the container (Fig. 15-20) to increase the surface area.

2. Apply high vacuum; the ice will sublime and leave the dried material behind.

   (a) Keep material frozen during sublimation. (Sublimation normally will maintain the frozen state.)

   (b) Use dilute solutions in preference to concentrated solutions.

   (c) Apply all safety procedures for working with high vacuum.

   (d) Protect the vacuum pump from water with a dry-ice trap, and insert chemical gas-washing towers to protect the pump from corrosive gases.

Advantages of Freeze Drying

1. Substances are locked in an ice matrix and cannot interact.

2. Oxidation is prevented because of the high vacuum.

3. The final product of freeze drying is identical to the original product minus the water removed.

Freeze Drying Corrosive Materials

The majority of freeze-drying operations involve solely the removal of water; however, many substances which are subjected to these procedures contain corrosive acids and bases. If these substances are to be freeze-dried, the unit must be protected against the corrosive vapors so that they do not attack the mechanical vacuum pump. *Always insert chemical gas towers to remove the corrosive vapors before they enter the pump.*

PREPARING THE SAMPLE FOR FINAL ANALYSIS

In order to complete many analyses, an aqueous solution of the sample is required; furthermore, the function to be determined must ordinarily be present in that solution in the form of a simple ion or molecule. Unfortunately many of the substances that are of interest can be converted to this form only by some treatment. For example, before the chlorine content of an organic compound can be determined, it is usually necessary to convert the element into a form that is amenable to analysis. Because this will require breaking of the carbon-chlorine bonds, the preliminary treatment of the sample is likely to be quite vigorous.

Various reagents and techniques exist for decomposing and dissolving analytical samples. Often the proper choice among these is critical to the success of an analysis, particularly where refractory substances are being dealt with.

Liquid Reagents Used for Dissolving or Decomposing Inorganic Samples

The most common reagents for attacking analytical samples are the mineral acids or their aqueous solutions. Solutions of sodium or potassium hydroxide also find occasional application.

Hydrochloric Acid

Concentrated hydrochloric acid is an excellent solvent for many metal oxides as well as those metals which lie above hydrogen in the electromotive series; it is often a better solvent for the oxides than the oxidizing acids. Concentrated hydrochloric acid is about 12 *N*, but upon heating hydrogen chloride is lost until a constant-boiling 6 *N* solution remains (boiling point about 110°C).

Nitric Acid

Concentrated nitric acid is an oxidizing solvent that finds wide use in attacking metals. It will dissolve most common metallic elements; aluminum and chromium, which become passive to the reagent, are exceptions. Many of the common alloys can also be decomposed by nitric acid. In this connection it should be mentioned that tin, antimony, and tungsten form insoluble acids when treated with concentrated nitric acid; this treatment is sometimes employed to separate these elements from others contained in alloys.

Sulfuric Acid

Hot concentrated sulfuric acid is often employed as a solvent. Part of its effectiveness arises from its high boiling point (about 340°C), at which temperature decomposition and solution of substances often proceed quite rapidly. Most organic compounds are dehydrated and oxidized under these conditions; the reagent thus serves to remove such components from a sample. Most metals and many alloys are attacked by the hot acid.

Perchloric Acid

Hot, concentrated perchloric acid is a potent oxidizing agent and solvent. It attacks a number of ferrous alloys and stainless steels that are intractable to the other mineral acids; it is frequently the solvent of choice. This acid also dehydrates and rapidly oxidizes organic materials.

CAUTION Violent explosions result when organic substances or easily oxidized inorganic compounds come in contact with the hot, concentrated acid; as a consequence, a good deal of care must be employed in the use of this reagent. For example, it should be heated only in hoods in which the ducts are clean and free of organic materials and where the possibility of contamination of the solution is absolutely nil.

Perchloric acid is marketed as the 60 or 72% acid. Upon heating, a constant-boiling mixture (72.4% $HClO_4$) is obtained at a temperature of 203°C. Cold, concentrated perchloric acid and hot, dilute solutions are quite stable with respect to reducing agents; it is only the hot, concentrated acid that constitutes a potential hazard. The reagent is a very valuable solvent and is widely used in analysis. *Before it is employed, however, the proper precautions for its use must be clearly understood.*

Oxidizing Mixtures

More rapid solvent action can sometimes be obtained by the use of mixtures of acids or by the addition of oxidizing agents to the mineral acids. *Aqua regia*, a mixture consisting of three volumes of concentrated hydrochloric acid and one of nitric acid, is well known. Addition of bromine or hydrogen peroxide to mineral acids often increases their solvent action and hastens the oxidation of organic materials in the sample. Mixtures of nitric and perchloric acid are also useful for this purpose, as are mixtures of fuming nitric and concentrated sulfuric acids.

Hydrofluoric Acid

The primary use for this acid is the decomposition of silicate rocks and minerals where silica is not to be determined; the silicon, of course, escapes as the tetrafluoride. After decomposition is complete, the excess hydrofluoric acid is driven off by evaporation with sulfuric acid or perchloric acid. Complete removal is often essential to the success of an analysis, because of the extraordinary stability of the fluoride complexes of several metal ions; the properties of some of these differ markedly from those of the parent cation. Thus, for example, precipitation of aluminum with ammonia is quite incomplete in the presence of small quantities of fluoride. Frequently, removal of the last traces of fluoride from a sample is so difficult and time-consuming as to negate the attractive features of this reagent as a solvent for silicates.

Hydrofluoric acid finds occasional use in conjunction with other acids in attacking some of the more difficultly soluble steels.

CAUTION Hydrofluoric acid can cause serious damage and painful injury when brought in contact with the skin; it must be handled with respect.

Decomposition of Samples by Fluxes

Quite a number of common substances—such as silicates, some of the mineral oxides, and a few of the iron alloys—are attacked slowly, if at all, by the usual liquid reagents. Recourse to more potent fused-salt media, or fluxes, is then called for. Fluxes will decompose most substances by virtue of the high temperature required for their use (300 to 1000°C) and the high concentration of reagent brought in contact with the sample.

Where possible, the employment of a flux is avoided, for several dangers and disadvantages attend its use. In the first place, a relatively large quantity of the flux is required to decompose most substances—often 10 times the sample weight. The possibility of significant contamination of the sample by impurities in the reagent thus becomes very real.

Furthermore, the aqueous solution resulting from the fusion will have a high salt content, and this may lead to difficulties in the subsequent steps of the analysis. The high temperatures required for a fusion increase the danger of loss of pertinent constituents by volatilization. Finally, the container in which the fusion is performed is almost inevitably attacked to some extent by the flux; this again can result in contamination of the sample.

In those cases where the bulk of the substance to be analyzed is soluble in a liquid reagent and only a small fraction requires decomposition with a flux, it is common practice to employ the liquid reagent first. The undecomposed residue is then isolated by filtration and fused with a relatively small quantity of a suitable flux. After cooling, the melt is dissolved and combined with the rest of the sample.

Method of Carrying Out a Fusion

In order to achieve a successful and complete decomposition of a sample with a flux, the solid must ordinarily be ground to a very fine powder; this will produce a high specific surface area. The sample must then be thoroughly mixed with the flux; this operation is often carried out in the crucible in which the fusion is to be done by careful stirring with a glass rod.

In general, the crucible used in a fusion should never be more than half-filled at the outset. The temperature is ordinarily raised slowly with a gas flame because the evolution of water and gases is a common occurrence at this point; unless care is taken there is the danger of loss by spattering. The crucible should be covered as an added precaution. The maximum temperature employed varies considerably depending upon the flux and the sample; it should be no greater than necessary, however, to minimize attack on the crucible and decomposition of the flux. The length of the fusion may range from a few minutes to one or two hours, depending upon the nature of the sample. It is frequently difficult to decide when the heating should be discontinued. In some cases, the production of a clear melt serves to indicate the completion of the decomposition. In others the condition is not obvious, and the analyst must base the heating time on previous experience with the type of material being analyzed. In any event, the aqueous solution from the fusion should be examined carefully for particles of unattacked sample.

When the fusion is judged complete, the mass is allowed to cool slowly; then just before solidification the crucible is rotated to distribute the solid around the walls of the crucible so that the thin layer can be readily detached.

Types of Fluxes

With few exceptions the common fluxes used in analysis (Table 15-4) are compounds of the alkali metals. Basic fluxes, employed for attack on acidic materials, include the carbonates, hydroxides, peroxides, and borates. The acidic fluxes are the pyrosulfates and the acid fluorides as well as boric oxide. If an oxidizing flux is required, sodium peroxide can be used. As an

TABLE 15-4
The Common Fluxes

Flux	*Melting point, °C*	*Type of crucible used for fusion*	*Type of substance decomposed*
Na_2CO_3	851	Pt	For silicates and silica-containing samples; alumina-containing sample; insoluble phosphates and sulfates
Na_2CO_3 + an oxidizing agent such as KNO_3, $KClO_3$ or Na_2O_2	Pt	Pt (not with Na_2O_2) Ni	For samples where an oxidizing agent is needed, that is, samples containing A, As, Sb, Cr, etc.
NaOH or KOH KOH	318–380	Au, Ag, Ni	Powerful basic fluxes for silicates, silicon carbide, and certain minerals; main limitation, purity of reagents
Na_2O_2	Decomposes	Fe, Ni	Powerful basic oxidizing flux for sulfides; acid-insoluble alloys of Fe, Ni, Cr, Mo, W, and Li; Pt alloys; Cr, Sn, Zn minerals
$K_2S_2O_7$	300	Pt porcelain	Acid flux for insoluble oxides and oxide-containing samples
B_2O_3	577	Pt	Acid flux for decomposition of silicates and oxides where alkali metals are to be determined
$CaCO_3$ + NH_4Cl		Ni	Upon heating of the flux, a mixture of CaO and $CaCl_2$ is produced; used for decomposing silicates for the determination of the alkali metals

alternative, small quantities of the alkali nitrates or chlorates are mixed with sodium carbonate.

Decomposition of Organic Compounds

Analysis of the elemental composition of an organic substance generally requires drastic treatment of the material in order to convert the elements of interest into a form susceptible to the common analytical techniques. These treatments are usually oxidative in nature, involving conversion of the carbon and hydrogen of the organic material to carbon dioxide and water; in some instances, however, heating the sample with a potent reducing agent is sufficient to rupture the covalent bonds in the compound and free the element to be determined from the carbonaceous residue.

Oxidation procedures are sometimes divided into two categories. Wet ashing (or oxidation) makes use of liquid oxidizing agents such as sulfuric or perchloric acids. Dry ashing (or oxidation) usually implies ignition of the organic compound in air or a stream of oxygen. In addition, oxidations can be carried out in certain fused-salt media, sodium peroxide being the most common flux for this purpose.

In the sections that follow we shall mention briefly some of the methods for decomposing organic substances prior to the analysis for the more common elements.

Combustion-Tube Methods

Several of the common and important elemental components of organic substances are converted to gaseous products when the material is oxidized. With suitable apparatus it is possible to trap these volatile compounds quantitatively and use them in analyzing for the element of interest. A common way to do this is to carry out the oxidation in a glass or quartz combustion tube through which is forced a stream of carrier gas. The stream serves to transport the volatile products to a part of the apparatus where they can be separated and retained for measurement; the stream may also serve as the oxidizing agent. The common elements susceptible to this type of treatment are carbon, hydrogen, oxygen, nitrogen, the halogens, and sulfur.

Figure 15-21 shows a typical combustion train for the determination of carbon and hydrogen in an organic substance. Figure 15-22 shows a movable combustion furnace. Oxygen is forced through the tube to oxidize the sample as well as to carry the products to the absorption part of the train. The sample is contained in a small platinum or porcelain boat that can be pushed into the proper position by means of a rod or wire. Ignition is initiated by slowly raising the temperature of that part of the tube which contains the sample. The sample undergoes partial combustion as well as

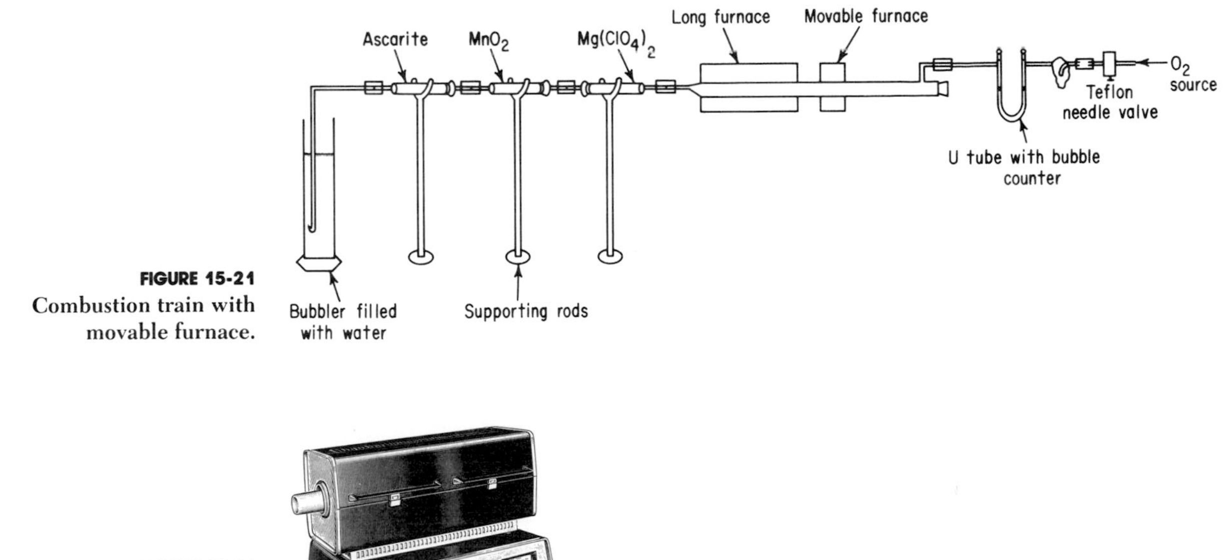

FIGURE 15-21 Combustion train with movable furnace.

FIGURE 15-22 Movable combustion furnace *(Fisher Scientific Company).*

thermal decomposition at this point, and the products are carried over a copper oxide packing that is maintained at a temperature of 700 to 900°C; this catalyzes the oxidation of the sample to carbon dioxide and water. Additional packing is often included in the tube to remove compounds that interfere with the determination of the carbon dioxide and water in the exit stream. Lead chromate and silver serve to remove halogen and sulfur compounds, while lead dioxide can be employed to retain the oxides of nitrogen.

The exit gases from the combustion tube are first passed through a massing tube packed with a desiccant that removes the water from the stream. The increase in mass of this tube gives a measure of the hydrogen content of the sample. The carbon dioxide in the gas stream is removed in the second massing tube packed with Ascarite® (sodium hydroxide held on asbestos). Because the absorption of carbon dioxide is accompanied by the formation of water, additional desiccant is contained in this tube.

Finally the gases are passed through a guard tube that protects the two massing tubes from contamination by the atmosphere.

Table 15-5 lists some of the applications of the combustion-tube method to other elements. A substance containing a halogen will yield the free element upon oxidation; this is frequently reduced to the corresponding

TABLE 15-5
Combustion-Tube Methods for the Elemental Analysis of Organic Substances

Element	*Name of method*	*Method of oxidation*	*Method of completion of analysis*
Halogens	Pregl	Sample combusted in a stream of O_2 gas over a red-hot Pt catalyst; halogens converted primarily to HX and X_2	Gas stream passed through a carbonate solution containing SO_3^{2-} (to reduce halogens and oxyhalogens to halides); product, the halide ion X^2, determined by usual procedures
	Grote	Sample combusted in a stream of air over a hot silica catalyst; products are HX and X_2	Same as above
S	Pregl	Similar to halogen determination; combustion products are SO_2 and SO_3	Gas stream passed through aqueous H_2O_2 which converts sulfur oxides to H_2SO_4, which can then be determined
	Grote	Similar to halogen determination; products are SO_2 and SO_3	Similar to above
N	Dumas	Sample oxidized by hot CuO to give CO_2, H_2O, and N_2	Gas stream passed through concentrated KOH solution leaving only N_2, which is measured volumetrically
C and H	Pregl	Similar to halogen analysis; products are CO_2 and H_2O	H_2O adsorbed on a desiccant and CO_2 on Ascarite®; determined gravimetrically
O	Unterzaucher	Sample pyrolized over C; O_2 converted to CO; H_2 used as carrier gas	Gas stream passed over I_2O_5 ($5CO + I_2O_5 \rightarrow 5CO_2 + I_2$); liberated I_2 titrated

halide prior to the analytical step. Sulfur finally yields sulfuric acid, which can be estimated by precipitation with barium ion or by alkalimetric titration.

Combustion with Oxygen in Sealed Containers

A relatively straightforward method for the decomposition of many organic substances involves oxidation with gaseous oxygen is a sealed container. The reaction products are absorbed in a suitable solvent before the reaction vessel is opened. Analysis of the solution by ordinary methods follows.

A remarkably simple apparatus for carrying out such oxidations has been suggested by Schöninger (see Fig. 15-23). It consists of a heavy-walled flask of 300- to 1000-mL capacity fitted with a ground-glass stopper. Attached to the stopper is a platinum-gauze basket which holds from 2 to 200 mg of sample. If the substance to be analyzed is a solid, it is wrapped

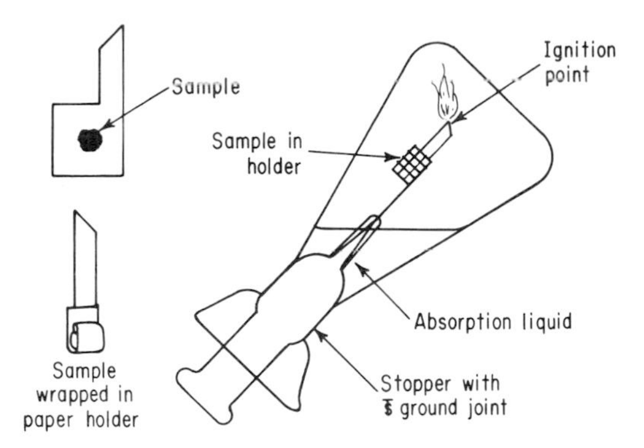

FIGURE 15-23
Apparatus for carrying out oxidation with gaseous oxygen in a sealed container, after Schöninger.

in a piece of low-ash filter paper cut in the shape shown in Fig. 15-23. Liquid samples can be massed in gelatin capsules which are then wrapped in a similar fashion. A tail is left on the paper and serves as an ignition point.

A small volume of an absorbing solution is placed in the flask, and the air in the container is then displaced by allowing tank oxygen to flow into it for a short period. The tail of the paper is ignited and the stopper is quickly fitted into the flask; the container is then inverted as shown in Fig. 15-23; this will prevent the escape of the volatile oxidation products. Ordinarily the reaction proceeds rapidly, being catalyzed by the platinum gauze surrounding the sample. During the combustion, the flask is shielded to avoid damage in case of explosion.

After cooling, the flask is shaken thoroughly and disassembled; then the inner surfaces are rinsed down. The analysis is then performed on the resulting solution. This procedure has been applied to the determination of halogens, sulfur, phosphorus, and various metals in organic compounds.

Peroxide Fusion

Sodium peroxide is a strong oxidizing reagent which, in the fused state, reacts rapidly and often violently with organic matter, converting carbon to the carbonate, sulfur to sulfate, phosphorus to phosphate, chlorine to chloride, and iodine and bromine to iodate and bromate. Under suitable conditions the oxidation is complete, and analysis for the various elements may be performed upon an aqueous solution of the fused mass.

Once started, the reaction between organic matter and sodium peroxide is so vigorous that a peroxide fusion must be carried out in a sealed, heavy-walled, steel bomb. Sufficient heat is evolved in the oxidation to keep the salt in the liquid state until the reaction is completed; ordinarily the oxidation is initiated by passage of current through a wire immersed in the flux or by momentary heating of the bomb with a flame. Bombs for peroxide fusions are available commercially.

One of the main disadvantages of the peroxide-bomb method is the rather large ratio of flux to sample needed for a clean and complete oxidation. Ordinarily an approximate 200-fold excess is used. The excess peroxide is subsequently decomposed to sodium hydroxide by heating in water; after neutralization, the solution necessarily has a high salt content. This may limit the accuracy of the method for completion of the analysis.

The maximum size for a sample that is to be fused is perhaps 100 mg. The method is more suited to semimicro quantities of about 5 mg.

Wet-Ashing Procedures

Solutions of a variety of strong oxidizing agents will decompose organic samples. The main problem associated with the use of these reagents is the prevention of volatility losses of the elements of interest.

One wet-ashing procedure is the Kjeldahl method for the determination of nitrogen in organic compounds. Here concentrated sulfuric acid is the oxidizing agent. This reagent is also frequently employed for decomposition of organic materials where metallic constituents are to be determined. Commonly, nitric acid is added to the solution periodically to hasten the rate at which oxidation occurs. A number of elements are volatilized at least partially by this procedure, particularly if the sample contains chlorine; these include arsenic, boron, germanium, mercury, antimony, selenium, tin, and the halogens.

An even more effective reagent than sulfuric–nitric acid mixtures is perchloric acid mixed with nitric acid. A good deal of care must be exercised in using this reagent, however, because of the tendency of hot, anhydrous perchloric acid to react explosively with organic material. Explosions can be avoided by starting with a solution in which the perchloric acid is well diluted with nitric acid and not allowing the mixture to become concentrated in perchloric acid until the oxidation is nearly complete. Properly carried out, oxidations with this mixture are rapid and losses of metallic ions negligible.

CAUTION *It cannot be too strongly emphasized that proper precautions must be taken in the use of perchloric acid to prevent violent explosions.*

Fuming nitric acid is another potent oxidizing reagent that is employed in the analysis of organic compounds. Its most important application is the analysis of the halogens and sulfur by the Carius method. The oxidation is carried out by heating the sample for several hours at 250 to 300°C in a heavy-walled sealed glass tube. Where halogens are to be determined, silver nitrate is added before the oxidation begins in order to retain them as the silver halides. Sulfur is converted to sulfate by the oxidation. A critical step in this procedure is that of forming a glass seal strong enough to withstand the rather high pressures that develop during the oxidation. Occasional explosions are almost inevitable, and a special tube furnace is ordinarily employed to minimize the effects of these.

Dry-Ashing Procedure

The simplest method for decomposing an organic sample is to heat it with a flame in an open dish or crucible until all the carbonaceous material has been oxidized by the air. A red heat is often required to complete the oxidation. Analysis of the nonvolatile components is then made after solution of the residual solid. Unfortunately a great deal of uncertainty always exists with respect to the recovery of supposedly nonvolatile elements when a sample is treated in this manner. Some losses probably arise from the mechanical entrainment of finely divided particulate matter in the hot convection currents around the crucible. In addition, volatile metallic compounds may be formed during the ignition. For example, copper, iron, and vanadium are appreciably volatilized when samples containing porphyrin compounds are heated.

In summary, the dry-ashing procedure is the simplest of all methods for decomposing organic compounds. It is often unreliable, however, and should not be employed unless tests have been performed that demonstrate its applicability to a given type of sample.

DETERMINING MASS OF SAMPLES

Store and dry samples in massing (weighing) bottles which have ground-glass contacting surfaces between the cover and the bottle (Fig. 15-24).

Procedure

1. Use a clean massing bottle fitted with a ground-glass cover.

2. Handle the bottle with suitable tongs or with a strip of lint-free paper (as illustrated in Fig. 15-25).

3. Do not touch the massing bottle with your fingers. Data will be significantly affected by the moisture and grease on your fingers.

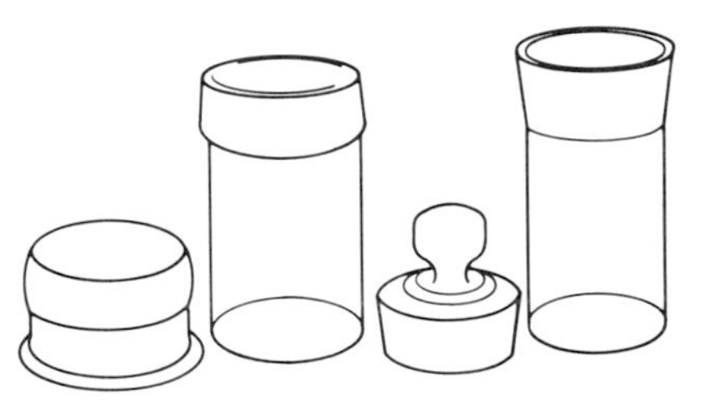

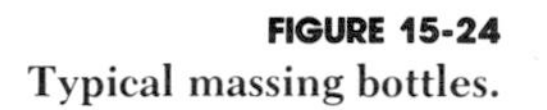

FIGURE 15-24
Typical massing bottles.

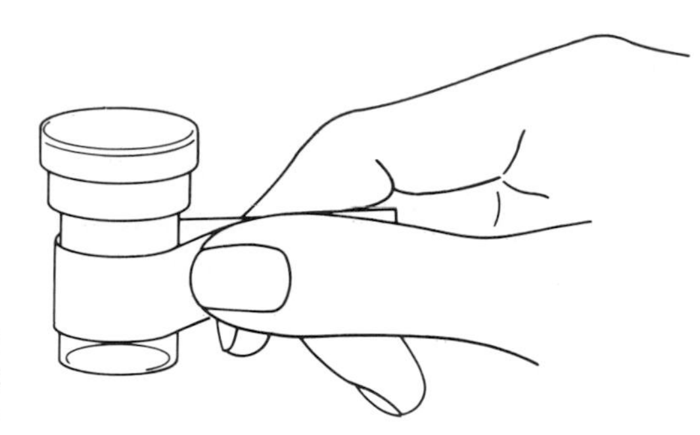

FIGURE 15-25
Method of handling massing bottle.

Direct Mass Determination

1. Mass a clean receiving vessel or dish or a piece of glazed paper.

2. Transfer the desired quantity of substance into the receiving container with a clean spatula or by gently tapping the tilted massing bottle.

3. Mass the substance and the glazed paper or massing dish.

4. Calculate the mass of the sample by subtracting the mass of the paper or dish from the mass of the material and dish found in step 3. The difference in these two masses is the mass of the substance.

Mass Determination by Difference

1. Mass the special tared bottle which contains the sample.

2. Quantitatively remove the desired amount of the substance to the receiving container by gently pouring the material out of the massing bottle.

3. Remass the bottle.

4. Subtract the mass found in step 3 from the mass found in step 1. The difference in these two masses is the amount of material transferred.

Use *direct mass determination* when an exact quantity of substance is needed.

Use *mass determination by difference* (Fig. 15-26) when several samples of the same material are to be massed. This method is preferable when determining the mass of hygroscopic substances.

NOTE Gently tapping the massing bottle or massing container enables you to better control the removal of the solid material without loss.

Gravimetric Calculations

The mass percent of a constituent is equal to the mass of the constituent divided by the sample mass and multiplied by 100.

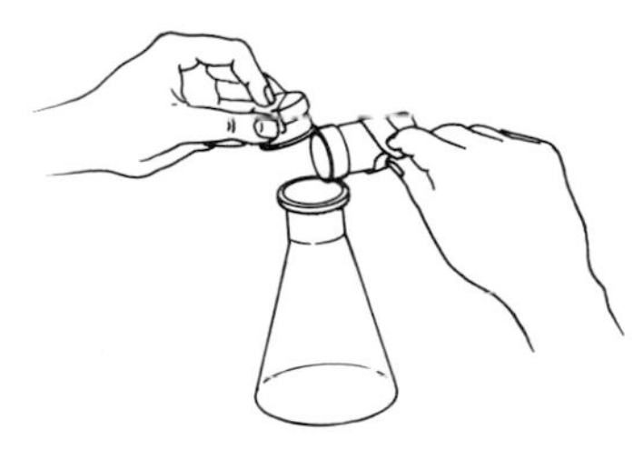

FIGURE 15-26
A convenient method to transfer a solid for massing by difference.

For example, if a 1.000-g sample of limestone is found to contain 0.3752 g of calcium, it has

$$\frac{0.3752}{1.000} \times 100 \text{ or } 37.52\% \text{ calcium}$$

$$\frac{\text{Mass of substance}}{\text{Mass of sample}} \times 100 = \text{percent by mass}$$

In most cases, however, the desired constituent is not massed directly but is precipitated and massed as some other compound. It is then necessary to convert the mass obtained to the mass in the desired form by using *a gravimetric factor.*

For example, a molecule of silver chloride is made up of one atom of silver and one atom of chlorine. The ratio of silver to chloride is as Ag/AgCl. Since the atomic mass of silver is 107.8 and that of chlorine is 35.5,

$$\frac{107.8 \text{ (atomic mass Ag)}}{107.8 \text{ (atomic mass Ag)} + 35.5 \text{ (atomic mass Cl)}} = 0.7526$$

This is called the *gravimetric factor,* and the percent of silver when weighed as AgCl is

$$\frac{\text{Mass of AgCl} \times 0.7526}{\text{Mass of sample}} \times 100$$

or $$\frac{\text{Mass of precipitate (AgCl)}}{\text{Mass of sample}} \times \text{gravimetric factor} \times 100 = \%\text{ Ag}$$

The factor for sodium when weighed as sodium sulfate is $2Na/Na_2SO_4$ since there are 2 sodium atoms in sodium sulfate.

A general equation for gravimetric calculation is

$$\frac{\text{Mass of precipitate} \times \text{gravimetric factor} \times 100}{\text{Mass of sample}} = \text{percent of constituent}$$

Gravimetric factors are given in Table 15-6.

TABLE 15-6
Table of Gravimetric Factors

Sought	*Massed*	*Factor*
Na	Na_2SO_4	0.3237
K	K_2SO_4	0.4487
Ba	$BaSO_4$	0.5885
Ca	$CaSO_4$	0.2944
Cu	CuO	0.7988
Fe	Fe_2O_3	0.6994
Pt	Pt	1.000
Au	Au	1.000
Ag	Ag	1.000
H	H_2O	0.1119
C	CO_2	0.2729
S	$BaSO_4$	0.1374

MICRO-DETERMINATION OF CARBON AND HYDROGEN

Since organic compounds are characterized by the fact that they contain carbon and usually hydrogen, it can be seen that the ability to measure these elements accurately is of extreme importance. In spite of its importance there is no universal test for carbon and hydrogen. There are almost as many modifications as there are people running the test. Since the time of Pregl, however, the basic microcombustion technique has been the principal means for determining carbon and hydrogen. The technique is threefold:

1. Combustion of the organic material
2. Removal of interfering elements
3. Measurement of the carbon dioxide and water formed

Any modification or variation is involved with one of these three categories.

Combustion of the Organic Material

The basic reactions in the combustion of organic material for the determination of carbon and hydrogen are

$$\text{Organic C} \xrightarrow[O_2]{\Delta} CO_2$$

$$\text{Organic H} \xrightarrow[O_2]{\Delta} H_2O$$

This combustion is usually carried out in a special tube. The tube packing, according to Pregl, contains silver, copper oxide, lead chromate, and lead dioxide. The combustion is carried out in oxygen; the copper oxide and lead chromate aid in the oxidation. Many investigators have described packings consisting of a variety of catalysts including cobaltic oxide, silver permanganate, silver vanadate, zirconium oxide, magnesium oxide, and silver tungstate.

Oxygen flow rates to carry out the oxidation have varied from 8 to >100 mL/min. At 50 mL/min a sample can be burned rapidly enough to effect complete combustion in 5 min and total sweep after 10 min, permitting the analyst enough time to prepare the next sample.

Aside from catalysts which are part of the combustion-tube packing, it is sometimes advantageous to cover the sample in the boat with catalyst to aid in the oxidation. This is especially true for organic compounds containing metals. Tungstic oxide appears to be a universal "clean" catalyst for most metal organics. A mixture of silver oxide and manganese dioxide has proved to be effective with highly chlorinated and sulfurized materials.

Removal of Interfering Substances

The interfering substances encountered in the determination of carbon and hydrogen are sulfur, the halogens, and nitrogen. Probably more has been published on removing these substances than on any other single area of carbon-hydrogen determination. Silver gauze or wire, maintained at 700°C, has been found effective for the removal of halogens (except fluoride) and sulfur products. The silver should be placed at the end of the combustion tube.

Removal of Sulfur

Sulfur in the carbon-hydrogen determination is converted to SO_2-SO_3 and is absorbed by silver, forming Ag_2SO_4. Manganese dioxide will absorb all the sulfur dioxide from the products, thereby preventing interference with the carbon analysis. This is not quantitative. Again if this is the only mechanism employed for the removal of sulfur dioxide, no condensate is permissible or sulfuric acid will be formed.

Removal of Halogens (Except Fluorine)

As previously mentioned, halogen can be removed by silver in the combustion tube. This is very effective. If halogen escapes from the tube, manganese dioxide placed in the train will absorb any chloride, and that chloride will not be absorbed on the magnesium perchlorate tube. However, it must be noted that no water may be permitted to condense prior to absorption.

Removal of Nitrogen Oxides

Nitrogen oxides formed during combustion of nitrogen compounds interfere with the carbon determination. Nitrogen oxides classically have been removed by absorption on lead dioxide maintained at 180°C. Since lead dioxide sets up an equilibrium with water and carbon dioxide (products of combustion), a weighed sample must be run prior to analysis of samples. When using this technique, one must not follow a high carbon-hydrogen analysis with a low one or vice versa. The analyst must also be absolutely sure of timing—one sample must follow the next in exactly the same time. Grades of lead dioxide are variable, and the analyst must exercise care in eliminating fines when packing the combustion tube.

Manganese dioxide is very effective when used for the absorption of nitrogen oxides. It is used externally between the magnesium perchlorate tube and the carbon dioxide absorption tube, normally at room temperature. It has a tendency to pick up carbon dioxide and release it; therefore, it must be dried prior to use. This material requires that no water condensation may take place prior to absorption: the formation of nitric acid will yield high hydrogen values. High flow rates and external heating problems alleviate these problems. A simple hair-dryer can be used to blow hot air on the connection between the combustion tube and absorption tube. This has proved to be very effective in preventing water condensation.

Measurement of Carbon Dioxide and Water

Water and carbon dioxide formed by the combustion of organic compounds have classically been collected on magnesium perchlorate and Ascarite® (NaOH on asbestos), respectively. This method is still extensively used today. One precaution when operating with a high flow rate is to have enough magnesium perchlorate at the exit end of the tube to remove all the water formed. Carbon dioxide is "absorbed" on Ascarite® as follows:

$$CO_2 + 2NaOH \rightarrow Na_2CO_3 + H_2O \tag{1}$$

Microprocedure for the Determination of Carbon and Hydrogen in Organic Compounds

Principle

In this procedure, a known amount of organic matter is combusted in an atmosphere of oxygen. The water and carbon dioxide formed are absorbed by magnesium perchlorate and Ascarite®, respectively. The difference in the mass of the tubes before and after the combustion is measured, and the amounts of carbon and hydrogen are calculated (see Fig. 15-27).

Scope

The method can be applied to all organic compounds containing carbon and hydrogen. Provision is made for the removal of interfering elements such as halogen, sulfur, and nitrogen. However, compounds containing large amounts of fluorine should not generally be analyzed by this method without prior treatment.

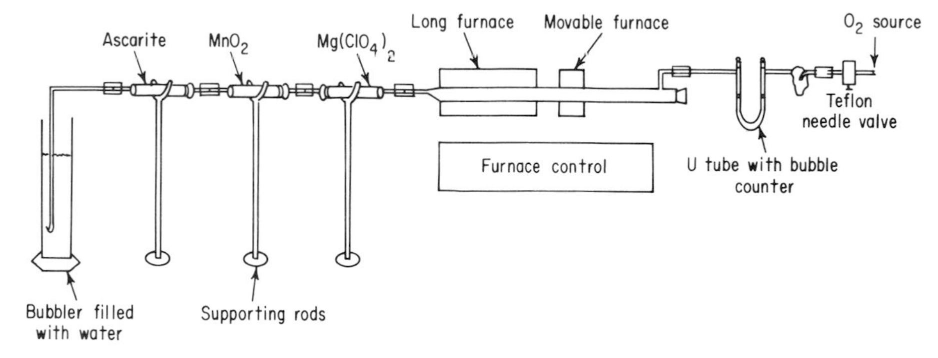

FIGURE 15-27 Schematic of equipment for analysis of carbon and hydrogen by combustion.

MICRO-DETERMINATION OF NITROGEN

The microdetermination of nitrogen in organic compounds has led to more investigation and publications than even the carbon-hydrogen test. Like other organic analyses it involves conversion of the nitrogen in the sample to a measurable form, and a means for measuring this nitrogen form quantitatively.

There are two basic methods for the determination of nitrogen which are used in one form or another almost exclusively. They are:

1. Kjeldahl
2. Dumas

Both methods have limitations, and any well-equipped laboratory will be capable of running both tests.

Kjeldahl Determination of Nitrogen

The method is based on the fact that digestion with sulfuric acid and various catalysts destroys the organic material and the nitrogen is converted to ammonium acid sulfate. When the reaction mixture is made alkaline, ammonia is liberated and removed by steam distillation, collected, and titrated. The equations are as follows:

$$\text{Organic N compound} \xrightarrow[\Delta \atop \text{catalysts}]{H_2SO_4} CO_2 + H_2O + NH_4HSO_4 \tag{2}$$

$$NH_4HSO_4 + 2NaOH \rightarrow NH_3 + Na_2SO_4 + 2H_2O \tag{3}$$

$$NH_3 + HCl \rightarrow NH_4Cl \tag{4}$$

A tremendous amount of work has been done on the selection of proper catalysts for the digestion. Various workers have used selenium, copper, mercury, and salts of each. Potassium sulfate added with the catalysts raises and controls the temperature of the reaction. Some workers have claimed that with the use of selenium too high a temperature is achieved, thereby yielding low nitrogen results. However, it has been shown that with $HClO_4$ in the digestion mixture, selenium can be used effectively. It must be noted that any refractory nitrogen compounds, pyridines, etc., must be digested at ~370°C but at no higher than 400°C.

In the Kjeldahl determination of nitrogen, compounds containing N—O or N—N linkages must be pretreated or subjected to reducing conditions prior to analysis. Numerous agents have been utilized to effect this reduction. The N—O linkages are much easier to reduce than the N—N, and zinc or iron in acid is suitable for this purpose. There is no such general technique for the N—N linkages. Samples containing very high concentrations of halide can in some instances cause trouble because of the formation of oxyacids known to oxidize ammonia to N_2.

For nitrate-containing compounds, salicylic acid is added to form nitrosalicylic acid which is reduced with thiosulfate. The ammonium hydrogen sulfate thus formed during digestion is reacted with NaOH to form free ammonia, which is distilled with steam. The reaction is as shown in Eq. (3). The ammonia is collected by passing it into boric acid solution and then is titrated with HCl. Some investigators have collected the ammonia in HCl and back-titrated with standard NaOH. The disadvantage of this technique is the need for two standard solutions, and the critical loss of two components on the condenser. The pH at which the ammonium chloride complex is formed is 5.2. A mixture of methyl red and methylene blue has been shown to be very effective. Some use methyl purple, which is a mixture of methyl red and a blue dye.

Automated Nitrogen Analysis (Kjeldahl)

The Technicon AutoAnalyzer® is presently being used to measure total nitrogen after Kjeldahl digestion. The sample is digested according to classical means using a digestion mixture containing selenium dioxide, sulfuric acid, and perchloric acid. The digest is then made up to a given volume and placed in the auto-analyzer. The ammonium acid sulfate is then automatically sampled and treated with caustic according to

$$NH_4HSO_4 + 2NaOH \rightarrow NH_3 + Na_2SO_4 + 2H_2O \qquad (5)$$

and the ammonia is liberated as a complex with the phenol–hypochlorite reagent according to the following proposed reaction:

$$NH_4^+ + OCl^- \longrightarrow NH_2Cl \xrightarrow{\text{phenol}} OC_6H_4NCl \xrightarrow{\text{phenol}}$$

$$OC_6H_4NC_6H_4OH \xrightarrow{\text{OH}} \underset{\text{(blue color)}}{OC_6H_4NC_6H_4O^-} \qquad (6)$$

The color formed is then measured with the spectrophotometer and the reading is recorded.

Dumas Determination of Nitrogen

The Dumas method for nitrogen was introduced in 1831. It was not practical, however, until after Pregl adapted it to the micro scale. The Dumas method is based on combustion of the nitrogen-containing organic material in the presence of a catalyst at 780°C. The oxidation products are then passed over a reducing medium and the NO_x . . . (various nitrogen oxides) are converted to N_2. The (various nitrogen oxides) following reactions show the process:

$$\text{Organic N compound} \xrightarrow[780^\circ C]{CuO} CO_2 + H_2O + NO \ldots + N_2 \qquad (7)$$

$$NO_x \ldots \xrightarrow[780^\circ C]{Cu^*} N_2 \qquad (8)$$

The nitrogen is then measured by bubbling the products into a solution of potassium hydroxide. The nitrogen displaces a corresponding volume of potassium hydroxide; the volume is measured, corrected for barometric pressure and temperature, and converted to mass. The percent nitrogen is then calculated. Samples containing a high concentration of methyl groups (CH_3), alkoxyl (OCH_3, etc.), and *N*-methyl (*N*-CH_3) compounds are known to release methane under normal conditions during the Dumas analysis. Cobaltic oxide has been shown to prevent the formation of methane.

Many investigators have attempted to automate the Dumas analysis. The Coleman analyzer programs the combustion and measures the nitrogen by means of a special nitrometer with a readout. Other techniques have applied thermal conductivity after a gas-chromatographic separation of the nitrogen. The automation of the Dumas analysis has been reasonably successful. The analyst must remember, however, that all samples are different and certain variables must be considered prior to analysis.

DETERMINATION OF HALOGENS

Many of the concepts covered later in discussing the determination of sulfur in organic compounds also apply to the determination of halogen. For simplicity, the term "halogen" will refer here only to chlorine, bromine, and iodine. Like most elemental analyses, the determination involves a

* Metallic copper as reducing medium.

three-step procedure:

1. Decomposition of the organic material and conversion of the halide to a measurable form
2. Removal of interfering ions
3. Measurement of the halogen-containing products

Decomposition of the Organic Material

The basic means for decomposing organic compounds in an effort to measure total halogen is oxidation.

Methods to carry out this procedure are:

1. Tube combustion (Carius method)
2. Bomb combustion, with peroxide (Parr method)
3. Flask combustion (Schöninger)
4. Pregl combustion methods (decomposition of the sample in an oxygen atmosphere at 680–700°C in the presence of a platinum catalyst)

The reaction for this oxidation is as follows:

$$\text{Organic } X \text{ compound} \xrightarrow[\Delta]{O_2} X_2 + CO_2 + H_2O \qquad (9)$$

X stands for Br, Cl, or I.

$$X_2 \xrightarrow[\text{absorbent}]{\text{suitable}} 2X^-$$

usually absorbed in a sodium carbonate solution (10)

Hydrazine is sometimes used in the absorbent to aid in the reduction. It is essential when X is iodine.

Removal of Interfering Ions

The primary ions interfering with the halogen determination are the other halides present. For example, bromide will interfere with chloride and vice versa.

Cyanides and thiols also interfere; however, they are not present after combustion. Their presence must be noted, however, when determining water-soluble fractions; in this case the organic material is not destroyed.

Ion Exchange

An ion-exchange separation on Dowex 1 separates chlorine from bromine from iodine when the sample is eluted with various concentrations of sodium nitrate. Review Ion-Exchange Chromatography in Chap. 36, "Chromatography."

Other Methods of Determining Halogen

1. Mohr method. Formation and determination of red silver chromate.

2. Volhard method. Back-titrate excess silver nitrate with standard thiosulfate; iron(III) is the indicator. Solubility of silver chloride is a real problem.

3. Colorimetry. One of the most sensitive techniques for the determination of chloride ion involves the reaction which displaces thiocyanate ion from mercury(II) thiocyanate by chloride ion. The color formed when this reaction is carried out in the presence of iron(III) ion to form iron(III) thiocyanate is stable and proportional to the chloride content of the material being measured. The reaction is as follows:

$$2Cl^- + Hg(SCN)_2 + 2FE^{3+} \rightarrow HgCl_2 + 2Fe(SCN)^{2+} \quad (11)$$

The technique can determine chloride down to 5 ppm.

DETERMINATION OF SULFUR

The determination of sulfur, like that of other elements in organic compounds, involves three basic steps:

1. Decomposition of the organic material and conversion of the sulfur to a measurable form
2. Removal of interfering ions
3. Measurement of the sulfur-containing products of combustion

Any modification or variation in a single method or between methods is involved with one of these three steps.

Decomposition of the Organic Material and Conversion of the Sulfur to a Measurable Form

Oxidation

The basic reaction for the determination of sulfur by an oxidative combustion is as follows:

$$\text{Organic S compound} \xrightarrow[O_2]{\Delta} CO_2 + H_2O + SO_3 + SO_2$$

Increase of temperature yields a mixture that is 95% SO_2. The resulting oxides ($SO_2 + SO_3$) are converted to SO_4^{2-} as H_2SO_4 as follows:

$$SO_2 + SO_3 + H_2O_2 + H_2O \rightarrow 2H_2SO_4 \quad (13)$$

or

$$SO_2 + SO_3 + Br_2 + 3H_2O \rightarrow 2H_2SO_4 + HBr \quad (14)$$

Tube Combustion (Dietert)

In this high-temperature combustion the sulfur is converted by oxidation to 95% theoretical SO_2. The sample is burned with the aid of a catalyst (V_2O_5) and pure oxygen in an alundum tube maintained at 1300°C.

Parr Bomb

This method is based on the fact that Na_2O_2 (sodium peroxide) is a powerful oxidizing agent. If the organic matter is placed in a closed vessel and heated in the presence of peroxide, all the sulfur will be converted to the corresponding sodium salt.

The danger to the operator is the pressure which is built up within the bomb during combustion. Remember 1 gram-molecular weight of a substance occupies 22.4 L in the vapor state at standard temperature and pressure.

Sugar or benzoic acid is added to increase the burning capacity of the bomb and decrease time for combustion. Excess peroxide is used up in this manner also.

The melt from the cooled products of combustion is put into solution and inspected, and the analysis is performed after all H_2O_2 has been removed by boiling.

Schöninger Combustion

In the Schöninger combustion technique (see Fig. 15-23) the sulfur is converted by oxidation to SO_2 and SO_3 and subsequently oxidized to $-SO_4^{2-}$ with H_2O_2. In this technique, a sample is weighed directly onto a piece of filter paper or a capsule and placed in a platinum basket. A flask is completely filled with oxygen, and the ignited paper is placed in the flask. Complete combustion takes place within 20 s, and the products are absorbed. This method is useful for nonvolatile compounds only.

Gravimetric Method

Barium chloride is added to a sample containing sulfate, and insoluble barium sulfate is formed. In this precipitation the acid concentration is critical because barium sulfate is soluble in excess hydrochloric acid.

Trivalent metals such as iron (Fe^{3+}) are known to coprecipitate, with a resultant loss in SO_3 when they are ignited to iron(III) oxide. One means of eliminating this as an interference is to precipitate the iron with ammonia and to dissolve the iron(III) hydroxide with dilute hydrochloric acid after the barium sulfate precipitation.

Nitrate and chlorate should be destroyed by heating with hydrochloric acid prior to the precipitation of barium sulfate.

Ion exchange is proving to be a very helpful tool in removing interfering ions from solutions.

16
LABORATORY GLASSWARE

INTRODUCTION

For a number of years manufacturers have been fabricating glass laboratory equipment with ground-glass joints having standard dimensions and designed to fit each other perfectly. This feature eliminates the use of rubber, plastic, and cork stoppers as connections between different pieces of equipment. Thus, by judicious use and choice of the multitude of items available, equipment for varied procedures can be assembled more quickly and with less effort.

Ground-glass assemblies (1) give perfect seals for vacuum and moderate pressures, (2) completely eliminate stopper contamination, (3) extend temperature limits upward for use of the equipment, and (4) provide neater and more professional assemblies.

The interjoint items are identified by their names and joint-size numbers. Size is designated by two figures. The first indicates the approximate diameter of the larger tube in millimeters. The second designates the length of the ground surface. A $^{19}/_{38}$ joint is about 19 mm in diameter and about 38 mm long. (See Table 16-1.) Reducing and expanding adapters permit the unlimited choice of joint designs and sizes. Interjoint glassware is designated as male or female connections, straight joint, ball joint, or any combination desired. (See Table 16-2.)

TABLE 16-1

Joint No.	*OD of tube, mm*	*Joint No.*	*OD of tube, mm*
5/20	5	34/45	32
7/25	6	40/50	37
10/30	8	45/50	42
12/30	10	50/50	47
14/35	12	55/50	52
19/38	17	60/50	57
24/40	22	71/60	68
29/42	27	103/60	100

TABLE 16-2

Ball or socket No.	*Ball or socket No.*
7/1	28/15
12/1	35/20
12/1.5	35/25
12/2	40/25
12/3	50/30
12/5	65/40
18/7	75/50
18/9	102/75
28/12	

Use the ground-joint glassware exactly as you would use regular glassware with flexible connections such as stoppers and tubing, except that precautions must be observed. (These precautions will be enumerated later in this section.)

Thousands of variations are available to the technician. Refer directly to the scientific glassware catalog to pick the exact item having the desired type and size of connector so that a perfect assembly can be made as desired.

Some of the advantages of ground-joint glassware are as follows.

1. Because no cork, rubber, or plastic stoppers or connections are used, contamination and discoloration of the chemicals are avoided.

2. Corks do not have to be selected, bored, or fitted—a timesaver.

3. Units can be assembled and reassembled again and again to carry out different operations.

4. Assembly of units is quick.

5. Broken parts are quickly replaced with duplicates which are standard and fit perfectly.

6. No impurities from broken cork, swollen rubber, or plastic enter the reaction.

7. Narrow tubes need not be used, as with cork and rubber stoppers, allowing full-width tubing.

GROUND-GLASS EQUIPMENT

Joints and Clamps

See Figs. 16-1 to 16-8.

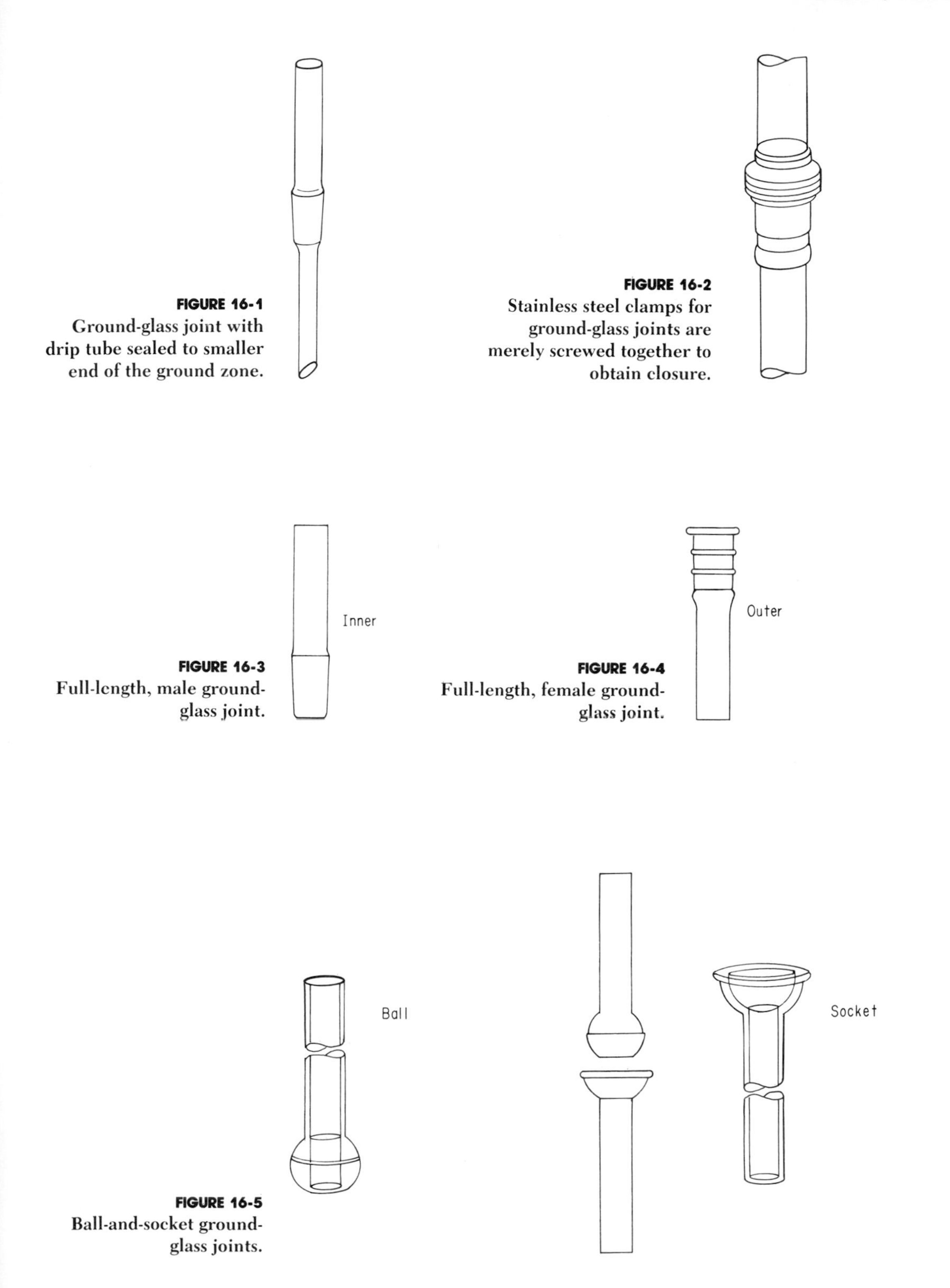

FIGURE 16-1
Ground-glass joint with drip tube sealed to smaller end of the ground zone.

FIGURE 16-2
Stainless steel clamps for ground-glass joints are merely screwed together to obtain closure.

FIGURE 16-3
Full-length, male ground-glass joint.

FIGURE 16-4
Full-length, female ground-glass joint.

FIGURE 16-5
Ball-and-socket ground-glass joints.

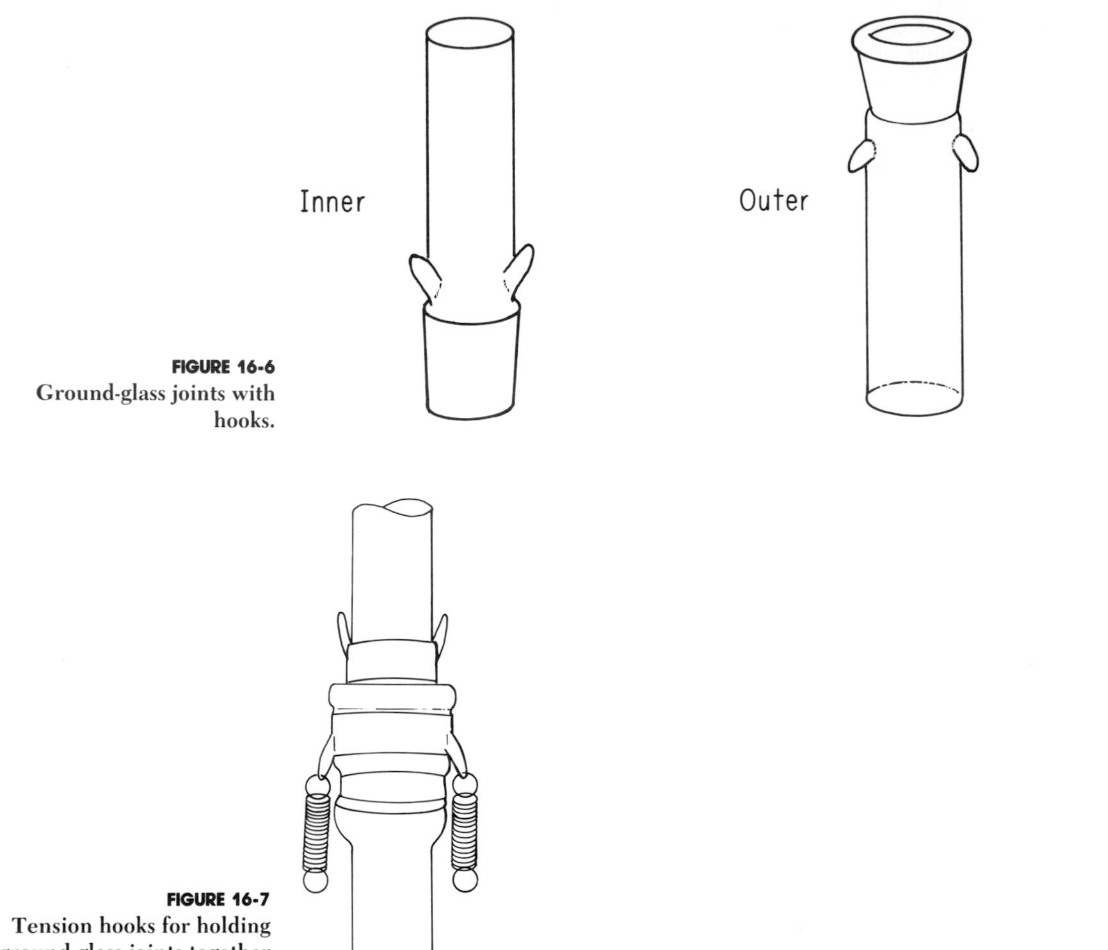

FIGURE 16-6
Ground-glass joints with hooks.

FIGURE 16-7
Tension hooks for holding ground-glass joints together securely (stainless steel tabs, hooks, and springs).

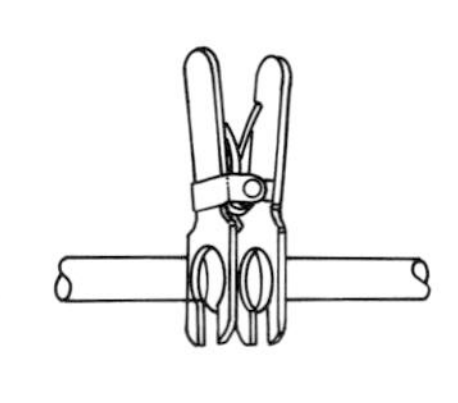

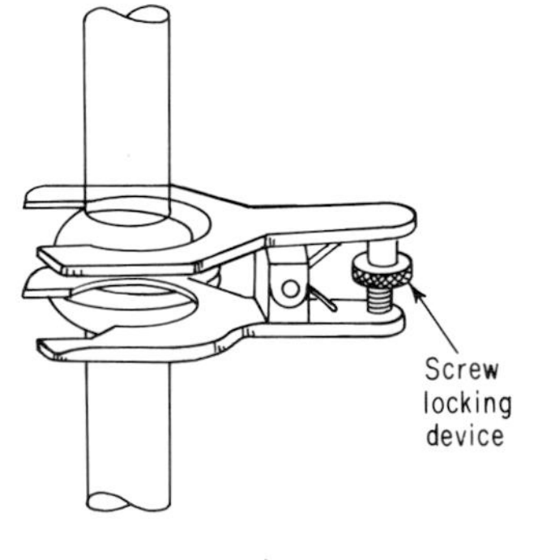

FIGURE 16-8
Clamps for ball-and-socket joints. (*a*) Spring-closed for smaller sizes. (*b*) Screw-locking device for larger sizes. The size number indicates the diameter, in millimeters, of the ball over which the clamp fits. (Size numbers: 7, 12, 12A, 18, 18A, 28, 35, 40, 50, 65, 75, 100.)

Components

See Figs. 16-9 to 16-23.

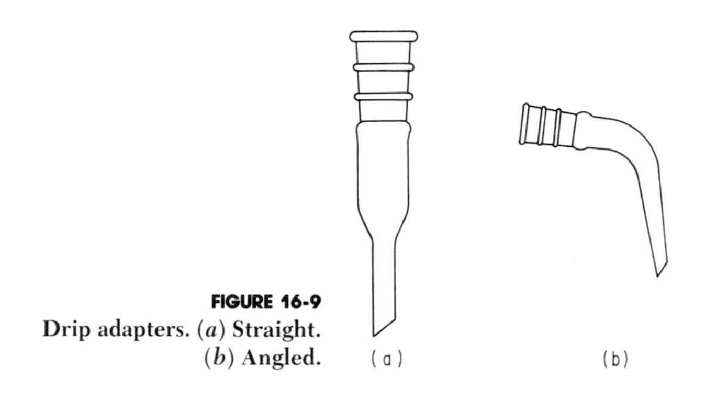

FIGURE 16-9
Drip adapters. (*a*) Straight. (*b*) Angled.

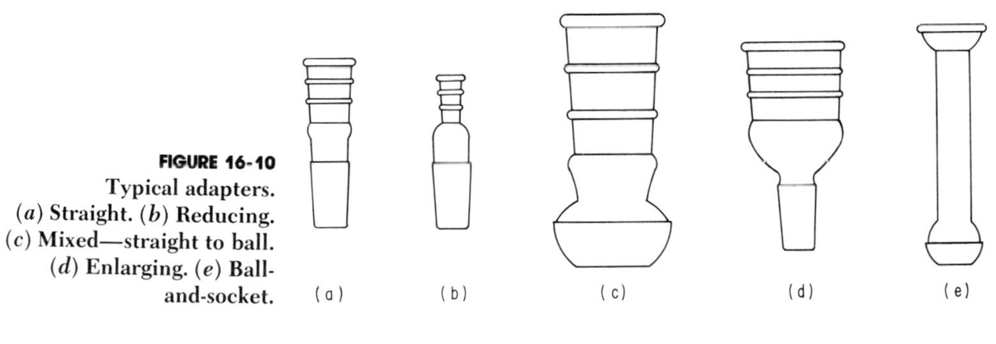

FIGURE 16-10
Typical adapters. (*a*) Straight. (*b*) Reducing. (*c*) Mixed—straight to ball. (*d*) Enlarging. (*e*) Ball-and-socket.

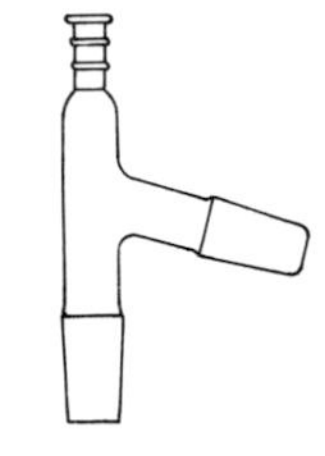

FIGURE 16-11
Adapters with simple distilling heads.

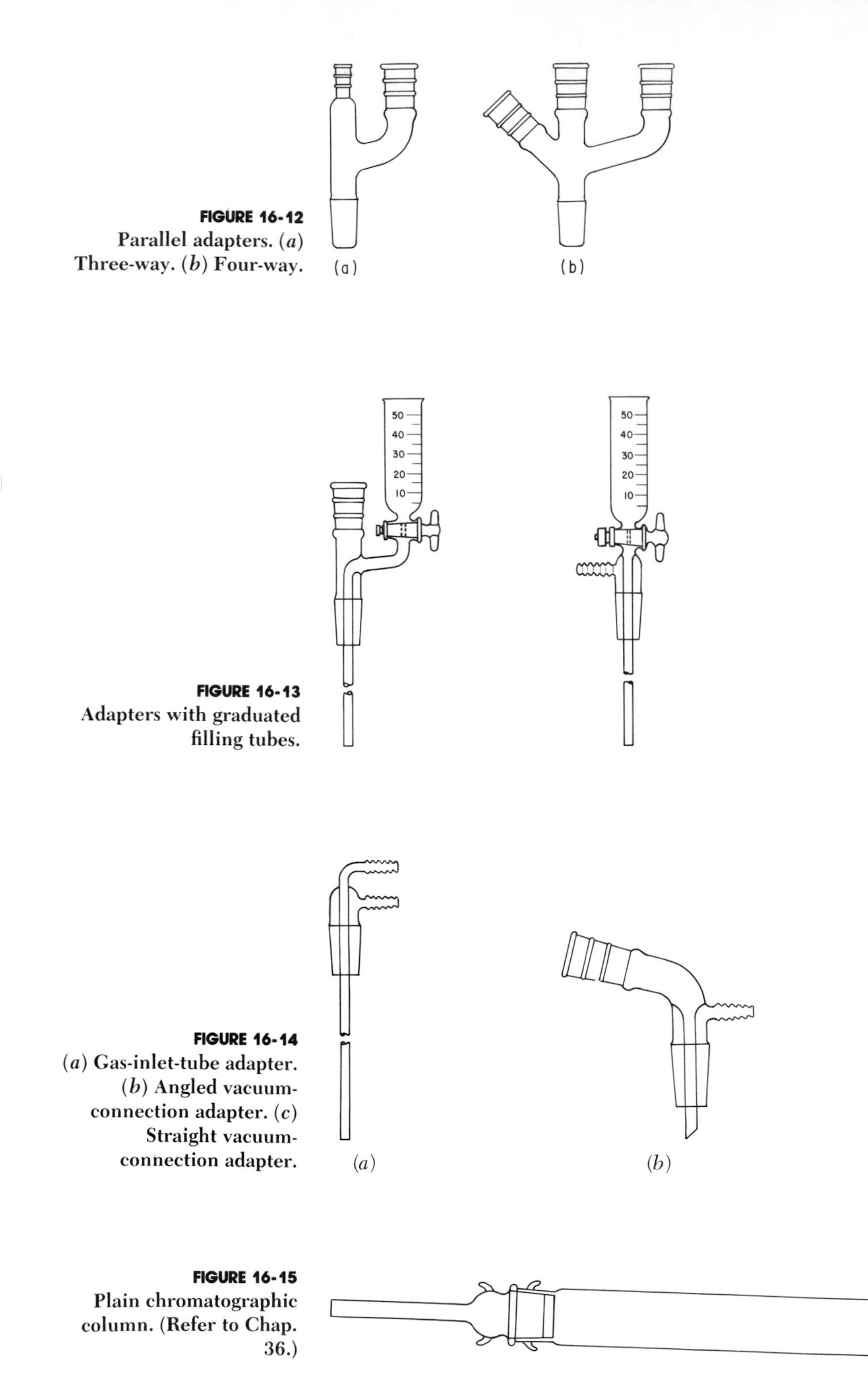

FIGURE 16-12
Parallel adapters. (*a*) Three-way. (*b*) Four-way.

FIGURE 16-13
Adapters with graduated filling tubes.

FIGURE 16-14
(*a*) Gas-inlet-tube adapter. (*b*) Angled vacuum-connection adapter. (*c*) Straight vacuum-connection adapter.

FIGURE 16-15
Plain chromatographic column. (Refer to Chap. 36.)

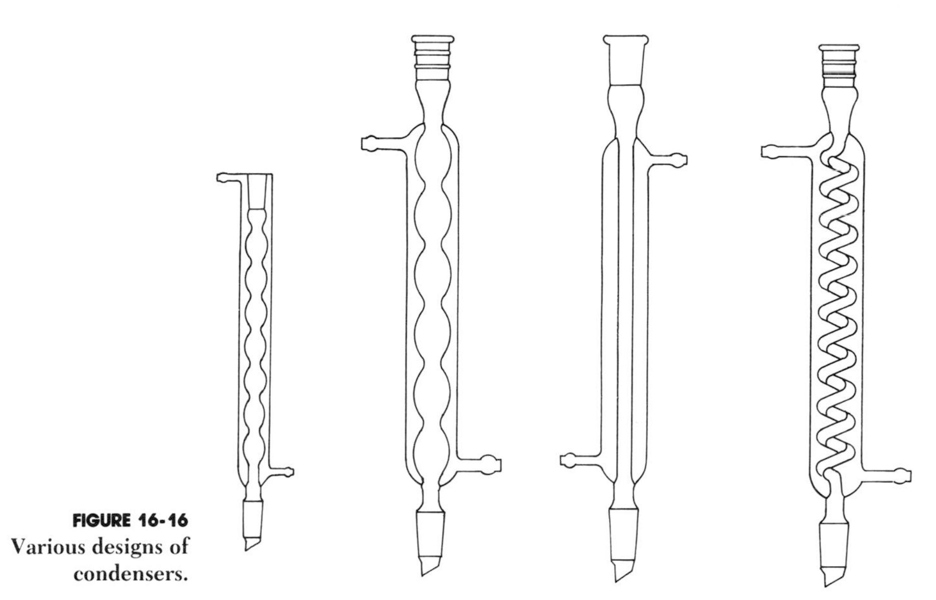

FIGURE 16-16 Various designs of condensers.

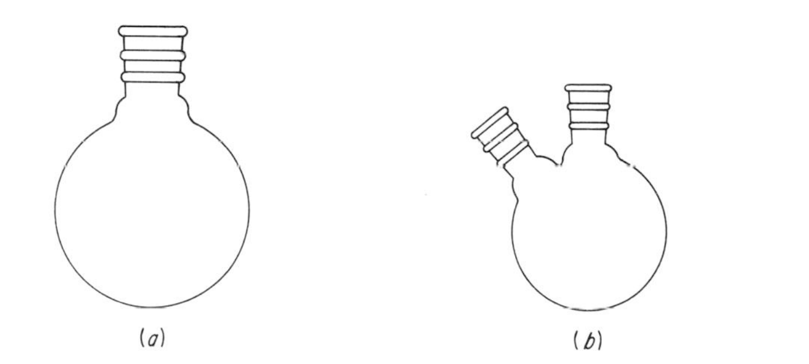

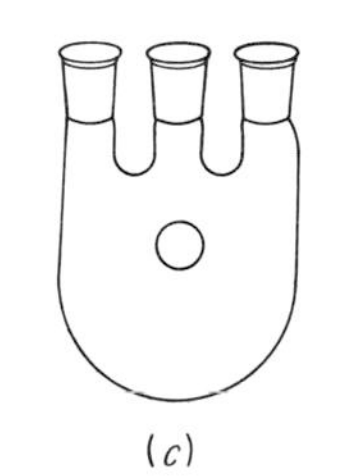

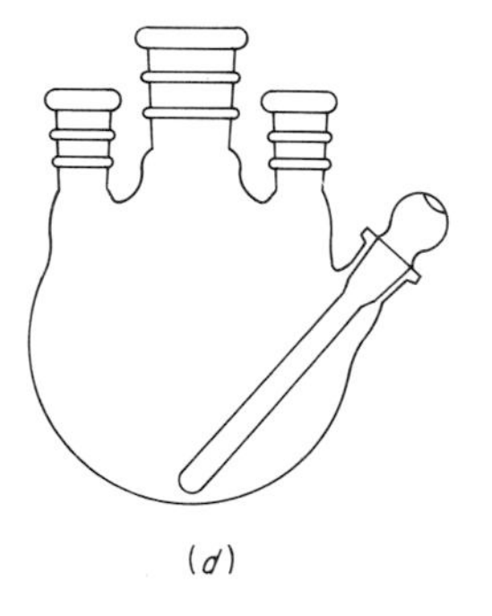

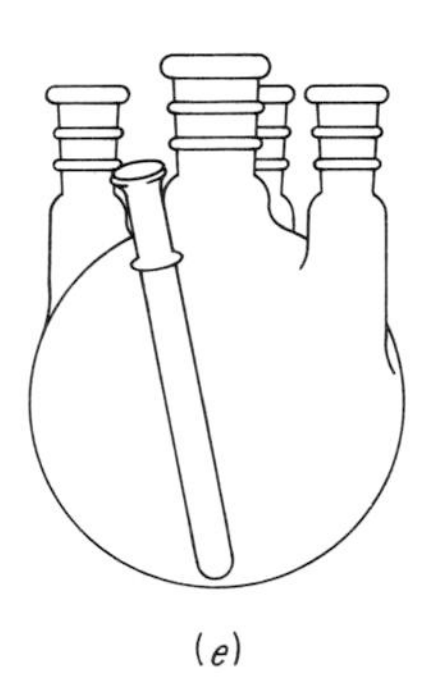

FIGURE 16-17 Round-bottomed reaction flasks fitted with different openings. (*a*) Standard single-neck. (*b*) Angled two-neck. (*c*) Parallel three-neck. (*d*) Three-neck and thermometer well. (*e*) Four-neck and thermometer well.

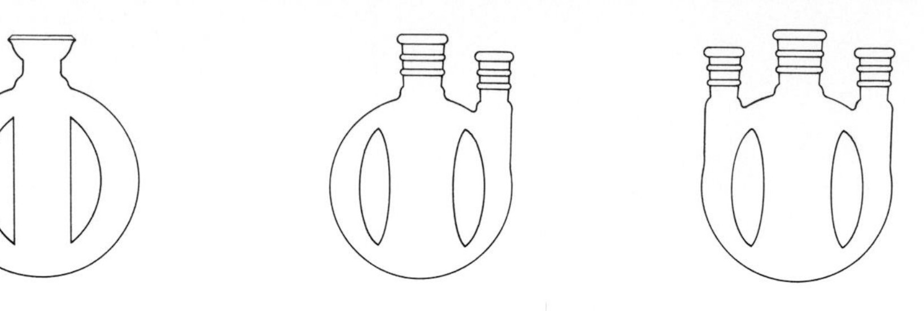

FIGURE 16-18
Reaction flasks with indentations to increase turbulence of mixing solution.

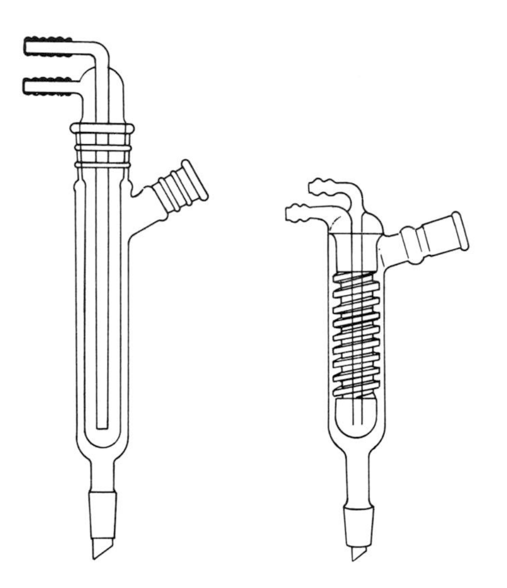

FIGURE 16-19
Reflux condensers come in various designs. These are two of the most common.

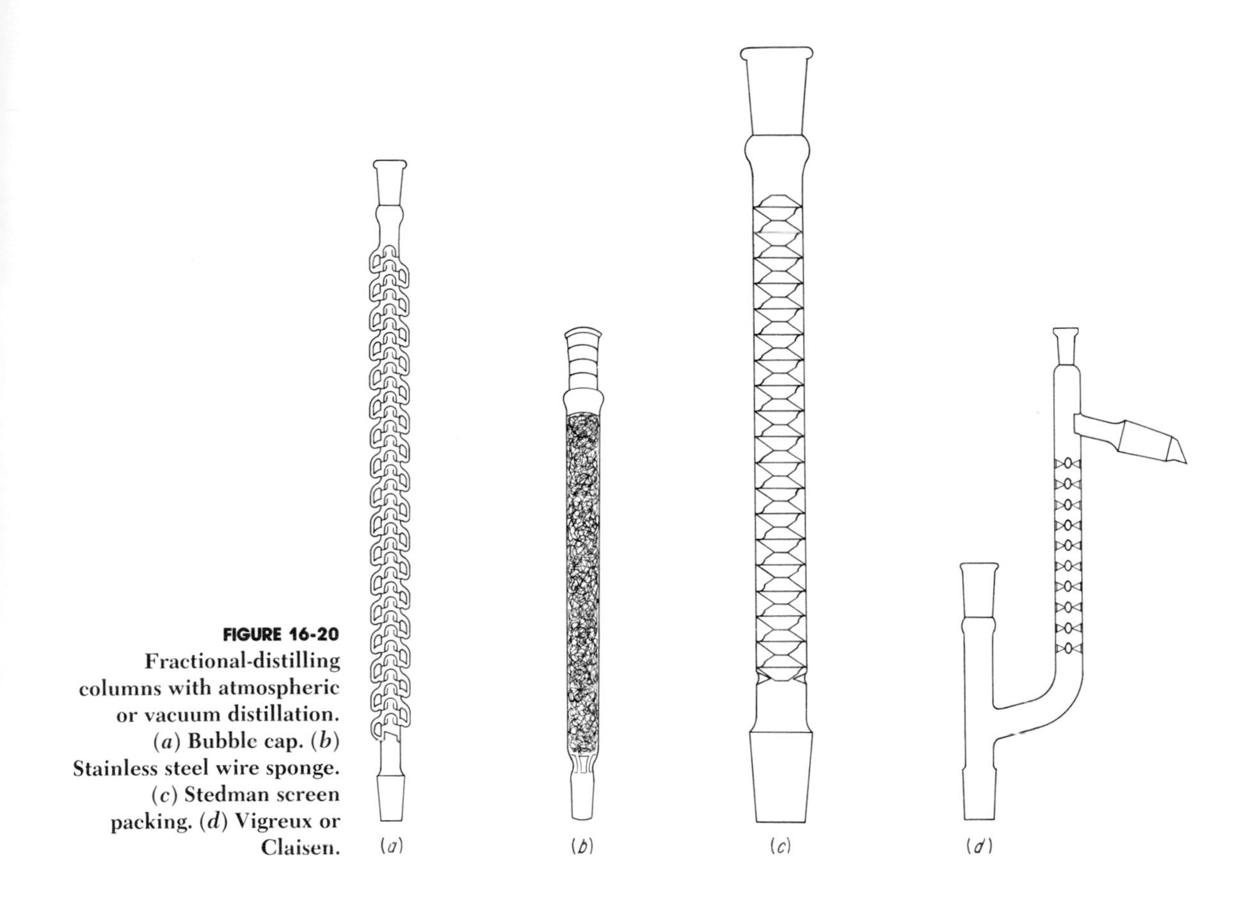

FIGURE 16-20
Fractional-distilling columns with atmospheric or vacuum distillation. (*a*) Bubble cap. (*b*) Stainless steel wire sponge. (*c*) Stedman screen packing. (*d*) Vigreux or Claisen.

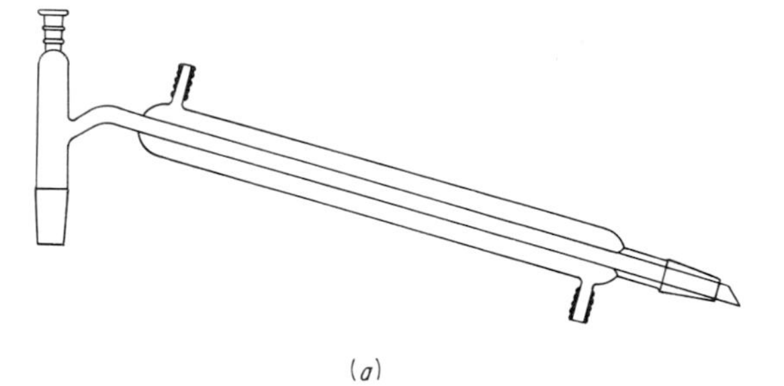

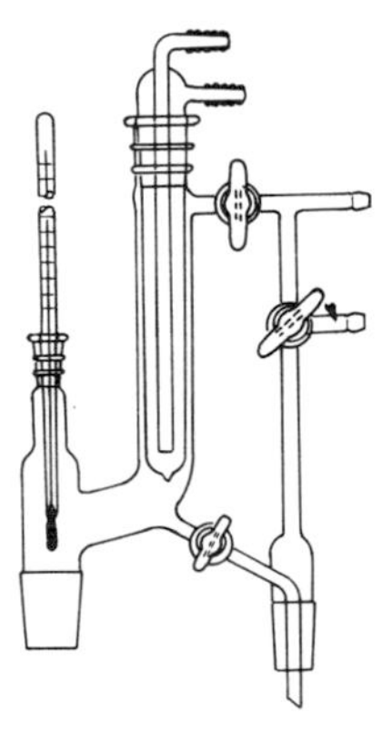

FIGURE 16-21
Distilling heads. (*a*) Atmospheric Liebig type. (*b*) Vacuum or atmospheric type with stopcock manifold allowing receiver distillate contents to be removed without breaking the vacuum. The reflux ratio is adjusted with the stopcock.

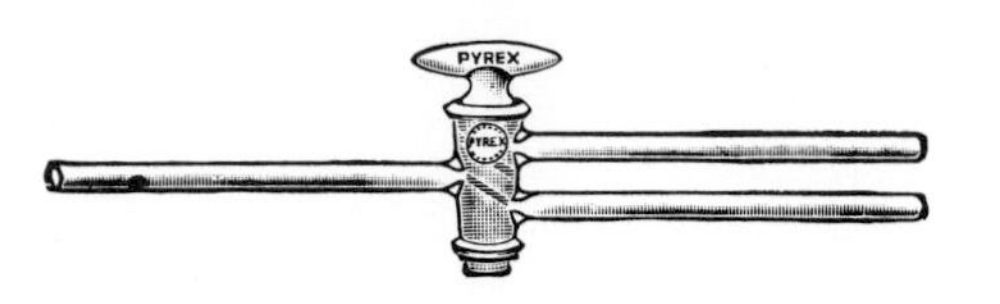

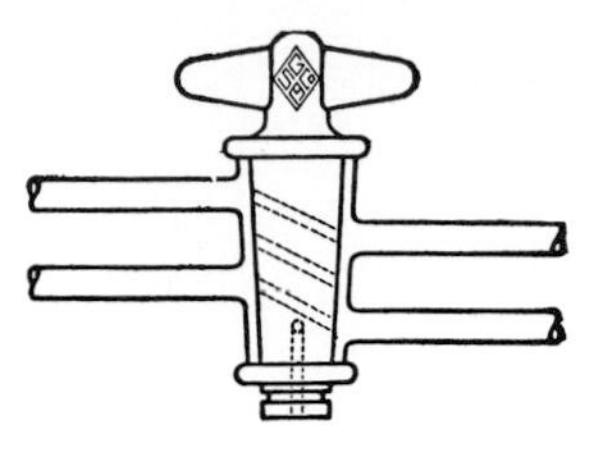

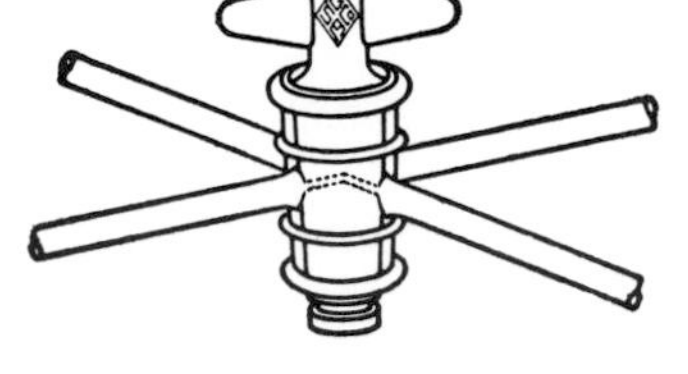

FIGURE 16-22
Stopcocks. (*a*) Three-way; allows fluids to be channeled as desired or to cut off flow completely. (*b*) Four-way oblique-bore with vent to bottom of plug. (*c*) Four-way V-bore.

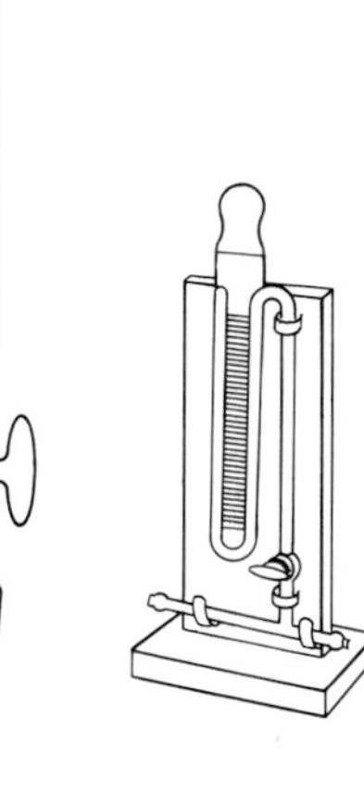

FIGURE 16-23
Manometer-type vacuum gauges used to indicate pressure in a closed system.

Assemblies

See Figs. 16-24 to 16-26.

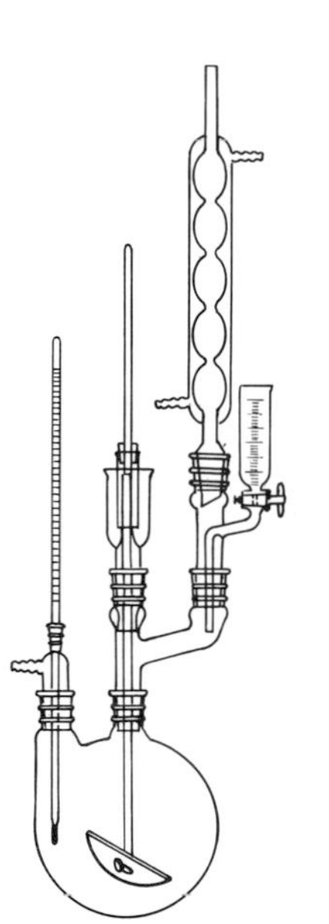

FIGURE 16-24
Reflux reaction apparatus with mixer, gas-inlet tube, thermometer, and addition funnel.

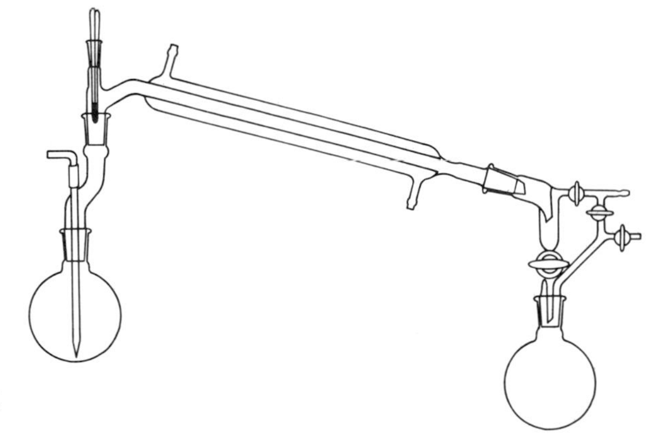

FIGURE 16-25
General distillation apparatus for use with vacuum or atmospheric pressure. Distillates are separable into fractions without interrupting vacuum distillation.

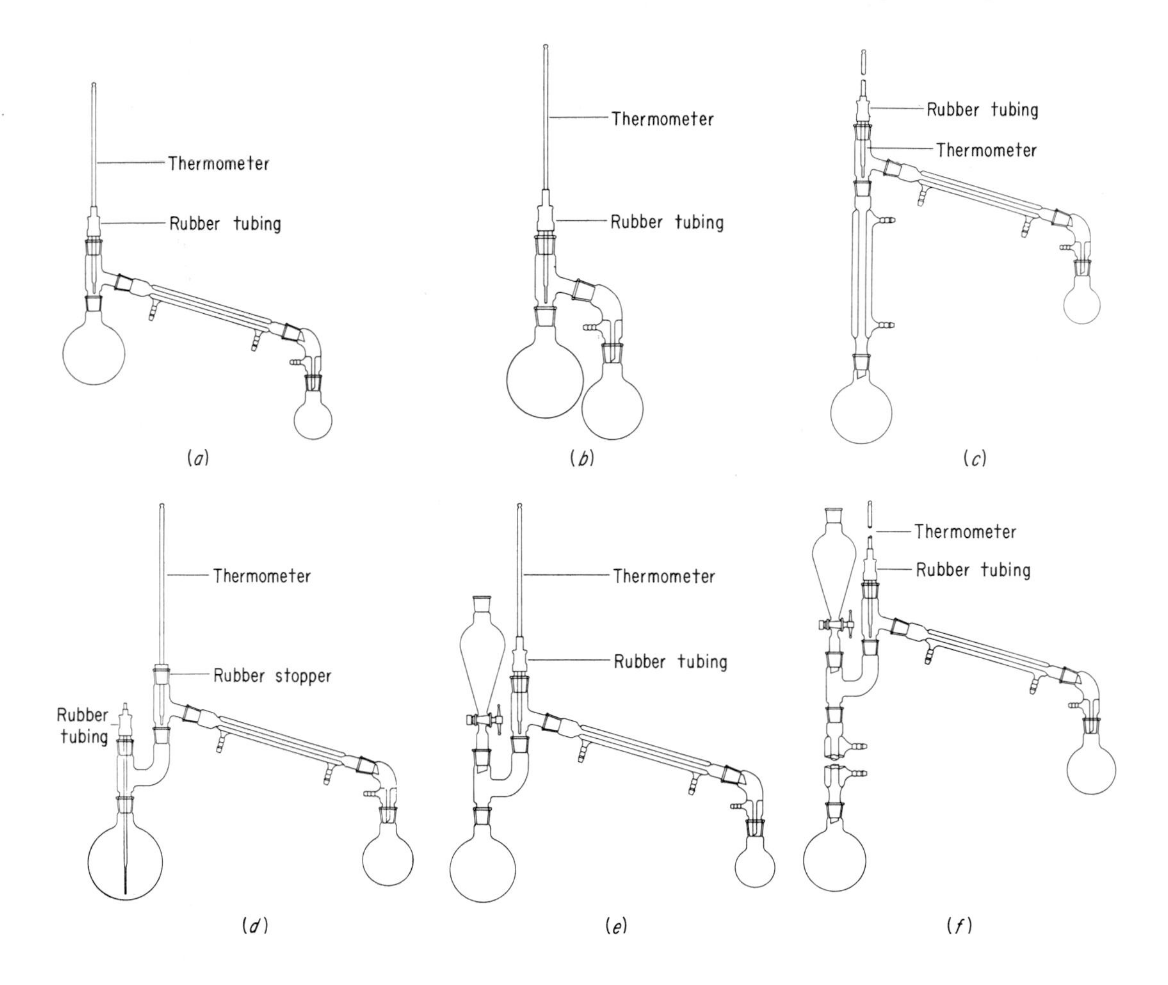

FIGURE 16-26
"How to" suggestions for using ground-glass-joint assemblies: (*a*) Simple distillation, atmospheric pressure, or vacuum. (*b*) Distillation of viscous or high-boiling liquids or solids. (*c*) Fractional distillation, vacuum or atmospheric pressure (column should be packed with stainless steel sponge). (*d*) Steam distillation, or vacuum distillation with gas capillary bubbler. (*e*) Concentration of a solvent or extract, vacuum or atmospheric pressure. (*f*) Concentration of a solvent or extract, with fractional distillation, vacuum, or atmospheric pressure (column should be packed with stainless steel sponge). (*g*) Simple reflux. (*h*) Simple reflux, dry atmosphere. (*i*) Reflux, inert atmosphere. (*j*) Addition and reflux, high-

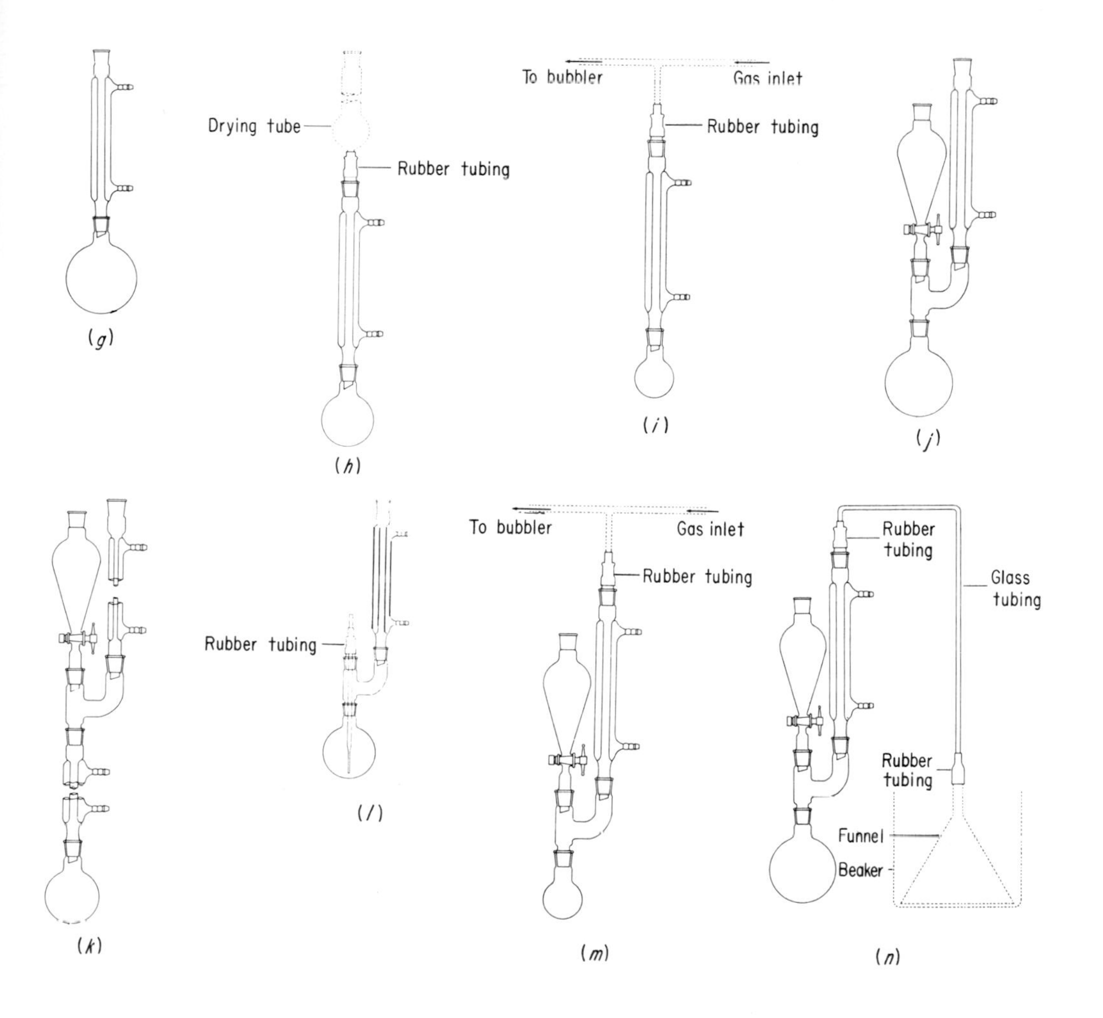

boiling liquids. (*k*) Addition and reflux for low-boiling liquids. (*l*) Gas inlet and reflux. (*m*) Addition and reflux, inert atmosphere. (*n*) Addition and reflux with trapping of evolved gases. Safety suggestions: (1) Remember to recheck setups, especially when using a closed system. (2) Provide adequate ventilation, particularly when working with solvents or materials which could produce toxic vapors. (3) By and large, materials being used are flammable and should be handled with caution. (4) In handling glassware, follow the safety tips given. (5) Learn the location of safety equipment and be familiar with its use in case of emergency.

CARE OF GROUND-GLASS SURFACES

Lubrication

Although ground-glass joints usually seal well without the use of lubricants, it is generally advisable to lubricate ground-glass joints to prevent sticking and, therefore, to prevent breakage. Under some reaction conditions, it may be necessary not to use a lubricant, but under most conditions a lubricant should be used. Lubrication makes it easy to separate ground-joint ware and prevents leakage. Ground-glass joints must be kept clean and must be cleaned prior to lubrication. Dust, dirt, and particulate matter may score the surfaces and cause leakage.

Choice of Lubricant

A lubricant must withstand high temperatures, high vacuum, and chemical reaction. Choice of a lubricant depends upon the conditions under which it is to operate.

Lubricate *only* the upper part of the joint with a small amount of grease. *Always* clean ground-glass surfaces thoroughly. *Avoid contamination by contact with grease* by cleaning joint of flask containing the liquid *prior* to pouring.

The following lubricants may be used:

Silicone grease for high temperatures and high vacuum. It is easily soluble in chlorinated solvents.

Glycerin for long-term reflux or extraction. It is a water-soluble, organic-insoluble greasing agent.

Hydrocarbon grease for general laboratory use. It is soluble in most organic laboratory solvents.

Lubricating Ground-Glass Joints

When lubricating ground-glass joints, use the following guidelines:

1. Lubricate only the upper part of the inner joint.

2. Avoid greasing any part of the joint which may come in contact with vapor or liquid and cause contamination.

3. The choice of the lubricant depends upon the materials used in the glassware, and the effect that the conditions of the reaction and the materials have on the lubricant.

4. A properly lubricated joint appears clear, without striations.

5. Lubrication is required when the joints must be airtight and when the glassware contains strong alkaline solutions.

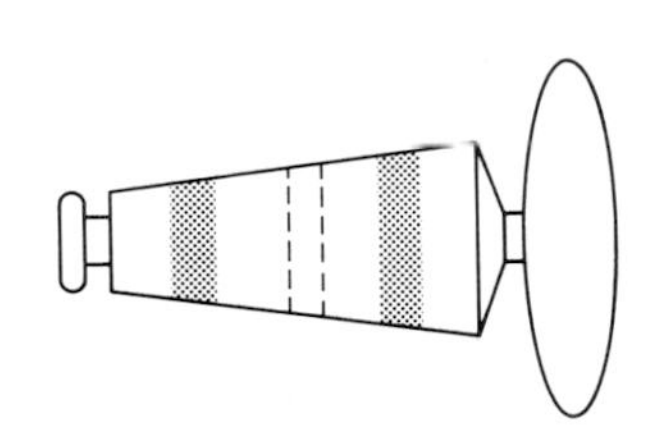

FIGURE 16-27
Greasing a stopcock.

Lubricating Stopcocks

A stopcock must be proberly lubricated with a suitable stopcock grease. Too much lubricant may plug the bore or the tip of a buret.

CAUTIONS Use recommended stopcock greases. Silicone greases should not be used. Ground surfaces *must be clean.*

Procedure

1. Spread two circular bands of grease around the stopcock (Fig. 16-27).
2. Insert stopcock in buret and twist several times. The grease will spread out and the joint will be completely transparent.

STORAGE OF GLASSWARE

In Drawers

Glassware may be stored in sliding drawers. To prevent glass breakage, take heed of the following:

1. If many glass items of various shapes and forms are crowded in sliding drawers, breakage may occur.
2. If the drawer is too shallow for the items, breakage can occur when the drawer is closed.
3. If the drawer sticks, a sudden jerk can cause the glass items to contact each other violently.
4. Round flasks, bulbs, etc., can roll in the drawer. They should be cushioned and secured with padding.
5. Care should be exercised when removing or placing glass items in drawers.

On Shelves and in Cabinets

When storing glassware on shelves and in cabinets:

1. There should be sufficient room.
2. No parts of the item should extend over the edge.

3. Items should be placed so that they cannot roll off or roll into other items.

4. Round-bottomed flasks should be seated on cork rings.

5. Heavy, bulky, or cumbersome items should be stored at low levels.

Preparation for Long-Term Storage

When lab glassware is not to be used for a long period of time, take apart buret stopcocks, ground-glass joints, and flask stoppers to prevent sticking. Remove grease from joints. Loosen Teflon® stoppers and stopcocks slightly to prolong the life of the sealant material.

For easy storage and reuse, put a strip of thin paper between the ground-glass surfaces. Otherwise the ground joints may stick together and may be extremely difficult to separate.

ASSEMBLY OF GROUND-JOINT GLASSWARE

1. Plan your assembly so that the working area will be uncluttered, with easy access to all components.

2. Use as few clamps as possible to support the apparatus firmly.

   (a) The precision of the ground-glass joints allows little room for misalignment.

   (b) The joints themselves provide mechanical support and rigidity.

3. Support all flasks with rings (Fig. 16-28) for stability, even though clamps may support the neck of the flask.

4. *Always* assemble the apparatus from the bottom up.

FIGURE 16-28
Support ring; use with suitable clamp holders to support round-bottomed vessels, funnels, and other apparatus on support stands or frames. Available in variety of sizes and rod lengths.

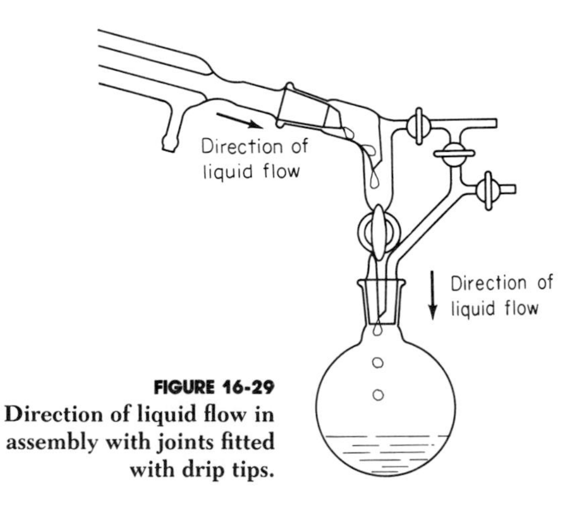

FIGURE 16-29
Direction of liquid flow in assembly with joints fitted with drip tips.

NOTE Assemble glassware so that any liquid flow always passes through the inner (male) joint. It should not flow into the joint (Fig. 16-29). This precaution keeps the joint surfaces free from liquid and prevents possible contamination from the lubricant.

(**a**) Fasten all clamps loosely at first, except the bottom clamp. *Use correct clamps of the proper size.* (See Figs. 16-30 and 16-31.)

(**b**) Gradually tighten the clamps as the apparatus goes into complete assembly, to accommodate the apparatus.

(**c**) *Be sure alignment is correct,* then finally tighten all clamps.

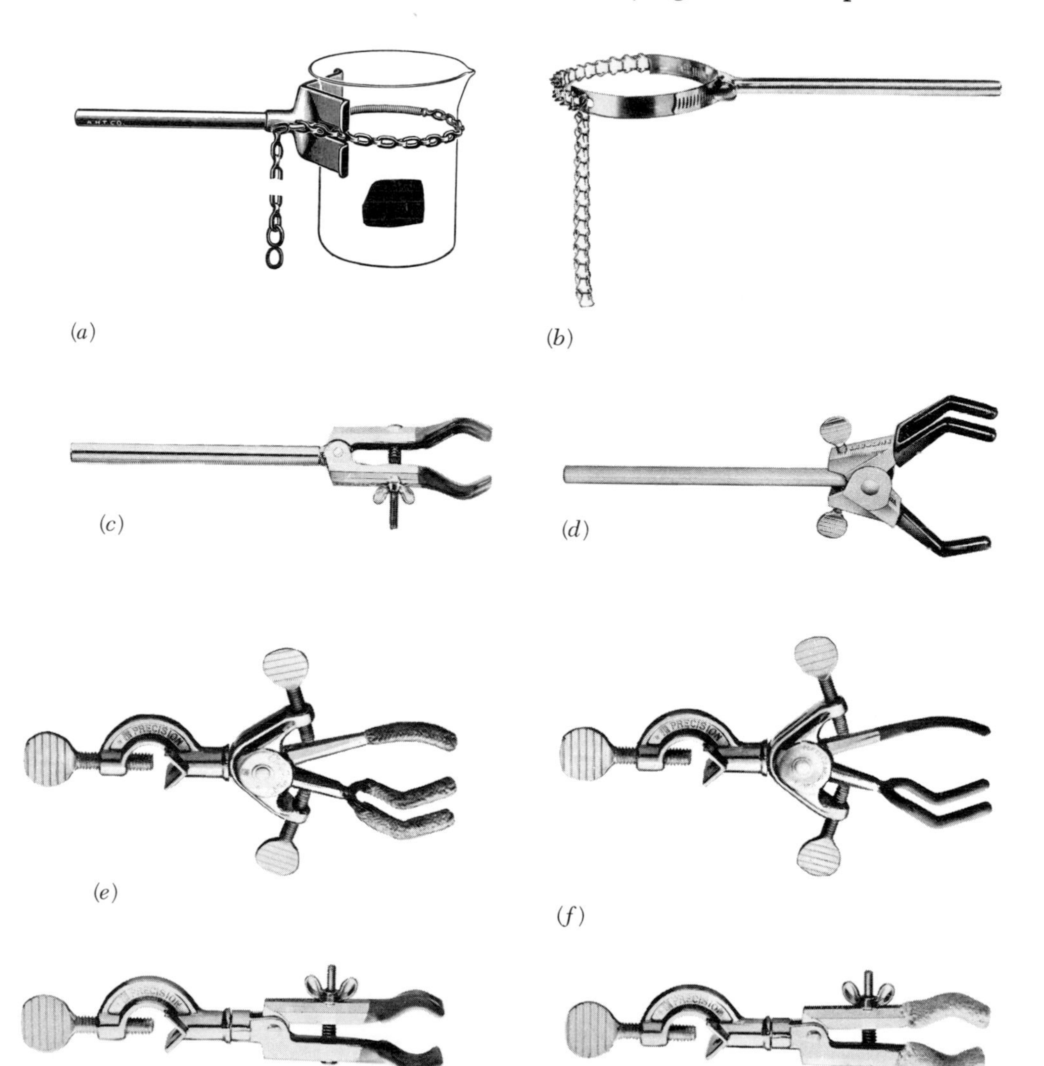

FIGURE 16-30
Clamps. (*a*) Beaker clamp, chain adjustable size, with spring tension. (*b*) Screw-ype clamp with adjustable tension. (*c*) Utility clamp with long handle. (*d*) Trigrip, double-jaw, vinyl-covered clamp with long andle to hold equipment. (*e*) Double-jaw, three-prong, asbestos-covered clamp. (*f*) Double-jaw, hree-prong, vinyl-covered clamp. (*g*) Fixed-position clamp with vinyl-covered jaws. (*h*) Fixed-position tility clamp with asbestos-covered jaws.

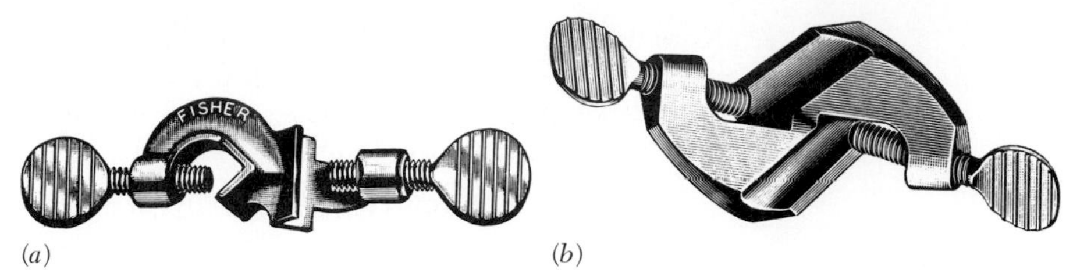

FIGURE 16-31
Clamp holders anchor rods securely to rods, support assemblies, and hold clamps to support assemblies. Available in parallel (*a*) or right-angle (*b*) direction.

CAUTION Always use stable ring stands, which *do not wobble,* or use a rigid frame network for support. (See Fig. 16-32.)

(**d**) Do not position your apparatus at an angle. Professional laboratory practices demand that vertically positioned items actually be exactly vertical.

(**e**) *Never force alignment* of ground-glass apparatus. This causes breakage, leakage, and improper function of the apparatus.

Safety

1. *Always recheck your systems,* especially when you are working under reduced pressure or with extremely hazardous materials.

2. *Always check your ventilation system,* especially when you are working with toxic gases. *Use hoods* if necessary.

3. *Never use broken or cracked glassware.* You may lose your material or you may be injured by contact with the material.

GLASSBLOWING

As a technician you will conduct experiments, tests, and analyses in glass equipment which can break or crack, or which may need modification or repair for current use. By becoming competent in the fundamentals of glassblowing, you can often avoid delays by repairing or modifying existing and available glass apparatus and save considerable money by doing so.

CAUTIONS 1. All glass is not the same; there are many different types of glass which vary in their softening point, strain point, annealing point, and linear coefficient of expansion. Identify the kind of glass in the apparatus to be worked, and be sure the glass you use for the repair or modification has the same characteristics.

2. If a soft glass is directly joined to a hard glass, strain will develop when they cool and the joint will crack or break.

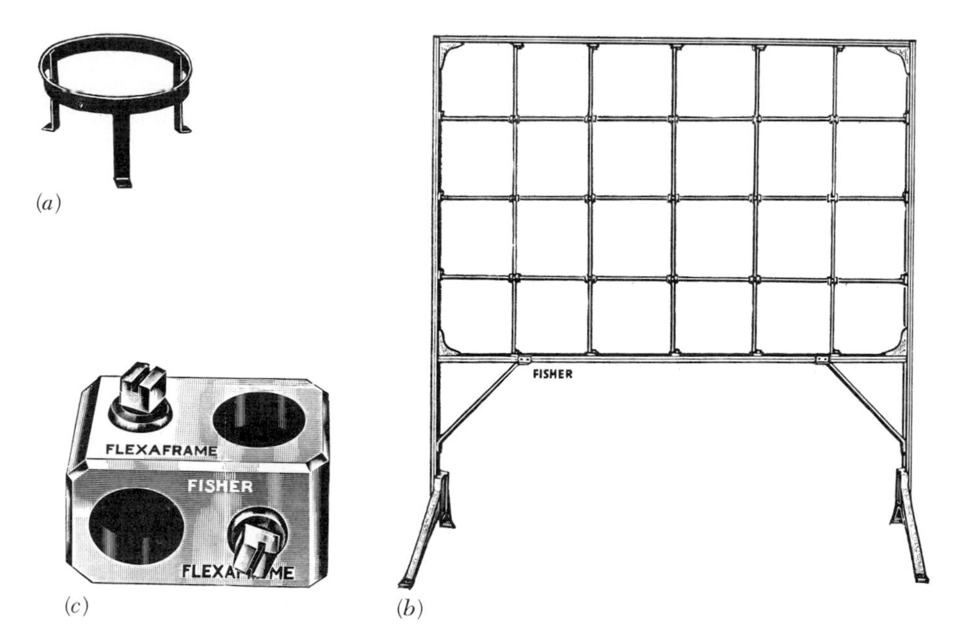

(*a*) (*c*) (*b*)

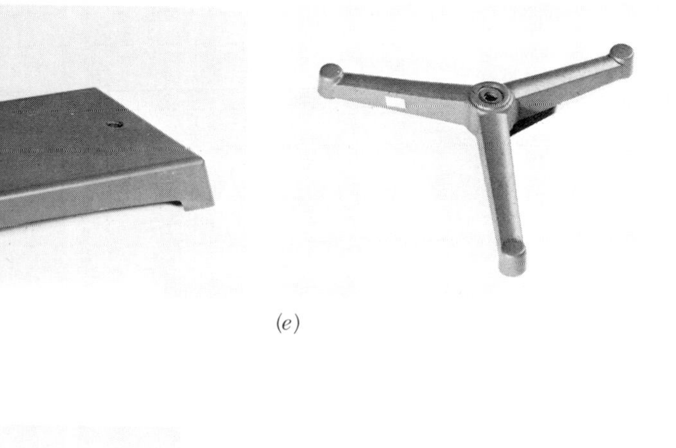

(*d*) (*e*)

(*f*) (*g*)

FIGURE 16-32
Supports. (*a*) Round, working-area-surface rest support base to support round-bottomed containers; available in a range of sizes. (*b*) Flexible frame can be assembled from rods of various lengths to provide the support requirements for the apparatus. Provides steady base without use of multiple support stands. (*c*) Right-angled rod clamps. (*d*) Rectangular single-rod support base. Fairly sturdy on level areas. (*e,f*) Tripods offer sturdy support for apparatus; may have two vertical rods. (*g*) Support plate clamped to support base or support frame with suitable clamp holder is used to hold flat-bottomed containers.

Glass is not a true crystalline solid. It does not have a true melting point, one that is sharp and distinct. It more nearly resembles a solid solution or an extremely viscous liquid, which gradually softens when heated. It is this property of glass which makes glassworking possible.

Soft Laboratory Glassware vs. Heat-Resistant Laboratory Glassware

Heat-resistant borosilicate laboratory glassware is more expensive than soft, soda-lime glassware which melts at a lower temperature, and working with it requires hotter burner flames. There is a place in the laboratory for both types of laboratory glassware, and the decision, of course, depends upon the conditions to which the glassware will be subjected.

Soft soda-lime glassware is not heat-resistant because it has a low melting point (600–800°C) and a high coefficient of expansion; when it is suddenly subjected to extreme temperature changes, it will break or crack. However, it can be used satisfactorily for such equipment as volumetric flasks for storing solutions, stirring rods, liquid- and gas-transport apparatus, and containers for normal-temperature mixing and reaction operations. It cannot be used over an open flame or with an electric heater. It can be worked with a Bunsen or Tirrill burner and is attacked by alkali.

Hard, high-temperature-melting (750–1100°C), heat-resistant borosilicate laboratory glassware (e.g., Pyrex brand glass) should be used wherever sudden changes in temperature may occur, e.g., for beakers which may be suddenly chilled, reaction flasks, and distilling columns and condensers. It should be used whenever the glassware is subjected to an open flame or to electric heating elements. It can be worked only with an oxygen torch and withstands attack by alkali. See Table 16-3 for properties of glasses. When choosing glassware, select the type that meets your needs.

How to Identify Types of Glass

CAUTION *Visual inspection is not reliable;* always test and compare glasses.

Method 1

Heat the two pieces of glass in the flame of a suitable torch, and compare the rates at which they soften and melt.

1. Notice if they soften and melt identically when you use samples of identical shape and size.

2. Join the molten ends of the samples, heating evenly.

3. Alternately push them together and pull them apart; notice if there is any difference in the way they handle. The softer glass will pull out more easily.

TABLE 16-3
Properties of Glasses

Property	*Type of glass*	
	*Pyrex**	*Kimble Flint*†
Coefficient of expansion (0–300°C)	3.25×10^{-8}/°C	9.3×10^{-8}/°C
Softening point	820°C	700°C
Annealing point	565°C	526°C
Strain point	515°C	486°C
Density	2.23 g/cm³	2.53 g/cm³
Refractive index	1.474‡	1.52

* Corning 7740 Glass (Pyrex) is a low-alkali-content, borosilicate glass free from magnesia-lime-zinc group elements, heavy metals, arsenic, and antimony. It possesses excellent chemical durability. Solubility is negligible under most conditions.
† Kimble Standard Flint (R-6) glass is a superior soda-lime glass developed primarily for vessels to be used at or near room temperature. In this temperature area, reaction is negligible between glass and liquids contained.
‡ Sodium D line.

Method 2

Borosilicate glass (Pyrex) is invisible when immersed in a liquid which has an identical refractive index.

1. Mix 16 parts of methyl alcohol and 84 parts of benzene in a closure tube jar.

2. Insert lengths of the tubing to be identified so that one end of each piece is submerged in the liquid.

3. If any sample of tubing is Pyrex 7740, its submerged portion will be practically invisible.

CAUTION This is not an infallible test because any glass which has the same refractive index will also be invisible.

Equipment Needed for Glassblowing

Glassblowing requires a permanent setup consisting of the following items:

1. Table or workbench, covered with a fire-resistant material or asbestos and of a height suitable for working the glass while standing or sitting.

2. Burner or torch (Fig. 16-33) mounted on a slide so that it can be pulled out over the edge of the table.

3. Compressed-air, oxygen, and gas outlets and controls located on the front or conveniently on the side of the table. Oxygen is usually obtained from a tank securely anchored to the table or wall.

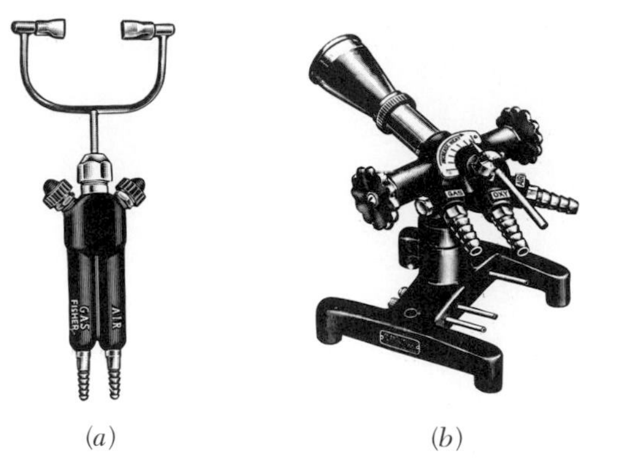

FIGURE 16-33 Glassworker's burners. (*a*) Sealing and tipping torch is used to seal ampuls and for glassblowing. It gives two fishtail flames opposite each other. (*b*) Triple-inlet burner for working borosilicate glass has an extremely hot flame suitable for working Pyrex. It uses gases with Btu ratings over 1000 together with oxygen from a tank.

4. Good lighting, located overhead or behind the glassblower.
5. Adequate glassworking tools.
6. Glassblower's goggles.

The Gas Burner or Torch

The gas burner or torch should be selected according to the type of gas that is to be used, providing all variations of heat from a sharp, intense flame to a soft, bushy flame by adjustments of the gas, air, and oxygen (Figs. 16-33*b* and 16-34). It may be fitted with a pilot light.

Glassblower's Goggles

Heated glass produces a blinding yellow sodium light which obscures the glassware being heated. Yellow-filtering glassblower's goggles (Fig. 16-

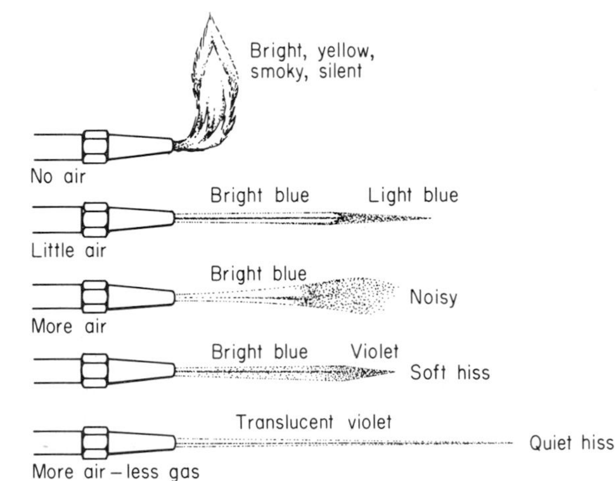

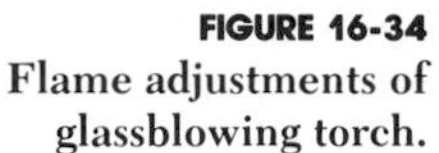

FIGURE 16-34 Flame adjustments of glassblowing torch.

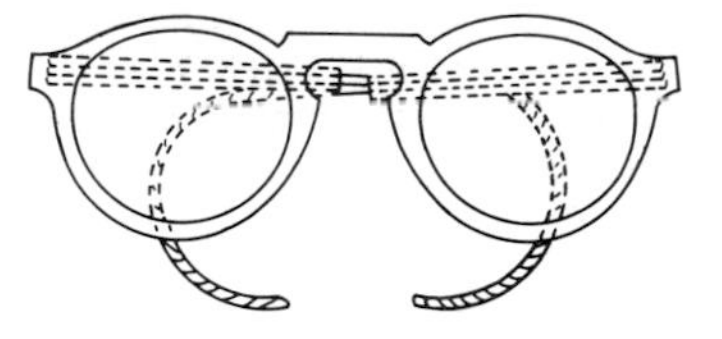

FIGURE 16-35
Glassblower's goggles.

35) are designed to eliminate this yellow light and permit a clear and unobscured view of the incandescent glass. Glassblower's goggles have lenses made of optical-quality didymium glass which is ground and polished, and they are fitted with safety frames. The goggles also provide a protective shield against glass particles.

Preparation of Glass

All glass surfaces must be clean, because foreign material or dust may cause the glass to vitrify rapidly when heated. Merely running water through tubing is not sufficient to clean it, because all particles of dust may not be removed. The best way to clean the inside of tubing is to use a wooden stick to force through snugly fitting pieces of wet paper, cloth, or cotton; metal or glass rods may scratch the glass. The outside of the tubing or apparatus must be washed clean, rinsed thoroughly, and dried.

CAUTION

1. If solvents are used to clean apparatus of grease or oil, do not heat the glass in a flame until all traces of the solvent and vapor have been removed. There is danger of fire and explosion.

2. Keep hands and fingers clean while working with glass. Finger marks readily burn into glass, and they are not easily removed once this has happened.

Review under Corks and Rubber Stoppers in Chap. 18, "Laboratory Tools and Hardware," for guidelines on handling glass tubing safely.

Cutting Glass Tubing

When glass tubing is cut, sharp edges result. These can cause cuts and serious injury. Always use care when cutting glass tubing and when handling the cut pieces of tubing.

There are many sizes of tubing and glass rod (see Fig. 16-36 and Tables 16-4 and 16-5). Methods used for cutting them vary with their size.

Cutting Small-Diameter Tubing (25-mm OD or less)

1. Scratch the tube or rod at the desired point with a three-cornered file or glasscutter. Use only one or two strokes. Use considerable pressure. Do *not* saw (Fig. 16-37).

FIGURE 16-36
Glass tubing. (*a*) Capillary wall thickness and diameter of hole vary. Bore varies from ¼ to 3 mm diameter, OD from 5 to 10 mm, and combinations of bore to OD are available. (*b*,*c*) Tubing is soft, Pyrex, or Vycor® glass, is available from 2 to 178 mm OD. It normally comes in 4-ft lengths. (*d*) Glass-tubing sizer.

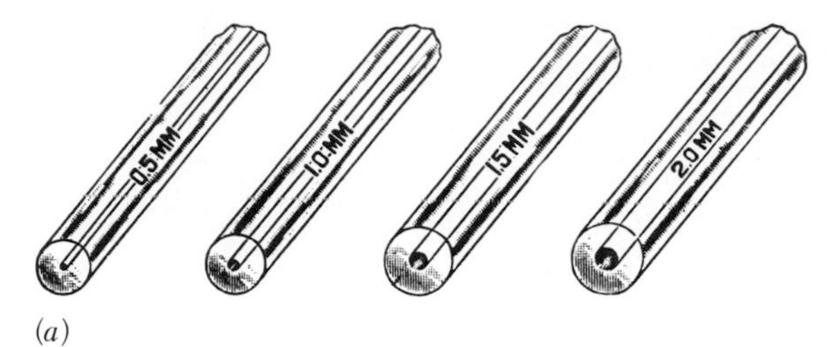

(*a*)

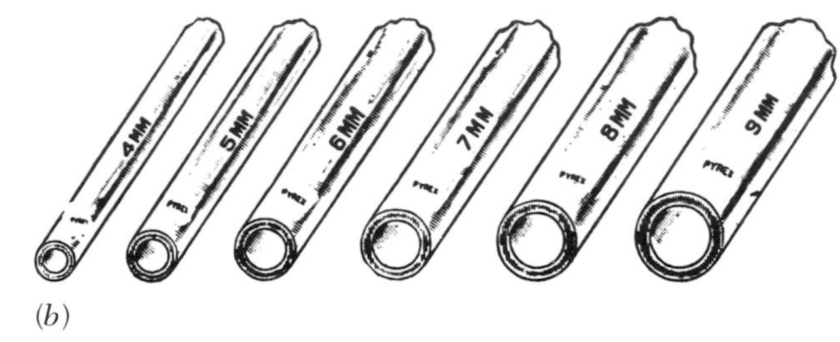

(*b*)

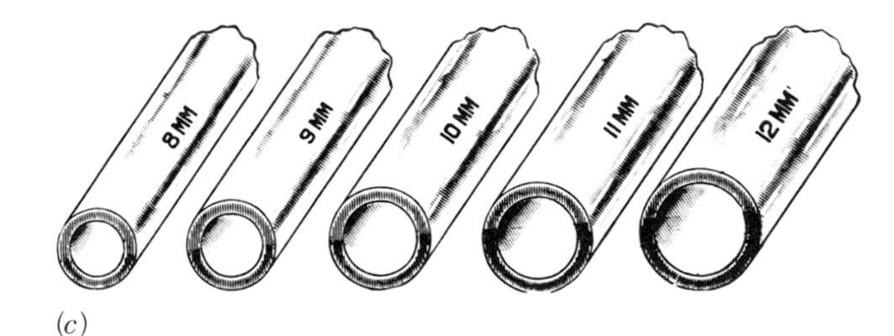

(*c*)

3	4	5	6	7	8	9	10	11	12	13	14	16	18	20

(*d*)

TABLE 16-4
Standard-Wall Glass Tubing

OD, *mm*	*Wall,* *mm*	*OD,* *mm*	*Wall,* *mm*
2	0.5	38	2.0
3	0.6	41	2.0
4	0.8	45	2.0
5	0.8	48	2.0
6	1.0	51	2.0
7	1.0	54	2.4
8	1.0	57	2.4
9	1.0	60	2.4
10	1.0	64	2.4
11	1.0	70	2.4
12	1.0	75	2.4
13	1.2	80	2.4
14	1.2	85	2.4
15	1.2	90	2.4
16	1.2	95	2.4
17	1.2	100	2.4
18	1.2	110	2.6
19	1.2	120	3.0
20	1.2	125	3.0
22	1.5		
25	1.5	130	3.0
28	1.5	140	3.5
30	1.8	150	3.5
32	1.8	178	3.5
35	2.0		

TABLE 16-5
Standard Glass Rod

OD, *mm*
2
3
4
5
6
7
8
9
10
11
13
16
19
25
32
38

2. Wrap the tubing in a protective cloth to avoid cutting your hands. Place thumbs together opposite scratch.

3. Using little force, pull back on the tube and push thumbs outward quickly to break the glass. A straight, clean break should result.

Dulling Sharp Glass Edges

Sharp edges, that result when glass tubing is cut, can be dulled in the following ways (see Fig. 16-38).

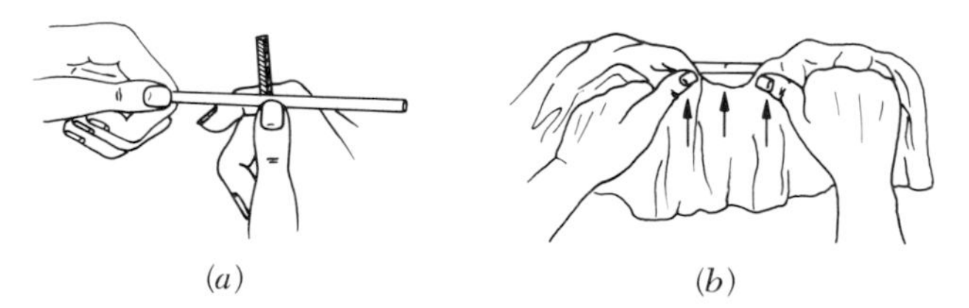

FIGURE 16-37
Proper method of (*a*) scratching, and (*b*) holding glass tubing for cutting.

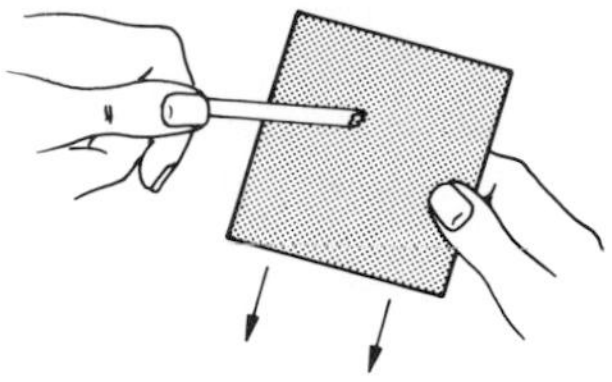

FIGURE 16-38
Dulling sharp edges with wire screen.

1. Hold the glass article in the left hand and a piece of clean, new wire gauze (without asbestos) in the right hand.

2. Gently stroke the broken end with wire gauze while rotating the glass object.

End Finishing (Fire Polishing)

Fire-polishing the ends of tubing eliminates the sharp edges, prevents cuts, and enables the technician to insert the fire-polished tubing easily into corks, stoppers, and rubber or plastic tubing.

1. Smooth the sharp edges to prevent cuts.

2. Insert the end into the hot nonluminous portion of burner flame and rotate smoothly and evenly. (See Figs. 16-39 and 16-40.)

CAUTION Wear gloves or keep hands well back from the end being heated. *It gets hot!*

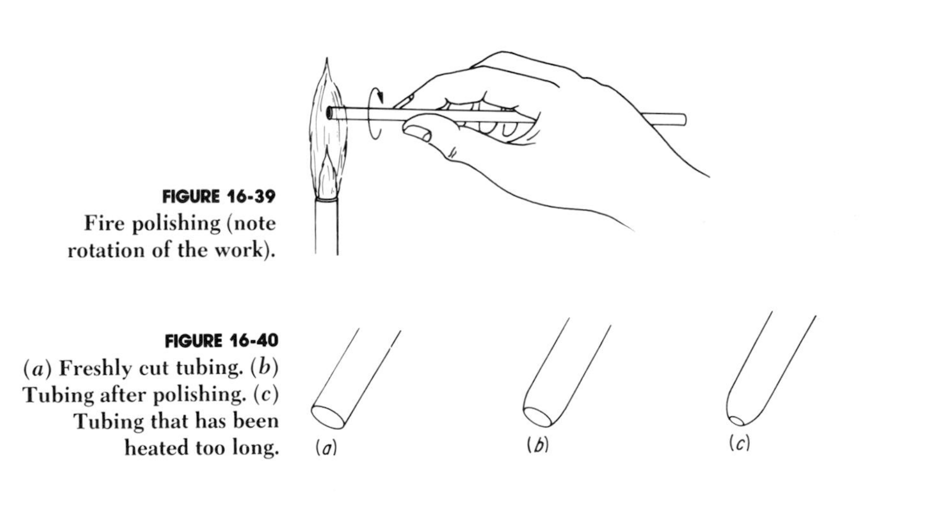

FIGURE 16-39
Fire polishing (note rotation of the work).

FIGURE 16-40
(*a*) Freshly cut tubing. (*b*) Tubing after polishing. (*c*) Tubing that has been heated too long.

Common Glassblowing Techniques

Making Points

Points are formed by pulling tubes to a smaller diameter (Fig. 16-41), and they can be used as handles for holding small pieces of tubing, and for closing them.

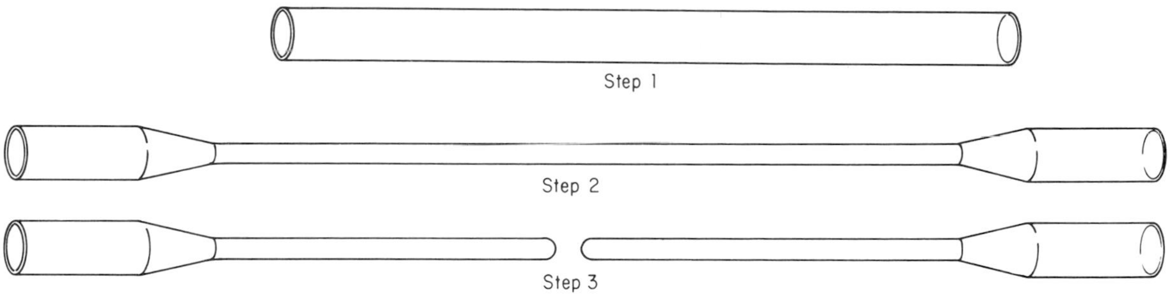

Step 1

Step 2

Step 3

FIGURE 16-41
Making a "point."

Procedure

1. Rotate the tube in the flame until it is pliable.

2. Remove it from the flame, while still rotating it, and slowly pull the ends about 8 in apart.

3. Melt the center of the smaller section until the ends are closed and sealed.

Points are useful when it is necessary to work a larger piece of glass tubing which is too short to handle. The smaller tube or rod is fused to the end of the large tubing, and it can be centered to form an effective handle similar to a point.

Sealing the Ends of Capillary Tubes

The end of a capillary tube can be sealed by gently heating its tip in a small flame (see Fig. 16-42).

CAUTION Use a small flame and heat only the tip until glass fuses.

Bending Glass Tubing

When bending glass, the diameter of the tubing and the distance between the tubes after the bend determine the length of tubing which must be heated. For small-bore tubing:

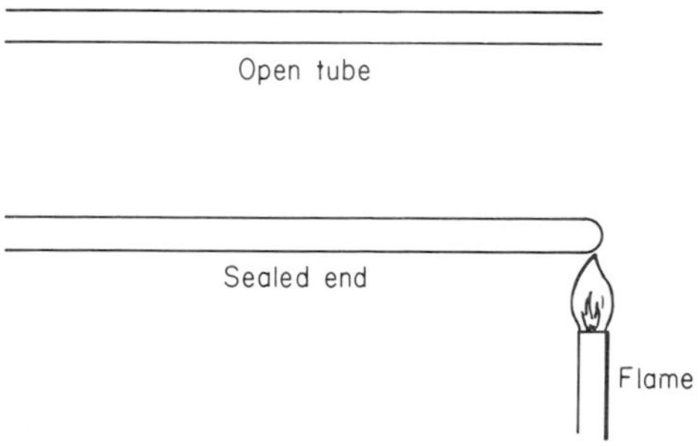

Open tube

Sealed end

Flame

FIGURE 16-42
Sealing capillary tubes.

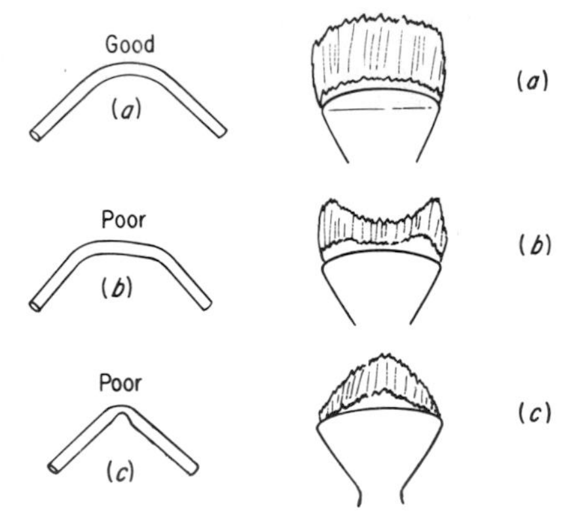

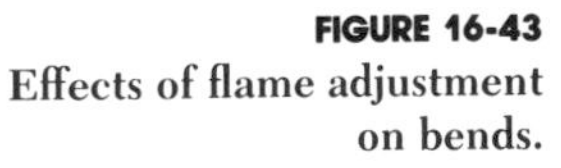

FIGURE 16-43
Effects of flame adjustment on bends.

1. Use a wing-tip Bunsen burner and rotate the tubing evenly with both hands for uniform heating. Adjust the burner to give a nonluminous flame and a well-defined blue cone (see Fig. 16-43).

2. Hold the tube lengthwise in the flame and rotate it evenly with a back-and-forth motion until the glass becomes soft (Fig. 16-44).

3. After the glass is soft (it bends under its own weight), remove it from the flame.

4. Bend the glass to the desired shape, holding it in position until it hardens.

CAUTION Wear gloves or keep hands well back from the end of the tube being heated. *It gets hot!*

HANDLING STOPPERS AND STOPCOCKS

Loosening "Frozen" Stoppers and Stopcocks

Laboratory glassware and reagent bottles have glass-to-glass connections which sometimes become "frozen." To loosen stopcocks, stoppers, or any glass-to-glass connections which are "frozen," you may use the following techniques.

Wing tip

FIGURE 16-44
Work must be constantly rotated.

CAUTION Remember *glass is fragile*. It will break under severe thermal or mechanical shock. *Use care and handle gently. Do not apply too much force or subject to rough treatment.*

1. *Gentle tapping.* Gently tap the frozen stopcock or stopper with the *wooden* handle of a spatula. Tap so that the direction of the force will cause the stopper to come out. If you tap *too hard*, you will break the stopper. When tapping, hold the item in the left hand and gently tap with the wooden handle of a spatula held in the right hand.* Always work directly over a desk top covered with soft cushioning material to prevent falling stopper from breaking.

2. *Heating.* Try immersing the frozen connection in hot water. If that doesn't work, warm the stopcock housing or the stopper housing *gently* in the smoky flame of a gas-flame burner such as a Bunsen burner. Heat causes the housing to expand. Rotate constantly for even heating. When it is hot, gently tap with handle of spatula. Repeat several times, allowing the housing to expand and contract to break the frozen seal.

3. *Soaking.* Soak the frozen assembly in hot dilute glycerin and water or chlorinated solvent (CCl_4) and grease.

4. *Stopcock-removing clamp.* Affix this clamp (Fig. 16-45) to the frozen stopcock and housing and *gently* twist; the knurled nut applies pressure to separate the components.

CAUTION Use care and use the stopcock-removing clamp with procedures 1 to 3 or combinations of these.

---

* Reverse if you are left-handed.

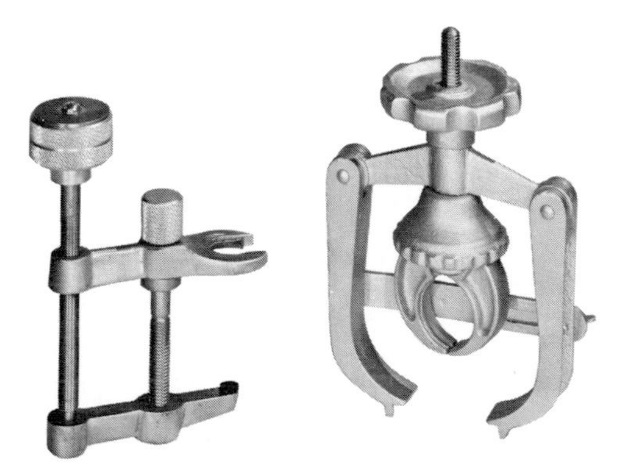

FIGURE 16-45 Clamps for removing frozen stopcocks and stoppers.

FIGURE 16-46
Devices for securing glass stoppers and stopcocks.

5. *Carbonated water method.* Frozen ground-glass joints are sometimes freed by immersing the frozen joint in a freshly opened container of carbonated water. The penetration of the liquid is evident by the appearance of the colored liquid between the joints. Allow to soak long enough to allow fullest penetration. Then remove, rinse thoroughly with tap water, and dry—leaving the inner surface wet. Gently warm the outer joint over a smoky Bunsen flame while continually rotating the joint. After a full deposit of carbon has covered the outer joint, remove from the flame and attempt to twist the components apart (using asbestos gloves for protection against the hot glass). *Do not use force.* Repeat if necessary.

CAUTION Never store alkaline reagents in ground-glass-stoppered flasks, bottles, or burets. Ground-glass stoppers or stopcocks may stick.

Stopcock-Securing Devices

When stopcocks are lost, broken, or chipped, the glassware may not be usable. To prevent such a mishap, always secure the stopcock in place with a securing device such as those shown in Fig. 16-46.

17

PLASTIC LABWARE

INTRODUCTION

Plastic labware offers the laboratory technician practically all the advantages of laboratory glassware, without the problem of breakage that is always present when glass is used.

Plastic labware made from a number of different plastic substances is available, each type offering specific advantages of clarity, flexibility, precision of manufacture, temperature stability, resistance to action or effect of chemical substances, resistance to heat, nonadhesive surfaces, resistance to the high gravities encountered during centrifugation, and ease of marking. The selection of the most suitable plastic labware depends upon an analysis of all the conditions to which the ware may be subjected, and manufacturers have provided guidelines for the judicious selection of the most appropriate material.

EFFECTS OF CHEMICALS ON PLASTICS

Chemicals can affect the strength, flexibility, surface appearance, color, dimensions, and weight of plastics. There are two basic modes of interaction which cause these changes. The first is chemical attack on the polymer chain, which can result in oxidation, reaction of functional groups in or on the chain, or depolymerization. The second is physical change, such as softening or swelling, resulting from absorption of a solvent, permeation by a solvent, or solution in a solvent. The plastic may also crack because of the interaction of a "stress-cracking agent" with its molded-in stresses.

The combination of compounds of two or more classes may be synergistic and increase the undesirable chemical effect. Other factors affecting chemical resistance include temperature, pressure, and other stresses (e.g., centrifugation), length of exposure, and concentration of the attacking chemical. As temperature increases, resistance to attack decreases.

INTERPRETATION OF CHEMICAL RESISTANCE

Table 17-1, a chemical-resistance chart, lists the resistance of plastic ware to various chemicals. It is intended to be used as a general guide only. Because so many factors can affect the chemical resistance of a given prod-

TABLE 17-1
Plastic Ware Chemical-Resistance Chart*

Chemical	Resins† CPE	LPE	PP	PMP	FEP/ ETFE	PC	PVC
Acetaldehyde	GN	GF	GN	GN	EE	FN	GN
Acetamine, sat.	EE	EE	EE	EE	EE	NN	NN
Acetic acid, 5%	EE	EE	EE	EE	EE	EG	EE
Acetic acid, 50%	EE	EE	EE	EE	EE	EG	EG
Acetone	EE	EE	EE	EE	EE	NN	FN
Adipic acid	EG‡	EE	EE	EE	EE	EE	EG
Alanine	EE	EE	EE	EE	EE	NN	NN
Allyl alcohol	EE	EE	EE	EG	EE	EG	GF
Aluminum hydroxide	EG	EE	EG	EG	EE	FN	EG
Aluminum salts	EE	EE	EE	EE	EE	EG	EE
Amino acids	EE	EE	EE	EE	EE	EE	EE
Ammonia	EE	EE	EE	EE	EE	NN	EG
Ammonium acetate, sat.	EE	EE	EE	EE	EE	EE	EE
Ammonium glycolate	EG	EE	EG	EG	EE	GF	EE
Ammonium hydroxide, 5%	EE	EE	EE	EE	EE	FN	EE
Ammonium hydroxide,	EG	EE	EG	EG	EE	NN	EG
Ammonium oxalate	EG	EE	EG	EG	EE	EE	EE
Ammonium salts	EE	EE	EE	EE	EE	EG	EG
n-Amyl acetate	GF	EG	GF	GF	EE	NN	FN
Amyl chloride	NN	FN	NN	NN	EE	NN	NN
Aniline	EG	EG	GF	GF	EE	FN	NN
Antimony salts	EE	EE	EE	EE	EE	EE	EE
Arsenic salts	EE	EE	EE	EE	EE	EE	EE
Barium salts	EE	EE	EE	EE	EE	EE	EG
Benzaldehyde	EG	EE	EG	EG	EE	FN	NN
Benzene	FN	GG	GF	GF	EE	NN	NN
Benzoic acid, sat.	EE	EE	EG	EG	EE	EG	EG
Benzyl acetate	EG	EE	EG	EG	EE	FN	FN
Benzyl alcohol	NN	FN	NN	NN	EE	GF	GF
Bismuth salts	EE	EE	EE	EE	EE	EE	EE
Boric acid	EE	EE	EE	EE	EE	EE	EE
Boron salts	EE	EE	EE	EE	EE	EE	EE
Brine	EE	EE	EE	EE	EE	EE	EE
Bromine	NN	FN	NN	NN	EE	FN	GN
Bromobenzene	NN	FN	NN	NN	EE	NN	NN
Bromoform	NN	NN	NN	NN	EE	NN	NN
Butadiene	NN	FN	NN	NN	EE	NN	FN
n-Butyl acetate	GF	EG	GF	GF	EE	NN	NN
n-Butyl alcohol	EE	EE	EE	EG	EE	GF	GF
sec-Butyl alcohol	EG	EE	EG	EG	EE	GF	GG
tert-Butyl alcohol	EG	EE	EG	EG	EE	GF	EG

TABLE 17-1 (Continued)

Chemical	*Resins*†						
	CPE	*LPE*	*PP*	*PMP*	*FEP/ ETFE*	*PC*	*PVC*
Butyric acid	NN	FN	NN	NN	EE	FN	GN
Cadmium salts	EE	EE	EE	EE	EE	EE	EE
Calcium hydroxide, conc.	EE	EE	EE	EE	EE	NN	EE
Calcium hypochlorite, sat.	EE	EE	EE	EG	EE	FN	GF
Carbazole	EE	EE	EE	EE	EE	NN	NN
Carbon bisulfide	NN	NN	EG	FN	EE	NN	NN
Castor oil	EE	EE	EE	EE	EE	EE	EE
Cedarwood oil	NN	FN	NN	NN	EE	GF	FN
Cellosolve acetate	EG	EE	EG	EG	EE	FN	FN
Cesium salts	EE	EE	EE	EE	EE	EE	EE
Chlorine, 10% in air	GN	EF	GN	GN	EE	EG	EE
Chlorine, 10% (moist)	GN	GF	GN	GN	EE	GF	EG
Chloroacetic acid	EE	EE	EG	EG	EE	FN	FN
p-Chloroacetophenone	EE	EE	EE	EE	EE	NN	NN
Chloroform	FN	GF	GF	FN	EE	NN	NN
Chromic acid, 10%	EE	EE	EE	EE	EE	EG	EG
Chromic acid, 50%	EE	EE	EG	EG	EE	EG	EF
Cinnamon oil	NN	FN	NN	NN	EE	GF	NN
Citric acid, 10%	EE	EE	EE	EE	EE	EG	GG
Citric acid, crystals	EE	EE	EE	EE	EE	EE	EG
Coconut oil	EE	EE	EE	EG	EE	EE	GF
Cresol	NN	FN	EG	NN	EE	NN	NN
Cyclohexane	GF	EG	GF	NN	EE	EG	GF
Decalin	GF	EG	GF	FN	EE	NN	EG
o-Dichlorobenzene	FN	FF	FN	FN	EE	NN	GN
p-Dichlorobenzene	FN	GF	EF	GF	EE	NN	NN
Diethyl benzene	NN	FN	NN	NN	EE	FN	NN
Diethyl ether	NN	FN	NN	NN	EE	NN	FN
Diethyl ketone	GF	GG	GG	GF	EE	NN	NN
Diethyl malonate	EE	EE	EE	EG	EE	FN	GN
Diethylene glycol	EE	EE	EE	EE	EE	GF	FN
Diethylene glycol ethyl ether	EE	EE	EE	EE	EE	FN	FN
Dimethyl formamide	EE	EE	EE	EE	EE	NN	FN
Dimethylsulfoxide	EE	EE	EE	EE	EE	NN	NN
1,4-Dioxane	GF	GG	GF	GF	EE	GF	FN
Dipropylene glycol	EE	EE	EE	EE	EE	GF	GF
Ether	NN	FN	NN	NN	EE	NN	FN
Ethyl acetate	EE	EE	EE	EG	EE	NN	FN
Ethyl alcohol	EG	EE	EG	EG	EE	EG	EG
Ethyl alcohol, 40%	EG	EE	EG	EG	EE	EG	EE
Ethyl benzene	FN	GF	FN	FN	EE	NN	NN

Chemical	Resins†						
	CPE	LPE	PP	PMP	FEP/ ETFE	PC	PVC
Ethyl benzoate	FF	GG	GF	GF	EE	NN	NN
Ethyl butyrate	GN	GF	GN	FN	EE	NN	NN
Ethyl chloride, liquid	FN	FF	FN	FN	EE	NN	NN
Ethyl cyanoacetate	EE	EE	EE	EE	EE	FN	FN
Ethyl lactate	EE	EE	EE	EE	EE	FN	FN
Ethylene chloride	GN	GF	FN	NN	EE	NN	NN
Ethylene glycol	EE	EE	EE	EE	EE	GF	EE
Ethylene glycol methyl ether	EE	EE	EE	EE	EE	FN	FN
Ethylene oxide	FF	GF	FF	FN	EE	FN	FN
Fluorides	EE	EE	EE	EE	EE	EE	EE
Fluorine	FN	GN	FN	FN	EG	GF	EG
Formaldehyde, 10%	EE	EE	EE	EG	EE	EG	GF
Formaldehyde, 40%	EG	EE	EG	EG	EE	EG	GF
Formic acid, 3%	EG	EE	EG	EG	EE	EG	GF
Formic acid, 50%	EG	EE	EG	EG	EE	EG	GF
Formic acid, 98–100%	EG	EE	EG	EF	EE	EF	FN
Fuel oil	FN	GF	EG	GF	EE	EG	EE
Gasoline	FN	GG	GF	GF	EE	FF	GN
Glacial acetic acid	EG	EE	EG	EG	EE	GF	EG
Glycerin	EE	EE	EE	EE	EE	EE	EE
n-Heptane	FN	GF	FF	FF	EE	EG	FN
Hexane	NN	GF	EF	FN	EE	FN	GN
Hydrochloric acid, 1–5%	EE	EE	EE	EG	EE	EE	EE
Hydrochloric acid, 20%	EE	EE	EE	EG	EE	EG	EG
Hydrochloric acid, 35%	EE	EE	EG	EG	EE	GF	GF
Hydrofluoric acid, 4%	EG	EE	EG	EG	EE	GF	GF
Hydrofluoric acid, 48%	EE	EE	EE	EE	EE	NN	GF
Hydrogen	EE	EE	EE	EE	EE	EE	EE
Hydrogen peroxide, 3%	EE	EE	EE	EE	EE	EE	EE
Hydrogen peroxide, 30%	EG	EE	EG	EG	EE	EE	EE
Hydrogen peroxide, 90%	EG	EE	EG	EG	EE	EE	EG
Isobutyl alcohol	EE	EE	EE	EG	EE	EG	EG
Isopropyl acetate	GF	EG	GF	GF	EE	NN	NN
Isopropyl alcohol	EE	EE	EE	EE	EE	EE	EG
Isopropyl benzene	FN	GF	FN	NN	EE	NN	NN
Kerosene	FN	GG	GF	GF	EE	GF	EE
Lactic acid, 3%	EG	EE	EG	EG	EE	EG	GF
Lactic acid, 85%	EE	EE	EG	EG	EE	EG	GF
Lead salts	EE	EE	EE	EE	EE	EE	EE
Lithium salts	EE	EE	EE	EE	EE	GF	EE
Magnesium salts	EE	EE	EE	EE	EE	EG	EE

TABLE 17-1 (Continued)

Chemical	Resins†						
	CPE	*LPE*	*PP*	*PMP*	*FEP/ETFE*	*PC*	*PVC*
Mercuric salts	EE	EE	EE	EE	EE	EE	EE
Mercurous salts	EE	EE	EE	EE	EE	EE	EE
Methoxyethyl oleate	EG	EE	EG	EG	EE	FN	NN
Methyl alcohol	EE	EE	EE	EE	EE	FN	EF
Methyl ethyl ketone	EG	EE	EG	EF	EE	NN	NN
Methyl isobutyl ketone	GF	EG	GF	FF	EE	NN	NN
Methyl propyl ketone	GF	EG	GF	FF	EE	NN	NN
Methylene chloride	FN	GF	FN	FN	EE	NN	NN
Mineral oil	GN	EE	EE	EG	EE	EG	EG
Nickel salts	EE	EE	EE	EE	EE	EE	EE
Nitric acid, 1–10%	EE	EE	EE	EE	EE	EG	EG
Nitric acid, 50%	EG	GN	GN	GN	EE	GF	GF
Nitric acid, 70%	EN	GN	GN	GN	EE	FN	FN
Nitrobenzene	NN	FN	NN	NN	EE	NN	NN
n-Octane	EE	EE	EE	EE	EE	GF	FN
Orange oil	FN	GF	GF	FF	EE	FF	FN
Ozone	EG	EE	EG	EE	EE	EG	EG
Perchloric acid	GN	GN	GN	GN	GF	NN	GN
Perchloroethylene	NN	NN	NN	NN	EE	NN	NN
Phenol, crystals	GN	GF	GN	FG	EE	EN	FN
Phosphoric acid, 1–5%	EE	EE	EE	EE	EE	EE	EE
Phosphoric acid, 85%	EE	EE	EG	EG	EE	EG	EG
Phosphorus salts	EE	EE	EE	EE	EE	EE	EE
Pine oil	GN	EG	EG	GF	EE	GF	FN
Potassium hydroxide, 1%	EE	EE	EE	EE	EE	FN	EE
Potassium hydroxide, conc.	EE	EE	EE	EE	EE	NN	EG
Propane gas	NN	FN	NN	NN	EE	FN	EG
Propylene glycol	EE	EE	EE	EE	EE	GF	FN
Propylene oxide	EG	EE	EG	EG	EE	GF	FN
Resorcinol, sat.	EE	EE	EE	EE	EE	GF	FN
Resorcinol, 5%	EE	EE	EE	EE	EE	GF	GN
Salicylaldehyde	EG	EE	EG	EG	EE	GF	FN
Salicylic acid, powder	EE	EE	EE	EG	EE	EG	GF
Salicylic acid, sat.	EE	EE	EE	EE	EE	EG	GF
Salt solutions	EE	EE	EE	EE	EE	EE	EE
Silver acetate	EE	EE	EE	EE	EE	EG	GG
Silver salts	EG	EE	EG	EE	EE	EE	EG
Sodium acetate, sat.	EE	EE	EE	EE	EE	EG	GF
Sodium benzoate, sat.	EE	EE	EE	EE	EE	EE	EE
Sodium hydroxide, 1%	EE	EE	EE	EE	EE	FN	EE
Sodium hydroxide, 50% to sat.	EE	EE	EE	EE	EE	NN	EG

TABLE 17-1 (Continued)

Chemical	*Resins*†						
	CPE	*LPE*	*PP*	*PMP*	*FEP/ ETFE*	*PC*	*PVC*
Sodium hypochlorite, 15%	EE	EE	EE	EE	EE	GF	EE
Stearic acid, crystals	EE	EE	EE	EE	EE	EG	EG
Sulfuric acid, 1–6%	EE	EE	EE	EE	EE	EE	EG
Sulfuric acid, 20%	EE	EE	EG	EG	EE	EG	EG
Sulfuric acid, 60%	EG	EE	EG	EG	EE	GF	EG
Sulfuric acid, 98%	EG	EE	EE	EE	EE	NN	NN
Sulfur dioxide, liq., 46 psi	NN	FN	NN	NN	EE	GN	FN
Sulfur dioxide, wet or dry	EE	EE	EE	EE	EE	EG	EG
Sulfur salts	FN	GF	FN	FN	EE	FN	NN
Tartaric acid	EE	EE	EE	EE	EE	EG	EG
Tetrachloromethane	FN	GF	GF	NN	EE	NN	GF
Tetrahydrofuran	FN	GF	GF	FF	EE	NN	NN
Thionyl chloride	NN	NN	NN	NN	EE	NN	NN
Titanium salts	EE	EE	EE	EE	EE	EE	EE
Toluene	FN	GG	GF	FF	EE	FN	FN
Tributyl citrate	GF	EG	GF	GF	EE	NN	FN
Trichloroethane	NN	FN	NN	NN	EE	NN	NN
Trichloroethylene	NN	FN	NN	NN	EE	NN	NN
Triethylene glycol	EE	EE	EE	EE	EE	EG	GF
Tripropylene glycol	EE	EE	EE	EE	EE	EG	GF
Turkey red oil	EE	EE	EE	EE	EE	EG	EG
Turpentine	FN	GG	GF	FF	EE	FN	GF
Undecyl alcohol	EF	EG	EG	EG	EE	GF	EF
Urea	EE	EE	EE	EG	EE	NN	GN
Vinylidene chloride	NN	FN	NN	NN	EE	NN	NN
Xylene	GN	GF	FN	FN	EE	NN	NN
Zinc salts	EE	EE	EE	EE	EE	EE	EE
Zinc stearate	EE	EE	EE	EE	EE	EE	EG

* Key to Classification Code:
E—30 days of constant exposure cause no damage. Plastic may even tolerate exposure for years.
G—Little or no damage after 30 days of constant exposure to the reagent.
F—Some signs of attack after 7 days of constant exposure to the reagent.
N—Not recommended; noticeable signs of attack occur within minutes to hours after exposure. (However, actual failure might take years.)
† Resins (Code):
CPE: Conventional (low-density) polyethylene
LPE: Linear (high-density) polyethylene
PP: Polypropylene
PMP: Polymethylpentene
FEP: Teflon FEP (fluorinated ethylene propylene). Teflon is a Du Pont registered trademark.
ETFE: Tefzel ethylene-tetrafluoroethylene copolymer. (For chemical resistance, see FEP ratings.) Tefzel is a Du Pont registered trademark.
PC: Polycarbonate
PVC: Rigid polyvinyl chloride
‡ 1st letter: at room temp. → EG ← 2d letter: at 52°C.
SOURCE: The Nalge Co.

uct, it is important that you test your own conditions. If any doubt exists about specific applications of plastic products, you should contact your supplier.

CAUTIONS Do not store strong oxidizing agents in plastic labware except that made of Teflon®. Prolonged exposure causes embrittlement and failure.

Do not place plastic labware in a direct flame or on a hot plate.

THE CHEMISTRY OF PLASTICS

This is a very complex subject and is largely beyond the scope of this book. However, technicians should know the names of the most common plastics as well as their chemical composition or structure so that they can follow the resistance chart and decide which kind of equipment is most suitable for their purposes. This information is summarized in Tables 17-2 and 17-3. As you can see, there are relatively few basic types, and you will soon be able to correlate trade names and types and thus choose your labware efficiently.

THE USE AND CARE OF PLASTIC LABWARE

Cleaning

Linear polyethylene, polypropylene, PMP, Teflon FEP®, Tefzel, and polycarbonate plastics can be cleaned in ordinary glassware-washing machines. Ultrasonic cleaners may also be used provided the product does not rest directly upon the transducer diaphragm. Abrasive cleaners, scouring pads, and strong oxidizing agents should be avoided.

Polyolefins

Polyolefins is the generic name for the family of plastics which includes conventional polyethylene, linear polyethylene, polypropylene, and polymethylpentene. Because of their adhesion resistance, washing in warm soap or detergent solution usually suffices. Autoclavable items should be washed free of detergent or other cleaning agents before being sterilized. Drying in hot air may shorten the life of polyolefins because of oxidation.

Occasionally, organic solvents such as alcohols or methylene chloride are preferred for removing materials which do not respond to aqueous solutions. Frequently a concentrated solution of sodium carbonate or bicarbonate removes acidic substances. For basic materials, a weakly acid solution is recommended. (It is generally wise to follow manufacturers' recommendations in this matter.) Polymethylpentene can be washed in a warm, soapy solution. Abrasives should not be used.

TABLE 17-2
Basic Molecular Structure of Polymeric Materials

Name	*Polymers*	*Structure compared to polyethylene*
Polyethylene	—	$\left[CH_2-CH_2-CH_2-CH_2-CH_2-CH_2 \right]_n$
Polypropylene	—	$\left[CH_2-CH(CH_3)-CH_2-CH(CH_3)-CH-CH(CH_3) \right]_n$
Teflon®	Polytetrafluoroethylene	$\left[CF_2-CF_2-CF_2-CF_2-CF_2-CF_2 \right]_n$
Vinyl	Polyvinyl chloride	$\left[CH_2-CH(Cl)-CH_2-CH(Cl)-CH_2-CH(Cl) \right]_n$
Polystyrene		$\left[CH_2-CH(C_6H_5)-CH_2-CH(C_6H_5)-CH_2-CH(C_6H_5) \right]_n$
Acrylic	Polymethyl methacrylate	$\left[CH_2-C(CH_3)(C(=O)-O-CH_3)-CH_2-C(CH_3)(C(=O)-O-CH_3)-CH_2-C(CH_3)(C(=O)-O-CH_3) \right]_n$
Nylon	Several polyamide types (6,6′ shown)	$\left[NH-(CH_2)_6-NH-C(=O)-(CH_2)_4-C(=O) \right]_n$
Mylar®	Polyethylene terephthalate	$\left[CH_2-CH_2-O-C(=O)-C_6H_4-C(=O)-O \right]_n$
Lexan®	Polycarbonate	$\left[C_6H_4-C(CH_3)_2-C_6H_4-O-C(=O)-O \right]_n$
Cellophane®	Cellulose	Contains CH, C—OH and C—O—C groups
Acetate	Cellulose acetate	Contains CH, C—OH, —C=O, and C—C(=O)—O—C groups
Butyrate	Cellulose butyrate	Contains CH, C—OH, —C=O, and C—C(=O)—O—C groups

TABLE 17-3
Trade Name, Composition, and Softening Point of Common Plastics

Trade name	*Composition*	*Softening point, °C*
Bakelite	Phenol-formaldehyde	135(d)*
Formica	Melamine-formaldehyde	——
Formvar	Polyvinyl formal	190
Geon	Vinyl chloride–vinylvinylidene copolymer	——
Kel-F	Polychlorotrifluoroethylene	——
Melmac	Melamine formaldehyde	125(d)
Nylon	Polyhexamethylene-adipamide	65(d)
Paraplex	Polyester; acrylic-modified polyester	——
Plexiglas	Polymethyl methacrylate	75(d)
polyethylene†	——	105(d)
polystyrene†	——	85(d)
PVC†	Polyvinyl chloride	54(d)
Saran	Vinylidene chloride—vinyl chloride copolymer	150
Styrofoam	Foamed polystyrene	85
Teflon	Polytetrafluoroethylene	300

* (d) = distortion.
† Not a trade name.

Polycarbonate

Strongly alkaline cleaning agents and organic solvents should not be used for cleaning plastic ware made from polycarbonate. Items should be rinsed with distilled or demineralized water before autoclaving.

Other Plastics

Soap or detergent solutions can be used when cleaning other plastics.

Sterilizing

Polypropylene, polymethylpentene, Tefzel, and Teflon FEP® may be autoclaved repeatedly under normal conditions. Linear polyethylene may be autoclaved with caution. Polycarbonate shows some loss of mechanical strength when autoclaved, and autoclaving should be limited to 20 min at 121°C. We recommend that items be cleaned and rinsed with distilled water before being sterilized because certain chemicals which have no appreciable effects on the plastic at room temperature may cause deterioration at autoclaving temperatures.

Polystyrene, polyvinyl chloride, styrene-acrylonitrile, and conventional polyethylene are not autoclavable. They may be gas-sterilized or chemically sterilized (by rinsing with benzalkonium chloride, formalin, or other agents).

None of the materials listed above, with the exception of Teflon FEP®, should be hot-air-sterilized because of the possibility of accelerated oxidative degradation.

Autoclaving Plastic Bottles

Be sure that closures are very loose during and after autoclaving until the bottles are fairly cool. If this is not done, a partial vacuum will form inside a bottle and cause some distortion or collapse of its walls.

Some transparent plastics and Nalgon® vinyl tubing (which may be autoclaved, too) may absorb minute quantities of water vapor and appear cloudy after autoclaving. This clouding is not cause for any concern and will disappear as the plastic dries. Clearing may be accelerated by placing the plastic ware in a drying oven at 110°C.

TUBING: SYNTHETIC AND NATURAL

Tygon® Tubing

Tygon® is a modified polyvinyl plastic. Tygon® tubing (Fig. 17-1) is one of the most useful plastic materials you will encounter in the laboratory. It is very clear and so resistant to chemical attack that it will handle practically any chemical. It can be used to transfer liquids, gases, and slurries. It comes in a variety of sizes (see Table 17-4) and is specified by inside diameter (ID), by outside diameter (OD), and by wall thickness.

Putting Tygon® Tubing into Service

Tygon® tubing is quickly and easily installed and is readily moved to meet changing needs. Because of its flexibility, it will curve around corners and snake around obstructions, occupying a minimum of space. Because it is

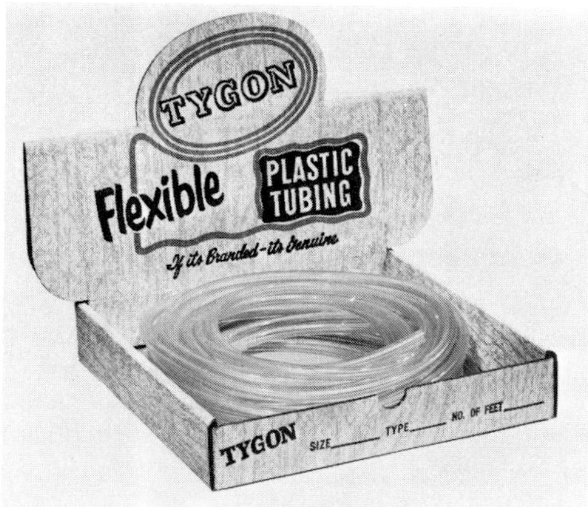

(a)

(b)

FIGURE 17-1 Tygon tubing.

TABLE 17-4
Specifications for Tygon® Tubing

Tubing bore or ID, in	*OD, in*	*Wall thickness, in*	*Maximum suggested working pressure at room temp., lb/in²*
1/16	3/16		75
3/32	7/32		70
1/8	1/4		65
5/32	9/32		65
3/16	5/16		55
1/4	3/8	1/16	55
5/16	7/16		50
3/8	1/2		45
7/16	9/16		40
1/2	5/8		30
3/16	3/8		65
1/4	7/16		60
5/16	1/2		60
3/8	9/16		50
7/16	5/8	3/32	45
1/2	11/16		40
9/16	3/4		40
5/8	13/16		35
11/16	7/8		30
3/16	7/16		75
1/4	1/2		70
5/16	9/16		70
3/8	5/8		60
7/16	11/16		50
1/2	3/4		45
9/16	13/16		45
5/8	7/8	1/8	40
11/16	15/16		40
3/4	1		35
7/8	1 1/8		30
1	1 1/4		25
1 1/8	1 3/8		25
1 1/4	1 1/2		20
1/2	13/16		60
9/16	7/8		55
5/8	15/16		50
11/16	1	5/32	45
3/4	1 1/16		35
7/8	1 3/16		35
1	1 5/16		30
3/4	1 1/8		50
1	1 3/8		40
1 1/8	1 1/2	3/16	35
1 1/4	1 5/8		35
1 1/2	1 7/8		30
3/4	1 1/4		55
1	1 1/2		45
1 1/4	1 3/4		45
1 1/2	2		40
1 3/4	2 1/4	1/4	40
2	2 1/2		35
2 1/4	2 3/4		35
2 1/2	3		30
3	3 1/2		25
2	2 3/4		40
2 1/2	3 1/4	3/8	35
3	3 3/4		30
2	3		50
3	4	1/2	40
4	5		30

available in continuous lengths, Tygon® tubing requires few joints or couplings.

Putting Tygon® tubing into service requires only the simplest of tools and equipment. It is readily cut with any sharp knife. If you are working with large-bore and/or thick-walled tubing, cutting can be done more easily if you lubricate the blade with a soapy water solution. A small amount of solvent or a means of softening the tubing ends with heat is also required if Tygon®-to-Tygon® couplings are necessary.

Couplings

Two lengths of Tygon® tubing are easily joined together by either heat sealing or solvent sealing.

In the first method, a soldering iron or heated knife can be used to soften the Tygon® at the ends to be coupled. A little experimentation will soon show the proper temperature (approximately 210°F): high enough to make a positive connection, yet not so hot that the Tygon® will become charred.

A good way of using the second method and making a tight Tygon®-to-Tygon® coupling is to utilize a sleeve of Tygon® tubing whose inside diameter is the same as the outside diameter of the two lengths to be joined. After making sure that the ends of the latter two are cut square, dip them into a solvent of the ketone type, such as Tygobond 50, leaving them immersed for about a minute to allow the tubing ends to soften. Then, slip these two ends inside the sleeve until they touch. After a few hours' drying time, the joint will be as strong as the tubing itself.

Connections to Metal or Glass Fittings

Tygon's natural tack will sometimes make it difficult to slip the tubing over metal, glass, or other fittings in making a connection. If this should happen, simply dip the end of the tubing into soap and water or one of the synthetic wetting agents. This will have the effect of lubricating the Tygon®, allowing it to slip on easily.

Y Fittings

For setups requiring branch lines and multiple connections, Y fittings made of Tygon® are available. Eight sizes are manufactured, with inside diameters approximating the outside diameters of standard tubing sizes. When a permanent connection is desired, Y fittings are easily and quickly joined to the tubing by using a ketone-type solvent such as Tygobond 50.

Clamps

Selecting a clamp for use with Tygon® tubing is as simple as picking one for use with rubber hose. Generally, a simple collar clamp will serve very satisfactorily. Stainless steel fittings are usually preferred in food-handling applications. An added convenience is afforded by clamps furnished with thumbscrews for hand tightening. Avoid clamps with sharp or ragged edges, since these may cut the tubing as the clamp is tightened.

Effect of Contact between Tygon® and Dissimilar Materials

With few exceptions, Tygon® tubing neither affects nor is affected by other materials, even with prolonged contact. It is compatible with all thermoplastics except polystyrene and acrylics. It can be used safely with all thermosetting plastics and resins. Although it is compatible with both natural and synthetic rubber, Tygon® may absorb some of the color from rubber stocks colored with organic dyes. However, this color transfer does not affect the physical or chemical properties of Tygon® in any way. Tygon® may become discolored when in contact at high temperatures with zinc, copper, or high-carbon steel. Because the plasticizers used in Tygon® tubing may soften some types of finishes, care should be exercised when Tygon® tubing is in contact with painted, lacquered, varnished, or enameled wood or metal surfaces.

Chemical-Resistance Characteristics of Standard Tygon® Tubing Formulations

Based on the results of both laboratory and field tests, the ratings contained in Tables 17-5 and 17-6 reflect the relative abilities of five standard Tygon® tubing formulations to withstand a specific corrosive. No guarantee is expressed or should be implied.

Various Tygon® tubings vary in their ability to resist chemical attack. Table 17-6 shows, for example, that whereas all formulations may be used with dilute sodium hydroxide, reports indicate that only formulations R-3603, R-2400, and R-3400 have proved satisfactory for use with concentrated sodium hydroxide. Again, formulations R-3603, R-2400, and R-3400 are shown as suitable for handling both wet and dry chlorine gas, while R-2807 will handle only wet chlorine gas.

Remember that all ratings are based on room temperature. Higher temperatures may affect the ability of Tygon® tubing to withstand the attack of some chemicals, particularly those rated C.

Teflon® Tubing

Teflon® tubing is chemically inert to synthetic lubricants, all acids, alkalis, organic solvents, and salt spray. It is also unaffected by fungi. It will withstand continuous temperatures to 500°F and will retain flexibility to −350°F. It does not absorb moisture and will not swell or disintegrate. It is available in four sizes (Table 17-7).

TABLE 17-5
Chemical-Resistance Ratings of Tygon® Tubing*

Chemicals	*Tygon® formulations*				
	B44-3 Crystal Clear	*R-3400 Black*	*R-2400 Black*	*R-3603 Clear*	*R-2807 Black*
Acetaldehyde	N	N	N	N	N
Acetates (low mol. wt.)	N	N	N	N	N
Acetic acid (less than 5%)	X	X	X	X	X
Acetic acid (more than 5%)	X	X	X	X	C
Acetic anhydride	C	C	C	N	N
Acetone	N	N	N	N	N
Acetyl bromide	N	N	N	N	N
Acetyl chloride	N	N	N	N	N
Air	X	X	X	X	X
Alcohols	X	X	X	X	X
Aliphatic hydrocarbons	X	X	X	X	X
Aluminum chloride	X	X	X	X	X
Aluminum sulfate	X	X	X	X	X
Alums	X	X	X	X	X
Ammonia (gas-liquid)	C	X	X	X	C
Ammonium acetate	X	X	X	X	X
Ammonium carbonate	X	X	X	X	X
Ammonium chloride	X	X	X	X	X
Ammonium hydroxide	C	X	X	X	C
Ammonium nitrate	X	X	X	X	X
Ammonium phosphate	X	X	X	X	X
Ammonium sulfate	X	X	X	X	X
Amyl acetate	N	N	N	N	N
Amyl alcohol	X	X	X	X	X
Amyl chloride	C	C	C	C	C
Aniline	C	C	C	N	N
Aniline hydrochloride	C	C	C	N	N
Animal oils	X	X	X	X	C
Antimony salts	X	X	X	X	X
Aqua regia	N	N	N	N	N
Aromatic hydrocarbons	Use R-4000†				
Arsenic salts	X	X	X	X	X
Barium salts	X	X	X	X	X
Benzaldehyde	N	N	N	N	N
Benzene	C	X	X	C	N
Benzenesulfonic acid	C	C	C	C	N
Benzoic acid	X	X	X	X	X
Benzyl alcohol	X	X	X	X	X
Bleaching liquors	X	X	X	X	X
Boric acid	X	X	X	X	X
Bromine‡	X	X	X	X	X
Butane	C	C	C	C	N
Butanol	X	X	X	X	X
Butyl acetate	N	N	N	N	N
Butyric acid	Use R-4000				
Calcium salts	X	X	X	X	X
Carbon bisulfide	N	N	N	N	N
Carbon dioxide	X	X	X	X	X
Chloracetic acid	N	N	N	N	N
Chlorbenzene	N	N	N	N	N
Chlorine (wet)	C	X	X	X	X
Chlorine (dry)	C	X	X	X	N
Chloroform	C	C	C	C	C
Chlorsulfonic acid	C	C	C	C	C
Chromic acid	X	X	X	X	X
Chromium salts	X	X	X	X	X
Copper salts	X	X	X	X	X
Cresol	C	N	N	N	N
Cyclohexanone	N	N	N	N	N
Essential oils	C	X	X	X	C
Ethers	C	C	C	C	C
Ethyl acetate	N	N	N	N	N
Ethyl alcohol	C	C	C	C	C
Ethyl bromide	N	N	N	N	N
Ethyl chloride	N	N	N	N	N
Ethylamine	N	N	N	N	N

TABLE 17-5 (Continued)

Chemicals	Tygon® formulations				
	B44-3 Crystal Clear	*R-3400 Black*	*R-2400 Black*	*R-3603 Clear*	*R-2807 Black*
Ethylene chlorohydrin	N	N	N	N	N
Ethylene dichloride	N	N	N	N	N
Ethylene glycol	X	X	X	X	X
Fatty acids	C	X	X	X	C
Ferric chloride	X	X	X	X	X
Ferric sulfate	X	X	X	X	X
Ferrous chloride	X	X	X	X	X
Ferrous sulfate	X	X	X	X	X
Fluoborate salts	X	X	X	X	X
Fluoboric acid	X	X	X	X	X
Fluosilicic acid	X	X	X	X	X
Formaldehyde	X	X	X	X	X
Formic acid	X	X	X	X	X
Freon	N	N	N	N	N
Gasoline (nonaromatic)	Use R-4000				
Gasoline (high aromaticity)	Use R-4000†				
Glucose	X	N	N	X	X
Glue	X	X	X	X	X
Glycerin	X	X	X	X	X
Hydriodic acid	X	X	X	X	X
Hydrobromic acid‡	X	X	X	X	X
Hydrochloric acid (dil.)	X	X	X	X	X
Hydrochloric acid (med. conc.)	X	X	X	X	X
Hydrochloric acid (conc.)	C	X	X	X	X
Hydrocyanic acid	X	X	X	X	X
Hydrofluoric acid	C	X	X	X	X
Hydrogen peroxide (dil.)	X	X	X	X	X
Hydrogen peroxide (conc.)§					
Hydrogen sulfide	X	X	X	X	X
Hypochlorous acid	X	X	X	X	X
Iodine‡ and solutions‡	X	X	X	X	X
Kerosene	Use R-4000				
Ketones	N	N	N	N	N
Lacquer solvents	N	N	N	N	N
Lactic acid	C	X	X	X	C
Lead acetate	X	X	X	X	X
Linseed oil	X	X	X	X	X
Magnesium chloride	X	X	X	X	X
Magnesium sulfate	X	X	X	X	X
Malic acid	X	X	X	X	X
Manganese salts	X	X	X	X	X
Mercury salts	X	X	X	X	X
Mixed acid (40% sulfuric, 15% nitric)	C	C	X	C	C
Naphtha	C	C	C	C	N
Natural gas	X	X	X	X	X
Nickel salts	X	X	X	X	X
Nitric acid (dil.)	X	X	X	X	X
Nitric acid (med. conc.)	C	X	X	X	X
Nitric acid (conc.)	C	C	X	C	X
Nitrobenzene	N	N	N	N	N
Nitrogen oxides	X	X	X	X	X
Nitrous acid	X	X	X	X	X
Oils, animal	Use R-4000				
Oils, mineral	Use R-4000				
Oils, vegetable	C	N	N	C	N
Oleic acid	N	X	X	C	C
Oxalic acid	X	X	X	X	X
Oxygen (gas)	X	X	X	X	X

TABLE 17-5 (Continued)

Chemicals	Tygon® formulations: B44-3 Crystal Clear	R-3400 Black	R-2400 Black	R-3603 Clear	R-2807 Black
Perchloric acid	N	N	N	N	N
Phenol	C	C	C	C	C
Phosphoric acid (ortho-)	X	X	X	X	X
Phthalic acid	X	X	X	X	C
Plating solutions	X	X	X	X	X
Potassium carbonate	X	X	X	X	X
Potassium chlorate	X	X	X	X	X
Potassium hydroxide	C	X	X	X	X
Potassium iodide‡	X	X	X	X	X
Pyridine	N	N	N	N	N
Silver nitrate	X	X	X	X	X
Soap solutions	X	X	X	X	X
Sodium bicarbonate	X	X	X	X	X
Sodium bisulfate	X	X	X	X	X
Sodium bisulfite	X	X	X	X	X
Sodium borate	X	X	X	X	X
Sodium carbonate	X	X	X	X	X
Sodium chloride	X	X	X	X	X
Sodium ferrocyanide	X	X	X	X	X
Sodium hydrosulfite	X	X	X	X	X
Sodium hydroxide (dil.)	X	X	X	X	X
Sodium hydroxide (med. conc.)	C	X	X	X	X
Sodium hydroxide (conc.)	C	X	X	X	C
Sodium hypochlorite	X	X	X	X	X
Sodium nitrate	X	X	X	X	X
Sodium silicate	X	X	X	X	X
Sodium sulfide	X	X	X	X	X
Sodium sulfite	X	X	X	X	X
Stearic acid	X	X	X	X	X
Sulfur chloride	C	C	C	C	C
Sulfur dioxide	X	X	X	X	X
Sulfur trioxide	X	X	X	X	X
Sulfuric acid (dil.)	X	X	X	X	X
Sulfuric acid (med. conc.)	X	X	X	X	X
Sulfuric acid (conc.)	C	C	C	C	C
Sulfurous acid	X	X	X	X	X
Tannic acid	X	X	X	X	X
Tanning extracts	X	X	X	X	X
Tartaric acid	X	X	X	X	X
Tetrachloromethane	C	C	C	C	C
Tin salts	X	X	X	X	X
Titanium salts	X	X	X	X	X
Toluol (Toluene)	N	N	N	N	N
Trichloracetic acid	N	N	N	C	C
Trisodium phosphate	X	X	X	X	X
Turpentine	X	X	X	X	C
Urea	X	X	X	X	X
Uric acid	X	X	X	X	X
Water	X	X	X	X	X
Water (brine)	X	X	X	X	X
Xylol	N	N	N	N	N
Zinc chloride	X	X	X	X	X

* Code: X—satisfactory; C—use only after test in your own plant; N—not suitable.
† Use only after test in your own plant.
‡ May absorb small quantities and impart characteristic color to tubing.
§ A special Tygon® formulation R-3604 is available for use with concentrated hydrogen peroxide.

TABLE 17-6
Chemical Resistance of Tygon® Tubing* to Food Products

Product	*Rating*	*Product*	*Rating*
Alcohol	C	Milk†	X
Beer	C	Milk of magnesia	X
Brandy	C	Molasses	X
Butter	X	Sauerkraut	X
Carrot	X	Sea foods	X
Certo	X	Sugar	X
Chocolate syrup	X	Tomato	X
Citric acid	X	Vegetable oil	X
Fish	X	Vinegar	X
Fruit juices	X	Whiskey	C
Karo syrup	X	Wines	C
Mayonnaise	X		

* Formulation B44-3 only.
† Use formulation B44-4X.

TABLE 17-7
Sizes of Teflon® Tubing

ID, in	*Wall thickness, in*
1/8	0.030
3/16	0.030
1/4	0.030
3/8	0.030

Rubber Tubing

Rubber tubing (Fig. 17-2) is available in pure latex, gum, red rubber, black rubber, black neoprene, and white rubber. It is available in various wall diameters and bores from ⅛-in ID and 3/16-in OD to 1-in ID and 1¼-in OD. (See Table 17-8.)

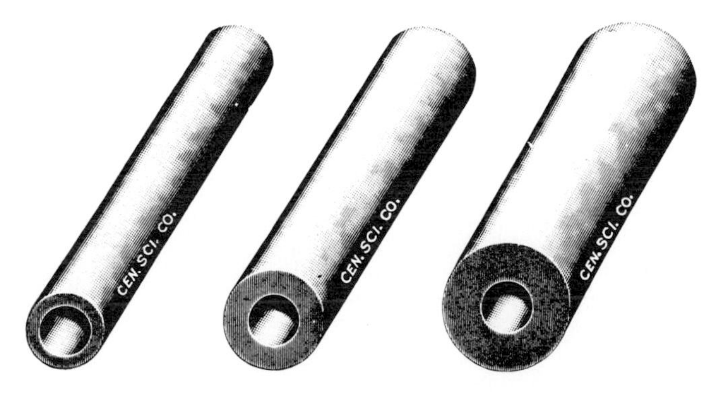

FIGURE 17-2
Rubber tubing.

TABLE 17-8
Rubber-Tubing Sizer

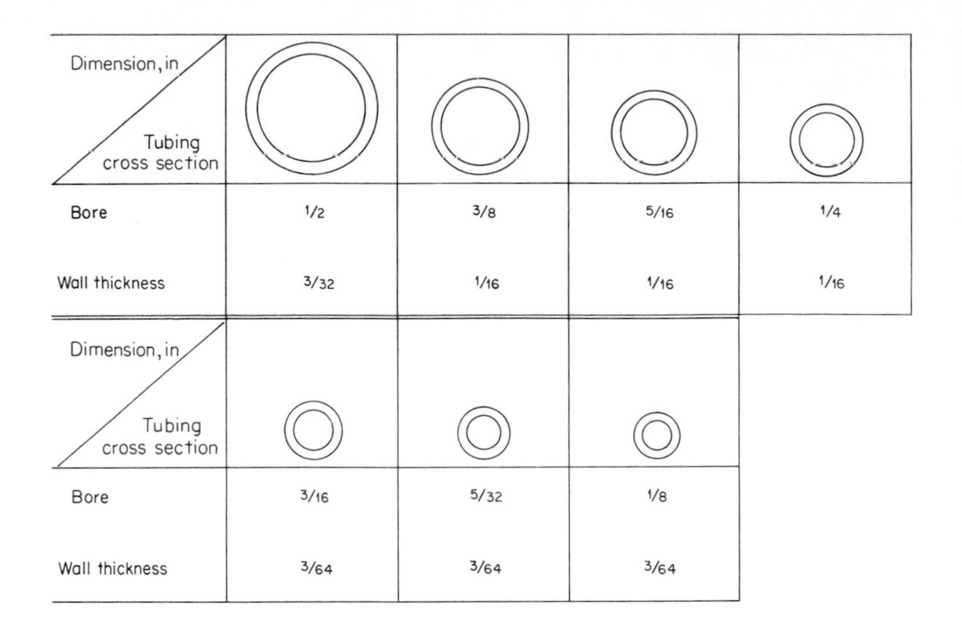

Dimension, in / Tubing cross section				
Bore	1/2	3/8	5/16	1/4
Wall thickness	3/32	1/16	1/16	1/16

Dimension, in / Tubing cross section			
Bore	3/16	5/32	1/8
Wall thickness	3/64	3/64	3/64

Amber latex rubber tubing is translucent and is recommended for all kinds of glass connections in which a high-quality clinging tubing, highly elastic and of long life, is desired. Special packing on reels in dispenser boxes (except sizes ¼ × ⅛ in, ⅜ × 3/32 in, ½ × 1/16 in, and ½ × ⅛ in) is available. (See Table17-9.)

TABLE 17-9
Sizes of Latex Tubing

ID, in	*Thickness of wall, in*	*Feet to reel or box*
1/8	1/32	100
3/16	1/16	100
3/16	3/32	50
1/4	1/16	50
1/4	1/8	50 (box)
5/16	1/16	50
5/16	3/32	50
3/8	3/32	50 (box)
1/2	1/16	50 (box)
1/2	1/8	25 (box)

Clamping Rubber Tubing

Use wire clamps or screw clamps to secure tubing to inlet and outlet fittings (Fig. 17-3*a* and *b*). Sudden changes in pressure or accidental compression of the tubing will cause it to slip off if it is not securely clamped. Never use old, cracked, or brittle tubing. Fluid-flow-control clamps are shown in Fig. 17-4.

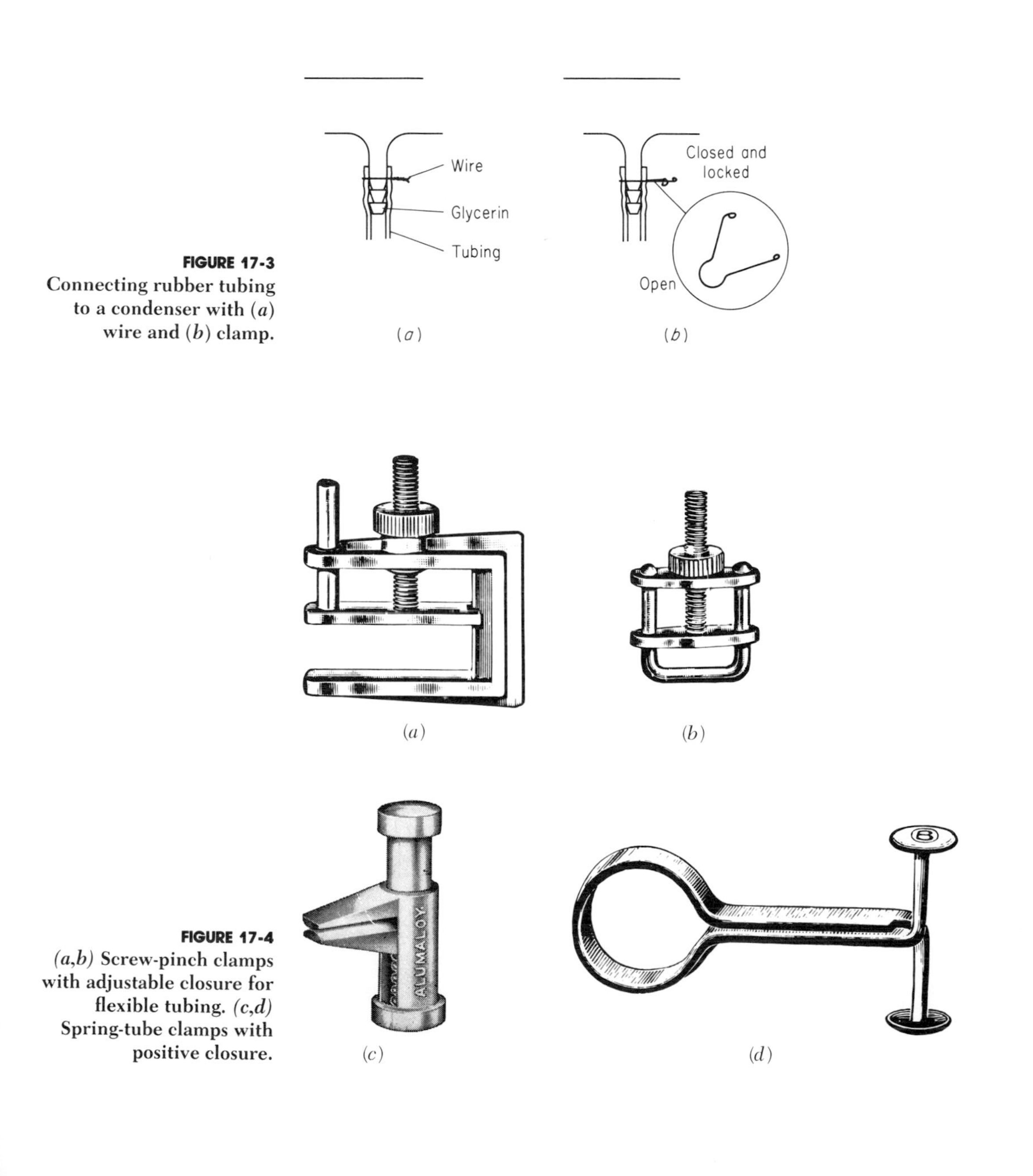

FIGURE 17-3 Connecting rubber tubing to a condenser with (*a*) wire and (*b*) clamp.

FIGURE 17-4 *(a,b)* Screw-pinch clamps with adjustable closure for flexible tubing. *(c,d)* Spring-tube clamps with positive closure.

Neoprene Tubing

Neoprene is a synthetic rubber. Neoprene tubing is soft and pliable with a smooth finish. It is not affected by oil, gasoline, alkalis, turpentine, grease, hot water, and many corrosive materials. It is not recommended for use with chlorinated or aromatic hydrocarbons. It is available in the sizes shown in Table 17-10.

TABLE 17-10
Sizes of Neoprene Tubing

ID, in	*Wall thickness, in*
1/8	1/4
1/4	3/8
3/8	5/8

18

LABORATORY TOOLS AND HARDWARE

CORKS AND RUBBER STOPPERS

These useful items form parts of many pieces of apparatus that you will use in the laboratory. Proper size of both the stopper and the hole in it are most important. It is often necessary to bore or enlarge a hole in a cork or rubber stopper (Figs. 18-1 and 18-2 and Table 18-1). If it is available, a cork-boring machine (Fig. 18-3) does the job quite easily. However, it is more likely that you will have to use a borer (Fig. 18-4). The correct procedure is outlined below.

1. Select a rubber or cork stopper of the correct size (Fig. 18-5).
2. Select a borer of slightly smaller size than the hole to be bored.
3. Sharpen the borer with a sharpener (Fig. 18-6).
4. Wet the borer with glycerin.
5. Hold the borer in the right hand and the stopper in the left hand; support the stopper with a cloth pad for palm protection. (Reverse hands if you are left-handed.)
6. Begin boring at the narrow end. Twist the borer and apply steady pressure.
7. Check the alignment of the borer after each twist.
8. Remove the borer when halfway through the stopper. Ream out any plug in the borer.
9. Begin boring the stopper from the other end; check alignment.
10. Complete the boring operation.

Inserting Glass Tubing into Stoppers

To insert glass tubing into the stopper:

1. Fire-polish the ends of the tubing. (Refer to Chap. 16, "Laboratory Glassware.")
2. Wet the tubing and stopper hole with water or glycerin.
3. Wrap hands in a towel for protection.

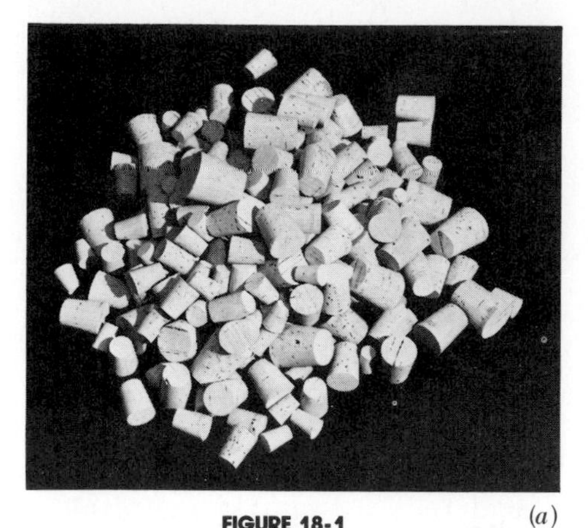

(*a*)

Cork Size No.

Size refers to top diameters

0 2 4 6 8 10 12 14 16 18 20 24

1 3 5 7 9 11 13 15 17 19 22 26

(*b*)

FIGURE 18-1
(*a*) Cork stoppers, various sizes. (*b*) Sizer for cork stoppers. Size refers to top diameter.

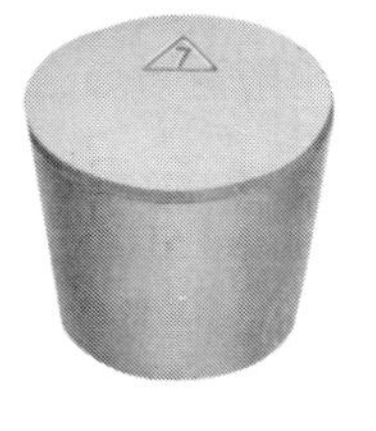

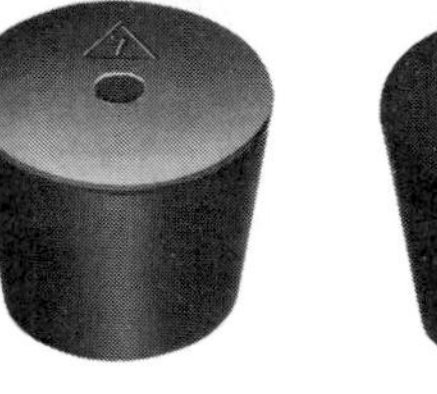

FIGURE 18-2
(*a*) Various sizes of one- or two-hole rubber stoppers. (*b*) Sizer for rubber stopper. Size refers to top diameter.

4. Grasp the tube near the point of insertion (Fig. 18-7).
5. Rotate the tube back and forth while gently pushing it into the stopper.
6. *Never attempt to push tubing through the stopper* (Fig. 18-8).

TABLE 18-1
Choosing the Correct Rubber Stopper

Stopper size No.	*Fits tubes, mm OD*	*or*	*Openings, mm ID*	*Top diameter, mm*	*Bottom diameter, mm*	*Length, mm*
00	12–15		10–13	15	10	26
0	16–18		13–15	17	13	26
1	19–20		15–17	19	15	26
2	20–21		16–18.5	20	16	26
3	22–24		18–21	24	18	26
4	25–26		20–23	26	20	26
5	27–28		23–25	27	23	26
5½	28–29		25–26	29	25	26
6	29–30		26–27	32	26	26
6½	30–34		27–31.5	34	27	26
7	35–38		30–34	37	30	26
8	38–41		33–37	41	33	26
9	41–45		37–41	45	37	26
10	45–50		42–46	50	42	26
10½	48–51		45–47	53	45	26
11	52–56		48–51.5	56	48	26
11½	57–61		51–56	60	51	26
12	62–64		54–59	64	54	26
13	64–68		58–63	67	58	26
13½	68–75		61–70	75	61	35
14	80–90		75–85	90	75	39
15	92–100		83–95	103	83	39

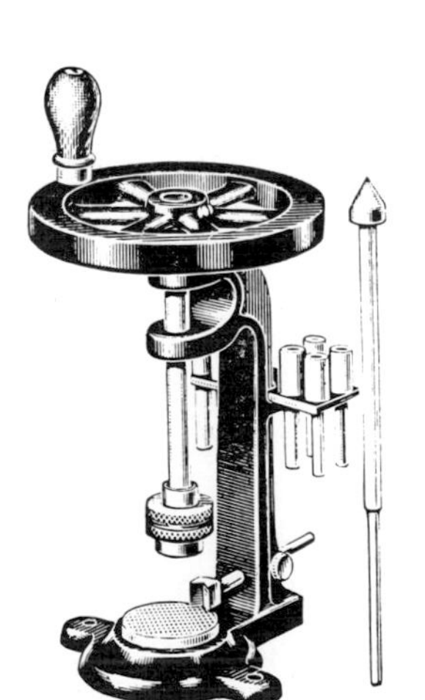

FIGURE 18-3
Cork-boring machine, supplied with different borers, gives straight-line borer holes.

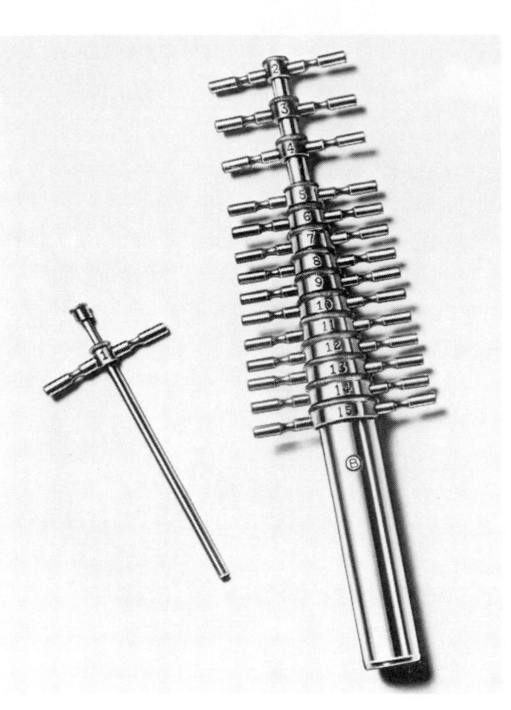

FIGURE 18-4
Set of cork borers. They are made of brass, and each set consists of borers of many sizes and a center plug punch.

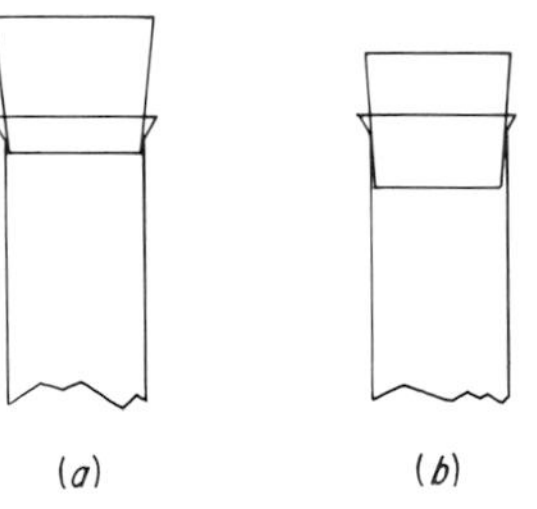

FIGURE 18-5
Rubber stoppers. (*a*) Correct size. (*b*) Too small.

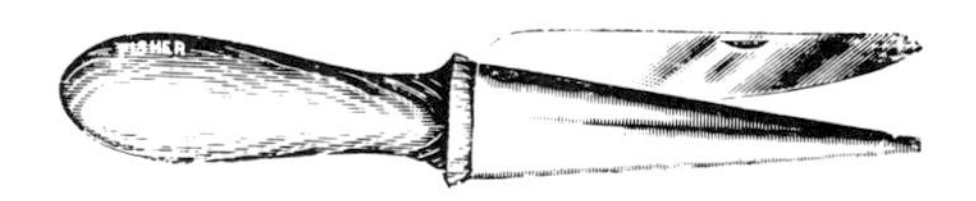

FIGURE 18-6
Brass-cone, steel-bladed cork-borer sharpener; used by rotating borer with gentle pressure on knife.

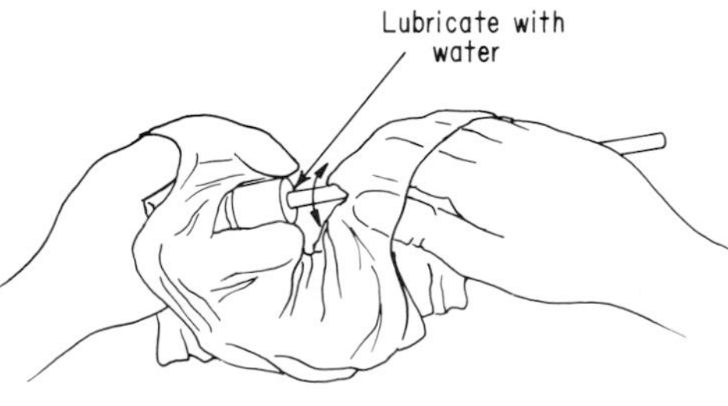

FIGURE 18-7
Proper way to insert glass tubing in a stopper.

FIGURE 18-8
Improper way to insert glass tubing in a stopper.

Separating Glass Tubing from Stoppers

To remove glass tubing and/or thermometers from stoppers:

1. Lubricate the tubing which is to be pulled from the stopper with water or glycerin.

2. Wrap the tubing with a towel.

3. Pull the tubing from the stopper with a gentle twisting motion.

4. If the tubing is stuck to the stopper, gently insert the end of a rat-tailed file between the tubing and the stopper and rotate gently, while lubricating with glycerin.

5. If step 4 fails, choose the smallest cork borer which fits over the tubing and gently work it through the stopper.

6. If all other efforts fail, cut the stopper away from the glass with a sharp knife (Figs. 18-9 and 18-10). Stoppers are inexpensive; thermometers are expensive.

Separating Glass Tubing from Rubber Tubing

When you must remove glass tubing that has frozen to rubber tubing:

1. Take a small file and gently insert the point of the handle into the glass-rubber junction, gently forcing the point to separate the rubber from the glass. Wrap the glass with a towel.

2. Lubricate by dropping some glycerin in the space created.

3. Rotate the file slowly while applying more pressure to force it in more deeply. Lubricate again after the file makes a complete revolution.

4. Remove the file and repeat steps 2 and 3 from the opposite end.

FIGURE 18-9
(Above) Sharp knife, with serrated blade, for cutting corks.

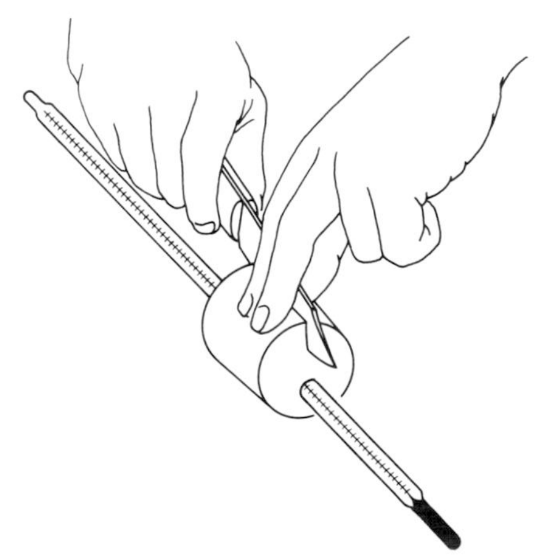

FIGURE 18-10
(Right) Cutting off a frozen stopper.

STAINLESS STEEL IN THE LABORATORY

Stainless steel is one of the best materials commercially available; it is very utilitarian, economical, and durable and has esthetic appeal. However, stainless steel is only corrosion-resistant. It is not corrosion-proof. Many tend to believe that products fabricated out of stainless steel are indestructible and require no preventive maintenance or service. Such is not the case. There are many different types of stainless steel, each designed to do a specific job well. Some are formulated for decorative purposes, some for special conditions of corrosion, others for combinations of machining properties and corrosion resistance. Each type of stainless steel formation is designated by a number, which is standard and which designates the characteristics of that particular stainless steel. For example, Type No. 304 is the corrosion-resistant stainless steel most widely used in the laboratory because it is readily obtainable, economical, and very practical.

Deterioration and Corrosion of Stainless Steel

Standard operating procedure requires that any object which is contaminated or splashed with corrosives be washed thoroughly with water. Even though this procedure may be followed scrupulously, we find that laboratory equipment fabricated with stainless steel tends to corrode and deteriorate. Even ordinary water can cause this to happen. Analyses of tap water throughout the country reveal that the composition and concentration of organic impurities, ions, pH, and dissolved minerals vary greatly. All halogens tend to pit stainless steel, and the concentration of chlorides, temperature, dissolved oxygen content, and time of contact affect the amount of deterioration and pitting. Even distilled water will have a deleterious effect upon stainless steel if it is left in contact with the metal for extended periods of time.

Corrosives may be spilled or splashed onto stainless steel and not be noticed because the water evaporates. Yet the residues may accumulate because of additional spills and, combined with the moisture in the air, gradually affect the surface. Stainless steel containers which are used for tap-water storage will accumulate calcium carbonate deposits, sometimes causing pitting beneath the surface of the steel, a phenomenon called *crevice corrosion.*

Corrosive vapors from reactions and industrial exhausts react with moisture in the air to form acidic conditions favorable to pitting and surface deterioration. Slight imperfections in the stainless steel, or microscopic pieces of the tool which was used in the fabrication of the item and which became imbedded in it, tend to form a surface rust. This rust can usually be removed by scrubbing with a stainless steel sponge, but it usually reappears.

Care of Stainless Steel Equipment

Observing the following rules will help to prolong the life of such equipment.

1. Periodically wash with a mild soap solution, then wipe clean with a clean, damp cloth and finally, dry with a soft, clean cloth.

2. Periodically scour the surface with a very mild abrasive cleanser that will not scratch or mar the surface (e.g., silver polish), follow by a thorough cleansing with a mild soap solution, and then wipe with clean, damp cloths, repeatedly rinsing out the cloths with water. Dry thoroughly.

3. Use solvent cleaners to remove grease, tar, oils, and fatty substances. Solvent degreasers are recommended.

4. *Never* use ordinary steel-wool cleaning pads to scrub stainless steel. They will scratch the surface and cause rust spots to appear. Use pads made out of stainless steel.

If you do not abuse or neglect your stainless steel equipment and if you maintain it properly, it can last indefinitely and continue to provide the esthetic appeal of shining stainless steel.

TOOLS USED IN THE LABORATORY

Materials of Construction

Steel is used to manufacture most tools, and the end application of the tool dictates the kind of steel needed. Other factors which are considered are hardness, toughness, strength, resistance to wear, and resistance to corrosion. Various formulations of steels include the addition of chromium, molybdenum, vanadium, nickel alloys, and different heat treatments to meet the need.

Normally, the more expensive tools are formulated with the hardest and longest-lasting steels. Naturally, they are harder to machine and to form or forge, and they do require more precise heat-treatment processing. Some of these steels are so extremely hard that they cannot be machined; they must be ground as needed. Although stainless steel is resistant to corrosion, stainless steel tools are not as durable, hard, or as long-lasting as tools made from other steels.

CAUTION Tools made of iron or steel tend to rust. *Keep all tools lightly oiled.*

Beryllium–copper alloys are nearly as hard as steel, but they are nonsparkling, nonmagnetic, and corrosion-resistant. These are to be used where there are operations using flammable and explosive solvents, and where dusts and hazardous gases are present. They are not quite a durable as steel tools, and they may be more expensive, but they should be used wherever sparks are dangerous.

Safety Rules for Using Hand Tools

1. Use the *correct tool* for the *correct job*. Never use wrenches as hammers or screwdrivers as chisels. Use the correct size of screwdriver to tighten or loosen screws. Too small a screwdriver blade distorts the screw opening. The screwdriver blade should fit the opening snugly.

Use the proper wrench to hold or to tighten nuts. *Do not use pipe wrenches on nuts* because the teeth of a pipe wrench will damage the face of a nut.

2. Use only tools that are in good order; discard those with broken handles, mutilated gripping surfaces, or broken sections. Do not use power tools with frayed, worn, or exposed conductors or with broken plugs.

3. Keep your tools clean, in good repair, oiled to prevent corrosion, and stored in their proper place.

4. Safety comes first, *always*, when using power tools. Wear safety glasses; use safety shields. Avoid breathing noxious fumes. Protect your hands with safety gloves—especially when applying force—to avoid injury because of slippage. Securely anchor objects when you must apply strong force to the tool to do a job.

5. Observe all electrical safety precautions, especially grounding (refer to Chap. 27, "Basic Electricity").

6. Plan your tool-use procedure. Do not work haphazardly. *Think:* anticipate all possible hazards and possible results. Keep your thoughts exclusively on the job. There is no room for horseplay or practical jokes when you are working with tools. Any distraction can result in serious injury or death.

7. To loosen frozen screws or nuts:

(**a**) Heat the outside part of the frozen assembly quickly, allow the heat to penetrate, and then apply pressure with a tool. If this is unsuccessful in loosening the assembly, apply penetrating oil.

(**b**) Spray the assembly with a *penetrating oil*, such as commercial Liquid Wrench or graphited oil. Tap gently and allow to stand. Repeat the operation several times, cautiously applying pressure with a tool.

(**c**) *When increasing the pressure applied to a tool by lengthening the lever with an extension, use caution. Think. Too much pressure causes breakage.*

Types of Laboratory Tools

Calipers

Calipers are often used (Fig. 18-11) to measure inside and outside diameters of openings, rods, shafts, and equipment. In measuring with calipers, adjust caliper to maximum diameter of cylinder (Fig. 18-11*a*). Measure opening with suitable scale or ruler (Fig. 18-11*b*). Micrometer calipers (Fig. 18-12) are also used for measurement.

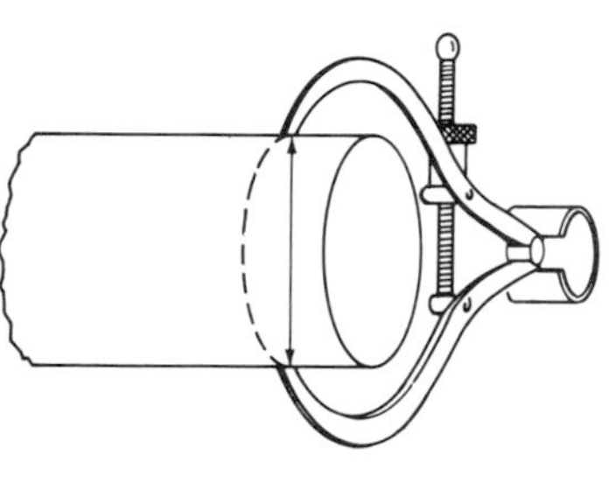

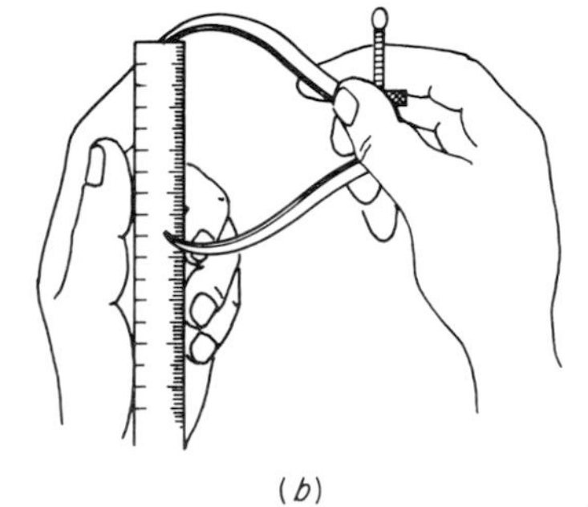

(*b*)

FIGURE 18-11 Measuring with calipers. (*a*) Calipers are set for size of cylinder. (*b*) Setting measured for distance between jaws.

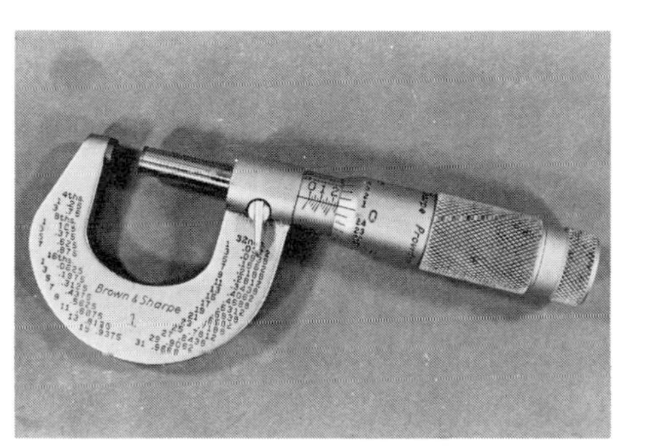

FIGURE 18-12 Micrometer caliper to measure thickness precisely.

Common micrometers measure diameters to 0.001 in. They are precision instruments and should be handled carefully.

Do not:

Drop them.

Apply excessive pressure; the frame may be sprung permanently.

Slide work between the tightened anvils; the surface will wear.

Always:

Open (release tension on) the anvils before removing work.

Treat with care and protect from corrosion.

Electric Drills

Electric drills (Fig. 18-13) are used with sharp bits to drill holes in steel, iron, plastic, wood, or other hard surfaces. Normally bits are available from $\frac{1}{64}$ to $\frac{1}{4}$ in (Tables 18-2 and 18-3). Exert moderate pressure when drilling to get "bite." *Do not* merely press lightly; you will only dull the point.

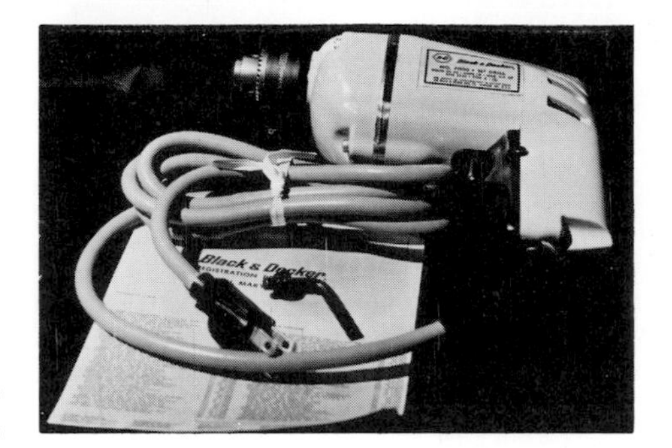

FIGURE 18-13
Electric drill.

TABLE 18-2
Bit Sizes

Bit no.	*Diam., in*	*Bit no.*	*Diam., in*	*Bit no.*	*Diam., in.*	*Bit no.*	*Diam., in*	*Bit no.*	*Diam., in*
1	0.2250	17	0.1730	33	0.1130	49	0.0730	65	0.0350
2	0.2210	18	0.1695	34	0.1110	50	0.0700	66	0.0330
3	0.2130	19	0.1660	35	0.1100	51	0.0670	67	0.0320
4	0.2090	20	0.1610	36	0.1065	52	0.0635	68	0.0310
5	0.2055	21	0.1590	37	0.1040	53	0.0595	69	0.0292
6	0.2040	22	0.1570	38	0.1015	54	0.0550	70	0.0280
7	0.2010	23	0.1540	39	0.0995	55	0.0520	71	0.0260
8	0.1990	24	0.1520	40	0.0980	56	0.0465	72	0.0250
9	0.1960	25	0.1495	41	0.0960	57	0.0430	73	0.0240
10	0.1935	26	0.1470	42	0.0935	58	0.0420	74	0.0225
11	0.1910	27	0.1440	43	0.0890	59	0.0410	75	0.0210
12	0.1890	28	0.1405	44	0.0860	60	0.0400	76	0.0200
13	0.1850	29	0.1360	45	0.0820	61	0.0390	77	0.0180
14	0.1820	30	0.1285	46	0.0810	62	0.0380	78	0.0160
15	0.1800	31	0.1200	47	0.0785	63	0.0370	79	0.0140
16	0.1770	32	0.1160	48	0.0760	64	0.0360	80	0.0135

Hacksaws

The hacksaw (Fig. 18-14) is used to cut iron or steel or other metals; blades can be fine to extremely coarse. Its sawtooth blade should point *forward*. Exert pressure *only* on the forward stoke, none on the reverse stroke.

TABLE 18-3
Conversion Chart: Millimeter, Fractional-Inch, and Decimal-Inch Equivalents

	In			*In*			*In*			*In*	
mm	*Frac.*	*Dec.*	*mm*	*Frac.*	*Dec.*	*mm*	*Frac.*	*Dec.*	*mm*	*Frac.*	*Dec.*
.91		0.0004	8.3344	21/64	0.3281	21.4312	27/32	0.8437	57		2.244
.02		0.0008	8.7312	11/32	0.3437	21.8281	55/64	0.8594	58		2.283
.03		0.0012	9.000		0.3543	22.000		0.8661	59		2.323
.04		0.0016	9.1387	23/64	0.3594	22.2250	7/8	0.875	60		2.362
.05		0.0020	9.505	3/8	0.375	22.6219	57/64	0.8906	61		2.402
.06		0.0024	9.9719	25/64	0.3906	23.000		0.9055	62		2.441
.07		0.0028	10.000		0.3937	23.0187	29/32	0.9062	63		2.480
.08		0.0032	10.3187	13/32	0.4062	23.4156	59/64	0.9219	64		2.520
.09		0.0035	10.7156	27/64	0.4219	23.8125	15/16	0.9375	65		2.559
.10		0.004	11.000		0.4331	24.000		0.9449	66		2.598
.20		0.008	11.1125	7/16	0.4375	24.2094	61/64	0.9531	67		2.638
.30		0.012	11.5094	29/64	0.4531	24.6062	31/32	0.9687	68		2.677
.3969	1/64	0.0156	11.9062	15/32	0.4687	25.000		0.9843	69		2.717
.40		0.0158	12.000		0.4724	25.0031	63/64	0.9844	70		2.756
.50		0.0197	12.3031	31/64	0.4844	25.400		1.000	71		2.795
.60		0.0236	12.700	1/2	0.500	26		1.024	72		2.835
.70		0.0276	13.000		0.5118	27		1.063	73		2.874
.7937	1/32	0.012	13.0968	33/64	0.5156	28		1.102	74		2.913
.80		0.0315	13.4937	17/32	0.5312	29		1.142	75		2.953
.90		0.0354	13.8906	35/64	0.5469	30		1.181	76		2.992
1.000		0.0394	14.000		0.5512	31		1.220	77		3.031
1.1906	3/64	0.0469	14.2875	9/16	0.5625	32		1.260	78		3.071
1.5875	1/16	0.0625	14.6844	37/64	0.5781	33		1.299	79		3.110
1.9844	5/64	0.0781	15.000		0.5906	34		1.339	80		3.150
2.000		0.0787	15.0812	19/32	0.5937	35		1.378	81		3.189
2.3812	3/32	0.0937	15.4781	39/64	0.6094	36		1.417	82		3.228
2.7781	7/64	0.1094	15.875	8/8	0.625	37		1.457	83		3.268
3.000		0.1181	16.000		0.6299	38		1.498	84		3.307
3.175	1/8	0.125	16.2719	41/64	0.6406	39		1.535	85		3.346
3.5719	9/64	0.1406	16.6687	21/32	0.6562	40		1.575	86		3.386
3.9637	5/32	0.1502	17.000		0.6693	41		1.614	87		3.425
4.000		0.1579	17.0656	43/64	0.6719	42		1.654	88		3.465
4.3656	11/64	0.1719	17.4625	11/16	0.6875	43		1.693	89		3.504
4.7625	3/16	0.1875	17.8594	45/64	0.7031	44		1.732	90		3.543
5.000		0.1969	18.000		0.7087	45		1.772	91		3.583
5.1594	13/64	0.2031	18.2562	23/32	0.7187	46		1.811	92		3.622
5.5562	7/32	0.2187	18.6532	47/64	0.7344	47		1.850	93		3.661
5.9531	13/64	0.2344	19.000		0.748	48		1.890	94		3.701
6.000		0.2362	19.050	3/4	0.750	49		1.929	95		3.740
6.3500	1/4	0.250	19.4369	49/64	0.7656	50		1.969	96		3.780
6.7459	11/64	0.2658	19.8433	15/32	0.7812	51		2.008	97		3.819
7.000		0.2756	20.000		0.7874	52		2.047	98		3.858
7.1437	9/32	0.2812	20.2402	51/64	0.7969	53		2.087	99		3.898
7.5406	19/64	0.2969	20.6375	13/16	0.8125	54		2.126	100		3.937
7.9375	5/16	0.3125	21.000		0.8268	55		2.165			
8.000		0.315	21.0344	53/64	0.8281	56		2.205			

FIGURE 18-14
Hacksaw.

Pliers

There are many types of pliers. Side-cutting pliers (Fig. 18-15) are used to cut wires. Needlenose pliers (Fig. 18-16) are used for bending wires, and utility pliers (Fig. 18-17) hold all kinds of objects securely.

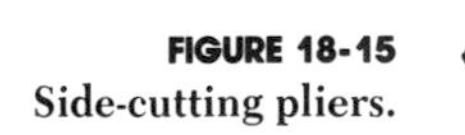

FIGURE 18-15
Side-cutting pliers.

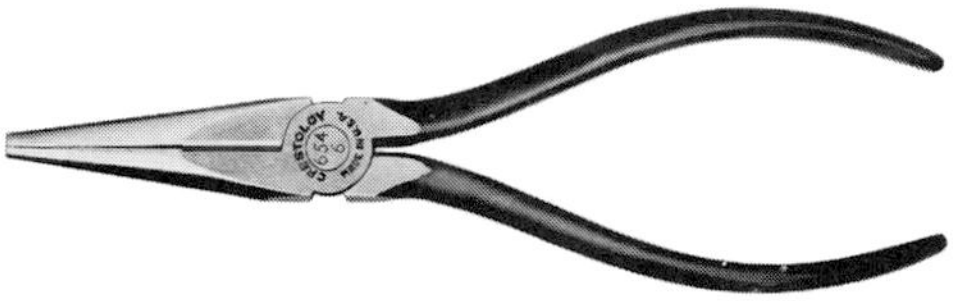

FIGURE 18-16
Needlenose pliers.

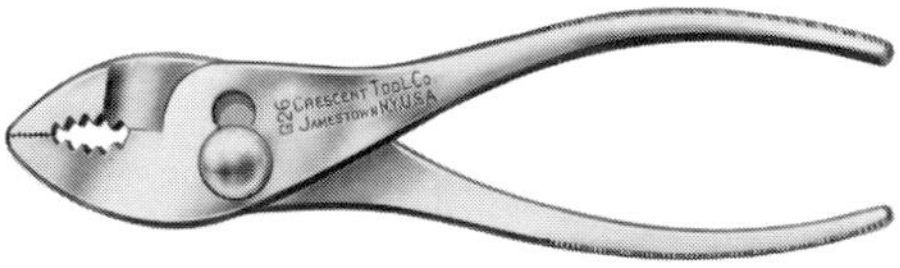

FIGURE 18-17
Utility pliers.

Heavy Shears

Heavy shears (Fig. 18-18) are used in the laboratory to cut paper, asbestos, rubber, or any other easily cut material.

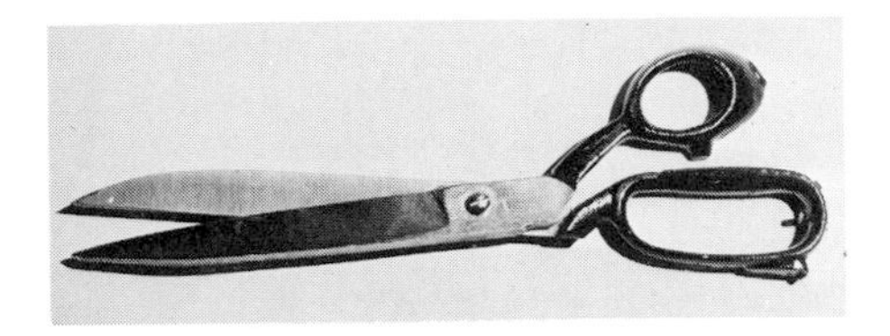

FIGURE 18-18
Heavy shears.

Screwdrivers

There are many kinds of screwdrivers (Fig. 18-19) and many kinds of tips (Fig. 18-20).

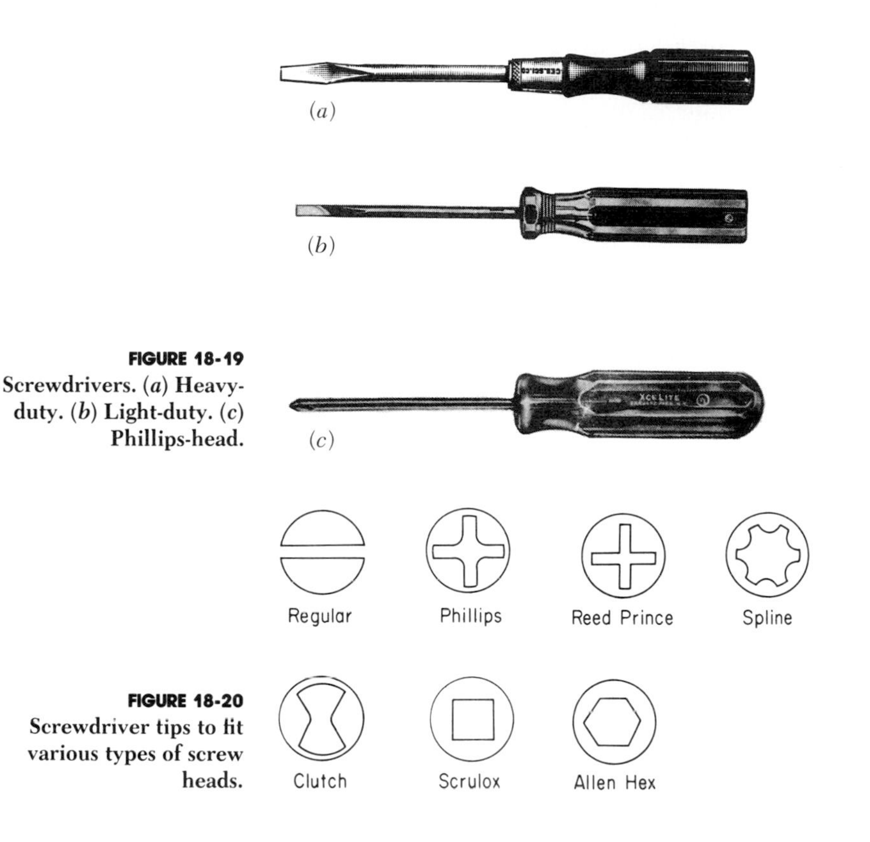

FIGURE 18-19
Screwdrivers. (*a*) Heavy-duty. (*b*) Light-duty. (*c*) Phillips-head.

FIGURE 18-20
Screwdriver tips to fit various types of screw heads.

Electric Soldering Irons

The electric soldering iron (Fig. 18-21) is used to heat surfaces, wires which are to be soldered to connections, or to seal. Refer to Chap. 27, "Basic Electricity." Tables 18-4 and 18-5 contain information about solders and solderability.

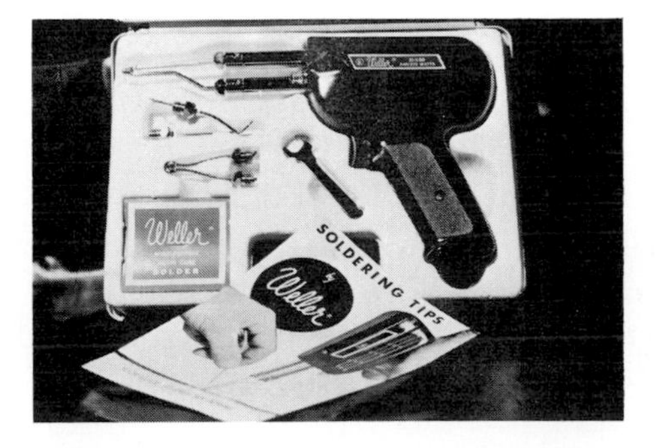

FIGURE 18-21
Electric soldering iron.

TABLE 18-4
Metals Solderability and Melting Point

Metal	*Solderability*	*mp, °F*
Magnesium	2	1204
Beryllium	3	2343
Aluminum	2	1215
Zinc	1	787
Chromium	3	2822
Iron	1	2795
Cadmium	1	610
Nickel	1	2645
Cobalt	3	2714
Tin	1	450
Lead	1	621
Stainless	2	2550
Bismuth	1	520
Copper	1	1981
Steel	1	2760
Silver	1	1761
Gold	1	1945
Antimony	1	1166

Key: 1, normally soldered; 2, soldered under special conditions; 3, not normally soldered.

TABLE 18-5
Common Solder Alloys

Identification	*% Composition*	*mp, °F*	*Comments*
Woods metal	12½-T, 25-L, 50B, 12½C	165	Low temp., nonelec.
Eutectic, T-L	63-T, 37-L	361	No pasty range
ASTM 60A	60-T, 40-L	370	Good electrical
ASTM 50A	50-T, 50-L	417	General purpose
Eutectic, T-S	96½-T, 3½-S	430	High temp., instruments
Indalloy-3	90-I, 10-S	448	Solders silver, ceramics
ASTM 40A	40-T, 60-L	460	Good wiping solder

Key: T, tin; L, lead; B, bismuth; I, indium; S, silver; C, cadmium.

Wrenches

There are many kinds of wrenches, some with fixed jaws and others with adjustable ones, depending upon the style of the wrench (Figs. 18-22 to 18-24). Many of them can be substituted for each other in various jobs; however, as with all tools, the correct wrench for the job should be used, if possible.

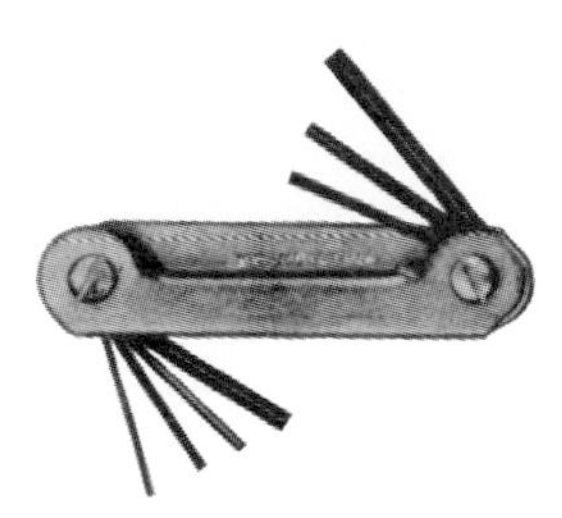

FIGURE 18-22
Allen wrenches for loosening and tightening hexagonal set screws.

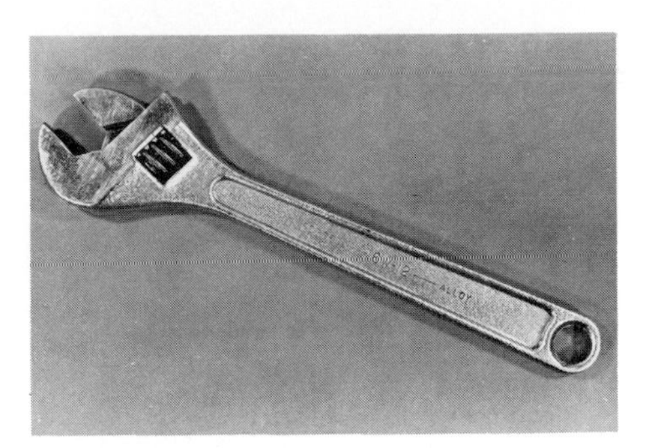

FIGURE 18-23
Adjustable-end wrench for loosening and tightening bolts and nuts.

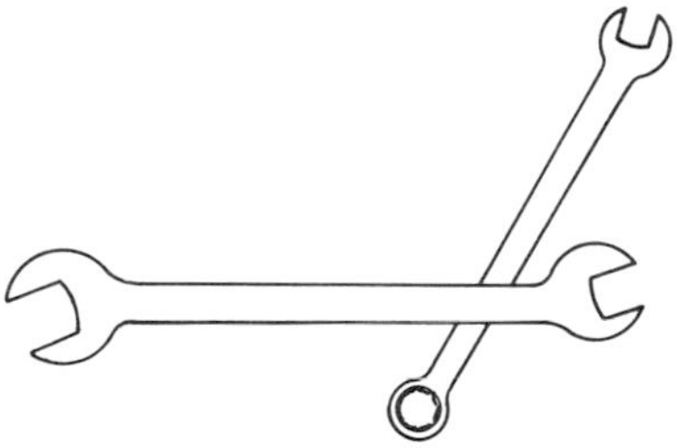

FIGURE 18-24
Open-end wrenches are available in various sizes, both USCS and SI (metric).

Allen wrenches, adjustable-end wrenches, and open-end wrenches are indispensable. Pipe wrenches (Fig. 18-25) are available in various sizes and the proper size should be used for the job. Pipe wrenches have movable jaws, and they should always be placed on the pipe so that the open jaws face the direction of turning. This makes the jaws tighten up and securely grip the pipe.

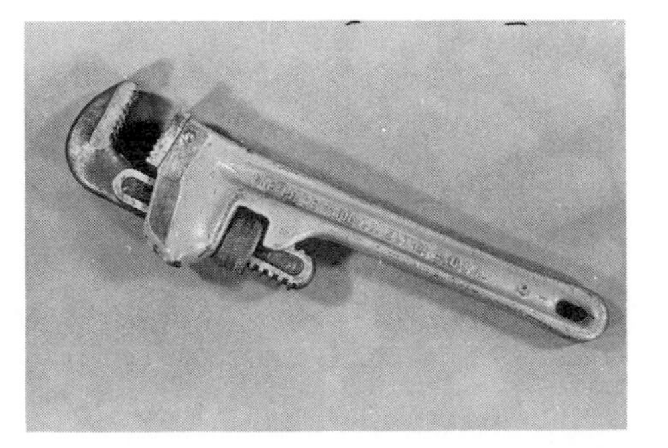

FIGURE 18-25
Pipe wrench with pivoted jaw to loosen or tighten rods or pipe. *Do not use on bolts and nuts.*

CAUTION Do not use pipe wrenches on nuts, bolts, or highly polished finishes, because the objects will get marred or distorted. For protection of your knuckles, apply force by pulling, not pushing, the wrench; if you must push, use the open palm of your hand. When using pipe wrenches to tighten pipes and fittings, always use two wrenches; one wrench to hold the pipe or fitting in a fixed position and the other to turn the pipe or fitting.

Socket wrenches (Fig. 18-26) are used to loosen and tighten small square and hexagonal nuts and bolts.

FIGURE 18-26
Socket wrenches.

A vise-grip wrench (Fig. 18-27) is a versatile tool, which can be used instead of certain other wrenches if necessary. For example, it can be used as a pipe wrench, pliers, clamp, portable vise, locking wrench, or as an adjustable wrench. The vise-grip wrench is available in several sizes.

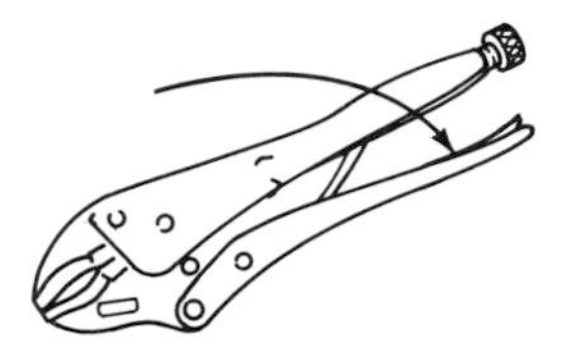

FIGURE 18-27
Vise-grip wrench.

Vises

A vise is used to hold objects securely between its jaws. It is anchored to a workbench. (See Fig. 18-28.)

FIGURE 18-28
Vise for bench use.

TUBING USED IN THE LABORATORY

Tubing is available in various sizes (Fig. 18-29) and materials of construction for use in the laboratory. Normally, copper tubing is used except where the corrosion resistance of stainless steel is needed (see Table 18-6). Tubing can be easily cut, bent, and formed to meet any need; and secure connections that do not leak can be made, provided proper tools, fittings, and working procedures are used. (Also refer to Chap. 17, "Plastic Labware.")

Stainless steel is available as standard or degreased and passivated tubing. The degreased and passivated tubing has been thoroughly cleaned and degreased, treated with an acid solution for passivation, and then rinsed in distilled water and dried.

TABLE 18-6
Types of Tubing

	OD, in	*Wall thickness, in*	*Maximum service pressure, lb/in²*
Annealed Copper			
Safety factor 5/1	1/8	0.030	3000
	1/4	0.030	3000
	1/4	0.049	2500
	3/8	0.032	1000
	1/2	0/049	1000
Stainless Steel			
Safety factor 5/1	1/8	0.020	5000
	1/4	0.020	3000
	3/8	0/020	2000
	1/2	0.028	2000

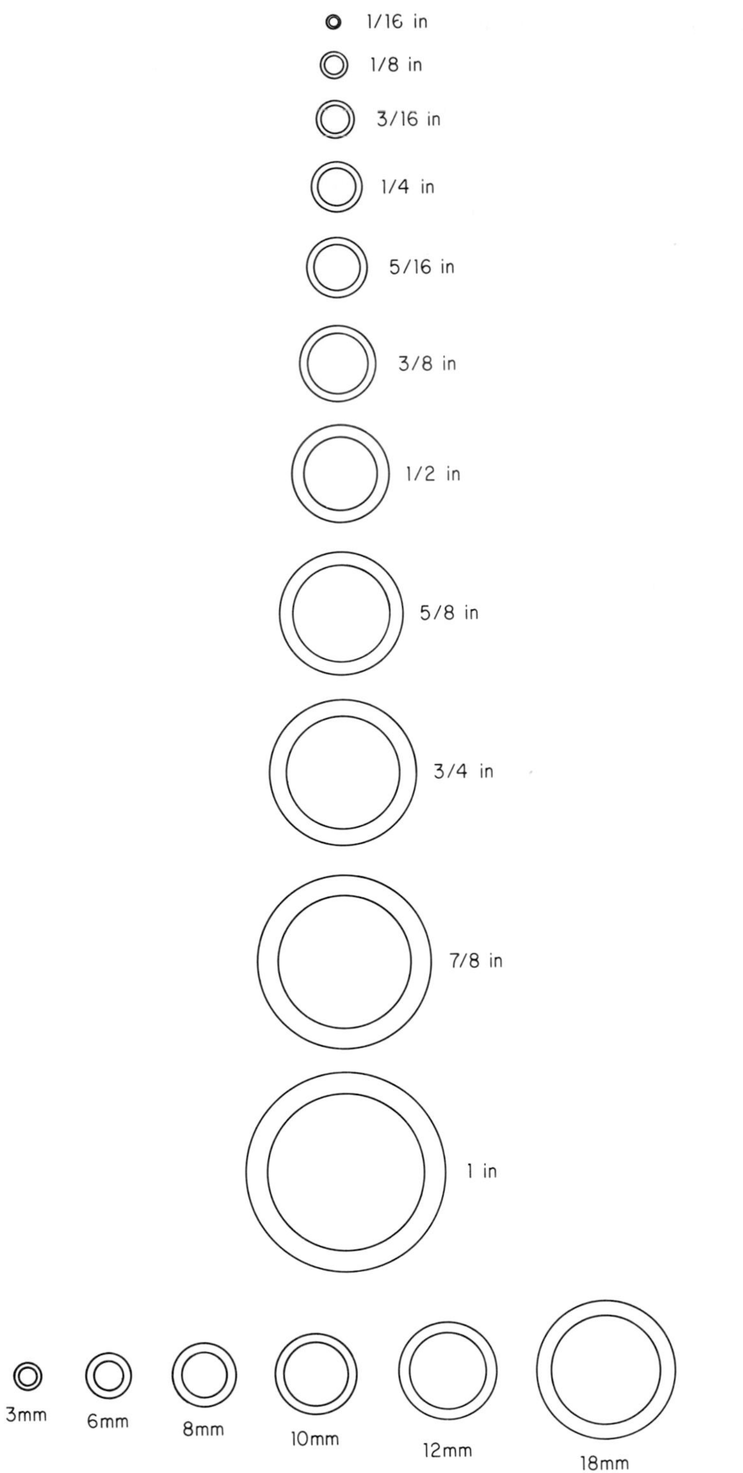

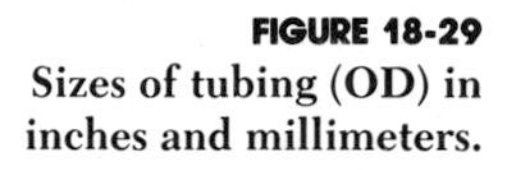

FIGURE 18-29
Sizes of tubing (OD) in inches and millimeters.

Uncoiling Tubing

Lay the tubing on a flat surface and slowly uncoil while holding the end flat against the surface (Fig. 18-30).

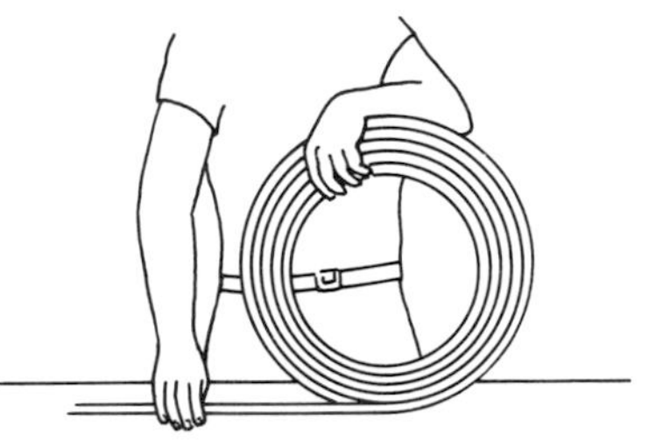

FIGURE 18-30
Uncoiling tubing.

Cutting Tubing

There are two methods to cut tubing:

1. Use a tube cutter (Fig. 18-31). Center the tubing between the two rollers and then rotate the cutter continuously around the tubing while *very slowly* applying pressure on the cutting wheel by twisting the knob screw (Fig. 18-32).

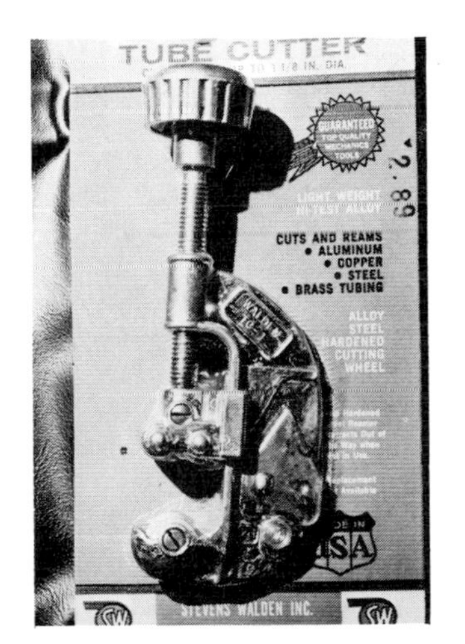

FIGURE 18-31
The tube cutter cuts brass and copper tubing cleanly without burs.

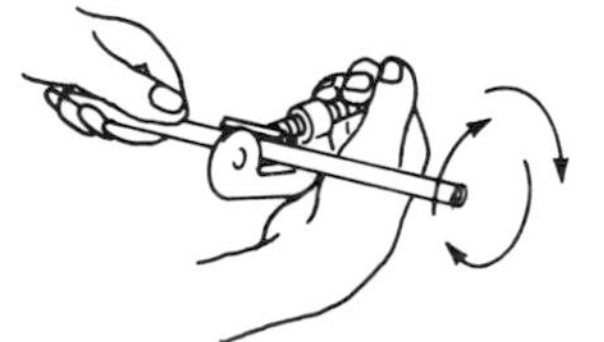

FIGURE 18-32
Cutting short pieces of tubing.

2. Use a hacksaw and guide blocks (Fig. 18-33).

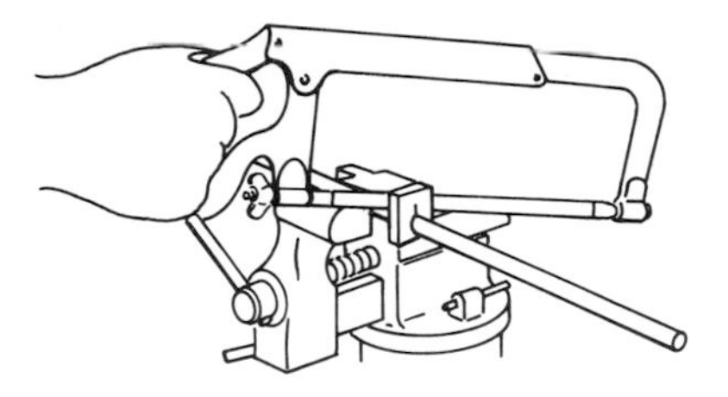

FIGURE 18-33
Hacksaw and guide blocks used to cut tubing.

The ends should be squared off at right angles to the tubing for secure connections.

Removing Burs from Cut Tubing

When tubing is cut either with a tube cutter or with a hacksaw, burring of the tubing occurs.

When a tube cutter is used, the burs are found inside (Fig. 18-34).

When a hacksaw is used, burs are found inside and outside (Fig. 18-35).

Burs can be easily removed by stroking the cut tubing gently with a flat file on the outside and with a rat-tailed file on the inside (Fig. 18-36).

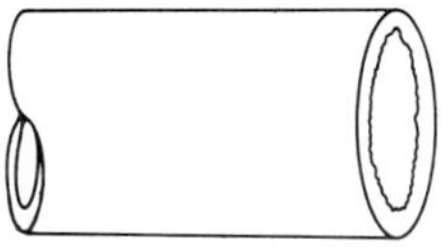

FIGURE 18-34
Burs formed inside tubing by action of cutter.

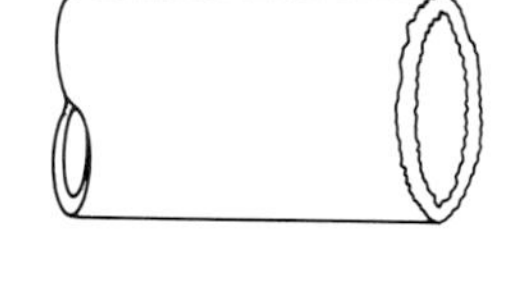

FIGURE 18-35
Hacksaw burs form both inside and outside the tubing.

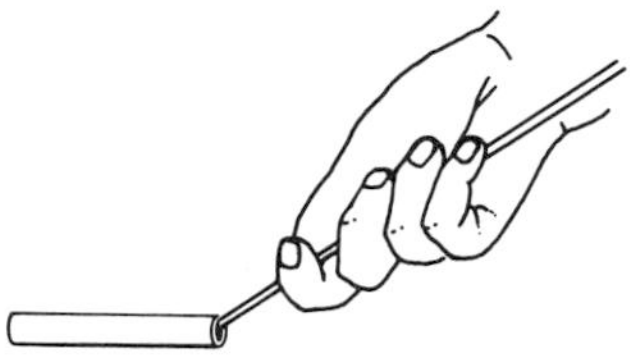

FIGURE 18-36
Removing inside burs with a file.

Bending Tubing

Tubing can be bent easily, and rounded bends without kinks can be obtained provided you do not attempt to bend the tubing too sharply. For large bends, pressure against the body, thighs, knees, compressed gas cylinders, drums, pails, or any cylindrical object works very well (Fig. 18-37).

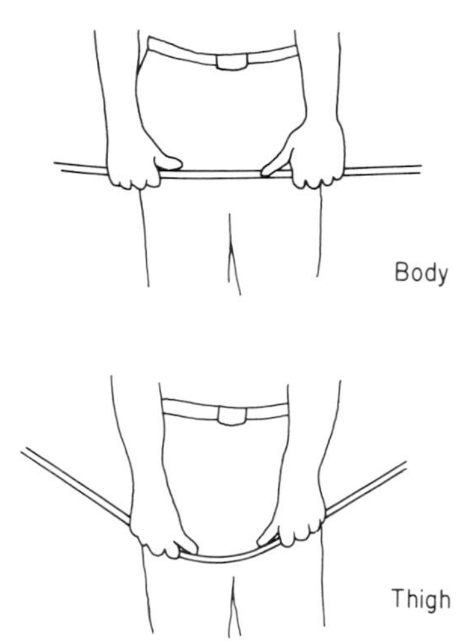

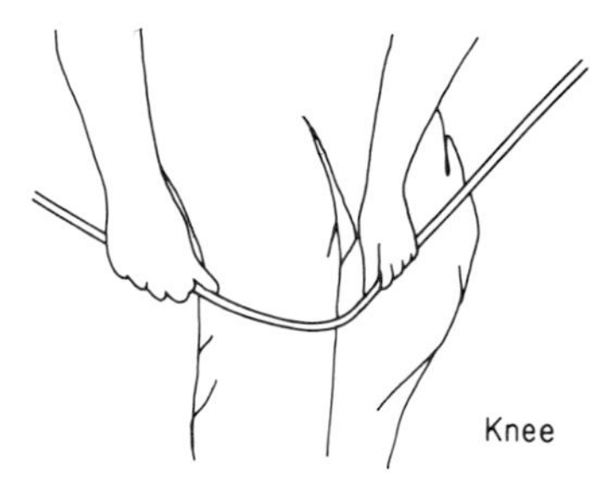

FIGURE 18-37
Bending tubing.

If very sharp bends are desired, a tool called a *tube bender* enables you to make those bends without kinking the tubing and thus restricting the fluid flow.

Connecting Copper Tubing

By Mechanical Means

Many laboratory procedures require the use of soft copper tubing securely connected to gas cylinders, gas-chromatography units, pressurized containers, cooling and heating containers, etc. Copper tubing can easily be connected to itself or to any unit that has a standard pipe-thread fitting, and there are all conceivable types of connections and interconnections available. It requires only thought and ingenuity to make the connection.

The preparation of the copper tubing end is most important. The cut must be squared off, must be cleaned of inside and outside burs, and must not be deformed in any way. Once the edge has been prepared, the fitting connection is made.

There are two basic types of connectors with which tubing can be mechanically connected together without the use of soldering equipment: one requires flaring of the tubing, the other does not. Both types provide secure connections that do not leak, if the proper procedures are followed.

Flaring Procedure: Method 1 (Fig. 18-38)

1. Cut the tubing to the proper lengths, making sure the ends are squared off and free of burs. Care must be taken that the end of the tubing to be connected is round and has not been deformed.

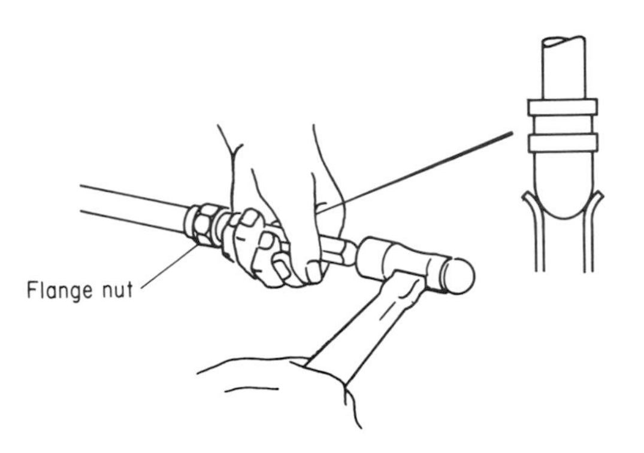

FIGURE 18-38
Flaring the end of tubing.

2. Slide the flange nut over the end of the tubing in such a way that the open end with the screw faces the connector.

3. Flare the open end of the tube by hitting the end of the flaring tool with a hammer repeatedly while securely holding the tubing with the other hand until the end is flared.

CAUTION The flared end of the tubing must not be too large, otherwise the flange nut will not slip over it to engage the male connector.

Flaring Procedure: Method 2 (Fig. 18-39)

1. Prepare ends of tubing properly; insert flange nut.

2. Place the tubing in the flaring holder of the vise grip and tighten securely so that the tubing will not slip. The open end should be flush with the face of the holder having the countersunk recesses.

3. Slide the flaring tool over the holder with the flaring screw facing the open end of the tubing.

4. Screw down the flaring tool knob tightly, flaring the tubing.

5. Remove flaring tool and tubing holder.

Connecting Flared Tubing

After the tubing has been flared (by method 1 or 2 above), make the connection as follows (see Fig. 18-40):

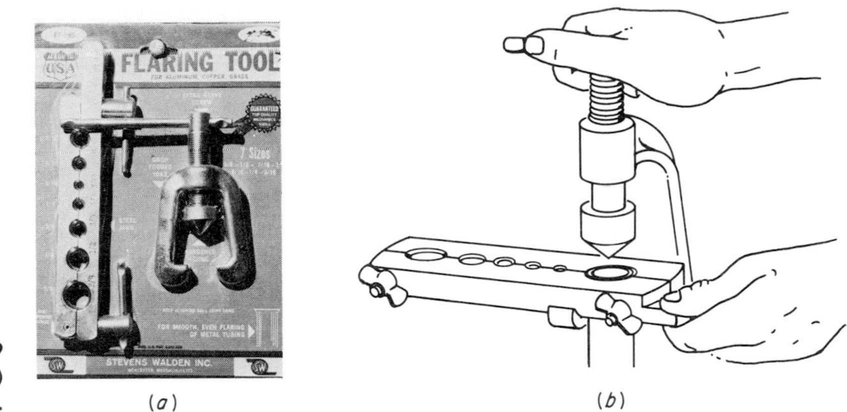

FIGURE 18-39
(*a*) Flaring tool. (*b*) Procedure for use.

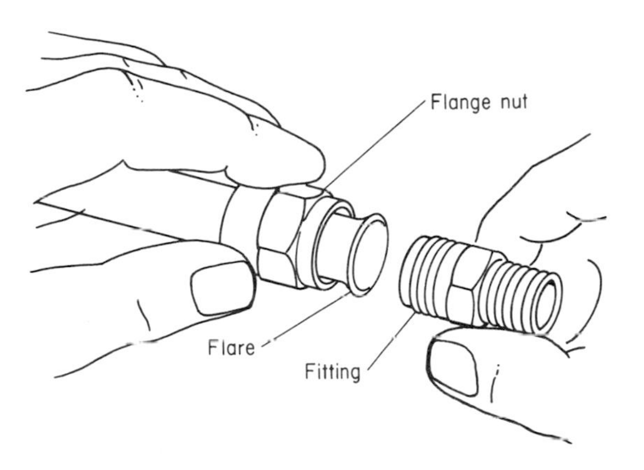

FIGURE 18-40
Assembling the flared tube and its fittings.

1. Select the desired connector.

2. Wipe clean the connector and flared tubing end.

3. Push tubing and connector together and slide flange nut to engage connector fitting.

4. Hand-tighten the flange nut as much as possible; it should screw on easily.

5. Select appropriate end wrenches—one for the connector, the other for the flange nut.

6. Tighten the flange nut securely by using both wrenches in a counterclockwise motion to each other.

CAUTION Do *not* exert excessive force.

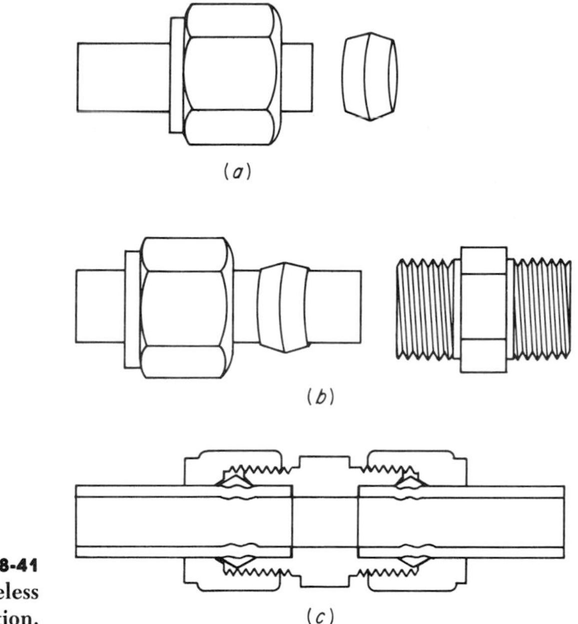

FIGURE 18-41 Procedure for flareless connection.

Compression Fittings

Flareless types of connectors do not require that the tubing be flared, because they have a self-contained method (an expanding ferrule) for providing leakproof connections. The tubing must be prepared as before, with the end clean; free from burs, and circular—not deformed. To make the connection:

1. Slide the compression nut over the end of the tubing, with the open threaded end facing the cut end of the tube (Fig. 18-41*a*).

2. Slide the compression ferrule over the tubing (Fig. 18-41*b*).

3. Insert the end of the copper tubing completely into the other part of the fitting (Fig. 18-41*c*).

4. While continuously exerting pressure to keep the tubing completely inserted into the fitting, tighten the compression nut firmly against the fitting, using an end wrench.

CAUTION Too much pressure will distort the fitting and cause breakage and leakage.

5. The connection can be disconnected and reconnected.

Other Connectors

Other types of connectors are shown in Fig. 18-42.

To connect tubing to a female pipe port use:

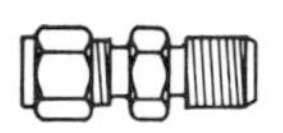

Male connector

Bulkhead male connector

Male elbow

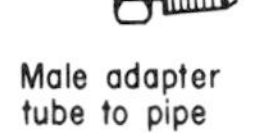

Male adapter tube to pipe

Male run tee

Male branch tee

To connect tubing to a male pipe stub use:

Female connector

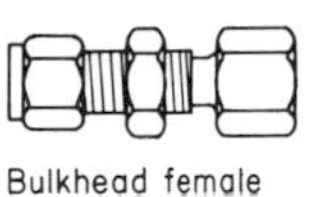

Bulkhead female connector

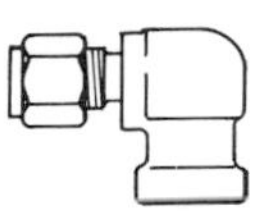

Female elbow

Female adapter tube to pipe

Female run tee

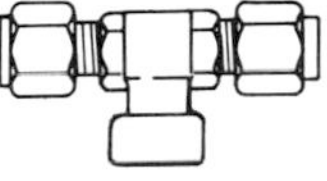

Female branch tee

To connect two or more tubes together use:

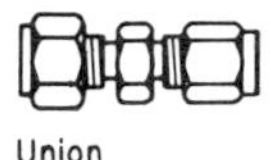

Union

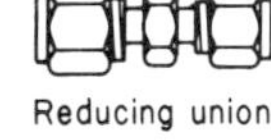

Bulkhead union

Reducing union

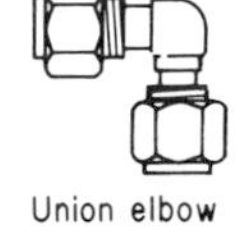

Union elbow

To connect two or more tube fittings together use:

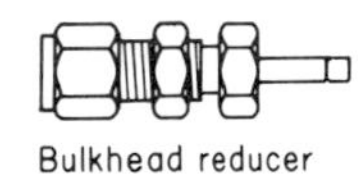

Bulkhead reducer

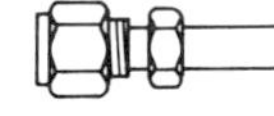

Reducer

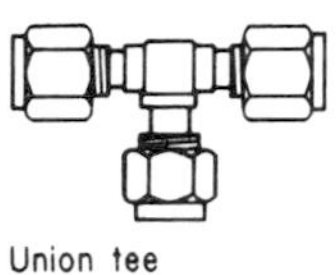

Union tee

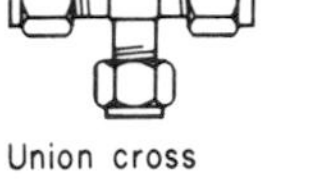

Union cross

Swagelok® to a union

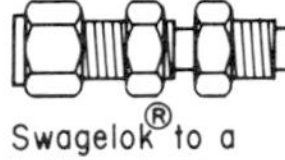

Swagelok® to a bulkhead union

Swagelok® to an adapter

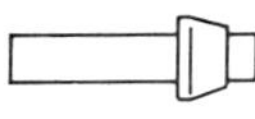

Port connector

To cap a tube or plug a fitting use:

Cap

Plug

To connect tubing to an all welded system use:

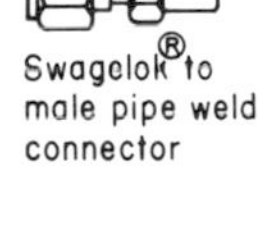

Swagelok® to male pipe weld connector

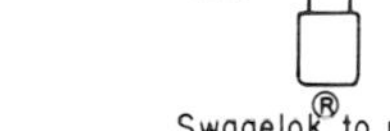

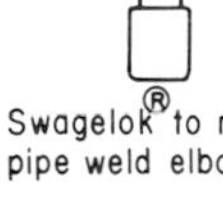

Swagelok® to male pipe weld elbow

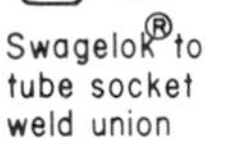

Swagelok® to tube socket weld union

Swagelok® to tube socket weld elbow

To connect tubing to pipe or straight thread ports using an o-ring seal use:

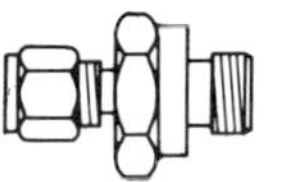

O-seal male connector pipe thread

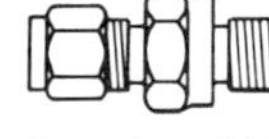

O-seal straight thread connector

To connect tubing to SAE straight thread ports use:

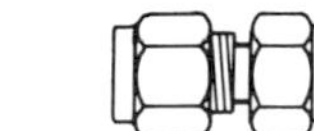

Male connector for straight thread boss

For special connections such as gas chromatographs, heat exchangers, or thermocouples use:

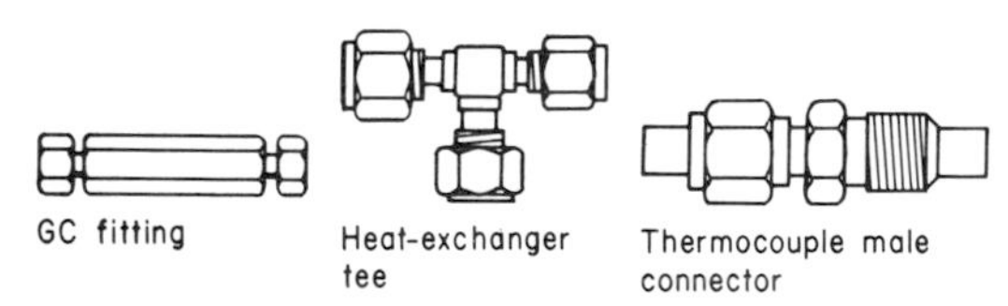

GC fitting

Heat-exchanger tee

Thermocouple male connector

As spare parts use:

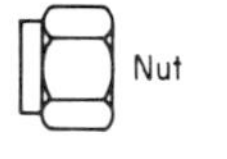

Nut

Back ferrule

Front ferrule

Knurled nut

Insert

Ferrule pak

FIGURE 18-42
Tube fittings, available in all machinable metals.

By Soldering

When permanent tubing installations are desired, the tubing can be interconnected and connected to standard fittings (Fig. 18-43) by soldering. Normal soldering procedures should be followed. (Refer to Chap. 27, "Basic Electricity.")

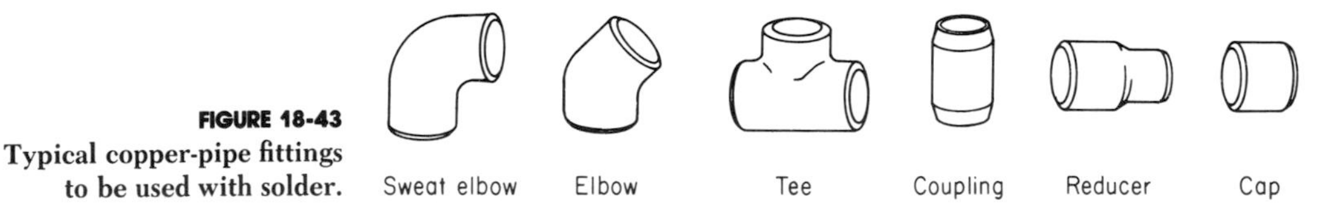

FIGURE 18-43
Typical copper-pipe fittings to be used with solder.

All surfaces should be cleaned to remove oxides and then lightly scraped with sandpaper to roughen the surface. Use a torch to heat the fitting (or the tube) hot enough so that the solder will melt and flow when solder wire is touched to the joint. *Do not heat the solder.*

Allow the melted solder to flow evenly throughout the heated surfaces to be connected, while maintaining secure contact by sufficient pressure. Remove the flame and wipe away excess solder with toweling while maintaining force to keep components together. Allow to cool and test for leaks.

Galvanized Pipe Assemblies

Galvanized pipe assemblies can be constructed by cutting the pipe to size, threading the ends, and then using any of the pipe fittings in Fig. 18-44

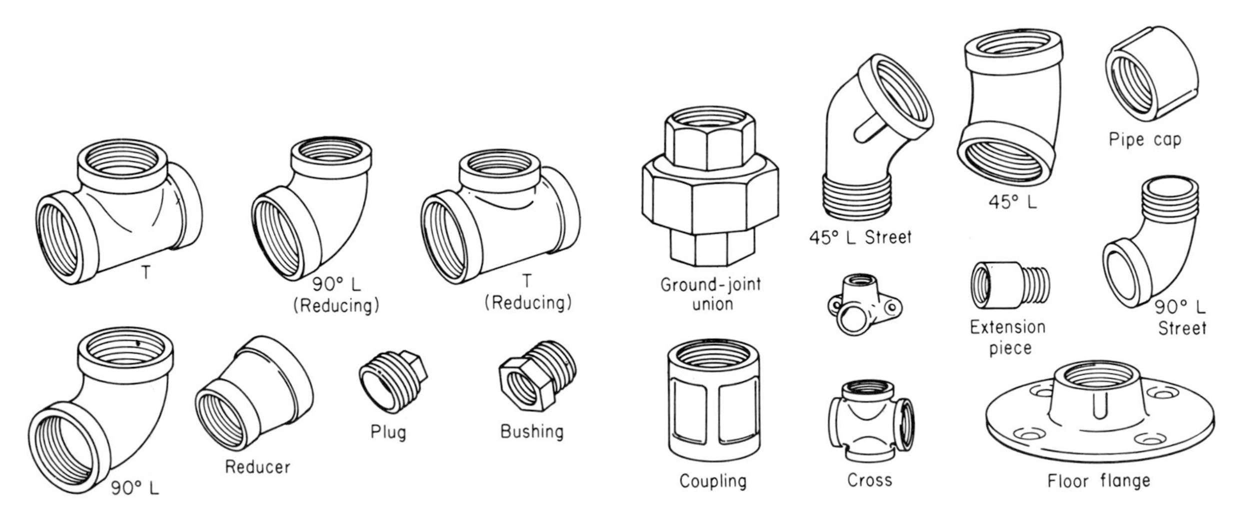

FIGURE 18-44
Typical malleable galvanized iron fittings.

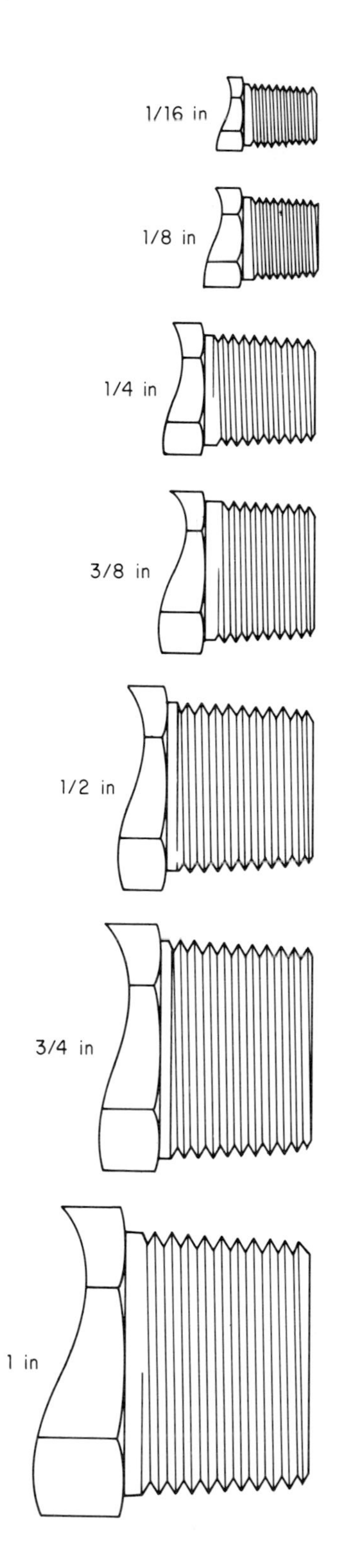

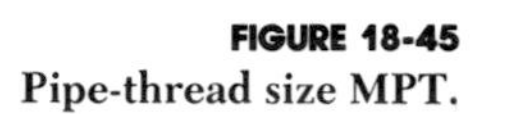

FIGURE 18-45
Pipe-thread size MPT.

or 18-45 to make the required assembly. The fittings are available in standard pipe sizes starting with 1/16 in ID.

MPT pipe thread (Fig. 18-45) is a *tapered thread.* It must be used with fittings that have the same thread. It cannot be used with parallel threaded machine fittings.

19

DETERMINATION OF PHYSICAL AND OTHER PROPERTIES

DENSITY

The *density* of any substance can be found by dividing the mass of that substance by the volume that it occupies. For example:

$$\text{Density} = \frac{\text{mass of the substance in grams}}{\text{volume of the substance in cubic centimeters}}$$

Density is expressed in the following units.

Grams per cubic centimeter	g/cm^3
Grams per milliliter	g/mL
Pounds per cubic foot	lb/ft^3

The density of water at 4°C is 1.000 g/cm^3 = 1.000 g/mL; therefore, the terms *milliliters* and *cubic centimeters* are usually interchangeable. (However, in the U.S. customary system the density of water at 4°C = 62.4 lb/ft^3.)

Regularly Shaped Solids

Procedure

1. Determine the mass of the object.
2. Measure the object and obtain relevant dimensions.
3. Calculate the volume, using mathematical formulas for box, sphere, or cylinder.
4. Divide the mass by the volume.

Irregularly Shaped Solids

Procedure

1. Determine the mass of the object.
2. Determine the volume by water displacement (Fig. 19-1).

   (a) Use a graduated cylinder containing a measured amount of water *(original volume)*.

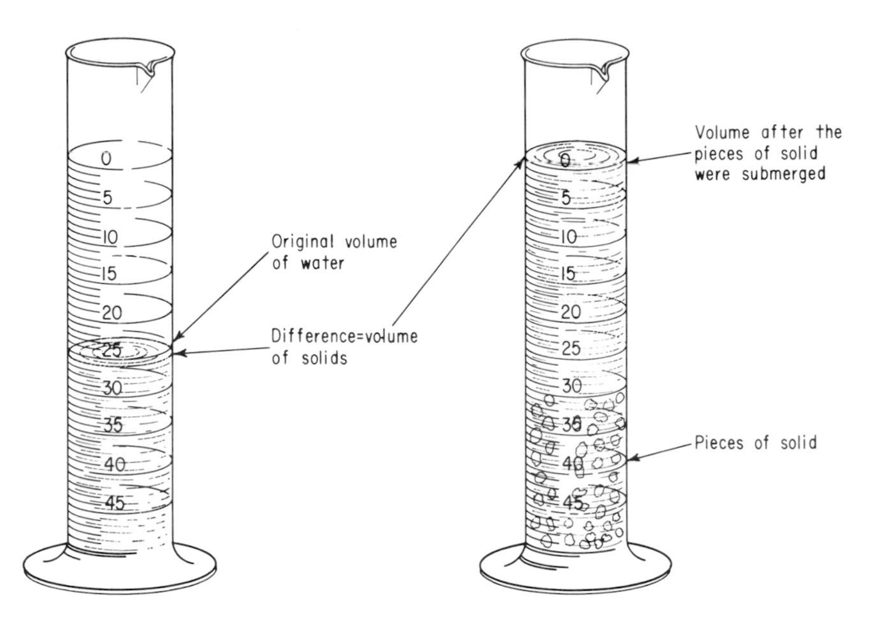

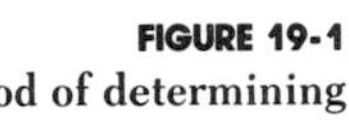

FIGURE 19-1
A method of determining the volume of an irregular solid.

(**b**) Submerge the weighed solid completely in the cylinder containing the water and record the larger volume reading *(final volume)*.

(**c**) Subtract the original volume from the final volume and obtain the volume of the object.

(**d**) Divide the mass of the object by the volume.

Liquids

Density-Bottle Method

Procedure

1. Use a calibrated-volume liquid container fitted with thermometer, or determine the volume of the container by using distilled water and reference tables listing the density of water at different temperatures. (See Figs. 19-2 and 19-6.)

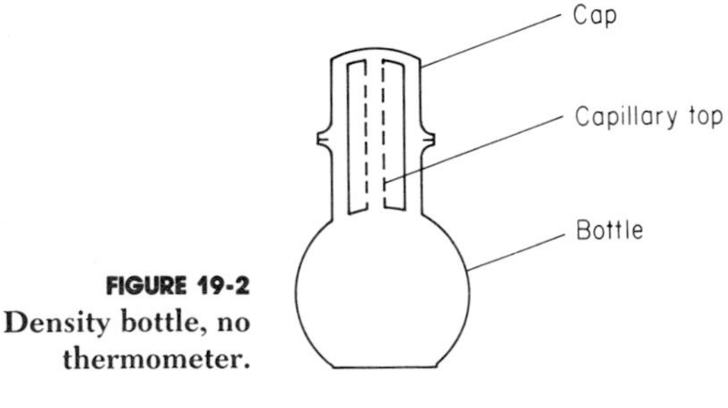

FIGURE 19-2
Density bottle, no thermometer.

2. Determine the mass of the empty container.

3. Determine the mass of the filled* container and record the temperature. The density bottle is placed in a thermostatically controlled bath. There is some loss of liquid from the density bottle during the time that it is removed from the bath, wiped dry and massed, especially with very volatile liquids, because of evaporation of the liquid.

4. Obtain the mass of the liquid by subtraction.

5. Divide the mass by the volume of the calibrated container.

CAUTION Take care to exclude air bubbles, and carry out operations as quickly as possible.

Fisher-Davidson Gravitometer Method

Principle

The height to which a liquid will rise is inversely proportional to its density. The Fisher-Davidson gravitometer is an instrument by which the density of an unknown liquid is determined by comparing the height to which an unknown sample will rise to that of a liquid of known density.

Procedure

1. Place known and unknown liquids in the instrument.

2. Apply a slight vacuum equally to both liquids.

3. Determine the ratio of the heights to which the two liquids will rise.

4. Calculate the density. The instrument is calibrated in grams per milliliter.

Westphal Balance Method

Principle

The Westphal balance (Fig. 19-3) is based upon two concepts:

1. The mass of a floating object is equal to the mass of the liquid that it displaces (Archimedes' principle).

2. If a body of constant mass is immersed in different liquids, then the corresponding apparent losses in weight of the body are proportional to the masses of the equal volumes of liquid that the body displaced. If one of the liquids is water, the density (specific gravity) of the liquid can be calculated.

---

* Filled means "filled to the top of the capillary."

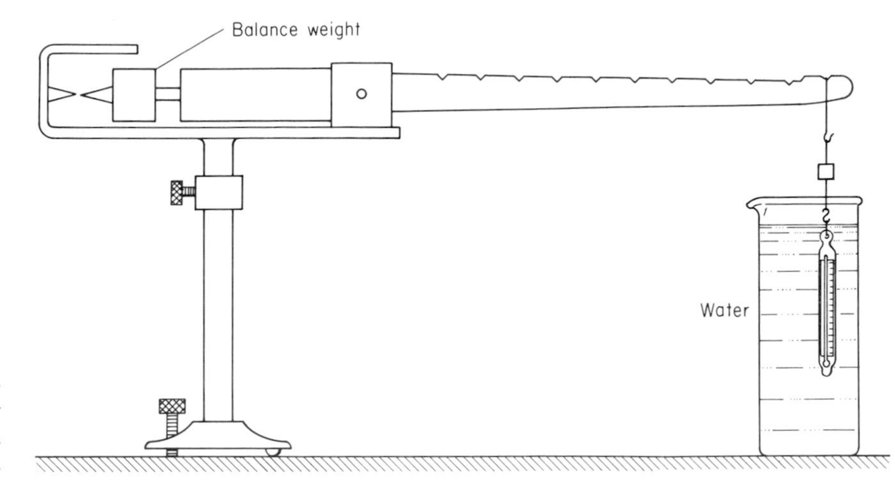

FIGURE 19-3
Westphal specific-gravity balance, standardized with water.

Procedure

1. The instrument is standardized in distilled water by fully immersing the plummet in the water and balancing the beam at the index end. When the balance is in equilibrium because it has been standardized with water (density 1.000 g/mL; sp gr 1.000), then the balance is ready for the unknown liquid.

2. The container of water is removed, and the plummet is carefully dried.

CAUTION Do not disturb the balance adjustment made in step 1.

3. Substitute the liquid to be tested. Fully immerse the plummet in the liquid.

4. Adjust the riders so that the indexes of the beam and frame are level.

NOTE The 5-g rider is left on the same hook as the plummet for liquids which are heavier than water. It is moved to index numbers on the beam for liquids which are lighter than water.

5. Read the density by the position of the riders.

Float Method

Principle

A group of weighted, hermetically sealed glass floats can be used to determine the density of a liquid. When such floats are dropped into a liquid and thoroughly wetted, they will float in a more dense liquid, sink in a less dense liquid, or remain suspended in a liquid which has equal density (Fig. 19-4). By experimentally determining which float remains suspended, the density of the liquid can be determined by reading the density of that float.

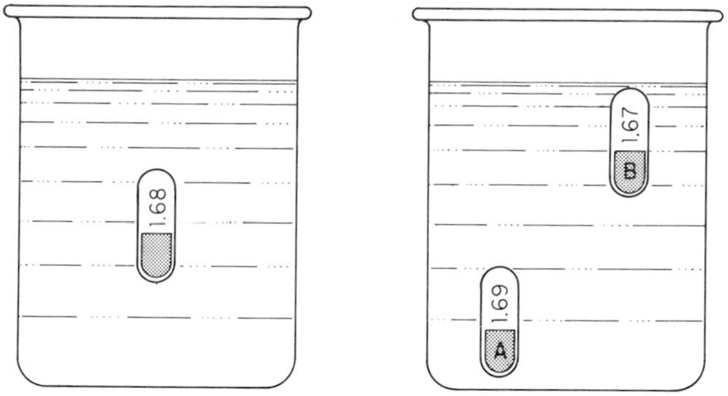

FIGURE 19-4
Float method of determining density of liquids. (*a*) Suspended float indicates that the densities of liquid and float match. (*b*) The density of float *A* is too great; that of float *B* is too low.

Procedure

1. Fill a beaker, graduated cylinder, or any glass container with the liquid to be tested.

2. Add floats and note which float remains suspended.

3. Read the density of the liquid from that float.

Gases

All gases have a much lower density than liquids, and the determination of the density of a gas is more difficult because of the extremely small mass involved.

Dumas Method

Principle

The Dumas method involves the direct determination of mass of a known volume of gas in a calibrated sphere of known volume (Fig. 19-5). The sphere of known volume is fitted with a capillary tube and stopcock, whereby the gas is introduced into the flask.

Procedure

1. The volume of the flask is verified by filling it with a liquid of known density and determining the mass of the filled sphere.

2. The flask is emptied, cleaned, dried, and evacuated.

3. The gas to be tested is introduced into the flask, and the mass of the flask and gas together is measured.

$$\text{Density of gas} = \frac{\text{mass of gas}}{\text{volume of gas}}$$

NOTE The density calculated is the density of the gas at the temperature at which the determination is made. A thermometer should be hung in the immediate vicinity and read several times during the procedure.

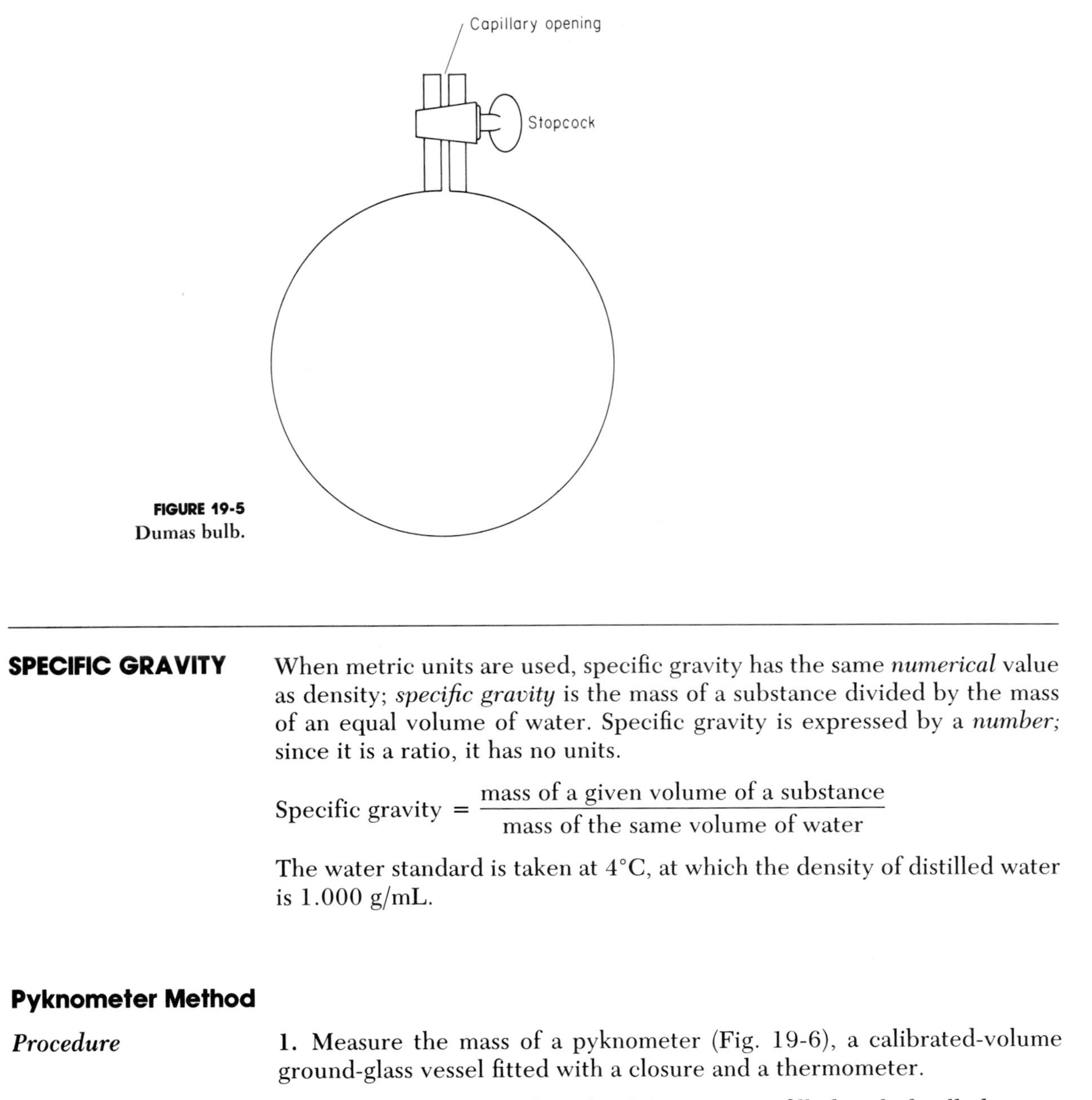

FIGURE 19-5
Dumas bulb.

SPECIFIC GRAVITY

When metric units are used, specific gravity has the same *numerical* value as density; *specific gravity* is the mass of a substance divided by the mass of an equal volume of water. Specific gravity is expressed by a *number;* since it is a ratio, it has no units.

$$\text{Specific gravity} = \frac{\text{mass of a given volume of a substance}}{\text{mass of the same volume of water}}$$

The water standard is taken at 4°C, at which the density of distilled water is 1.000 g/mL.

Pyknometer Method

Procedure

1. Measure the mass of a pyknometer (Fig. 19-6), a calibrated-volume ground-glass vessel fitted with a closure and a thermometer.

2. Measure the mass when the pyknometer is filled with distilled water. Subtract the mass obtained in step 1 from this value. This gives the mass of the water.

3. Repeat the procedure when the pyknometer is filled with the unknown liquid. Subtract the mass obtained in step 1 from this quantity. This gives the mass of the equal volume of the unknown liquid.

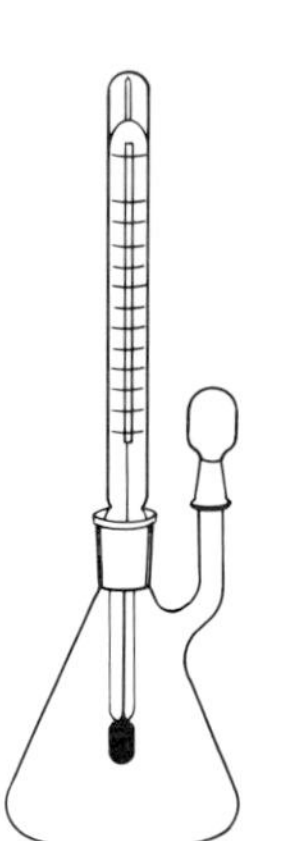

FIGURE 19-6
Pyknometer (can also be used as a density bottle).

4. Divide the mass obtained in step 3 by the mass obtained in step 2 to get the specific gravity. (All mass measurements should be made at the same temperature.)

Hydrometer Method

A hydrometer is a glass container, weighted at the bottom, having a slender stem calibrated to a standard (Fig. 19-7). The depth to which the container will sink in a liquid is a measure of the specific gravity of the liquid. Specific gravity is read directly from the calibrated scale on the stem of the container (Fig. 19-8).

FIGURE 19-7
Glass hydrometer for specific-gravity determinations. It has a weighted tip and a graduated, direct-reading tube. Such hydrometers are available in a variety of sizes and ranges.

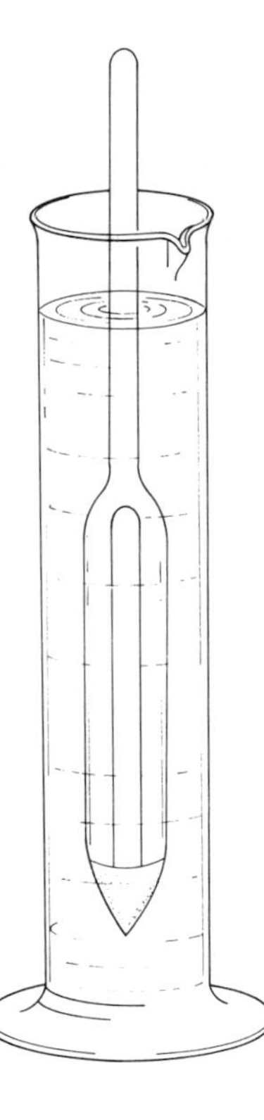

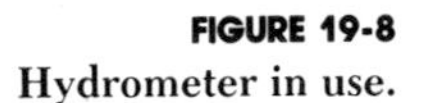

FIGURE 19-8
Hydrometer in use.

Hydrometers are calibrated in specific gravities (the mass of a liquid divided by the mass of an equal volume of water taken at certain temperatures, such as 60°F or 70°F) or in arbitrary units as degrees Baumé (Bé), degrees A.P.I. (American Petroleum Institute), or degrees Brix (also called Fisher). Degrees Brix are arbitrarily graduated so that 1° Brix = 1% sugar in solution. Hydrometer readings given in specific gravities can be converted to these units by the following formulas:

Liquids lighter than water:

$$°\text{Bé} = \frac{140}{\text{sp gr } 60°\text{F}/60°\text{F}} - 130$$

$$°\text{A.P.I.} = \frac{141.5}{\text{sp gr } 70°\text{F}/60°\text{F}} - 131.5$$

$$°\text{Brix} = \frac{400}{\text{sp gr } 60°\text{F}/60°\text{F}} - 400$$

Liquids heavier than water:

$$°\text{Bé} = 145 - \frac{145}{\text{sp gr } 60°\text{F}/60°\text{F}}$$

TABLE 19-1 Conversion Degrees Baumé American to Specific Gravity at 60°F (15.55°C)

°Bé	0°	1°	2°	3°	4°	5°	6°	7°	8°	9°
For liquids *lighter than water*, degrees Baumé $= \frac{140}{\text{sp gr}} - 130$										
10	1.000	0.993	0.986	0.979	0.972	0.966	0.959	0.952	0.946	0.940
20	0.933	0.927	0.921	0.915	0.909	0.903	0.897	0.892	0.886	0.880
30	0.875	0.870	0.864	0.859	0.854	0.848	0.843	0.838	0.833	0.828
40	0.824	0.819	0.814	0.809	0.804	0.800	0.795	0.791	0.786	0.782
50	0.778	0.773	0.769	0.765	0.761	0.757	0.753	0.749	0.745	0.741
60	0.737	0.733	0.729	0.725	0.722	0.718	0.714	0.711	0.707	0.704
70	0.700	0.696	0.693	0.690	0.686	0.683	0.680	0.676	0.673	0.670
80	0.667	0.664	0.660	0.657	0.654	0.651	0.648	0.645	0.642	0.639
90	0.636	0.633	0.631	0.628	0.625	0.622	0.619	0.617	0.614	0.611
For liquids *heavier than water*, degrees Baumé $= 145 - \frac{145}{\text{sp gr}}$										
0	1.000	1.007	1.014	1.021	1.028	1.036	1.043	1.051	1.058	1.066
10	1.074	1.082	1.090	1.098	1.107	1.115	1.124	1.133	1.142	1.151
20	1.160	1.169	1.179	1.188	1.198	1.208	1.218	1.229	1.239	1.250
30	1.261	1.272	1.283	1.295	1.306	1.318	1.330	1.343	1.355	1.368
40	1.381	1.394	1.408	1.422	1.436	1.450	1.465	1.480	1.495	1.510
50	1.526	1.543	1.559	1.576	1.593	1.611	1.629	1.648	1.667	1.686
60	1.706	1.726	1.747	1.768	1.790	1.812	1.835	1.859	1.883	1.908

SOURCE: Fisher Scientific Co.

See Table 19-1 for conversion of degrees Baumé to specific gravity.

MELTING POINT

The *melting point* of a crystalline solid is the temperature at which the solid substance begins to change into a liquid. Pure organic compounds have sharp melting points. Contaminants usually lower the melting point and extend it over a long range. The temperature of the melting point and the sharpness of the melting point are criteria of purity. The melting-point range is the temperature range between which the crystals begin to collapse and melt and the material becomes completely liquid.

The majority of organic compounds melt at convenient temperatures which range from about 50° to 300°C, and their melting points are useful aids in identifying the compounds as well as indicating their purity. Many compounds have the same melting point, yet mixtures of different compounds having the same melting point will melt at lower temperatures. This depression is a characteristic feature of mixed melting points and is extremely useful when one is trying to identify a compound; the melting point may be as much as 50°C lower than that of the pure compound.

NOTES

1. As a rule, samples which melt at the same temperature and whose melting point is not depressed by admixture are usually considered to be the same compound.

2. Narrow-range melting points are indicative of relative purity of a compound. Acceptably pure compounds have a 1°C range; normal commercially available compounds have a 2 to 3°C range. Extremely pure compounds have a 0.1 to 0.3°C range.

3. A wide melting-point range indicates that the compound is impure and contaminated.

Aberrant Behaviors

Some substances will decompose, discolor, soften, and shrink as they are being heated, and the technician must be able to recognize and to distinguish such behavior from that at the true melting point of a compound.

Shrinking and softening: Some substances tend to shrink and soften prior to reaching their melting point. Others may release solvents of crystallization, but the melting point is reached only when the solid substance begins to change to a liquid, that is, when the first drop of liquid becomes visible.

Decomposition and discoloration: When substances decompose upon melting, discoloration and/or charring usually takes place. The decomposition point is usually taken as the reliable temperature, and the value listed is followed by the letter "d" to indicate this, for example: 269°d. (The scale—°C, °F, or K—is specified in the table.)

There are several procedures for determining melting points.

Capillary-Tube Methods for Determining Melting Point

Procedure 1

Determine melting points by introducing a tiny amount of the compound into a small capillary tube attached to the stem of a thermometer which is centered in a hot-oil bath.

Setup of Apparatus

1. Obtain commercially available capillary tubes (Fig. 19-9) or make them by drawing out 12-mm soft-glass tubing.

2. Fill the capillary tube with the powdered compound to a height of 3 to 4 mm:

   (a) Scrape the powder into a pile.

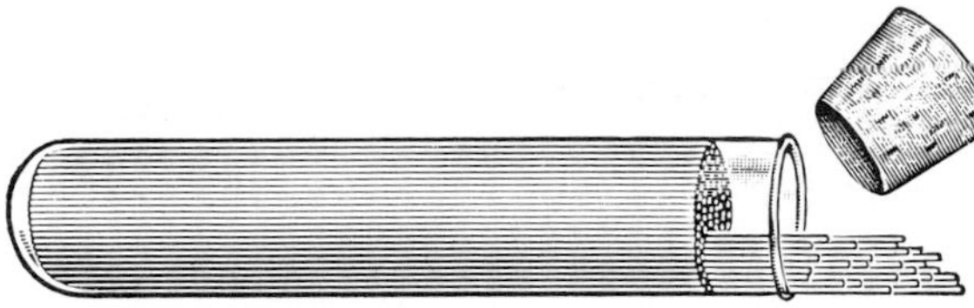

FIGURE 19-9
Capillary tubes with sealed ends for use in melting-point and capillary boiling-point determinations.

(b) Push the powder into the open end of the capillary tube.

(c) Shake the powder to the bottom of the tube by tamping lightly against the desk top or by gently scraping the tube with a file. *Pack tightly.*

3. Attach the capillary to the thermometer with a rubber band, and immerse in an oil bath (Fig. 19-10).*

4. Heat the oil bath quickly to about 5°C below the melting point, stirring continuously.

5. Now heat slowly; raise the temperature about 1°C/min, mixing continuously.

6. Record the temperature when fusion is observed, and record the melting-point range.

7. Discard the capillary after the determination has been made.

---

* Heated oil expands. Hot oil will swell and loosen the rubber band, causing the capillary tubing to fall off the thermometer. Be sure the rubber band is placed well above the oil.

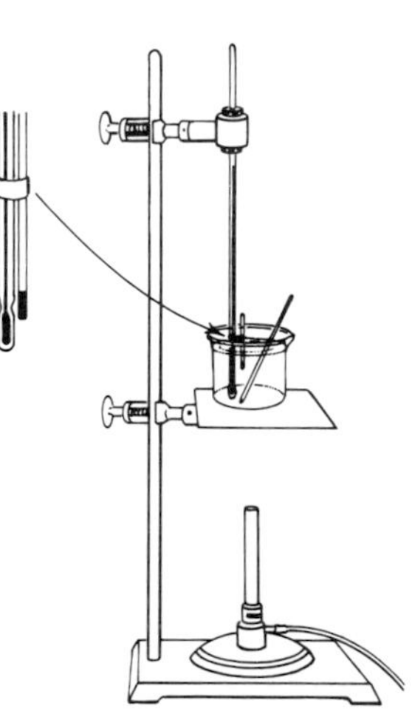

FIGURE 19-10
Laboratory setup for determining melting points.

CAUTIONS

1. Step 4 enables you to save a great deal of time. Therefore prepare two samples of the compound. Determine the approximate melting point first. Then allow the bath to cool about 15°C below that point and insert the second tube. Reheat slowly to obtain the melting point.

2. Stir vigorously and constantly so that the temperature reading will not lag behind the actual temperature of the heating fluid.

Procedure 2

An alternate method makes use of the Thiele melting-point apparatus (Fig. 19-11). The Thiele tube is a glass tube so shaped that when heat is applied by a microburner or Bunsen burner to the side arm, that heat is distributed to all parts of the vessel by convection currents in the heating fluid. No stirring is required.

CAUTIONS

1. Never heat a closed system. Always vent the tube.

2. Do not heat the bath too fast; the thermometer reading will lag behind the actual temperature of the heating fluid.

3. See footnote to procedure 1 on placing the rubber band.

4. Never determine the melting point by observing the temperature at which the melted substance solidifies when the bath cools. The substance may have decomposed, forming a new substance with a different melting

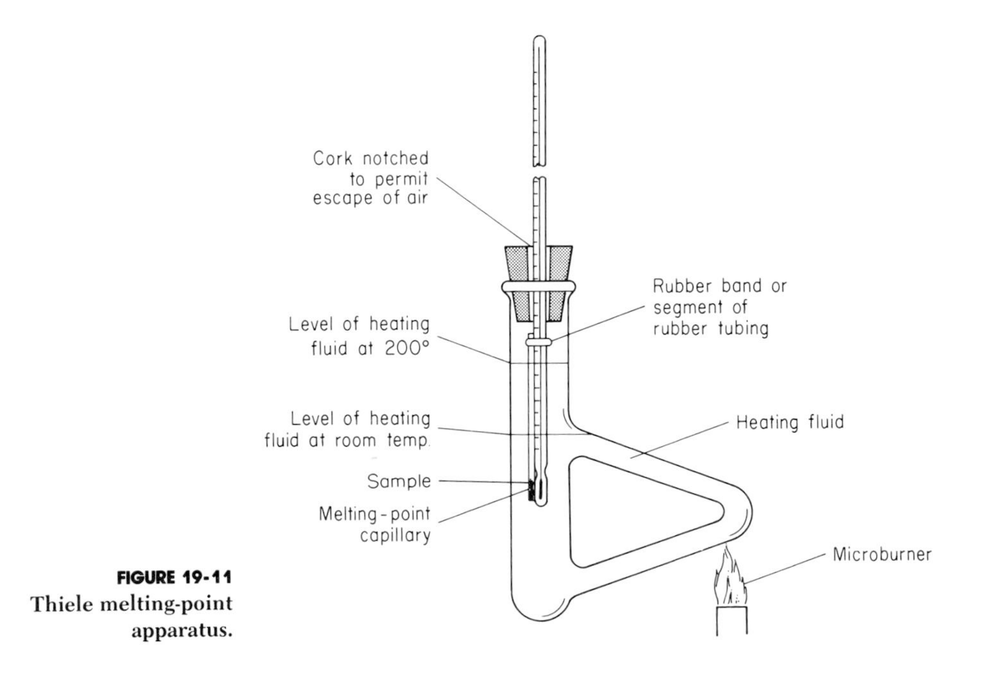

FIGURE 19-11 Thiele melting-point apparatus.

point, or the substance may have changed into another crystalline form having a different melting point. Multiple melting points may be run simultaneously if the melting points of the different substances differ by 10°C. *Identify tubes to avoid mistakes.*

Substances which sublime: Seal both ends of the capillary tube. (Refer to the section titled Sublimation in Chap. 21, "Distillation.")

Substances which tend to decompose: Insert the capillary in the heating bath when the temperature is only a few degrees below the melting point.

Electric Melting-Point Apparatus

This is a metal block equipped with a thermometer inserted into a close-fitting hole bored into the block, which is heated by electricity controlled by a variable transformer or a rheostat (Fig. 19-12).

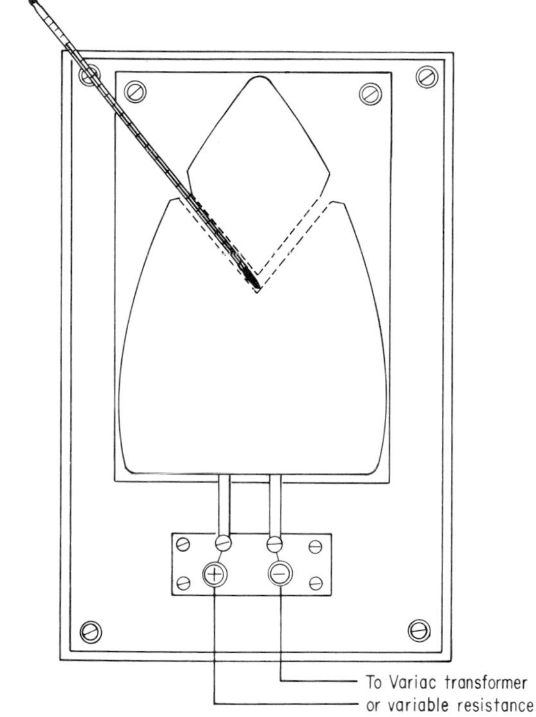

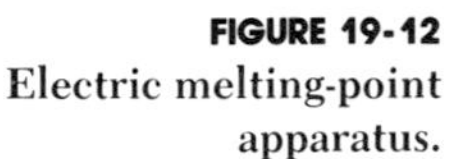

FIGURE 19-12
Electric melting-point apparatus.

Principle

The metal block is so constructed that the temperature reading of the thermometer indicates the temperature of the metal block on which the solid melts.

Procedure

1. Clean the surface of the block.

2. Place a very small quantity of finely powdered material on the proper area.

3. Follow the heating procedure detailed under Capillary-Tube Methods for Determining Melting Point, raising the heat quickly to about 5°C below the melting point of the substance, and then increasing the heat slowly.

4. When the determination is complete, turn off the electricity.

5. Multiple melting points may be determined simultaneously.

6. Always clean the metal surface scrupulously after each use.

Automatic Melting-Point Determination

Melting points which are taken by visual methods give a range, rather than a point, because the values determined are based upon the judgment of the observer who makes the decision as to which point within the observed range is the melting point.

The exact melting point can be determined automatically with instruments which use a beam of light to survey the process and a photocell to signal the instant that the sample melts. Readout indicators give the measured value of the melting point. As the substance melts, the light transmission increases, and resistance changes in the photocells are produced by the variations in the transmission. Some instruments permit the simultaneous determination of the melting point of three samples, each in its own capillary tube; the operator merely reads the individual melting points, each of which is indicated clearly on the readout panel. Mixed melting points, used in the identification of organic compounds, can be measured simultaneously, and each melting point, the knowns and the mixture, can be readily determined and compared.

Melting Points of Common Metals

The melting points shown in Fig. 19-13 are approximate. Factors such as type of processing for purification, mounting, composition of the atmosphere while the metal is being melted, etc., affect the values.

BOILING POINT

The *boiling point* of a liquid is indicated when bubbles of its vapor arise in all parts of the volume. This is the temperature at which the pressure of the saturated vapor of the liquid is equal to the pressure of the atmosphere under which the liquid boils. Normally, boiling points are determined at standard atmospheric pressure: 760 mmHg (torr) or 1 atm.

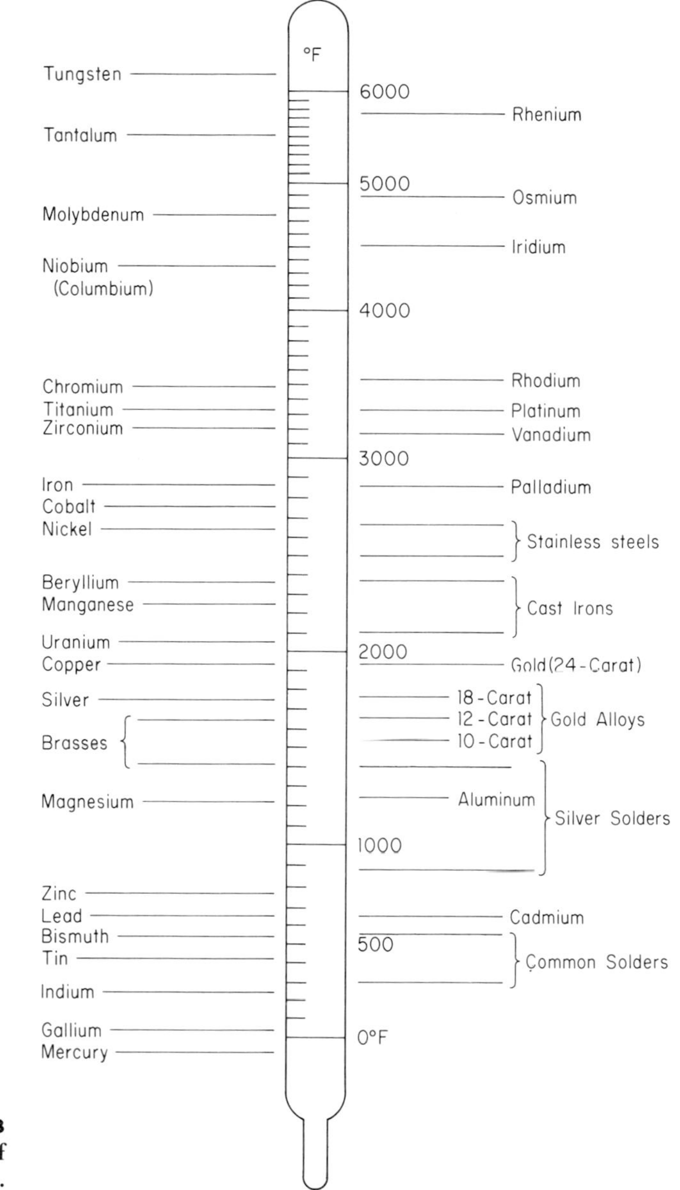

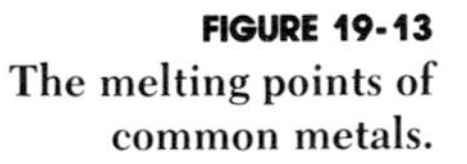

FIGURE 19-13
The melting points of common metals.

The boiling point of a liquid is sensitive to atmospheric pressure, and varies with it. As the pressure decreases, the boiling point will drop; at approximately normal pressure it will drop about 0.5°C for each 10-mm drop in pressure. At much lower pressures, close to 10 mmHg, the temperature will drop about 10°C when the pressure is halved.

As with melting points, there are several methods for determining boiling points.

Boiling-Point Determination during Distillation

When a liquid is distilled (refer to Chap. 21, "Distillation"), the boiling point of the distilling liquid can be read from the thermometer in the distilling head, which is constantly in contact with the vapors.

Test-Tube Method

Equipment

Ring stand, clamps, thermometer, test tube, and burner. (See Fig. 19-14.)

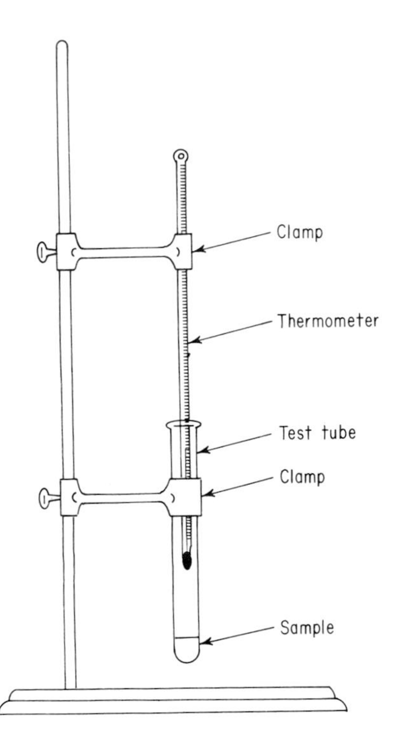

FIGURE 19-14
Laboratory setup for determining boiling points.

Procedure

1. Clamp a test tube containing 2 to 3 mL of the compound on a stand.
2. Suspend a thermometer with the bulb of the thermometer 1 in above the surface of the liquid.
3. Apply heat gently until the condensation ring of the boiling liquid is 1 in above the bulb of the thermometer.
4. Record the temperature when the reading is constant.

Capillary-Tube Method

Boiling points of liquids can be determined in micro quantities by the capillary-tube method.

Procedure

1. Seal one end of a piece of 5-mm glass tubing.

2. Attach to a thermometer with a rubber band.

3. Use a pipet to introduce the liquid (a few milliliters) whose boiling point is to be determined.

4. Drop in a short piece of capillary tubing (sealed at one end) so that the open end is down.

5. Begin heating in a Thiele tube filled with oil. A beaker half-filled with mineral oil (Fig. 19-15) may be substituted for the Thiele tube (see Capillary-Tube Methods for Determining Melting Point, pages 400–403).

6. Heat until a continuous, rapid, and steady flow of vapor bubbles emerges from the open end of the capillary tube, then stop heating.

7. The flow of bubbles will stop and the liquid will start to enter the capillary tube. Record this temperature. It is the boiling point.

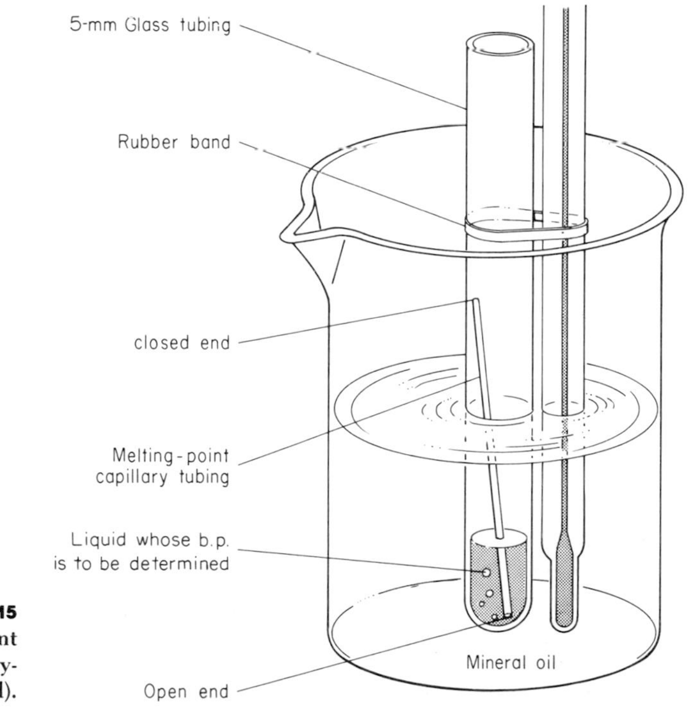

FIGURE 19-15 Micro boiling-point determination (capillary-tube method).

CAUTIONS

1. Heating cannot be stopped before step 6 is reached. If heating is stopped below the boiling point, the liquid will enter the capillary tube immediately because of atmospheric pressure. If this happens, discard the capillary tube, add more liquid, and insert a new capillary tube before restarting procedure.

2. Do not boil away all the liquid by heating too rapidly.

3. Stem corrections for the true boiling point must be made. (Refer to Chap. 9, "Measuring Temperature," beginning on p. 169.)

4. Heated oil expands. Hot oil will swell and loosen the rubber band, causing the capillary tubing to fall off the thermometer. Be sure that the rubber band is placed well above the oil.

Electronic Methods

The boiling points of very small amounts of liquid can be determined accurately and speedily with electronic sensing equipment using photocells. The substance is put into a special boiling-point sample tube, which is designed to prevent superheating and achieve smooth and continuous boiling. The tube is illuminated from the bottom by dark-field illumination, and as long as no bubbles are present (the compound is not boiling), no light passes through the liquid to reach the photocell sensor. When the boiling point is reached and bubbles begin to rise, the bubbles reflect light to the photocell. Light is reflected to the photocell with sufficient intensity and frequency to trigger the readout indicator when the true boiling point is reached. The initial outgassing bubbles (if any) do not actuate the sensor. Boiling points are determined with extreme accuracy, to ±0.3°C.

VISCOSITY

Viscosity is the internal friction or resistance to flow that exists within a fluid, either liquid or gas. Fluid flow in a line has a greater velocity at the center than next to the metal surfaces, partly because of the friction between the fluid and the boundary surfaces. This causes the adjacent layers to move more slowly. The slower-moving layers in turn retard the motion of adjacent layers. Viscosity or internal friction is a very important characteristic of fluids, both liquids and gases.

In the study of oils and organic liquids, viscosity is very significant, because, in industry, "heavier" oils and liquids have higher viscosities, not greater densities.

The unit of viscosity is the poise (cgs system), equal to 1 g/(cm) (s). Viscosities are usually tabulated in centipoises (cp).

Fluidity is the reciprocal of viscosity, and in the cgs system the unit is called the *rhe*. 1 rhe = 1/1 poise (P) or 1 poise^{-1}.

Newtonian and Non-Newtonian Fluids

If there is a linear relationship between the magnitude of the applied shear stress and the resulting rate of deformation, the fluid is classified as a *newtonian liquid*. Most oils fall in this category. If there is a nonlinear relationship between the magnitude of the applied shear stress and the resulting rate of deformation, it is classified as a *non-Newtonian liquid*.

Newtonian liquids	*Non-Newtonian liquids*
Oils	Synthetic oils
Solvents	Thermosetting resins
	Latex paints (thixotropic)

Viscosity Standards

References in handbooks and the literature provide viscosity data for many pure solvents and chemical substances. However, extremely high viscosity measurements require that the unit be standardized against materials of certified viscosity. These viscosity standards are available in a very wide range of viscosities, and they are certified as permanent viscosity standards. Most of them are fluid silicones and are accurate to within 1% of the stated viscosity value. Viscosities range from 5 to over 100,000 cp.

There are various experimental methods for determining viscosity.

Small-Bore-Tube Method

The flow of the liquid through a small-bore tube, such as is found in a thermometer, can be measured with a graduated cylinder and a stopwatch. (See Fig. 19-16.) A constant hydrostatic pressure (head) is main-

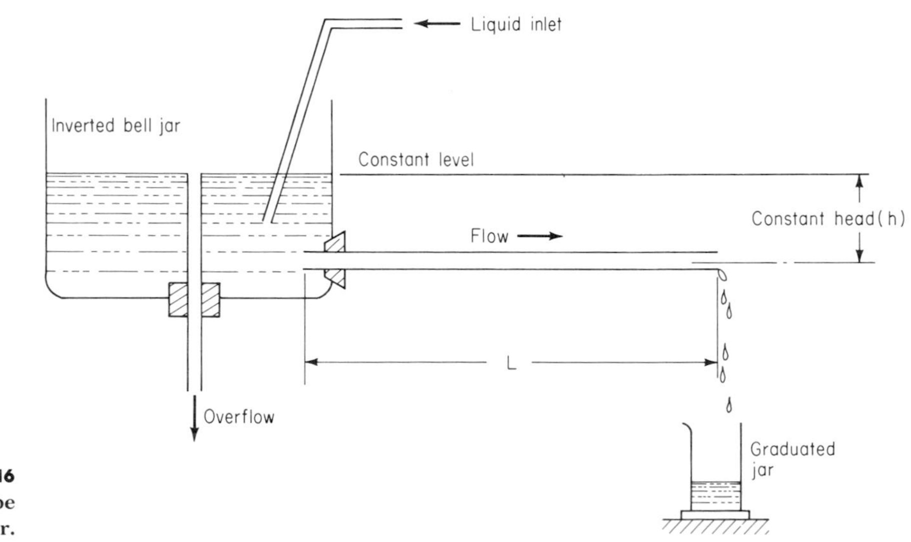

FIGURE 19-16 Small-bore-tube viscometer.

tained by constant feed and overflow. The volume of liquid that passes through the capillary tube is collected in a graduated cylinder, and the time required is measured with the stopwatch.

$$\text{Coefficient of absolute viscosity} = \frac{\text{volume collected}}{\text{time}}$$

Saybolt Viscometer Method

The Saybolt viscometer (Fig. 19-17) has a container for liquids with a capacity of 60 mL, fitted with a short capillary tube of special length and diameter. The liquid flows through the tube, under a falling head, and the time required for the liquid to pass through is measured in seconds. If temperature is a critical factor, the viscometer is kept at constant temperature in a temperature-controlled bath.

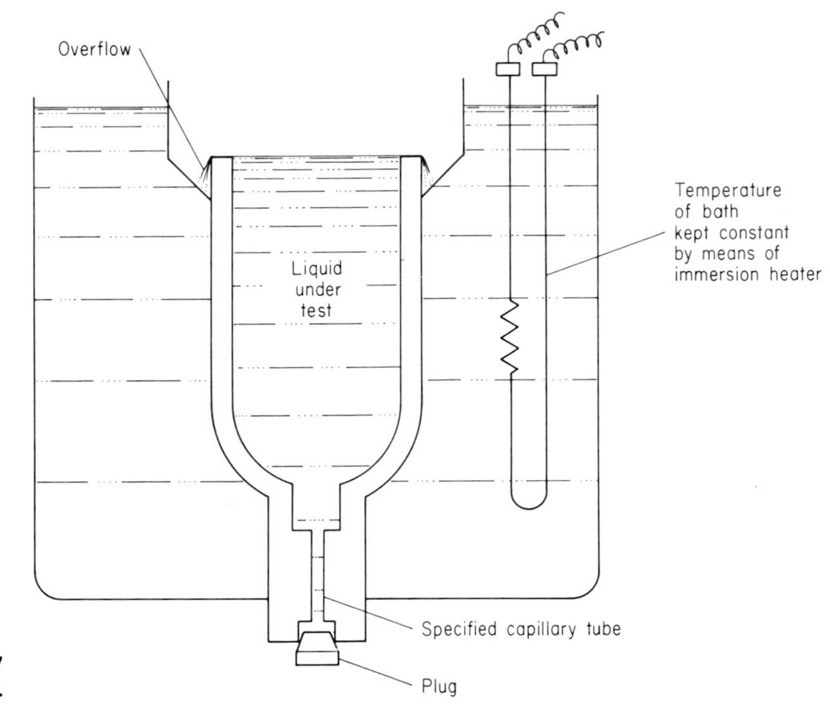

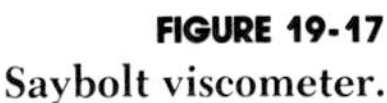
FIGURE 19-17
Saybolt viscometer.

The Falling-Piston-Viscometer Method

The liquid to be tested is placed in the test cylinder, and the falling piston is raised to a fixed, measured height. (See Fig. 19-18.) The time that is required for the piston to fall is a measure of the viscosity. The higher the viscosity, the more time it takes for the piston to drop to the bottom.

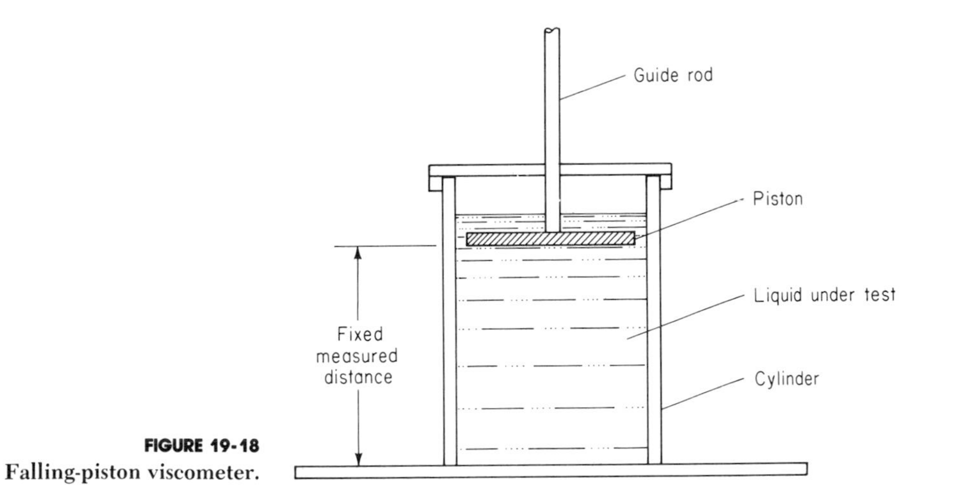

FIGURE 19-18
Falling-piston viscometer.

Rotating-Concentric-Cylinder-Viscometer Method

Two concentric cylinders, which are separated by a small annular space, are immersed in the liquid to be tested (Fig. 19-19). One cylinder rotates with respect to the other, and liquid in the space rotates in layers. A viscous force tends to retard the rotation of the cylinder when the viscome-

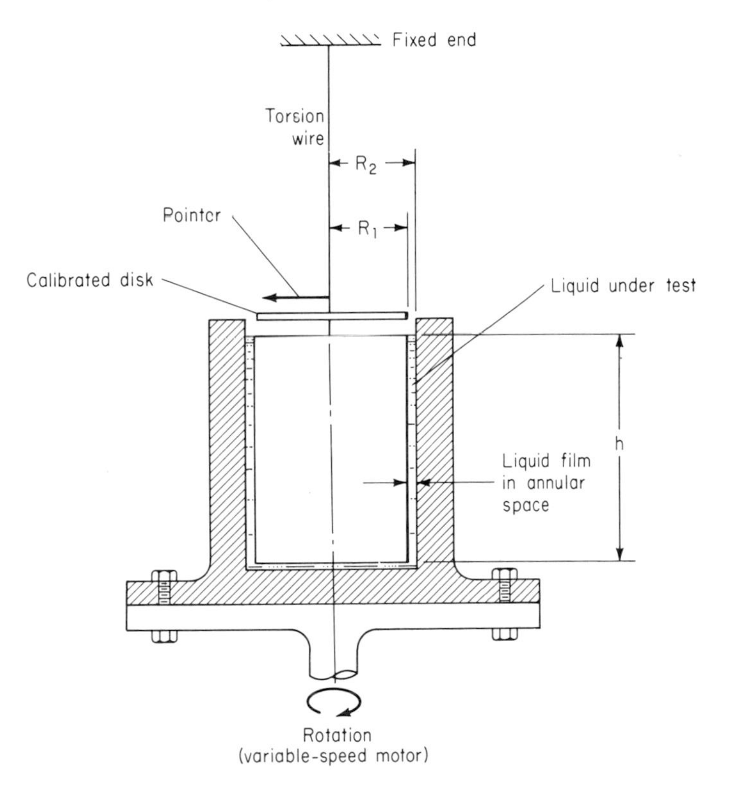

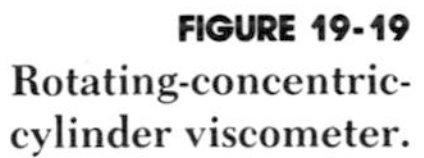

FIGURE 19-19
Rotating-concentric-cylinder viscometer.

ter is in motion. The torque on the inner cylinder, which is caused by the viscous force retarding its rotation, is measured by a torsion wire from which the cylinder is suspended.

Fixed-Outer-Cylinder-Viscometer Method

The inner cylinder rotates at a constant speed, actuated by a wire wrapped around a drum on the shaft. Force is exerted by weights of different mass at each end of the wire (Fig. 19-20). The number of revolutions of the cylinder is counted by a revolution indicator. The viscosity of liquids is determined by timing a definite number of revolutions; the time is proportional to the viscosity of the liquid under test and is the basis of comparison.

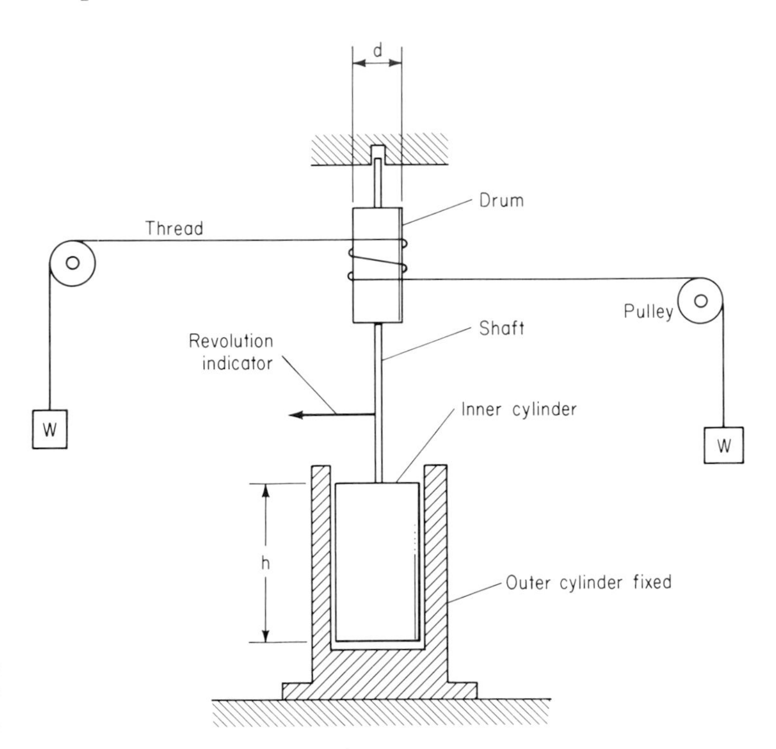

FIGURE 19-20 Fixed-outer-cylinder viscometer.

Falling-Ball-Viscometer Method

A ball will fall slowly through a viscous liquid. At first the ball accelerates, but then it will fall with a constant velocity. The technician measures the time required for the ball to fall a known distance, after the condition of uniform viscosity has been achieved. The cylinder (Fig. 19-21) must have a large enough diameter so that (1) no eddy currents are set up and (2) the cylinder surface will not affect the fall of the ball.

Commercial falling-ball viscometers are fitted with a ball-release device, and the time required for the ball to descend is measured with a stopwatch (see Fig. 19-22).

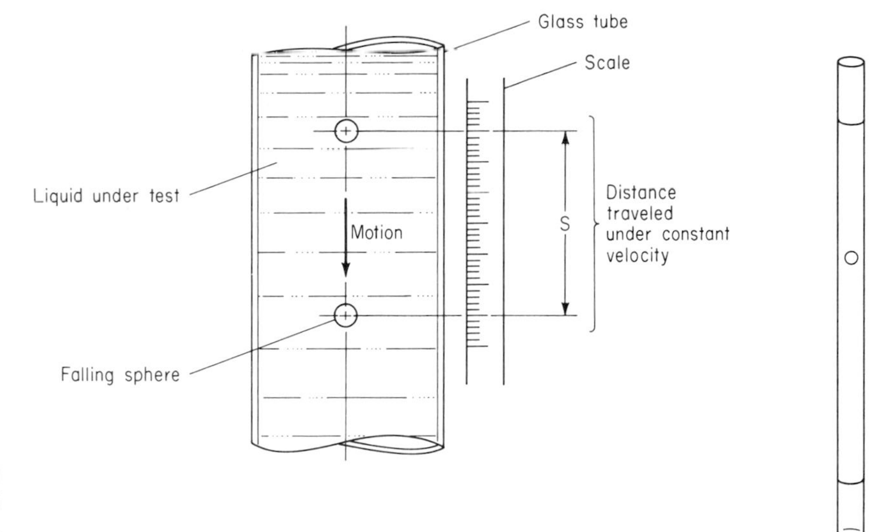

FIGURE 19-21
Schematic of falling-ball viscometer.

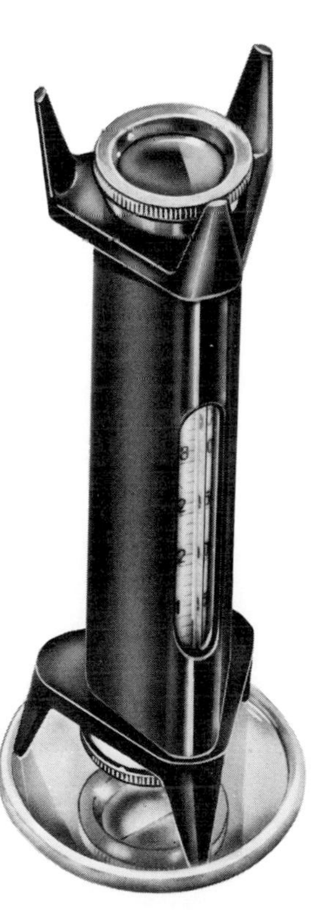

FIGURE 19-22
Commercial falling-ball viscometer is used for rapid and accurate determinations of viscosities of liquids. Range is to 40,000 cP. This instrument measures the time required for a steel ball to drop through a precision-bore brass tube. It requires about 25 mL of liquid.

Ostwald Viscometer Method

Procedure

1. Wash the viscometer and make sure it is absolutely clean.

   (a) Wash thoroughly with soap and water.

   (b) Rinse at least five times with distilled water.

   (c) Finally, introduce sufficient distilled water into the large round bulb, or reservoir. (See Fig. 19-23).

2. Allow the distilled water to come to thermal equilibrium in a constant-temperature bath.

3. Apply suction with a rubber tube to the upper part of the viscometer. This is best done by inverting the viscometer. Draw the liquid up into the tube with the two bulbs to a level above the second bulb (Fig. 19-24).

4. Clock the time needed for the level of the water to pass the signal markings. Make several determinations, using a stopwatch.

5. Drain the viscometer and dry it completely. (Refer to Drying Laboratory Glassware in Chap. 25, "Volumetric Analysis.")

6. With a pipet, add an appropriate volume of the solution to be tested to the reservoir, that is, the same volume as in step 1(c).

7. Clock the time needed for the level of the liquid under test to pass the signal markings.

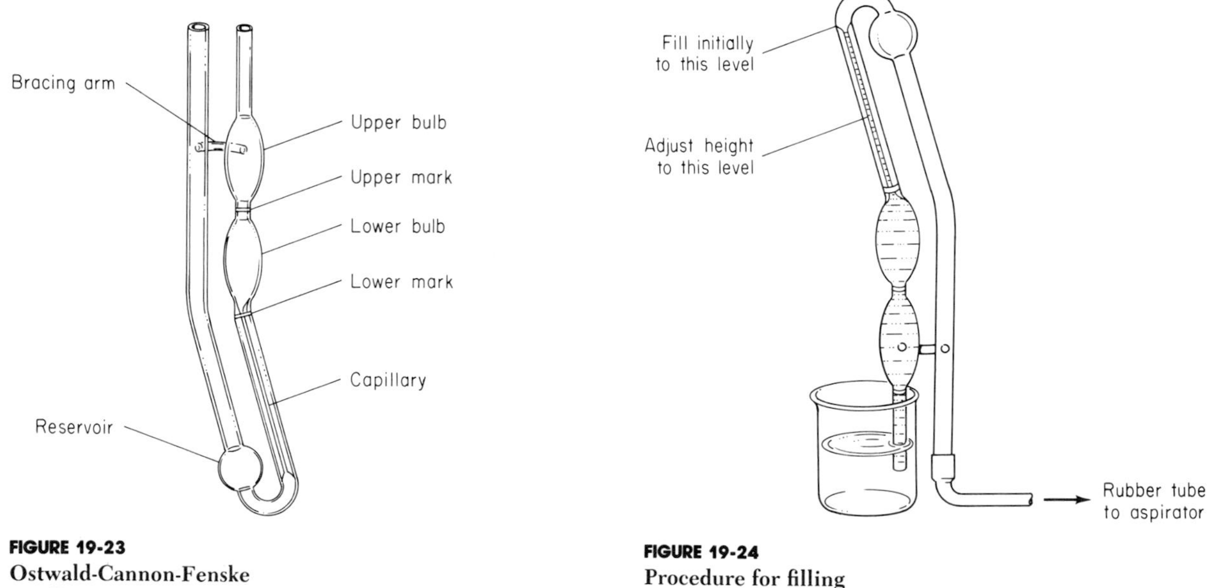

FIGURE 19-23
Ostwald-Cannon-Fenske capillary viscometer.

FIGURE 19-24
Procedure for filling viscometer.

8. Calculate the relative viscosity of the test liquid by comparing the average time required for its flow against that of water at 25°C (reference standard).

Method for Determining Kinematic Viscosity

Kinematic viscosity is equal to the absolute viscosity divided by the mass density, and most laboratories measure kinematic viscosity. The unit of kinematic viscosity is the stokes (St) and is equal to 1 mL/s. A viscometer used for this purpose is shown in Fig. 19-25.

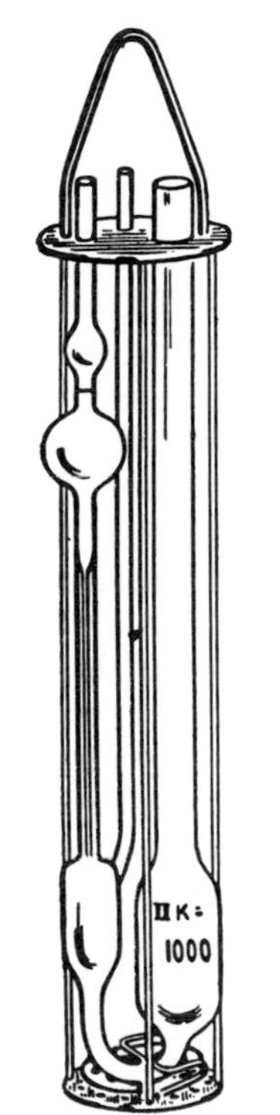

FIGURE 19-25 Frame-supported viscometer tube is used to determine kinematic viscosities. It is available for a wide range of viscosities and is calibrated in kinematic centistokes (2 to 10,000) and Saybolt universal seconds (33 to 46,000).

SURFACE TENSION

The cohesion of the molecules of a liquid is manifested in the phenomenon called *surface tension.* The molecules in the interior of the liquid are subjected to balanced forces between them. The molecules at the surface of the liquid are subjected to unbalanced forces, because they are attracted to the molecules below them. As a result, the surface of the liquid appears to resemble that of an elastic membrane, causing liquid surfaces to contract. This inward pull on those surface molecules results in surface tension, the tendency of a liquid to form drops and the resistance to expansion of the surface area. As a result, in the absence of any external influence, a small liquid sample will assume the spherical shape of a drop.

Liquid molecules also have attraction for other substances; this property is called *adhesion*. When there is an adhesive force between liquids and the surfaces of containers (tubes, beakers, etc.), the liquid is said to "wet" the surface. The property is called *capillarity* and is related to surface tension.

Capillary-Rise Method

Principle

Liquids will rise in a capillary tube until the gravitational force or pull on the column of liquid is exactly equal to the wetting force (Fig. 19-26). Surface tensions of liquids can be calculated by measuring the capillary rise in tubes and by comparing them with the rise of known standards, such as water.

The following formula can be used for calculation.

$$T = \frac{rhdg}{2}$$

where T = surface tension, dyn/cm
d = the density of the liquid, g/mL
h = the height of the column, cm
r = the internal radius of the tube, cm
g = the acceleration due to gravity, 980 cm/s^2

Procedure

1. Set up experimental apparatus as shown in Fig. 19-26.

2. Standardize the apparatus with water as the test liquid. (A drop of water-soluble dye facilitates reading the height of the liquid.)

3. Apply gentle suction to raise water to the top of the capillary tube, release suction, and allow the water to fall to equilibrium. Measure the height of the water column; repeat this step and your measurements to obtain reliable data.

4. Clean and dry the equipment. Substituting the liquid under test for water, repeat step 3, and obtain liquid-height data.

CAUTIONS

1. Temperature rise decreases surface tension, and therefore reliable results require thermostated procedures.

2. Tubing must be absolutely clean, and the column of liquid must not contain any air bubbles; otherwise the experimental results will not be reproducible and will be incorrect.

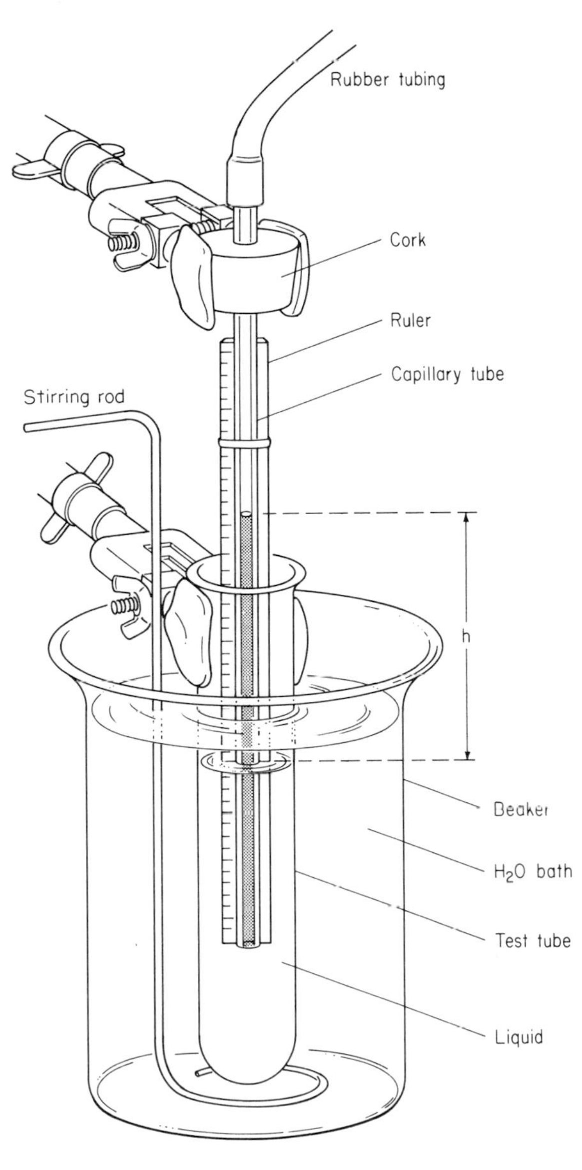

FIGURE 19-26
Apparatus for measuring surface tension by the capillary-rise methods (h is the height to which the liquid rises).

Angle-of-Contact Method

Principle

The angle of contact between the liquid and the capillary tube (Fig. 19-27) can be used to calculate surface tension by using the formula

$$h = \frac{2\ T \cos \theta}{rdF}$$

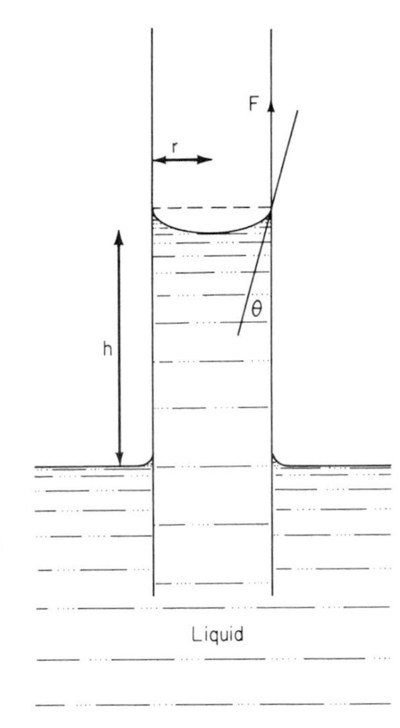

FIGURE 19-27
Capillary rise and angle of contact. Key: h = height, r = radius of tube, F = force upward = force downward, θ = contact angle.

where h = height of liquid column
r = internal radius of tubing
F = gravitational force = force upward
θ = angle of contact

Procedure

1. Immerse a clean microscope slide in a beaker of water which is filled to the brim; then carefully adjust the angle of the slide until you have obtained a flat meniscus. (See Fig. 19-28.)

2. Measure the angle with a common protractor. Make several determinations and measurements.

3. Measure the angle of other liquids to obtain the contact angle.

4. Calculate the surface tension by using the formula on page 416.

DuNouy Torsion-Wire Tensiometer Method

A stainless steel torsion wire which is attached to a torsion head extends through the scale and carries a vernier readable to 0.1 dyn. The torsion ring is lowered into the liquid, and, as the torsion knob is turned, the torsion is transmitted to the head. The scale is read at the point when the ring breaks the surface of the sample, giving the interfacial tension (surface tension) directly.

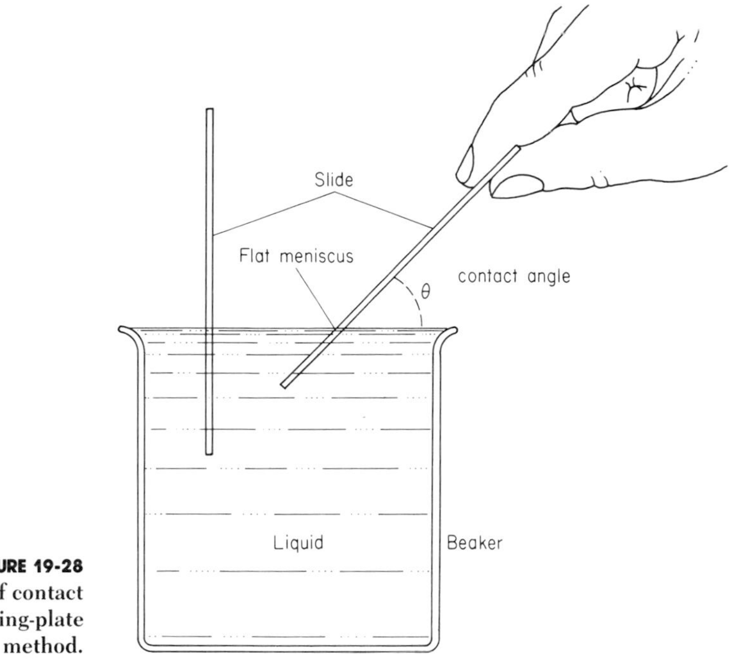

FIGURE 19-28 Measurement of contact angle by the tilting-plate method.

OPTICAL ROTATION

Polarimetry

Ordinary white light vibrates in all possible planes which are perpendicular to the direction of propagation, and it consists of many different wavelengths. Sodium light is monochromatic light, having only one frequency, but it still vibrates in all possible planes. (See Fig. 19-29*a*.)

When light which is vibrating in all possible planes is passed through a Nicol prism, two polarized beams of light (Fig. 19-29*b*) are generated (see Fig. 19-30). One of these beams passes through the prism, while the other beam is reflected so that it does not interfere with the plane-polarized transmitted beam.

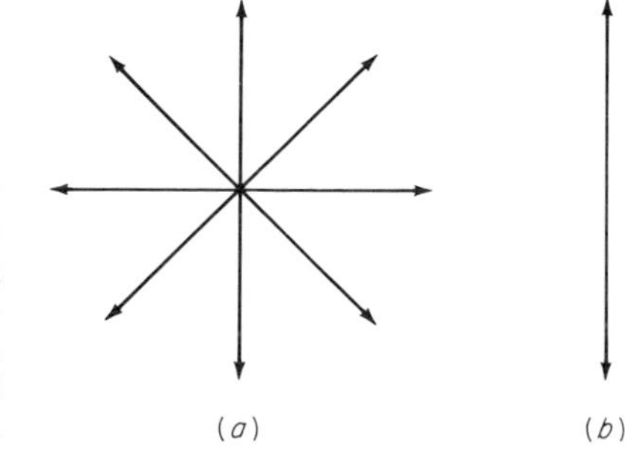

FIGURE 19-29 Light polarization. (*a*) Unpolarized light vibrating in all possible planes. (*b*) Light wave vibrating in only one plane (polarized light).

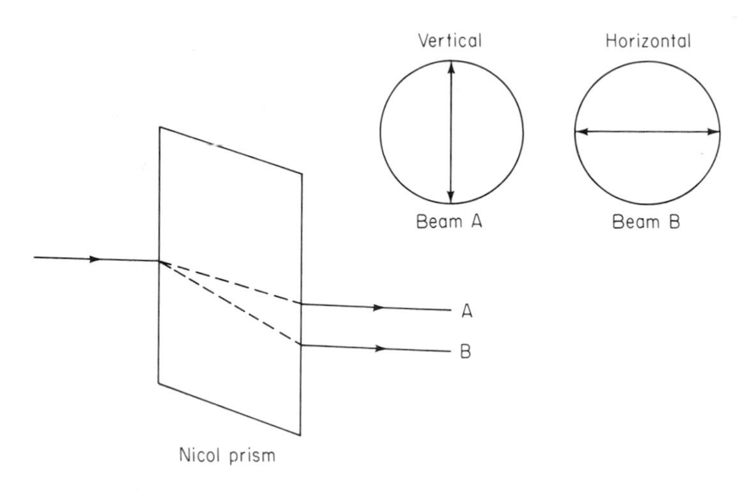

FIGURE 19-30
Generation of two beams of polarized light through a Nicol prism.

If the beam of plane-polarized light is also passed through another Nicol prism, it can pass through only if the second Nicol prism has its axis oriented so that it is parallel to the plane-polarized light. If its axis is perpendicular to that of the plane-polarized light, the light will be absorbed and will not pass through.

Some substances have the ability to bend or deflect a plane of polarized light, and that phenomenon is called *optical rotation.* Only asymmetric molecules can do this, and the magnitude and direction of the deflection of the plane of polarized light which passes through an asymmetric substance in solution can be measured in a polarimeter.

Polarimeter

Principle

Unpolarized light from the light source passes through a polarizer. Only polarized light is transmitted. This polarized light passes through the sample cell. If the sample does not deflect the plane of light, its angle is unchanged. If an optically active substance is in the sample tube, the light is deflected. The analyzer prism is rotated to permit maximum passage of light and is then said to be lined up. The degree (angle) of rotation α is measured. (See Fig. 19-31.) Factors that affect the angle of rotation α are:

1. Concentration—the greater the concentration, the greater the angle of rotation.
2. Solvent
3. Temperature

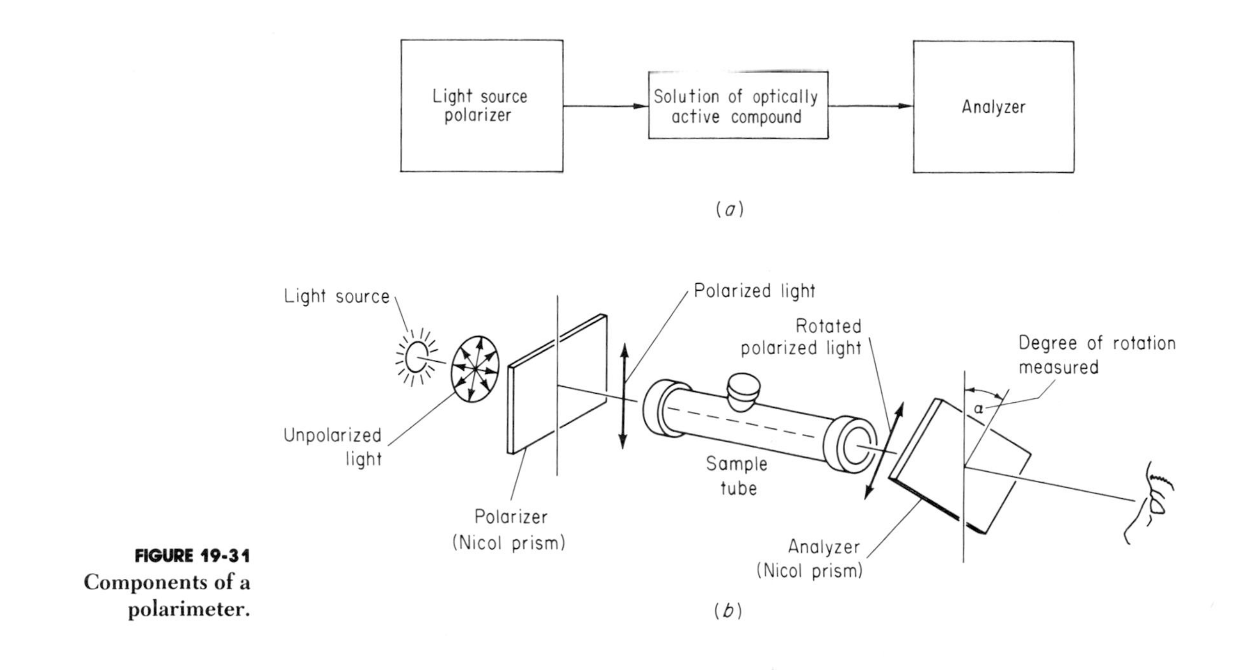

FIGURE 19-31 Components of a polarimeter.

4. Wavelength of the polarized light
5. Nature of the substance
6. Length of sample tube

Direction of Rotation of Polarized Light

The direction of rotation is indicated by a (+) for dextrorotation (to the right) and a (−) for levorotation (to the left). To avoid confusion and to correlate structure and nomenclature, the symbols D and L are used *without regard to the direction of optical rotation.* Substances are labeled D or L with regard to the *configuration* of the asymmetric carbon atom.

Specific Rotation

The angle of rotation can be combined in a formula with the sample concentration and length of the sample tube to provide a value called *specific rotation* $[\alpha]_D^{25}$, which includes notation for the temperature and wavelength of the polarized light.

$$[\alpha]_D^{25} = \frac{\text{angle of rotation } \alpha}{\text{concentration} \times \text{length of sample tube}}$$

where concentration is expressed in grams of solute per milliliter of solvent, length is expressed in decimeters, D = D line of sodium, and 25 = 25°C.

Calculation of the Specific Rotation

If the angle of rotation is plotted on the y axis of a graph against the concentration in grams per milliliter on the x axis, the specific rotation can be calculated from the slope of the line obtained (see Fig. 19-32).

$$[\alpha]_D^{25} = \text{slope} = \frac{\text{change in angle of rotation}}{\text{change in concentration}}$$

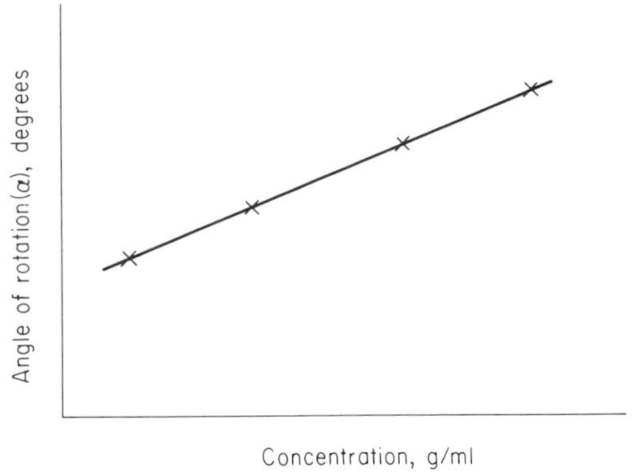

Angle of rotation (α), degrees

Concentration, g/ml

FIGURE 19-32 Graphing to obtain specific rotation.

Using the Polarimeter

Equipment

Polarized light source, analyzer, polarimeter tube

Procedure

1. In a 10-mL volumetric flask, prepare a solution* of the compound whose optical activity is to be determined.

2. Clean, dry, and partially assemble a polarimeter tube (Fig. 19-33).

3. With a dropper fill the tube completely to overflow (Fig. 19-34).

4. Slide an end glass on the tube so that no air is entrapped. Close the end of the polarizer tube. Do not screw on too tightly.

5. Position the polarimeter tube in the polarimeter.

6. Turn on the light source and allow it to warm up.

7. Rotate the analyzer tube until the two halves of the image which is viewed through the eyepiece match exactly.

---

* Temperature change causes a change in solution concentration due to the contraction or expansion of the solution. Compensate for this effect by making up the solution at room temperature (the temperature of the polarimeter) or use a polarimeter with a thermostat.

FIGURE 19-33
Polarimeter sample tube.

FIGURE 19-34
Filling a polarimeter tube.

8. Read the dial and record the magnitude of rotation.

9. Obtain *blank solvent reading* by repeating steps 2 through 8 with pure solvent blank in a clean polarimeter tube.

10. The difference between readings 8 and 9 is α (see Fig. 19-35).

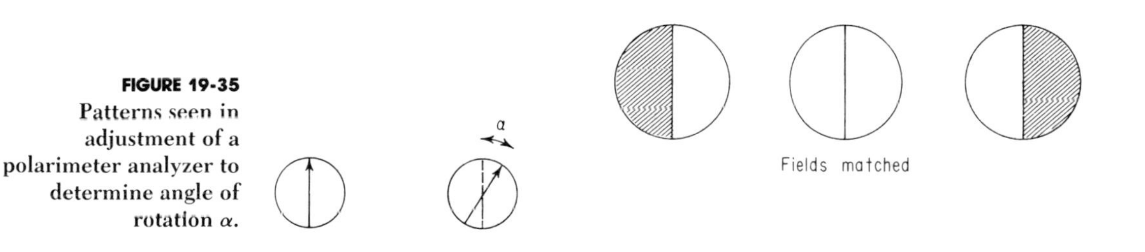

FIGURE 19-35
Patterns seen in adjustment of a polarimeter analyzer to determine angle of rotation α.

Double-Field Matching

Newer types of polarimeters use a double-field optical system. The split-field image enables the operator to match the light intensity more accurately. As the plane of polarized light is rotated, the split field changes as the analyzer is rotated (see Fig. 19-36).

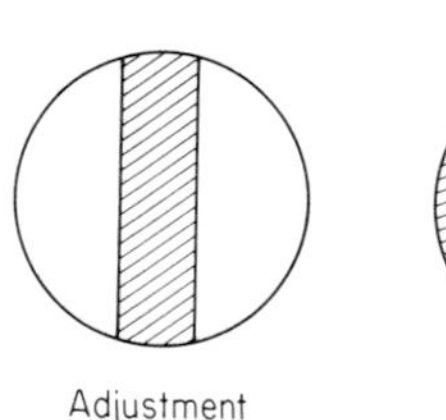

Adjustment
correct

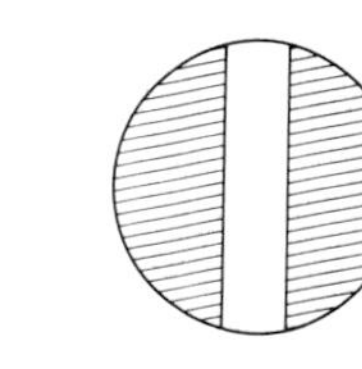

FIGURE 19-36
Split-field image changes in the polarimeter as the analyzer is rotated.

Causes of Inaccurate Measurements

1. Solid particles suspended in the solution. (Filter solution if dust particles are present.)

2. Entrapped air bubbles

3. Strains (improperly annealed or strained glass) in the glass end plates

Avoidance of Inaccuracy

1. Filter solutions properly.

2. Do not create strains in the glass end plates by tightening them with excessive pressure.

REFRACTIVE INDEX

The *refractive index* of a liquid is the ratio of the velocity of light in a vacuum to the velocity of light in the liquid. The refractive index of a liquid is a constant for that liquid, and its determination furnishes us with both a method of identifying a substance and a method for determining the purity of substances. Since the angle of refraction varies with the wavelength of the light, the measurement of refractive index requires that light of a known wavelength be used, usually that of the yellow sodium D line with a wavelength of 5890 Å.

A typical refractive index would be

$$n_D^{20} = 1.4567$$

where the superscript indicates the temperature and the subscript indicates that the sodium D line was used as a reference.

The commonly encountered refractive index is reported to four decimal places, and since refractive indices can easily be experimentally determined to a few parts in 10,000, it is a very accurate physical constant. Small amounts of impurities have a significant effect on the experimental value, and, in order to match the established reported refractive indices, substances must be very carefully purified.

Specific Refraction

The refractive index decreases with temperature because the density decreases, resulting in fewer molecules per unit volume. However, a quantity called the specific refraction is independent of the temperature and may be calculated by the equation:

$$r = \frac{(1)(n^2 - 1)}{(d)(n^2 + 2)}$$

where r = specific refraction
d = density
n = refractive index

Molar Refraction

The refraction of light is an additive property of atoms and also partly a constitutive one of molecules. Numerical values of atomic refraction have been assigned to functional groups which are found in organic compounds. Experimental values of molar refraction can be compared with the molar refraction calculated on the basis of assigned structure.

White Light Substituted for Yellow Sodium Light

Commercial refractometers have built-in compensations which enable the technician to use ordinary white light instead of yellow sodium light for illumination. The index of refraction that is obtained from the instrument, therefore, is that which would have been obtained if the yellow sodium light had been used.

Abbé Refractometer

Principle

The refractive index is easily determined with an Abbé refractometer. The refractometer compares the angles at which light from an effective point source passes through the test liquid and into a prism whose refractive index is known. The refractive index of the liquid is read from the dial. (See Fig. 19-37.)

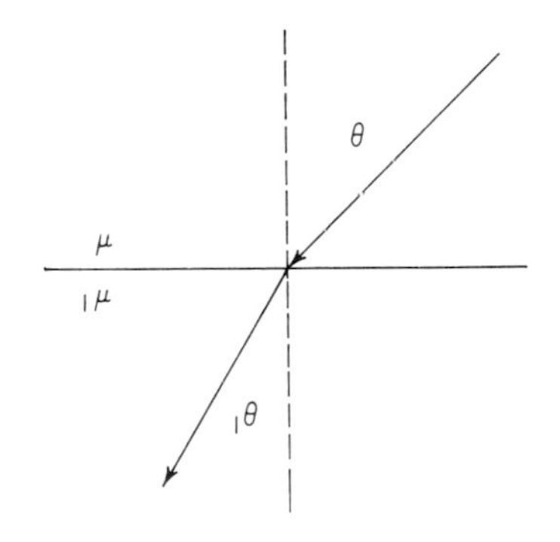

FIGURE 19-37 Defining angles and refractive indices used.

General Precautions and Techniques

1. The prisms can be temperature-controlled to meet the need.

2. Only a few drops of sample are required.

3. Free-flowing liquids can be introduced with an eyedropper into the channel which is alongside the prisms. When a mixture to be analyzed contains a volatile component, this technique is particularly useful because it minimizes the loss by evaporation of the volatile component and thus affects the accuracy of the readout.

4. *Never touch the prisms with any hard object.* Always clean with soft tissues moistened with alcohol or petroleum ether; the prisms, after being cleaned, should always be left in a locked position so that dirt and dust will not collect on them.

Equipment

Abbé refractometer (Figs. 19-38 and 19-39), constant-temperature bath, standard solution of known refractive index

Procedure

1. Adjust the constant-temperature bath to the desired temperature. Connect to the Abbé refractometer with a small circulating pump.

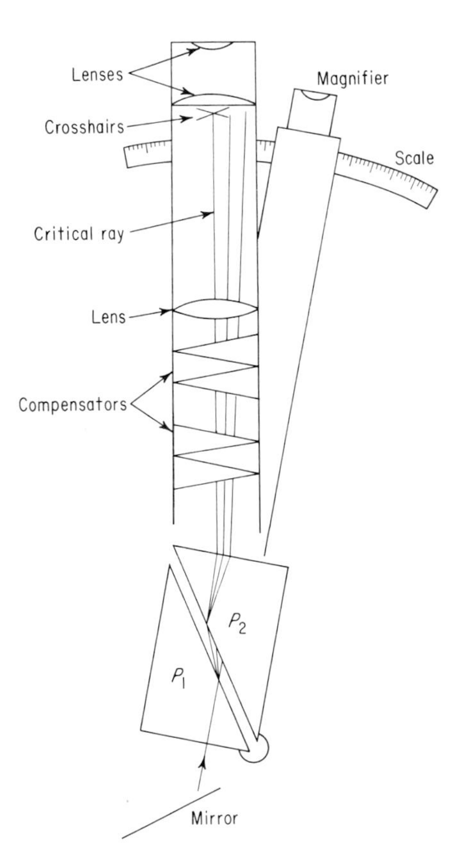

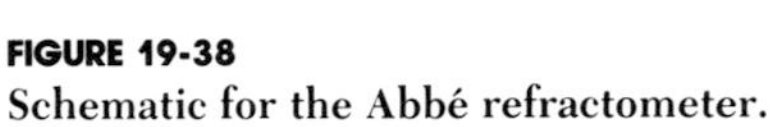

FIGURE 19-38
Schematic for the Abbé refractometer.

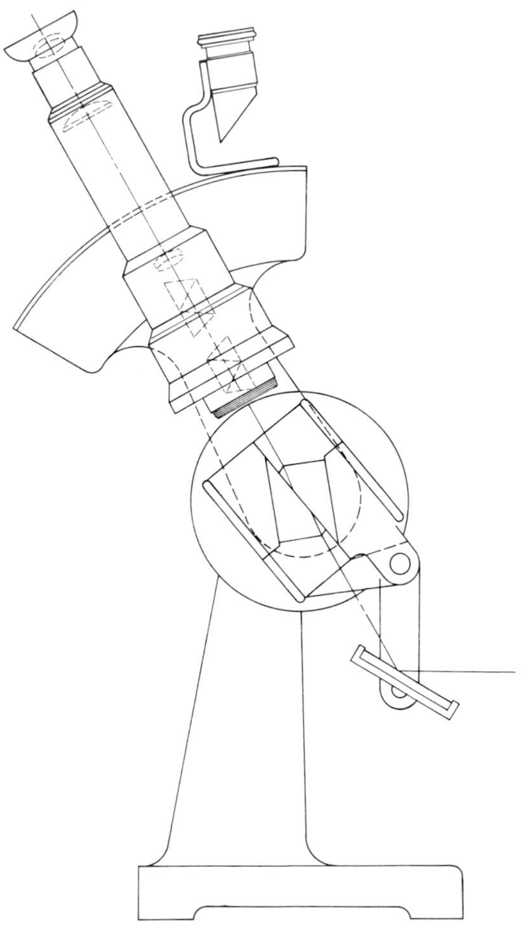

FIGURE 19-39
Drawing of the Abbé refractometer.

2. Twist the knurled locking screw counterclockwise to unlock the prism assembly. Lower the bottom part of the hinged prism until it is parallel with the desk top.

3. Clean the upper and lower prisms with soft, nonabrasive, absorbent, lint-free cotton wetted with alcohol; allow them to dry.

4. Place a drop of the standard test solution of known refractive index at that temperature on the prism. Check the thermometer in the well of the prism assembly.

5. Close the prism assembly. Lock by twisting the knurled knob.

6. Set the magnifier index on the scale to correspond with the known refractive index of the standard.

7. Look through the eyepiece and turn the compensator knob until the colored, indistinct boundary seen between the light and dark fields becomes a sharp line.

8. Adjust the knurled knob at the bottom of the magnifier arm until the sharp line exactly intersects the midpoint of the cross hairs in the image (Fig. 19-40).

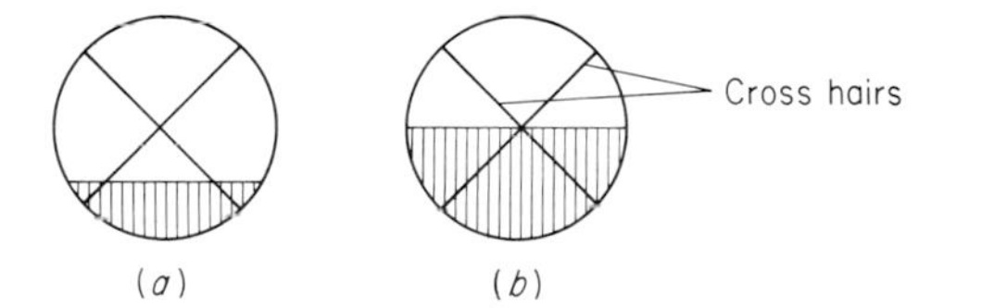

FIGURE 19-40 Adjustment of the refractometer. (*a*) Incorrect. (*b*) Correct.

CAUTION If necessary, recheck and reposition the index on the magnifier arm to read as in step 6.

9. The refractometer is standardized.

10. Repeat steps 2 through 5.

(**a**) Open the prism assembly.

(**b**) Clean the prisms.

(**c**) Place a drop of the sample on the prisms.

(**d**) Close the prism assembly.

11. Look through the eyepiece and move the magnifier arm until the sharp line exactly intersects the midpoint of the cross hairs on the image.

12. Read the refractive index from the magnifier-index pointer.

13. Clean the prisms and lock them together.

MOLECULAR WEIGHT DETERMINATION

Vapor-Density Method

Principle

Equal volumes of gases contain the same number of molecules at the same temperature and pressure. This enables one to determine the relative masses (weights) of gases and of other substances which can be vaporized. The mass of a unit volume of a gas, its vapor density,* can be substituted in the general gas equation, and its molecular weight can be determined by using the following formula:

$$M = \frac{gRT}{VP}$$

where M = molecular weight
g = acceleration due to gravity
R = Avogadro's number
T = temperature in kelvins
V = the volume of the sample
P = pressure at which volume is measured

Procedure (for a Volatile Liquid)

1. Clean and dry suitable flask (125-mL Erlenmeyer).

2. Cover mouth of flask with aluminum foil secured with thin copper wire. Puncture the foil with a fine needle.

CAUTION The hole should be almost invisible to the eye. A large hole introduces appreciable error.

3. Determine mass of flask, foil, and copper wire. Record data.

4. Remove wire and foil and place in the flask about 3 to 5 mL of the volatile liquid which is to be analyzed. Replace wire and foil. (The liquid may also be injected through the needle hole with a hypodermic syringe.)

5. Immerse flask (including foil, wire, and sample) in a beaker of boiling water. Clamp the flask securely so that it is immersed to the neck (Fig. 19-41).

6. Keep the flask in the boiling water until all the liquid has volatilized. No vapor should emerge from the hole.

---

* Refer to the section on determination of gas densities, page 395.

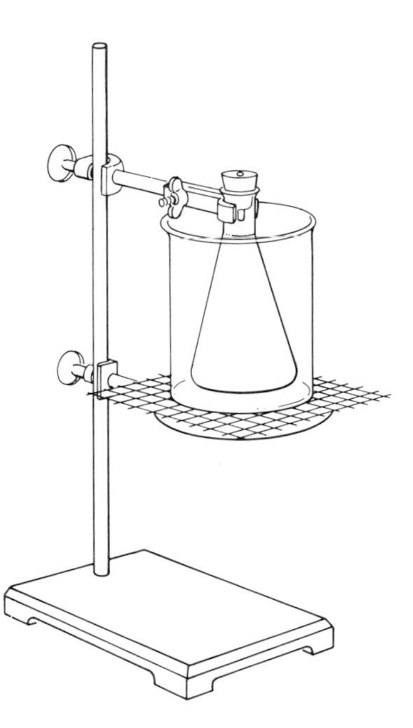

FIGURE 19-41
Apparatus used to determine molecular weight by the vapor-density method.

7. Remove the flask and allow it to cool; wipe it dry. Record the barometric pressure and boiling point of the water.

CAUTION *Use calibrated thermometer.* Then determine the mass of the flask.

8. Determine the volume of the flask by the standard procedure:

Clean and dry the flask; determine the mass of the empty flask; fill the flask to its lip with water; remass; calculate mass of water; refer to density tables for the density of water; calculate the volume of the water and, thus, the flask.

9. Calculate the molecular weight, substituting the proper values in the formula.

Alternate Procedure Substitute a Dumas bulb for the flask and follow the same procedure as above.

Boiling-Point-Elevation Method

Principle A nonvolatile solute raises the boiling point of a solvent by an amount proportional only to the concentration of the solute particles; the chemical nature and masses of the particles are immaterial. Thus, the boiling-

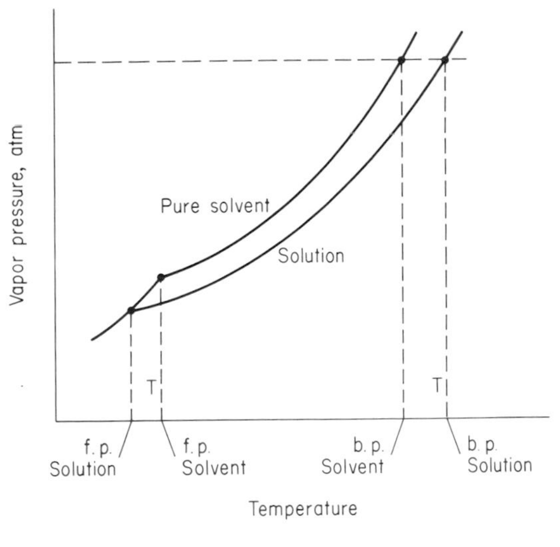

FIGURE 19-42 Vapor-pressure curve of a solution containing a nonvolatile solute compared with that of the pure solvent. The boiling-point elevation is equal to the boiling point of the pure solvent.

point elevation, the freezing-point depression, and the vapor pressure of such solutions are the *colligative properties* of solutions, because the amount of change depends primarily upon the number of particles involved (see Fig. 19-42). Each solvent has its own *molal boiling-point constant,* that is, the temperature change in degrees Celsius that is caused by dissolving one mole of a solute in one kilogram of the solvent. This boiling-point elevation is a direct consequence of the vapor-pressure depression, governed by Raoult's law, which states that the vapor pressure of a solvent above a solution (p) is equal to the product of the vapor pressure of the pure solvent (p_0) times the mole fraction of the solute, or

$$\frac{p_0 - p}{p_0} = \frac{aM}{bm}$$

where a = mass in grams of solute
m = molecular weight of solute
b = mass in grams of solvent
M = molecular weight of solvent

The molal boiling-point constant for each solvent is obtained when one mole of a nonvolatile solute is dissolved in one kilogram of pure solvent. Fractional moles of a solute will produce proportional increases in the boiling point. Some of these constants are given in Table 19-2.

Procedure 1

1. Arrange the apparatus as shown in Fig. 19-43.

CAUTION If flammable organic solvents are used, be extremely careful. Use emergency hood and alternate source of heat, without an open flame.

TABLE 19-2
Molal Boiling-Point Constants

Substance	*Boiling point, °C*
Bromobenzene	6.26
Camphor	5.95
Chlorobenzene	4.15
Diphenyl	7.08
Naphthalene	5.65
Tetrachloromethane	5.03
Water	0.512

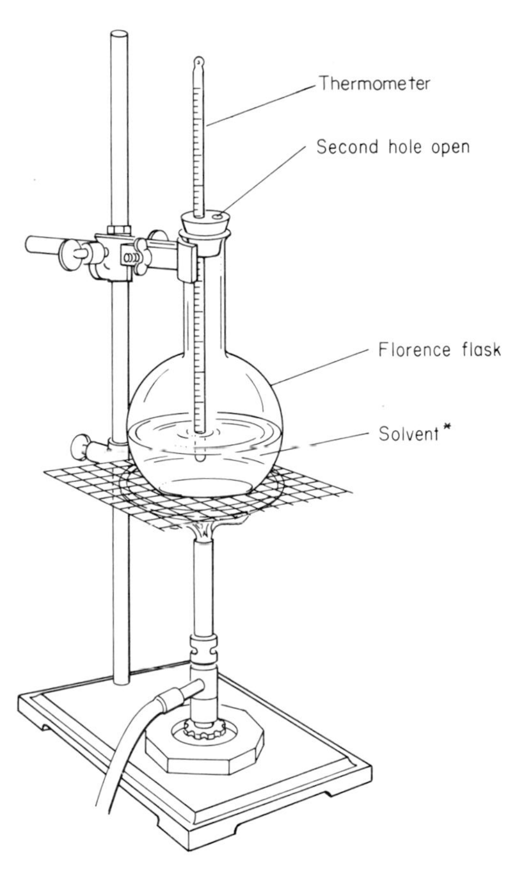

FIGURE 19-43
Boiling-point-elevation apparatus.

2. Place 250 mL of solvent into the flask; and add a few boiling stones.
3. Heat gently (thermometer submerged) to boiling.

CAUTION Do not superheat. Do not boil too vigorously or for too long a time; solvent will be lost.

4. Record the temperature of the boiling solvent to 0.1°C. For precision work use special thermometers capable of reading to 0.001°C.*

5. Discontinue heating; add a sample of known mass of the nonvolatile solute (about 25.0 ± 0.1 g) to the solvent.

6. Reheat gently and record the boiling point of the solution. Observe all safety rules.

7. Subtract the boiling point of the solvent from that of the solution. The difference in the boiling points is the boiling-point elevation.

Calculations

1. Calculate the number of grams of solute per kilogram of solvent. Use the ratio

$$\frac{\text{Mass (actual) solute}}{\text{Mass (actual) solvent}} = \frac{\text{mass solute}}{\text{1000 g solvent}}$$

2. Divide the measured boiling-point elevation by the molal boiling-point-elevation constant to obtain the mole fraction:

$$\text{Mole fraction} = \frac{\text{measured boiling-point elevation}}{\text{molal boiling-point elevation}}$$

3. Divide step 1 by step 2 to obtain the number of grams of solute per mole of solute, which is the molecular weight.

Procedure 2

This procedure uses the micro-boiling-point technique which is described more fully in the section on boiling points in this chapter. The boiling points of the solvent and solution are determined by the capillary-tube method.

1. Arrange the experimental setup as shown in Fig. 19-44.

2. Select the solvent and carefully weigh out about 20 g (±0.1 g).

3. Heat the bath gently until a steady stream of bubbles issues from the capillary tube.†

---

* The boiling point of the pure solvent as determined may not agree with the value in the literature because of variations in atmospheric pressure. This does not affect the determination, because the *difference* in the boiling points is the important factor.

† The selection of the heating-bath fluid is determined by the boiling point of the solvent. Use mineral oil for high-boiling solvents.

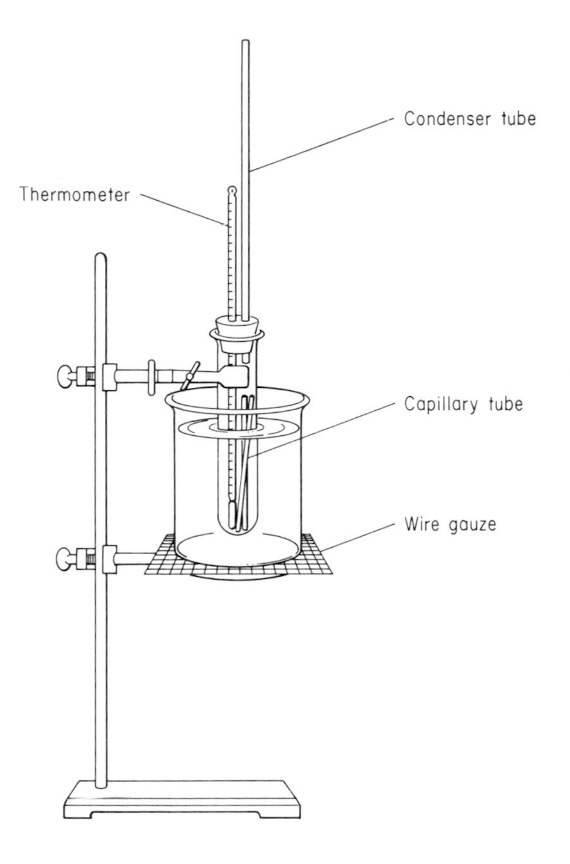

FIGURE 19-44
Apparatus for determining molecular weight by boiling-point elevation (micro method).

CAUTION Observe safety precautions for flammable and toxic vapors.

Record the temperature (the boiling point of the pure solvent). Do not boil for a long time; solvent will escape and affect the accuracy of the determination.

4. Remove the test tube from the bath and cool it. Discard the capillary tube.

5. Measure accurately to 0.1 g the mass of substance to be analyzed; add it to the solvent. The substance must dissolve in the solvent. Use reference tables or ascertain its complete solubility experimentally before the determination.

6. Add a new, clean capillary tube, with its open end pointed down, to the test tube.

7. Reimmerse the test tube in the heating bath and determine the boiling point of the solution.

8. Subtract the boiling point of the solvent from that of the solution; this is the boiling-point elevation for that concentration of solute.

9. Calculate the molecular weight of the solute, using the method detailed in procedure 1.

Vapor-Pressure-Equilibrium Method

Principle

As previously stated, the colligative properties of solutions depend solely upon the number of molecules, ions, atoms, or other dissolved particles of nonvolatile solute per unit mass of the solvent. They *do not* depend upon their nature or type. The vapor-pressure-equilibrium (isopiestic) method is used to determine molecular weights up to around 125,000. A schematic of the apparatus used is shown in Fig. 19-45.

A chamber containing two carefully matched thermistors is carefully saturated with solvent vapor. Both thermistors assume the same temperature when solvent is placed on both of them. A drop of a solution of known concentration, prepared for the determination, is placed on one of the thermistors. That drop of solution, which has a lower vapor pressure than the pure solvent because of the dissolved solute, acts as a site for the condensation of solvent vapor. This condensation heats the sensitive thermistor until the vapor pressure is raised to that of the pure solvent. The change in temperature causes a change in the resistance of the thermistor, and that change in resistance of the thermistor is reflected in the sensitive

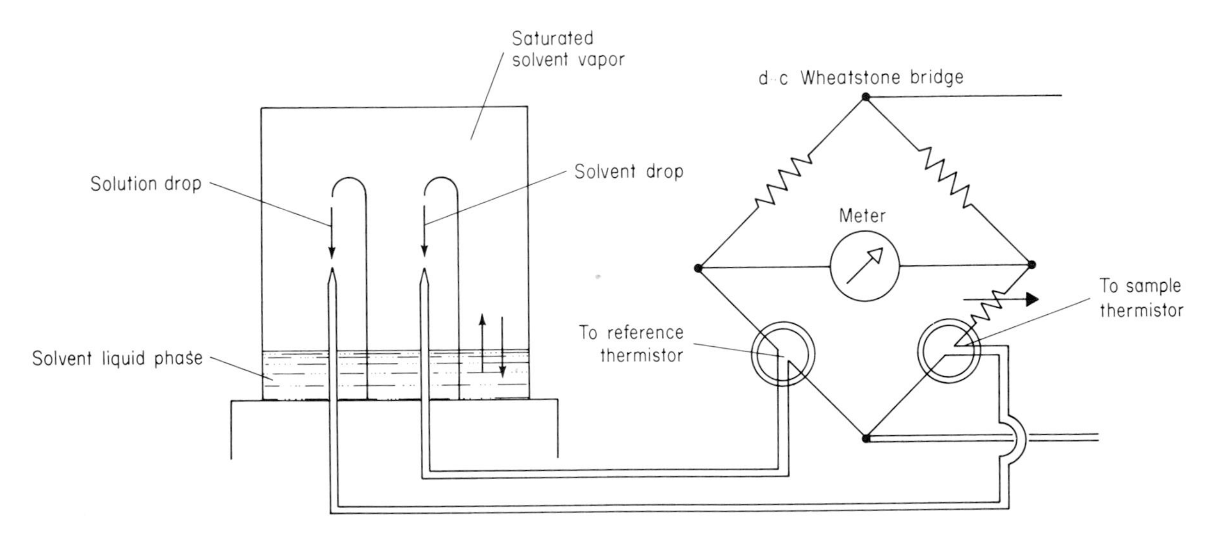

FIGURE 19-45
Schematic of sample chamber and measuring circuit illustrating operating principle of vapor-pressure-equilibrium molecular-weight apparatus.

Wheatstone bridge circuit. The bridge circuit is calibrated in units of molarity, a measure of the molecular weight, and can be coupled to a digital readout unit and/or recorder.

Solvent Considerations

The sensitivity of this type of apparatus varies inversely with the heat of vaporization of the solvent. Organic solvents, such as toluene, have a low heat of vaporization and require high sensitivity; water, with a high heat of vaporization, requires less sensitive equipment.

Calibration of the Apparatus

Each equipment unit must be calibrated; a constant K is determined for each unit because of heat losses due to conduction, convection, and radiation. Calibration is accomplished by using a drop of a standardized solution. If the readout is in microvolts (μV), then the calibration constant (K) can be calculated:

$$K = (E/c) \times M$$

where K = constant
E = readout of the meter, μV
c = concentration of the standard, g/L
M = molecular weight of the standard

The final readout is then multiplied by K.

Freezing-Point-Depression Method

Principle

The molecular weight of a substance can be found by observing the lowering of the freezing point of a solvent, the freezing-point depression, caused by the presence of a solute. This colligative property again depends solely upon the number of solute particles dissolved in the solvent, not upon the kind of particles.

The extent of the change in the freezing point that one mole of nonvolatile solute in one kilogram of solvent will produce depends upon the solvent (Fig. 19-46). Each solvent has its own characteristic molal freezing-point constant, that is, the depression in degrees Celsius of the freezing point of the solvent when one mole of solute particles is dissolved in one kilogram of the solvent. Some of these values are given in Table 19-3.

When a solvent is heated above the melting point and then allowed to cool, the temperature will gradually fall until the melting point of the solvent is reached. At equilibrium the temperature will remain constant until the solvent has solidified completely, and then it will drop as it cools. (Refer to the section on melting points.) The melting point of the pure solvent is determined, and then the melting point of a solution of known

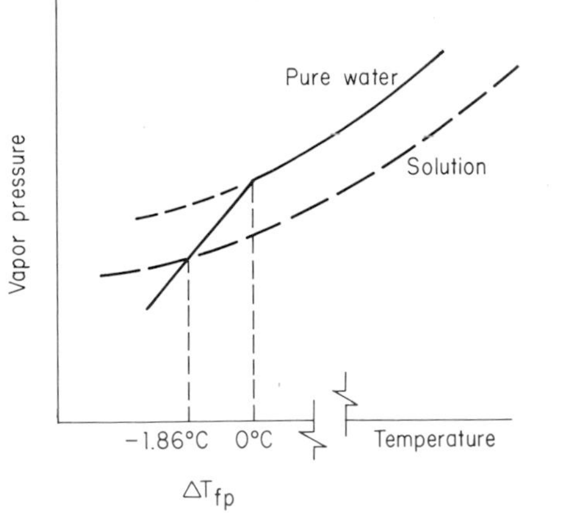

FIGURE 19-46 Freezing-point-depression curve for a 1 *M* solution of a nonvolatile (nonionic) solute in water. The freezing point is lowered by 1.86°C.

TABLE 19-3 Molal Freezing-Point Constants

Solvent	*Freezing-point depression, °C*
Water	−1.86
Benzene	−5.1
Naphthalene	−6.9
p-Dichlorobenzene	−7.1
Cyclohexanol	−39.3

concentration of the solute in the solvent is determined. The melting-point depression can then be calculated, and the mole fraction and corresponding molecular weight of the solute can be found.

CAUTION Solvents may supercool before crystallization takes place. The temperature may fall lower than the freezing point, due to supercooling, but will rise to the true melting point when equilibrium is established.

Procedure

1. Set up the apparatus as shown in Fig. 19-47. The bath may be a cooling or heating bath.*

2. Measure out about 10 g of selected solvent (±0.05 g) and transfer it quantitatively to the cleaned and dried test tube.

3. Heat the selected solvent (if a solid) to about 10°C above its melting point.*

---

* If the solvent is a solid at room temperature, the bath must be heated. If the solvent is a liquid (benzene, water), the bath must be a cooling bath (NaCl, ice, water) to freeze the solvent. If the solvent is frozen, removing the cooling bath allows it to warm up slowly and melt, the temperature at equilibrium being the melting point.

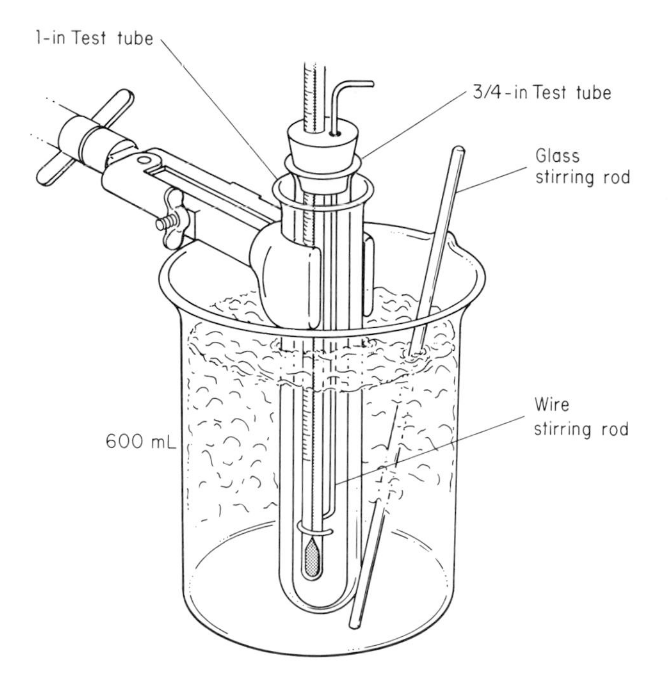

FIGURE 19-47
Freezing-point apparatus for molecular-weight determination.

4. Raise the test tube out of the bath; allow it to cool slowly while mixing, and record the freezing point of the solute (when the temperature remains constant). Record the temperature to 0.1°C, T_2. Special thermometers, such as the Beckmann thermometer (Refer to Chap. 9, "Measuring Temperature") permit readings to be made to 0.001°C.

5. Measure out accurately (to 0.01 g) about 1 g of the powdered solute; add it to the test tube containing the crystallized solvent.

6. Immerse the mixture in the heated bath and mix until all the solute has dissolved in the solvent and a clear solution results.

7. Raise the solution out of the heating bath and determine the melting point of the solution as in step 4. Stir continuously. Record the melting point, T_1.

Calculations

1. Calculate the number of grams of solute per kilogram of solvent. Use the ratio

$$\frac{\text{Mass (actual) solute}}{\text{Mass (actual) solvent}} = \frac{\text{mass solute}}{\text{1000 g solvent}}$$

2. Divide the measured freezing-point depression by the molal freezing-point depression to find the mole fraction:

$$\text{Mole fraction} = \frac{\text{measured freezing-point depression}}{\text{molal freezing-point constant}}$$

3. Divide step 1 by step 2 to obtain the number of grams of solute per mole of solute, which is the molecular weight.

Method for Determining Molecular Weights by Low-Angle-Laser Light-Scattering Photometry

The term *laser* is an acronym for *l*ight-*a*mplified *s*timulated-*e*mission *r*adiation. Lasers emit monochromatic radiation in the visible and near-visible ultraviolet and infrared spectra. The electromagnetic waves comprising these radiations originate from changes within atoms or molecules. Moreover, they are coherent; that is, they are all in phase. Laser beams spread out much less than other, noncoherent, sources of radiation, and therefore extremely large amounts of energy can be concentrated in a very small area. Laser emission can be continuous or pulsed; however, for analytical work, it is usually pulsed.

Lasers produce only one wavelength if conditions remain constant, and they can be tuned to any desired working wavelength. Their energy output per pulse can range from milliwatts to megawatts. It is therefore important to know the energy output of the equipment in your laboratory. A very common laser makes use of a ruby rod as the emitter of monochromatic radiation. Figure 19-48 is a schematic of such a laser.

Other laser systems use organic dyes to obtain the desired frequencies. Certain organic dyes such as fluorescein can be used to cause changes in laser emission. Lasers incorporating these dyes are being used in high-resolution spectrophotometry. Organic-dye lasers have a frequency output that can be tuned over a range of about 400 Å, and there is no significant loss of output energy if their normally broad output range is narrowed. Xanthene dyes are available commercially in an extremely high state of purity; this is a very desirable situation because the purer the dye, the greater the output energy.

Dye lasers are used for several purposes; among them are the production of laser activity in metal vapors, the detection of atmospheric pollution,

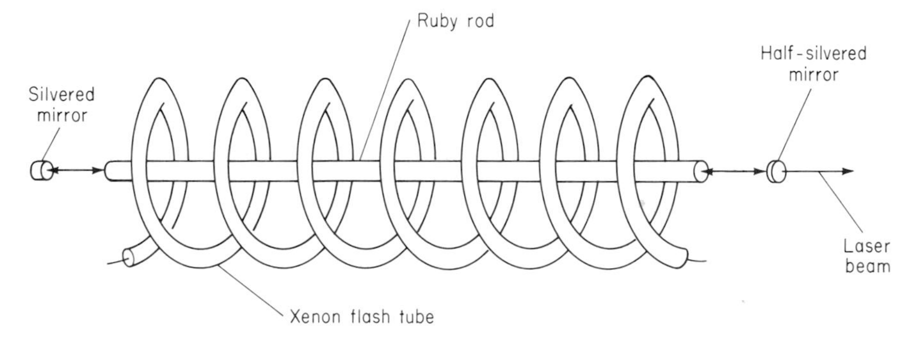

FIGURE 19-48
Schematic of a ruby laser.

the measurement of the wavelength dependence of laser-induced gas breakdown, and the determination of molecular weights by means of light-scattering photometers.

Principle

Light-scattering photometry provides a measure of average molecular weight, chain size and stiffness, long-chain branching, and interaction parameters. The intensity of light scattered by a solution is a function of its polarizability as well as the concentration and size of the solute particles as compared to those of the medium.

Low-angle-laser light-scattering photometers utilize the spatial characteristics of a small, single, transverse-mode laser beam to allow measurements of scattering at forward angles less than 2° from the incident beam. This measurement permits the determination of molecular weights (average) in several minutes, and it is especially sensitive to solutes of high molecular weight, which can be determined with a single concentration and angle.

Safety Rules for a Laser Laboratory

1. Post the recommended warning signs and use the required caution and danger labels wherever they are called for. Remember to take precautions for both types of danger, electrical and optical.

2. Use properly marked eye protection wherever eye exposure is possible.

3. Always discharge the laser beam into a background that is nonreflecting and fire-resistant.

4. Design the laser setups to avoid accidental pulsing of the laser and to minimize the possibility of electric shocks from storage capacitors, power supplies, and other equipment.

5. Keep laboratory personnel a reasonable distance away from *all sides* of the anticipated path of the laser beam.

6. Keep the level of general illumination high in areas where lasers are operated. Darkened rooms cause the pupils of the eyes to dilate, and thus increase the amount of energy that may inadvertently enter the eye.

7. Do not count on goggles or safety glasses to protect your eyes. No glasses that are now available offer protection from *all* wavelengths emitted by existing lasers.

8. *Never* look into the primary beam. Avoid looking at specular reflections of the beam, including those from the lens surface.

9. *Never* look directly at the pump source.

10. *Never* aim the laser with the eye. Looking along the axis of the laser beam increases the hazard from reflections.

11. Be especially cautious around lasers that operate in the ultraviolet and infrared. These beams are invisible to the human eye. Consequently, they can cause many small lesions in the eye without the victim's being aware that damage has occurred until the loss of sight becomes noticeable.

12. Do not let the laser beam strike exposed skin surfaces.

13. Anyone who is working in laser test procedures, and anyone who may be frequently exposed to laser discharges, should have thorough general ophthalmological examinations at regular intervals. These examinations should include funduscopic studies and mapping of visual fields.

CHANGES IN THERMAL ENERGY

Thermochemistry

Heat, a form of energy sometimes called thermal energy, passes spontaneously from objects at higher temperatures to objects at lower temperatures. Many physical and chemical changes take place with the evolution or absorption of heat. This heat change can be measured in a calorimeter, an instrument which is insulated to restrict the heat exchange to the unit and to minimize any heat exchange to the surroundings. Such a system is called an *adiabatic system.* Heat which is absorbed by a reaction equals the heat lost by the calorimeter and its contents. That heat which is evolved by a reaction equals the heat gained by the calorimeter and its contents, because energy cannot be created or destroyed in ordinary chemical reactions, only transferred or changed into another form. When reactions are carried out at constant pressure—for example, atmospheric pressure—the heat flow is known as enthalpy, or ΔH. The values of ΔH are expressed in calories per mole (cal/mol) or kilocalories per mole (kcal/mol).* When heat is absorbed during a reaction, the reaction is an *endothermic* one, and the value of ΔH is positive. When heat is evolved during a reaction, the reaction is an *exothermic* one, and the value of ΔH is negative.

Thermochemistry is concerned with the energy changes brought about by a reaction; these can be classified as follows:

1. *Heat of formation of a substance.* The amount of heat which is required to form one mole of the substance directly from its elements in their standard states; also known as the standard heat of formation.

$$H_2(g) + \tfrac{1}{2}O_2\,(g) \rightarrow H_2O\,(l) \qquad \Delta H_f^\circ = -68.32 \text{ kcal/mol}$$

---

* In SI, which is currently accepted in much of industry, the calorie has gone into disuse. The joule (J) is the accepted unit, and 1 cal = 4.184 J.

2. *Heat(s) of solution, vaporization, sublimation, and fusion.* Heat involved in changes in state of substances or the hydration of ions or molecules. When a solid and a liquid (both at the same temperature) are mixed and the solid dissolves in the liquid, the temperature of the solution is different from that of the initial system. The heat flow from the solution of one mole of the solute is called the *molar heat of solution,* expressed in kilocalories per gram-mole (kcal/g·mol).

NaOH (solid) → NaOH (solution)

The heat of vaporization is reported in gram-calories per gram (g·cal/g); the heat of fusion is also reported in gram-calories per gram. To calculate the number of calories required to fuse (melt) a solid, multiply the number of grams by the heat of fusion. To calculate the number of calories required to vaporize a liquid (at its boiling point), multiply the number of grams by the heat of vaporization. (See Fig. 19-49.)

3. *Heat of combustion.* The amount of heat which is evolved when one mole of a substance reacts with oxygen.

$$CH_4 + 2O_2 \rightarrow CO_2 + 2H_2O \qquad \Delta H = -211 \text{ kcal/mol}$$

Determination of heats of combustion must be carried out in special steel bombs; some are called Parr bombs.

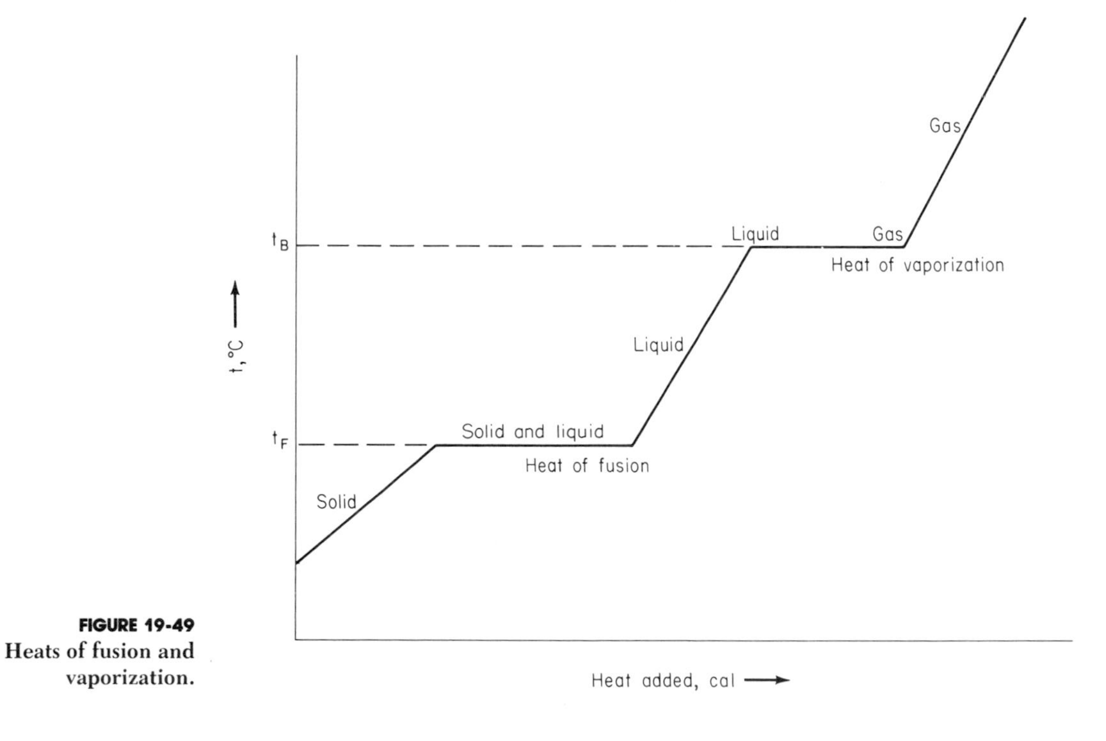

FIGURE 19-49
Heats of fusion and vaporization.

4. *Heat of neutralization*. The heat evolved when one mole of water is produced by the reaction of an acid and a base. The reaction may be represented by:

$$H^+ (aq) + OH^- (aq) \rightarrow H_2O \qquad \Delta H = -13.8 \text{ kcal/mol}$$

5. *Specific heat*. The quantity of heat required to raise the temperature of one gram of a substance one degree Celsius, expressed in calories per gram per degree Celsius. In a calorimetric determination of the specific heat of a substance, the heat gain or loss is equal to the mass of the substance times its specific heat times the change in temperature.

$$\text{Heat gain or loss} = \text{mass} \times \text{specific heat} \times \Delta T$$

The specific heat of metals is related to their atomic mass according to the law of Dulong and Petit, which states that six calories (approximately) are required to raise one gram-atomic mass of many metals one degree Celsius. Knowing the specific heat enables you to calculate the approximate atomic mass.

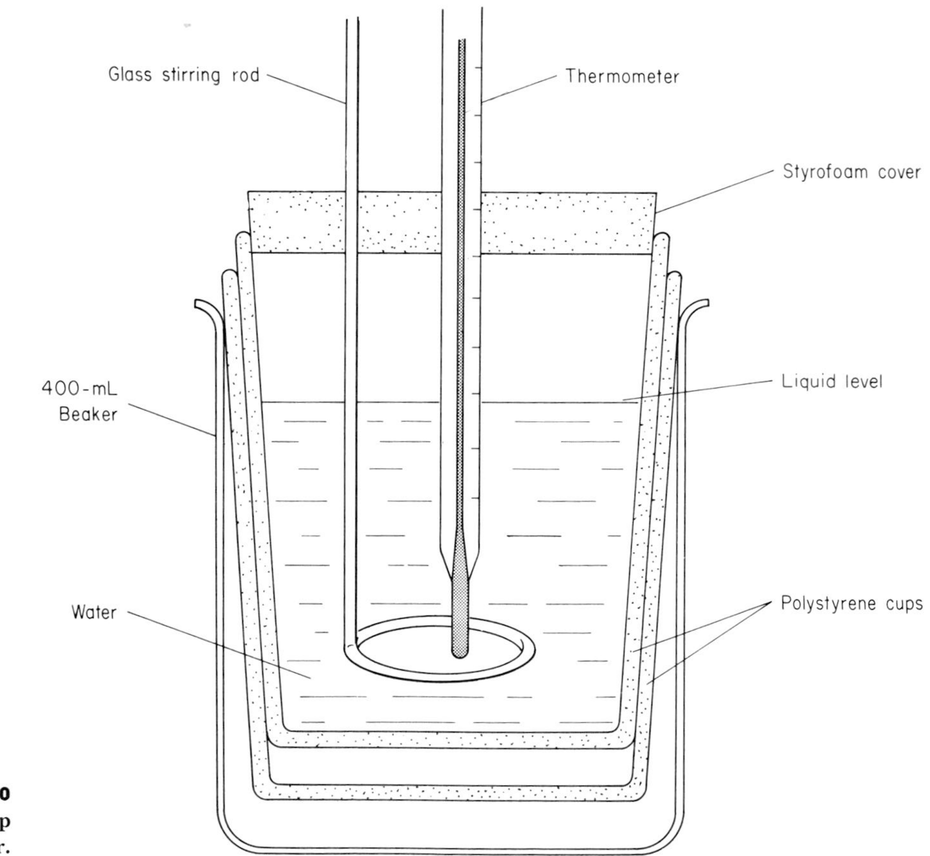

FIGURE 19-50 A polystyrene-cup calorimeter.

The Calorimeter

A calorimeter consists of a reaction vessel which is surrounded by a large volume of water contained in a well-insulated vessel (Fig. 19-50). When a reaction takes place in the calorimeter, the temperature of the water reservoir either increases or decreases, according to whether the reaction is exothermic or endothermic. The heat liberated or absorbed is measured in calories and is calculated by multiplying the change in temperature by the heat capacity of the calorimeter. Usually, the heat capacity of the calorimeter is known or can be experimentally determined by conducting a reaction with an accurately known enthalpy in the calorimeter. The heat capacity or the water equivalent can be calculated.

Heat Leakage in Calorimeters

No calorimeter is perfectly insulated, or perfectly adiabatic, and therefore, all determinations are complicated by the loss of heat from the calorimeter to its surroundings or the absorption of heat by the calorimeter from its surroundings. Correction can be made for this heat leakage by taking a number of temperature measurements after the reaction has taken place, and then extrapolating back to zero time (Fig. 19-51).

Thermal Analysis

Principle

Thermal analysis is the measurement of changes in the physical or chemical properties of materials as a function of temperature, relating heat transfer to either temperature or time. This can provide both qualitative

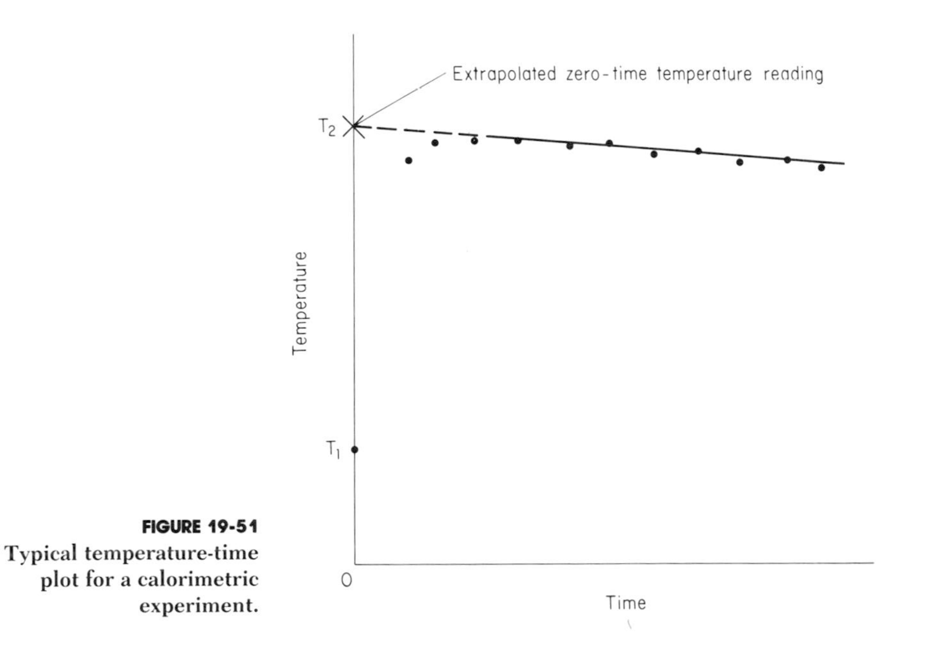

FIGURE 19-51
Typical temperature-time plot for a calorimetric experiment.

and quantitative insight into the basic process. It is used for the measurement and study of processes which involve changes in thermal energy.

Changes of state: vaporization, fusion, and sublimation

Reaction studies: decomposition, oxidation-reduction, hydration-dehydration, polymerization, depolymerization

Changes in the crystalline structures of substances

Specific heat

Thermogravimetric Analysis

Thermogravimetric analysis (TGA) is used to determine physical and chemical changes which result in changes in mass when a material is heated.

Materials in the solid or liquid state can be examined, with minimum sample sizes about 1 mg.

Various environments and heating rates are available for temperatures ranging from ambient to 1200°C.

TGA is a useful technique for investigating reactions accompanied by changes in mass occurring while heating a solid or a liquid. The sample temperature may be held constant for a given time or may be increased at a constant rate. Sample amounts normally used are between 5 and 500 mg.

Principle

TGA is a technique which measures and automatically records changes in mass as a function of temperature, or as a function of time at a constant temperature (see Fig. 19-52). The resulting mass change gives information concerning the thermal stability, composition, and other properties of the original sample.

Differential Thermal Analysis and Differential Scanning Calorimetry

The differential-thermal-analysis (DTA) technique and the differential-scanning-calorimetric (DSC) technique can be used to investigate the thermal properties of inorganic and organic materials. The DTA technique can be used to detect the physical and chemical changes which are accompanied by a gain or loss of heat in a substance as the temperature of the substance is raised. The DSC technique can provide quantitative information about these heat changes; this technique produces data which provide information concerning the total amount of heat and also the rate of heat transfer dQ/dt, that is, the change in heat at a given time (see Fig. 19-53).

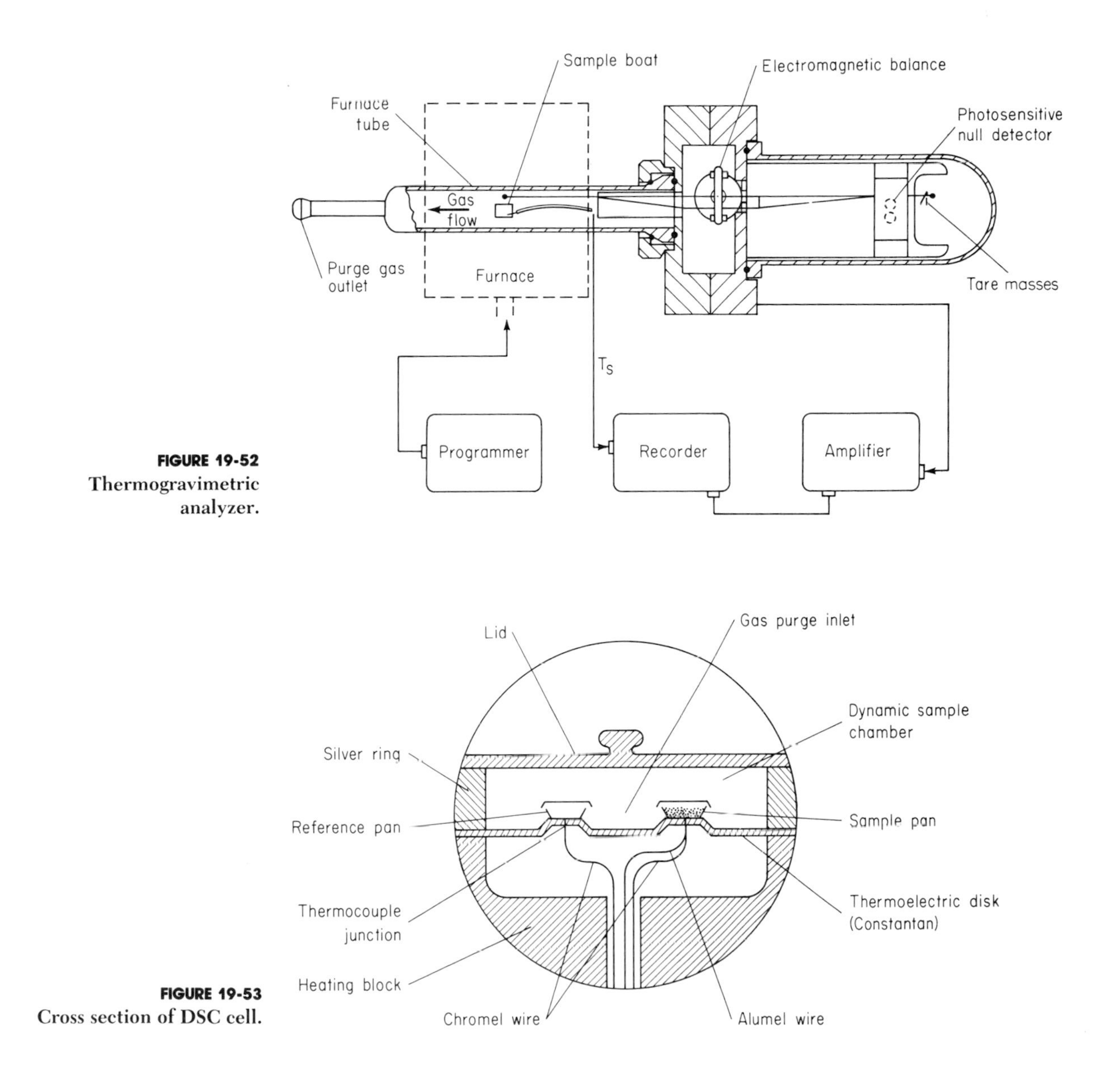

FIGURE 19-52
Thermogravimetric analyzer.

FIGURE 19-53
Cross section of DSC cell.

Samples may be examined in the solid or liquid state. Some information may be obtained with samples as small as 0.1 mg. However, quantitative studies usually require samples of at least 1 mg.

Principle

DTA and DSC are techniques for studying the thermal behavior of substances as they undergo physical and chemical changes during heat treatment. When a substance is heated, various chemical and physical transformations occur involving the absorption of heat (endothermic) or evolution of heat (exothermic). DTA provides information by measuring the temperature difference arising between the sample and a thermostable inert reference material as both are heated at a constant rate in the same environment. The DSC technique, on the other hand, measures the amount of heat that is involved as a material undergoes either an endothermic or exothermic heat change.

Various environments (vacuum, inert, or controlled gas composition) and heating rates (from 0.5°C/min to 100°C/min) are available for temperatures ranging from −100 to 1400°C.

Thermomechanical Analysis

Thermomechanical analysis (TMA) measures changes in linear dimensions, expansion, contraction, etc. Transducers are used to sense movement, and they are called *linear variable differential transformers*, or LVDTs (see Fig. 19-54). Probes transmit any movement of the sample under test into the LVDT, which generates the corresponding electrical signal; this is then recorded as a function of time or heating rate.

Four different interchangeable probes are available to measure penetration, expansion, and shrinkage as a function of temperature. These measurements can also be made in an isothermal mode as a function of time. Predetermined force can also be applied to study changes under load; the sensitivity of this method is useful even in the range of micro-inches (μm).

Samples of materials can be examined as films (from 0.1-mil thickness), fibers, and solids in other forms up to ½-in thick. Various heating and cooling rates are available for temperatures ranging from −150 to +1200°C.

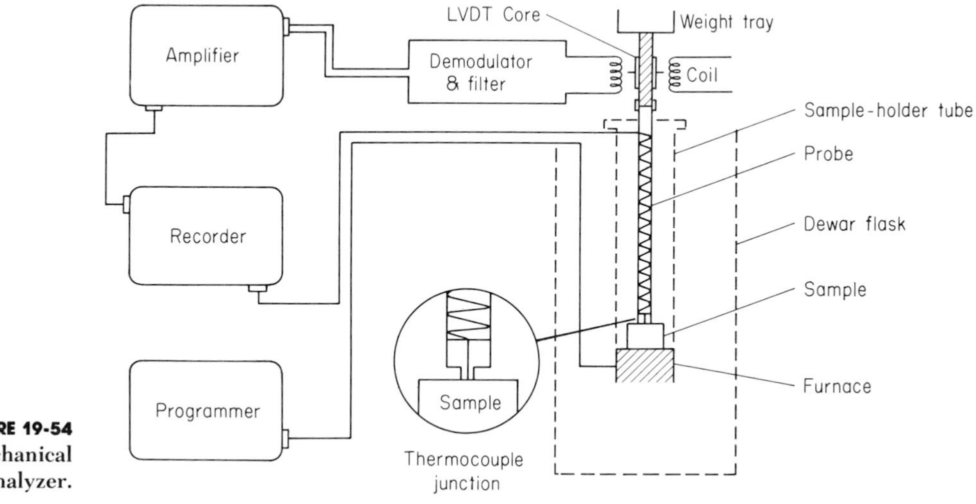

FIGURE 19-54 Thermomechanical analyzer.

Dilatometry

Dilatometric techniques can be used to measure coefficients of thermal expansion, first- and second-order transition temperatures, and rates of crystallization.

Solid samples of approximately 20 g are required. Smaller quantities of material can be analyzed, but the accuracy of the measurements will be reduced. The time necessary for the determination of coefficients of thermal expansion and first- and second-order transition temperatures may range from several hours to approximately 1 day. Evaluations of rates of crystallization will require more extended investigations.

Principle

Volume changes in solids or semicrystalline materials caused by the onset of molecular motion, thermal expansion, or phase transformations are measured as a function of temperature. The property frequently used to analyze these transitions is the specific volume. Discontinuities and changes in the slope of the specific volume–temperature curve correspond to first- and second-order transitions, respectively. Such plots give information on nucleation and on both rates and mode of crystallization.

PHYSICAL CHARACTERISTICS

Optical Microscopy

Optical microscopy is used for the identification and characterization of matter in the visible region of the electromagnetic spectrum. Any solid material and many liquids may be studied by means of optical microscopy. In optical microscopy, information is obtained by the interaction of light with matter. Optical microscopes consist of a light source, a condenser, two systems of lenses, and other accessories. See Figs. 19-55 to 19-59. They are capable of producing enlarged images of small objects, thereby revealing details too small to be seen with the naked eye. The magnification ranges from 40 × to 1400 ×. The limit of resolution is approximately 0.18 μm. Cameras may be attached to microscopes to permit photographing the image.

Studies can be carried out using transmitted and reflected light, polarizing optics, and bright-field, dark-field, and phase-contrast illumination.

Special Precautions for Using a Microscope

1. Never touch lens with fingers. Wipe only with lens tissue, moistened with distilled water, in a circular motion. If film persists, repeat with lens paper moistened with mild detergent.

2. Never force movement of mechanical adjustments.

3. Remove immersion oil immediately after use. It will harden on standing. Any hardened oil can be removed with lens tissue and toluene.

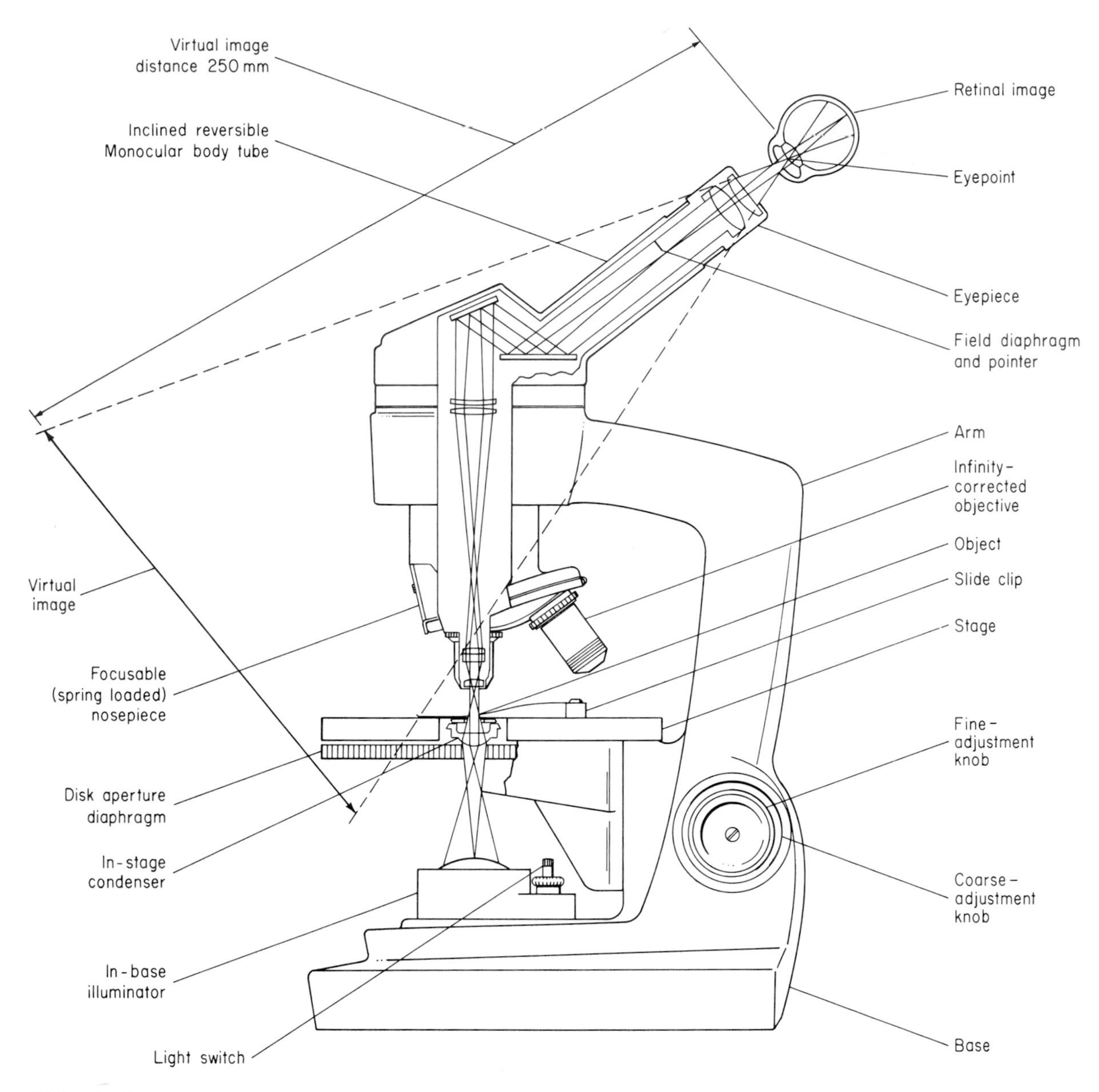

FIGURE 19-55
Monocular microscope. All important parts are labeled.

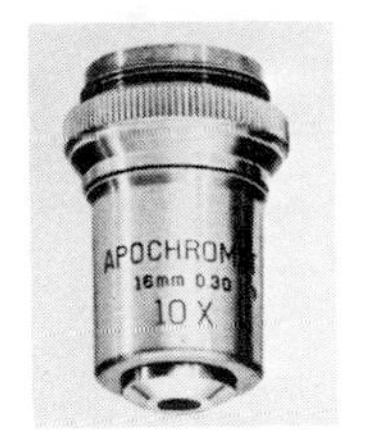

FIGURE 19-56
Microscope objective 10 × power. Handle with care.

FIGURE 19-57
Microscope oculars of varied power. Handle with care.

4. Never clean optical lenses unnecessarily.

5. Never tamper with or remove objectives, oculars, condensers, or mirrors unless you are experienced in microscope maintainance.

6. *Never* switch optical components from one microscope to another.

7. Always handle with care. Use two hands when lifting and moving the microscope.

8. When finished using the microscope, adjust objective lenses so that they are off target and will not hit against anything.

General Procedure for Using the Microscope

1. Place slide on stage, securing it with the spring levers (clamps).

2. Adjust mirror (or microscope light) for maximum light through specimen, using low power.

3. Position low-power objective about ¼ in from slide.

4. Slowly raise the objective to focus. *Never lower it while looking through eyepiece.*

5. Adjust iris diaphragm and substage condenser for optimum light, neither dull nor glaring.

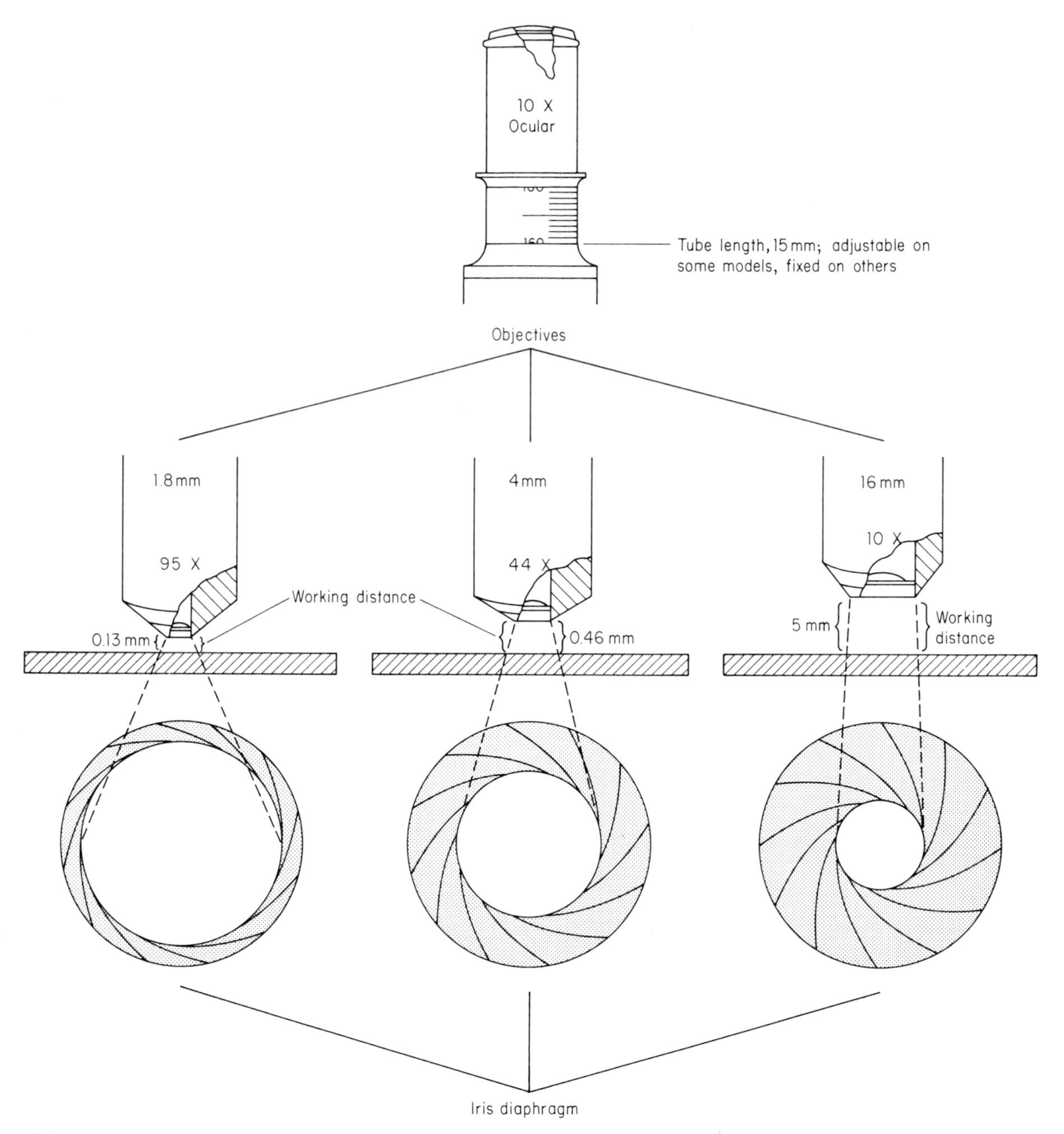

FIGURE 19-58

Relationship between the working distance of the objective lens of the microscope and the optimum adjustment of the iris diaphragm. The shorter the working distance, the greater the diaphragm opening.

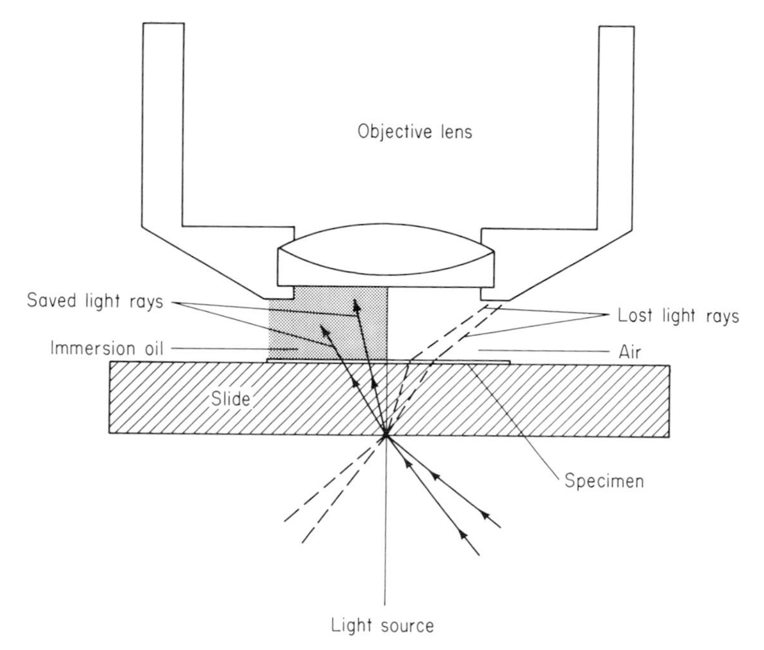

FIGURE 19-59
The oil used with the immersion objective increases the amount of light passing from the specimen into the objective lens.

6. Shift to higher power objective by rotating nosepiece, and fine-adjust as needed. (Remember, never *lower* objective while looking through eyepiece.)

7. Oil-immersion objectives (see Fig. 19-59) require additional care:
   (a) Position slide as desired.
   (b) Raise objective (oil-immersion).
   (c) Place drop of immersion oil on slide.
   (d) Carefully lower objective into oil.
   (e) Fine-adjust until image appears.
   (f) Clean thoroughly after using immersion oil.

CENTRIFUGATION

The principle of centrifugation is that the rate of settling of a precipitate or the rate of separation of two immiscible liquids is increased manyfold by the application of centrifugal force thousands of times that of gravity. The force exerted on the liquid is directly proportional to the speed of rotation, the radius of rotation, and the mass (weight) of the object.

Centrifuges rotate at high speeds, and therefore the head and tubes containing the liquid must be balanced to avoid excessive vibration which can damage the equipment. Special centrifuge tubes must be used because the tubes must be able to withstand the very high centrifugal forces. *Never* use ordinary test tubes in a high-speed centrifuge; they will break.

Centrifugal Force

The force exerted on rotating particles in a liquid is described in terms of *relative centrifugal force,* RCF, which is another term for "number of *g*'s," i.e., multiples of the force of gravity. Centrifuges are rated by their RCF's, which range from less than 10 to almost 750,000. Since centrifuges have heads of fixed radius, and since the mass of the particles of the liquid is independent of the centrifuge used, the only variable at our disposal is the speed of rotation. This means that by varying the speed of the rotor (head), the number of *g* forces exerted on the particles can be varied.

Effect of *g* on Particles of Differing Mass

All particles in a liquid slurry do not have the same mass; therefore the same force will not be exerted upon each one. This allows the technician to selectively centrifuge the slurry to effect separation. If lightweight particles are desired, centrifuge the slurry and save the decanted supernatant liquid containing the lighter-weight ones. The heavier particles will be in the pellet at the bottom of the tube.

Centrifuges

The common centrifuge (Fig. 19-60), a small, portable table-top model, can hold up to eight 15-mL centrifuge tubes, and it has a maximum speed

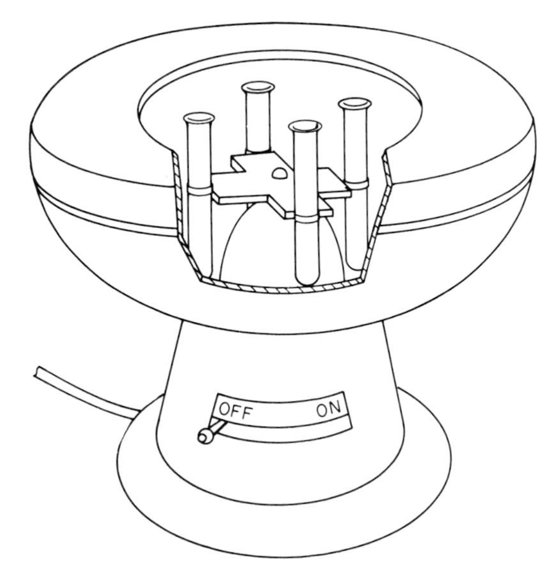

FIGURE 19-60
A common laboratory centrifuge.

of about 1800 rpm. This provides about 2000 *g* of force, and most solid-liquid mixtures can easily be separated with this model.

Special centrifuges rotate at high speeds—up to 70,000 rpm—and may have an RCF of 750,000. They may be refrigerated because the rotating head generates heat as a result of friction with air. *Ultracentrifuges* operate in a vacuum and are used for differential centrifugation.

Use centrifuges to:

1. Clarify nonsettling solutions containing finely divided solids.
2. Break emulsions of two immiscible liquids.
3. Collect solids from a slurry.
4. Obtain a clear solution from a liquid-solid mixture.
5. Separate two immiscible liquids from each other.

Use of Batch-Type Centrifuge

Procedure

1. Divide the liquid to be centrifuged among the minimum number of centrifuge tubes. (See Fig. 19-61.)

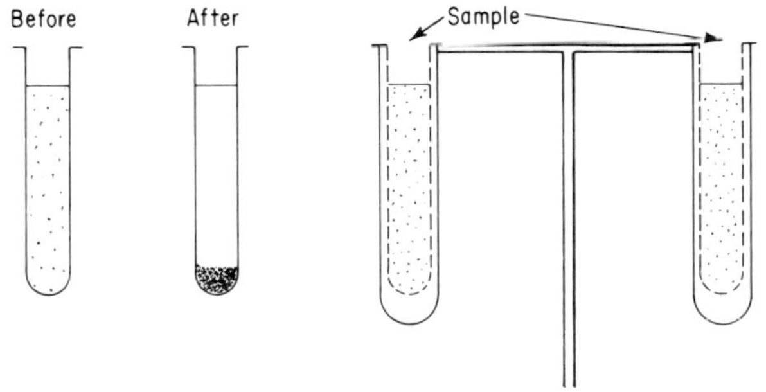

FIGURE 19-61 Solid-liquid separation through centrifugation.

CAUTION Use an equal number of tubes or fill one with a counterbalancing solution. *All tubes and contained liquid should be massed and adjusted to the same mass.*

2. Insert the tubes, equally spaced from each other.

3. Start the motor and run the centrifuge until the objective has been achieved.

4. Turn off the switch and *allow the rotating centrifuge assembly to come to rest.*

CAUTION Do not attempt to stop the rotation manually when the centrifuge is rotating at high speed. *Use the brake* if one is available.

5. If the solid is firmly packed against the bottom of the tube, the liquid can be decanted from the solid by pouring off the liquid (Fig. 19-62). Alternatively, you may withdraw the supernatant liquid with a pipet fitted with a rubber bulb (Fig. 19-63).

FIGURE 19-62
Decanting off supernatant liquid.

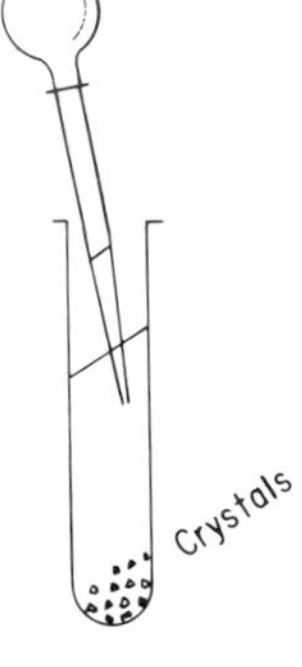

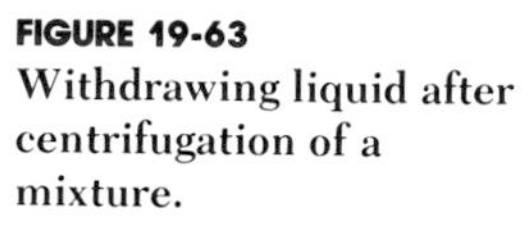

FIGURE 19-63
Withdrawing liquid after centrifugation of a mixture.

6. To wash centrifuged precipitates, add sufficient water, mix well (Fig. 19-64), centrifuge again, and discard the washings. Several small-volume washes are more effective than one large wash.

FIGURE 19-64
Washing a centrifuged precipitate.

CAUTIONS

1. *Always* use balanced pairs of centrifuge tubes which are placed opposite each other on the rotating head. If there is only one tube, counterbalance it with another tube containing water.
2. *Always* use rubber cushions to prevent tube breakage.
3. *Always* use *only* centrifuge tubes (Fig. 19-65); others may break.

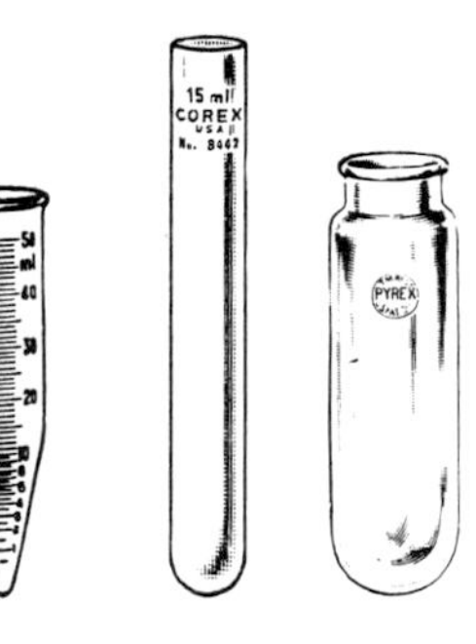

FIGURE 19-65
Centrifuge tubes of general and special designs.

4. *Always* close the centrifuge cover before you start the motor, and open it *only* after the assembly has stopped.
5. *Always* turn off the centrifuge if vibration develops. Tubes may have broken, their contents may be lost, and the rotor head may have become unbalanced. Use the brake (if available); never use your hands to stop the machine.

20

EXTRACTION

INTRODUCTION

Solutes have different solubilities in different solvents, and the process of selectively removing a solute from a mixture with a solvent is called *extraction.*

The solute to be extracted may be in a solid or in a liquid medium, and the solvent used for the extraction process may be water, a water-miscible solvent, or a water-immiscible solvent. The selection of the solvent to be uesd depends upon the solute and upon the requirements of the experimental procedure.

SOXHLET EXTRACTION

A Soxhlet extractor (Fig. 20-1) can be used to extract solutes from solids, using any desired volatile solvent, which can be water-miscible or water-immiscible. The solvent is vaporized. When it condenses, it drops on the solid substance contained in a thimble and extracts soluble compounds. When the liquid level fills the body of the extractor, it automatically siphons into the flask. This process continues repeatedly as the solvent in the flask is vaporized and condensed.

Procedure

Set up apparatus as illustrated in Fig. 20-1.

1. Put the solid substance in the porous thimble, and place the partially filled thimble in the Soxhlet inner tube.
2. Fill the flask one-half full of extracting solvent.
3. Assemble the unit.
4. Turn on the cooling water.
5. Heat.
6. When the extraction is complete, turn off the heat and cooling water.
7. Dismantle the apparatus, and pour the extraction solvent containing the solute into a beaker. Isolate the extracted component by evaporation.

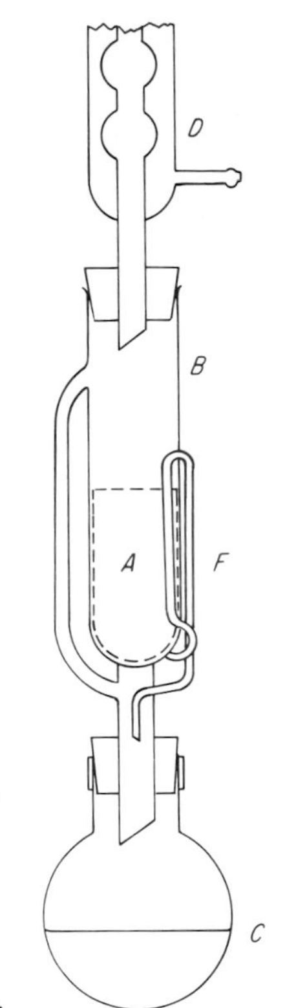

FIGURE 20-1
Soxhlet extractor.
D = reflux condenser,
B = body of extractor,
A = extraction thimble,
F = liquid return siphon,
C = extracting solvent.

EXTRACTION OF A SOLUTE USING IMMISCIBLE SOLVENTS

Principle

A solute may be soluble in many solvents which are immiscible. When a solution of that solute in one of two immiscible solvents is shaken vigorously with the other immiscible solvent, the solute will be distributed between the two solvents in such a manner that the ratio of the concentrations (in moles per liter) of the solute is constant. This ratio is called the *distribution coefficient,* and it is independent of the volumes of the two solvents and the total concentration of the solute.

This type of extraction transfers a solute from one solvent to another. It can be used to separate reaction products from reactants and to separate desired substances from others in solution. The separatory funnel is used for this purpose. *Immiscible solvents,* which are incapable of mixing with each other to attain homogeneity and will separate from each other into separate phases, *must be used. Miscible solvents,* which are capable of being mixed in any ratio without separation into two phases, *cannot be used.*

NOTE *Multiple extractions with smaller portions of the extraction solvent are more effective than one extraction with a large volume.*

The choice of the extraction solvent determines whether the solute remains in the separatory funnel or is in the solvent which is drawn off. *The solvent which has the greater density will be the bottom layer.* Thus the less dense extraction solvent remains in the separatory funnel, and the more dense extraction solvent is drawn off.

Procedure Using Extraction Solvent of Higher Density

Use a separatory funnel with a ring stand or other support. (See Fig. 20-2.)

1. Use a clean separatory funnel, lubricating the barrel and plug of the

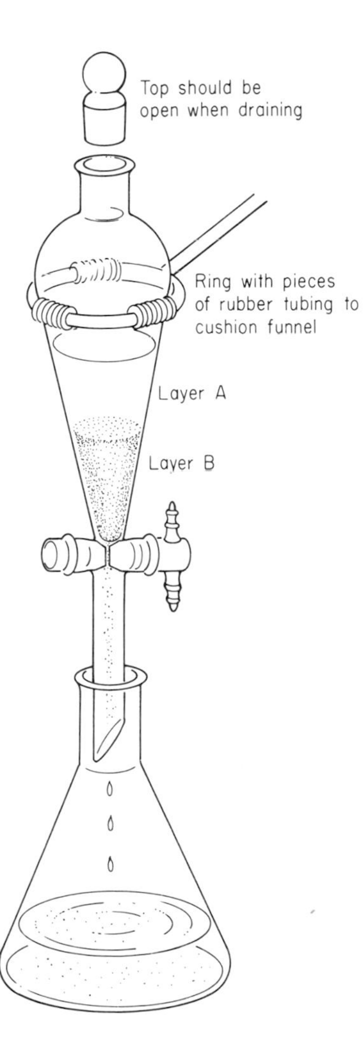

FIGURE 20-2
The extraction procedure: separating immiscible liquids by means of a separatory funnel.

funnel with a suitable lubricant. (Refer to procedures for the lubrication of a stopcock, Chap. 16.)

2. Pour the solution to be extracted into the funnel, which should be large enough to hold at least twice the total volume of the solution and the extraction solvent.

3. Pour in the extraction solvent; close with the stopper.

4. Shake the funnel *gently*.

5. Invert the funnel and open the stopcock slowly to relieve the pressure built up.

6. Close the stopcock while the funnel is inverted, and shake again.

7. Repeat steps 5 and 6.

8. Place the funnel in a ring-stand support and allow the two layers of liquid to separate. Remove stopper closure.

9. Open the stopcock slowly and drain off the bottom layer.

10. Repeat operation, starting at step 3, with fresh extraction solvent as many times as desired.

11. Combine the lower layers which have been drawn off.

Cautions and Techniques

1. Separatory funnels are very fragile and expensive (especially if they are graduated). If one is to be supported in an iron ring stand, pad the ring with rubber tubing (cut longitudinally). This prevents breakage if the funnel bumps against the ring or the ring support when it is removed or inserted.

2. Always be certain that the stopcock is in a closed position before the funnel is returned to the normal vertical position and that it is securely seated, not floating loosely.

3. Always hold the stopper securely seated in the funnel. When the funnel is shaken vigorously to mix the two solvents, pressure inside the funnel increases as a result of the additive effect of the two partial vapor pressures of the two immiscible solvents. If $NaHCO_3$ is used and there are acids present, the pressure may also be increased by the resultant evolution of CO_2 gas. The high pressure is reduced to atmospheric pressure by holding the inverted funnel in both hands, holding the stopper securely in place with one hand, and opening the stopcock with the other (Fig. 20-3). The venting procedure is repeated as many times as is necessary, until no further pressure buildup can be detected. This state is signaled by the disappearance of the audible "whoosh" of the escaping vapors.

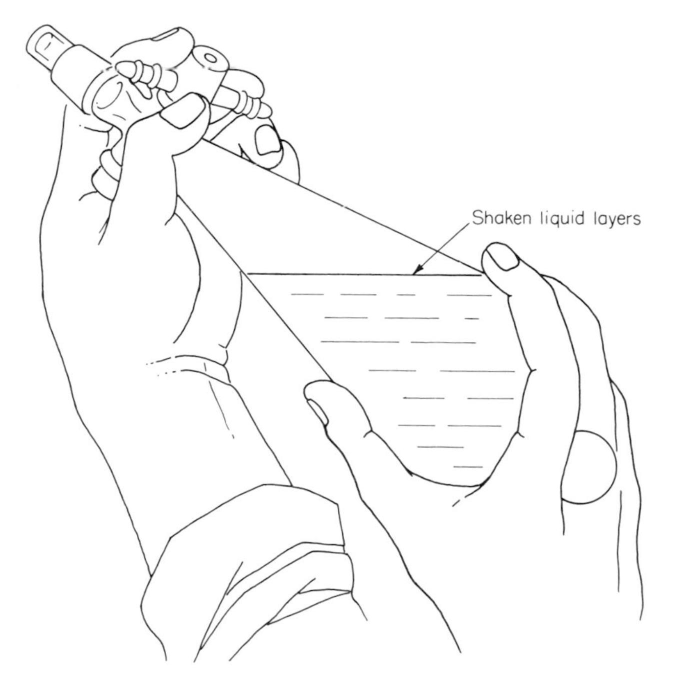

FIGURE 20-3
How to hold, shake, and vent a separatory funnel.

4. Traces of insoluble material often collect at the interfaces between the two insoluble liquids. It is extremely difficult to separate the layers without taking some of this material along, but it can be easily removed by filtration when the extraction is completed, or at an even later stage, while washing or drying the extract.

EXTRACTION PROCEDURES IN THE LABORATORY

Extraction procedures are used to separate, purify, and analyze substances; three principal methods follow:

Using Water

Water is a polar solvent, and polar substances are soluble in it; examples of such substances are inorganic salts, salts of organic acids, strong acids and bases, low-molecular-weight compounds, carboxylic acids, alcohols, polyhydroxy compounds, and amines. Water will extract these compounds from any immiscible organic solvents which contain them.

Using Dilute Aqueous Acid Solution

Dilute hydrochloric acid (between 5 and 10% HCl) will extract basic substances such as organic amines, cyclic nitrogen-containing ring compounds, alkaloids, etc. The basic compound is converted to the corresponding hydrochloride, which is soluble in the aqueous solution, and is therefore extracted from the immiscible organic solvent. After the acid extraction is completed, the organic solvent is extracted with water to remove any acid that might be left in the organic solvent.

Using Dilute Aqueous Basic Solution

Dilute NaOH or 5% $NaHCO_3$ will extract acidic solutes from an immiscible organic solvent by converting the acidic solute to the corresponding sodium salt, which is soluble in water. After the basic extraction is completed, the organic solvent is extracted with water to remove any base that might be left in the organic solvent.

Selective Extraction

Extraction procedures may be used to separate phenols from carboxylic acids because, even though phenols are acidic, the carboxylic acids are about 10^5 times more acidic than phenols. Phenols are not converted to the corresponding salt by $NaHCO_3$, whereas carboxylic acids are converted to the corresponding salt. Therefore an extraction with $NaHCO_3$ solution will extract the carboxylic acid as the salt. An NaOH extraction will convert both phenols and carboxylic acids to the corresponding sodium salts, and NaOH will extract both into the aqueous phase. The NaOH is a strong enough base to deprotonate the phenol, whereas the $NaHCO_3$ is too weak.

CONSIDERATIONS IN THE CHOICE OF A SOLVENT

1. Like substances tend to dissolve like substances.

2. Organic solvents tend to dissolve organic solutes (see Table 20-1).

3. Water tends to dissolve inorganic compounds and salts of organic acids and bases.

4. Organic acids, soluble in organic solvents, can be extracted into water solutions by using bases (NaOH, Na_2CO_3, or $NaHCO_3$).

See Table 20-2 for information regarding various solvents.

TABLE 20-1
Common Organic Solvents

Lighter than water	*Heavier than water*
Diethyl ether	Chloroform
Benzene	Ethylene dichloride
Petroleum ether	Methylene chloride
Ligroin	Tetrachloromethane
Hexane	

TABLE 20-2 Characteristics of Various Solvents Used for Extraction of Aqueous Solutions

Compound	*Characteristics*
Diethyl ether	Is generally a good solvent; absorbs 1.5% water; has a strong tendency to form peroxides.
Methylene chloride	May form emulsions; is easily dried.
Petroleum ethers (pentanes, hexanes, etc.)	Are easily dried; are poor solvents for polar compounds.
Benzene	Tends to form emulsions.
Ethyl acetate	Is good for polar compounds; absorbs large amounts of water.
2-Butanol	Is good for highly polar compounds; dries easily.
Tetrachloromethane	Is good for nonpolar compounds; is easily dried.
Chloroform	Is easily dried; tends to form emulsions.
Diisopropyl ether	Tends to form peroxides.

Diethyl Ether

Diethyl ether is a commonly used organic solvent because:

1. It is easily removed from solutes because of its high volatility.
2. It is cheap.
3. It is an excellent high-power solvent.

However, it has disadvantages:

1. It is a fire hazard.
2. It is toxic.
3. It is highly soluble in water.
4. It is poorly recoverable because of its high volatility.
5. It is an explosion hazard because peroxides form.

PEROXIDES IN ETHER

The safety of using diethyl ether can be increased by detecting and removing any peroxides that form.

Detection of Peroxides in Ethers

Ethers tend to form peroxides on standing, and these peroxides can cause severe and destructive explosions. *Always* test ethers for peroxides before distilling them, either in concentrating solutions or purifying the ethers. Any one of three tests can be used.

Test 1

Colorless ferrothiocyanate changes to red ferrithiocyanate.

1. Prepare reagent.

   (a) Dissolve 9 g $FeSO_4 \cdot 7H_2O$ in 50 mL 18% HCl.

   (b) Add a little granular zinc.

   (c) Add 5 g sodium thiocyanate. When the red color fades add an additional 12 g of sodium thiocyanate.

   (d) Decant clear supernatant from excess zinc into fresh storage bottle.

2. Add ether to be tested dropwise to reagent in a test tube. The solution will turn red if peroxides are present. This test is sensitive to 0.001%.

Test 2

Method 1

Colorless iodide changes to yellow (brown) iodine.

1. Prepare 10% aqueous KI solution.

2. Add 1 mL of KI solution to a sample of the ether to be tested. Let the mixture stand 1 min.

3. Appearance of yellow color indicates peroxides.

Method 2

1. Prepare 10% KI solution in glacial acetic acid (100 mg KI/mL).

2. Add 1 mL of liquid to be tested to 1 mL of the solution.

3. A yellow color indicates a low peroxide concentration; a brown color indicates a high peroxide concentration.

Test 3

1. Prepare a test solution containing the following proportion of substances (a 0.1% solution):

   1 mg $Na_2Cr_2O_7 \cdot 2H_2O$

   1 mL H_2O

   1 drop dilute H_2SO_4

2. Add a few drops of the solution containing peroxides to the test solution. The development of a blue color indicating presence of perchromate ion in the organic layer is a positive test for peroxides.

Removal of Peroxides from Ethers

There are several convenient methods which can be used:

1. Pass the ether through a column containing activated alumina.

CAUTION Do not allow alumina to dry out. Elute or wash the alumina with 5% aqueous $FeSO_4$.

2. Store the ether over activated alumina.

CAUTION See above.

3. Shake ether with a concentrated $FeSO_4$ solution (100 g $FeSO_4$ + 42 mL conc. HCl + 85 mL H_2O).

NOTE Some ethers produce aldehydes when so treated. Remove them by washing with 1% $KMnO_4$ followed by 5% aqueous NaOH extraction to remove any acids formed; follow again with water wash.

4. Wash ethers with sodium metabisulfite solution: sodium pyrosulfite ($Na_2S_2O_5$). This substance reacts stoichiometrically with ethers.

5. Wash ethers with cold triethylenctetramine (make the mixture 25% by mass of ether).

6. In cases of water-soluble ethers, reflux with 0.5% (by mass) CuCl and follow by distillation of the ether.

RECOVERY OF THE DESIRED SOLUTE FROM THE EXTRACTION SOLVENT

After extraction the extraction solvent and the solute which it now contains are processed to recover the *solute* wanted and, if practical, to reclaim the extraction solvent for economic reasons. (Refer to Chap. 21, "Distillation and Evaporation.")

Emulsions

Frequently, when aqueous solutions are extracted with organic solvents, or organic solutions are extracted with aqueous solutions, emulsions form instead of two separate and distinct phases. *Emulsions* are colloidal suspensions of the organic solvent in the aqueous solvent or suspensions of the aqueous solvent in the organic solvent as minute droplets. This situation occurs when the solute may act as a detergent or soap or when viscous and gummy solutes are present. It also may happen if the separatory funnel containing the two solvents is shaken especially vigorously. Once emulsions form, it may be a very long time before the components separate.

Breaking Emulsions

Emulsion formation can be minimized by gently swirling the separatory funnel instead of shaking it vigorously; or the funnel may be gently inverted many times to achieve extraction.

Emulsions caused by too small a difference in the densities of the water and organic layer can be broken by the addition of a high-density organic solvent such as tetrachloromethane (carbon tetrachloride). Pentane can be added to reduce the density of the organic layer, if so desired, especially when the aqueous layer has a high density because of dissolved salts. Saturated NaCl or Na_2SO_4 salt solutions will increase the density of aqueous layers.

"Salting out" also helps to separate the layers. In simple extractions, the distribution coefficient of the extracting solvent may be increased by the addition of a soluble inorganic salt (NaCl or Na_2SO_4) to the water layer. The salt dissolves in the water layer and decreases the solubility of organic liquids in it. The mixture of immiscible liquids in the separatory funnel may form a *homogeneous solution* on shaking and may not separate to form two separate and distinct layers. This condition may be caused by the presence of a mutual solvent, such as alcohol or dioxane. In these cases, add NaCl or Na_2SO_4 crystals or small increments of a saturated solution of these salts, and shake again. Normally, two layers will begin to separate.

One or more of the following techniques may also be of value in breaking emulsions.

1. Add a few drops of silicone defoamer.
2. Add a few drops of dilute acid (if permissible).
3. Draw a stream of air over the surface with a tube connected to a water pump.
4. Place the emulsion in a suitable centrifuge tube and centrifuge until the emulsion is broken.
5. Filter by gravity or with a Büchner funnel (using an aspirator or pump, for higher vacuum).
6. Add a few drops of a detergent solution.
7. Allow the emulsion to stand for a time.
8. Place the emulsion in a freezer.

EXTRACTION BY DISPOSABLE PHASE SEPARATORS

When many extractions must be made, speed is important. A new method utilizes the 1PS® separator manufactured by Whatman, Inc. 1PS® is a high-speed, disposable medium providing complete separation of immiscible aqueous solutions from organic solvents. The 1PS® form of separatory funnel is so effective and yet so low-cost that it can be discarded freely after use. An important advantage of 1PS® (over conventional, fragile, and expensive separatory funnels) is that 1PS® simultaneously, with phase separation, filters out most if not all solids in the organic phase.

The 1PS® can be used with solvents that are either lighter or heavier than water. If heavier than water, the solvent passes directly through the apex of the 1PS® cone; if lighter than water, the solvent passes through the walls of the cone.

It is not necessary when using 1PS® to wait until the two phases have settled out into separate layers; 1PS® will separate drops of one phase suspended in the other just as well as if settling out had occurred. It will not, however, separate the components of a stable emulsion.

1PS® can be used *flat* under suction to separate solvents heavier than water, provided that the pressure differential does not exceed 70 mm Hg.

1PS® is unaffected by mineral acids to 4 *N*. It will tolerate alkalies to 0.4 *M*.

Successful separations have been achieved at 90°C. However, surface tension is inversely proportional to temperature; temperature limits for a given separation, therefore, must be experimentally determined.

Phase separation *will not occur* if the interfacial surface tension is lowered by addition of polar compounds to the system. Surface tension is also lowered in the presence of surface-active agents, i.e., surfactants, above certain proportions.

Procedure

Disposable Phase Separator 1PS® separates aqueous and organic solvent phases in a simple conical filter funnel. (See Fig. 20-4.)

FIGURE 20-4
Extraction with the Disposable Phase Separator®.

1. Fold the paper in the normal way and place it in a conical filter funnel.

2. Pour the mixed phases directly into the funnel. It is not normally necessary to allow the phases to separate cleanly before pouring.

3. Allow the solvent phase to filter completely through the paper.

4. If required, wash the retained aqueous phase with a small volume of clean solvent. This is normally necessary when separating lighter-than-water solvents in order to clean the meniscus of the aqueous phase.

Do not allow the aqueous phase to remain in the funnel for a long time after phase separation, or it will begin to seep through. This effect is caused by evaporation of the solvent from the pores of the paper, creating a local vacuum which in turn draws the water through.

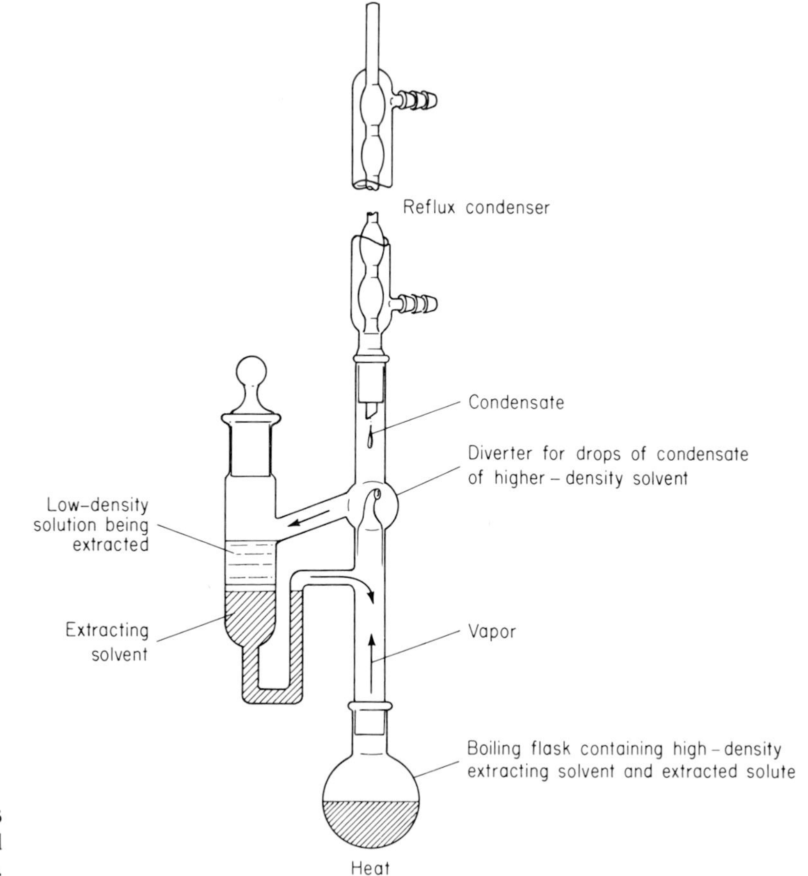

FIGURE 20-5
High-density liquid extractor.

CONTINUOUS LIQUID-LIQUID EXTRACTIONS

When a solute is to be transferred from one solvent into another, a procedure which requires many extractions with large volumes of solvent because of its relative insolubility, continuous liquid-liquid extractions may be used. Specially designed laboratory glassware is required, because the extracting solvent is continually reused as the condensate from a total reflux. Choose the solvent in which the solubility of the desired solute to be extracted is most favorable, and one in which the impurities are least soluble. The solvents must be immiscible.

Higher-Density-Solvent Extraction

In this method the extracting solvent has a higher density than the immiscible solution being extracted. The condensate from the total reflux of the extracting heavier solvent is diverted through the solution to be extracted, passes through that solution, extracting the solute, and siphons back into the boiling flask (Fig. 20-5). Continuous heating vaporizes the higher-density solvent, and the process is continued as long as is necessary.

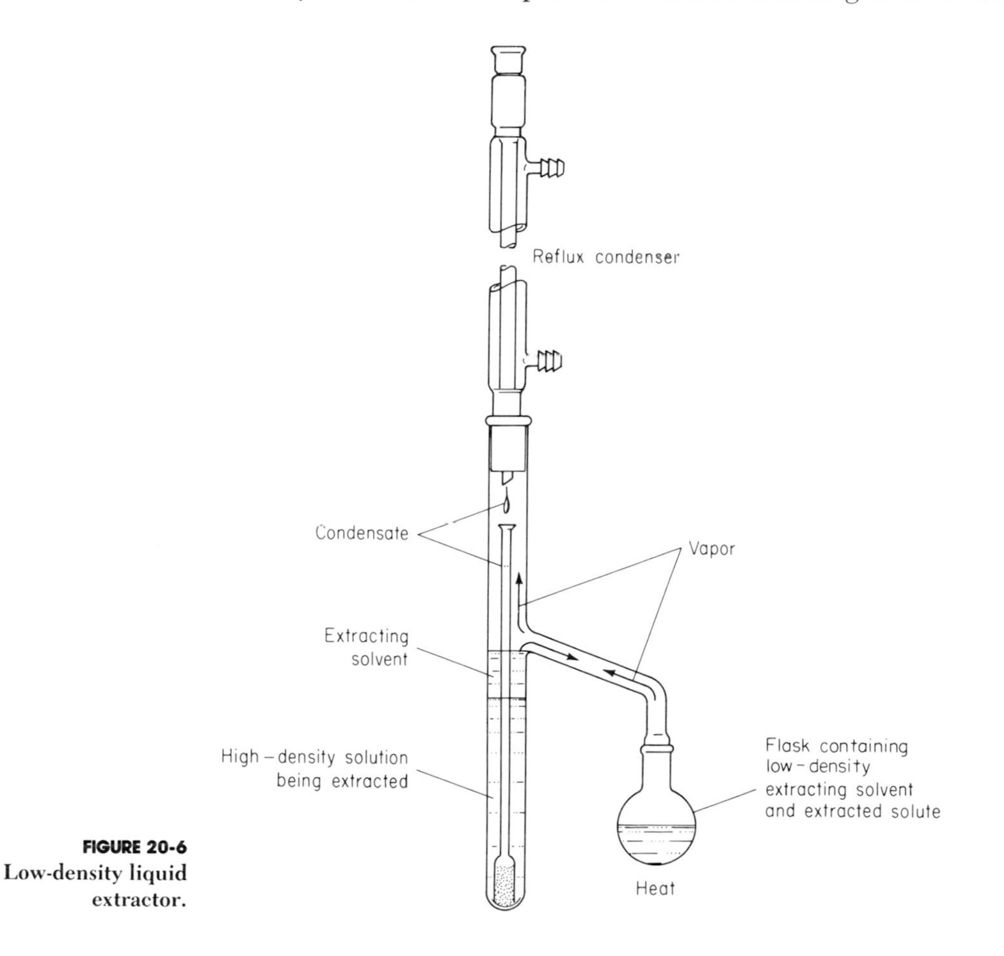

FIGURE 20-6 Low-density liquid extractor.

Lower-Density-Solvent Extraction

The extracting solvent has a lower density than the immiscible solution being extracted. The condensate from the total reflux of the extracting lower-density solvent is caught in a tube (Fig. 20-6). As the tube fills, the increased pressure forces some of the lower-density solvent out through the bottom. It rises through the higher-density solvent, extracting the solute, and flows back to the boiling flask. Continuous heating vaporizes the low-density solvent, and the process is continued as long as is necessary.

ULTRAFILTRATION

Dialysis of small sample volumes for buffer exchange and desalting is time-consuming; solutes diffuse at different rates, and often enormous quantities of dialysate are needed. Ultrafiltration with the Immersible Molecular Separator® is a better way.

The Immersible Molecular Separator® consists of a Pelicon membrane mounted on a cylindrical plastic core. The membrane can handle substances with molecular weights up to 10,000. To use the equipment, simply dip it in the sample solution, attach the tubing to a suitable vacuum source, and process. Solutes are removed automatically through the membrane into the vacuum trap while macromolecules (with molecular

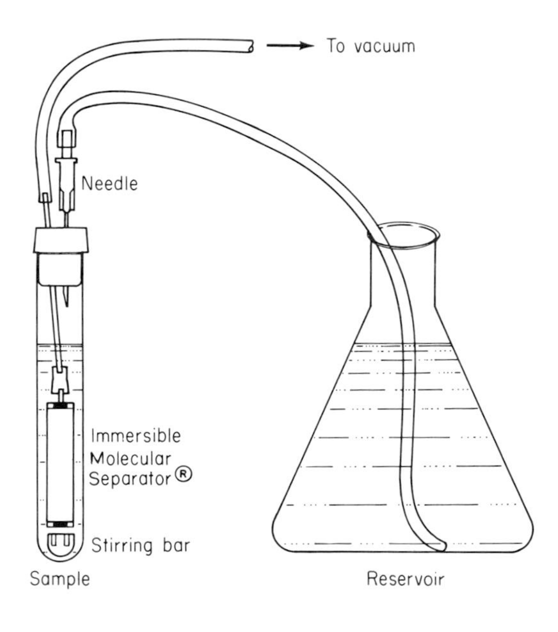

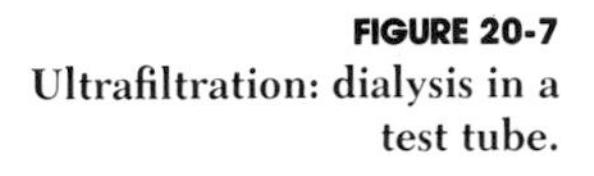

FIGURE 20-7
Ultrafiltration: dialysis in a test tube.

The membrane tube filter

weights over 10,000) are retained in the original vessel. Adsorption of protein to the membrane is negligible.

Figure 20-7 depicts a typical setup for constant-volume buffer exchange. The Immersible Molecular Separator® is placed in a stoppered test tube or vessel containing the sample. The immersible tubing is attached to a vacuum source with a trap, while the line from the replacement solution reservoir is connected to the sample container. During filtration, the vacuum created by the departing filtrate draws in an equal amount of new solution to mix with the sample.

21
DISTILLATION AND EVAPORATION

INTRODUCTION

Distillation is a process in which the liquid is vaporized, recondensed, and collected in a receiver. The liquid which has not vaporized is called the *residue*. The resultant liquid, the condensed vapor, is called the *condensate* or *distillate*.

Distillation is used to purify liquids and to separate one liquid from another. It is based on the difference in the physical property of liquids called *volatility*. Volatility is a general term used to describe the relative ease with which the molecules may escape from the surface of a pure liquid or a pure solid. The vapor pressure of a substance at a given temperature expresses this property. (See Fig. 21-1.)

Vapor Pressure

A volatile substance is one which exerts a relatively high vapor pressure at room temperature. A nonvolatile substance is one which exerts a low vapor pressure. The more volatile a substance, the higher its vapor pressure and the lower its boiling point. The less volatile a substance, the lower its vapor pressure and the higher its boiling point.

All liquids and solids have a tendency to vaporize at all temperatures, and this tendency varies with temperature and the external pressure which is

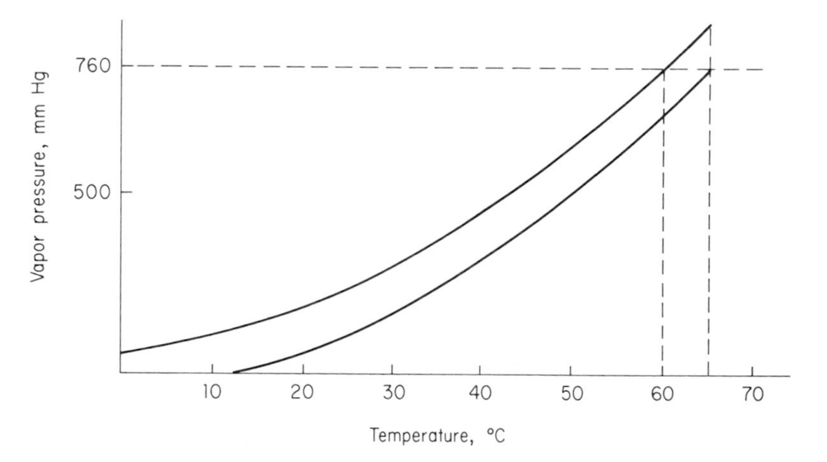

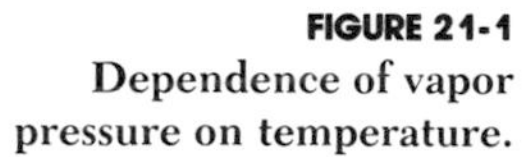
FIGURE 21-1
Dependence of vapor pressure on temperature.

applied. When a solvent is enclosed, vaporization will take place until the partial pressure of the vapor above the liquid has reached the vapor pressure at that temperature. Further evaporation of the liquid can be accomplished by removing some of the vapor above it, which in turn reduces the vapor pressure over the liquid.

SIMPLE DISTILLATION

An experimental setup for simple distillation is shown in Figs. 21-2 and 21-3. The glass equipment may be standard and require corks or may have ground-glass fitted joints. To be sure your setup is correct, follow the checklist below:

1. The distilling flask should accommodate twice the volume of the liquid to be distilled.

2. The thermometer bulb should be slightly below the sidearm opening of the flask. The boiling point of the corresponding distillate is normally accepted as the temperature of the vapor. If the thermometer is not positioned correctly, the temperature reading will not be accurate. If the entire bulb of the thermometer is placed too high, above the side arm leading to the condenser, the entire bulb will not be heated by the vapor of the distillate and the temperature reading will be too low. If the bulb is placed too low, too near the surface of the boiling liquid, there may be

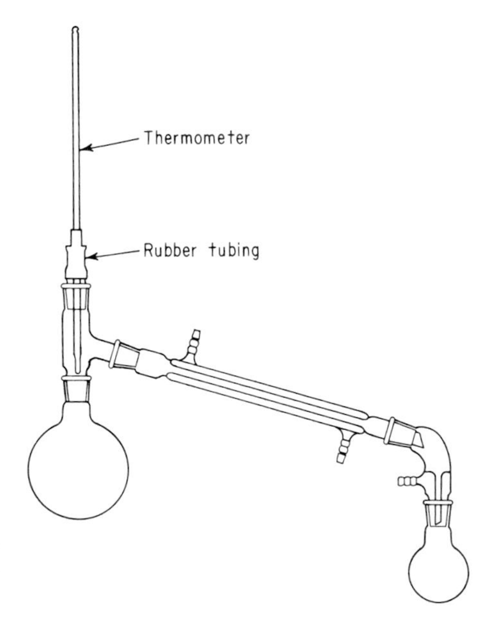

FIGURE 21-2 Apparatus for simple distillation at atmospheric pressure.

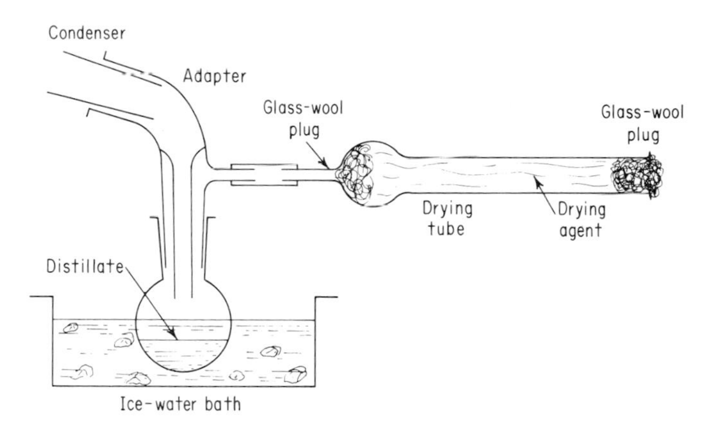

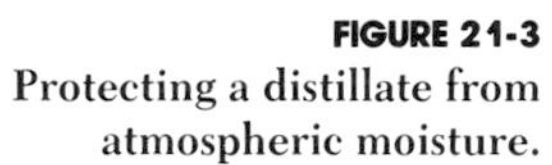
FIGURE 21-3
Protecting a distillate from atmospheric moisture.

a condition of superheating, and the thermometer will show too high a temperature.

3. All glass-to-glass or glass-to-cork connections should be firm and tight.

4. The flask, condenser, and receiver should be clamped independently in their proper relative positions on a steady base.

5. The upper outlet for the cooling water exiting from the condenser should point upward to keep the condenser full of water.

Procedure

1. Pour the liquid into the distilling flask with a funnel which extends below the side arm.

2. Add a few boiling stones to prevent bumping.

3. Insert the thermometer.

4. Open the water valve for condenser cooling.

5. Heat the distilling flask until boiling begins; adjust the heat input so that the rate of distillate is a steady 2 to 3 drops per second.

6. Collect the distillate in the receiver.

7. Continue distillation until only a small residue remains. Do not distill to dryness.

Distillation of Pure Liquids

Distillation can be used to test the purity of liquids or to remove the solvent from a solution.

The experimental setup for pure liquids is the same as that shown in Figs. 21-2 and 21-3.

1. The composition of the condensate is necessarily the same as the original liquid and is the same as the residue.

2. The composition does not change (see Fig. 21-4).

3. The boiling temperature remains constant throughout the distillation.

4. Distillation establishes only the purity and boiling point of the pure liquid.

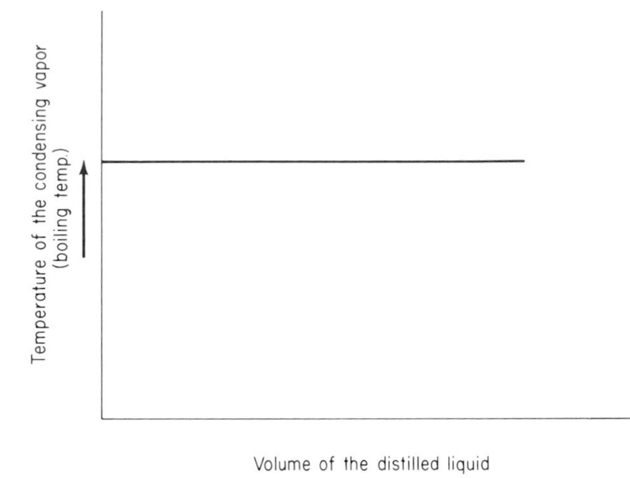

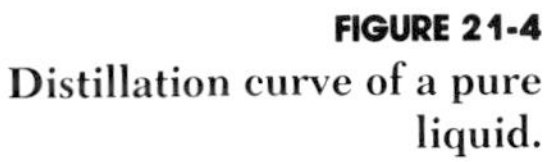

FIGURE 21-4
Distillation curve of a pure liquid.

Distillation of a Solution

This process effects the separation of the nonvolatile dissolved solids because they remain in the residue and the volatile liquid is distilled, condensed, and collected.

1. The temperature of the distillate is constant throughout because it is pure.

2. The temperature of the boiling solution increases gradually throughout the distillation because the boiling solution becomes saturated with the nonvolatile solids.

When a nonvolatile substance is in the liquid being distilled, the temperature of the distilling liquid (the *head temperature*) will be the same as that of the pure liquid, since the vapor being condensed is uncontaminated by the impurity. The temperature of the pot liquid will be higher, because of the decreased vapor pressure of the solution containing the nonvolatile solute. The temperature of the pot liquid will continue to increase as the volatile component distills away, further lowering the vapor pressure of the solution and increasing the concentration of the solute.

CAUTION When evaporating a solution to recover the solute or when using electric heat or burners to distill off large volumes of solvent to recover the solute, do not evaporate completely to dryness. The residue may be superheated and begin to decompose.

Distillation of a Mixture of Two Liquids

Principle

Simple distillation of a mixture of two liquids will not effect a complete separation. If both are volatile, both will vaporize when the solution boils and both will appear in the condensate. The more volatile of the two liquids will vaporize and escape more rapidly and will form a larger proportion of the distillate. The less volatile constituent will concentrate in the liquid which remains in the distilling flask, and the temperature of the boiling liquid will rise. The more volatile of the two liquids will appear first in the distillate. When the difference in the volatilities of the two liquids is large enough, the first distillate may be almost pure. The last of the distillate collected will be richer in the less volatile component. If there is sufficient difference in the volatilities of the two liquids, the last of the distillate may be almost pure. The distillate collected between the first portion and the last portion will contain varying amounts of the two liquids. (See Fig. 21-5.)

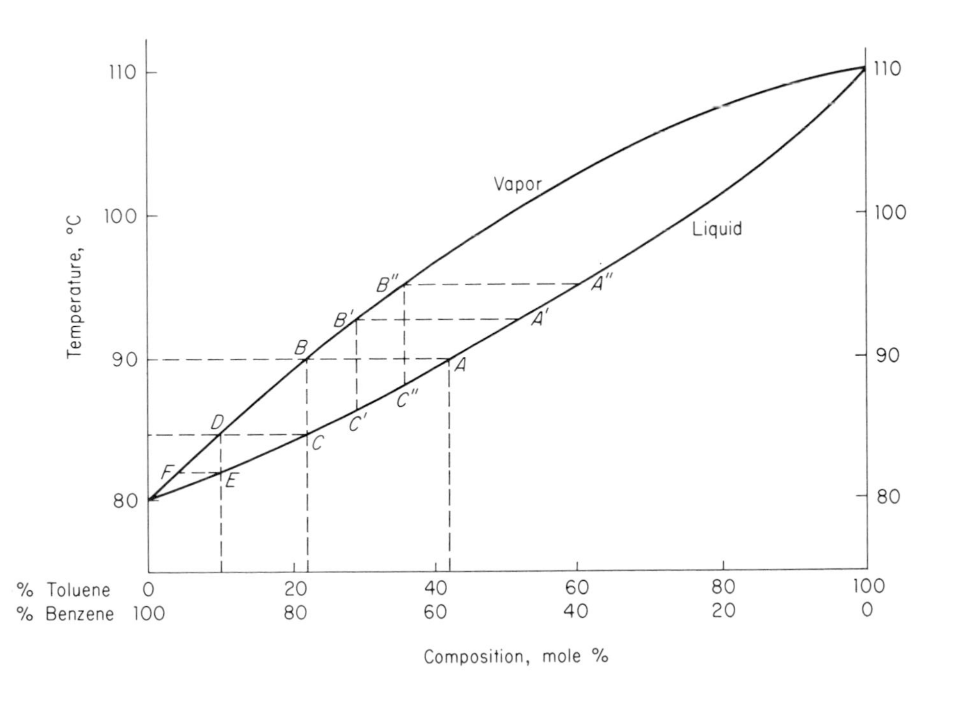

FIGURE 21-5 Boiling-point-composition diagram for the system benzene–toluene.

You may separate two liquids with different volatilities by *changing the receiver* several times during the distillation, thus collecting several portions of distillate.

1. The first portion collected while the boiling temperature is near that of the more volatile liquid may contain that liquid with little impurity.

2. The last portion collected when the distillation temperature is nearly equal to the boiling point of the less volatile liquid may contain that liquid and little of the other. Intermediate portions will contain *both liquids in varying proportions.* You may redistill each of the intermediate portions collected in the receiver to separate them further into their pure components.

The lower curve of the diagram in Fig. 21-5 gives the boiling points of all mixtures of these compounds. The upper curve gives the composition of the vapor *in equilibrium* with the boiling liquid phase. The vapor phase is much richer in the more volatile component, benzene, than the liquid phase with which it is in equilibrium. Therefore, the first few drops of vapor which condense will be richer in the more volatile component than in the less volatile one, toluene.

Rate of Distillation (No Fractionation)

The rate of distillation is controlled by the rate of the input of heat. In normal distillations a rate of about 3 to 10 mL/min, which corresponds roughly to 1 to 3 drops per second, is average. When liquids are to be separated by fractional distillation (see Fractional Distillation, this chapter), the rate may be a great deal smaller, depending upon the difficulty of the separation.

Concentration of Large Volumes of Solutions

When it becomes necessary to distill off large volumes of solvent to recover very small quantities of the solute, it is advisable to use a large distilling flask at first. (Never fill any distilling flask over one-half full.) When the volume has decreased, transfer the material to a smaller flask and continue the distillation. This minimizes losses caused by the large surface area of large flasks. If the solute is a high-boiling substance, the walls of the flask act as a condenser, making it difficult to drive the material over.

AZEOTROPIC DISTILLATION

Azeotropic mixtures distill at constant temperature without change in composition. Obviously, one cannot separate azeotropic mixtures by normal distillation methods.

TABLE 21-1
Maximum-Boiling-Point Azeotropic Mixtures

Component A		*Component B*		*bp of Azeotropic mixture, °C*	*% of B (by mass) in mixture*
Substance	*bp, °C*	*Substance*	*bp, °C*		
Water	100.0	Formic acid	100.8	107.1	77.5
Water	100.0	Hydrofluoric acid	19.4	120.0	37
Water	100.0	Hydrochloric acid	−84.0	108.6	20.22
Water	100.0	Hydrobromic acid	−73	126	47.6
Water	100.0	Hydriodic acid	−35	127	57.0
Water	100.0	Nitric acid	86.0	120.5	68
Water	100.0	Sulfuric acid	10.5 (mp)	338	98.3
Water	100.0	Perchloric acid	110.0	203	71.6
Acetone	56.4	Chloroform	61.2	64.7	80
Acetic acid	118.5	Pyridine	115.5	130.7	65
Chloroform	61.2	Methyl acetate	57.0	64.8	23
Phenol	181.5	Aniline	184.4	186.2	58

Azeotropic solutions are nonideal solutions. Some display a greater vapor pressure than expected; these are said to exhibit *positive deviation*. Within a certain composition range such mixtures boil at temperatures higher than the boiling temperature of either component; these are *maximum-boiling azeotropes* (see Table 21-1 and Fig. 21-6).

Mixtures which have boiling temperatures much lower than the boiling temperature of either component exhibit *negative deviation;* when such

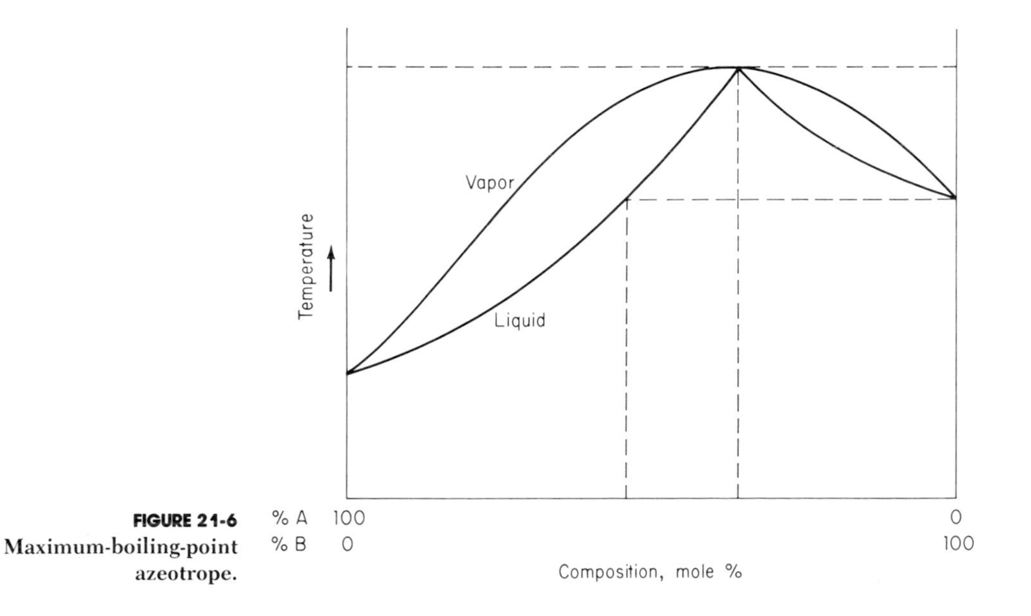

FIGURE 21-6 Maximum-boiling-point azeotrope.

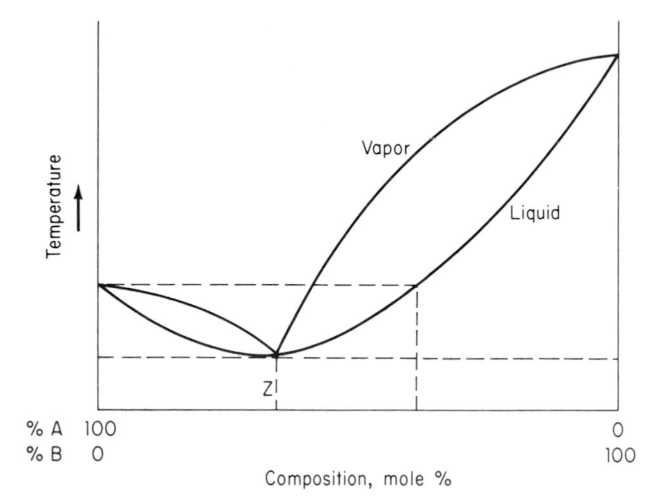

FIGURE 21-7 Minimum-boiling-point azeotrope.

TABLE 21-2 Minimum-Boiling-Point Azeotropic Mixtures

Component A		*Component B*		*bp of Azeotropic mixture, °C*	*% of A (by mass) in mixture*
Substance	*bp, °C*	*Substance*	*bp, °C*		
Water	100.0	Ethyl alcohol	78.3	78.15	4.4
Water	100.0	Isopropyl alcohol	82.4	80.4	12.1
Water	100.0	*n*-Propyl alcohol	97.2	87.7	28.3
Water	100.0	*tert*-Butyl alcohol	82.6	79.9	11.8
Water	100.0	Pyridine	115.5	92.6	43.0
Methyl alcohol	64.7	Methyl iodide	44.5	39.0	7.2
Ethyl alcohol	78.3	Ethyl iodide	72.3	63.0	13
Methyl alcohol	64.7	Methyl acetate	57.0	54.0	19
Ethyl alcohol	78.3	Ethyl acetate	77.2	71.8	31
Water	100.0	Butyric acid	163.5	99.4	18.4
Water	100.0	Propionic acid	140.7	100.0	17.7
Benzene	80.2	Cyclohexane	80.8	77.5	55
Ethyl alcohol	78.3	Benzene	80.2	68.2	32.4
Ethyl alcohol	78.3	Toluene	110.6	76.7	68
Methyl alcohol	64.7	Chloroform	61.2	53.5	12.5
Ethyl alcohol	78.3	Chloroform	61.2	59.4	7.0
Ethyl alcohol	78.3	Methyl ethyl ketone	79.6	74.8	40
Methyl alcohol	64.7	Methylal	42.2	41.8	18.2
Acetic acid	118.5	Toluene	110.6	105.4	28

mixtures have a particular composition range, they act as though a third component were present. In Fig. 21-7, the minimum boiling point at Z is a constant boiling point because the vapor is in equilibrium with the liquid and has the same composition as the liquid does. Pure ethanol (bp 78.4°C) cannot be obtained by fractional distillation of aqueous solutions which contain less than 95.57% of ethanol because this is the azeotropic composition; the boiling point of this azeotropic mixture is 0.3° lower than that of pure ethanol.

Other examples are:

1. A mixture of 32.4% ethyl alcohol and 67.6% benzene (bp 80.1°C) boils at 68.2°C.

2. A ternary azeotrope (bp 64.9°C) is composed of 74.1% benzene, 18.5% ethyl alcohol, and 7.4% water. (See also Table 21-2.)

Absolute ethyl alcohol can be obtained by distilling azeotropic 95.5% ethyl alcohol with benzene. The water is removed in the volatile azeotrope formed.

The procedure is described above under Distillation of Pure Liquids.

FRACTIONAL DISTILLATION

Principle

The separation and purification of a mixture of two or more liquids, present in appreciable amounts, into various fractions by distillation is *fractional distillation.* It consists essentially in the systematic redistillation of distillates (fractions of increasing purity). Figures 21-8, 21-9, and 21-10 are distillation curves showing how two liquids separate. Fractionations can be carried out with an ordinary distilling flask; but, where the components do not have widely separated boiling points, it is a very tedious process. A fractionating column (Fig. 21-11) is essentially an apparatus for performing a large number of successive distillations without the necessity of actually collecting and redistilling the various fractions. The glass column is filled with pieces of glass, glass beads, metal screening, or glass helices (Fig. 21-12). Some columns are more efficient than others.

The separation of mixtures by this means is a refinement of ordinary separation by distillation. Thus a series of distillations involving partial vaporization and condensation concentrates the more volatile component in the first fraction of distillate and leaves the less volatile component in the last fraction or in the residual liquid. The vapor leaves the surface of the liquid and passes up through the packing of the column. There it condenses on the cooler surfaces and redistills many times before entering the con-

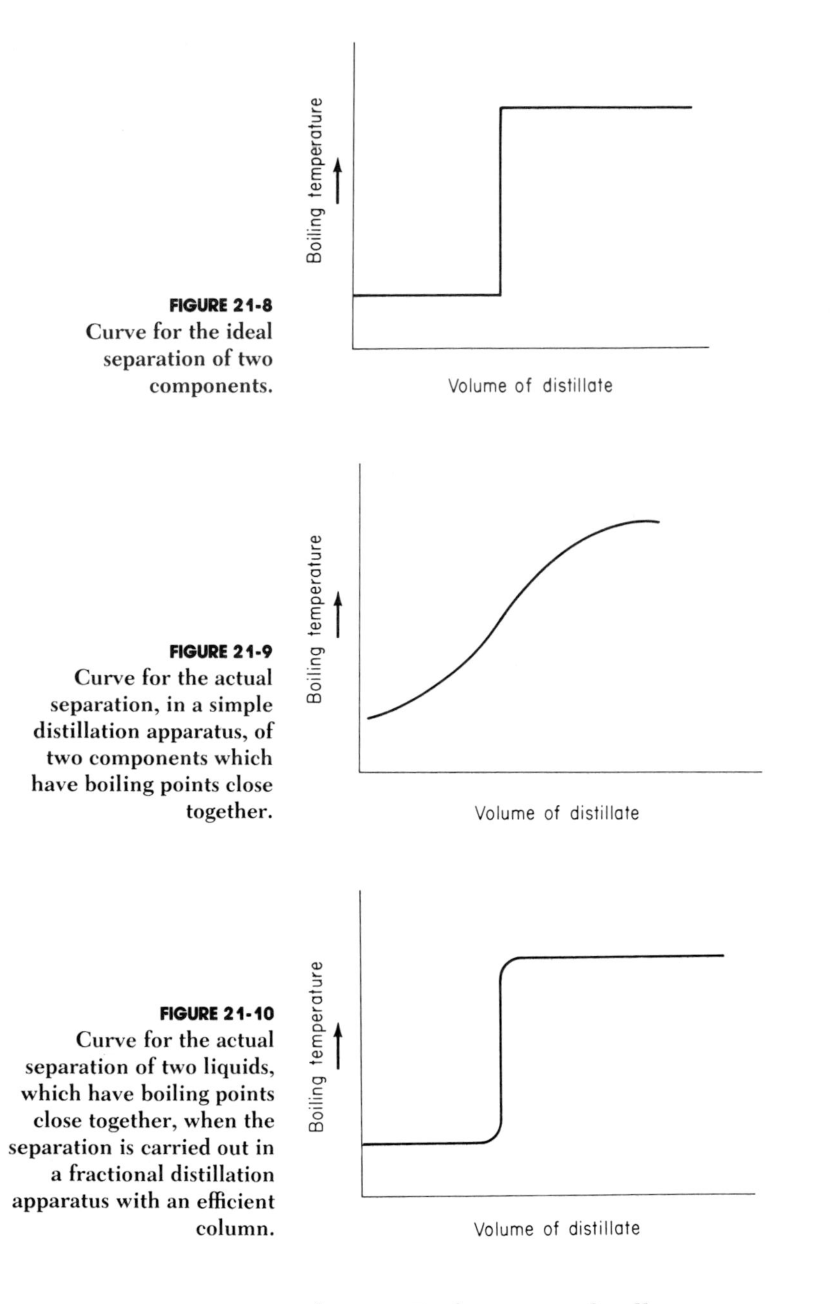

FIGURE 21-8
Curve for the ideal separation of two components.

FIGURE 21-9
Curve for the actual separation, in a simple distillation apparatus, of two components which have boiling points close together.

FIGURE 21-10
Curve for the actual separation of two liquids, which have boiling points close together, when the separation is carried out in a fractional distillation apparatus with an efficient column.

denser. Each minute distillation causes a greater concentration of the more volatile liquid in the rising vapor and an enrichment of the residue which drips down through the column in the less volatile components.

By means of long and efficient distillation columns (see Fig. 21-11), two liquids may be completely separated.

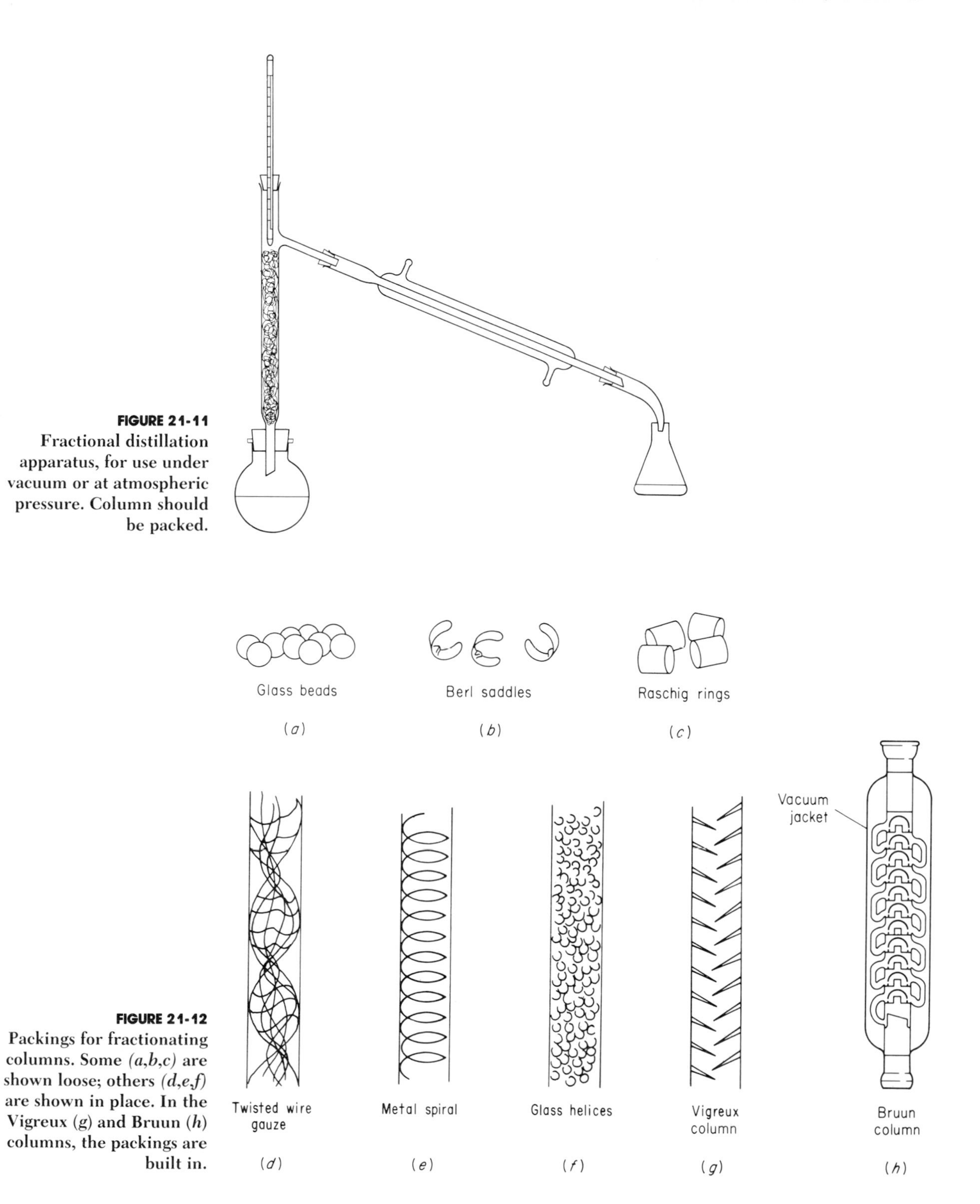

FIGURE 21-11
Fractional distillation apparatus, for use under vacuum or at atmospheric pressure. Column should be packed.

FIGURE 21-12
Packings for fractionating columns. Some (*a,b,c*) are shown loose; others (*d,e,f*) are shown in place. In the Vigreux (*g*) and Bruun (*h*) columns, the packings are built in.

Procedure

1. Select the type of fractionating column to be used, one that offers a large surface contact inside (see Fig. 21-12).
2. Set up the equipment.
3. Open the inlet cooling-water valve.
4. Apply heat.
5. Keep a large volume of liquid condensate continually returning through the column.
6. *Distill slowly* to effect efficient separation (See *caution*, page 485.)

Heating Fractionating Columns

Sometimes a fractionating column must be heated in order to achieve the most efficient fractionation of distillates. This may be accomplished by wrapping the column with heating tape or a resistance wire, such as Nichrome wire, controlled by variable transformers (see Fig. 21-13). (Refer to the section on heating mantles in Chap. 10, "Heating and Cooling.")

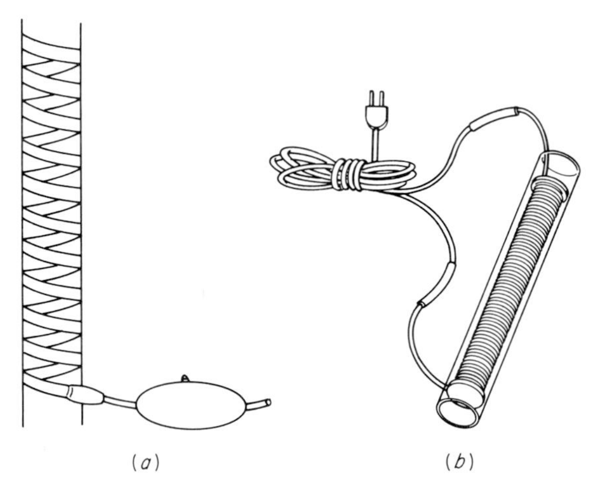

FIGURE 21-13 Methods of heating fractionating columns. (*a*) Using heating tape. (*b*) Nichrome wire heater.

Efficiency of Fractionating Columns

We measure the efficiency of a fractionating column in terms of the number of theoretical plates that it contains. A column with one theoretical plate is one in which the initial distillate has a vapor composition which is at equilibrium with the original solution. It is impossible to operate fractionating columns at equilibrium.

Bubble-Cap Fractionating Columns

Bubble-cap fractionating columns (Fig. 21-14) have definite numbers of trays or plates and are fitted with either bubble caps, or sieve perforations—or modifications of these two—to enable the achievement of inti-

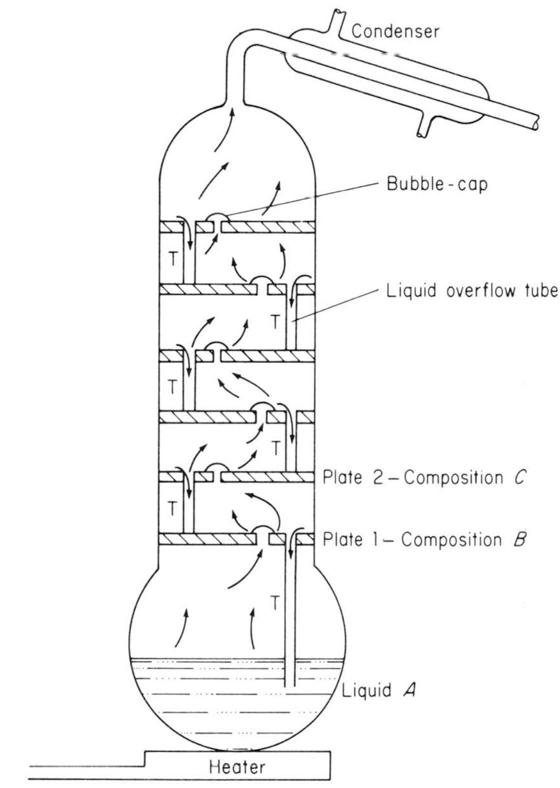

FIGURE 21-14 Bubble-cap fractionating column.

mate vapor-liquid dispersion. However, because fractionating columns cannot be operated at equilibrium (practically), the number of theoretical plates is always lower than the number of actual plates, depending upon the rate of distillation, reflux ratio, and other factors.

CAUTION If you heat the pot too vigorously and remove the condensed vapor too quickly, the whole column will heat up uniformly and there will be no fractionation. The fractionating column will become flooded by the returning condensate.

Total-Reflux–Partial-Takeoff Distilling Heads

Exercise good judgment in the control of the amount of heat applied, and, for truly effective fractionation, use a total-reflux–partial-takeoff distilling head as shown in Fig. 21-15. With the stopcock *S* completely closed, all condensed vapors are returned to the distilling column, a total reflux condition. With the stopcock partially opened, the number of drops of condensate falling from the condenser which returns to the fractionating column can be adjusted. The ratio of the number of drops of distillate allowed to pass through stopcock *S* into the receiver to the number of

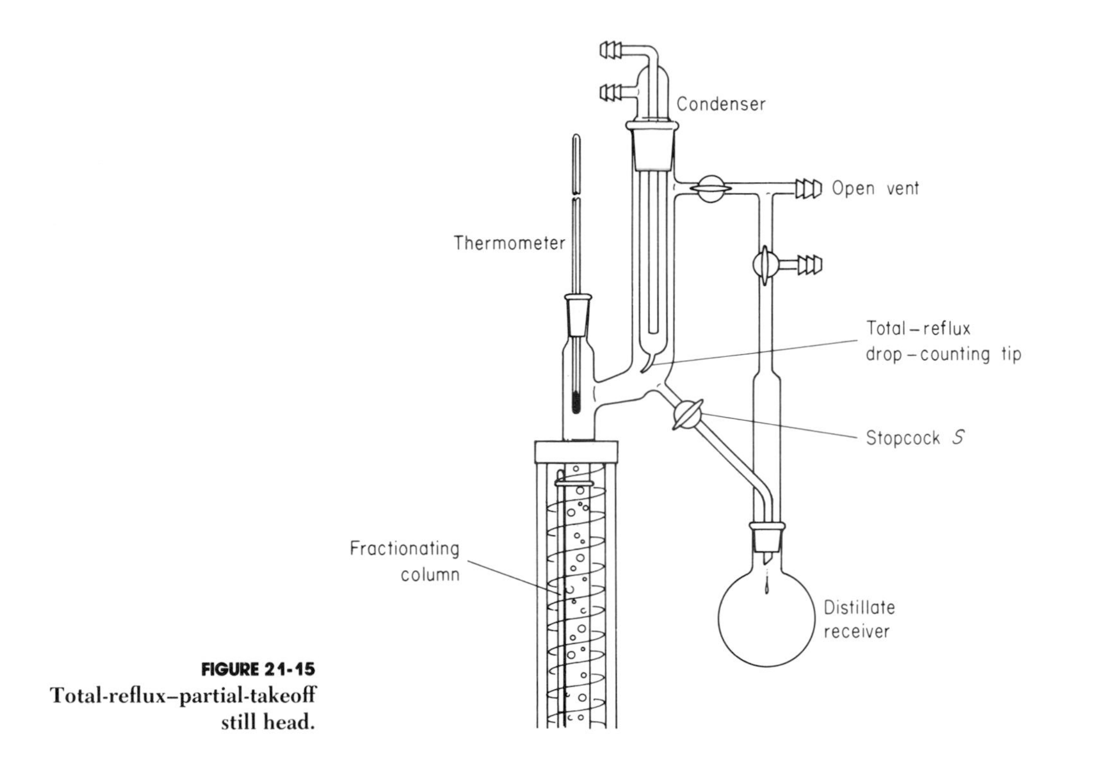

FIGURE 21-15
Total-reflux–partial-takeoff still head.

drops of reflux is called the *reflux ratio.* With an efficient column, reflux ratios as high as 100 to 1 can be used to effectively separate compounds which have very close boiling points.

VACUUM DISTILLATION

Principle

Many substances cannot be distilled satisfactorily at atmospheric pressure because they are sensitive to heat and decompose before the boiling point is reached. Vacuum distillation, distillation under reduced pressure, makes it possible to distill at much lower temperatures. The boiling point of the material is affected by the pressure in the system. The lower the pressure, the lower the boiling point; the higher the pressure, the higher the boiling point.

Nomographs

Nomographs are special graphs which enable the technician to determine more accurately the boiling points at different pressures; they also provide a method of converting boiling points to the pressure desired.

Procedure for using the nomograph (see Fig. 21-16):

1. Select the desired boiling point at the reduced pressure.

2. Use a transparent plastic ruler, and connect that boiling point (column A) with the given corresponding pressure (column C). It will intersect column B at a definite point. Record that point.

3. Using the point obtained from column B in step 2, select the new pressure desired on column C. Align the plastic ruler with these two points and read the corresponding temperature for the boiling point at the new pressure where the ruler intersects column A.

EXAMPLE A liquid boils at 200°C at 10 torr pressure. What would be (1) the normal boiling point at 760 torr and (2) the boiling point at 1 torr pressure?

Solution Given: bp = 200°C (column A) at 10 torr (column C).

1. Intersection point in column B is 350°C, the boiling point at 760 torr pressure (corrected).

2. The line connecting the points 350°C on column B, and 1 torr pressure

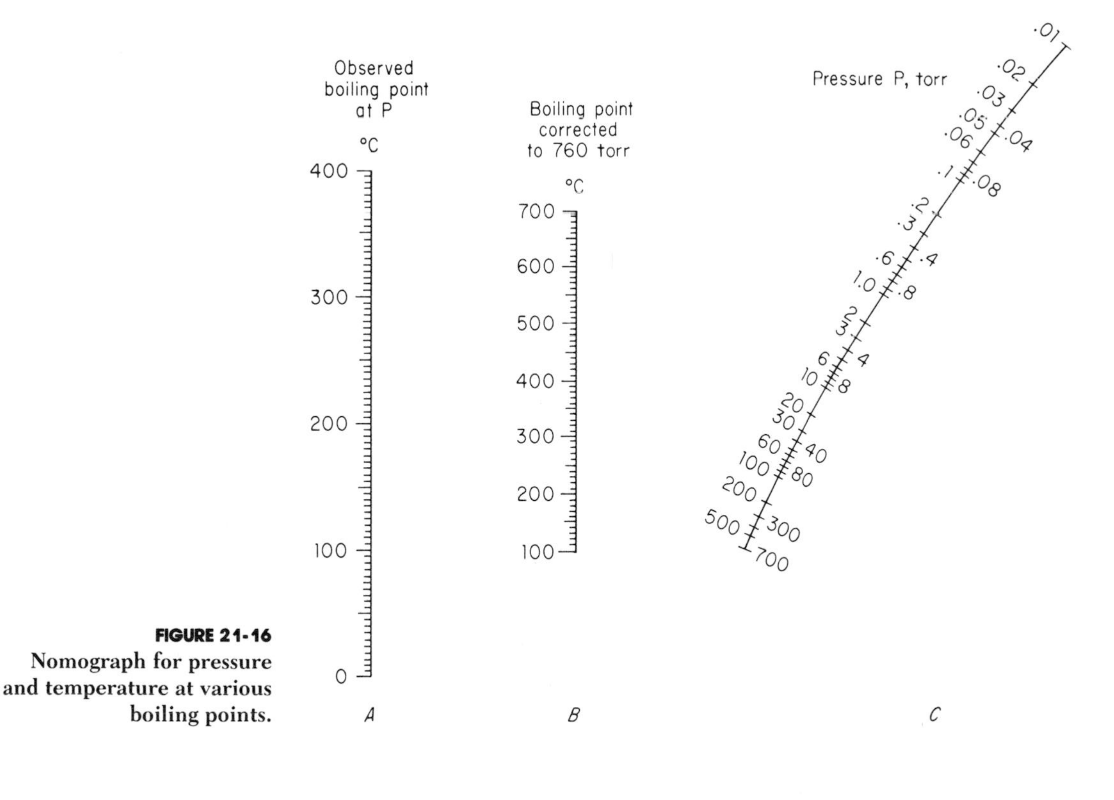

FIGURE 21-16
Nomograph for pressure and temperature at various boiling points.

on column C intersects column A at 150°C. Thus 150°C is the boiling point at 1 torr pressure.

CAUTION Glass equipment may collapse under reduced pressure. Use *safety glasses* and a *safety shield* as well as special equipment shown in Fig. 21-17.

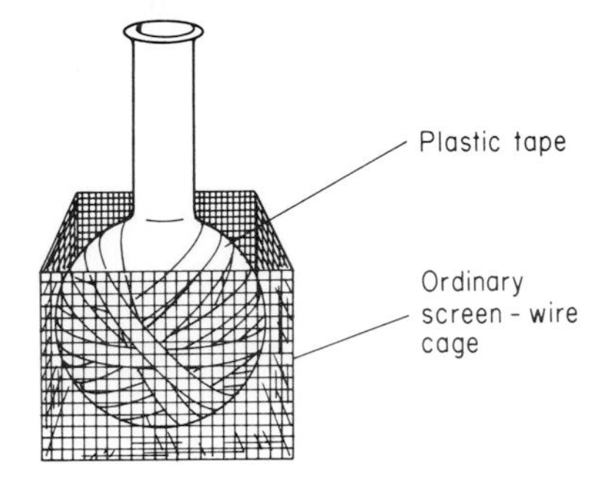

FIGURE 21-17
Safety precautions for a large flask in a vacuum system.

General Requirements

The experimental setup is as shown in Fig. 21-18.

1. A source of vacuum. Efficient water pumps, aspirators, will theoretically reduce the pressure in the system to the vapor pressure of the water passing through the pump. In practice, the pressure is usually about 10

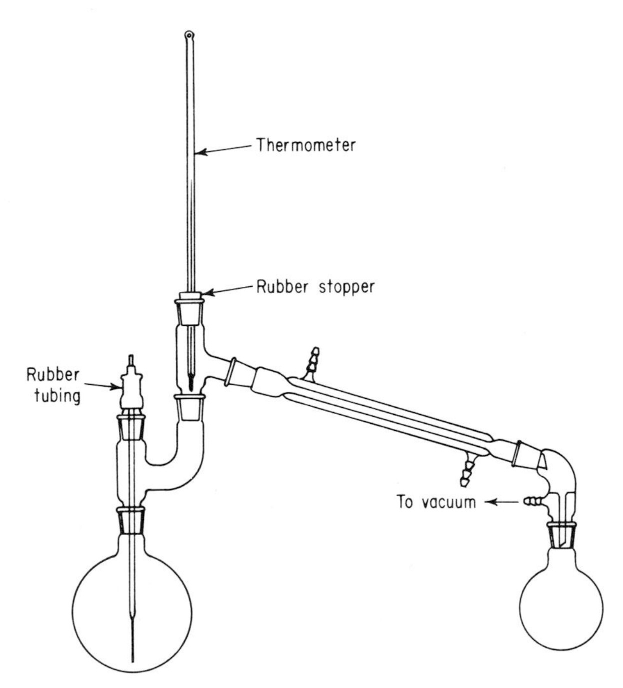

FIGURE 21-18
Vacuum distillation with gas-capillary bubbler.

mm higher. (*a*) Oil mechanical vacuum pumps. (*b*) Use rubber pressure tubing. (*c*) The entire distillation system should be airtight, free from leaks. (*d*) Lubricate all joints and connections.

2. Safety trap to protect manometer and vacuum source from overflow-liquid contamination (Fig. 21-19).

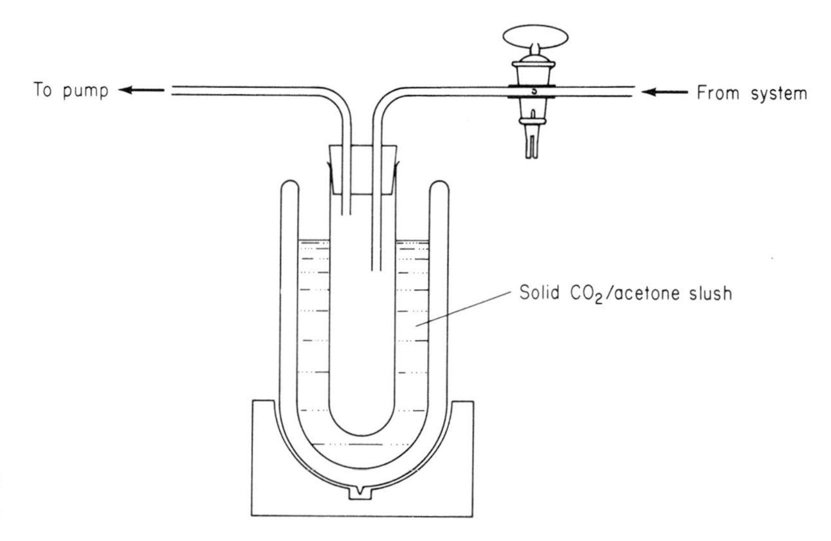

FIGURE 21-19
Dry-ice vapor trap.

The trap must be correctly connected. Vapors condense on the sides of the trap and fall to the bottom if they are not solidified.

Dry ice will freeze your skin; *do not handle with bare fingers or hands.*

(a) Crush dry ice in a cloth towel with a hammer.

(b) Use a scoop to fill the Dewar flask after the trap has been inserted.

(c) Add solvent, acetone, or isopropanol in small increments until the trap is filled and the liquid level is near the top.

3. Pressure gauge (manometer). Exercise great care when allowing air into the evacuated system. It must be done slowly to avoid breakage when the mercury column rises to the top of the closed tube. (Refer to the section on manometers in Chap. 5, "Pressure and Vacuum.")

4. Manostat (pressure regulator). To maintain constant pressure in the system, it automatically opens and closes needle valves, permitting air to enter or keeping the system airtight because of vacuum variations. Refer to Chap. 5, "Pressure and Vacuum."

5. Capillary air inlet (Fig. 21-20).

FIGURE 21-20
Capillary gas- or air-inlet tube.

6. Special vacuum distillation flasks to minimize contamination of the distillate caused by frothing of the boiling solution.

7. Heating baths, electric mantles, and fusible alloy or sand baths.

8. Special distilling heads (Fig. 21-21) to permit removal of distillate fractions without interrupting the distillation.

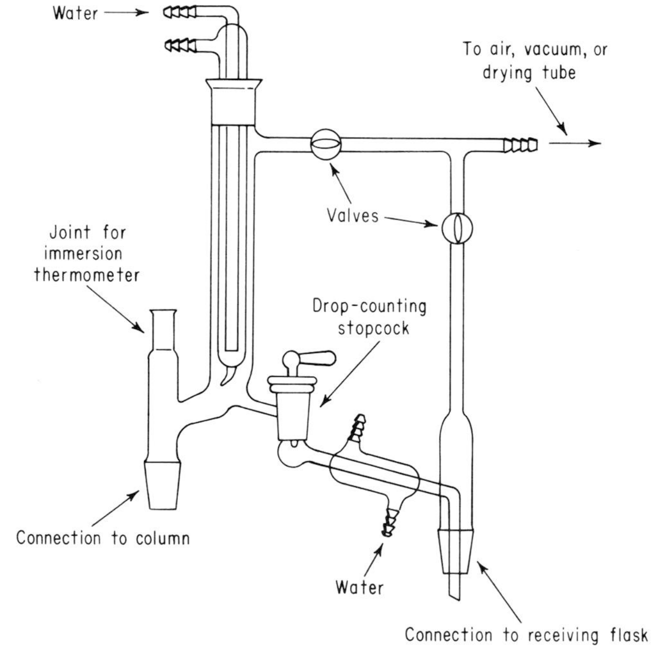

FIGURE 21-21
Partial-takeoff distillation head and its component parts.

Assemblies for Simple Vacuum Distillation and Fractionation

Simple Vacuum Distillation

Procedure

A *Claisen flask* for use in this process is shown in Fig. 21-22.

1. Fill the Claisen flask (*A*) one-third full.
2. Apply vacuum; adjust the capillary air inlet (*C*) with the pinch clamp (*D*).
3. Heat bath to about 20°C higher than the temperature at which the material will distill.
4. Cooling water flowing over the receiver (*B*) condenses the vapors to give a distillate.
5. A safety trap prevents any condensate from contaminating the suction pump or manometer.
6. When distillation is completed:
   (a) Remove the heating bath; allow the flask to cool.
   (b) Remove the capillary pinch clamp.
   (c) Cut off the cooling water.
   (d) Turn off the vacuum pump.

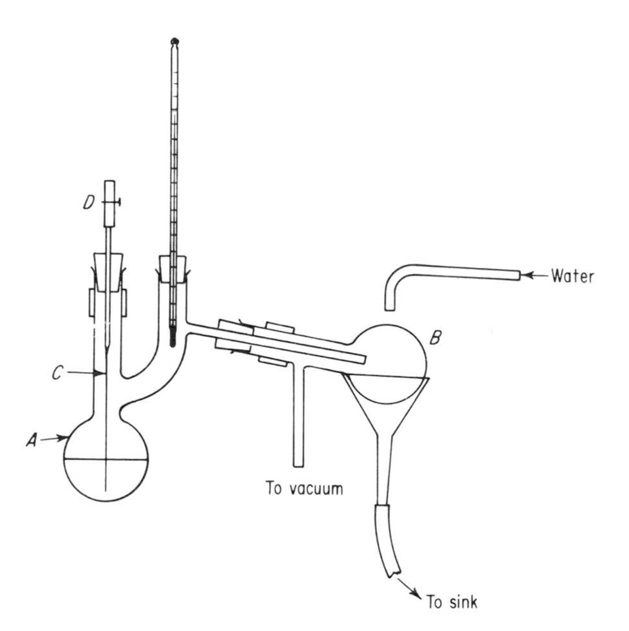

FIGURE 21-22
Claisen flask setup for vacuum distillation.

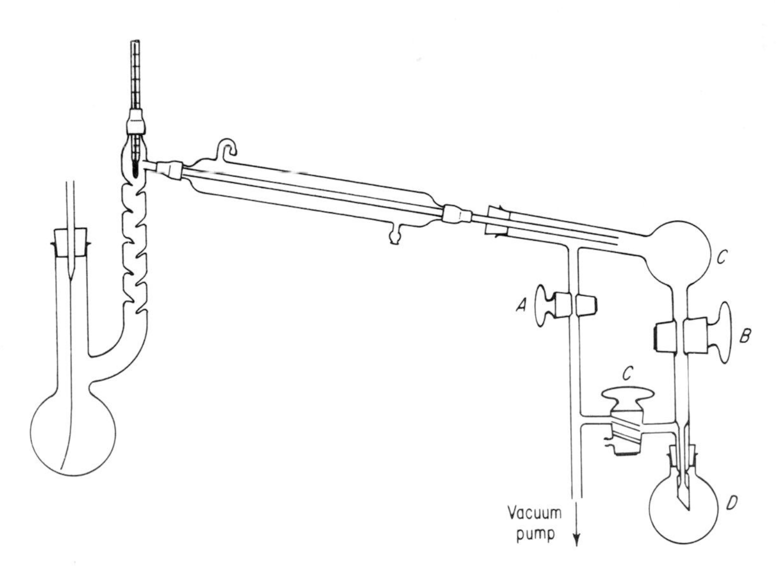

FIGURE 21-23
Claisen apparatus for fractional distillation in a vacuum.

Vacuum Fractionation

A *Claisen fractionating column* is shown in the apparatus in Fig. 21-23. The distillation neck of the Claisen flask serves as a fractionating column because of the indentations. The receiver makes it possible to remove distillate fractions without interrupting the distillation or breaking the vacuum.

Procedure

Follow the steps and observe the cautions previously listed. To remove distillate fractions during distillation:

1. Close stopcock B.
2. Close stopcock C.
3. Reverse rotation stopcock C so that air flows into flask D.
4. Gently remove flask D, empty into bottle, and replace.
5. Rotate stopcock C 180° so that D is now under vacuum.
6. Open stopcock B to allow collected distillate to drain into D.

Complete the fractional distillation and disconnect the equipment as described above.

Modified Distillate Receiver

Collecting individual distillate fractions without removing the receiver (as in the procedure above) can be accomplished with the modified receiver. The receiver is adjusted to collect the particular fraction desired and then rotated to meet the need. This method does not introduce any possible problems which might occur in the procedure if the vacuum distilling pressure is affected when the receiver is removed to isolate the distillate fraction. (See Fig. 21-24.)

A more complex apparatus is shown in Fig. 21-25.

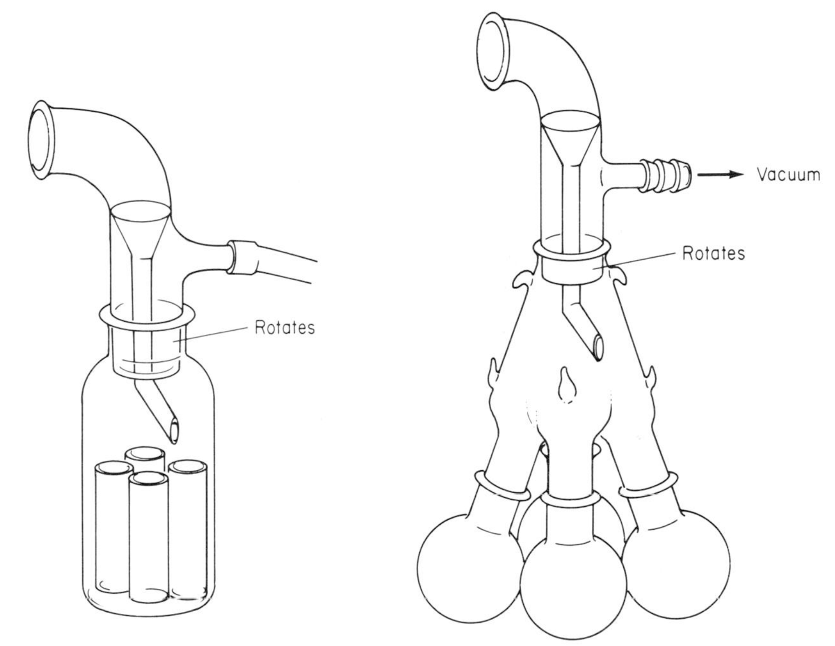

FIGURE 21-24
Modified receivers for collection of distillate fractions.

STEAM DISTILLATION

Principle

Steam distillation is a means of separating and purifying organic compounds by volatilization. The organic compound must be insoluble or slightly soluble in water. When steam is passed into a mixture of the compound and water, the compound will distill with the steam. In the distillate, this distilled compound separates from the condensed water because it is insoluble in water.

Most compounds, regardless of their normal boiling point, will distill by steam distillation at a temperature below that of pure boiling water. For example, naphthalene is a solid with a boiling point of 218°C. It will distill with steam and boiling water at a temperature less than 100°C.

Some high-boiling compounds decompose at their boiling point. Such substances can be successfully distilled at low temperature by steam distillation.

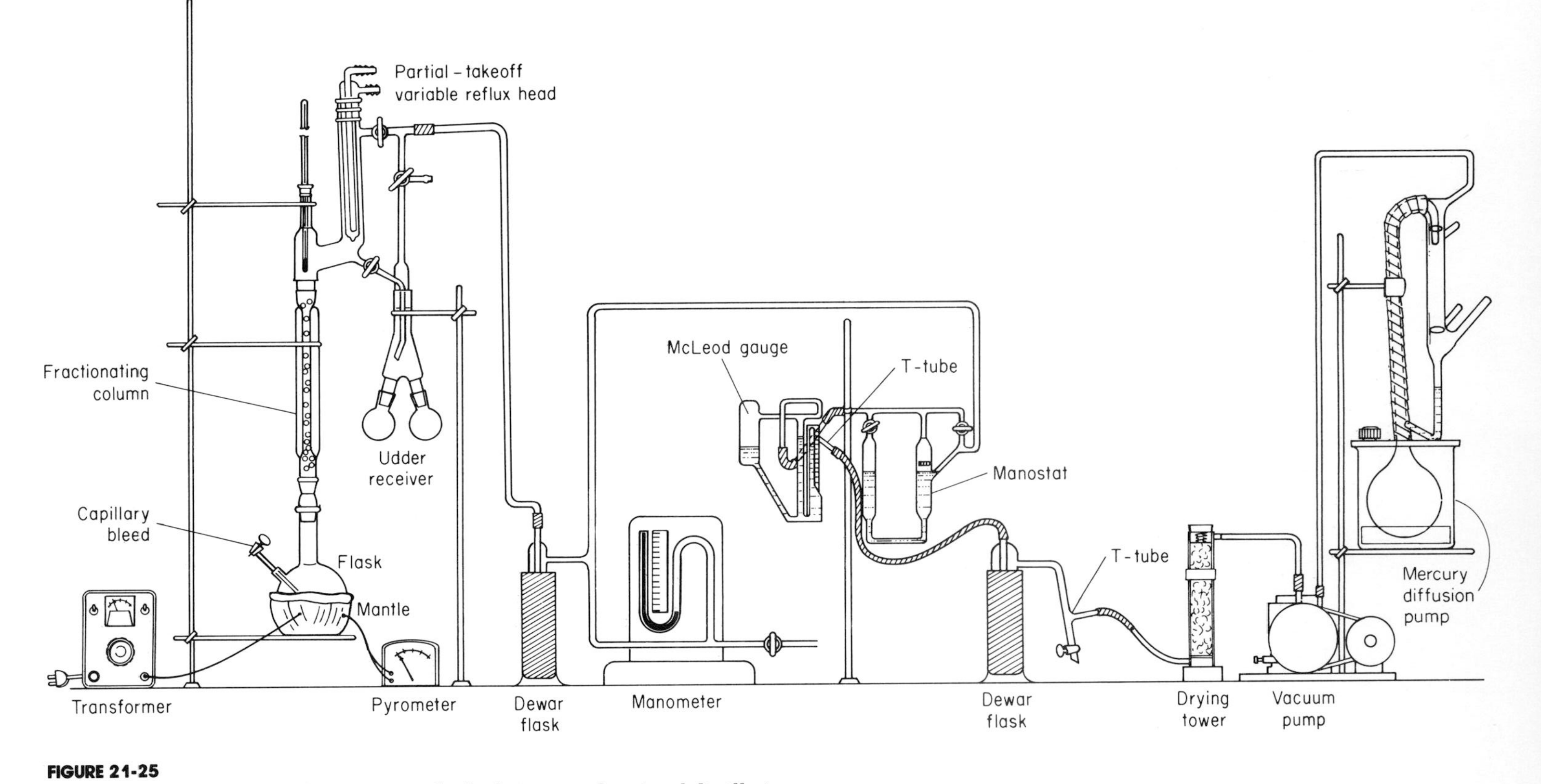

FIGURE 21-25
Apparatus setup using a diffusion pump for high-vacuum fractional distillation.

Steam distillation can be used to rid substances of contaminants because some water-insoluble substances are steam-volatile and others are not.

When it is desirable to separate nonvolatile solids from high-boiling organic solvents, steam distillation will remove all solvents (water-insoluble).

Procedure

1. Place the compound or mixture in the distilling flask with a little water. Pass cooling water through the condenser (see Fig. 21-26). A Claisen flask may be substituted for the round-bottomed flask. The Claisen stillhead helps to prevent any contamination of the distillate caused by spattering of the steam-distilled mixture.

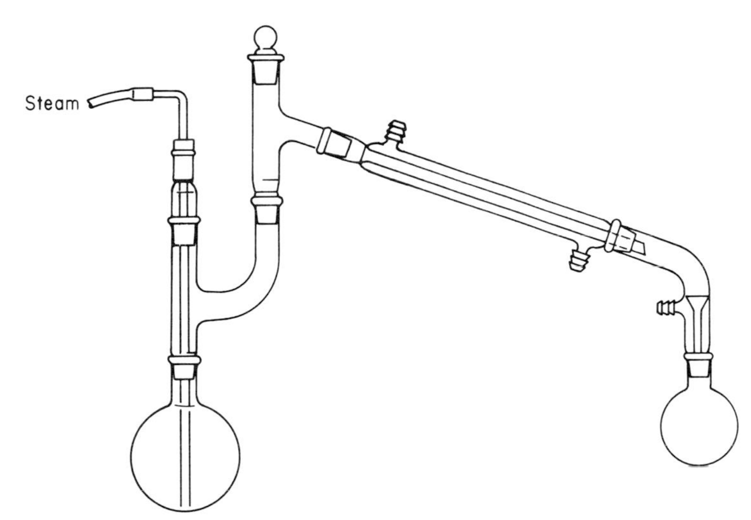

FIGURE 21-26 Apparatus for steam distillation.

If there is no readily available source of piped steam, the steam can be generated in an external steam generator (Fig. 21-27) and then passed into the mixture to be steam-distilled.

CAUTION Always equip the steam generator with a safety tube to prevent explosions.

NOTE If only a small amount of steam is needed, for instance, to rid a substance of a small amount of steam-volatile impurity, water can be combined with the material directly in the distilling flask; the flask is then directly heated with a Bunsen flame or any other suitable heat source. (Refer to Chap. 10, "Heating and Cooling.") The long steam-inlet tube is replaced with a stopper.

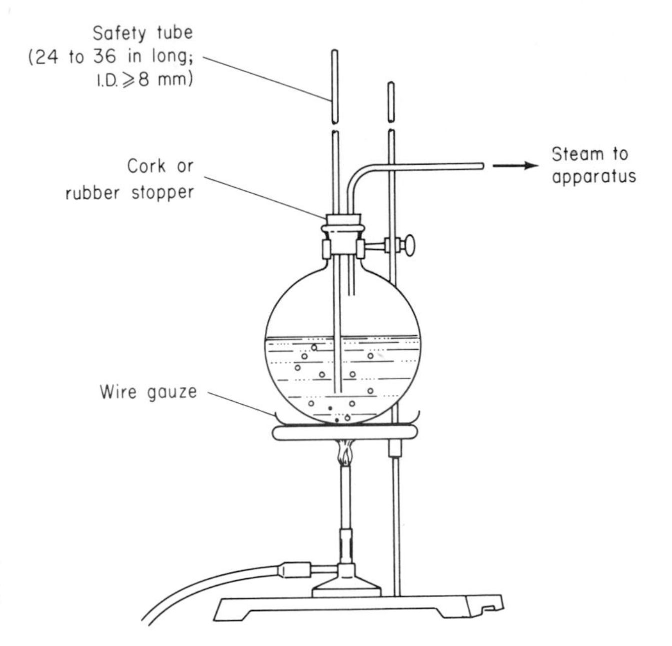

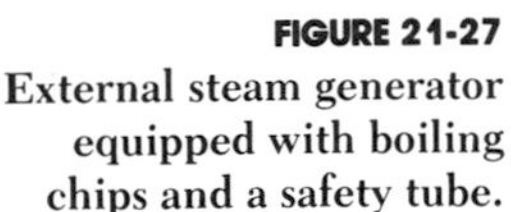
FIGURE 21-27
External steam generator equipped with boiling chips and a safety tube.

2. Pass steam into the distilling flask, with the steam outlet below the surface of the liquid. The distilling flask *itself* may be heated gently with a burner. If steam is available from a laboratory steam line, insert a water trap (Fig. 21-28) in the entering steam line to trap condensed water. Otherwise, the condensed water may fill up the distilling flask.

3. Continue passing steam into the flask until no appreciable amount of water-insoluble material appears in the condensate.

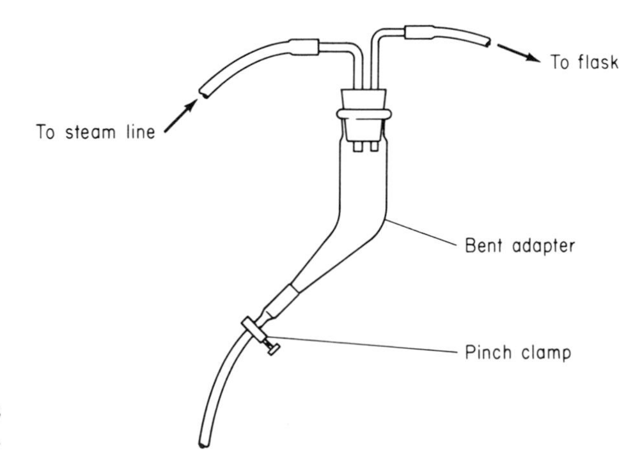

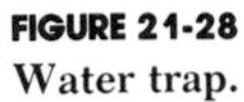
FIGURE 21-28
Water trap.

CAUTION Steam will cause severe burns. Handle with care!

DANGER If the substance crystallizes in the condenser, it will close the tube. Steam pressure could build up when the tube closes and cause an explosion. *Use care! Drain the condenser of cooling water. The crystals will melt and pass into the distillate.* When the tube is clear, slowly pass the cooling water through the condenser.

4. Always disconnect the steam-inlet tube from the flask.

Steam Distillation with Superheated Steam

The use of superheated steam can increase the proportion of the low-vapor-pressure component in the distillate, and, at the same time, reduce the amount of steam condensate in the distilling flask. The distilling flask is surrounded by a heating bath (refer to Chap. 10, "Heating and Cooling") which is heated to the same temperature as the superheated steam; this minimizes any cooling of the steam which could occur before it enters the flask. (See Fig. 21-29.)

A commercially available metal superheater, heated by a Meker burner, is shown in Fig. 21-30.

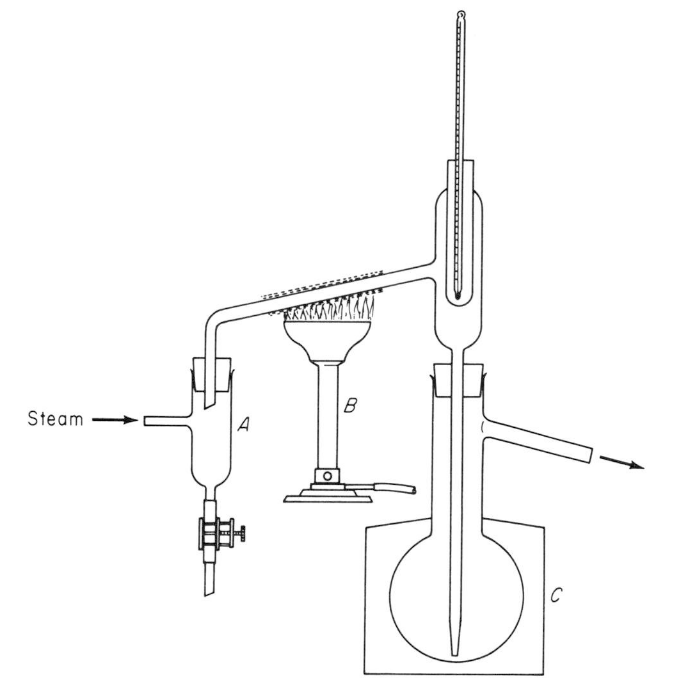

FIGURE 21-29 Distillation apparatus for use with superheated steam.

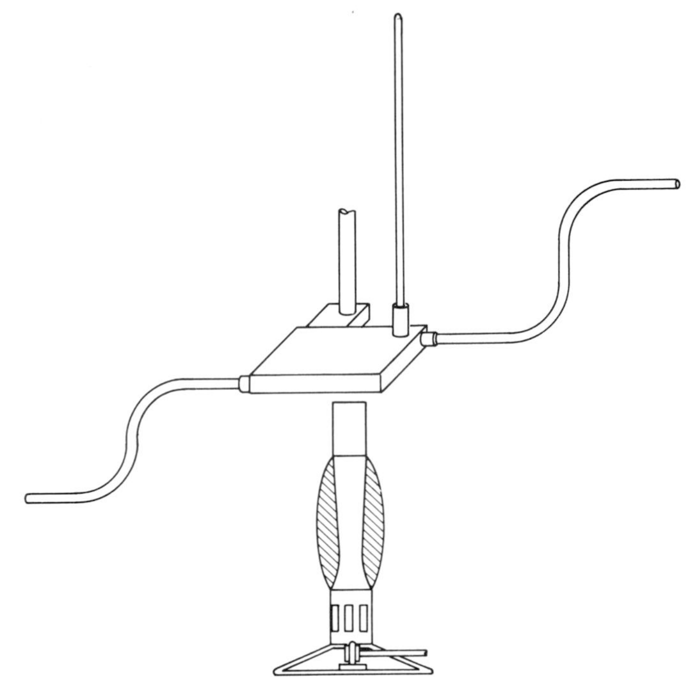

FIGURE 21-30
Commercially available metal superheater, heated with a Meker burner.

REFLUXING

The reflux procedure allows you to heat a reaction mixture for an extended period of time without loss of solvent. The condenser, which is fixed in a vertical position directly above the heated flask, condenses all vapors to liquid. Because none of the vapors escape, the volume of liquid remains constant.

Reflux procedures are carried out in neutral, acid, or basic solution, depending upon the reaction.

Typical operations include hydrolysis-saponification of acid amides, esters, fats, nitriles, substituted amides, and sulfonamides. Hydrolysis-saponification is used to split organic molecules (which were made by combination of two or more compounds) into the original compounds.

Experimental setups are shown in Figs. 21-31 to 21-34.

Procedure

1. The water inlet to the condenser is the lower one. The water outlet to the condenser is the upper one.
2. Fill the heating flask no more than half full; add a few boiling stones.
3. Turn on the cooling water.
4. Heat to reflux for the desired period of time.

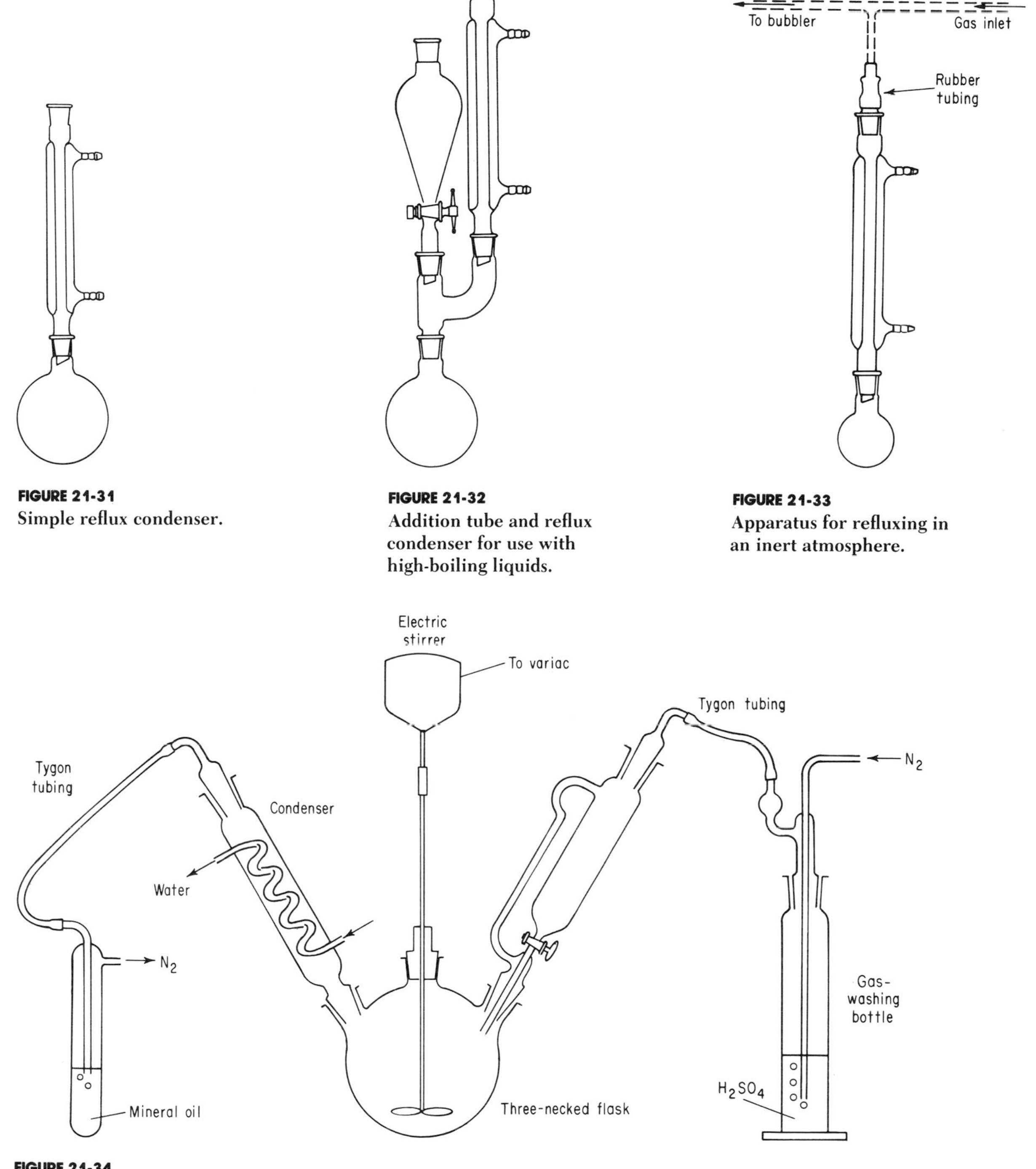

FIGURE 21-31
Simple reflux condenser.

FIGURE 21-32
Addition tube and reflux condenser for use with high-boiling liquids.

FIGURE 21-33
Apparatus for refluxing in an inert atmosphere.

FIGURE 21-34
Apparatus with mechanical stirrer for reflux reaction in an inert atmosphere.

MOLECULAR STILLS

Many organic substances cannot be distilled by any of the ordinary distilling methods because: (1) They are extremely viscous, and any condensed vapors would plug up the distilling column, side arm, or condenser. (2) Their vapors are extremely susceptible to condensation.

To distill high-molecular-weight substances (molecular weights around 1300 for hydrocarbons and around 5000 for silicones and halocarbons) molecular stills are used. Molecular distillations differ from other distillations because:

1. All condensed vapor flows to the distillate receiver or collector.

2. Very low pressure (high vacuum) in the system favors vaporized molecules reaching the condensing surface without collision with other molecules to condense prematurely.

3. There is a very short distance between the surface of the evaporating liquid and the condenser surface. (See Fig. 21-35.)

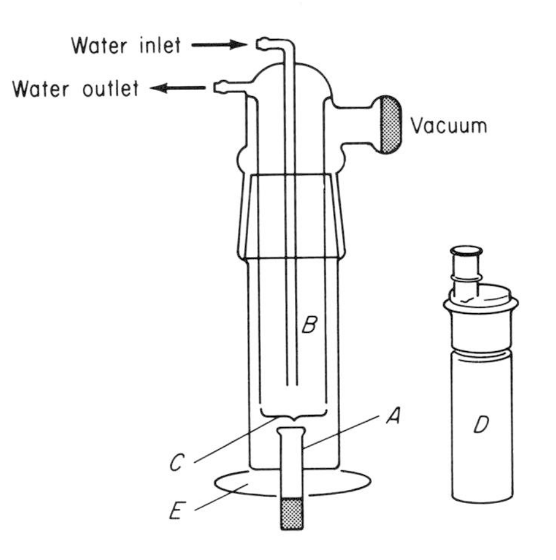

FIGURE 21-35 Simple molecular still made of glass. *A*, sample to be distilled; *B*, cold-finger condenser; *C*, tip of condenser to guide condensate into *D*; *D*, distillate collector; *E*, source of heat.

Laboratory models include a rotating still. Materials are fed at slightly above their melting point into the rotating still. They are distributed evenly and thinly over the heated evaporating surface. The sample distills, requiring a very short time, and its vapors condense to run into the collector. The degree of vacuum is controlled to collect the distillate effectively at the condenser, and the pressure can be as low as 1 μm Hg.

EVAPORATION OF LIQUIDS

Small Volumes

Evaporation of solvents is necessary at times to concentrate solutions and to obtain crystallization of solutes.

Method 1

1. Pour the small volume of solution into the watch glass placed over a beaker of water (Fig. 21-36). (Refer to Chap. 10, "Heating and Cooling.")

2. Boil the water. The heat transfer through the steam formed evaporates the solvent of the solution.

Method 2

Use an evaporating dish instead of a watch glass (Fig. 21-37). Evaporating dishes come in various sizes and are made of various materials (Fig. 21-38). Use the appropriate one.

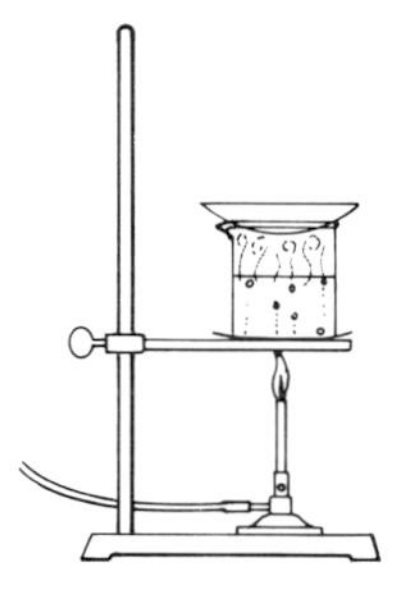

FIGURE 21-36
Evaporation over a water bath.

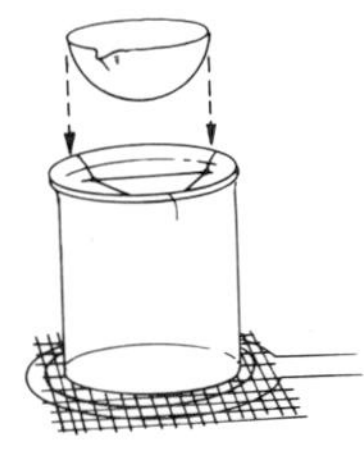

FIGURE 21-37
Alternate procedure using evaporating dish instead of watch glass.

FIGURE 21-38
Evaporating dishes, (*a*) Porcelain with heavy rim. (*b*) Nickel. (*c*) Crystallizing dish to hold or contain solutions from which solids are expected to crystallize.

Direct Heating of Evaporating Dishes

You can speed up the evaporation of water by directly heating the evaporating dish with a Bunsen burner, diffusing the heat with a wire gauze.

When the volume is very low, transfer the evaporating dish to the top of a beaker of boiling water. The steam acts as the heating agent. (See Fig. 21-38.)

FIGURE 21-39
Watch glass is rinsed with a wash bottle using a back-and-forth motion.

Transferring Residues from Watch Glasses or Evaporating Dishes

When the water has been evaporated to the desired volume, or the desired concentration, the material can be transferred to an appropriate container, such as a smaller evaporating dish, by rinsing with distilled water. The watch glass or evaporating dish is rinsed with a wash bottle, using a back-and-forth motion (Fig. 21-39).

Large Volumes

Concentrate solutions by boiling off the desired volume of solvent. Refer to the sections on heating flammable liquids and heating organic liquids in Chap. 10, "Heating and Cooling."

Method 1

1. Pour the solution which is to be concentrated by boiling off solvent into a suitably sized beaker which is covered by a watch glass resting on glass hooks (Fig. 21-40).
2. Heat the solution to evaporate the solvent.

FIGURE 21-40
Arrangement for the evaporation of liquids.

Method 2

When you need to concentrate a solution by evaporating the solvent, you can accelerate the process by directing a stream of air (or nitrogen gas, for easily oxidized substance) gently toward the surface of the liquid. This stream of gas will remove the vapors which are in equilibrium with the solution.

CAUTION If toxic or flammable solvent vapors should be involved, conduct the evaporation in a hood.

Compressed air sometimes contains oil and water droplets. Filter the air by passing it through a cotton-filled tube and a tube filled with a drying agent, such as anhydrous calcium chloride.

Evaporation under Reduced Pressured

Solvents can be evaporated more quickly by evaporating them at reduced pressure and gently heating (Fig. 21-41). Refer to Fig. 21-42 for trap bottle.

Procedure

1. Place the solution to be concentrated in a round-bottomed flask or suction flask.

CAUTION Do not use Erlenmeyer flasks having volumes larger than 125 mL. There is a danger of implosion and collapse.

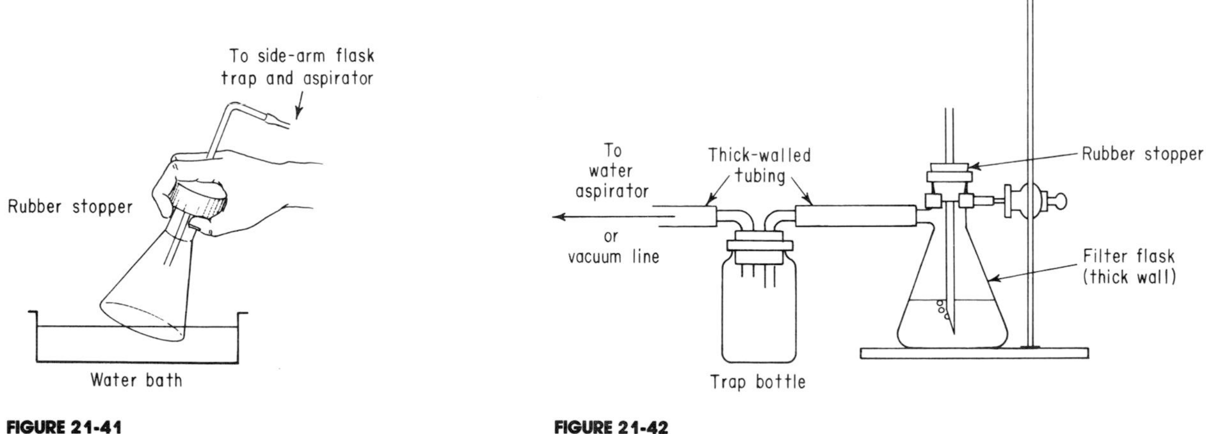

FIGURE 21-41
Evaporation of solvent under reduced pressure at elevated temperature.

FIGURE 21-42
Evaporation of solvent under reduced pressure at room temperature.

2. Connect with rubber tubing to a safety trap which in turn is connected to a water aspirator.

3. Apply vacuum by turning on water.

4. *Swirl* the flask to expose large areas of the liquid and speed evaporation.

NOTE *Swirling* technique helps suppress bumping.

5. The flask cools as the solvent evaporates.

6. Heat the flask by immersing in a warm-water bath.

Evaporation under Vacuum

Water Aspirator

1. Place liquid in a flask (fitted with a capillary air-inlet tube, Fig. 21-42) which is connected to a water aspirator with tubing.

2. Turn on the water aspirator to apply vacuum; adjust the capillary in the flask.

3. Gently apply heat with a warm-water bath.

4. When evaporation is completed, disconnect the tubing from the water aspirator *before* turning off the water.

Mechanical Vacuum Pump

1. Place liquid in a flask (fitted with a capillary air-inlet tube) connected to a dry-ice trap by tubing.

2. Connect the outlet of the trap to a vacuum pump.

3. Turn on vacuum pump; adjust capillary on flask to a fine-air-bubble stream.

4. Gently apply heat with a warm-water bath.

5. Disconnect the tubing, connecting the flask to the vapor trap *before* turning off the vacuum pump.

ROTARY EVAPORATOR

Rotary evaporators (Fig. 21-43) provide a very rapid means to evaporate solvents and concentrate solutions. The flask rotates while the system is under vacuum (see section on the water aspirator in Chap. 5, "Pressure and Vacuum") providing a very large surface area for evaporation. The walls of the flask are constantly rewetted as the flask rotates, minimizing superheating and bumping. Heat is supplied to the flask by steam bath, oil bath, heating mantle, or other heat source to meet the need.

FIGURE 21-43
Rotary evaporator.

Rotary evaporators can be used for evaporation and vacuum drying of powders and solids, and for low-temperature distillation of heat-sensitive substances. Substances can be degassed and distilled under inert atmospheres. The rotating flask ensures good mixing and good heat transfer from the heating bath.

SUBLIMATION

The vapor pressure of solids increases as the temperature increases, and because of this some solids can go from the solid to the vapor state without passing through the liquid state. This phenomenon is called *sublimation*, and the vapor can be resolidified by reducing its temperature. The process can be used for purifying solids if the impurities in the solids have a much lower vapor pressure than that of the desired compound.

The advantages of sublimation as a purification method are:

1. No solvent is used.

2. It is a faster method of purification than crystallization.

3. More volatile impurities can be removed from the desired substance by subliming them off.

4. Substances which contain occluded molecules of solvation can sublime to form the nonsolvated product, losing the water or other solvent during the sublimation process.

5. Nonvolatile or less volatile solids can be separated from the more volatile ones.

The disadvantage of sublimation as a purification method is that the process is not as selective as crystallization because the vapor pressures of the sublimable solids may be very close together.

Atmospheric and Vacuum Sublimation

A sublimation point is a point at which the vapor pressure of a solid equals the applied pressure: it is a constant property like a melting point or a boiling point. Many liquids evaporate at temperatures below their boiling points, and some solids sublime (evaporate) below their melting points. If a substance sublimes below its melting point at atmospheric pressure, its melting point must be determined under pressure, in a sealed capillary tube.

Many solids do not develop enough vapor pressure at 760 mm Hg (atmospheric pressure) to sublime, but they do develop enough vapor pressure to sublime at reduced pressure. For this reason, most sublimation equipment is constructed with fittings making it adaptable for vacuum connections. Furthermore, the use of vacuum is advantageous because the lower temperatures required reduce thermal decomposition.

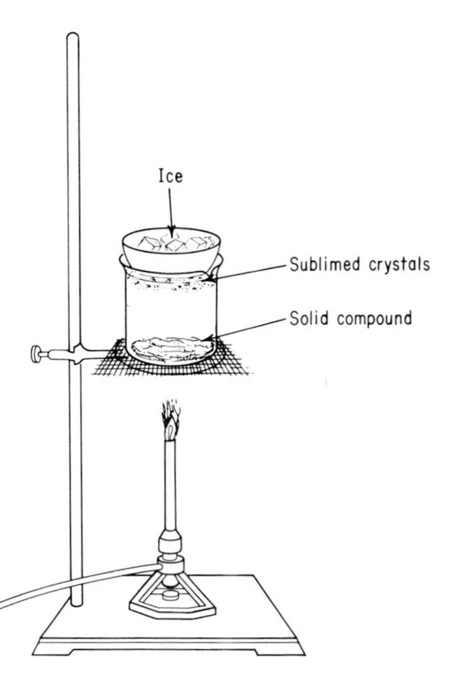

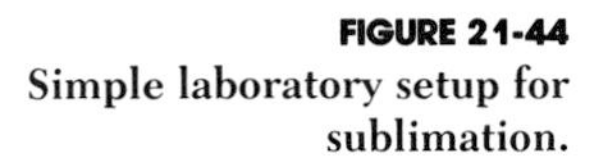
FIGURE 21-44
Simple laboratory setup for sublimation.

Methods of Sublimation

Simple Laboratory Procedure at Atmospheric Pressure

Gently heat the sublimable compound in a container which has a loosely fitting cover that is chilled with cold water or ice (Fig. 21-44).

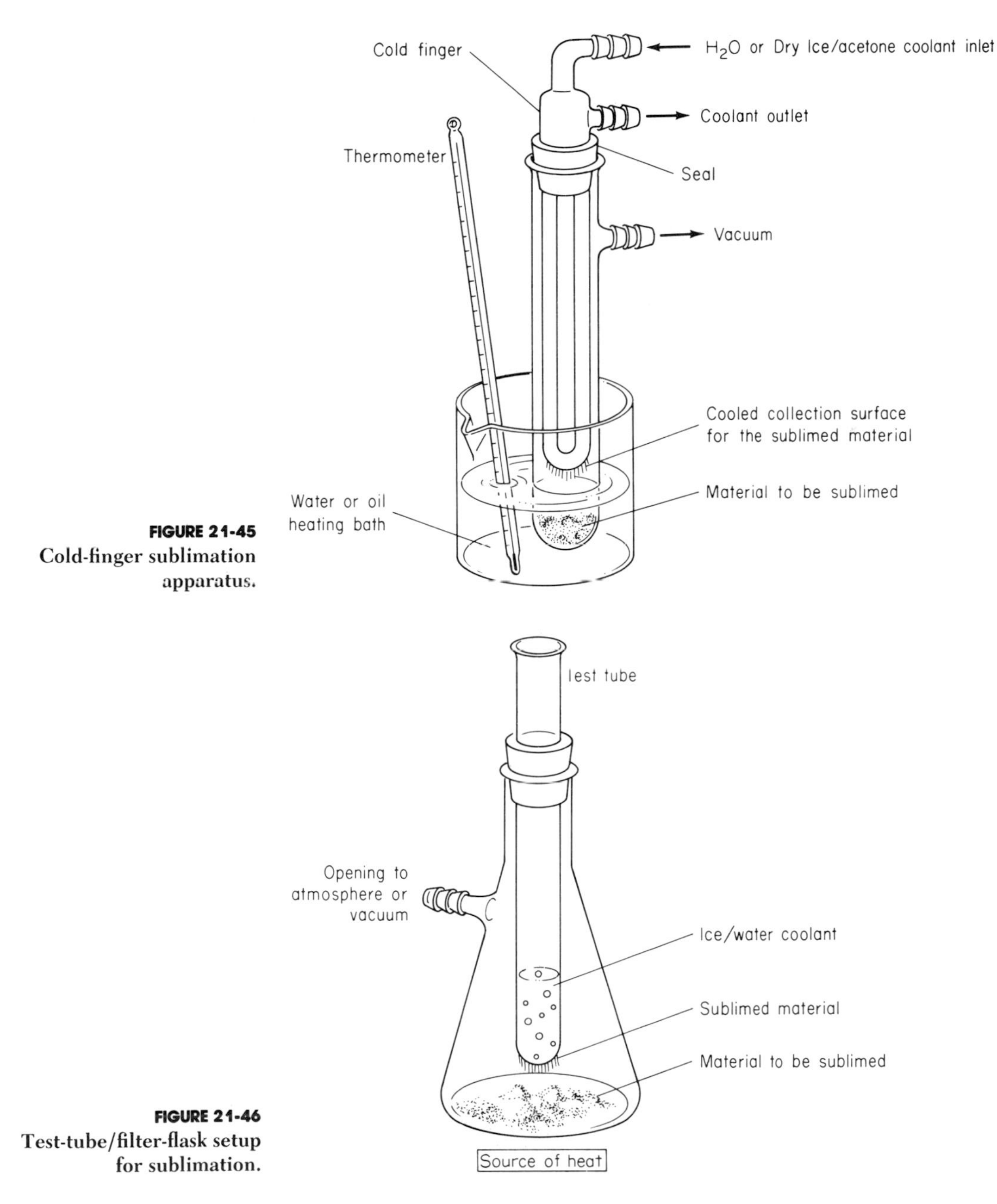

FIGURE 21-45 Cold-finger sublimation apparatus.

FIGURE 21-46 Test-tube/filter-flask setup for sublimation.

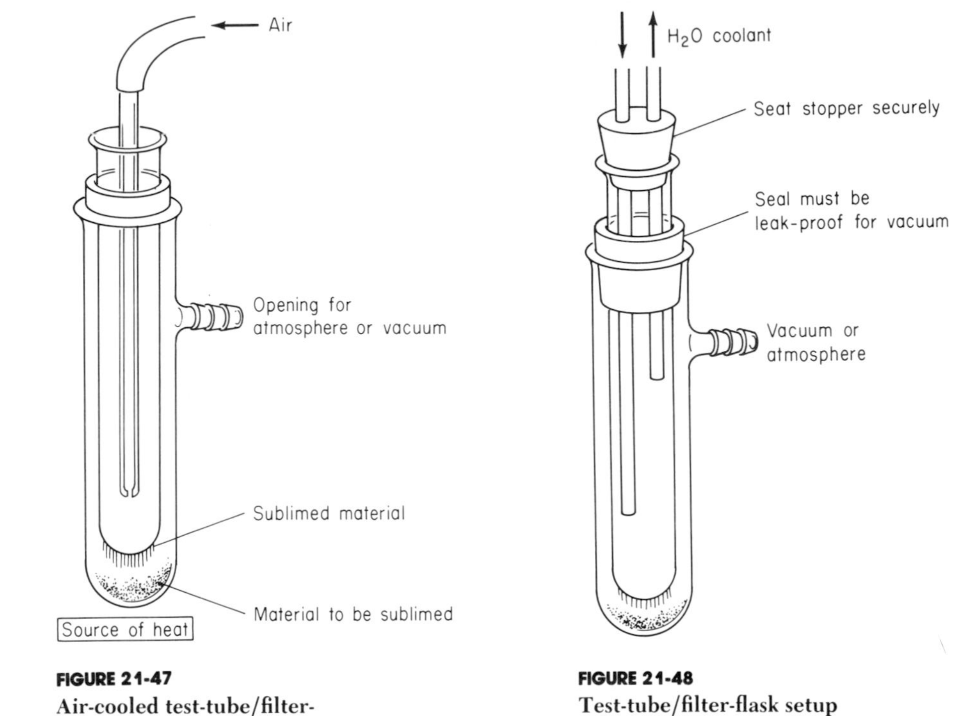

FIGURE 21-47
Air-cooled test-tube/filter-flask setup.

FIGURE 21-48
Test-tube/filter-flask setup with continuous water coolant.

Methods Useful at Atmospheric or Reduced Pressure

The sublimation equipment illustrated in Figs. 21-45 to 21-48 can be easily constructed in the laboratory and can be used for sublimation procedures at normal atmospheric pressure or at reduced pressure.

The apparatus in Fig. 21-45 is based on the use of a cold finger as a condenser. Figure 21-46 demonstrates the use of a coolant-filled test tube as a condenser; the coolant here is a mixture of ice and water. In Fig. 21-47 the condenser is an air-cooled test tube, and in Fig. 21-48 the coolant for the test-tube condenser is constantly circulating water.

22
FUNDAMENTALS OF CHEMISTRY

INTRODUCTION

Matter is anything that occupies space (has volume) and has mass.* There are three states of matter: *gases, liquids,* and *solids,* and the simplest form of matter is an element. A so-called fourth state, *plasma,* exists in powerful magnetic fields.

Elements are substances that are homogeneous; have a definite composition; and cannot be broken down, changed, or decomposed by ordinary physical or chemical means into simpler forms. Elements have a constant composition, because each element is composed of one specific type of the indestructible† particles called *atoms,* the smallest particle of an element that retains the chemical identity of the element. Elements may be classified into two main types: metals and nonmetals. Few elements are found in the pure state; examples are mercury, lead, gold, silver, and copper. Most of the others are found in various combinations with other elements in forms called *compounds.* There are 89 naturally occurring elements and 16 synthetic elements. Each element is designated by a symbol, such as Fe for iron, Pb for lead, Ag for silver, Na for sodium, to identify that element as being distinct from other elements.

Elements are composed of atoms; each atom contains an equal number of positive charges, protons, in its nucleus and negative charges, electrons, in orbit around the nucleus. The nucleus also contains neutral particles, neutrons, which have no electrical charge. The nuclear charge is equal to the number of positive charges, the number of protons, and this number is called the *atomic number* of the element. The neutrons in the nucleus have the same mass as that of the protons, and the mass of the atom equals the sum of the masses of the protons and neutrons. The mass of the electrons is so small and contributes so little to the mass of the atom that the sum of the masses of the protons and neutrons is called the *atomic mass.**

---

* The terms "mass" and "weight" are often used (loosely) interchangeably. However, in this and the following chapters we will use the term "mass" for a quantity without a gravity component.

† Atoms are indestructible by *ordinary* physical or chemical means.

Compounds are complex substances which are made up of two or more elements which are held together by chemical bonds, chemically combined in a constant ratio with respect to the number and kind of atoms. Compounds have a constant composition by mass. Like elements, compounds are pure substances having a definite composition.

Mixtures are of indefinite (variable) composition. They are not chemically combined, are composed of two or more substances, and they have no definite ratio by mass. They may be composed of different atoms and/or compounds (combination of atoms) that are only physically mixed. They may be separated by physical means such as varying the degrees of solubility of the components, by differences in boiling point, by magnetic properties, and so forth.

PHYSICAL AND CHEMICAL CHANGES

Substances, either elements or compounds, can undergo two types of change. They can undergo *physical change,* where the identity of the substance is not lost but the shape or form has been changed by the application of heat or pressure. The original shape, form, or state of matter always can be attained after infinite physical changes; the process is reversible. Examples are the melting of ice, freezing of water, melting of iron, vaporization of water, and so forth.

A *chemical change* occurs when new substances are formed which are entirely different in composition and properties than the original substances. Chemical bonds have been made or broken; there is a change in energy and also a change in composition. Examples are the burning of fuel, the rusting of iron, the cooking of foods, and so forth. When a piece of iron rusts, undergoing a chemical change, it forms flaky brown rust (iron oxide), which does not conduct electricity, is not attracted by a magnet, is not shiny, and has no tensile strength.

Physical Properties of Substances

Each substance is characterized by its own physical properties. Comparison of the properties of a newly formed substance with the properties of the original substance determines whether or not a new substance is formed by chemical change, or whether there has been a physical change. Physical properties of substances are:

Odor	Solubility	Tensile strength
Color	Density	Thermal conductivity
Melting point	Luster	Ductility
Boiling point	Electrical conductivity	Malleability
Freezing point	Hardness	Refractive index
Taste	Crystal structure	Optical rotation
Specific heat		

Evidence of Chemical Change

When a chemical reaction takes place, one or more of the following phenomena occurs (these are indications only, see *cautions* below):

1. The evolution or absorption of heat
2. The formation of a precipitate
3. The evolution of a gas
4. The change in color of a solution
5. The emission of an odor
6. The development or loss of fluorescence
7. The emission of light (chemiluminescence)

CAUTIONS 1. The evolution of a gas may be a physical change; for example, the loss of CO_2 gas from carbonated water.

2. The formation of a precipitate may be the change of a supersaturated solution to a saturated solution, or the loss of solvent and the crystallization of the excess solute.

Figure 22-1 shows the relationship of elements, compounds, and mixtures to uniformity and composition.

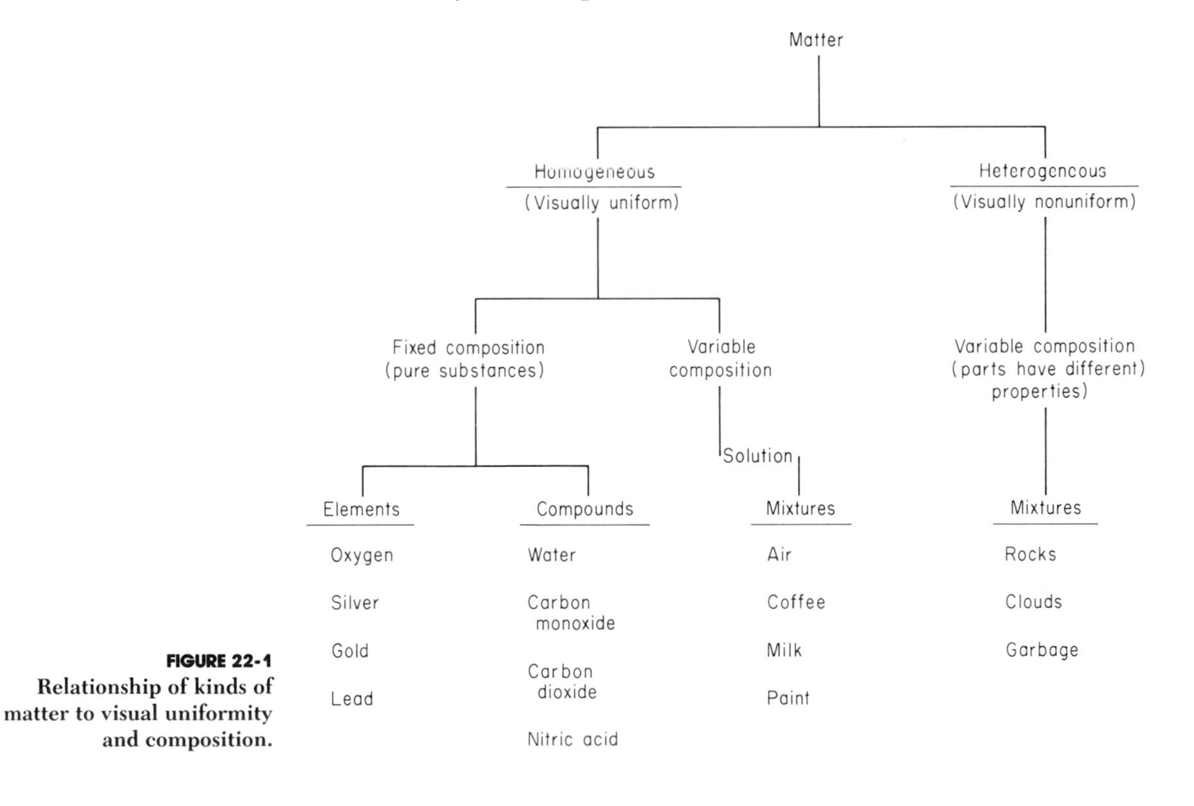

FIGURE 22-1
Relationship of kinds of matter to visual uniformity and composition.

CHEMICAL SYMBOLS AND FUNDAMENTAL UNITS

The periodic chart (Table 22-1) provides the symbol of the element, the atomic number of the element, and the atomic mass of the element. See also Table 22-2.

Mole

Individual atoms have extremely minute masses. Chemists use a fundamental unit called a *mole.* The mole is the amount of substance of a system which contains as many elementary particles as there are atoms in 0.012 kg of carbon-12 (^{12}C). When the mole is used, the elementary entities must be specified and may be atoms, molecules, ions, electrons, etc. (See Fig. 22-2.)

Avogadro's Number

A mole of atoms of any element thus contains the same number of atoms as exactly 12 g of pure ^{12}C. This number is Avogadro's number; it equals 6.02×10^{23}.

The formula mass (molecular mass) of any compound in grams (gram-molecular mass) contains the same number of molecules, Avogadro's number.

A gram-atomic mass of an element, the atomic mass of the element expressed in grams, contains Avogadro's number of atoms.

CHEMICAL BONDING

Chemical compounds are formed when atoms interact in such a way that the energy of the system is decreased, giving rise to forces of attraction between the atoms; these forces are called chemical bonds.

There are two generalized models of chemical bonds, the ionic bond and the covalent bond; a bond is classified as ionic if it is reasonably similar to the ionic model. If it is reasonably similar to the covalent model, it is classified as a covalent bond.

Ionic Bonds

When an electropositive element reacts with an electronegative element by the lending and borrowing of electrons, ionic bonds are formed. The element which lent (donated) the electrons is left with a positive charge (a deficiency of electrons) and the element which borrowed (accepted) the electrons has a negative charge (excess of electrons). The result of the transfer of the electrons results in the formation of oppositely charged particles (ions) which are held together by their mutual electrostatic attractions. Ionic compounds generally have high melting points, exist in

TABLE 22-1
Periodic Table of the Elements

PERIODIC TABLE OF THE ELEMENTS

KEY

Oxidation States	+1 +3
Atomic Number	79
Symbol	Au
Name	Gold
Atomic Weight	196.9665
Electron Configuration	-32-18-1

Filled Shells	I	II	Transition Elements										III	IV	V	VI	VII	0
	-1 1 H Hydrogen 1.0079 1																	0 2 He Helium 4.00260 2
	+1 3 Li Lithium 6.941 2-1	+2 4 Be Beryllium 9.01218 2-2											+3 5 B Boron 10.81 2-3	+2 ±4 6 C Carbon 12.011 2-4	±1 ±2 ±3 +4 +5 7 N Nitrogen 14.0067 2-5	-2 8 O Oxygen 15.9994 2-6	-1 9 F Fluorine 18.99840 2-7	0 10 Ne Neon 20.179 2-8
	+1 11 Na Sodium 22.98977 2-8-1	+2 12 Mg Magnesium 24.305 2-8-2											+3 13 Al Aluminum 26.98154 2-8-3	+2 ±4 14 Si Silicon 28.086 2-8-4	±3 +5 15 P Phosphorus 30.97376 2-8-5	+4 +6 -2 16 S Sulfur 32.06 2-8-6	±1 +5 +7 17 Cl Chlorine 35.453 2-8-7	0 18 Ar Argon 39.948 2-8-8
2	+1 19 K Potassium 39.098 -8-8-1	+2 20 Ca Calcium 40.08 -8-8-2	+3 21 Sc Scandium 44.9559 -8-9-2	+2 +3 +4 22 Ti Titanium 47.90 -8-10-2	+2 +3 +4 +5 23 V Vanadium 50.9414 -8-11-2	+2 +3 +6 24 Cr Chromium 51.996 -8-13-1	+2 +3 +4 +7 25 Mn Manganese 54.9380 -8-13-2	+2 +3 26 Fe Iron 55.847 -8-14-2	+2 +3 27 Co Cobalt 58.9332 -8-15-2	+2 +3 28 Ni Nickel 58.70 -8-16-2	+1 +2 29 Cu Copper 63.546 -8-18-1	+2 30 Zn Zinc 65.38 -8-18-2	+3 31 Ga Gallium 69.72 -8-18-3	+2 +4 32 Ge Germanium 72.59 -8-18-4	±3 +5 33 As Arsenic 74.9216 -8-18-5	+4 +6 -2 34 Se Selenium 78.96 -8-18-6	±1 +5 35 Br Bromine 79.904 -8-18-7	0 36 Kr Krypton 83.80 -8-18-8
2-8	+1 37 Rb Rubidium 85.4678 -18-8-1	+2 38 Sr Strontium 87.62 -18-8-2	+3 39 Y Yttrium 88.9059 -18-9-2	+4 40 Zr Zirconium 91.22 -18-10-2	+3 +5 41 Nb Niobium 92.9064 -18-12-1	+6 42 Mo Molybdenum 95.94 -18-13-1	+4 +6 +7 43 Tc Technetium (97) -18-13-2	+3 44 Ru Ruthenium 101.07 -18-15-1	+3 45 Rh Rhodium 102.9055 -18-16-1	+2 +4 46 Pd Palladium 106.4 -18-18-0	+1 47 Ag Silver 107.868 -18-18-1	+2 48 Cd Cadmium 112.40 -18-18-2	+3 49 In Indium 114.82 -18-18-3	+2 +4 50 Sn Tin 118.69 -18-18-4	±3 +5 51 Sb Antimony 121.75 -18-18-5	+4 +6 -2 52 Te Tellurium 127.60 -18-18-6	±1 +5 +7 53 I Iodine 126.9045 -18-18-7	0 54 Xe Xenon 131.30 -18-18-8
2-8-18	+1 55 Cs Cesium 132.9054 -18-8-1	+2 56 Ba Barium 137.34 -18-8-2	57-71 See Lantha-nides	+4 72 Hf Hafnium 178.49 -32-10-2	+5 73 Ta Tantalum 180.9479 -32-11-2	+6 74 W Tungsten 183.85 -32-12-2	+4 +6 +7 75 Re Rhenium 186.207 -32-13-2	+3 +4 76 Os Osmium 190.2 -32-14-2	+3 +4 77 Ir Iridium 192.22 -32-15-2	+2 +4 78 Pt Platinum 195.09 -32-17-1	+1 +3 79 Au Gold 196.9665 -32-18-1	+1 +2 80 Hg Mercury 200.59 -32-18-2	+1 +3 81 Tl Thallium 204.37 -32-18-3	+2 +4 82 Pb Lead 207.2 -32-18-4	+3 +5 83 Bi Bismuth 208.9804 -32-18-5	+2 +4 84 Po Polonium (209) -32-18-6	±1 +5 +7 85 At Astatine (210) -32-18-7	0 86 Rn Radon (222) -32-18-8
2-8-18-32	+1 87 Fr Francium (223) -18-8-1	+2 88 Ra Radium 226.0254 -18-8-2	89-103 See Acti-nides	+4 104 Rf-Ku (Rutherfordium; Kurchatovium) (261) -32-10-2	105 Ha Hahnium (262) -32-11-2	106 (263) -32-12-2	107 (262) -32-13-2	108 (265) -32-14-2	109 (266) -32-15-2	(110)	(111)	(112)	(113)	(114)	(115)	(116)	(117)	(118)

Lanthanides	+3 57 La Lanthanum 138.9055 -18-9-2	+3 +4 58 Ce Cerium 140.12 -19-9-2	+3 59 Pr Praseodymium 140.9077 -21-8-2	+3 60 Nd Neodymium 144.24 -22-8-2	+3 61 Pm Promethium (145) -23-8-2	+2 +3 62 Sm Samarium 150.4 -24-8-2	+2 +3 63 Eu Europium 151.96 -25-8-2	+3 64 Gd Gadolinium 157.25 -25-9-2	+3 65 Tb Terbium 158.9254 -26-9-2	+3 66 Dy Dysprosium 162.50 -28-8-2	+3 67 Ho Holmium 164.9304 -29-8-2	+3 68 Er Erbium 167.26 -30-8-2	+3 69 Tm Thulium 168.9342 -31-8-2	+2 +3 70 Yb Ytterbium 173.04 -32-8-2	+3 71 Lu Lutetium 174.97 -32-9-2
Actinides	+3 89 Ac Actinium (227) -18-9-2	+4 90 Th Thorium 232.0381 -18-10-2	+4 +5 91 Pa Protactinium 231.0359 -20-9-2	+3 +4 +5 +6 92 U Uranium 238.029 -21-9-2	+3 +4 +5 +6 93 Np Neptunium 237.0482 -22-9-2	+3 +4 +5 +6 94 Pu Plutonium (244) -24-8-2	+3 +4 +5 +6 95 Am Americium (243) -25-8-2	+3 96 Cm Curium (247) -25-9-2	+3 +4 97 Bk Berkelium (247) -27-8-2	+3 98 Cf Californium (251) -28-8-2	+3 99 Es Einsteinium (252) -29-8-2	+3 100 Fm Fermium (257) -30-8-2	+2 +3 101 Md Mendelevium (258) -31-8-2	+2 +3 102 No Nobelium (259) -32-8-2	+3 103 Lr Lawrencium (262) -32-9-2

Note:
Atomic weights are those of the most commonly available long-lived isotopes on the 1973 IUPAC Atomic Weights of the Elements. A value given in parentheses denotes the mass number of the longest-lived isotope.

TABLE 22-2
Atomic Masses of Elements—Based on ^{12}C = 12

	Symbol	*Atomic no.*	*Atomic mass*		*Symbol*	*Atomic no.*	*Atomic mass*
Actinium	Ac	89	[227]	Erbium	Er	68	167.26
Aluminum	Al	13	26.98	Europium	Eu	63	151.96
Americium	Am	95	[243]	Fermium	Fm	100	[253]
Antimony	Sb	51	121.75	Fluorine	F	9	19.00
Argon	Ar	18	39.95	Francium	Fr	87	[223]
Arsenic	As	33	74.92	Gadolinium	Gd	64	157.25
Astatine	At	85	[210]	Gallium	Ga	31	69.72
Barium	Ba	56	137.34	Germanium	Ge	32	72.59
Berkelium	Bk	97	[249]	Gold	Au	79	196.97
Beryllium	Be	4	9.01	Hafnium	Hf	72	178.49
Bismuth	Bi	83	208.98	Helium	He	2	4.00
Boron	B	5	10.81	Holmium	Ho	67	164.93
Bromine	Br	35	79.91	Hydrogen	H	1	1.01
Cadmium	Cd	48	112.40	Indium	In	49	114.82
Calcium	Ca	20	40.08	Iodine	I	53	126.90
Californium	Cf	98	249	Iridium	Ir	77	192.2
Carbon	C	6	12.01	Iron	Fe	26	55.85
Cerium	Ce	58	140.12	Krypton	Kr	36	83.80
Cesium	Cs	55	132.91	Kurchatovium*	Ku	104	257
Chlorine	Cl	17	35.45	Lanthanum	La	57	138.91
Chromium	Cr	24	52.00	Lawrencium	Lr	103	257
Cobalt	Co	27	58.93	Lead	Pb	82	207.19
Copper	Cu	29	63.54	Lithium	Li	3	6.94
Curium	Cm	96	[247]	Lutetium	Lu	71	174.97
Dysprosium	Dy	66	162.50	Magnesium	Mg	12	24.31
Einsteinium	Es	99	[254]	Manganese	Mn	25	54.94

* This name is not accepted in the United States; in this country the accepted name is rutherfordium. Elements 105, 106, 108, and 109 have also been discovered, but no names or symbols have been officially assigned to them.

	Symbol	Atomic no.	Atomic mass		Symbol	Atomic no.	Atomic mass
Mendelevium	Md	101	[256]	Ruthenium	Ru	44	101.07
Mercury	Hg	80	200.59	Samarium	Sm	62	150.35
Molybdenum	Mo	42	95.94	Scandium	Sc	21	44.96
Neodymium	Nd	60	144.24	Selenium	Se	34	78.96
Neon	Ne	10	20.18	Silicon	Si	14	28.09
Neptunium	Np	93	[237]	Silver	Ag	47	107.87
Nickel	Ni	28	58.71	Sodium	Na	11	22.99
Niobium	Nb	41	92.91	Strontium	Sr	38	87.62
Nitrogen	N	7	14.01	Sulfur	S	16	32.06
Nobelium	No	102	[256]	Tantalum	Ta	73	180.95
Osmium	Os	76	190.2	Technetium	Tc	43	[99]
Oxygen	O	8	16.00	Tellurium	Te	52	127.60
Palladium	Pd	46	106.4	Terbium	Tb	65	158.92
Phosphorus	P	15	30.97	Thallium	Tl	81	204.37
Platinum	Pt	78	195.09	Thorium	Th	90	232.04
Plutonium	Pu	94	[242]	Thulium	Tm	69	168.93
Polonium	Po	84	[210]	Tin	Sn	50	118.69
Potassium	K	19	39.10	Titanium	Ti	22	47.90
Praseodymium	Pr	59	140.91	Tungsten	W	74	183.85
Promethium	Pm	61	[147]	Uranium	U	92	238.03
Protactinium	Pa	91	[231]	Vanadium	V	23	50.94
Radium	Ra	88	[226]	Xenon	Xe	54	131.30
Radon	Rn	86	[222]	Ytterbium	Yb	70	173.04
Rhenium	Re	75	186.2	Yttrium	Y	39	88.91
Rhodium	Rh	45	102.91	Zinc	Zn	30	65.37
Rubidium	Rb	37	85.47	Zirconium	Zr	40	91.22

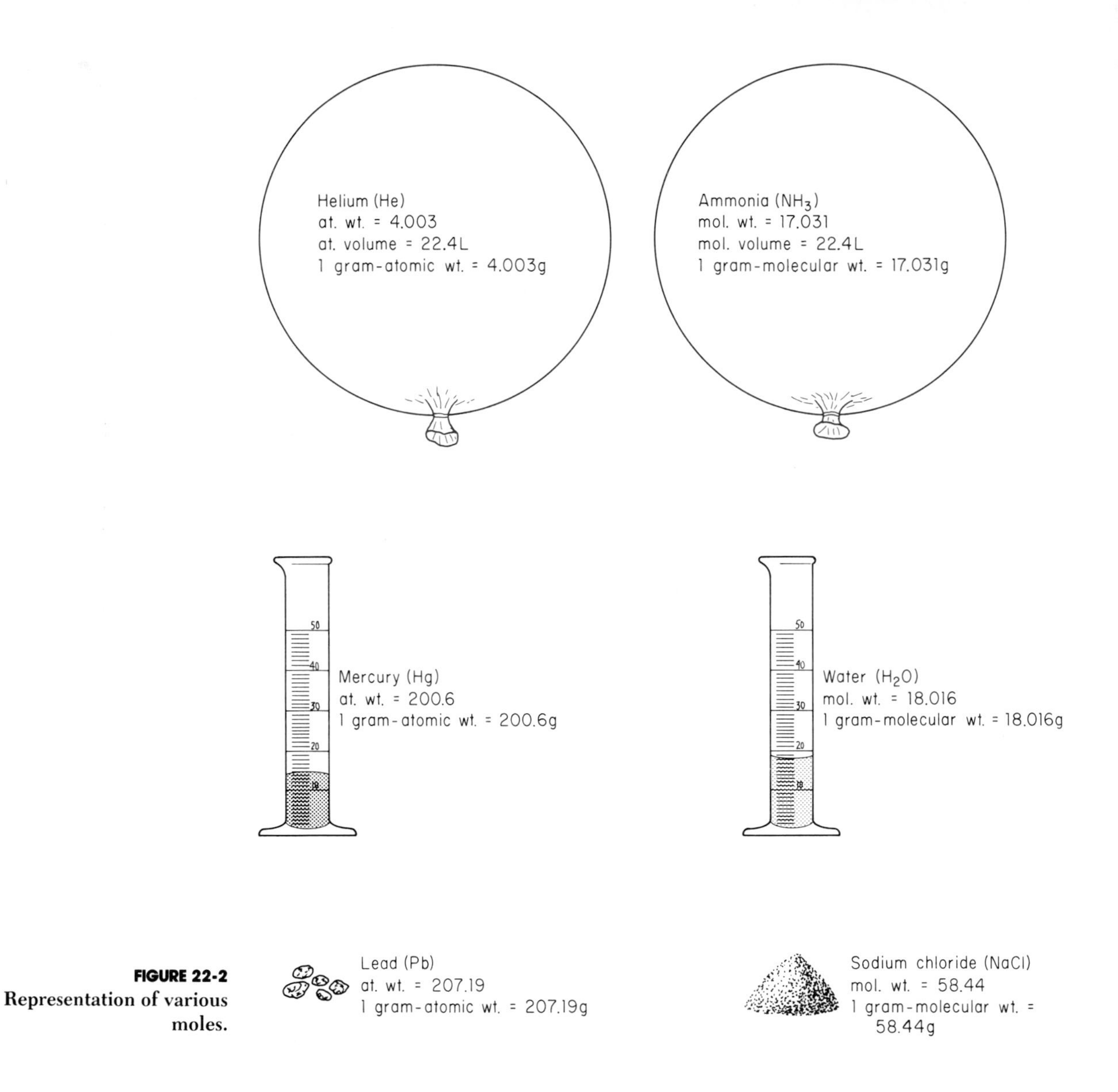

FIGURE 22-2 Representation of various moles.

crystalline form, and are, in general, the compounds normally classified as inorganic compounds.

Covalent Bonds

When elements react with each other by sharing electrons, structural units (molecules) are formed which are termed covalent. The nucleus of each atom which shares electrons reaps the benefit of those electrons. No electrons have been transferred; there are no excesses or deficiencies of electrons, and the result is an electrically neutral molecule, generally composed of nonmetallic elements. The organic compounds, whose basic atoms are the carbon and hydrogen atoms, are covalently formed molecules. These compounds decompose on application of heat, have relatively low melting and boiling points, and are the substances of which all life forms are composed.

Valence of Ions

The combining capacity or *valence* of various ions is given in Table 22-3.

Radicals

Sometimes a group of two or more atoms act as a unit in a reaction. These polyatomic charged particles are called *radicals.* They are treated as a unit and are named as a unit. Examples are the sulfate radical SO_4^{2-} in H_2SO_4, the nitrate radical NO_3^- in HNO_3, and the ammonium radical NH_4^+ in ammonium hydroxide, NH_4OH.

COMPOUNDS

Compounds are made of two or more identical or different atoms. All atoms have definite atomic masses; therefore, all compounds have definite masses because of the masses of the atoms. Every pure compound has a definite number of atoms arranged in a definite structure; the mathematical representation of this is called the formula.

Empirical Formula

An empirical formula expresses the relative number of atoms of the different elements in a compound *expressed with the smallest possible set of integers.*

EXAMPLE Benzene, C_6H_6. A molecule of benzene contains 6 atoms of carbon and 6 atoms of hydrogen. The ratio of the elements carbon to hydrogen is therefore 1 : 1, and the *empirical formula is* CH.

Molecular Formula

A molecular formula not only indicates the *relative* number of atoms of each element in a molecule, but also gives the *actual* number of atoms of each element in one molecule of the compound.

Arithmetical Meaning of a Molecular Formula

A formula represents one unit of a compound. Ethane gas has the molecular formula C_2H_6, which can denote the following:

1. 1 molecule of ethane, C_2H_6
2. 1 mole of C_2H_6 molecules, which is equal to 6.02×10^{23} molecules

TABLE 22-3
Valences of Ions

Name	*Formula*	*Valence*
Positive ions:		
Aluminum	Al^{3+}	+3
Ammonium	NH_4^+	+1
Antimony	Sb^{3+}	+3
Arsenic(V) (arsenic)	As^{5+}	+5
Arsenic(III) (arsenious)	As^{3+}	+3
Barium	Ba^{2+}	+2
Bismuth	Bi^{3+}	+3
Cadmium	Cd^{2+}	+2
Calcium	Ca^{2+}	+2
Chromium(III) (chromic)	Cr^{3+}	+3
Chromium(II) (chromous)	Cr^{2+}	+2
Cobalt(III) (cobaltic)	Co^{3+}	+3
Cobalt(II) (cobaltous)	Co^{2+}	+2
Copper(II) (cupric)	Cu^{2+}	+2
Copper(I) (cuprous)	Cu^{+}	+1
Hydrogen	H^{+}	+1
Iron(III) (ferric)	Fe^{3+}	+3
Iron(II) (ferrous)	Fe^{2+}	+2
Lead(IV) (plumbic)	Pb^{4+}	+4
Lead(II) (plumbous)	Pb^{2+}	+2
Lithium	Li^{+}	+1
Magnesium	Mg^{2+}	+2
Manganese(III) (manganic)	Mn^{3+}	+3
Manganese(II) (manganous)	Mn^{2+}	+2
Mercury(II) (mercuric)	Hg^{2+}	+2
Mercury(I) (mercurous)	Hg^{+}	+1
Nickel(III) (nickelic)	Ni^{3+}	+4
Nickel(II) (nickelous)	Ni^{2+}	+2
Potassium	K^{+}	+1
Silver	Ag^{+}	+1
Sodium	Na^{+}	+1
Tin(IV) (stannic)	Sn^{4+}	+4
Tin(II) (stannous)	Sn^{2+}	+2
Strontium	Sr^{2+}	+2
Zinc	Zn^{2+}	+2

Name	*Formula*	*Valence*
Negative ions:		
Acetate	$C_2H_3O_2^-$	−1
Bicarbonate	HCO_3^-	−1
Binoxalate	$HC_2O_4^-$	−1
Bisulfate	HSO_4^-	−1
Bisulfide	HS^-	−1
Bisulfite	HSO_3^-	−1
Bromate	BrO_3^-	−1
Bromide	Br^-	−1
Carbonate	CO_3^{2-}	−2
Chlorate	ClO_3^-	−1
Chloride	Cl^-	−1
Chlorite	ClO_2^-	−1
Chromate	CrO_4^{2-}	−2
Dichromate	$Cr_2O_7^{2-}$	−2
Ferricyanide	$Fe(CN)_6^{3-}$	−3
Ferrocyanide	$Fe(CN)_6^{4-}$	−4
Fluoride	F^-	−1
Hydroxide	OH^-	−1
Hypochlorite	ClO^-	−1
Iodide	I^-	−1
Nitrate	NO_3^-	−1
Nitrite	NO_2^-	−1
Oxalate	$C_2O_4^{2-}$	−2
Perchlorate	ClO_4^-	−1
Phosphate	PO_4^{3-}	−3
Phosphate (meta)	PO_3^-	−1
Phosphite	HPO_3^{2-}	−2
Sulfate	SO_4^{2-}	−2
Sulfide	S^{2-}	−2
Sulfite	SO_3^{2-}	−2
Thiosulfate	$S_2O_3^{2-}$	−2

3. 6.02×10^{23} molecules of C_2H_6 (Avogadro's number of molecules)

4. $8 \times 6.02 \times 10^{23}$ atoms (8 atoms make up one molecule of C_2H_6)

5. 22.4 L of C_2H_6 because 1 mole of any gas occupies 22.4 L at standard temperature and pressure (0°C and 760 mmHg pressure)

6. 30.06 g of C_2H_6 (1 molecule of C_2H_6 consists of 2 atoms of carbon and 6 atoms of hydrogen, and can be considered to contain 2 moles of carbon atoms and 6 moles of hydrogen atoms)

1 mole of carbon atoms	=	12.00 g	
2 moles of carbon atoms	=	2 × 12.00 g	= 24.00 g
1 mole of hydrogen atoms	=	1.01 g	
6 moles of hydrogen atoms	=	6 × 1.01 g	= 6.06 g
Total for each mole of C_2H_6			= 30.06 g

Calculating Molecular Masses*

EXAMPLE Calculate the molecular mass of sulfuric acid, H_2SO_4. The molecule contains 2 hydrogen atoms, 1 sulfur atom, and 4 oxygen atoms.

Solution The molecular formula indicates the actual number of atoms of each element:

1. Multiply the number of atoms of each element by the atomic mass.

2. Add the weights of all the atoms to get the molecular mass.

2 hydrogen atoms	2 × 1.008 =	2.016	(2 × atomic mass hydrogen)
1 sulfur atom	1 × 32.064 =	32.064	(1 × atomic mass sulfur)
4 oxygen atoms	4 × 16.000 =	64.000	(4 × atomic mass oxygen)
		98.08	(molecular mass H_2SO_4)

Molecular Masses of Compounds Containing Radicals

When a radical group is present in a compound it is treated as a unit. When more than one of these radicals is a part of the formula, the radical is enclosed in parentheses, and the number of radicals is designated by a subscript to the right of the parentheses.

EXAMPLES $Ca_3(PO_4)_2$ contains 2 phosphate radicals.

$Al(OH)_3$ contains 3 hydroxide radicals.

* "Molecular mass" was called "molecular weight" in Chap. 19 because the determinations involved a gravity factor.

$(NH_4)_3PO_4$ contains 3 ammonium radicals.

$Fe_2(SO_4)_3$ contains 3 sulfate radicals.

$Al(NO_3)_3$ contains 3 nitrate radicals.

Name of compound	*Skeletal formula*	*Correct formula*	*+ charges = − charges*
Ferric sulfate	Fe^{3+} SO_4^{2-}	$Fe_2(SO_4)_3$	$2 \times 3+ = 3 \times 2-$
Aluminum nitrate	Al^{3+} NO_3^-	$Al(NO_3)_3$	$1 \times 3+ = 3 \times 1-$
Ammonium phosphate	NH_4^+ PO_4^{3-}	$(NH_4)_3PO_4$	$3 \times 1+ = 1 \times 3-$

To calculate the molecular mass of such compounds, the mass of each and every atom in the molecule must be accounted for. There are two methods for doing this:

Method 1

Calculate the molecular mass of the whole radical. Then multiply by the number of radicals present in the formula. Then add the masses of the other atoms.

EXAMPLE To calculate the molecular mass of $Ca_3(PO_4)_2$ by this method:

1. Calculate the molecular mass of PO_4^{3-}:

1 phosphorus atom = $1 \times 30.97 = 30.97$
4 oxygen atoms = $4 \times 16.00 = 64.00$
94.97

2. Since there are two radicals, multiply by 2:

$2 \times 94.97 = 189.94$

3. Then add the atomic masses of the three calcium atoms:

3×40.08 120.24
189.94
310.18 = molecular mass of $Ca_3(PO_4)_2$

Method 2

Determine the total number of each kind of atom in the molecule, then multiply the number of each one by its atomic mass. To determine the number of atoms in the radical part of the molecule, multiply the *subscript of the entire radical by the subscript of each atom in the radical.* If there is no subscript, the number is 1.

EXAMPLE To calculate the molecular mass of $Ca_3(PO_4)_2$ by this method:

1. Determine the total number of each kind of atom:

3 calcium atoms

2 phosphorus atoms

2 × 4 or 8 oxygen atoms

2. Multiply the number of each atom by its atomic mass:

3 Ca atoms = 3 × 40.08 = 120.24
2 P atoms = 2 × 30.97 = 61.94
8 O atoms = 8 × 16.00 = 128.00

310.18 = molecular mass of $Ca_3(PO_4)_2$

NOMENCLATURE OF INORGANIC COMPOUNDS

Binary Acids

Hydrogen reacts with the elements of groups VIA and VIIA to form binary compounds called acids, which dissolve in water to yield both positive hydrogen ions and negative nonmetallic ions.

Aqueous solutions of HF, HCl, HBr, HI, H_2S, and H_2Se are acid solutions. These binary acids are distinguished by the prefix "hydro-" and the ending "-ic":

HCN(aq) HYDROcyanIC acid

HCl(aq) HYDROchlorIC acid

HI(aq) HYDROiodIC acid

H_2S(aq) HYDROsulfurIC acid

Salts

Salts are ionic-type compounds. They consist of an electropositive radical or element (excluding hydrogen) combined with an electronegative radical or element (excluding the hydroxyl ion). They may be formed by any of the representative following reactions (the salt is underlined):

1. Base + acid

$NaOH + HCl \rightarrow \underline{NaCl} + H_2O$

2. Basic oxide + acid

$Na_2O + 2HCl \rightarrow \underline{2NaCl} + H_2O$

3. Metal + nonmetal

$2Na + Cl_2 \rightarrow \underline{2NaCl}$

4. Metal + acid

$2Na + 2HCl \rightarrow \underline{2NaCl} + Cl_2$

5. Basic oxide + acid nonmetal hydroxide

$Na_2O + 2HOCl \rightarrow \underline{2NaOCl} + H_2O$

6. Base + acid oxide

$NaOH + CO_2 \rightarrow \underline{NaHCO_3}$

7. Basic oxide + acid oxide

$Na_2O + CO_2 \rightarrow \underline{Na_2CO_3}$

To name a salt, name the metal, abbreviate the nonmetal, and end the name with "-ide":

BaF_2 barium fluorIDE

NaCl sodium chlorIDE

MgI_2 magnesium iodIDE

K_2S potassium sulfIDE

The prefixes "mono-," "di-," "tri-," etc., are not used with metals in groups IA and IIA because they do not show variable valence (combining power with other elements or groups of elements) (see Table 22-2).

Metals with Variable Valence

Various metal salts can be distinguished by one of the following systems:

1. The *Stock system* (IUPAC system), which designates the oxidation state of the metal with a roman numeral.

$FeBr_2$ iron(II) bromide

$FeBr_3$ iron(III) bromide

HgCl mercury(I) chloride

$HgCl_2$ mercury(II) chloride

VCl_2 vanadium(II) chloride

VCl_3 vanadium(III) chloride

VCl_4 vanadium(IV) chloride

VCl_5 vanadium(V) chloride

2. The *"ous-ic"* system. In this nearly obsolete system, the ending "-ous" denotes the lower oxidation state (valence or combining power) of the metal, "-ic" the higher oxidation state (valence or combining power of the element):

HgO mercurIC oxIDE [one mercury (combining power of 2) requiring one oxygen (combining power of 2)]

Hg_2O mercurOUS oxIDE [two mercuries (combining power of 1) requiring one oxygen (combining power of 2)]

$FeCl_3$ ferrIC chlorIDE [one iron (combining power of 3) requiring 3 chlorines (combining power of 1)]

$FeCl_2$ ferrOUS chlorIDE [one iron (combining power of 2) requiring 2 chlorines (combining power of 1)]

Covalent Compounds

Covalent compounds involve two nonmetallic elements. To name these compounds, use the appropriate Greek prefixes "mono-," "di-," "tri-," "tetra-," "penta-," etc., to indicate the number of atoms of each element involved and use the ending "-ide."

$H_2S(g)$ hydrogen sulfIDE

CO_2 carbon DIoxIDE

$HCl(g)$ hydrogen chlorIDE

PCl_3 phosphorus TRIchlorIDE

Cl_2O DIchlorine MONoxIDE

N_2O_4 DInitrogen TETRoxIDE

Compounds Containing Radicals

When a polyatomic particle, a radical, is present in a compound, it is named as a unit.

$Fe_2\underline{(SO_4)_3}$ ferric sulfate

$Al\underline{(NO_3)_3}$ aluminum nitrate

$\underline{(NH_4)_3}\underline{PO_4}$ ammonium phosphate

$Ca_3\underline{(PO_4)_2}$ calcium phosphate

$Al\underline{(OH)_3}$ aluminum hydroxide

Use the ending "-ide" for compounds containing hydroxyl, ammonium, or cyanide radicals.

$NH_4\underline{CN}$ ammonium cyanIDE

$NH_4\underline{Cl}$ ammonium chlorIDE

$Na\underline{OH}$ sodium hydroxIDE

$Ca\underline{(OH)_2}$ calcium hydroxIDE

$K\underline{CN}$ potassium cyanIDE

$Fe\underline{(OH)_2}$ ferric hydroxIDE

Common Names or Trivial Names

The common or trivial names for some ordinary compounds are:

H_2O	water
NH_3	ammonia
PH_3	phosphine
$Ca(OH)_2$	lime
Fe_3O_4	iron rust
HCl	muriatic acid
H_2SO_4	battery acid
$NaOH$	caustic soda
KOH	caustic potash
$K(SbO)C_4H_4O_6 \cdot \frac{1}{2}H_2O$	tartar emetic
$Pb(C_2H_3O_2)_2 \cdot 3H_2O$	sugar of lead
Hg_2Cl_2	calomel
$HgCl_2$	corrosive sublimate
$NaNH_4HPO_4 \cdot 4H_2O$	microcosmic salt
$Na_2B_4O_7 \cdot 10H_2O$	borax
$Na_2CO_3 \cdot 10H_2O$	washing soda
$NaHCO_3$	baking soda
$Na_2S_2O_3 \cdot 5H_2O$	hypo

Various groups in the periodic chart are often referred to by common names:

Group	*Members of group*	*Common name for group*
Group IA	Li, Na, K, Rb, Cs	Alkali metals
Group IIA	Mg, Ca, Sr, Ba	Alkaline earth metals
Group 0	He, Ne, Kr, Ar, Xe, Rn	Rare gases, inert elements, or noble gases
Heaviest members of Groups VIII and IB	Pt, Au, Ir, Rh, Pd	Noble metals
Group VIIA	F, Cl, Br, I	Halogens
	Salts or ions of the halogens	Halides

Peroxides

Metals containing two oxygen atoms with a single covalent bond between them (binding force which prevents the separation of the atoms) are named *peroxides.* Normal dioxides, such as MnO_2, TiO_2, and SiO_2, are not peroxides.

Oxygen, as O_2^{2-}, is considered to be a radical liberating oxygen, and it has a valence of -2. Peroxides decompose to form normal oxides:

$$2H_2O_2 \rightarrow 2H_2O + O_2$$

The prefix "per-" combined with the name of an oxide indicates that there is additional oxygen in the molecule and that it is therefore a peroxide.

Ternary Acids and Salts

Oxy acids containing the nonmetal in different oxidation states and their respective salts are named as follows:

Oxidation state of Cl	*Acid*	*Name*
+1	$HClO$	HYPOchlorOUS acid
+3	$HClO_2$	chlorOUS acid
+5	$HClO_3$	chlorIC acid
+7	$HClO_4$	PERchlorIC acid
Oxidation state of Cl	*Salt*	*Name*
+1	$NaClO$	sodium HYPOchlorITE
+3	$NaClO_2$	sodium chlorITE
+5	$NaClO_3$	sodium chlorATE
+7	$NaClO_4$	sodium PERchlorATE

When acids end in "-ous," the corresponding salt ends in "-ite." For acids that end in "-ic," the salt ends in "-ate." The prefix "hypo-," Greek for *under,* denotes the lowest oxidation state of the nonmetal. The prefix "per-," from the Greek *hyper,* denotes the highest oxidation state of the nonmetal. The "-ous" acids are usually two oxidation states lower than the "-ic" acids. The oxidation state measures the relative combining force of the elements. Formulas of "-ous" or "-ic" acids are:

-ous acid	*Oxidation state*	*-ic acid*	*Oxidation state*
Sulfurous acid H_2SO_3	+4	Sulfuric acid H_2SO_4	+6
Nitrous acid HNO_2	+3	Nitric acid HNO_3	+5

Polyprotic Acids

Polyprotic acids contain more than one replaceable hydrogen. Anhydrides react with water to produce the "ortho-" acids on hydration. The "ortho-" acids form "pyro-" or "meta-" acids by dehydration (removal of water).

$P_4O_{10} + 6H_2O \rightarrow 4H_3PO_4$ ORTHOphosphoric acid (hydration of anhydride)

$2H_3PO_4 \rightarrow H_2O + H_4P_2O_7$ PYROphosphoric acid [dehydration of ortho acid (2 molecules)]

$H_3PO_4 \rightarrow H_2O + HPO_3$ METAphosphoric acid [dehydration of ortho acid (1 molecule)]

"Ortho-" refers to the normal acid. "Meta-" refers to the normal acid *less* 1 molecule of water. "Pyro-" refers to the acid formed by removing 1 molecule of water from 2 molecules of the "ortho-" acid. (See above.)

Formulas of the acids of phosphorus are:

ORTHOphosphoric acid = H_3PO_4

METAphosphoric acid = HPO_3

PYROphosphoric acid = H_4PO_7

ORTHOphosphorous acid = H_3PO_3

METAphosphorous acid = HPO_2

PYROphosphorous acid = $H_4P_2O_5$

Salts of Polyprotic Acids

Salts of polyprotic acids such as H_2SO_4 or H_3PO_4 are formed when one or more of the hydrogen ions are replaced by metal ions (charged particles). The first name given is usually preferred:

$NaHCO_3$	sodium bicarbonate, sodium acid carbonate, sodium hydrogen carbonate
Na_2SO_4	sodium sulfate, normal sodium sulfate
NaH_2PO_4	monosodium phosphate, primary sodium phosphate
Na_2HPO_4	disodium phosphate, secondary sodium phosphate
Na_3PO_4	trisodium phosphate, tertiary sodium phosphate

Summary

See Table 22-4 for a summary of the nomenclature rules.

FORMATION OF COORDINATION COMPOUNDS AND COMPLEX IONS

When metal ions are bonded by attraction to oppositely charged ions, the bonding is electrostatic. Metal ions can also bond covalently, when the metal ion acts as an electron-pair acceptor. The donor, called the ligand, must therefore have one pair of unshared electrons available. The donor can be a molecule or an ion. If it is a molecule, the resulting complex is called a *coordination compound.* Since the donor molecule is electrically neutral, the resulting complex has the same valence as the metal ion.

EXAMPLES

$$Cu^{2+} + 4NH_3 \rightarrow Cu(NH_3)_4^{2+}$$

$$Co^{3+} + 6NH_3 \rightarrow CO(NH_3)_6^{3+}$$

$$Ag^{+} + 2NH_3 \rightarrow Ag(NH_3)_2^{+}$$

If an ion is the donor, the resulting complex is called a *complex ion,* and the resulting valence of the complex is the algebraic sum of the charges of the component ions:

$$Ag^{+} + 2CN^{-} \rightarrow Ag(CN)_2^{-}$$
$$Fe^{3+} + 6CN^{-} \rightarrow Fe(CN)_6^{3-}$$

The formation of complex ions and coordination compounds is extremely important in analytical procedures and in medicine. Metal ions which react in certain ways no longer do so when they have been complexed. The chemical properties of the complex ion are different from those of the metal ion, and the number of molecules or ions which are grouped around the central metal ion is called the *coordination number.* The most common coordination numbers found in coordination compounds are 2, 4, and 6.

TABLE 22-4
Summary: How to Write Chemical Formulas and How to Name Compounds

	Compounds	Type of bonding	Naming rules	Typical compounds — Names	Typical compounds — Formulas
Binary and ternary compounds ending in "-ide"	Metals + nonmetals	Ionic	Name positive metal ion then negative non-metal ion	Potassium chloride	KCl
	Metals + nonmetals, variable valence	Ionic	Name metal; use Stock (Roman numeral) method or use -ic or -ous ending*	Iron(II) chloride or ferrous chloride	$FeCl_2$
	OH^-; NH_4^+; CN^-; CO_3^{2-}	Ionic generally	Include name of radical as part of the compound	Sodium hydroxide Ammonium chloride Sodium cyanide Potassium carbonate	NaOH NH_4Cl NaCN K_2CO_3
Ternary compounds, ionic	Compounds of different nonmetallic elements	Covalent	Individualized names; prefixes used	Carbon monoxide Carbon dioxide Dinitrogen pentoxide	CO CO_2 N_2O_5
	Compounds ending in -ate	Ionic	Name ions; -ate ions have 3 to 4 oxygen atoms	Sodium nitrate Sodium phosphate Sodium chlorate	$NaNO_3$ Na_3PO_4 $NaClO_3$
	Compounds ending in -ite	Ionic	Name ions; -ite ions have 1 less oxygen atom than the -ate ion	Sodium nitrite Sodium phosphite Sodium chlorite	$NaNO_2$ Na_2HPO_3 $NaClO_2$

* The -ous and -ic endings are found in older literature.

AMPHOTERISM

Hydroxides of alkali metals are basic, for example, NaOH and KOH. However, hydroxides of nonmetals are acidic; for example, $P(OH)_3$ is acidic and is usually written as H_3PO_3. Some of the transition elements can act as either acids or bases, depending upon the environment to which they are subjected. In the presence of strong acids, they act as bases; in the presence of strong bases they act as acids. The solubility of the transition hydroxides in strong bases is due to the formation of a complex ion:

$$Al(OH)_3 + 3NaOH \rightarrow 3Na^+ + AlO_3^{3-}$$

Other transition-element amphoteric hydroxides are $Cr(OH)_3$, $Sn(OH)_2$, and $Zn(OH)_2$.

MOLECULAR RELATIONSHIPS FROM EQUATIONS

Equations are used to represent what is reacting and what is being formed in a chemical reaction. The relative numbers of reactant particles and resulting particles are indicated by the coefficients of the formulas which represent these particles. The particles may be atoms, molecules, or ions.

EXAMPLE Ammonia burns in oxygen to yield nitrogen and water.

Balanced chemical equation:

$$4NH_3 + 3O_2 \rightarrow 2N_2 + 6H_2O$$

Analysis: 4 molecules of ammonia react with 3 molecules of oxygen to give 2 molecules of nitrogen and 6 molecules of water.

When ammonia reacts with oxygen as in this equation, 3 molecules of oxygen are *always* consumed for every 4 molecules of NH_3 and *always* yield 2 molecules of nitrogen and 6 molecules of water.

The coefficients are 4, 3, 2, and 6, and these indicate the relative number of the *reactant and resultant* particles formed.

The *law of conservation of mass* is observed: All atoms are accounted for—every nitrogen, hydrogen, and oxygen atom is counted in the reactants and the products.

The *number of atoms* of any element in a molecule of a substance can be found as follows: Multiply the coefficient of the molecule by the subscript of the elements in the molecule. When there is no subscript for an element, the value for that subscript is always 1.

EXAMPLE

$$4NH_3 + 3O_2 \rightarrow 2N_2 + 6H_2O$$

Reactants:

4×1 nitrogen gives 4 nitrogens

4×3 hydrogens gives 12 hydrogens

3×2 oxygens gives 6 oxygens

Products:

2×2 nitrogens gives 4 nitrogens

6×2 hydrogens gives 12 hydrogens

6×1 oxygen gives 6 oxygens

All reactant atoms are found in the products.

MASS RELATIONSHIPS FROM EQUATIONS

Balanced chemical equations tell what the reactants are and what the products are and indicate the relative masses of the reactants and the products. All atoms and all molecules have definite masses, and the actual masses of the atoms and the molecules are proportional to their atomic and molecular masses.

EXAMPLE Phosphoric acid reacts with sodium hydroxide to yield sodium phosphate and water.

Equation:

H_3PO_4	$+ 3\ NaOH$	$= Na_3PO_4$	$+ 3H_2O$
1 molecule	+ 3 molecules	= 1 molecule	+ 3 molecules
1 molecular mass	+ 3 molecular mass	= 1 molecular mass	+3 molecular mass
98 g	+ 3 × 40 g	= 164 g	+ 3 × 18 g
98 g	+ 120 g	= 164 g	+54 g
	218 g	= 218 g	

The equation tells us: Every molecule of phosphoric acid that reacts consumes 3 molecules of sodium hydroxide, and 1 molecule of sodium phosphate and 3 molecules of water are formed.

For every 98 g of H_3PO_4 by weight, 120 g NaOH are used up, and 164 g of Na_3PO_4 and 54 g H_2O are formed. (Refer to Table 22-2 for atomic masses.)

EXAMPLES Molecular mass of H_3PO_4

3 hydrogens	= 3 × 1.008 =	3.024
1 phosphorus	= 1 × 30.98 =	30.98
4 oxygens	= 4 × 16.00 =	64.00
Molecular mass		98.00

Molecular mass of NaOH:

1 sodium	= 1 × 23.00 =	23.00
1 oxygen	= 1 × 16.00 =	16.00
1 hydrogen	= 1 × 1.008 =	1.008
Molecular mass		40.01

Molecular mass of Na_3PO_4:

3 sodiums	= 3 × 23.00 =	69.00
1 phosphorus	= 1 × 30.98 =	30.98
4 oxygens	= 4 × 16.00 =	64.00
Molecular mass		163.98

Molecular mass of H_2O:

2 hydrogens	= 2 × 1.008 =	2.016
1 oxygen	= 1 × 16.00 =	16.00
Molecular mass		18.02

Total mass of the reactants = total mass of the products

98 g H_3PO_4 + 120 g NaOH = 218 g

164 g Na_3PO_4 + 54 g H_2O = 218 g

This is the *conservation of mass:* In normal chemical reactions:

Mass of the reactants = mass of the products

Mass is neither created nor destroyed.

VOLUME RELATIONSHIPS FROM EQUATIONS

When volatile or gaseous compounds are formed, volume relationships are determined by the coefficients of the reactants and products.

1. Every gram-molecular mass of gas that reacts or is formed occupies 22.4 L at standard conditions of pressure (760 mmHg) and temperature (0°C).

2. For every fraction of a mole of gas that reacts or is formed as in rule 1, that same fraction of 22.4 L of gas is involved.

EXAMPLES

Hydrogen + oxygen → water
$2H_2 + O_2 \rightarrow 2H_2O$

Observations: 2 volumes of hydrogen react with 1 volume of oxygen to give 2 volumes of water vapor.

Zinc + hydrochloric acid → zinc chloride and hydrogen gas
$Zn + 2HCl \rightarrow ZnCl_2 + H_2$

Observations: 1 gram-atomic mass of zinc will yield 22.4 L of hydrogen (1 gram mole); 2 gram moles of HCl were needed. The use of ratio and proportion will provide the answer of reactant or product mass or volume relationships.

23
ORGANIC CHEMISTRY NOMENCLATURE

INTRODUCTION

This chapter is designed to review the rules for naming organic compounds by the International Union of Pure and Applied Chemistry (IUPAC) system. The IUPAC names will be routinely given for each compound, and common names, if used, will be in stated in parentheses. An *organic* compound can be defined as any substance that contains carbons. Carbon has four electrons in its valence shell and needs to share four additional electrons to achieve stability. Therefore, many organic compounds will contain carbon atoms with four single covalent (sharing) bonds. However, carbon atoms are capable of forming double and triple bonds. Molecules having only carbon and hydrogen atoms are known as *hydrocarbons*. Those hydrocarbon molecules having single, double, and triple bonds are known as the *alkane, alkene, and alkyne* families respectively. Most hydrocarbons fall into one of the following categories: (1) compounds with a linear structure and varying numbers of single, double, and triple bonds, called *aliphatic hydrocarbons*; (2) *cyclic hydrocarbons* that form ring structures with primarily single bonds; (3) hydrocarbons called *aromatic hydrocarbons* which have alternating single and double bonds in a circular structure. This relationship of the hydrocarbons is outlined in Fig. 23-1.

ALKANES

The simplest hydrocarbon molecules, known as *alkanes*, have carbon and hydrogen atoms bonded to each other by only single bonds. A hydrocarbon containing all single bonds between the carbon and hydrogen atoms is referred to as *saturated* (meaning all four possible bonding sites are filled by hydrogen atoms). The general formula for an alkane is C_nH_{2n+2} where n represents the number of carbon atoms and $2n + 2$ represents the number of hydrogen atoms in the hydrocarbon chain.

Three basic IUPAC rules for naming alkanes are as follows:

IUPAC Rule 1: *Name the alkane by selecting the name of the longest continuous chain of carbons found in the structure using Table 23-1.*

IUPAC Rule 2: *Each carbon-hydrogen group (called carbon branches) not counted in the continuous chain is named as an alkyl group. An alkyl group*

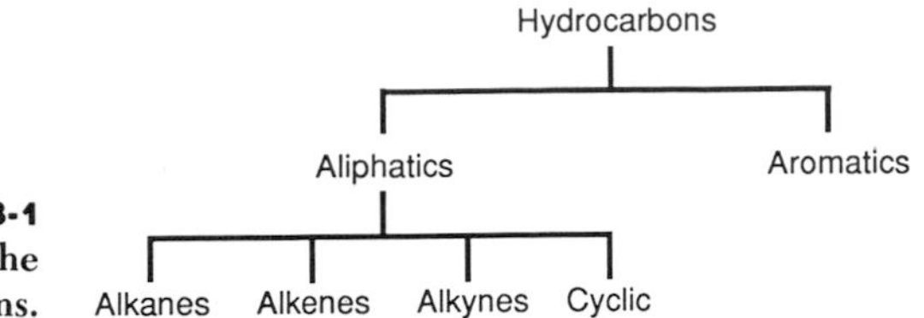

FIGURE 23-1
Relationship of the hydrocarbons.

TABLE 23-1
The IUPAC Names of the First Ten Simple Alkanes, Number of Carbons, and Their Formulas

Name	*Number of carbons*	*Formula*
Methane	1	CH_4
Ethane	2	C_2H_6
Propane	3	C_3H_8
Butane	4	C_4H_{10}
Pentane	5	C_5H_{12}
Hexane	6	C_6H_{14}
Heptane	7	C_7H_{16}
Octane	8	C_8H_{18}
Nonane	9	C_9H_{20}
Decane	10	$C_{10}H_{22}$

(al′kĭl) is simply a hydrocarbon with one hydrogen atom removed as shown in Table 23-2. The name is derived by dropping the "ane" ending on the parent hydrocarbon and adding "yl."

The alkyl group name should precede the name of the longest continuous chain. If similar alkyl groups are attached (branched) to the same parent chain, then the prefixes shown in Table 23-3 should be used to indicate the number of groups.

IUPAC Rule 3: *The carbon atoms on the longest continuous chain are numbered in an order which will locate the attached alkyl groups using the smallest numbers possible.* These alkyl branches are normally listed in

TABLE 23-2
Naming Alkyl Groups

Parent	*Alkyl name*	*Formula*
Methane	Methyl	$-CH_3$
Ethane	Ethyl	$-CH_2CH_3$
Propane	Propyl	$-CH_2CH_2CH_3$
Butane	Butyl	$-CH_2CH_2CH_2CH_3$
Pentane	Pentyl	$-CH_2CH_2CH_2CH_2CH_3$

TABLE 23-3
IUPAC Prefixes for Multiple Alkyl Branches

Number of similar groups	*Prefix*
2	di
3	tri
4	tetra
5	penta
6	hexa
7	hepta
8	octa
9	nona
10	deca

alphabetical order (the alkyl prefixes as shown in Table 23-3 are normally not considered in alphabetizing this order). If two of the same alkyl groups are bonded to the same carbon, the number is listed twice. For example, see Fig. 23-2.

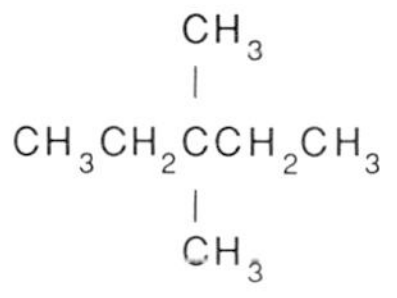

$$\begin{array}{c} CH_3 \\ | \\ CH_3CH_2CCH_2CH_3 \\ | \\ CH_3 \end{array}$$

2,2-Dimethylpentane

FIGURE 23-2
Identical alkyl groups bonded to the same carbon atom.

ALKENES

Hydrocarbon molecules containing one or more carbon-carbon double bonds are known as *alkenes*. The general formula for an alkene is C_nH_{2n}. Hydrocarbons which contain double and/or triple bonds are sometimes referred to as *unsaturated*. The IUPAC rules for naming alkenes are similar to those for naming alkanes:

1. Determine the base name by selecting the longest continuous chain that contains the double bond. Change the ending of the alkane name which corresponds to the number of carbon atoms in the chain from "ane" to "ene." For instance, if the longest chain has five carbon atoms (pentane), the base name for the alkene would be "pentene."

2. Number the chain so as to include both carbons of the double bond and begin numbering at the end of the chain nearest the double bond.

Designate the location of the double bond by using the number of the first atom of the double bond as a prefix (Fig. 23-3):

$CH_3-CH_2-HC=CH-CH_3$

2-Pentene
(not 3-Pentene)

$H_2C=CH-CH_2-CH_3$

1-Butene
(not 3-Butene)

FIGURE 23-3
How to designate the location of a double bond.

3. The locations of alkyl groups are indicated by the number of the carbon atom to which they are attached (Fig. 23-4):

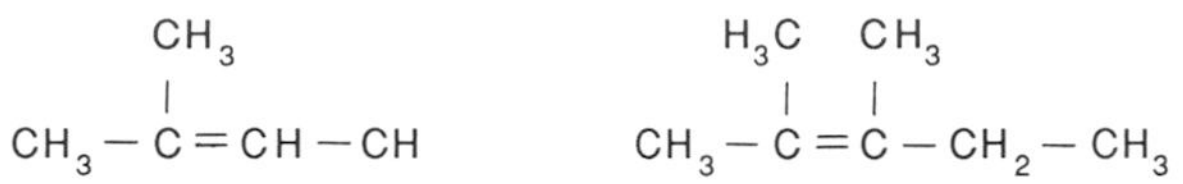

$CH_3-C(CH_3)=CH-CH$

2-Methyl-2-butene

$CH_3-C(CH_3)=C(CH_3)-CH_2-CH_3$

2,3-Dimethyl-2-pentene

FIGURE 23-4
How to designate the locations of alkyl groups.

The naming of organic compounds is complicated because of isomers. *Isomers* are compounds with the same formula, but different structures. Table 23-4 indicates how the number of possible isomeric structures increases rapidly as the number of carbons and multiple bonds increases. Butane (formula: C_4H_{10}) is the first saturated hydrocarbon to have two isomeric structures as shown in Fig. 23-5.

TABLE 23-4
Possible Number of Isomers with Increasing Number of Carbons in a Compound

Formula	*Number of isomers*
CH_4	1
C_2H_6	1
C_3H_8	1
C_4H_{10}	2
C_5H_{12}	3
C_8H_{18}	18
$C_{10}H_{22}$	75

$CH_3CH_2CH_2CH_3$

Butane

$CH_3CH(CH_3)CH_3$

2-Methylpropane
(Isobutane)

FIGURE 23-5
Isomers of butane.

Unlike the single-bonded alkanes, there is no rotation about the double bond of the alkenes. This loss of rotation about the double bond creates *stereoisomers* of some alkenes. *Stereoisomers* are defined as compounds with identical groups on the same side *(cis)* of the double bond and *(trans)* if the identical groups are on opposite sides of the double bond. In order for there to be *cis-trans* stereoisomers, both double-bonded carbons must be connected to two different groups. If either carbon is connected to two identical groups, there are no stereoisomers. 2-Butene is an example of a compound that can form stereoisomers (Fig. 23-6), and like all *cis-trans* isomers, they have different physical (boiling points, freezing points, etc.) and chemical properties.

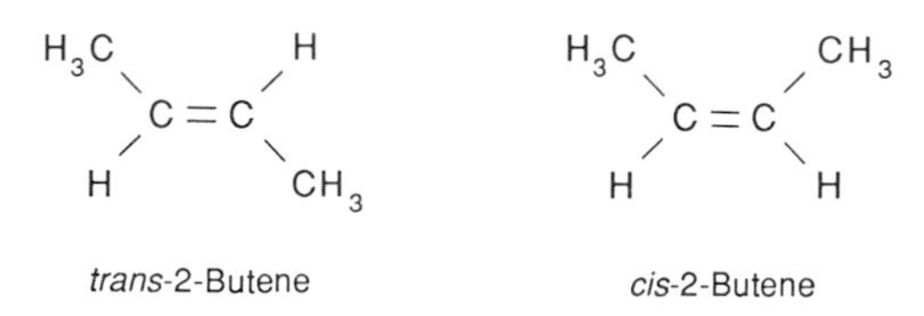

FIGURE 23-6
Stereoisomers of 2-butene.

ALKYNES

Hydrocarbon molecules that contain triple bond(s) between the carbon atoms are called *alkynes.* The general formula for alkynes is C_nH_{2n-2}. The IUPAC rules for naming alkynes are similar to those for alkenes, except that the parent alkane (longest continuous chain containing the triple bond) name ending is changed from "ane" to "yne." The numbering system is exactly the same as for alkanes. The triple bond is designated by the lowest possible number; this determines which end of the chain the numbering begins from, and the constituents or alkyl groups are then numbered accordingly. (See Fig. 23-7, for example.)

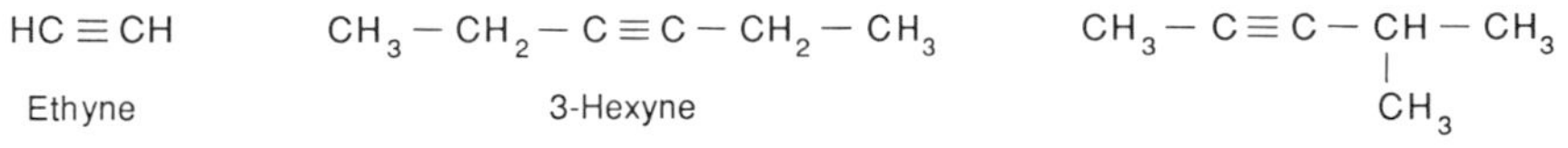

$HC \equiv CH$ — Ethyne

$CH_3-CH_2-C \equiv C-CH_2-CH_3$ — 3-Hexyne

$CH_3-C \equiv C-CH(CH_3)-CH_3$ — 4-Methyl-2-pentyne

FIGURE 23-7
How to designate compounds containing a triple bond.

CYCLIC HYDROCARBONS

Cycloalkanes

Several of the organic molecules discussed thus far are capable of forming cyclic molecules. The carbon atoms normally associated in a straight chain can form a ring by connecting the end carbons. *Cycloalkanes* have two fewer hydrogen atoms than the same straight-chained alkane. Therefore the general formula for a cycloalkane is C_nH_{2n}. Cyclic molecules are named using the IUPAC system by attaching the prefix "cyclo-" to the name of the organic molecule (see Fig. 23-8).

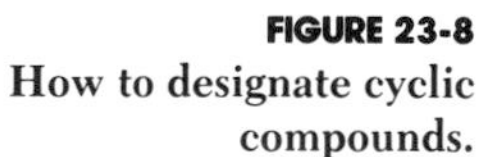

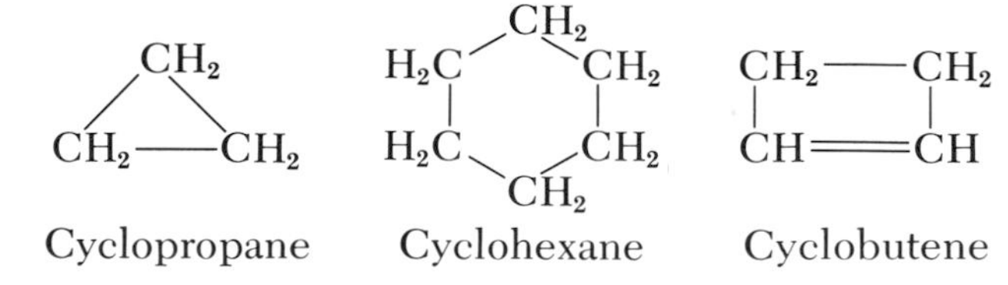

Cyclopropane Cyclohexane Cyclobutene

FIGURE 23-8 How to designate cyclic compounds.

The substituents on the ring are named, and their positions are indicated by numbers, the lowest combination of numbers being used. In simple cycloalkenes and cycloalkynes, the double- and triple-bonded carbons are considered to occupy positions 1 and 2. For example, see Fig. 23-9.

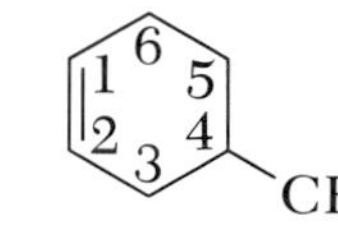

4-Methylcyclohexene

can also be written

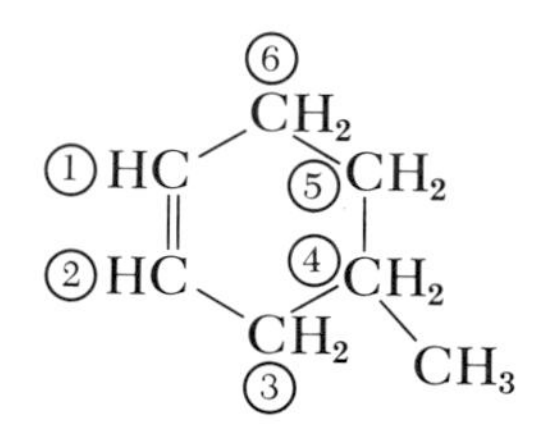

FIGURE 23-9 How to designate substituents attached to cyclic compounds.

Aromatics

An *aromatic* compound is one which contains a benzene molecule (C_6H_6). Benzene is unique because it is a cyclic molecule with alternating double and single bonds. This alternating single- and double-bond configuration gives the benzene ring a great deal of stability, and many compounds are found in nature with this structure. Because of its abundance, benzene is used as the IUPAC parent name for most molecules containing a six-membered (that is, six-carbon) alternating double- and single-bonded ring as shown in Fig. 23-10. A shorthand method of writing the benzene structure is given in Fig. 23-10; a circle inserted into the hexagon structure represents the alternating bonds. When benzene is a substituent on another molecule, it is called *phenyl* (fen′uhl). Again, the lowest numbers possible are used to locate two or more substituents bonded to the ring.

Many important compounds are not just aliphatic (alkanes or saturated hydrocarbons) or just aromatic (benzene derivative), but contain both aliphatic and aromatic units. Ethylbenzene, for example, contains a benzene ring and an aliphatic side chain (Fig. 23-11).

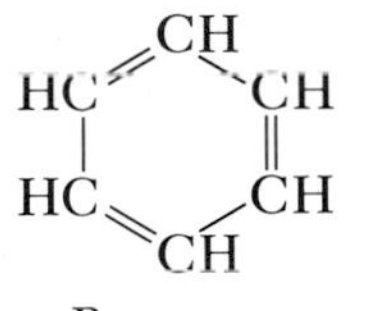

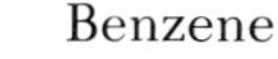

A B

FIGURE 23-10 The benzene ring. A is preferred over B to represent benzene

(The hexagon with the three double bonds or the inserted circle represents the six-carbon ring, the six hydrogen atoms, and the three double bonds.)

CH_2CH_3

FIGURE 23-11 Ethylbenzene. Ethylbenzene

The simplest of the alkyl benzenes is given the special name of toluene (methylbenzene). Compounds containing longer side chains are named by prefixing the name of the alkyl group to the word benzene, as, for example, in ethylbenzene, *n*-propylbenzene, and isobutylbenzene (Fig. 23-12).

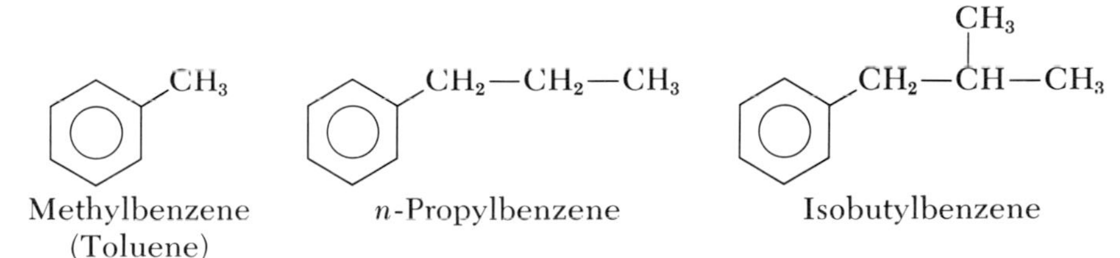

FIGURE 23-12 How to designate the alkyl benzenes.

The relative positions of the groups on the benzene ring are designated by the prefixes, "ortho-," "meta-," and "para-" (usually abbreviated *o, m,* and *p*). (See Fig. 23-13.)

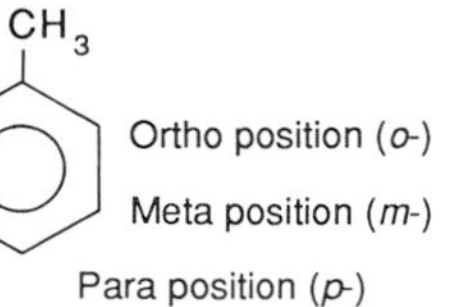

FIGURE 23-13 Relative positions of attachments on the benzene ring.

The simplest of the dialkylbenzenes, the dimethylbenzenes, are given the special name of xylene; we have, then, *o*-xylene, *m*-xylene, and *p*-xylene. Dialkylbenzenes containing the 1-methyl group are named as derivatives of toluene, while others are named by prefixing the names of both alkyl groups to the word benzene. (See Fig. 23-14.)

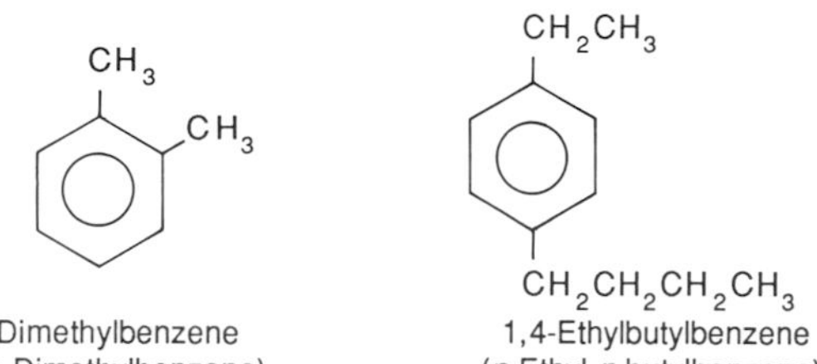

1,2-Dimethylbenzene
(*ortho*-Dimethylbenzene)
(*o*-Xylene)

1,4-Ethylbutylbenzene
(*p*-Ethyl-*n*-butylbenzene)

FIGURE 23-14 How to designate dialkylbenzenes.

ORGANIC FAMILIES AND THEIR FUNCTIONAL GROUPS

Although carbon-hydrogen bonds are predominant in organic molecules, carbon can also covalently bond to other atoms such as oxygen, nitrogen, and chlorine. These different constituents are known as *functional groups*. All organic compounds are classified into *families* (see Table 23-5) by their functional groups. All organic compounds in this chapter will be named using the IUPAC system, and common names will be indicated in parentheses.

Alkyl Halides

Alkanes with a halogen atom as a substituent are named, using the IUPAC nomenclature system, as *haloalkanes*. The general formula for haloalkanes is RH_2X. (See Fig. 23-15.) The parent alkane name is preceded by the name of the halogen which is present and the number of the carbon to which the halogen is bonded. (See Fig. 23-16.)

```
    H
    |
R — C — X        where: X = F, Cl, Br, or I
    |                   R = any alkyl or aryl group
    H
```

FIGURE 23-15 The general formula for alkyl halides.

```
Cl — CH3          CH3CH2 — CHCH3          CH3 — C = CH2
                          |                     |
                          Br                    Cl

Chloromethane     2-Bromobutane           2-Chloro-1-propene
```

FIGURE 23-16 How to designate alkanes with a halogen atom as substituent.

TABLE 23-5
Organic Families and Their Functional Groups

Formula	*Functional Group*	*Family*
C_nH_{2n+2}	Single bond	Alkane
$C_nH_{n\,2}$	Double bond	Alkane
C_nH_{2n-2}	Triple bond	Alkyne
C_6H_6	Benzene	Aromatic
$R-X$	Halogen	Alkyl halide
$R-OH$	Hydroxyl	Alcohol
$R-C-NH_2$	Amino	Amine
$R-O-R'$	Oxygen	Ether
$R-C-O$ $\vert$ H	Carbonyl (terminal)	Aldehyde
$R-C-O$ $\vert$ R	Carbonyl (nonterminal)	Ketone
$R-C-O$ $\vert$ OH	Carbonyl + hydroxyl	Carboxylic acid
$R-C-O$ $\vert$ NH_2	Carboxyl + amine	Amide
$R-C-O$ $\vert$ OR'	Carboxyl + carbinol	Ester

Alcohols

An *alcohol* is any compound with the general formula $R-OH$ where R represents any alkyl or aryl group and $-OH$ a hydroxyl group. This alcoholic functional group is referred to as a *carbinol* ($C-OH$) group. If the hydroxyl group is attached to an aryl group (benzene ring), the resulting compound is called a *phenol.* Consider alcohols, phenols, and ethers as derivatives of water. "Absolute alcohol" is another name for pure, water-free ethanol. *Denatured alcohol* is ethanol that has been made unfit for human consumption by adding small portions of methanol, benzene, and/or aviation gasoline. The contaminants do not change the chemical properties of the ethanol and no alcohol tax is required.

Follow these rules to arrive at the IUPAC name for an alcohol:

1. Select the longest continuous carbon chain that contains the $-OH$ group.
2. Drop the "e" ending and replace it with the suffix "ol."
3. The "longest" chain must include the carbon bearing the $-OH$ group.

TABLE 23-6
The Common and IUPAC Names for Some Alcohols

IUPAC name	*Structure*	*Common name*
Methanol	CH_3OH	Methyl alcohol, wood alcohol
Ethanol	CH_3CH_2OH	Ethyl alcohol, grain alcohol
1-Propanol	$CH_3CH_2CH_2OH$	Propyl alcohol
2-Propanol	$CH_3CHOHCH_3$	Isopropyl alcohol, rubbing alcohol

Table 23-6 gives the IUPAC names and common names of some simple alcohols.

Alcohols are classified as primary, secondary, and tertiary:

Primary alcohol: One carbon atom is directly attached to the carbinol carbon (Fig. 23-17).

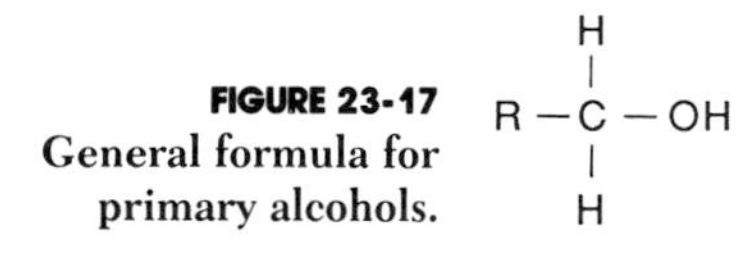

FIGURE 23-17 General formula for primary alcohols.

where R is any H or carbon-containing group. The R groups can be similar or different as indicated by R′ and R″.

Secondary alcohol: Two carbon atoms are directly attached to the carbinol carbon (Fig. 23-18).

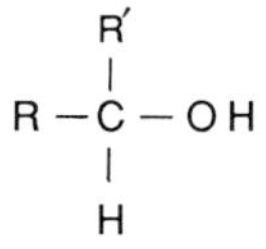

R′
|
R — C — OH
|
H

FIGURE 23-18 General formula for secondary alcohols.

Tertiary alcohol: Three carbon atoms are directly attached to the carbinol carbon (Fig. 23-19).

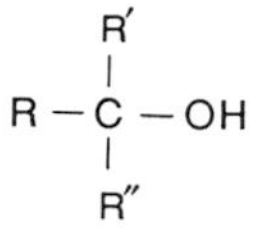

R′
|
R — C — OH
|
R″

FIGURE 23-19 General formula for tertiary alcohols.

Alcohols with two OH groups are called *glycols* or *diols*, and those with three OH groups are called *triols*. (See Fig. 23-20.)

$HO-CH_2CH_2-OH$ 1,2-Ethanediol (Ethylene glycol) (Automobile antifreeze)

$CH_3CH(OH)CH_2-OH$ 1,2-Propanediol (Propylene glycol)

$HO-CH_2CH(OH)CH_2-OH$ 1,2,3-Propanetriol (Glycerol) (Glycerine)

FIGURE 23-20 How to designate alcohols with multiple hydroxyl groups.

Amines

Amines (uh meen′) are compounds with organic groups attached to a central nitrogen atom. Consider amines as derivatives of ammonia (NH_3). The general formula for amines is $R-NH_2$. Most amines have a strong decaying-fish odor. (See Fig. 23-21.)

H—N(—H)—H	R—N(—H)—H	R—N(—R)—H	R—N(—R)—R
Ammonia	Primary amine	Secondary amine	Tertiary amine

FIGURE 23-21 Ammonia and general formulas for the amines.

Notice that the classification of amines is based on the number of organic groups attached to the nitrogen atom, not those attached to a carbon as was the case with primary, secondary, and tertiary alcohols. A *quaternary amine* is defined as an amine which has four alkyl or aryl groups attached to the central nitrogen atom, for example, diethyldimethylammonium chloride.

Amines are very often called by common names as alkyl derivatives of ammonia with the word "ammonia" being changed to "amine." A capital *N* preceding the name of an amine indicates that the branches are located on the nitrogen atom, not the carbon chain. (See Fig. 23-22.)

CH_3-NH_2 (Methylamine)

$CH_3CH_2-N(H)-CH_2CH_3$ (N, N-Diethylamine)

FIGURE 23-22 Common names for alkyl derivatives of ammonia.

The IUPAC name for the amine functional group is *amino* ($-NH_2$, uh meen′ oh). It is considered the substituent and its position on the chain is indicated by the lowest numbers (Fig. 23-23).

```
                CH3
                |
CH3CH2CHCH2CHCH3
       |
       NH2
```

5-Methyl-3-aminohexane

$H_2N-CH_2CH_2CH_2CH_2CH_2CH_2-NH_2$

1,6-Diaminohexane

FIGURE 23-23 How to designate amino substituents.

Ethers

An *ether* (ee′thr) is defined as any compound with an oxygen atom linked by single bonds to two carbon-containing (R) groups. The general formula for ethers is R—O—R′.

Symmetrical ether: The R and R′ groups are identical.

Mixed ether: The R and R′ groups are different (thus unsymmetrical).

CH_3-O-CH_3 (Dimethyl ether)

$CH_3-O-CH_2CH_3$ (Methyl ethyl ether)

Common names are assigned to ethers by naming the two alkyl groups bonded to the oxygen and adding the word "ether." The IUPAC system names ethers as alkoxyalkanes, alkoxyalkenes, and alkoxyalkynes by selecting the longest continuous hydrocarbon chain as the parent and adding the *alkoxy* (RO—) prefix. Table 23-7 gives the common and IUPAC names and the structures of some ethers.

TABLE 23-7 The Common and IUPAC Names for Some Ethers

IUPAC name	*Structure*	*Common name*
Methoxymethane	CH_3OCH_3	Dimethyl ether
Methoxyethane	$CH_3OCH_2CH_3$	Methyl ethyl ether
Ethyoxyethane	$CH_3CH_2OCH_2CH_3$	Diethyl ether
Ethyoxypropane	$CH_3CH_2OCH_2CH_2CH_3$	Ethyl propyl ether

CAUTION *Use of ethers:* Ethers tend to be chemically stable; however, aliphatic ethers react slowly with air to produce peroxides. All ether containers should be dated when received. Ethers should be purchased only in iron containers to reduce production of peroxides, and the excess discarded properly after opening the container. Ethers tend to be volatile, have low solubility in water, and have very low flash points. Do not discard excess ethers down drain; consult supervisor and/or Material Safety Data Sheets (MSDS) for proper disposal.

Aldehydes

An *aldehyde* (al′duh hīd) is defined as any compound that contains a *carbonyl group* (C=O, kar′buh neel) with one or more hydrogen atoms. The general formula for aldehydes is shown in Fig. 23-24.

FIGURE 23-24 General formula for aldehydes.

```
    H
    |
R — C = O
```

The IUPAC nomenclature rules state that the "e" ending on the parent hydrocarbon's name be deleted and the "al" suffix be added. The word "aldehyde" starts with the prefix letters "al." Since it will always be the first carbon in the chain, no locator numbers are needed to designate the carbon to which the carbonyl is bonded. Table 23-8 gives both the common and IUPAC names for some aldehydes.

TABLE 23-8 The Common and IUPAC Names for Some Aldehydes

IUPAC name	*Structure*	*Common name*
Methanal	H_2CO	Formaldehyde
Ethanal	H_3CHCO	Acetaldehyde
Propanal	H_3CCH_2HCO	Propionaldehyde
Butanal	$H_3CCH_2CH_2HCO$	Butylaldehyde

CAUTION *Use of aldehydes:* The simplest aldehyde is formaldehyde and is a gas (bp −21°C) at room temperature. It is very soluble in water and is sold as a 37% aqueous solution called *formalin.* Formaldehyde has been used as a very effective preservative and bactericide, but is a carcinogen and should not be used without proper precautions.

Ketones

Ketones (kee′tōn) are structurally very similar to aldehydes. Ketones also have a carbonyl functional group, but the difference is that it is located on a carbon other than the first carbon of the chain. The general formula for ketones is R—CO—R′.

Thus, the carbon to which the carbonyl is bonded must be assigned the lowest possible number. The parent alkane's ending is changed by dropping the "e" and adding the suffix "one." The word "ketone" ends in the letters "one." (See Fig. 23-25.)

$CH_3-CO-CH_3$ 2-Propanone (Dimethyl ketone) (Acetone)

$CH_3-CO-CH_2-CH_3$ 2-Butanone (Methyl ethyl ketone) (MEK)

FIGURE 23-25
How to designate ketones.

Carboxylic Acids

Carboxylic acids (kar box′ĭl ĭc) are classified as compounds that contain both the carboxyl group and the hydroxyl group (carbonyl + hydroxyl). These organic acids tend to be weak acids and have unpleasant odors. The general formula for carboxylic acids is R—CO—OH.

The IUPAC rules state that the "e" should be deleted from the parent hydrocarbon's name and the suffix "oic" and word "acid" be added. No location numbers are needed to designate the carbon containing the carboxylic functional group since it will always be the first carbon. Table 23-9 gives the IUPAC name of several carboxylic acids, the structure, and the common name and its derivation.

TABLE 23-9
The Common and IUPAC Names for Some Carboxylic Acids

IUPAC name	*Structure*	*Common name*	*Derivation of name*
Methanoic acid	H—COOH	Formic acid	Formica: *ants* (Latin)
Ethanoic acid	CH_3-COOH	Acetic acid	Acetum *(vinegar)*
Propanoic acid	CH_3CH_2COOH	Propionic acid	Proto *(first)* pion *(fat)*
Butanoic acid	$CH_3(CH_2)_2COOH$	Butyric acid	Butyrum *(butter)*
Pentanoic acid	$CH_3(CH_2)_3COOH$	Valeric acid	Valerian root
Hexanoic acid	$CH_3(CH_2)_4COOH$	Caproic acid	Caper *(goat)*

Amides

Amides (am′eed) are carboxylic acid derivatives in which the —OH group has been replaced by an amine (NH_2) group. The general formula for amides is $R-CONH_2$. Location numbers are used to locate branches on carbon atoms of the parent hydrocarbon, and a capital N prefix is used to indicate branches on the nitrogen atom of the amine group. The IUPAC system requires: Select the longest carbon chain as the parent carboxylic acid, drop the "oic" ending, and add suffix "amide."

$CH_3CH_2CH_2CH_2CONH_2$	Pentanamide
$CH_3CONHCH_2CH_3$	(*N*-Ethylacetamide)

Esters

Esters are compounds produced by the reaction of an alcohol and a carboxylic acid. The ester's common name is derived by naming the alkyl group of the alcohol followed by the organic anion of the acid molecule. The general formula for esters is R—CO—OR.

$$CH_3CO{-}OH + CH_3OH \rightarrow CH_3CO{-}O{-}CH_3 + H_2O$$

(Acetic acid) + (Methyl alcohol) → (Methyl acetate) + Water

The IUPAC nomenclature rules require that the acid's "ic" ending be changed to "ate" and that this name be preceded by the name of the alkyl or aryl (R) group of the alcohol. For example:

Pentyl ethanoate	(pentanol + ethanoic acid)
Ethyl propanoate	(ethanol + propanoic acid)
Methyl butanoate	(methanol + butanoic acid)

REFERENCES

Dean, J. A. (ed.), *Lange's Handbook of Chemistry,* 13th ed., McGraw-Hill Book Co., New York, 1985.

Handbook of Chemistry and Physics, Chemical Rubber Publishing Company, Cleveland, Ohio, various annual editions.

Morrison, R. T., and Boyd, R. N., *Organic Chemistry,* 4th ed., Allyn and Bacon, Inc., Boston, 1983.

Rodd, E. H. (ed.), *Chemistry of Carbon Compounds,* Elsevier Publishers, New York, 1951 to present.

24
CHEMICALS AND PREPARATION OF SOLUTIONS

INTRODUCTION

Part of a technician's job is to prepare the solutions needed for the various procedures performed in the laboratory. It is most important that directions be followed exactly and that all caution be observed.

CAUTION Always recheck the label of the chemical that you are using. Use of the wrong chemical can cause an explosion or ruin a determination.

GRADES OF PURITY OF CHEMICALS

Chemicals are manufactured in varying degrees of purity. Select the grade of chemical that meets the need of the work to be done. It is wasteful to use costly reagent grades when technical grades would be satisfactory. The various grades are listed and explained below.

Commercial or Technical Grade

This grade is used industrially, but is generally unsuitable for laboratory reagents because of the presence of many impurities.

Practical Grade

This grade does contain impurities, but it is usually pure enough for most organic preparations. It may contain some of the intermediates resulting from its preparation.

USP

USP grade chemicals are pure enough to pass certain tests prescribed in the U.S. Pharmacopoeia and are acceptable for drug use, although there may be some impurities which have not been tested for. This grade is generally acceptable for most laboratory purposes.

CP

CP stands for chemically pure. Chemicals of this grade are almost as pure as reagent-grade chemicals, but the intended use for the chemicals determines whether or not the purity is adequate for the purpose. The classification is an ambiguous one; read the label and use caution when substituting for reagent-grade chemicals.

Spectroscopic Grade

Solvents of special purity are required for spectrophotometry in the uv, ir, or near ir ranges as well as in nmr spectrometry and fluorometry. Specifications of the highest order in terms of absorbance characteristics, water, and evaporation residues are given. The principal requirement of a solvent for these procedures is that the background absorption be as low as possible. Most of these chemicals are accompanied by a label which states the minimum transmission at given wavelengths. Residual absorption within certain wavelengths is mainly due to the structure of the molecule.

Chromatography Grade

These chemicals have a minimum purity level of 99+ mol % as determined by gas chromatography; each is accompanied by its own chromatogram indicating the column and parameters of the analysis. No individual impurity should exceed 0.2%.

Reagent Analyzed (Reagent Grade)

Reagent-analyzed, or reagent-grade, chemicals are those which have been certified to contain impurities in concentrations below the specifications of the Committee on Analytical Reagents of the American Chemical Society (ACS). Each bottle is identified by batch number. Use only reagent-analyzed chemicals in chemical analysis.

CAUTION Be certain that the bottle has not been contaminated in previous use. Impurities may not have been tested for, and the manufacturer's analysis may possibly have been in error.

Primary Standard

Substances of this grade are sufficiently pure that they may serve as reference standards in analytical procedures. You may use them directly to prepare standard solutions by dissolving massed amounts in solvents and then diluting them to known volumes. Primary standards must satisfy extremely high requirements of purity; they usually contain less than 0.05% impurities.

CAUTION Many compounds form hydrates with water. Some form more than one hydrate with water, combining with a different number of water molecules to form definite compounds. When using tables from reference handbooks, check *carefully* to confirm that the formula is the one that you seek and not another hydrate.

COMMON HAZARDOUS CHEMICALS

Consider all chemicals, reagents, and solutions as *toxic* substances. Many of the hazards of chemicals are not obvious or evident by smell, odor, appearance, or immediately detectable by the organs of the body. See Fig. 24-1 for appropriate labels used on hazardous materials. Tables 24-1

Please turn to page 566.

FIGURE 24-1 Warning labels for hazardous materials.

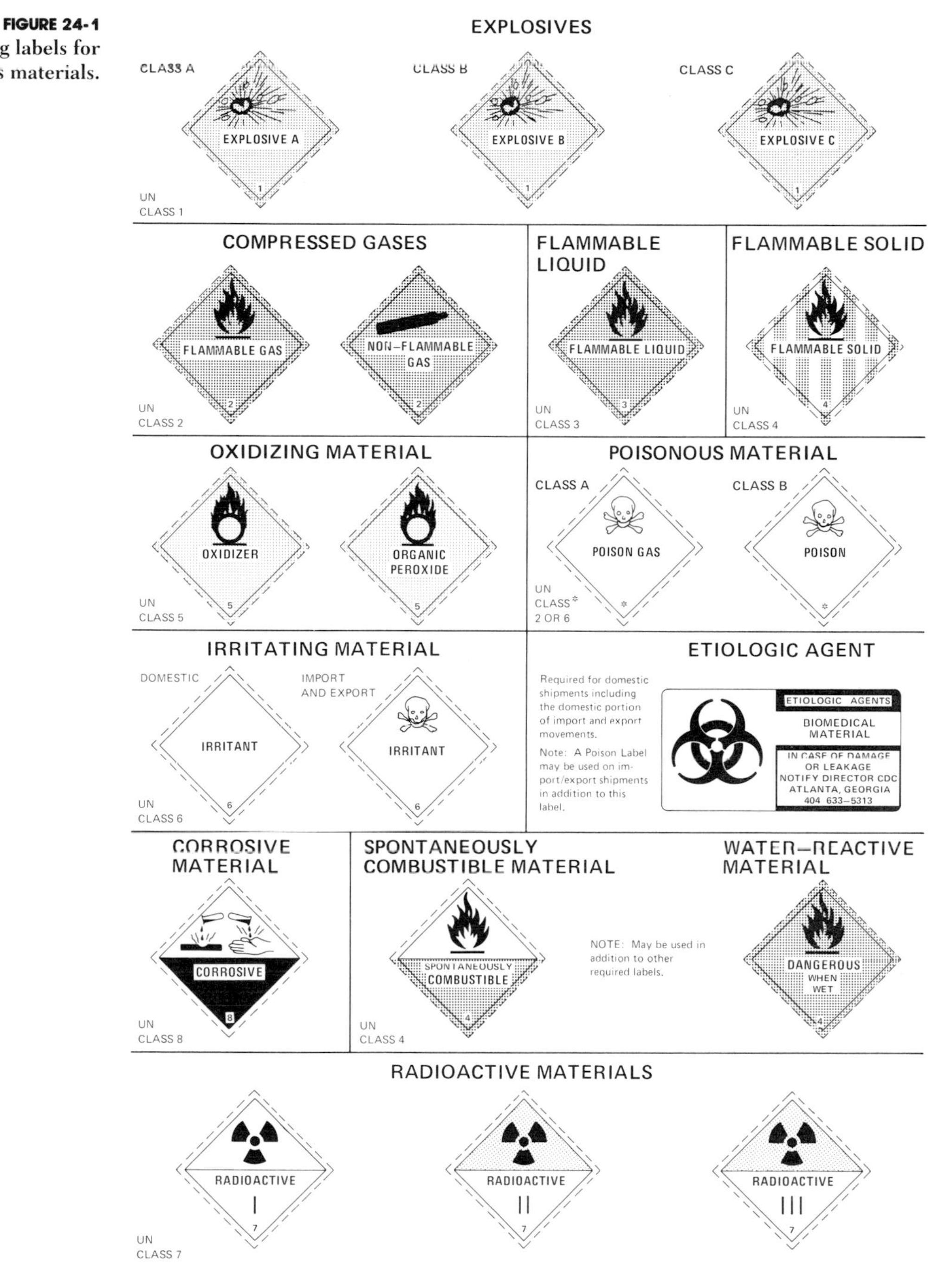

TABLE 24-1
A Table of Common Hazardous Chemicals

Chemical	*Usual shipping container*	*Life hazard*	*Fire-fighting phases*	*Fire hazard*	*Storage*
Acetic acid (glacial)[a]	Glass carboys and barrels.	May cause painful burns of skin.	Extinguishing agent is water.	Dangerous in contact with chromic acid, sodium peroxide, or nitric acid; yields moderately flammable vapors above flash point 40°C (104°F).	Safeguard against mechanical injury. Isolate from oxidizing materials as noted under Fire Hazard.
Acetone	Carboys, steel drums, tank cars.	Toxicity of a comparatively low order.	Lighter than water (sp gr 0.792) but soluble in it in all proportions. Water, particularly in the form of spray, is the best extinguisher. Carbon dioxide may also be used. Automatic sprinkler systems or total-flooding carbon dioxide systems may be employed for protection in storage rooms.	A volatile liquid. Gives off vapors which form with air flammable and explosive mixtures. Flash point −16°C (3°F), Explosive range 2.55% to 12.8% (upward propagation). The ignition temperature is comparatively high, being within the range 538 to 566°C (1000° to 1050°F). The vapors are heavier than air (vapor density 2). Fire hazard slightly less than that of gasoline.	Safeguard containers against mechanical injury. Only electric equipment of the explosionproof type, Group D classification, permitted in atmospheres containing acetone vapor in flammable proportions.
Aluminum dust	Barrels or boxes.		Smother with sand, ashes, or rock dust. Do not use water, which may cause explosion.	Forms flammable and explosive mixtures with air.	Keep in dry place. In case of fire do not use water; it may cause an explosion.
Aluminum resinate	Wooden barrels.			Combustible.	Storage should be ventilated and safeguarded as for oil-storage building.
Ammonia (anhydrous)	Steel cylinders or steel tank cars.	Irritant. An outstanding serious effect produced by ammonia in concentrations of the order of ½% by volume for duration of exposure of the order of ½ h is blindness. A concentration of 0.03% of ammonia in air for duration of exposure of the order of ½ to 1 h, according to Lehmann, does not cause serious effects.	Soluble in water. Hose streams comparatively effective in removing the gas from the atmosphere.	Gas density 0.60 (air = 1). Not flammable in air except in comparatively high concentration, which is seldom encountered under practical conditions, the low limit of the flammable or explosive range being about 15 to 16% and the upper limit about 25 to 26% by volume (horizontal propagation). The presence of oil will increase the fire hazard. Ammonia aqua does not burn.	Safeguard against mechanical injury and excessive heating of cylinders of tanks. Fire-resistive storage recommended. In combustible buildings or if near combustibles, sprinklered storage recommended. Isolate from other chemicals, particularly chlorine, bromine, iodine, and mineral acids.
Ammonium perchlorate	Wooden barrels or kegs and glass bottles.			Oxidizing material. May explode in a fire. Hazard classes with potassium chlorate.	Safeguard against mechanical injury. Isolate from mineral acids, also from combustibles.

Antimony pentasulfide (golden antimony sulfide) Sb_2S_3[b]	Fiber drums or tins.	Gaseous products of combustion contain sulfur dioxide and are irritating and corrosive.	Practically insoluble in water. Use water.	Combustible. Readily ignited by a small flame. Hazardous in contact with oxidizing material. Yields flammable hydrogen sulfide on contact with mineral acid.	Safeguard against mechanical injury. Isolate from acids, chlorates, nitrates, and other oxidizing agents.
Barium chlorate	Wooden boxes, barrels, or kegs.		See potassium chlorate.	Oxidizing material. Hazard classes with potassium chlorate.	Isolate.
Barium nitrate	Wooden boxes; barrels.	Soluble in water. Poisonous when taken internally.	See sodium nitrate.	Oxidizing material. Hazard in class with sodium nitrate.	Do not store with combustible materials.
Barium peroxide	Tightly closed metal containers packed in wooden boxes or barrels, or in bulk in metal barrels or drums.		Smother with sand, ashes, or rock dust. Do not use water.	Oxidizing material. Hazard in class with sodium peroxide.	Do not store with combustible materials.
Benzoyl peroxide (dry granular or powdered wet)	Dry granular material in individual 1-lb containers inside wooden boxes; finely powdered material shipped wet (30% water by weight) in glass containers placed inside tightly sealed metal containers in wooden boxes or in aluminum drums.	Dust irritating to eyes and lungs. Use goggles and dust respirator in dusty atmospheres.	Water, carbon dioxide, foam, sand, soda ash, or rock dust may be used as extinguishing agents.	Highly flammable in the dry state. Strong supporter of combustion. Do not subject dry material to heat of friction or grinding. Not miscible with water.	Store in a cool ventilated place. Powder should be stored with not less than 30% water by weight. Keep away from all sources of heat and separate from all combustible materials and acids.
Bleaching powder; calcium hypochlorite; chlorinated lime; chloride of lime (incorrect name)[c]	Airtight tin containers, wooden barrels, and steel drums.	Corrosive. Irritating to skin, eyes, and lungs. See chlorine.	Fires where the compound is present may be fought with water, preferably spray. Protect eyes and skin, using gas mask of a type approved by Bureau of Mines.	Not combustible but evolves chlorine and at higher temperatures oxygen. With acids or moisture evolves chlorine freely at ordinary temperatures. See chlorine.	Store in cool, dry, well-ventilated place away from combustibles. See chlorine. Rupture of drums containing bleaching powder, particularly if the chlorine content is high, may result from exposure to heat.
Borneol	Barrels, kegs, boxes, and tins.		See camphor.	Combustible. Hazard similar to camphor.	Store in well-ventilated compartment or building.
Bromine	Glass bottles; earthen jugs. (Bottles should be surrounded by incombustible packing.)	Corrosive; at ordinary temperatures gives off poisonous suffocating vapors.		Causes oxidizing effect, resulting in heating, and may cause fire when in contact when organic material.	Isolate; safeguard against mechanical injury.
Bronze dust	Barrels or boxes.		Smother with sand or ashes.	When aluminum is present forms flammable and explosive mixtures with air. Composition usually free from aluminum.	Bronze dust free from aluminum not considered dangerous.
Butane	Steel cylinders.			Flammable gas under pressure. Classes with gasoline vapor in fire hazard.	Safeguard against mechanical injury. Keep cool.

TABLE 24-1 (Continued)

Chemical	*Usual shipping container*	*Life hazard*	*Fire-fighting phases*	*Fire hazard*	*Storage*
Calcium carbide	Iron drums and tin cans.	Serious under fire conditions.	Smother with sand or ashes. Do not use water.	Gives off acetylene gas on contact with water or moisture.	Store in dry, well-ventilated place in accordance with NFPA Standards.
Calcium oxide; quick lime	Wooden barrels and bags.			Heats upon contact with water or moisture and may cause ignition of organic material. Swells when moist and may burst container.	Isolate; store in dry place away from water or moisture.
Camphene	Tins.		Smother with sand or ashes. Avoid water.	When heated gives off flammable vapors. Classes with turpentine.	Isolate; keep in unheated compartment away from fire or heat.
Camphor	Tins and wooden kegs.		Smother with sand or ashes. Chemical streams. Avoid water.	Flammable; gives off flammable vapors when heated which may form explosive mixture with air. Flash point 82°C (180°F).	Detach from other storage. Keep in well-ventilated room remote from fire.
Carbolic acid—*See Phenol.*					
Carbon disulfide[d]	Small glass, earthenware, or metal containers packed in outside barrels or boxes (see I.C.C. Regulations). Steel drums, steel tank cars.	Toxic. 3200 to 3850 parts of vapor per million (0.32 to 0.385% by volume) may cause dangerous illness in ½ to 1 h. Direct contact with the skin should be avoided. Products of combustion contain sulfur dioxide, which in concentrations of 0.2% by volume in air may cause serious injury in ½ h or less. Often poisonous carbon monoxide is present in the products of combustion.	Heavier than water (sp. gr. 1.29) and sparingly soluble in it. Use sand, carbon dioxide, or other inert gas as extinguishing agents. Cooling and blanketing action of water may be utilized in case of fires in metal containers or tanks. Total-flooding carbon dioxide systems may be employed for protection in storage rooms. Foam not effective. Do not use tetrachloromethane (carbon tetrachloride). Use of gas masks or oxygen helmets of the type approved for the purpose by U.S. Bureau of Mines recommended.	A highly volatile liquid with an offensive odor, giving off even at comparatively low temperatures vapors which form with air flammable and explosive mixtures. Flash point −30°C (−22°F). Flammable range 1 to 50% (upward propagation). The ignition temperature is dangerously low, being about 100 to 106°C (212 to 223°F). It is endothermic, and the vapor may be ignited by a heavy blow. The vapors are heavier than air (vapor density 2.62), and may travel a considerable distance to a source of ignition and flash back. More hazardous than gasoline.	Isolate and safeguard containers against mechanical injury and metallic blows, and keep in unheated compartment away from sunlight and any source of ignition, including electric lighting fixtures and other electric equipment. Storage tanks should be constructed over concrete basins containing water, and the carbon disulfide kept blanketed with water or inert gas at all times.
Charcoal (wood)	Boxes, barrels, or bulk.	There is danger from carbon monoxide poisoning during burning unless adequate ventilation is provided.	Use water, completely extinguishing the fire, after which the storage pile should be moved.	Spontaneously ignitable when freshly calcined and exposed to air, or when wet; hazardous when freshly ground and tightly packed.	Isolate, prevent dust accumulations; ventilate well; make daily inspections.

Chinese wax	Burlap bags and wooden barrels.			Combustible.	Detach from other storage.
Chlorine	Steel cylinders and tank cars.	Corrosive. Irritating to eyes and mucous membrane. Toxic. 0.004% to 0.006% by volume in air causes dangerous illness in ½ to 1 h.	Use gas masks on entering atmospheres containing chlorine gas. If, however, concentration is high, or there is doubt as to the degree of concentration, use oxygen helmet of a type approved for such use by U.S. Bureau of Mines.	Is not combustible in air but reacts chemically with many common substances and may cause fire or explosion when in contact with them. See remarks under Storage. Dangerous to neutralize chlorine in a room with ammonia.	Isolate from turpentine, ether, ammonia gas, illuminating gas, hydrocarbons, hydrogen, and finely divided metals. Safeguard against mechanical injury of containers.
Chromium trioxide or chromium anhydride CrO_3 (often called chromic acid)[e]	Iron drums and glass bottles.	Irritating to skin. Poisonous.	Use water, completely extinguishing the fire, after which the storage pile should be removed.	Oxidizing material; will ignite on contact with acetic acid and alcohol. Chars organic material such as wood, sawdust, paper, or cotton, and may cause ignition. Combustible material in presence of chromium trioxide when ignited burns with great intensity. May cause explosion in fire.	Isolate.
Cobaltous nitrate	Wooden barrels.			Oxidizing material. Classes with sodium nitrate.	See sodium nitrate.
Colophony; rosin	Barrels.			Combustible; gives off flammable vapors when heated.	Ventilate storage; avoid dust; keep away from fire or heat.
Copper nitrate	Wooden barrels and kegs.	Poisonous when taken internally. Soluble in water.		Oxidizing material. Hazard classes with sodium nitrate.	Safeguard against mechanical injury; isolate. See sodium nitrate.
Cyclopropane	Steel cylinders.	Anesthetic.	Use water to cool cylinders not on fire. If gas is burning at valves or safety releases, usually the best course to follow is not to disturb or attempt to extinguish flame. To do so will cause the release of unburned gas and quickly create highly dangerous explosive atmospheres. If cylinder is mounted on an anesthetic machine or truck, it may be possible to move it to a safe place. Carbon dioxide or tetrachloromethane (carbon tetrachloride) are best extinguishing agents.	Highly flammable gas. Forms flammable and explosive mixtures with air or oxygen. Explosive range 2.40 to 10.4% (upward propagation). Only electric equipment of the explosion-proof type. Group C classification tentative pending further tests, permitted in atmospheres containing cyclopropane in flammable proportions.	Isolate from oxygen cylinders and store in a cool, well-ventilated storeroom.
Didymium nitrate	Wooden kegs.			Oxidizing material. Classes with sodium nitrate.	Isolate. See sodium nitrate.

TABLE 24-1 (Continued)

Chemical	*Usual shipping container*	*Life hazard*	*Fire-fighting phases*	*Fire hazard*	*Storage*
Dioxane	Glass bottles, metal cans, and metal drums.	Irritant and toxic in high concentrations.	Water best extinguishing agent. Slightly heavier than and completely soluble in water (sp gr 1.03).	Moderately volatile flammable liquid. Flash point 12°C (54°F). Explosive range 1.97 to 22.25% (upward propagation). Vapors are heavier than air (vapor density 3.03). Capable of forming peroxides under certain conditions, and there may be danger of explosion if redistilled, unless certain precautions are taken.	Isolate and safeguard against mechanical injury.
Ethyl ether	Glass bottles or tin cans in boxes, steel drums.	Anesthetic. See National Board of Fire Underwriters' Recommended Safeguards for the Installation and Operation of Anesthetical Apparatus Employing Combustible Anesthetics.	Lighter than water (sp gr 0.7135). Soluble in about ten times its own volume of water. Water may be utilized only to cool metal containers. Best extinguishing agents are carbon dioxide and sand, also tetrachloromethane (carbon tetrachloride) in case of fires involving limited amounts of ether. Total-flooding carbon dioxide systems may be employed for protection in storage rooms.	A highly volatile liquid, giving off even at comparatively low temperatures vapors which form with air or oxygen flammable and explosive mixtures. Explosive range 1.85 go 36.5% (upward propagation). The ignition temperature is comparatively low, being approximately 180°C (356°F). Spontaneously explosive peroxides sometimes form on long standing or exposure in bottles to sunlight. The vapors are heavier than air (vapor density 2.6) and may travel a considerable distance to a source of ignition and flash back. More hazardous than gasoline.	Safeguard containers against mechanical injury. Isolate and keep in unheated compartment away from sunlight and any source of ignition. Only electric equipment of the explosionproof type, Group C classification, permitted in atmospheres containing ether vapor in flammable proportions.
Ethylene	Steel cylinders.	Anesthetic.	Use water to cool cylinders not on fire. If gas is burning at valves or safety releases, usually the best course to follow is not to disturb or attempt to extinguish flame. To do so will cause the release of unburned gas and quickly create highly dangerous explosive atmospheres. If cylinder is mounted on an anesthetic machine or truck, it may be possible to move it to a safe place.	Highly flammable gas. Forms flammable and explosive mixtures with air or oxygen. Explosive range 2.75 to 28.6% (upward propagation). Ignition temperature about 450°C (842°F). Slightly lighter than air (density 0.97). The gas is spontaneously explosive in sunlight with chlorine. Only electric equipment of the explosion-proof type, Group C clas-	Isolate from oxygen cylinders and store in a cool, well-ventilated storeroom.

			In a closed storeroom carbon dioxide or carbon tetrachloride are best extinguishing agents.	sification, permitted in atmospheres containing ethylene in flammable proportions.	
Ferric nitrate	Wooden barrels.			Oxidizing material.	Isolate. Safeguard against mechanical injury.
Formic acid	Barrels and carboys.	Corrosive; has caustic effect on the skin.		Flammable; gives off flammable vapors which may form explosive mixtures with air.	Safeguard against mechanical injury.
Fulminate of mercury				High explosive.	Explosive restrictions.
Fulminate of silver				High explosive.	Explosive restrictions.
Hydrochloric acid (muriatic acid)	Tank cars (rubber-lined), carboys, and glass bottles.	Aqueous solution is corrosive, irritating, and poisonous. Fumes are corrosive and irritating to mucous membranes.	Use water or chemically basic substances such as soda ash or slaked lime.	Not combustible (in air) but if allowed to come in contact with common metals, hydrogen is evolved, which may form explosive mixtures with air.	Safeguard containers against mechanical injury. Keep away from oxidizing agents, particularly nitric acid and chlorates. Avoid contact by leakage or otherwise with all common metals.
Hydrocyanic acid (prussic acid)	Cylinders, or when completely absorbed in inert material in metal cans with outside wooden boxes.	Poisonous. Few breaths may cause unconsciousness and death. Avoid contact with the skin.	The gas is slightly lighter than air. Soluble in water. Water is the best extinguisher. When entering premises where used or stored during a fire, use oxygen helmet or gas mask equipped with canister of a type approved by the Bureau of Mines for hydrocyanic acid.	Forms flammable and explosive mixtures with air. Explosive range 6 to about 40% by volume (horizontal propagation). Concentrations of the gas ordinarily employed for fumigation (1% or less) are considerably below the lower limit of flammability (6% by volume). Some methods of fumigation, however, are employed which temporarily yield flammable mixtures even though the final concentration is low.	Isolate. Keep away from any source of heat. Safeguard containers against mechanical injury.
Hydrofluoric acid (HF)	Aqueous solution in lead carboys and wax or gutta-percha bottles.	Acid and its vapors highly toxic and irritating to skin, eyes, and respiratory tract. Fumes produced by contact with ammonia and many metals are poisonous. May be neutralized with chalk. Bicarbonate of soda solution may be immediately applied to burns as first-aid and used as gargle.	Use water in case of fires involving hydrofluoric acid. Use oxygen helmet of a type approved for such use by the U.S. Bureau of Mines on entering atmospheres known to contain hydrofluoric acid vapors.	Colorless, volatile liquid. Not combustible but reacts with glass and most substances, platinum being an exception. Aqueous solution also attacks glass and several metals.	Isolate. Ventilate. Safeguard against mechanical injury. Encountered in glass works and chemical laboratories. Used to remove sand from castings and in the manufacture of filter paper. Vapors have been known to cause serious corrosion of sprinkler piping and heads.
Hydrofluosilicic acid	Lead carboys, hard rubber or paraffin bottles.	Corrosive.		None.	Safeguard against mechanical injury.

TABLE 24-1 (Continued)

Chemical	*Usual shipping container*	*Life hazard*	*Fire-fighting phases*	*Fire hazard*	*Storage*
Hydrogen peroxide (27.5% by weight)	Glass carboys, aluminum drums, aluminum tank cars (all containers must be vented).	Prolonged exposure to vapor irritating to eyes and lungs. Causes skin irritation. Use goggles to protect eyes from splash.	Use water.	Oxidizing liquid. May cause ignition of combustible material if left standing in contact with it. May decompose violently if contaminated with iron, copper, chromium, and most metals or their salts.	Store in a cool place in ventilated containers remote from combustible material and catalytic metals such as iron, copper, chromium.
Hydrogen sulfide (sulfuretted hydrogen)	Steel cylinders.	Toxic. 0.05 to 0.07% by volume in air causes dangerous illness in ½ to 1 h. Should be used under hoods in chemical laboratories to avoid danger of breathing dangerous concentrations.	Use gas masks in entering atmospheres containing hydrogen sulfide. If, however, concentration is high, use oxygen helmet of a type approved for such use by U.S. Bureau of Mines.	Flammable gas. Forms flammable and explosive mixture with air or oxygen. Explosive range in air (upward propagation) 4.3 (low limit) to 46. Heavier than air. Specific gravity 1.19 (air = 1). Ignition temperature 346–379°C (655–714°F)	Store in ventilated place away from fuming nitric acid and oxidizing materials. Encountered in chemical laboratories, metallurgical and smelting works, gas works, sewers.
Lead nitrate	Wooden barrels.	Poisonous.		Oxidizing material. Classes with sodium nitrate.	Isolate; safeguard against mechanical injury.
Lime (unslaked)—*See Calcium oxide.*					
Magnesium	Shavings or powder in tightly closed metal or metal-lined containers. Ingots and bars in ordinary boxes.	Serious under fire conditions. Danger of explosion and from flying particles. Do not attempt to smother unless at a safe distance, or else protection is provided for eyes and face.	Smother with an excess of dry graphite. Dry sand may be used on small fires. Not advisable to use sand around machinery. Do not use water, foam, tetrachloromethane (carbon tetrachloride), or carbon dioxide.	Combustible, particularly in the form of powder, shavings, or thin sheets. When powder is disseminated in the air, explodes by spark. In finely divided form liberates hydrogen in contact with water. In massive form (ingots or blocks) comparatively difficult to ignite.	Store remote from water or moisture, oxidizing materials, chlorine, bromine, iodine, acids, and alkalies.
Magnesium alloys (high percentage of magnesium)		Protect eyes and skin from flying particles in case of fire.	Smother with an excess of dry graphite. Dry sand may be used on small fires. Not advisable to use sand around machinery. Do not use water, foam, tetrachloromethane (carbon tetrachloride), or carbon dioxide.	In compact or bulk form (castings, plates, etc.) difficult to ignite. Readily combustible in form of dust, turnings, and hazardous in such form with chlorine, bromine, iodine, oxidizing agents, acids, and alkalies.	Store dust, shavings, and turnings in metal containers in detached building or fire-resistive room. Detailed safety precautions for handling are usually supplied by manufacturers of magnesium alloys.
Magnesium nitrate	Wooden boxes.		See sodium nitrate.	Oxidizing material. Classes with sodium nitrate.	See sodium nitrate.
Muriatic acid—*See Hydrochloric acid.*					

Naphthalene	Tins, barrels, and burlap bags.	Irritant. Slight narcotic effect on the skin. The hot vapors produce itching, pain, and eczema.	Water is the best extinguishing agent. Foam or water applied to molten naphthalene at temperatures over 110°C (230°F) will cause foaming.	Gives off flammable vapors when heated. Flash point 80°C (176°F). Naphthalene dust forms explosive mixtures with air. Ignition temperature 559°C (1038°F).	Isolate; keep away from fire or heat.
Nickel nitrate	Wooden kegs.	Poisonous when taken internally.		Oxidizing material. Classes with sodium nitrate.	Safeguard against mechanical injury.
Nitraniline or nitroaniline	Wooden kegs.	Poisonous.		In presence of moisture causes nitration of organic materials and may result in spontaneous ignition.	Store in dry place; safeguard against mechanical injury.
Nitric acid	Carboys and glass bottles.	Corrosive; causes severe burns by contact, deadly if inhaled.		May cause ignition when in contact with combustible materials; corrodes iron or steel; may cause explosion when in contact with hydrogen sulfide and certain other chemicals.	Safeguard against mechanical injury of containers; isolate from turpentine, combustible materials, carbides, metallic powders, fulminates, picrates or chlorates.
Nitrochlorobenzene	Wooden kegs.	Serious under fire conditions.		Gives off flammable vapors when heated, which may form explosive mixtures with air.	Isolate, preferably in the open; if inside should be in unheated compartment or building.
Phenol		Poisonous.		When heated yields flammable vapors. Flash point 78°C (172.4°F).	Soluble in water. Never store with or above food products
Phosphorus, red	Hermetically sealed tin cans inside of wooden boxes.	Yields toxic fumes when burning.	Flood with water, and when fire is extinguished cover with wet sand or dirt. Under certain conditions at high temperatures reverts to white phosphorus.	Flammable. Explosive when mixed with oxidizing materials. Not as dangerous to handle as white phosphorus, and when afire, more readily extinguished.	Isolate from other chemicals, safeguard against mechanical injury of container.
Phosphorus, white (or yellow)	Under water usually in hermetically sealed cans enclosed in other hermetically sealed cans with outside wooden boxes, or in drums or tank cars.	Poisonous. Serious under fire conditions. Yields highly toxic fumes when burning. Contact of phosphorus with the skin causes severe burns.	Deluge with water until fire is extinguished and phosphorus solidified; then cover with wet sand or dirt.	Highly flammable. Explosive in contact with oxidizing material. Ignites spontaneously on contact with air.	Isolate from chemicals. Store large quantities under water in underground iron or concrete tanks.
Phosphorus sesquisulfide, P_2S_5	Wooden boxes, iron drums, glass bottles.	Fumes in fire toxic.		Highly flammable. Ignites by friction.	Isolate from chemicals. Safeguard container against shock.
Picric acid	Wooden kegs, boxes, bottles.	Classes with high explosives in respect to life hazard.		Flammable, explosive. Oxidizing material.	Isolate or store under water, keep away from other material, including metals, with which it forms sensitive and explosive picrates.

TABLE 24-1 (Continued)

Chemical	*Usual shipping container*	*Life hazard*	*Fire-fighting phases*	*Fire hazard*	*Storage*
Potassium (metallic potassium)	Hermetically sealed steel drums, tin cans, and tank cars.	Strong caustic reaction. Dangerous.	Smother with an excess of dry graphite or dry sand. Do not use water. It is difficult to extinguish fires in large quantities of potassium.	Oxidizes rapidly on exposure to atmosphere, igniting spontaneously if warm enough. Water is decomposed suddenly by contact with potassium, sufficient heat being generated to ignite spontaneously the evolved hydrogen (in the presence of air). Its reaction with water is more violent than that of sodium.	Do not tier if it can be avoided. Keep away from water, avoiding sprinkler systems. Safeguard against mechanical injury of containers.
Potassium chlorate	Wooden barrels or kegs.	Dangerous under fire conditions.	Water is the best extinguishing agent.	Oxidizing material; explosive when in contact with combustible material.	Isolate from combustible material, acids, and sulfur.
Potassium cyanide	Tightly closed glass, earthenware, or metal containers; wooden boxes with inside metal containers, or with hermetically sealed metal lining; metal barrels or drums.	Highly poisonous when taken internally. Evolves hydrocyanic acid gas (poisonous) on contact with acids or moisture.		Cyanides not flammable but evolve hydrocyanic acid (see) on contact with acids or moisture.	Isolate. Safeguard containers against mechanical injury.
Potassium hydroxide	Wooden barrels, glass bottles.			Generates heat on contact with water. Classes with calcium oxide (lime) in hazard.	Store in dry place, keep remote from water or moisture.
Potassium nitrate	Bags, tins, and glass bottles.		See sodium nitrate.	In contact with organic materials causes violent combustion on ignition. Classes with sodium nitrate.	Store in dry place; prevent contact with organic material.
Potassium perchlorate	Paper-lined metal containers.		Use water to prevent the spread of fire.	Oxidizing material. Combustible in contact with organic materials. More stable than chlorates. Explosive in contact with concentrated sulfuric acid.	Store in dry place away from acids and combustible material.
Potassium permanganate	Tins.			Oxidizing material. Explosive when treated with sulfuric acid, and in contact with alcohol, ether, flammable gases, and combustible materials.	Isolate from other chemicals, especially those noted under fire hazard.

Potassium peroxide	Tins and steel drums.	Strong caustic reaction and dangerous under fire conditions. Avoid breathing dust in handling, and wear goggles to protect eyes.	Smother with dry sand, soda ash, or rock dust. Do not use water.	Does not burn or explode per se but mixtures of potassium peroxide and combustible substances are explosive and ignite easily even by friction or on contact with a small amount of water. Reacts vigorously with water, and in large quantities this reaction may be explosive.	Store remote from organic substances and water. Do not expose to sprinkler systems.
Potassium persulfate	Glass bottles and stone jars.			Oxidizing material. May cause explosion in a fire.	Keep dry; safeguard against mechanical injury of containers.
Potassium sulfide	Iron drums, cans, glass bottles.	Yields irritating and corrosive gases when burning.		Moderately flammable, yields flammable hydrogen sulfide on contact with mineral acids and sulfur dioxide when burning.	Safeguard against mechanical injury of containers.
Salicyclic acid[f]	Bottles, cartons, kegs and barrels.	There have been explosions in sublimation chambers.	Slightly soluble in water. Extinguish with water or smother with carbon dioxide or sand.	Combustible solid. Flash point 157°C (315°F). Ignition temperature 545°C (1013°F). Salicyclic dust forms explosive mixtures with air.	Store in dry place.
Saltpeter—*See Potassium nitrate.*					
Silver nitrate	Amber or black glass bottles.	Corrosive and poisonous.		Oxidizing material.	Store in dark place; keep cool and away from combustible material.
Soda, caustic—*See Sodium hydroxide.*					
Sodium	Hermetically sealed steel drums, tin cans, and tank cars.	Strong caustic reaction. Dangerous.	Smother with excess of dry graphite or dry sand. Do not use water. It is difficult to extinguish fires in large quantities of sodium.	Water is suddenly decomposed by contact with sodium with the evolution of hydrogen, which may ignite spontaneously (in the presence of air). Classes with potassium in respect to fire hazard but its reaction with water is less violent than that of potassium.	Do not tier if it can be avoided. Keep away from water, avoiding sprinkler systems. Safeguard against mechanical injury of containers.
Sodium chlorate	Wooden barrels, glass bottles.	See potassium chlorate.	See potassium chlorate.	Oxidizing material. Classes with potassium chlorate. See potassium chlorate.	See potassium chlorate.

TABLE 24-1 (Continued)

Chemical	*Usual shipping container*	*Life hazard*	*Fire-fighting phases*	*Fire hazard*	*Storage*
Sodium chlorite ($NaClO_2$)	Wooden boxes with inside containers which must be glass or earthenware not over 2¼-lb capacity, or metal not over 5-lb capacity each. *Used in bleaching textiles and paper.*	Poisonous when taken internally. Dangerous under fire conditions.	Soluble in water. Water is best extinguishing agent.	Strong oxidizing material. Decomposes with evolution of heat at about 175°C (347°F). Explosive in contact with combustible material. See potassium chlorate. In contact with strong acid liberates chlorine dioxide, an extrahazardous gas.	Isolate from combustible material, sulfur, and acids.
Sodium cyanide—*See Potassium cyanide.*					
Sodium hydrosulfite[g]	Wooden barrels, kegs, or boxes with inside glass bottles of capacity not exceeding 5 lb each, or metal containers.		Smother with sand or foam.	Combustible. Heats spontaneously in contact with moisture and air, and may ignite nearby combustible material.	Store in dry place away from combustible materials.
Sodium hydroxide	Iron drums.			Classes with potassium hydroxide and calcium oxide.	Isolate from heat and water. See calcium oxide and potassium hydroxide.
Sodium nitrate	Bags, tins, and glass bottles.		Most fires involving sodium nitrate can safely be fought with water in the early stages; at such times the fire should be flooded with water. When large quantities are involved in the fire, the sodium nitrate may fuse or melt, in which condition application of water may result in extensive scattering of the molten material, and, therefore, care should be taken in applying water to the material after fire has been burning for some time.	Oxidizing material. Bags or barrels may become impregnated with nitrate, in which condition they are readily ignitable. In contact with organic or other readily oxidizable (combustible) substances it will cause violent combustion on ignition.	Store in dry place; prevent contact with organic or combustible material. Fire hazard less if removed from bags and stored in noncombustible bins.
Sodium perchlorate	Paper-lined metal containers.		See potassium perchlorate.	See potassium perchlorate.	See potassium perchlorate. Sodium perchlorate in anhydrous form is very hygroscopic and not used much in industries.

Sodium peroxide—*See Potassium peroxide.*					
Sodium sulfide	Iron drums and bottles.	See potassium sulfide.		Moderately flammable. Classes with potassium sulfide.	See potassium sulfide.
Strontium nitrate	Barrels and boxes.		See sodium nitrate.	Oxidizing material. Classes with sodium nitrate.	Safeguard against mechanical injury; keep away from other materials.
Strontium peroxide (SrO_2)	Metal cans.		Smother with sand, ashes, or rock dust. Do not use water.	Oxidizing material. Hazards in class with barium peroxide.	Store in dry place away from combustible materials.
Sulfur	Sacks, boxes, barrels, and box cars.	When burning forms sulfur dioxide, which in concentrations of 0.2% by volume in air may cause serious injury in ½ h or less.	Water in form of spray best extinguisher. Small fires may be smothered with sand or additional sulfur. (Sulfur dioxide does not support combustion.) Avoid use of pressure hose (solid) streams, and do not scatter sulfur dust. If spray nozzle is not available, water may be allowed to flow out of hose (without nozzle) on to a burning pool of sulfur, or saturated steam may be used. See NFPA Code for the Prevention of Sulfur Dust Explosions and Fires.	Flammable. Dust or vapor forms explosive mixtures with air. Hazardous in contact with oxidizing material.	Provide good ventilation. Isolate from chlorates, nitrates, and other oxidizing materials.
Sulfuric acid	Carboys, iron drums, glass bottles, and tank cars.	Corrosive, dangerous fumes under fire conditions.	Smother with sand, ashes, or rock dust, but avoid water.	May cause ignition by contact with combustible materials. Corrodes metal.	Safeguard against mechanical injury, isolate from saltpeter, metallic powders, carbides, picrates, fulminates, chlorates, and combustible materials.
Thorium nitrate	Wooden kegs.	Possibly radioactive.	See sodium nitrate.	Oxidizing material. Classes with sodium nitrate.	Store in dry place, remote from water or moisture.
Uranium nitrate	Glass bottles, boxes.	Possibly radioactive.		Oxidizing material. Classes with sodium nitrate.	See sodium nitrate.

TABLE 24-1 (Continued)

Chemical	*Usual shipping container*	*Life hazard*	*Fire-fighting phases*	*Fire hazard*	*Storage*
Vinyl ether	Glass bottles and metal cans.	Anesthetic.	Lighter than water (sp. gr. 0.774). Not soluble in water. Water may be used only to cool metal containers. Best extinguishing agents are carbon dioxide, sand, and tetrachloromethane (carbon tetrachloride).	Highly volatile flammable liquid. Flash point below −30°C (−22°F). Gives off even at comparatively low temperatures vapors which form flammable mixtures with air or oxygen. Explosive range 1.70 to 27.0% (upward propagation). Hazard in a class with ethyl ether. Only electric equipment of the explosion-proof type, Group C classification tentative pending further tests, permitted in atmospheres containing vinyl ether vapor in flammable proportions.	Isolate. Safeguard containers against mechanical injury. Store in a cool, well-ventilated storeroom.
Zinc chlorate	Glass bottles, iron drums.	Serious under fire conditions.		When in contact with organic material explodes by slight friction, percussion, or shock. Classes with potassium chlorate.	Safeguard against mechanical injury; avoid tiering; isolate.
Zinc powder or dust	Cartons, wooden barrels, or steel drums.	Zinc is comparatively volatile at elevated temperatures. Under fire conditions precautions should be taken to avoid breathing fumes, which may cause metal fume fever.	Smother with sand, ashes, or rock dust. Do not use water.	Hydrogen is evolved when commercial zinc is in contact with acids, sodium hydroxide, or potassium hydroxide. Hydrogen also evolved by acid-forming combinations containing zinc, such as zinc chloride and moisture. Dust may form explosive mixtures with air. Zinc dust in bulk in a damp state may heat and ignite spontaneously on exposure to the air.	Store in dry, ventilated place away from water or moisture. Isolate from acids.

Zirconium	Wooden kegs, glass bottles.		Available data indicate fires can be controlled by foam or sand. Tetrachloromethane (carbon tetrachloride), carbon dioxide, soda and acid extinguishers ineffective.	Has comparatively low ignition temperature. Highly flammable in dry state. Burns with intensely brilliant flame. Explosive in contact with oxidizing agents. Powder very susceptible to ignition by static electricity, and explosion may be caused when disbursed into a cloud in air by static charges generated.	Encountered in granular, finely divided powder; also in form of small, friable, spongy lumps. Store only in wet condition and in small quantities. Isolate from oxidizing materials.

[a] Expands on solidification and may burst container unless kept at a temperature above 16°C (60.8°F).

[b] Used in manufacture of matches, ammunition, and fireworks, and of certain rubber compounds.

[c] Encountered in paper, textile, disinfectant, and alkali industries; also where water-purification processes are employed.

[d] Carbon disulfide should never be transferred by means of air. Use inert gas, water, or pump. Use a wood measuring stick for measuring contents of storage tanks or tank cars. Tank cars when being loaded or unloaded should be well grounded. Do not dispose of carbon disulfide by pouring it on the ground. Provide a safe place for burning it.

[e] Used in chromium plating, in electric batteries, and in photography.

[f] Used in manufacture of aspirin, salol, and methyl salicylate; also in manufacture of azo dyes. Preservative.

[g] Bleaching agent for removing dyes.

SOURCE: *Prepared by Committees of the National Fire Protection Association and American Chemical Society. Tentatively adopted* 1928, *adopted* 1929, *amended* 1931, 1935, 1938, 1939, 1941, 1942.

NOTE: These columns are informative only; it is not considered necessary that the material be kept or stored only in the containers as listed, nor that each package be labeled. The requirements in the table on the storage of containers refer to chemicals in usual containers, and are not intended to apply to small bottles of chemicals such as are found in drug stores and chemical laboratories.

TABLE 24-2
Examples of Incompatible Chemicals

Chemical	*Keep out of contact with:*
Acetic acid	Chromic acid, nitric acid, hydroxyl compounds, ethylene glycol, perchloric acid, peroxides, permanganates
Acetylene	Chlorine, bromine, copper, fluorine, silver, mercury
Alkaline metals, such as powdered aluminum or magnesium, sodium, potassium	Water, tetrachloromethane or other chlorinated hydrocarbons, carbon dioxide, the halogens
Ammonia, anhydrous	Mercury (in manometers, for instance), chlorine, calcium hypochlorite, iodine, bromine, hydrofluoric acid (anhydrous)
Ammonium nitrate	Acids, metals powders, flammable liquids, chlorates, nitrites, sulfur, finely divided organic or combustible materials
Aniline	Nitric acid, hydrogen peroxide
Bromine	Same as for chlorine
Carbon, activated	Calcium hypochlorite, all oxidizing agents
Chlorates	Ammonium salts, acids, metals powders, sulfur, finely divided organic or combustible materials
Chromic acid	Acetic acid, naphthalene, camphor, glycerin, turpentine, alcohol, flammable liquids in general
Chlorine	Ammonia, acetylene, butadiene, butane, methane, propane (or other petroleum gases), hydrogen, sodium carbide, turpentine, benzene, finely divided metals
Chlorine dioxide	Ammonia, methane, phosphine, hydrogen sulfide
Copper	Acetylene, hydrogen peroxide
Cumene hydroperoxide	Acids, organic or inorganic
Flammable liquids	Ammonium nitrate, chromic acid, hydrogen peroxide, nitric acid, sodium peroxide, the halogens
Fluorine	Isolate from everything
Hydrocarbons (butane, propane, benzene, gasoline, turpentine, etc.)	Fluorine, chlorine, bromine, chromic acid, sodium peroxide

(pages 552 to 565) and 24-2 (above) contain valuable information about safe handling of chemicals and should be consulted before solutions are prepared.

CAUTION *Never fill a receptacle with a material other than that called for by the label. Label all containers before filling them. Throw away contents of all unlabeled containers.*

Chemical	*Keep out of contact with:*
Hydrocyanic acid	Nitric acid, alkali
Hydrofluoric acid, anhydrous	Ammonia, aqueous or anhydrous
Hydrogen peroxide	Copper, chromium, iron, most metals or their salts, alcohols, acetone, organic materials, aniline, nitromethane, flammable liquids, combustible materials
Hydrogen sulfide	Fuming nitric acid, oxidizing gases
Iodine	Acetylene, ammonia (aqueous or anhydrous), hydrogen
Mercury	Acetylene, fulminic acid, ammonia
Nitric acid (concentrated)	Acetic acid, aniline, chromic acid, hydrocyanic acid, hydrogen sulfide, flammable liquids, flammable gases
Oxalic acid	Silver, mercury
Perchloric acid	Acetic anhydride, bismuth and its alloys, alcohol, paper, wood
Potassium	Tetrachloromethane, carbon dioxide, water
Potassium chlorate	Sulfuric and other acids
Potassium perchlorate (see also chlorates)	Sulfuric and other acids
Potassium permanganate	Glycerin, ethylene glycol, benzaldehyde, sulfuric acid
Silver	Acetylene, oxalic acid, tartaric acid, ammonium compounds
Sodium	Tetrachloromethane, carbon dioxide, water
Sodium peroxide	Ethyl or methyl alcohol, glacial acetic acid, acetic anhydride, benzaldehyde, carbon disulfide, glycerin, ethylene glycol, ethyl acetate, methyl acetate, furfural
Sulfuric acid	Potassium chlorate, potassium perchlorate, potassium permanganate (or compounds with similar light metals, such as sodium, lithium)

SOURCE: Chemical Manufacturers Association, *Waste Disposal Manual*.

Incompatible Chemicals

Certain chemicals may react and create a hazardous condition. Separate storage areas should be provided for such *incompatible chemicals*. (Refer to Chap. 2, "Laboratory Safety," for information regarding storage of volatile and flammable chemicals.)

WATER FOR LABORATORY USE

Water is needed in the laboratory in various grades of purity depending upon the requirements of the procedures for which it is to be used.

Water-Purity Specifications

Solutions which conduct electric current are said to have electric conductivity, and the current is carried solely by the ions in solution. This electrical conductivity is also called the specific conductance, and it is measured in units of siemens per centimeter (S/cm), which is the reciprocal of the resistance in ohms per centimeter (Ω/cm). This characteristic is used to specify the purity of the water. A microsiemens is one-millionth of a siemens. The resistance measurements require the use of an ac Wheatstone bridge, a conductivity cell, and a null indicator. Alternating current is normally used to avoid polarization of the electrodes.

Contrary to folk belief, rainwater is not absolutely pure water because it contains dissolved gases: O_2, N_2, CO_2, and oxides of nitrogen. As it falls on the ground and flows to city reservoirs, it dissolves minerals and other soluble substances. As a result, surface water can contain such positive ions as Na^+, K^+, Ca^{2+}, Mg^{2+}, Fe^{2+}, and Fe^{3+}, as well as such negative ions as CO_3^{2-}, HCO_3^-, SO_4^{2-}, NO_3^-, and Cl^-. Rainwater, after appropriate treatment, is often the source of city water. Depending upon the particular processes used in its purification and the efficiency of those processes, average city water normally contains dissolved inorganic substances, microorganisms, dissolved organic compounds of vegetable and plant origin, and particulate matter. This tap water contains sufficiently high levels of these impurities so that it cannot be used for testing procedures and analytical evaluations. It is also frequently "hard." Hard water contains appreciable quantities of Ca^{2+}, Mg^{2+}, Fe^{2+}, and Fe^{3+}; these minerals react with soap to form insoluble curds. Thus hard water does not lather readily. Soft water contains little mineral matter, lathers well, and is needed for laboratory use.

Softening Hard Water

Temporarily Hard Water

Some waters are temporarily hard, because they contain the bicarbonates of calcium, magnesium, and/or iron. These bicarbonates can be removed simply by heating, thus converting the soluble bicarbonate to the insoluble carbonate, which can be filtered out. Therefore, temporarily hard water can be converted to soft water by boiling.

Permanently Hard Water

When water contains sulfates or chlorides of calcium, magnesium, or iron, it is permanently hard water and is unsuitable for most laboratory work. It can be made soft in a number of ways. The first consists of adding substances which convert the undesirable soluble calcium, magnesium, and iron salts to precipitates which can be removed by suitable filtration. The most common water-softening substances are listed below:

Borax (sodium tetraborate), $Na_2B_4O_7$

Ammonium hydroxide, NH_4OH

Sodium carbonate, Na_2CO_3

Sodium hydroxide, NaOH

Potassium hydroxide, KOH

Trisodium phosphate, Na_3PO_4

Mixture of calcium hydroxide and sodium carbonate, $Ca(OH)_2$ and Na_2CO_3

Ion-Exchange Resins

Soft water can also be obtained by passing hard water through a bed of special synthetic resins. These resins exchange their "soft" sodium ions for the water's "hard" calcium, magnesium, and iron ions. As long as the resins will exchange their sodium ions for calcium ions, the water will be softened. However, when the resins will no longer effect this exchange because all the sodium ions are gone, the resin is said to be exhausted and it must be recharged. In the recharging process, sodium ions are replaced in the resin and the calcium ions are discarded. Water obtained in this way is called *deionized water*. For additional information on ion exchange, see Chap. 36, "Chromatography," p. 783.

Purifying Water by Reverse Osmosis

High-quality water of consistent purity can be obtained by the process of reverse osmosis whereby the dissolved solids are separated from feedwater by applying a pressure differential across a semipermeable membrane. This semipermeable membrane allows the water to flow through it, but prevents dissolved ions, molecules, and solids from passing through. The differential pressure forces the water through the membrane, leaving the dissolved particles behind, thus producing laboratory-grade water.

Distilled Water

Water purified by distillation is called *distilled water*. Electrically heated or steam-heated water stills (Fig. 24-2) distill raw water to give high-grade distilled water, venting volatile impurities. High-quality distilled water (Fig. 24-3) has a specific conductance of less than one microsiemens* per

---

* A more familiar name for this unit may be the micromho; the siemens is the SI unit of conductance.

FIGURE 24-2
Still for water. Flexible tube conducts distilled water to holding container.

centimeter (μS/cm), which corresponds to about 0.5 ppm of a dissolved salt. Absolutely pure water has a conductivity of 0.55 μS/cm at 25°C. As mentioned before, this characteristic is used to specify the purity of the water.

Demineralized Water

In the laboratory, the presence of even sodium ions may not be desirable or tolerated, and water which is free of all inorganic ions must be supplied. Demineralizers (Fig. 24-4) are used to remove mineral ions from water. Demineralizers contain several beds of resins and are packaged in

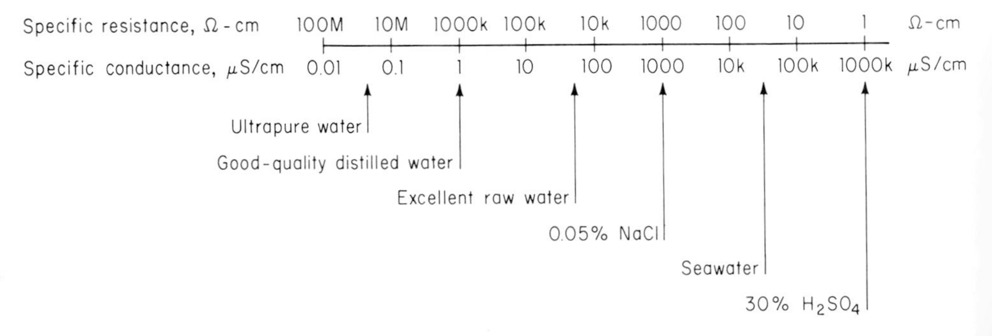

FIGURE 24-3
Specific conductances of various grades of water.

FIGURE 24-4
Demineralizer with dial to indicate the hardness of the water.

cartridges; the water that is obtained this way is free of both mineral cations and anions.*

CAUTION Demineralizers will not remove organic matter or nonelectrolytes from water. These must first be removed by distillation. You cannot substitute demineralized water for distilled water in every case. Check solution requirements.

Reagent-Grade Water

Reagent-grade water is water of the highest purity that is available from a practical standpoint; it is even purer than triple-distilled water, which can contain significant amounts of dissolved inorganic substances, organic substances, suspended particulates, and bacteria. These substances can interfere with or adversely affect analytical procedures employing atomic absorption photometry, chromatography, tissue culturing, etc.

Reagent-grade water is obtained from pretreated water, that is, water which has been distilled, deionized, or subjected to reverse osmosis; it is then passed through an activated carbon cartridge to remove the dissolved organic materials, through two deionizing cartridges to remove any dissolved inorganic substances, and through membrane filters to remove microorganisms and any particulate matter with a diameter larger than 0.22 micrometers (μm).

* The terms "deionized water" and "demineralized water" are sometimes used interchangeably. Make sure that the water you use meets the specifications named for conductivity and purity.

SOLUTIONS

Most chemical compounds found in the laboratory are not pure, but as previously explained, contain various percentages of impurities, depending on their grade. They are often used in solution form.

Solutions are mixtures which are characterized by homogeneity, absence of settling, and a molecular or ionic state of subdivision of the components. There are many kinds of solutions possible, because of the differences in the state of the solute (substance dissolved) and the solvent (the dissolving medium).

The Dissolution Process

When a solute is added to a solvent and enters the dissolved state, it will rapidly diffuse throughout the solvent, dispersing into separate atoms, ions, or molecules.

Solubility

Solvents have limited capacity to dissolve solutes, and that limit defines the *solubility* of the solute in the solvent (the maximum amount of solute that will dissolve in a fixed amount of solvent at a definite temperature). Pressure has an effect only upon gases, otherwise its effect is unimportant. When a two-component system contains the maximum quantity of solute in a solvent, the system is said to be *saturated.* An equilibrium is established between the pure solute and the dissolved solute. At the equilibrium point, the rate at which the pure solute enters the solution equals the rate at which the dissolved solute crystallizes out of the solution to return to the pure state. When a solution retains more than the equilibrium concentration of the dissolved solute (which happens under certain conditions) the solution is said to be *supersaturated.* Supersaturation is an unstable condition which can revert back to a stable one as a result of physical shock, decreased temperature, merely standing for a period of time, or some indeterminate factor.

Solubility Rules for Common Inorganic Compounds

Following are general solubility rules for inorganic compounds. See also Tables 24-3 and 24-4.

NO_3^- — All nitrates are soluble.

$C_2H_3O_2^-$ — All acetates are soluble; $AgC_2H_3O_2$ is moderately soluble.

Cl^- — All chlorides are soluble except AgCl, $PbCl_2$, and HgCl. $PbCl_2$ is soluble in hot water, slightly soluble in cold water.

TABLE 24-3
Soluble Complex Ions

Cation	NH_3	CNS^-	Cl^-	CN^-	OH^-
Al^{3+}					$Al(OH)_4^-$
Ag^+	$Ag(NH_3)_2^+$			$Ag(CN)_2^-$	
Cd^{2+}	$Cd(NH_3)_4^{2+}$		$[CdCl_4]^{2-}$	$[Cd(CN)_4]^{2-}$	
Co^{2+}	$Co(NH_3)_6^{3+}$			$Co(CN)_4^{3+}$	
Cr^{2+}					$Cr(OH)_4^-$
Cu^{2+}	$Cu(NH_3)_4^{2+}$		$CuCl_4^{2-}$	$Cu(CN)_2^-$	
Fe^{2+}				$Fe(CN)_6^{4-}$	
Fe^{3+}		$Fe(CNS)^{2+}$	$FeCl_4^-$	$Fe(CN)_6^{3-}$	
Hg^{2+}	$Hg(NH_3)_4^{2+}$	$Hg(CNS)_4^{2-}$	$HgCl_4^{2-}$	$Hg(CN)_4^{2-}$	$HHgO_2^-$
Ni^{2+}	$Ni(NH_3)_6^{2+}$			$Ni(CN)_4^{2-}$	
Pb^{2+}			$PbCl_4^{2-}$		$Pb(OH)_3^-$
Zn^{2+}	$Zn(NH_3)_4^{2+}$			$Zn(CN)_4^{2-}$	$Zn(OH)_4^{2-}$

Arsenic (+3 or +5), antimony (+3 or +5), and tin (+2 or +4) react with Na_2S_x or $(NH_4)_2S_x$ to yield soluble AsS_4^{3-}, SnS_2^{2-}, and SbS_4^{4+}. Mercury will form HgS_2^{2-}. Concentrated NaOH or KOH yields AsO_2^-, AsO_4^{3+}, SnO_2^2, SnO_3^{2-}, SbO_4^{4+}.

SO_4^{2-}	All sulfates are soluble except $BaSO_4$ and $PbSO_4$. Ag_2SO_4, Hg_2SO_4, and $CaSO_4$ are slightly soluble.
HSO_4^-	The bisulfates are more soluble than the sulfates.
CO_3^{2-}, PO_4^{3-}, CrO_4^{2-}, SiO_4^{2-}	All carbonates, phosphates, chromates, and silicates are insoluble, except those of sodium, potassium, and ammonium. An exception is $MgCrO_4$, which is soluble.
OH^-	All hydroxides (except sodium, potassium, and ammonium) are insoluble; $Ba(OH)_2$ is moderately soluble; $Ca(OH)_2$ and $Sr(OH)_2$ are slightly soluble.
S^{2-}	All sulfides (except sodium, potassium, ammonium, magnesium, calcium, and barium) are insoluble. Aluminum and chromium sulfides are hydrolyzed and precipitate as hydroxides.
Na^+, K^+, NH_4^+	All sodium, potassium, and ammonium salts are soluble. Exceptions: $Na_4Sb_2O_7$, $K_2NaCo(NO_2)_6$, K_2PtCl_6, $(NH_4)_2NaCo(NO_2)_6$, and $(NH_4)_2PtCl_6$.
Ag^+	All silver salts are insoluble. Exceptions: $AgNO_3$ and $AgClO_4$; $AgC_2H_3O_2$ and Ag_2SO_4 are moderately soluble.

TABLE 24-4
Solubilities of Inorganic Compounds in Water and Other Aqueous Solvents

Cations	*Acetate*	*Borate*	*Bromide*	*Carbonate*	*Chlorate*	*Chloride*	*Chromate*	*Ferri-cyanide*
Aluminum	W	A	W		W	W	A	
Ammonium	W	W	W	W	W	W	W	W
Barium	W	A	W	A	W	W	A	W
Bismuth	W	A	3	A	W	5	A	
Cadmium	W		W	A	W		A	
Calcium	W	A	W	A	W	W	W	W
Chromium(III)	W	A	W		W	W	A	
Cobalt(II)	W	A	W	A	W	W	A	i
Copper(II)	W	A	W	A	W	W	W	i
Iron(II)	W	A	W	A	W	W		i
Iron(III)	W	A	W		W	W	W	W
Lead(II)	W	2	1	2	W	19	2	1
Magnesium	W	A	W	A	W	W	W	W
Mercury(I)	1A		2	2	W	2	A	A
Mercury(II)	W		A	A	W	W	A	
Nickel	W	A	W	A	W	W	A	i
Potassium	W	W	W	W	W	W	W	W
Silver	1	A	4	2	W	6	A	6
Sodium	W	W	W	W	W	W	W	W
Tin(II)	W	A	W		W	A		i
Tin(IV)	W		W			W	W	
Zinc	W	A	W	A	W	W	A	A

1 = only slightly soluble
2 = soluble in nitric acid, HNO_3
3 = soluble in hydrobromic acid, HBr
4 = soluble in potassium cyanide, KCN
5 = soluble in hydrochloric acid, HCl
6 = soluble in ammonium hydroxide, NH_4OH
7 = hydrolyzes
8 = soluble in sulfuric acid
9 = soluble in hot water
W = soluble in water
A = soluble in acid
B = soluble in base
d = decomposes in water
i = insoluble in water

Increasing the Rate of Solution of a Solute in a Liquid

A solvent will only dissolve a limited quantity of solute at a definite temperature. (Refer to Chap. 13, "Recrystallization.") However, the rate at which the solute dissolves can be speeded up by the following methods:

1. Pulverizing, or grinding up the solid to increase the surface area of the solid on contact with the liquid.

2. Heating the solvent. This will increase the rate of solution because the molecules of both the solvent and the solute move faster.

3. Stirring.

Combinations of all three methods, when practical, enable one to dissolve solids more quickly.

Ferro-cyanide	*Fluoride*	*Hydroxide*	*Iodide*	*Nitrate*	*Oxide*	*Phosphate*	*Sulfate*	*Sulfide*	*Sulfite*
A	W	AB	W	W	A	A	W	7	A
W	W	W	W	W		W	W	W	W
W	2	1	W	W	1	A	i	7	A
A	A	A	A	2	A	A	A	2	
	W	A	W	W	A	A	W	A	
W	i	1	W	W	1	A	1	7	A
	i	A		W	A	A	W	7	
i	1	AB	W	W	A	A	W	2	A
i	1	AB		W	A	A	W	2	A
i	1	A	W	W	A	A	W	A	A
i	1	A	W	W	A	A	W	A	
i	1	2B	1	W	A	2	i	2	2
W	A	A	W	W	A	A	W	7	W
	W		2	2	A	2	1	i	
A	W	A	i	W	A	A	A	i	A
i	i	AB	W	W	A	A	W	2	A
W	W	W	W	W	d	W	W	W	W
6	W		4	W	2	2	W	2	A
W	W	W	W	W	d	W	W	W	W
i	W	AB	W	W	A	A	1–8	A	A
i	W	AB	7d	W	A	A	W	A	
i	A	AB	W	W	A	A	W	A	A

COLLOIDAL DISPERSIONS

Colloidal dispersions are dispersions of particles which are hundreds and thousands of times larger than molecules or ions. Yet these particles are not large enough to settle rapidly out of solution, nor can they be seen under an ordinary microscope. These particles, which make up the dispersed phase in colloidal systems, are intermediate in size between those of a true solution and those of a coarse suspension. Thus, colloidal systems are somewhere between true solutions and coarse suspensions. However, they exhibit significant phenomena which are associated with adsorbed electric charges and enormous surface area. There are three types of colloidal systems: emulsions, gels, and sols.

Every colloidal system consists of two phases or portions:

1. The colloidal particles (the *dispersed phase*) are small aggregates or conglomerates of materials which are larger than the usual ion or molecule. These particles will remain suspended. They cannot be seen under the ordinary light microscope.

2. The dispersing medium *(dispersing phase)* is a continuous phase. It corresponds to the solvent of a solution.

Electrical Behavior

Since colloidal dispersions are either positively or negatively charged, they will migrate to an electrode of opposite charge. They can be coagulated by addition of highly charged ions of opposite sign: negatively charged particles are coagulated by positively charged ions, and positively charged particles, by negatively charged ions.

Distinguishing Between Colloids and True Solutions

Method 1

Colloidal dispersions, like true solutions, may be perfectly clear to the naked eye, but when they are examined at right angles to a beam of light (Fig. 24-5), they will appear turbid. This phenomenon is called the *Tyndall effect,* and it is caused by the scattering of light by the colloidal particles.

Method 2

The dispersed phase of a colloidal dispersion will pass through ordinary filter paper, but only the molecules and ions of a true solution will pass through semipermeable membranes. This selective passage of molecules and ions and retention of the particles of colloidal dispersions is called *dialysis*.

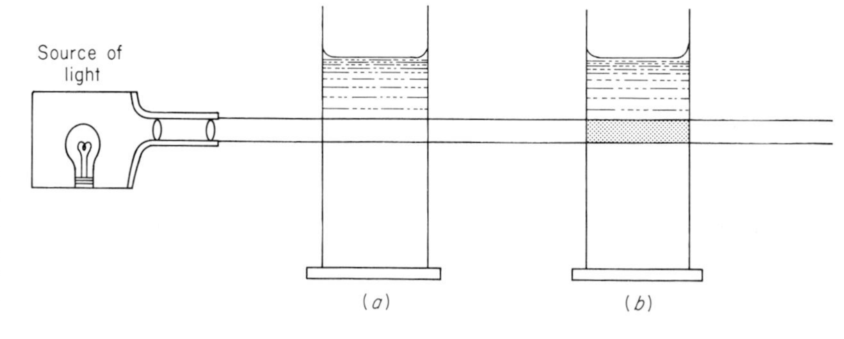

FIGURE 24-5
The Tyndall effect. (*a*) In a true solution a light beam is invisible. (*b*) In a colloidal solution a light beam is visible.

Distinguishing Between Colloids and Ordinary Suspensions

1. Carefully examine the solution visually, or if possible, determine the transmission of light through the sample in a spectrophotometer.

CAUTION *Do not shake the sample.*

2. Shake the sample and compare results with those of step 1.

3. Ordinary suspensions become turbid and opaque on shaking. Particles tend to settle. Colloidal dispersions appear unchanged.

LABORATORY SOLUTIONS

The solutions of chemical reagents used in the laboratory are prepared so that their concentrations or compositions are known and can be used in appropriate calculations. Some general definitions follow:

Solute That substance which is dissolved or has gone into solution (sugar into water) (less than 50 percent).

Solvent That substance which does the dissolving (water dissolving sugar) (more than 50 percent).

Solution Homogeneous system of two or more substances which may be present in varying amounts.

Composition Mass of solute per unit mass of solvent.

Concentration Amount of solute per unit volume of solvent.

Chemical Calculations and Computations for Preparation of Laboratory Solutions

Mass Percent

Grams of solute per 100 grams of solution

EXAMPLE 25 g of NaCl in 100 g of H_2O is a *20% by mass solution.*

Solution Mass NaCl = 25 g (solute)

Mass NaCl + H_2O = 25 g + 100 g = 125 g (mass of solution)

$$\text{Mass percent} = \frac{\text{mass of solute}}{\text{mass of total solution}}$$

25 g NaCl/125 g solution = ⅕ = 20% = 20 g NaCl/100 g solution

Procedure

Preparing percent by mass solutions:

1. Determine the percent and quantity of solution wanted.

2. Change the percent to a fraction or a decimal.

3. Multiply the total mass of the solution (step 1) by the fraction obtained in step 2 to get the mass of the material needed to make the solution.

4. Subtract the mass of the material obtained in step 3 from the total mass of the solution (step 1) to get the mass of the solvent needed to make the desired solution.

5. Dissolve the amount of material found in step 3 in the amount of solvent found in step 4 to make the desired solution percentage.

EXAMPLE Make 1000 g of a 10% NaCl solution by mass.

1. Total mass of solution wanted = 1000 g.

2. 10% = $\frac{1}{10}$.

3. $\frac{1}{10} \times 1000 = 100$ g NaCl or 0.1×1000 g = 100 g NaCl needed.

4. 1000 g − 100 g = 900 g of water needed.

5. Dissolve 100 g NaCl (step 3) in 900 g water (step 4) to get 1000 g of the 10% NaCl solution by mass.

Volume Percent

Milliliters of solute per 100 milliliters of *solution*

EXAMPLE 10 mL of ethyl alcohol plus 90 mL of H_2O is a *10% by volume solution.*

NOTE This rule generally holds true, but the volume percent *cannot always* be calculated directly from the volumes of the components mixed because the final volume *may not equal the sum* of the separate volumes.

Procedure

Preparing volume percent solutions:

1. Decide how much solution you want to make and of what strength (volume percent).

2. Express the strength wanted as a fraction or a decimal instead of a percent.

3. Multiply the total volume of the solution (step 1) by the fraction or decimal obtained in step 2 to get the volume of the material needed (step 3).

4. Subtract the volume of material needed in step 3 from the total volume of the solution (step 1) to get the volume of the other material needed.

5. Dissolve the volume of material found in step 3 in the volume found in step 4.

EXAMPLE Make 1000 mL of a 5% by volume solution of ethylene glycol in water.

1. Total volume wanted = 1000 mL.

2. 5% = $\frac{1}{20}$ or 5% = 0.05.

3. $\frac{1}{20} \times 1000$ mL = 50 mL or 0.05×1000 mL = 50 mL needed.

4. 1000 mL − 50 mL = 950 mL water needed.

5. Dissolve 50 mL ethylene glycol in 950 mL water to get the 1000 mL of the 5% ethylene glycol solution wanted. (See note above.) Use slightly less than 950 mL of water to start and adjust the volume if necessary.

Conversion of Mass Percent to Volume Percent

EXAMPLE A 10% by mass solution of ethyl alcohol in water contains 10 g of ethyl alcohol for each 90 g of water. What is the *volume percent* of this solution?

1. Determine the volume of the component by using the formula

$$\text{Volume} = \frac{\text{mass of component}}{\text{density of component}}$$

2. Calculate the volume of the total solution by dividing the mass of the solution by the density of solution.

3. Calculate the *percent by volume* by dividing the volume of the component by the volume of the solution.

Solution **1.** Mass of ethyl alcohol = 10 g (given)

Density of ethyl alcohol = 0.794 (obtained from reference tables)

$$\text{Volume} = \frac{\text{mass}}{\text{density}}$$

$$\text{Volume of ethyl alcohol} = \frac{10\text{ g}}{0.794\text{ g/mL}} = 12.6\text{ mL}$$

2. $\text{Volume of solution} = \dfrac{\text{mass of solution}}{\text{density of solution}}$

Mass of solution = 100 g (given)

Density of solution (10% ethyl alcohol) = 0.983 g/mL (obtained from reference tables)

$$\text{Volume of solution} = \frac{100 \text{ g}}{0.983 \text{ g/mL}} = 101.8 \text{ mL}$$

[See previous note on adding volumes: notice that, if the volume of alcohol (12.6 mL) is added to the volume of water (90 mL), the final volume will equal 102.6 mL.]

3. Volume percent of solution:

$$\text{Percent} = \frac{\text{volume of ethyl alcohol}}{\text{total volume of solution}} = \frac{12.6}{101.8} = 12.4\%$$

Reverse this process to convert volume percent to mass percent.

Molar Solutions and Molal Solutions

A *gram-atom* or *gram-atomic mass* is the atomic mass of an element expressed in grams. This is a *mass* term.

EXAMPLE The atomic mass of iron is 55.8. Therefore, 55.8 g of iron equals 1 gram-atom. Thus ½ of 55.8 g is 27.9 g, and 27.9 g is equal to 0.5 gram-atom.

Gram-molecular mass is the molecular mass of a molecule expressed in grams. This is also a mass term.

EXAMPLE The molecular mass of NaCl is 58. Fifty-eight grams of NaCl equals 1 gram-molecular mass of NaCl. Therefore, 116 g of NaCl (2 × 58 g) equals 2 gram-molecular masses.*

Molality is the number of gram-molecular masses of solute per *1000 grams of solvent.* This is a *composition* term.†

EXAMPLE A solution containing 58 g of NaCl per 1000 g of water is a 1 *molal* (1 *m*) solution. A solution which contained 29 g of NaCl in 1000 g of water would be a 0.5 *m* solution.

Molarity is the number of gram-molecular masses of solute per *liter or 1000 milliliters of solution.* This is a *concentration* term.

---

* These numbers have been rounded for convenience. The molecular mass of NaCl is actually 22.99 + 35.45 or 58.44, and is usually taken as 58.5.

† A similar composition term, *formality*, is expressed in the number of gram-formula masses of solute per kilogram of solvent. A solution which contained 58 g of NaCl in 1 kg of H_2O would be a 1 *F* solution.

EXAMPLE If 58 g of NaCl (1 gram-molecular mass) is dissolved in water and the *volume of the solution is 1000 mL,* it is a 1 molar (1 *M*) solution. If it is dissolved in 1000 g of water, it is a 1 molal (1 *m*) solution.

In setting up a calibration, molality is used rather than molarity when:

Reference data are given as mass functions (certain constants).

Small amounts of solvent are being utilized, and measured mass is more accurate than measured volume.

The *equivalent mass* of a substance is the mass of that substance that will combine with 1 gram-atom of hydrogen. It is based on the *reactivity* of the substance in a chemical reaction. *It depends strictly upon the reaction in which the substance participates.*

Gram-equivalent mass is the equivalent mass expressed in grams. This is a mass term.

EXAMPLE In neutralization reactions, acid-base reactions, obtain the equivalent mass of the acid by dividing the molecular mass of the acid by the number of replaceable hydrogen ions, and the equivalent mass of the base by the number of replaceable hydroxyl ions.

For acids and bases containing only one reactive hydrogen or hydroxyl the equivalent mass is equal to the molecular mass. The equivalent mass of HCl is 36.5. The molecular mass of HCl is 36.5.

For acids or bases containing more than one hydrogen and/or hydroxyl ion the completeness of the reaction must be determined.

In the case of sulfuric acid (H_2SO_4) both hydrogens enter into any neutralization reaction, and the equivalent mass is therefore one-half the molecular mass. The molecular mass of H_2SO_4 is 98. The equivalent mass of H_2SO_4 is 49, which is

$$\frac{98 \text{ (molecular mass)}}{2 \text{ (number of replaceable hydrogen ions)}} = 49 \text{ (equivalent mass)}$$

The case of phosphoric acid, however, is quite different. Depending on the reaction, either one, two, or all three of the hydrogens in phosphoric acid may react, and then the equivalent mass is equal to half or even one-third the molecular mass based on the reaction taking place.

For systems other than neutralization reactions the determination of equivalents can be confusing. The equivalent mass is obtained by dividing the molecular mass by the *change in oxidation state* or valence of the species in the reaction.

$Fe^{2+} \rightarrow Fe^{3+}$

The equivalent mass of Fe in this system is

$$\frac{55.8 \text{ (atomic mass)}}{1} \text{ or } 55.8 \text{ (equivalent mass Fe)}$$

EXAMPLES To make 1000 mL of solutions of different molarity of sodium chloride and sodium sulfate.

1. First calculate the molecular masses of the compounds.

NaCl:
- 1 sodium = 1 × 23 = 23
- 1 chlorine = 1 × 35.5 = 35.5
- Molecular mass = 58.5 g

$Na_2SO_4 \cdot 10H_2O$:
- 2 sodiums = 2 × 23 = 46
- 1 sulfur = 1 × 32 = 32
- 4 oxygens = 4 × 16 = 64
- 10 waters = 10 × 18 = 180
- Molecular mass = 322 g

2. Decide the molarity of the solution you want to make, 0.10 *M*, 0.5 *M*, 1 *M*, etc.

3. Multiply the molarity you want by the molecular mass of the compound to find out how many grams you will need to make 1000 mL of solution.
For NaCl: Dilute the mass of salt to 1000 mL with water.

0.10 *M* solution requires 0.1 × 58.5 g = 5.85 g

0.5 *M* solution requires 0.5 × 58.5 g = 29.25 g

1.0 *M* solution requires 1.0 × 58.5 g = 58.5 g

2.0 *M* solution requires 2.0 × 58.5 g = 117.0 g

For $Na_2SO_4 \cdot 10H_2O$: Dilute the mass of salt to 1000 mL with water.

0.10 *M* solution requires 0.10 × 322 g = 32.2 g

0.5 *M* solution requires 0.5 × 322 g = 161 g

1.0 *M* solution requires 1.0 × 322 g = 322 g

2.0 *M* solution requires 2.0 × 322 g = 644 g

Molar Solutions: Volumes Other than 1000 mL

1. Calculate the mass of material needed to prepare 1000 mL of the solution of the desired molarity. (See previous example.)

2. Decide how much solution is to be made: 100 mL, 250 mL, 500 mL, etc.

3. Change the volume in milliliters to liters as follows:

$$\frac{100\ \text{mL}}{1000\ \text{mL/L}} = 0.1\ \text{L}$$

$$\frac{250\ \text{mL}}{1000\ \text{mL/L}} = 0.25\ \text{L}$$

$$\frac{500\ \text{mL}}{1000\ \text{mL/L}} = 0.50\ \text{L}$$

$$\frac{750\ \text{mL}}{1000\ \text{mL/L}} = 0.75\ \text{L}$$

4. Multiply the mass calculated in step 1 by the decimal obtained in step 3 to get the mass needed.

5. Dissolve the mass obtained in step 4 in water and add enough water to make up the desired volume.

You have prepared the solution having the molarity and volume you wanted.

EXAMPLE Exactly 40.0 g NaOH is required to make 1 L of a 1 *M* solution. How many grams are required to make 250 mL of a 1 *M* solution?

1. 40.0 g is needed to make 1 L of a 1 *M* solution.

2. 250 mL is the volume wanted.

3. 250 mL = 0.25 L.

4. 40.0 g/L × 0.25 L = 10.0 g.

5. Dissolve 10.0 g NaOH in water and dilute to the desired volume: 250 mL.

EXAMPLE How much sodium sulfate should be dissolved to make 5000 mL of a 0.1 *M* solution?

1. The molecular mass of sodium sulfate is 322 g ($Na_2SO_4 \cdot 10H_2O$):
A 1 *M* solution requires 322 g/L.
A 0.1 *M* solution requires 32.2 g/L.

2. 5000 mL is the volume desired.

3. 5000 mL = 5.0 L.

4. 32.2 g/L × 5.0 L = 161.0 g for the 5 L.

5. Dissolve the 161.0 g of sodium sulfate in water and dilute to 5 L.

Method for Preparing Solutions of Definite Molarity

Use this formula:

Molecular mass of compound × molarity wanted × number of liters = grams of compound needed to make up the solution.

EXAMPLE Prepare 50 mL of a 0.5 *M* NaCl solution. The molecular mass of NaCl is 58.5. The molarity wanted is 0.5 *M*. The number of liters is 0.05. Multiply as above: $58.5 \times 0.5 \times 0.05 = 1.46$ g NaCl. Dissolve in water and add water until the volume is 50 mL.

Preparing Molar Solutions by Diluting More Concentrated Solutions

1. Decide the molarity of the solution you wish to prepare, and decide how much you want to prepare.

2. Find out what solutions are available, which are more concentrated, and the molarity of those solutions.

3. Use the following formula to calculate how much of the more concentrated solution you need:

Volume (concentrated solution) × molarity (concentrated solution)
= volume (final solution) × molarity (final solution)

EXAMPLE **1.** Prepare 100 mL of a 1 *M* H_2SO_4 solution from stock 3 *M* H_2SO_4:

2. Concentrated solution: molarity = 3 *M*, volume = ?

Final solution: molarity = 1 *M*, volume = 100 mL

3. Substitute in equation and solve:

$? \times 3 = 100 \times 1$
$? \times 3 = 100$
$? = 33.3$ mL of concentrated 3 *M* acid needed

4. Subtract the volume of concentrated acid obtained in step 3 from the final volume wanted to find out how much water is needed:

100 mL − 33.3 mL = 66.7 mL water

5. Mix the volume of acid obtained in step 3 with the volume of water obtained in step 4 to get the final solution having the volume and molarity wanted. (Refer to note on mixing volumes of liquids. *Always pour the acid into the water.*)

Calculating Molarity from Gram Concentrations

1. Determine the mass of solution which you have by multiplying the volume of the solution by the density of the solution. Mass = volume × density.

2. You know the concentration in percent by mass of the compound in solution (given). *Change to the decimal equivalent.*

3. Calculate the molecular mass of the compound.

4. Multiply step 1 by step 2 and divide by step 3 to find the number of moles present in the whole solution.

5. Divide the number of moles (obtained in step 4) by the volume in liters of the solution to find the molarity of the solution.

EXAMPLE What is the molarity of 2000 mL of sulfuric acid solution—density 1.18 g/mL, 20% by mass of acid?

Solution 1. The mass of the solution equals volume times density:

$$2000 \text{ mL} \times 1.18 \text{ g/mL} = 2360 \text{ g}$$

2. The concentration of the sulfuric acid by mass is 20%:

$$20\% = 0.20$$

3. The molecular mass of sulfuric acid equals 98.

4. Multiply 2360 by 0.20 and divide by 98 to find the number of gram-moles present:

$$\frac{2360 \times 0.20}{98} = 4.8 \text{ gram-moles in 2 L}$$

5. The number of gram-moles in 1 liter is the *molarity*. Divide the number of gram-moles by the number of liters to get the *molarity:*

$$\frac{4.8 \text{ gram-moles}}{2 \text{ L}} = 2.4 \text{ gram-moles/L} = 2.4\ M$$

Equivalent Masses

In chemical reactions, equivalent amounts of reactants will combine with each other to form products. For example, $NaOH + HCl \rightarrow NaCl + H_2O$. In this reaction, one equivalent of NaOH reacts with one equivalent of HCl to yield one equivalent of NaCl and one equivalent of H_2O.

The equivalent mass of a compound is equal to the gram-molecular mass of the compound divided by the number of hydrogen ions (or hydroxyl ions) accepted or donated for each molecular unit (see Table 24-5):

$$\text{Equivalent mass} = \frac{\text{gram-molecular mass}}{\text{number of } H^+ \text{ ions (or } OH^- \text{ ions) accepted or donated per molecular unit}}$$

TABLE 24-5
Molecular and Equivalent Masses (Acids and Bases)

Acids or bases	*No. of H^+ or OH^- ions*	*Gram-molecular mass, g*	*Gram-molecular mass divided by H^+ ions or OH^- ions*	*Equivalent mass, g*
HCl	1	36.5	36.5/1	36.5
HNO_3	1	63.1	60.1/1	60.1
H_2SO_4	2	98.1	98.1/2	49.05
$HC_2H_3O_2$	1	60	60/1	60
H_3PO_4	3	98	98/3	32.7
NaOH	1	40	40/1	40
NH_4OH	1	35	35/1	35
$Ca(OH)_2$	2	74	74/2	37

Normality

Normal solutions are solutions which have a specific number of equivalent masses of the acid or base dissolved in the solution per liter. A 1 *N* solution (a 1 normal solution) contains 1 equivalent mass per liter, a 2 *N* solution contains 2 equivalent masses per liter, and so on.

General Formula Relating the Terms: Normality, Equivalent Mass, Grams of Substance, and Volume

Most calculations involving normality, equivalent mass, grams, and volume can be most easily handled by the following formula:

$$\text{Grams of substance} = N \times \text{equivalent mass} \times V$$

where N = normality
V = volume in *liters*

EXAMPLE How many grams of H_2SO_4 are required to prepare 3000 mL of a 0.1 *N* solution?

Solution We have

$N = 0.1$

Equivalent mass = 49 g

Volume = 3000 mL = 3 L

Substituting:

$\text{Mass} = 0.1 \times 49 \times 3 = 14.7 \text{ g}$

Using Normality to Express Concentrations in Liters and Milliliters

1. Divide the number of gram-equivalent masses of solute by the number of liters of solution to get the *normality.* This is a concentration term.

$$\frac{\text{Number of gram-equivalent masses of solute}}{\text{Number of liters}} = \text{normality}$$

2. Divide the equivalent mass by 1000 to get the milliequivalent mass:

Compound	*Equivalent mass, g*	*Milliequivalent mass, g*
NaOH	40	0.040
HCl	36.5	0.0365
H_2SO_4	49.05	0.04904
H_3PO_4	32.7	0.0327

One equivalent mass dissolved in one liter yields a solution of the same concentration or normality as one milliequivalent mass dissolved in one milliliter.

3. Normality is expressed in milliequivalents per milliliter or equivalents per liter.

Calculations with Normality

One gram-equivalent mass of any acid will neutralize (react with) one gram-equivalent mass of any base. One milliequivalent of any acid will neutralize (react with) one milliequivalent of any base.

1. Number of gram-equivalent masses = volume in liters × normality
2. Number of milliequivalents = volume in milliliters × normality
3. When two solutions react exactly with each other (acid-base reactions):

Equivalents of acid = equivalents of base

Volume (acid) × normality (acid) = volume (base) × normality (base)

Both volumes are in liters.

4. Milliequivalents of acid = milliequivalents of base

Volume of acid in milliliters × normality = volume of base in milliliters × normality of base

EXAMPLE Exactly 40.0 mL of 0.1 *N* HCl are required to neutralize 10.0 mL of NaOH. What is the normality of the base?

Solution 1. Start with step 2 above.

Milliequivalents of acid = 40.0 × 0.1 *N* = 4.

2. Milliequivalents of base needed = 4 (step 3 above).

10.0 mL × normality = 4: Solve for normality.

$N = 4/10 = 0.4$ *N* NaOH

Alternative solution Equivalents of acids = equivalents of base

1. Equivalents of acid = liters × normality.

Equivalents = $0.040 \times 0.1\ N = 0.0040$ equivalents

2. Equivalents of base needed = 0.0040.

3. Equivalents of base = liters of base × normality.

Equivalents of base = 0.0040 = 0.010 liter × normality of base

or $0.01 \times N = 0.004\ N$

Solve for normality (0.4 *N* NaOH).

Gravimetric Calculations

The mass percent of a constituent is equal to the mass of the constituent divided by the sample mass and multiplied by 100.

For example, if a 1.000-g sample of limestone is found to contain 0.3752 g of calcium, it has

$$\frac{0.3752}{1.000} \times 100 \text{ or } 37.52\% \text{ calcium}$$

$$\frac{\text{Mass of substance}}{\text{Mass of sample}} \times 100 = \text{percent by mass}$$

In most cases, however, the desired constituent is not massed directly but is precipitated and massed as some other compound. It is then necessary to convert the mass obtained to the mass in the desired form by using *a gravimetric factor.*

For example, a molecule of silver chloride is made up of one atom of silver and one atom of chlorine. The ratio of silver to chloride is as Ag/AgCl. Since the atomic mass of silver is 107.8 and that of chlorine is 35.5,

$$\frac{107.8 \text{ (atomic mass Ag)}}{107.8 \text{ (atomic mass Ag)} + 35.5 \text{ (atomic mass Cl)}} = 0.7526$$

This is called the *gravimetric factor,* and the percent of silver when weighed as AgCl is

$$\frac{\text{Mass of AgCl} \times 0.7526}{\text{Mass of sample}} \times 100$$

$$\text{or } \frac{\text{Mass of precipitate (AgCl)}}{\text{Mass of sample}} \times \text{gravimetric factor} \times 100 = \%\ \text{Ag}$$

The factor for sodium when weighed as sodium sulfate is $2Na/Na_2SO_4$ since there are 2 sodium atoms in sodium sulfate.

A general equation for gravimetric calculation is

$$\frac{\text{Mass of precipitate} \times \text{gravimetric factor} \times 100}{\text{Mass of sample}} = \text{percent of constituent}$$

Gravimetric factors are given in Table 24-6.

Volumetric Calculations

The basic formula for all volumetric calculations is

Mass percent of constituent

$$= \frac{V \times N \times \text{milliequivalent mass (g) of titrant} \times 100}{\text{sample mass (g)}}$$

where V = volume of titrant, mL
N = normality of titrant or equivalents per liter or milliequivalents per milliliter

The equation then becomes

$$\frac{\text{mL} \times \text{milliequivalents/mL} \times \text{milliequivalent mass of constituent} \times 100}{\text{sample mass (g)}}$$

$$= \frac{\text{milliequivalents} \times \text{milliequivalent mass (constituent)} \times 100}{\text{sample mass (g)}}$$

$$= \frac{\text{mass of constituent}}{\text{sample mass (g)}} \times 100 = \text{weight percent constituent}$$

TABLE 24-6
Table of Gravimetric Factors

Sought	*Massed*	*Factor*
Na	Na_2SO_4	0.3237
K	K_2SO_4	0.4487
Ba	$BaSO_4$	0.5885
Ca	$CaSO_4$	0.2944
Cu	CuO	0.7988
Fe	Fe_2O_3	0.6994
Pt	Pt	1.000
Au	Au	1.000
Ag	Ag	1.000
H	H_2O	0.1119
C	CO_2	0.2729
S	$BaSO_4$	0.1374

EXAMPLE What is the percent of Cl in a sample if 1 g of sample requires 10 mL of 0.1 N $AgNO_3$ for complete titration?

Solution

$$\frac{V \times N \times \text{milliequivalent mass (Cl)} \times 100}{\text{sample mass (g)}} = \%\ \text{Cl}$$

$$\frac{10 \times 0.1 \times 0.35 \times 100}{1} = 3.5\%\ \text{Cl}$$

Preparation of Standard Laboratory Solutions

Table 24-7 lists the concentration of acids and bases used in the laboratory and their preparation. Table 24-8 lists the preparation of standard laboratory solutions. These tables will save you calculation time.

Special instructions are required for the preparation of a carbonate-free NaOH solution.

Text continues on page 600.

TABLE 24-7
Concentration of Desk Acids and Bases Used in the Laboratory

Reagent	*Formula*	*Molecular mass*	*Molarity*	*Density, g/mL*	*% solute*	*Preparation*°
Acetic acid, glacial	$HC_2H_3O_2$	60	17	1.05	99.5	
Acetic acid, dil.			6	1.04	34	Dilute 333 mL 17 *M* to 1 L
Hydrochloric acid, conc.	HCl	36.4	12	1.18	36	
Hydrochloric acid, dil.			6	1.10	20	Dilute 500 mL 12 *M* to 1 L
Nitric acid, conc.	HNO_3	63.0	16	1.42	72	
Nitric acid, dil.			6	1.19	32	Dilute 375 mL 16 *M* to 1 L
Sulfuric acid, conc.	H_2SO_4	98.1	18	1.84	96	
Sulfuric acid, dil.			3	1.18	25	Dilute 165 mL 18 *M* to 1 L
Ammonium hydroxide, conc.	NH_4OH	35.05	15	0.90	58	
Ammonium hydroxide, dil.			6	0.96	23	Dilute 400 mL 15 *M* to 1 L
Sodium hydroxide, dil.	NaOH	40.0	6	1.22	20	Dissolve 240 g in water, dilute to 1 L

° Caution: Always pour acid into water slowly with careful mixing.

TABLE 24-8
Preparation of Standard Laboratory Solutions

Name	*Formula*	*Molecular mass*	*Concentration*	*Preparation*
Acetic acid	CH_3COOH	60.05	1 *M*	Dilute 58 mL of glacial acetic acid to 1 L with distilled water
Aluminum chloride	$AlCl_3 \cdot 6H_2O$	241.43	0.05 *M*	Dissolve 12.1 g in distilled water and dilute to 1 L
Aluminum nitrate	$Al(NO_3)_3 \cdot 9H_2O$	375.13	0.1 *M*	Dissolve 37.5 g $Al(NO_3)_3 \cdot 9H_2O$ in distilled water and dilute to 1 L
			0.2 *F**	Dissolve 75 g of $Al(NO_3)_3 \cdot 9H_2O$ in distilled water and dilute to 1 L
Aluminum sulfate	$Al_2(SO_4)_3 \cdot 18H_2O$	666.42	0.083 *M*	Dissolve 55 g of $Al_2(SO_4)_3 \cdot 18H_2O$ in distilled water and dilute to 1 L
			0.1 *M*	Dissolve 66.6 g $Al_2(SO_4)_3 \cdot 18\ H_2O$ in distilled water and dilute to 1 L
Ammonium acetate	$NH_4C_2H_3O_2$	77.08	1 *M*	Dissolve 77 g in distilled water and dilute to 1 L
Ammonium carbonate	$(NH_4)_2CO_3$	96.10	0.5 *M*	Dissolve 48 g $(NH_4)_2CO_3$ in distilled water and dilute to 1 L
Ammonium chloride	NH_4Cl	53.49	0.01 *M*	Dissolve 0.54 g NH_4Cl in distilled water and dilute to 1 L
Ammonium hydroxide	NH_4OH	35.05	1 *M*	Dilute 67 ml conc. NH_4OH (15 *M*) to 1 L with distilled water
Ammonium molybdate	$(NH_4)_2MoO_4$	196.01	0.5 *M*, 1 *N*	Dissolve 72 g MoO_3 in 200 mL of water, add 60 mL conc. NH_4OH, filter into 270 mL conc. HNO_3, add 400 mL water, and dilute to 1 L. Dissolve 87 g $(NH_4)_6Mo_7O_{24} \cdot 4H_2O$ and 240 g NH_4NO_3 in about 800 mL of warm distilled water and 40 mL conc. (15 *M*) NH_4OH, and dilute to 1L
Ammonium nitrate	NH_4NO_3	80.04	0.1 *M*	Dissolve 8 g NH_4NO_3 in distilled water and dilute to 1 L
Ammonium oxalate	$(NH_4)_2C_2O_4 \cdot H_2O$	142.11	0.25 *M*	Dissolve 35.5 g $(NH_4)_2C_2O_4 \cdot H_2O$ in distilled water and dilute to 1 L
Ammonium persulfate	$(NH_4)_2S_2O_8$	228	0.1 *M*	Dissolve 22.8 g $(NH_4)_2S_2O_8$ in distilled water and dilute to 1 L
Ammonium sulfate	$(NH_4)_2SO_4$	132.14	0.1 *M*	Dissolve 13 g $(NH_4)_2SO_4$ in distilled water and dilute to 1 L
Ammonium sulfide	$(NH_4)_2S$	68.14	1.0 *M*	Dilute 250 mL $(NH_4)_2S$ (22%) to 1 L with distilled water. Pass H_2S into 500 mL ice cold 1 *M* NH_4OH until saturated; then add 500 mL 1 *M* NH_4OH (use hood)

TABLE 24-8 (Continued)

Name	*Formula*	*Molecular mass*	*Concentration*	*Preparation*
Ammonium sulfide, colorless	$(NH_4)_2S$	68.14	3 *M*	Saturate 200 mL NH_4OH with H_2S, add 200 mL NH_4OH, and dilute to 1 L with distilled water (use hood)
Ammonium thiocyanate	NH_4CNS	76.12	0.1 *M*	Dissolve 7.6 g NH_4CNS in distilled water and dilute to 1 L
Antimony trichloride	$SbCl_3$	228.11	0.1 *F**	Dissolve 22.7 g $SbCl_3$ in 200 mL conc. HCl and dilute to 1 L with distilled water
Aqua regia				Mix 1 part conc. HNO_3 with 3 parts conc. HCl
Barium acetate	$Ba(C_2H_3O_2)_2 \cdot H_2O$	273.45	0.1 *M*	Dissolve 27.3 g $Ba(C_2H_3O_2)_2 \cdot H_2O$ in distilled water and dilute to 1 L (use hood)
Barium chloride	$BaCl_2 \cdot 2H_2O$	244.28	0.1 *M*	Dissolve 24.4 g $BaCl_2 \cdot 2H_2O$ in distilled water and dilute to 1 L
Barium hydroxide	$Ba(OH)_2 \cdot 8H_2O$	315.48	0.02 *N*	Dissolve 3.1 g $Ba(OH)_2 \cdot 8H_2O$ in distilled water and dilute to 1 L
			0.1 *M*, 0.2 *N*	Dissolve 32 g $Ba(OH)_2 \cdot 8H_2O$ in distilled water and dilute to 1 L
Barium nitrate	$Ba(NO_3)_2$	261.35	0.1 *M*	Dissolve 26 g $Ba(NO_3)_2$ in distilled water and dilute to 1 L
Bismuth chloride	$BiCl_3$	315.34	0.1 *M*	Dissolve 31.5 g $BiCl_3$ in 200 mL conc. HCl and dilute to 1 L with distilled water
Bismuth nitrate	$Bi(NO_3)_3 \cdot 5H_2O$	485.07	0.1 *M*	Dissolve 49 g of $Bi(NO_3)_3 \cdot 5H_2O$ in 50 mL conc. HNO_3 and dilute to 1 L with distilled water
Boric acid	H_3BO_3	61.83	0.1 *M*	Dissolve 6.2 g of H_3BO_3 in 1 L distilled water
Bromine in carbon tetrachloride	Br_2	159.82	0.1 *M*	Dissolve 1 g liquid bromine in 100 g (63 ml) CCl_4 (use hood)
Bromine water	Br_2	159.82		Saturate 1 L distilled water with 10–15 g liquid bromine (use hood)
Cadmium chloride	$CdCl_2$	183.31	0.1 *M*	Dissolve 18.3 g $CdCl_2$ in distilled water and dilute to 1 L
Cadium nitrate	$Cd(NO_3)_2 \cdot 4H_2O$	308.47	0.2 *M*	Dissolve 62 g $Cd(NO_3)_2 \cdot 4H_2O$ in distilled water and dilute to 1 L
Cadmium sulfate	$CdSO_4 \cdot 4H_2O$	280.5	0.25 *M*	Dissolve 70 g $CdSO_4 \cdot 4H_2O$ in distilled water and dilute to 1 L
Calcium acetate	$Ca(C_3H_3O_2)_2 \cdot H_2O$	176.2	0.5 *M*	Dissolve 88 g $Ca(C_2H_3O_2)_2 \cdot H_2O$ in distilled water and dilute to 1 L

TABLE 24-8 (Continued)

Name	*Formula*	*Molecular mass*	*Concentration*	*Preparation*
Calcium chloride	$CaCl_2$	111	0.1 *M*	Dissolve 11 g $CaCl_2$ in distilled water and dilute to 1 L
	$CaCl_2 \cdot 6H_2O$	219.1	0.1 *M*	Dissolve 22 g $CaCl_2 \cdot 6H_2O$ in distilled water and dilute to 1 L
Calcium hydroxide	$Ca(OH)_2$	74.1		Saturated: Shake 3–4 g reagent-grade CaO with distilled water and dilute to 1L, then filter
				Saturated solution (prepared every few days): Saturate 1 L distilled water with solid $Ca(OH)_2$; filter
Calcium nitrate	$Ca(NO_3)_2 \cdot 4H_2O$	236.2	0.1 *M*	Dissolve 23.6 g $Ca(NO_3)_2 \cdot 4H_2O$ in distilled water and dilute to 1 L
Chlorine water	Cl_2	70.91		Slightly acidify 225 mL 3% NaClO or 175 mL 5% NaClO with 6 mL acetic acid and dilute 1 L with distilled water. Chlorine water (prepared every few days): saturate 1 L distilled water with chlorine gas (use hood)
Chloroplatinic acid	$H_2PtCl_6 \cdot 6H_2O$	517.92	0.512 *M*, 0.102 *N*	Dissolve 26.53 g $H_2PtCl_6 \cdot 6H_2O$ in distilled water and dilute to 1 L
Chromium(III) chloride	$CrCl_3$	158.36	0.167 *M*	Dissolve 26 g $CrCl_3$ in distilled water and dilute to 1 L
Chromium(III) nitrate	$Cr(NO_3)_3 \cdot 9H_2O$	400.15	0.1 *M*	Dissolve 40 g $Cr(NO_3)_3 \cdot 9H_2O$ in distilled water and dilute to 1 L
Chromium(III) sulfate	$Cr_2(SO_4)_3 \cdot 18H_2O$	716.45	0.083 *M*	Dissolve 60 g $Cr(SO_4)_3 \cdot 18H_2O$ in distilled water and dilute to 1 L
Cobalt(II) chloride	$CoCl_2 \cdot 6H_2O$	237.93	0.1 *M*	Dissolve 24 g $CoCl_2 \cdot 6H_2O$ in distilled water and dilute to 1 L
Cobalt(II) nitrate	$Co(NO_3)_2 \cdot 6H_2O$	291.04	0.25 *M*	Dissolve 73 g $Co(NO_3)_2 \cdot 6H_2O$ in distilled water and dilute to 1 L
Cobalt(II) sulfate	$CoSO_4 \cdot 7H_2O$	281.10	0.25 *M*	Dissolve 70 g $CoSO_4 \cdot 7H_2O$ in distilled water and dilute to 1 L
Copper(II) chloride	$CuCl_2 \cdot 2H_2O$	170.48	0.1 *M*	Dissolve 17 g $CuCl_2 \cdot 2H_2O$ in distilled water and dilute to 1 L
Copper(II) nitrate	$Cu(NO_3)_2 \cdot 3H_2O$	241.6	0.1 *M*	Dissolve 24 g $Cu(NO_3)_2 \cdot 3H_2O$ in distilled water and dilute to 1 L
	$Cu(NO_3)_2 \cdot 6H_2O$	295.64	0.1 *M*	Dissolve 29.5 g $Cu(NO_3)_2 \cdot 6H_2O$ in distilled water and dilute to 1 L
			0.25 *M*	Dissolve 74 g $Cu(NO_3)_2 \cdot 6H_2O$ in distilled water and dilute to 1 L
Copper(II) sulfate	$CuSO_4 \cdot 5H_2O$	249.68	0.10 *M*	Dissolve 24.97 g $CuSO_4 \cdot 5H_2O$ in distilled water, add 3 drops conc. H_2SO_4, and dilute to 1 L

TABLE 24-8 (Continued)

Name	*Formula*	*Molecular mass*	*Concentration*	*Preparation*
Disodium phosphate	$Na_2HPO_4 \cdot 12H_2O$	358.14	0.1 *M*	Dissolve 35.8 g $Na_2HPO_4 \cdot 12H_2O$ in distilled water and dilute to 1 L
Fehling's solution A				Dissolve 35 g $CuSO_4 \cdot 5H_2O$ in 500 mL distilled water
Fehling's solution B				Dissolve 173 g $KNaC_4H_4O_6 \cdot 4H_2O$ (Rochelle salt) in 200 mL water, add 50 g solid NaOH in 200 mL water, and dilute to 500 mL with distilled water
Hydrochloric acid	HCl	36.46	0.05 *M*	Dilute 4.1 mL of conc. HCl (37%) to 1 L with distilled water
			0.1 *M*	Dilute 8.2 mL of conc. HCl (37%) to 1 L with distilled water
			1.0 *M*	Dilute 82 mL of conc. HCl (37%) to 1 L with distilled water
			3 *M*	Dilute 246 mL HCl (37%) to 1 L with distilled water
			6.0 *M*	Dilute 492 mL of conc. HCl (37%) to 1 L with distilled water
Hydrogen peroxide	H_2O_2	34.01	0.2 *M*	Dilute 23.3 mL 30% H_2O_2 to 1 L with distilled water. Hydrogen peroxide 3% is standard concentration commercially available
Iodine in potassium iodide				Dissolve 5 g I_2 and 15 g KI in distilled water and dilute to 250 mL
Iron(III) ammonium sulfate, saturated	$Fe_2(SO_4)_3(NH_4)_2SO_4 \cdot 24H_2O$	964.4		Dissolve about 1240 g $Fe_2(SO_4)_3(NH_4)_2SO_4 \cdot 24H_2O$ in distilled water and dilute to 1 L
Iron(III) chloride	$FeCl_3 \cdot 6H_2O$	270.3	0.1 *M*	Dissolve 27.0 g $FeCl_3 \cdot 6H_2O$ in water containing 20 mL conc. HCl in distilled water and dilute to 1 L
Iron(III) nitrate	$Fe(NO_3)_3 \cdot 6H_2O$	349.95	0.1 *M*	Dissolve 35 g $Fe(NO_3)_3 \cdot 6H_2O$ in distilled water and dilute to 1 L
Iron(III) sulfate	$Fe_2(SO_4)_3 \cdot 9H_2O$	562.02	0.25 *M*	Dissolve 140.5 g $Fe_2(SO_4)_3 \cdot 9H_2O$ in 100 mL 3 *M* H_2SO_4 and dilute to 1 L with distilled water
Iron(II) ammonium sulfate	$FeSO_4(NH_4)_2SO_4 \cdot 6H_2O$	392.14	0.5 *M*	Dissolve 196 g $FeSO_4(NH_4)SO_4 \cdot 6H_2O$ in water containing 10 mL conc. H_2SO_4 in distilled water and dilute to 1 L

TABLE 24-8 (Continued)

Name	*Formula*	*Molecular mass*	*Concentration*	*Preparation*
Iron(II) sulfate	$FeSO_4 \cdot 7H_2O$	278.02	0.1 *M*	Dissolve 28 g $FeSO_4 \cdot 7H_2O$ in water containing 5 mL conc. H_2SO_4, and dilute to 1 L with distilled water
Lead(II) acetate	$Pb(C_2H_3O_2)_2 \cdot 3H_2O$	379.33	0.1 *M*	Dissolve 38 g $Pb(C_2H_3O_2)_2 \cdot 3H_2O$ in distilled water and dilute to 1 L
Lead nitrate	$Pb(NO_3)_2)$	331.2	0.05 *M*	Dissolve 16.6 g $Pb(NO_3)_2$ in distilled water and dilute to 1 L
			0.1 *M*	Dissolve 33 g $Pb(NO_3)_2$ in distilled water and dilute to 1 L
Lead nitrate (basic)	$Pb(OH)NO_3$	286.19	0.1 *M*	Dissolve 28.6 g of $Pb(OH)NO_3$ in distilled water and dilute to 1 L
Lime water				See Calcium hydroxide
Magnesium chloride	$MgCl_2 \cdot 6H_2O$	203.31	0.25 *M*	Dissolve 51 g $MgCl_2 \cdot 6H_2O$ in distilled water and dilute to 1 L
Magnesium nitrate	$Mg(NO_3)_2 \cdot 6H_2O$	256.41	0.1 *M*	Dissolve 25.6 g of $Mg(NO_3)_2 \cdot 6H_2O$ in distilled water and dilute to 1 L
Magnesium sulfate	$MgSO_4 \cdot 7H_2O$	246.48	0.25 *M*	Dissolve 62 g $MgSO_4 \cdot 7H_2O$ in distilled water and dilute to 1 L
Manganese(II) chloride	$MnCl_2 \cdot 4H_2O$	197.91	0.25 *M*	Dissolve 50 g $MnCl_2 \cdot 4H_2O$ in distilled water and dilute to 1 L
Manganese(II) nitrate	$Mn(NO_3)_2 \cdot 6H_2O$	287.04	0.25 *M*	Dissolve 72 g $Mn(NO_3)_2 \cdot 6H_2O$ in distilled water and dilute to 1 L
Manganese(II) sulfate	$MnSO_4 \cdot 7H_2O$	277.11	0.25 *M*	Dissolve 69 g $MnSO_4 \cdot 7H_2O$ in distilled water and dilute to 1 L
Mercury(II) chloride	$HgCl_2$	271.5	0.25 *M*	Dissolve 68 g $HgCl_2$ in distilled water and dilute to 1 L
Mercury(I) nitrate	$HgNO_3 \cdot H_2O$	280.61	0.1 *M*	Dissolve 28 g $HgNO_3 \cdot H_2O$ in 100 mL 6 *F* HNO_3 and dilute to 1 L with distilled water
Mercury(II) nitrate	$Hg(NO_3)_2 \cdot \frac{1}{2}H_2O$	332	0.02 *M*	Dissolve 6.6 g of $Hg(NO_3)_2 \cdot \frac{1}{2}H_2O$ in distilled water and dilute to 1 L
Nickel(II) chloride	$NiCl_2 \cdot 6H_2O$	237.71	0.25 *M*	Dissolve 59 g $NiCl_2 \cdot 6H_2O$ in distilled water, dilute to 1 L
Nickel(II) nitrate	$Ni(NO_3)_2 \cdot 6H_2O$	290.81	0.02 *M*	Dissolve 5.8 g $Ni(NO_3)_2 \cdot 6H_2O$ in distilled water, dilute to 1 L
			0.1 *F**	Dissolve 29.1 g $Ni(NO_3)_2 \cdot 6H_2O$ in distilled water and dilute to 1 L
Nickel(II) sulfate	$NiSO_4 \cdot 6H_2O$	262.86	0.25 *M*	Dissolve 66 g $NiSO_4 \cdot 6H_2O$ in distilled water, dilute to 1 L

TABLE 24-8 (Continued)

Name	*Formula*	*Molecular mass*	*Concentration*	*Preparation*
Nitric acid	HNO_3	63.01	1 *M*	Add 63 mL conc. HNO_3 (16 *M*) to distilled water and dilute to 1 L
			3 *M*	Add 189 mL conc. HNO_3 (16 *M*) to distilled water to make 1 L
			6.0 *M*	Add 378 mL conc. HNO_3 to distilled water to make 1 L
Oxalic acid	$C_2H_2O_4 \cdot 2H_2O$	126.07	0.1 *M*	Dissolve 12.6 g $C_2H_2O_4 \cdot 2H_2O$ in distilled water and dilute to 1 L
Potassium bromide	KBr	119.01	0.1 *M*	Dissolve 11.9 g KBr in distilled water and dilute to 1 L
Potassium carbonate	K_2CO_3	138.21	1.5 *M*	Dissolve 207 g K_2CO_3 in distilled water and dilute to 1 L
Potassium chloride	KCl	74.56	0.1 *M*	Dissolve 7.45 g KCl in distilled water and dilute to 1 L
Potassium chromate	K_2CrO_4	194.20	0.1 *F**	Dissolve 19.4 g K_2CrO_4 in distilled water and dilute to 1 L
Potassium cyanide	KCN	65.12	0.5 *M*	Dissolve 33 g KCN in distilled water and dilute to 1 L (use hood)
Potassium dichromate	$K_2Cr_2O_7$	294.19	0.1 *F**	Dissolve 29.4 g $K_2Cr_2O_7$ in distilled water and dilute to 1 L
Potassium dihydrogen phosphate	KH_2PO_4	136.09	0.1 *M*	Dissolve 13.6 g of KH_2PO_4 in distilled water and dilute to 1 L
Potassium ferricyanide	$K_3Fe(CN)_6$	329.26	0.167 *M*	Dissolve 55 g of $K_3Fe(CN)_6$ in distilled water and dilute to 1 L
Potassium ferrocyanide	$K_4Fe(CN)_6 \cdot 3H_2O$	422.41	0.1 *M*	Dissolve 42.3 g of $K_4Fe(CN)_6 \cdot 3H_2O$ in distilled water and dilute to 1 L
Potassium hydrogen phthalate	$KHO_4C_8H_4$	204	0.1 *M*	Dissolve 20.4 g $KHO_4C_8H_4$ in distilled water and dilute to 1 L
Potassium hydrogen sulfate	$KHSO_4$	136.17	0.1 *F**	Dissolve 13.6 g $KHSO_4$ in distilled water and dilute to 1 L
Potassium hydrogen sulfite	$KHSO_3$	120.17	0.20 *M*	Dissolve 24.0 g $KHSO_3$ in distilled water and dilute to 1 L
Potassium hydroxide	KOH	56.11	0.1 *F**	Dissolve 5.6 g KOH in distilled water and dilute to 1 L
			1.0 *F**	Dissolve 56 g KOH in distilled water and dilute to 1 L
			20%	Dissolve 250 g of KOH in distilled water and dilute to 800 mL
			3*M*, 3 *N*	Dissolve 168 g KOH in distilled water and dilute to 1 L

TABLE 24-8 (Continued)

Name	*Formula*	*Molecular mass*	*Concentration*	*Preparation*
Potassium iodate	KIO_3	214	0.10 *M*	(Acidified) Dissolve 21.4 g KIO_3 in distilled water, add 5 mL H_2SO_4, and dilute with distilled water to 1 L
Potassium iodide	KI	166.01	0.1 *M*	Dissolve 16.6 g KI in distilled water and dilute to 1 L
Potassium nitrate	KNO_3	101.11	01. *M*	Dissolve 10 g KNO_3 in distilled water and dilute to l L
Potassium permanganate	$KMnO_4$	158.04	0.1 *N*	Dissolve 3.2 g $KMnO_4$ in distilled water, add 2 mL (18 *M*) H_2SO_4, and dilute to 1 L with distilled water
			0.1 *M*	(Acidified) Dissolve 15.8 g $KMnO_4$ in distilled water, add 10 mL (18 *M*) H_2SO_4, and dilute to 1 L
			0.1 *M*	Dissolve 15.8 g $KMnO_4$ in distilled water and dilute to 1 L
Potassium phosphate dibasic	K_2HPO_4	174.18	0.1 *M*	Dissolve 17.4 g K_2HPO_4 in distilled water and dilute to 1 L
Potassium sulfate	K_2SO_4	174.27	0.1 *M*	Dissolve 17.4 g K_2SO_4 in distilled water and dilute to 1 L
			0.25 *M*	Dissolve 44 g K_2SO_4 in distilled water and dilute to 1 L
Potassium thiocyanate	KSCN	97.18	0.1 *M*	Dissolve 9.7 g KSCN in distilled water and dilute to 1 L
Silver nitrate	$AgNO_3$	169.87	0.05 *M*	Dissolve 8.50 g $AgNO_3$ in distilled water and dilute to 1 L
			0.10 *M*	Dissolve 17.0 g $AgNO_3$ in distilled water and dilute to 1 L
			0.20 *M*	Dissolve 34.0 g $AgNO_3$ in distilled water and dilute to 1 L
			0.50 *M*	Dissolve 85 g $AgNO_3$ in distilled water and dilute to 1 L
			1.0 *M*	Dissolve 170 g $AgNO_3$ in distilled water and dilute to 1 L
Sodium acetate	$NaC_2H_3O_2 \cdot 3H_2O$	136.08	0.50 *M*	Dissolve 68.0 g $NaC_2H_3O_2 \cdot 3H_2O$ in distilled water and dilute to 1 L
			1 *M*	Dissolve 136 g $NaC_2H_3O_2 \cdot 3H_2O$ in distilled water and dilute to 1 L
			1 *M*	Dissolve 82.0 g $NaC_2H_3O_2 \cdot 3H_2O$ in distilled water and dilute to 1 L
			3 *M*, 3 *N*	Dissolve 408 g $NaC_2H_3O_2 \cdot 3H_2O$ in distilled water and dilute to 1 L

TABLE 24-8 (Continued)

Name	*Formula*	*Molecular mass*	*Concentration*	*Preparation*
Sodium arsenate	$Na_3AsO_4 \cdot 12H_2O$	424.07	0.1 *M*	Dissolve 42.4 g $Na_3AsO_4 \cdot 12H_2O$ in distilled water and dilute to 1 L
Sodium bicarbonate	$NaHCO_3$	84.01	0.1 *F**	Dissolve 8.4 g $NaHCO_3$ in distilled water and dilute to 1 L
Sodium bromide	NaBr	102.9	0.1 *F**	Dissolve 10.3 g NaBr in distilled water and dilute to 1 L
Sodium carbonate	Na_2CO_3	105.99	1 *N*	Dissolve 53 g Na_2CO_3 in distilled water and dilute to 1 L
Sodium chloride	NaCl	58.44	0.1 *M*	Dissolve 5.9 g NaCl in distilled water and dilute to 1 L
Sodium cobaltinitrite	$Na_3Co(NO_2)_6$	404	0.08 *M*	Dissolve 25 g $NaNO_2$ in 75 mL water, add 2 mL glacial acetic acid, then 2.5 g $Co(NO_2)_2 \cdot 6H_2O$, and dilute with distilled water to 100 mL
Sodium hydrogen phosphate	$Na_2HPO_4 \cdot 7H_2O$	268.1	0.05 *M*	Dissolve 13.4 g $Na_2HPO_4 \cdot 7H_2O$ in distilled water and dilute to 1 L
Sodium hydrogen sulfate	$NaHSO_4$	120.0	1 *M*	Dissolve 120 g $NaHSO_4$ in distilled water and dilute to 1 L
Sodium hydroxide	NaOH	40.0	0.1 *M*	Dissolve 4 g NaOH in distilled water and dilute to 1 L
			1 *M*	Dissolve 40.0 g NaOH pellets in distilled water and dilute to 1 L (Caution: heat evolved; use goggles)
			6.0 *M*	Dissolve 240 g NaOH pellets in distilled water and dilute to 1 L (Caution: heat evolved; use goggles)
Sodium iodide	NaI	149.9	0.1 *F**	Dissolve 15 g NaI in distilled water and dilute to 1 L
Sodium nitrate	$NaNO_3$	85.00	0.2 *M*	Dissolve 17 g $NaNO_3$ in distilled water and dilute to 1 L
Sodium nitroprusside†	$Na_2Fe(CN)_5NO \cdot 2H_2O$	297.95	10%	Dissolve 100 g $Na_2Fe(CN)_5NO \cdot 2H_2O$ in 900 mL distilled water
Sodium polysulfide	Na_2S_x			Dissolve 480 g $Na_2S \cdot 9H_2O$ in 500 mL water, add 40 g NaOH, 16 mg, powdered sulfur, mix well, and dilute to 1 L
Sodium sulfate	Na_2SO_4	142.04	0.1 *F**	Dissolve 14.2 g Na_2SO_4 in distilled water and dilute to 1 L

TABLE 24-8 (Continued)

Name	*Formula*	*Molecular mass*	*Concentration*	*Preparation*
Sodium sulfide	$Na_2S \cdot 9H_2O$	240.1	0.1 *M*	(Fresh) Dissolve 24 g $Na_2S \cdot 9H_2O$ in distilled water and dilute to 1 L
			0.2 *M*	Dissolve 48 g $Na_2S \cdot 9H_2O$ in distilled water and dilute to 1 L
Sodium sulfite	Na_2SO_3	126.04	0.1 *M*	Dissolve 12.6 g anhydrous Na_2SO_3 in distilled water and dilute to 1 L
Sodium tetraborate (borax)	$Na_2B_4O_7 \cdot 10H_2O$	381.37	0.0025 *M*	Dissolve 9.5 g $Na_2B_4O_7 \cdot 10H_2O$ in distilled water and dilute to 1 L
Sodium thiosulfate	$Na_2S_2O_3 \cdot 5H_2O$	248.18	0.1 *M*	Dissolve 24.8 g $Na_2S_2O_3 \cdot 5H_2O$ in distilled water and dilute to 1 L
Strontium chloride	$SrCl_2 \cdot 6H_2O$	266.6	0.25 *M*	Dissolve 66.6 g $SrCl_2 \cdot 6H_2O$ in distilled water and dilute to 1 L
Sulfuric acid‡	H_2SO_4	98.08	0.1 *M*	Add 5.6 mL conc. H_2SO_4 slowly to 500 mL distilled water and dilute to 1 L
			1 *M*	Add 56 mL conc. H_2SO_4 slowly to 500 mL distilled water and dilute to 1 L
			3.0 *M*, 6.0 *N*	Add 168 mL conc. H_2SO_4 to at least 700 mL distilled water and dilute to 1 L
			6 *M*	Add 336 mL conc. H_2SO_4 to at least 600 mL distilled water and dilute to 1 L
Tin(II) chloride	$SnCl_2 \cdot 2H_2O$	225.63	0.15 *M*	Dissolve 33.8 g $SnCl_2 \cdot 2H_2O$ in 75 mL conc. HCl and dilute to 1 L
Tin(IV) chloride	$SnCl_4 \cdot 5H_2O$	350.58	0.1 *M*	Dissolve 35.1 g $SnCl_4 \cdot 5H_2O$ in 167 mL 12 *F* HCl and dilute to 1 L with distilled water
Zinc chloride	$ZnCl_2$	136.3	0.1 *M*	Dissolve 13.6 g $ZnCl_2$ in distilled water and dilute to 1 L
Zinc nitrate	$Zn(NO_3)_2 \cdot 6H_2O$	297.5	0.1 *M*	Dissolve 30 g $Zn(NO_3)_2 \cdot 6H_2O$ in distilled water and dilute to 1 L
Zinc sulfate	$ZnSO_4 \cdot 7H_2O$	287.6	0.25 *M*, 0.5 *N*	Dissolve 72 g $ZnSO_4 \cdot 7H_2O$ in distilled water and dilute to 1 L

* *F* denotes the *formality* of a solution; see footnote on page 580.
† Also called (more properly) sodium nitroferricyanide.
‡ Caution: heat evolved, pour acid slowly with mixing.

Preparation and Storage of Carbonate-Free NaOH (0.1*N*)

Reagent-grade NaOH usually contains a considerable amount of Na_2CO_3, because it reacts with the CO_2 in the atmosphere. Carbonate-free NaOH (0.1*N*) can be prepared as follows:

Procedure

1. Boil 1 L distilled water in a 2-L flask to free it of dissolved CO_2. Allow to cool (while covered with a watch glass), then transfer to a polyethylene bottle (or a rubber-stoppered hard-glass bottle).

2. Dissolve about 8 g reagent-grade NaOH in 8 mL distilled water.*

CAUTION This reaction is highly exothermic and evolves considerable heat. The solution is highly caustic and corrosive.

3. Filter the 50% NaOH solution through a Gooch crucible seated on a suction flask and catch the filtrate in a clean, hard-glass test tube. (See Fig. 24-6, below.)

NOTE Highly caustic 50% NaOH solution will attack paper. The filter medium should be asbestos fiber or glass-fiber mat. To prepare the mat, stir acid-washed asbestos fibers in distilled water to make a suspension. With vacuum off, pour a small amount of the suspension through the crucible. Turn on suction and continue to pour the suspension into the crucible until a

---

* Only 4 g of NaOH are actually needed to prepare 1 L of 0.1 *N* NaOH, but we use 8 g because of the hygroscopic nature of NaOH and our uncertainty as to the water content of the NaOH.

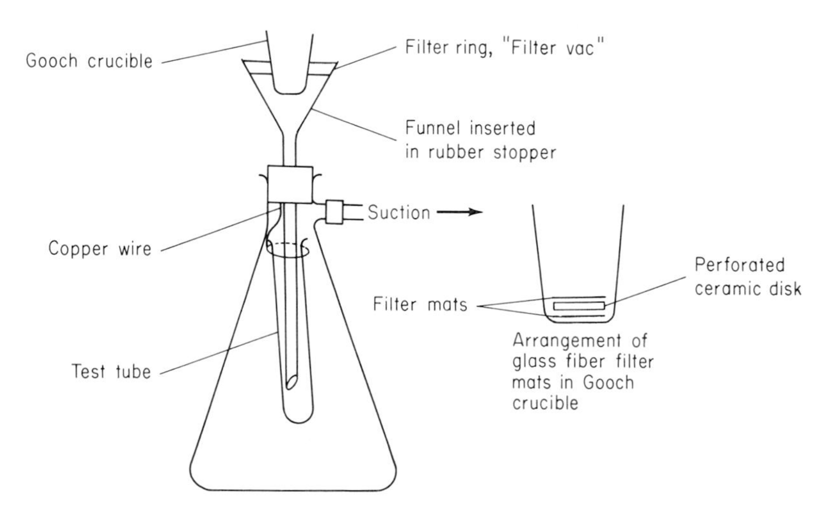

FIGURE 24-6
Assembly incorporating a Gooch crucible for filtering 50% NaOH.

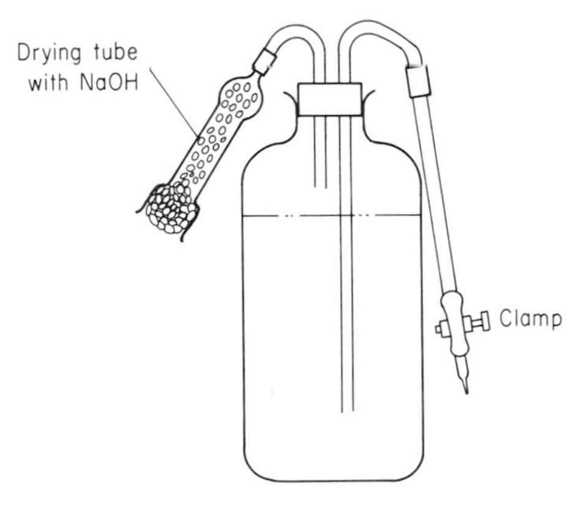

FIGURE 24-7
Method of storing carbonate-free standard NaOH solutions.

1-mm-thick mat has been formed. When the crucible is held to the light, the perforations should be visible, but no light should pass through the openings.

4. Transfer the filtered 50% NaOH to a clean 10-mL graduated cylinder and add 6 mL of the solution to the prepared liter of cool boiled water. Shake well to ensure a uniform concentration of the resulting solution. The solution should be approximately 0.1 *N*, and when stored against the CO_2 in the atmosphere should be protected with a drying tube of NaOH. (See Fig. 24-7.)

5. Standardize the solution by standard methods of analysis. (Refer to Chap. 25, "Volumetric Analysis.")

Commercially Packaged Standard Laboratory Solutions

Table 24-9 contains some useful information about commercially packaged solutions.

The following instructions will help you use standard packaged solutions. (See Fig. 24-8 and Table 24-10.)

DILUT-IT™ VOLUMETRIC CONCENTRATES*

Special Dilution Technique

Two techniques are useful for the preparation of standard solutions from DILUT-IT™ Volumetric Concentrates:

Where the concentration desired is an *exact* multiple or fraction of one normal, the instructions provided with each package are followed, employing an appropriate volumetric flask, 500-mL or larger.

---

* Reprinted by permission of J. T. Baker Chemical Company.

TABLE 24-9
DILUT-IT® Volumetric Concentrates

Reagent and concentration of solution after exact dilution to 1 L		*Equivalents*	*Formula and formula mass*		*Packaging*
Ammonium thiocyanate	0.1 *N*	0.1 equiv = 7.612 g	NH_4SCN	76.12	Plastic ampul
Disodium (ethylenedinitrilo) tetraacetate	0.2 *M*	0.2 mol = 6.724 g	$C_{10}H_{14}N_2Na_2O_8$	336.21	Plastic ampul
Disodium (ethylenedinitrilo) tetraacetate	0.1 *M*	0.1 mol = 33.62 g	$C_{10}H_{14}N_2Na_2O_8$	336.21	Plastic bottle
Hydrochloric acid	0.1 *N*	0.1 equiv = 3.646 g	HCl	36.46	Plastic ampul
Hydrochloric acid	1 *N*	1 equiv = 36.46 g	HCl	36.46	Plastic ampul
Hydrochloric acid	5 *N*	5 equiv = 182.3 g	HCl	36.46	Plastic bottle
Iodine (iodine-iodide)	0.01 *N*	0.01 equiv = 1.269 g	I_2	253.81	Glass ampul
Iodine (iodine-iodide)	0.1 *N*	0.1 equiv = 12.69 g	I_2	253.81	Glass ampul
Oxalic acid	0.1 *N*	0.1 equiv = 4.502 g	HOCOCOOH	90.04	Plastic ampul
Potassium bromate	0.1 *N*	0.1 equiv = 2.783 g	$KBrO_3$	167.01	Plastic ampul
Potassium dichromate	0.1 *N*	0.1 equiv = 4.903 g	$K_2Cr_2O_7$	294.19	Plastic ampul
Potassium hydroxide, carbonate-free	0.1 *N*	0.1 equiv = 5.611 g	KOH	56.11	Plastic ampul
Potassium hydroxide, carbonate-free	1 *N*	1 equiv = 56.11 g	KOH	56.11	Plastic ampul
Potassium iodate	0.1 *N*	0.1 equiv = 3.567 g	KIO_3	214.00	Plastic ampul
Potassium permanganate	0.1 *N*	0.1 equiv = 3.161 g	$KMnO_4$	158.04	Glass ampul
Silver nitrate	0.1 *N*	0.1 equiv = 16.99 g	$AgNO_3$	169.87	Plastic ampul
Sodium chloride	0.1 *N*	0.1 equiv = 5.844 g	NaCl	58.44	Plastic ampul
Sodium hydroxide, carbonate-free	0.1 *N*	0.1 equiv = 4.000 g	NaOH	40.00	Plastic ampul
Sodium hydroxide, carbonate free	1 *N*	1 equiv = 40.00 g	NaOH	40.00	Plastic ampul
Sodium hydroxide, carbonate-free	5 *N*	5 equiv = 200.0 g	NaOH	40.00	Plastic bottle
Sodium thiosulfate	0.01 *N*	0.01 equiv = 1.581 g	$Na_2S_2O_3$	158.11	Plastic ampul
Sodium thiosulfate	0.1 *N*	0.1 equiv = 15.81 g	$Na_2S_2O_3$	158.11	Plastic ampul
Sulfuric acid	0.1 *N*	0.1 equiv = 4.904 g	H_2SO_4	98.08	Plastic ampul
Sulfuric acid	1 *N*	1 equiv = 49.04 g	H_2SO_4	98.08	Plastic ampul
Sulfuric acid	5 *N*	1 equiv = 245.2 g	H_2SO_4	98.08	Plastic bottle

SOURCE: J. T. Baker Chemical Company.

Where the concentration desired is *not* an exact multiple or fraction of one normal, the weighing technique given below is applicable. Such solutions are often used where 1 mL is to equal a desired number of milligrams of a sought-for substance or a certain percentage of it for a fixed sample weight.

Calculate the final weight of the desired solution using the following equation:

$$W = \frac{1000\ E\ k}{N}$$

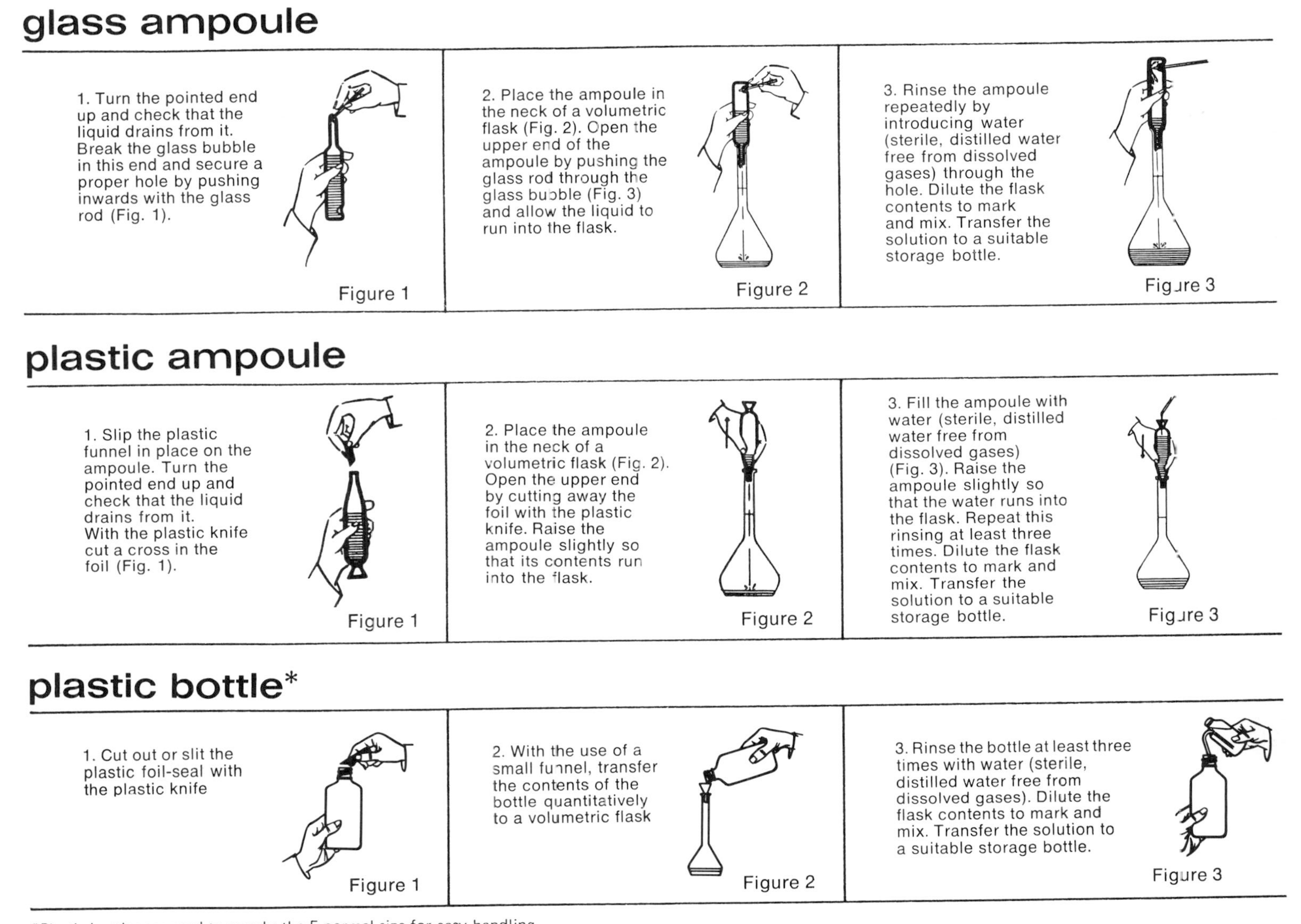

FIGURE 20-8 General directions for preparing standard solutions from commercial concentrates.

TABLE 24-10
DILUT-IT® Ionic Standard Concentrates

Ion	*Each ampul contains*	*Solute*	*Packaging*
Aluminum	1 g Al^{3+} = 37.06 mmol	$AlCl_3$	Plastic ampul
Ammonium	0.1 g NH_4^+ = 5.544 mmol	NH_4Cl	Plastic ampul
Arsenic	0.01 g As^{3+} = 0.1335 mmol	$HAsO_2 + nNa_2SO_4$	Plastic ampul
Calcium	1 g Ca^{2+} = 24.95 mmol	$CaCl_2$	Plastic ampul
Chloride	1 g Cl^- = 28.21 mmol	NaCl	Plastic ampul
Chromium	1 g Cr^{6+} = 19.23 mmol	K_2CrO_4	Plastic ampul
Copper	1 g Cu^{2+} = 15.74 mmol	$CuSO_4$	Plastic ampul
Hardness	1 g $CaCO_3$ = 9.991 mmol	$CaCl_2$	Plastic ampul
Iron	1 g Fe^{3+} = 17.91 mmol	$FeCl_3$	Plastic ampul
Lead	1 g Pb^{2+} = 4.826 mmol	$Pb(NO_3)_2$	Glass ampul
Manganese	1 g Mn^{2+} = 18.20 mmol	$MnSO_4$	Plastic ampul
Mercury	0.1 g Hg^{2+} = 0.4985 mmol	$Hg(NO_3)_2$	Glass ampul
Nickel	1 g Ni^{2+} = 17.03 mmol	$NiSO_4$	Plastic ampul
Phosphate	1 g PO_4^{3-} = 10.529 mmol	KH_2PO_4	Plastic ampul
Potassium	1 g K^+ = 25.57 mmol	KCl	Plastic ampul
Silica	1 g SiO_2 = 16.64 mmol	$Na_2SiO_3 + nNaOH$	Plastic ampul
Silver	1 g Ag^+ = 9.270 mmol	$AgNO_3$	Plastic ampul
Sodium	1 g Na^+ = 43.50 mmol	NaCl	Plastic ampul
Sulfate	1 g SO_4^{2-} = 10.409 mmol	Na_2SO_4	Plastic ampul
Zinc	1 g Zn^{2+} = 15.30 mmol	$ZnSO_4$	Plastic ampul

SOURCE: J. T. Baker Chemical Company.

In this equation, W = the final weight in grams of the solution, N = the desired normality at 20°C, E = the number of equivalents provided by the one or more DILUT-IT™ Volumetric Concentrates used, and k = a constant read from the accompanying curves for the reagent and the value of N.

To prepare the desired solution, tare a dry flask or bottle and transfer to it the Volumetric Concentrate (following the general directions), then dilute with water to the calculated weight (±0.05%) and mix thoroughly. Water used in the dilution should be sterile, distilled, and free of dissolved gases.

For best accuracy, the temperature of the final solution should be brought to 20 ± 2°C just before use.

25
VOLUMETRIC ANALYSIS

INTRODUCTION

A volumetric method is one in which the analysis is completed by measuring the volume of a solution of established concentration needed to react completely with the substance being determined. A review of Chap. 24, "Chemicals and Preparation of Solutions," and Chap. 26, "pH Measurement," is recommended.

Definitions of Terms

Units of Volume

Liter (L): The volume occupied by one cubic dekameter.

Millileter (mL): One-thousandth of a liter.

Cubic centimeter (cm^3): Can be used interchangeably with milliliter without effect.

Titration

A process by which a substance to be measured is combined with a reagent and quantitatively measured. Ordinarily, this is accomplished by the controlled addition of a reagent of known concentration to a solution of the substance until reaction between the two is judged to be complete; the volume of reagent is then measured.

Back Titration

A process by which an excess of the reagent is added to the sample solution and this excess is then determined with a second reagent of known concentration.

Standard Solution

A reagent of known composition used in a titration. The accuracy with which the concentration of a standard solution is known sets a limit on the accuracy of the test.

Standard solutions are prepared by one of the following methods.

- Carefully measuring a quantity of a pure compound and calculating the concentration from the mass and volume measurements

- Carefully dissolving a massed quantity of the pure reagent itself in the solvent and diluting it to an exact volume
- Using a prestandardized commercially available standard solution

Equivalence Point

The point at which the standard solution is chemically equivalent to the substance being titrated. The equivalence point is a theoretical concept. We estimate its position by observing physical changes associated with it in the solution.

End Point

The point where physical changes arising from alterations in concentration of one of the reactants at the equivalence point become apparent.

Titration Error

The inadequacies in the physical changes at the equivalence point and our ability to observe them.

Meniscus

The curvature that is exhibited at the surface of a liquid which is confined in a narrow tube such as a buret or a pipet.

Typical Physical Changes during Volumetric Analysis

- Appearance or change of color due to the reagent, the substance being determined, or an indicator substance
- Turbidity formation resulting from the formation or disappearance of an insoluble phase
- Conductivity changes in a solution
- Potential changes across a pair of electrodes
- Refractive index changes
- Temperature changes

Reading the Meniscus

Volumetric flasks, burets, pipets, and graduated cylinders are calibrated to measure volumes of liquids. When a liquid is confined in a narrow tube such as a buret or a pipet, the surface is bound to exhibit a marked curvature, called a meniscus. It is common practice to use the bottom of the meniscus in calibrating and using volumetric ware. Special care must be used in reading this meniscus. Render the bottom of the meniscus, which is transparent, more distinct by positioning a black-striped white card behind the glass. (See Fig. 25-1.)

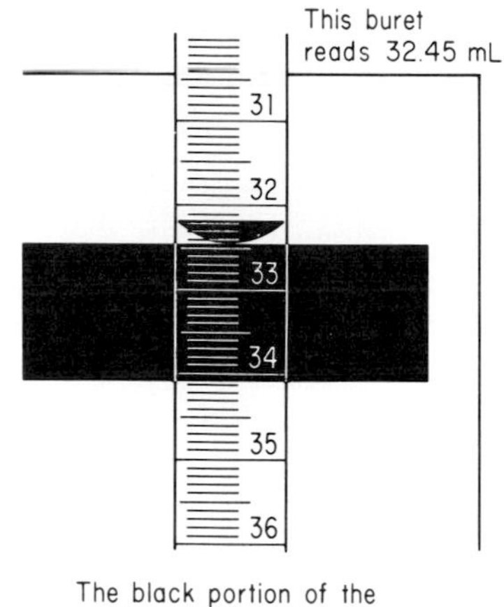

FIGURE 25-1
Useful technique for reading a meniscus.

Procedure

Location of the eye in reading any graduated tube is important:

1. With the eye above the meniscus, too small a volume is observed.
2. With the eye at the same level as the meniscus, the correct volume is observed.
3. With the eye below the meniscus, too large a volume is observed.

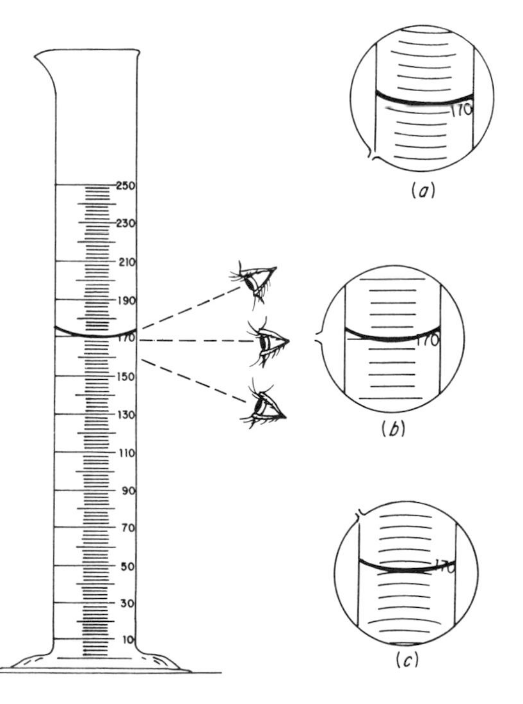

FIGURE 25-2
Avoiding parallax error in reading a meniscus. (*a*) Eye level too high; volume too high. (*b*) Eye level correct; volume correct. (*c*) Eye level too low; volume too low.

The eye must be level with the meniscus of the liquid to eliminate parallax errors (Fig. 25-2). Read the top of the black part of the card with respect to the graduations on a buret.

Washing and Cleaning Laboratory Glassware

General Rules

1. Always clean your apparatus immediately after use, if possible. It is much easier to clean the glassware *before* the residues in them become dry and hard. If dirty glassware cannot be washed immediately, put it in water to soak.

2. Handle glassware carefully while cleaning it, especially bulky flasks and long, slender columns.

CAUTION Round-bottomed flasks are especially fragile. When they are full, never set them down on a hard, flat surface while holding them by the neck. Even gentle pressure can generate a great deal of pressure and cause a star crack, which is difficult to see in a wet flask. Cracked glassware is a hazard. Check each round-bottomed flask carefully before it is used. Support round-bottomed flasks *only* on ring stands or on cork rings.

3. Rinse off all soap or detergent residue after washing glassware to prevent any possible contamination later.

Most pieces of laboratory glassware can be cleaned by washing and brushing with a detergent or a special laboratory cleaning product called Alconox. After they have been thoroughly cleaned, they are rinsed with tap water and finally with a spray of distilled water. If the surface is clean, the water will wet the surface uniformly. On soiled glass the water stands in droplets (Fig. 25-3).

Use brushes carefully and be certain that the brush has no exposed sharp metal points which can scratch the glass and cause it to break. Such

FIGURE 25-3 Water spreads out smoothly and evenly on clean glass (*a*), but stands in droplets on soiled glass (*b*).

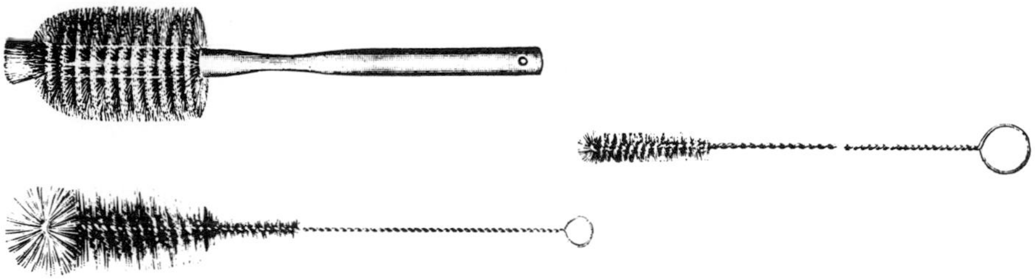

FIGURE 25-4
Brushes in a varied selection of sizes and designs for cleaning laboratory equipment.

scratches are a frequent and unexpected cause of breakage when the glassware is heated. Brushes come in all shapes and sizes (Fig. 25-4). Using the correct brush makes the cleaning job much easier.

CAUTION Never exert excessive force on the brush. You may force the metal through the bristles to scratch the glassware, causing breakage and injury to yourself.

Cleaning Volumetric Glassware

1. Always rinse volumetric glassware equipment three times with distilled water after you have emptied and drained it. This prevents solutions from drying on the glassware, causing difficulty in cleaning.

2. Dry volumetric glassware at room temperature, *never in a hot oven.* Expansion and contraction may change the calibration.

3. The glass surfaces should be wetted *evenly.* Spotting is caused by grease and dirt. *Removing grease:* Rinse and scrub with hot detergent solution followed by adequate distilled-water rinses. *Removing dirt:* Fill or rinse with dichromate cleaning solution. Allow to stand for several hours, if necessary. Follow with multiple distilled-water rinses.

Pipets

Pipets may be cleaned with a warm solution of detergent or with cleaning solution. Draw in sufficient liquid to fill the bulb to about one-third of its capacity. While holding it nearly horizontal, carefully rotate the pipet so that all interior surfaces are covered. Drain inverted, and rinse thoroughly with distilled water. Inspect for water breaks, and repeat the cleaning cycle as often as necessary.

Burets

Thoroughly clean the tube with detergent and a long brush. If water breaks persist after rinsing, clamp the buret in an inverted position with the end dipped in a beaker of cleaning solution. Connect a hose from the buret tip to a vacuum line. Gently pull the cleaning solution into the buret, stopping well short of the stopcock. Allow to stand for 10 to 15 min, and then drain. Rinse thoroughly with distilled water and again inspect for water breaks. Repeat the treatment if necessary.

After some use, the grease lubricant on stopcocks tends to harden; small particles of grease can break off, flow to the tips of burets, and clog them. The grease may be cleaned out by inserting a fine, flexible wire into the buret tip and breaking up the plug. Flush out the particles with water.

CAUTION The wire must be flexible and not too thick. Otherwise, you may break or split the tip.

Alternative method of cleaning buret tips:

1. Carefully heat the plugged tip of the buret with a match (Fig. 25-5), warming the grease and melting it; use several matches if necessary. Water pressure or air (blowing through the cleaned top of the buret) will push the melted grease out.

CAUTION Heat cautiously; the tip may break.

2. Wipe surface soot from tip.

CAUTION Handle burets and pipets very carefully. They are longer than you think. Their tips are especially fragile and are easily broken on contact with sinks or water faucets. Protective rubber mats help prevent breakage and chipping.

Automatic Pipet and Tube Cleaners

Apparatus designed to clean and dry many pipets or tubes at one time is available. It consists of a stainless steel or high-density polyethylene cylinder which utilizes intermittent siphon action, setting up a turbulent "fill-and-empty" action cycle. Either hot or cold water can be used, and some units are fitted with an electric heater to provide heated air through the cylinder after the washing cycles.

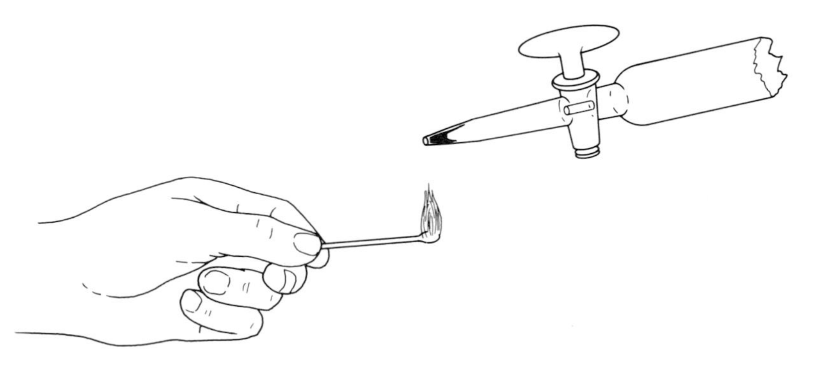

FIGURE 25-5
Cleaning a buret tip.

Cleaning Glassware Soiled with Stubborn Films and Residues

When you cannot completely clean glassware by scrubbing it with a detergent solution, more drastic cleaning methods must be used.

Dichromate–Sulfuric Acid Cleaning Solution

CAUTION Prepare and handle this cleaning solution with extreme care. Avoid contact with clothing or skin.

Dissolve 92 g $Na_2Cr_2O_7 \cdot 2\ H_2O$ (sodium dichromate) in 458 mL H_2O, and cautiously add with stirring 800 mL concentrated H_2SO_4.* The contents of the flask will get very hot and become a semisolid red mass. When this happens, add just enough sulfuric acid to bring the mass into solution. Allow the solution to cool before attempting to transfer it to a soft-glass bottle. After the glassware has been cleaned with a detergent and rinsed carefully, pour a small quantity of the chromate solution into the glassware, allowing it to flow down all parts of the glass surface. Pour the solution back into its stock bottle. Then rinse the glassware, first with tap water and then with distilled water, until the glass surface looks clean.†

The solution may be reused until it acquires the green color of the chromium(III) ion. Once this happens, it should be discarded.

Dilute Nitric Acid Cleaning Solution

Films which adhere to the inside of flasks and bottles may often be removed by wetting the surface with dilute nitric acid, followed by multiple rinses with distilled water.

Aqua Regia Cleaning Solution

Aqua regia is made up of three parts of concentrated HCl and one part of concentrated HNO_3. This is a very powerful, but extremely dangerous and corrosive, cleaning solution. *Use in a hood with extreme care.*

Alcoholic Potassium Hydroxide or Sodium Hydroxide Cleaning Solution

Add about 1 L ethanol (95%) to 120 mL H_2O containing 120 g NaOH or 105 g KOH.

This is a very good cleaning solution. Avoid prolonged contact with ground-glass joints on interjoint glassware because the solution will etch glassware and damage will result. This solution is excellent for removing carbonaceous materials.

---

* Potassium dichromate may also be used, but it is less soluble than the sodium compound.

† As many as 10 washings with demineralized or distilled water are necessary to remove traces of the chromium ions.

Trisodium Phosphate Cleaning Solution

Add 57 g Na_3PO_4 and 28.5 g sodium oleate to 470 mL H_2O.

This solution is good for removing carbon residues. Soak glassware for a short time in the solution, and then brush vigorously to remove the incrustations.

Nochromix Cleaning Solution

This is a commercial oxidizer solution, but it contains no metallic ions. The powder is dissolved in concentrated sulfuric acid, yielding a clear solution. The solution turns orange as the oxidizer is used up. *Use with care.*

Metal Decontamination of Analytical Glassware

Where metal decontamination is desired, in chelometric titration, treat acid-washed glassware by soaking it in a solution containing 2% NaOH and 1% disodium ethylenediamine tetraacetate for about 2 h; follow this bath with a number of rinses with distilled water.

Ultrasonic Cleaning

Ultrasonic cleaning units generate high-frequency sound waves (sound waves of higher frequencies than those detectable by the human ear) that penetrate deep recesses, turn corners, and pass through barriers. Laboratory glassware and optical equipment as well as narrow-bore pipets, manometers, and similar items are easily cleaned with such units. The items are immersed in the cleaning solution, and the power is turned on.

The cleaners are usually fitted with stainless steel tanks and come in various sizes. They can clean a load of badly soiled pipets and assorted glassware faster and better than any other method of cleaning. The principle on which ultrasonic cleaners work is *cavitation*, the formation and collapse of submicron bubbles (bubbles smaller than 1 μm in diameter). These bubbles form and collapse about 25,000 times each second with a violent microscopic intensity which produces a scrubbing action. This action effectively frees every surface of the glassware of contaminants, because the glassware is immersed in the solution and the sound energy penetrates wherever the solution reaches.

Fluidized-Bath Method

A fluidized bath may be used for cleaning polymeric and carbonaceous residues from laboratory glassware and metallic apparatus. (For discussion on fluidized baths refer to Chap. 10, "Heating and Cooling.")

1. Establish a temperature in the fluidized medium of approximately 500°C (932°F).

2. Place the workpieces and hardware to be cleaned in a basket, and lower the basket into the fluidized medium.

3. Adjust heater control. Heat is transmitted to the immersed parts, and thermal breakdown of the plastic polymer commences almost immediately. Cleaning is effected rapidly. Be sure the fumes which are emitted from the surface of the bath are removed from the work area by means of a fume hood and extractor fan.

4. Remove the basket and parts from the bath, and allow them to cool to room temperature. Metals can be quenched in water.

5. Any noncombustible residues may be easily removed by blowing with an air jet or by brushing lightly.

Drying Laboratory Glassware

After glassware has been thoroughly cleaned and rinsed, it must be dried.

Drainboards and Drain Racks

Drainboards and drain racks are used for draining and drying various sizes and shapes of glassware. The supports have pins and pegs anchored in an inclined position to ensure drainage. Some drainboards are fitted with hollow pegs and a hot-air blower to speed up the drying process. Place glassware securely on rack. Do not allow pieces to touch each other and cause accidental breakage.

Dryer Ovens

Dryer ovens are designed for efficient high-speed drying of glassware. They come in various sizes and power ratings; some of them are fitted with timers.

Quick Drying

If it is necessary to dry the insides of flasks or similar vessels, gently warm them over a Bunsen flame and then gently pass a stream of compressed air through a glass tube which leads to the bottom of the flask until the item is dry.

Rinsing Wet Glassware with Acetone

Water-wet glassware can be dried more quickly by rinsing the item with several small portions of acetone, discarding the acetone rinse after each time. Then place the item in a safety oven or gently pull air through the glassware by connecting a pipet with rubber tubing to an aspirator and inserting the pipet in the glassware.

CAUTION *Never heat thick-walled glassware in a flame.* It will break (Fig. 25-6).

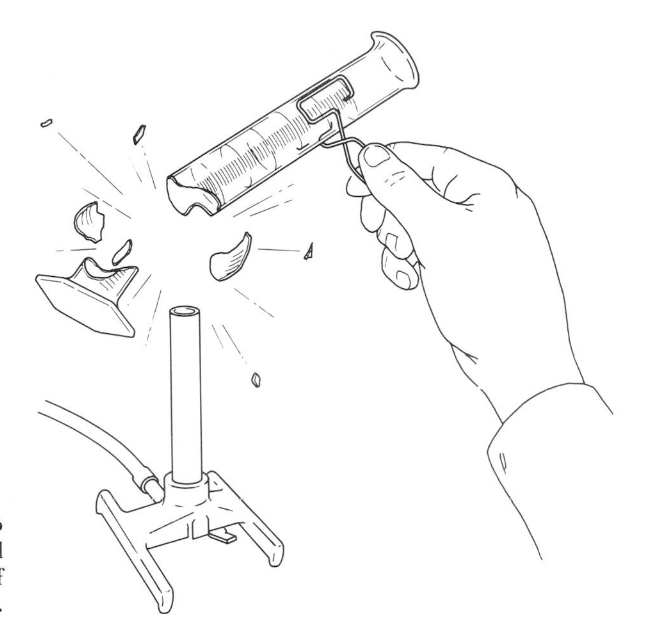

FIGURE 25-6
Heavy, thick-walled glassware will break if heated in a flame.

TOOLS OF VOLUMETRIC ANALYSIS

Pipets, burets, and volumetric flasks are standard volumetric equipment. Volumetric apparatus calibrated to contain a specified volume is designated TC, and apparatus calibrated to deliver a specified amount, TD.

Only clean glass surfaces will support a uniform film of liquid; the presence of dirt or oil will tend to cause breaks in this film. The appearance of water breaks is a sure indication of an unclean surface. Volumetric glassware is carefully cleansed by the manufacturer before being supplied with markings, and in order for these to have meaning, the equipment must be kept equally clean when in use.

As a general rule, the heating of calibrated glass equipment *should be avoided.* Too rapid cooling can permanently distort the glass and cause a change in volume.

Volumetric Flasks

Volumetric flasks are calibrated to contain a specified volume when filled to the line etched on the neck.

Directions for the Use of a Volumetric Flask

Before use, volumetric flasks should be washed with detergent and, if necessary, cleaning solution. Then they should be carefully and repeatedly rinsed in distilled water; only rarely need they be dried. Should drying be required, however, it is best accomplished by clamping the flasks in an inverted position and employing a mild vacuum to circulate air through them.

Introducing Standard Directly into a Volumetric Flask

Direct preparation of a standard solution requires that a known mass of solute be introduced into a volumetric flask. In order to minimize the possibility of loss during transfer, insert a funnel into the neck of the flask. The funnel is subsequently washed free of solid.

Dilution to the Mark

After introducing the solute, fill the flask about half full and swirl the contents to achieve solution. Add more solvent, and again mix well. Bring the liquid level almost to the mark, and allow time for drainage. Then use a medicine dropper to make such final additions of solvent as are necessary. Firmly stopper the flask and invert repeatedly to assure uniform mixing.

NOTES

1. If, as sometimes happens, the liquid level accidentally exceeds the calibration mark, the solution can be saved by correcting for the excess volume. Use a gummed label to mark the actual position of the meniscus. After the flask has been emptied, carefully refill to the mark with water. Then, using a buret, measure the volume needed to duplicate the actual volume of the solution. This volume, of course, should be added to the nominal value for the flask when the concentration of the solution is calculated.

2. For exacting work, the flask should be maintained at the temperature indicated on the flask.

Pipets

Pipets are designed for the transfer of known volumes of liquid from one container to another. Pipets which deliver a fixed volume are called *volumetric* or *transfer pipets* (Fig. 25-7). Other pipets, known as measuring pipets, are calibrated in convenient units so that any volume up to maximum capacity can be delivered (Fig. 25-8). Table 25-1 gives these units for some measuring pipets.

Certain accessories are useful in working with pipets. See Fig. 25-9.

Directions for the Use of a Pipet

The following instructions pertain specifically to the manipulation of transfer pipets, but with minor modifications they may be used for other types as well. Liquids are usually drawn into pipets through the application of a slight vacuum. Use mechanical means, such as a rubber suction bulb or a rubber tube connected to a vacuum pump. See Fig. 25-9*c* and *d*.

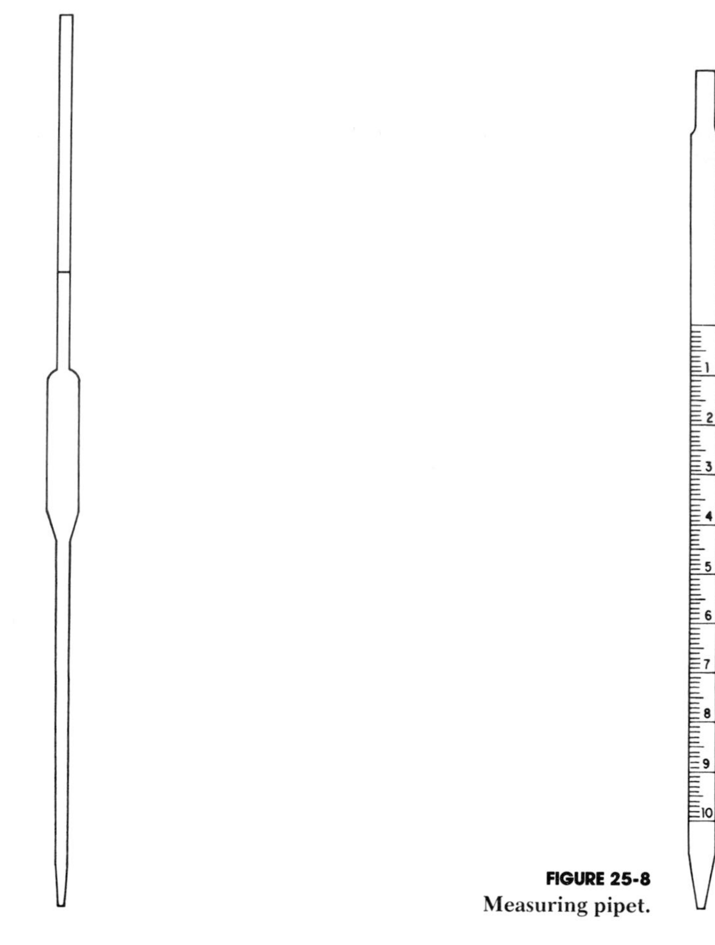

FIGURE 25-7
Volumetric, or transfer, pipet.

FIGURE 25-8
Measuring pipet.

Procedure

1. Clean pipet thoroughly and rinse with distilled water.

2. Drain completely, leaving no rinse-water drops inside. If the pipet is wet with water, rinse three times with the solution to be used in the analysis.

3. Keep the tip of the pipet below the surface of the liquid.

TABLE 25-1
Calibration Units of Measuring Pipets

Capacity, mL	*Divisions, mL*	*Capacity, mL*	*Divisions, mL*
0.1	0.01	5	0.1
0.2	0.01	10	0.1
1	0.01	25	0.1
1	0.1	50	0.2
2	0.1		

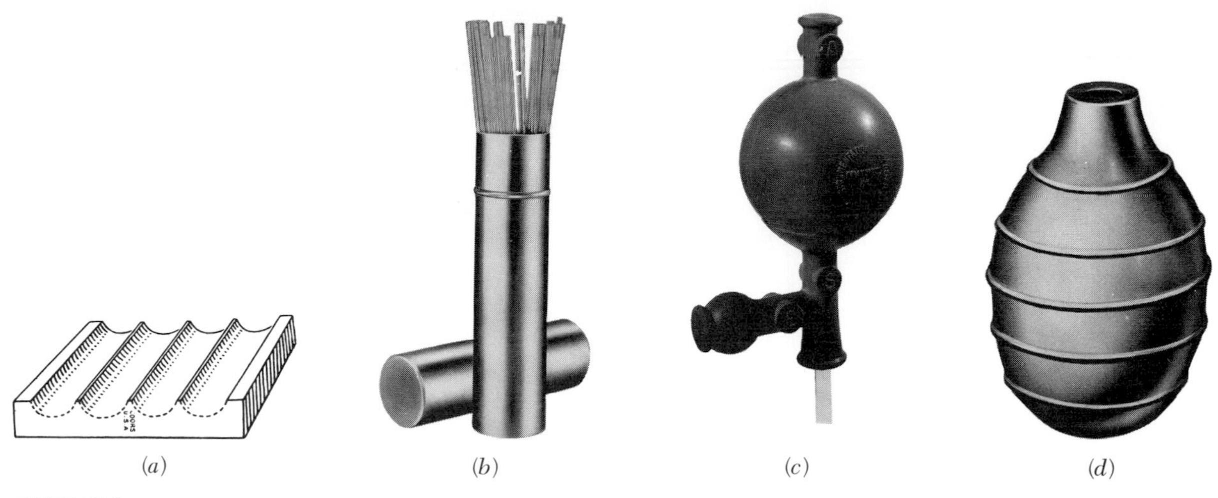

(a) (b) (c) (d)

FIGURE 25-9

Pipet accessories, (*a*) Porcelain pipet rest to prevent contamination and to prevent pipets from rolling and breaking. (*b*) Cylindrical can, color-coded top and bottom, for storage of pipets after cleaning and use. (*c*) Pipet filler used to transfer sterile, corrosive, and toxic liquids safely. Easily controlled by squeezing, it delivers quickly, precisely, and safely. (*d*) Pipet filler. Squeeze first, then immerse in liquid; release pressure gradually as needed.

4. Draw the liquid up in the pipet using the pipet bulb or suction (mouth or aspirator) (Fig. 25-10).

CAUTION *Always* use an aspirator, pipet bulb, and/or safety flask when working with liquids (Figs. 25-9*c* and *d* and 25-10). *Never pipet any solution by mouth.*

5. Disconnect the suction when the liquid is above the calibration mark. *Quickly remove the suction unit and immediately place the index finger of the hand holding the pipet over the exposed end of the pipet to the closed end.*

6. Release pressure on the index finger to allow the meniscus to approach the calibration mark.

7. At the mark, apply pressure to stop the liquid flow, and drain the drop on the tip by touching it to the wall of the liquid-holding container.

8. Transfer the pipet to the container to be used and release pressure on the index finger. Allow the solution to drain completely; allow a time lapse of 10 s, or the period specified on the pipet. Remove the last drop by touching the wall of the container (Figs. 25-11*c* and *d* and 25-12).

9. The calibrated amount of liquid has been transferred. *Do not blow out the pipet.* In the case of color-coded pipets, a frosted ring indicates complete blowout.

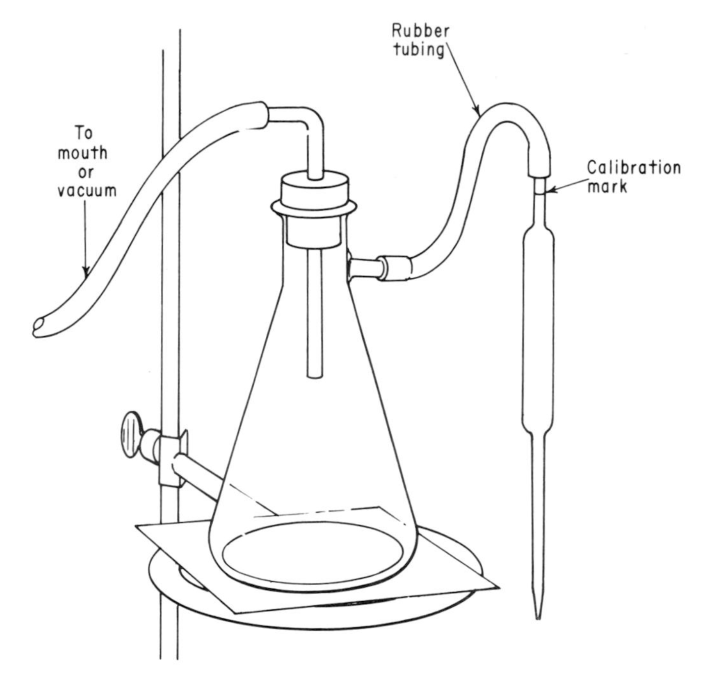

FIGURE 25-10
Using a pipet with a safety flask and water aspirator.

NOTES

1. The liquid can best be held at a constant level in the pipet if the forefinger is slighly moist (use distilled water). Too much moisture, however, makes control difficult.
2. It is good practice to avoid handling the pipet by the bulb.
3. Pipets should be thoroughly rinsed with distilled water after use.

Burets

Burets, like measuring pipets, deliver any volume up to their maximum capacity. Burets of the conventional type (Fig. 25-13*a*) must be manually

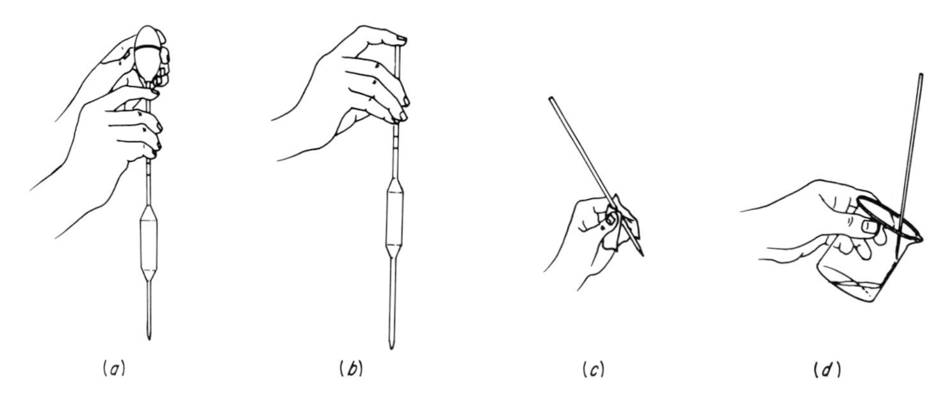

FIGURE 25-11
Technique for using a volumetric pipet. (*a*) Draw liquid past the graduation mark. (*b*) Use forefinger to maintain liquid level above the graduation mark. (*c*) Tilt pipet slightly and wipe away any drops on the outside surface. (*d*) Allow pipet to drain freely.

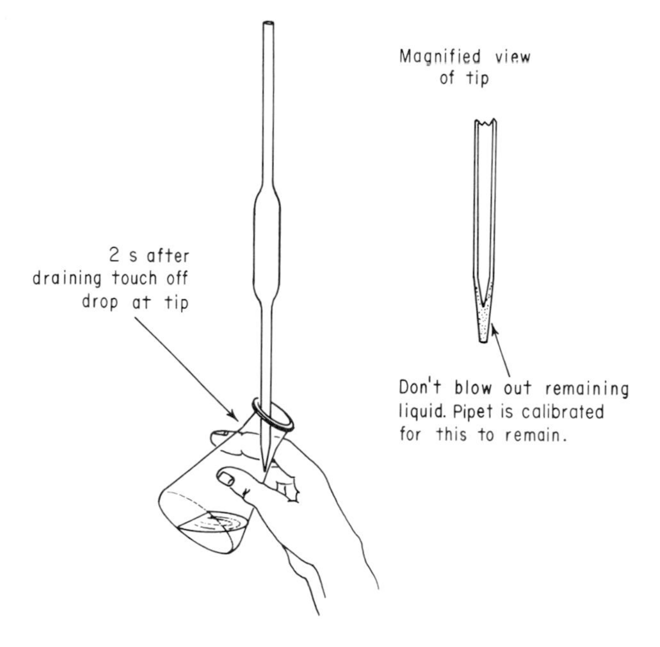

FIGURE 25-12
Correct technique for draining a pipet.

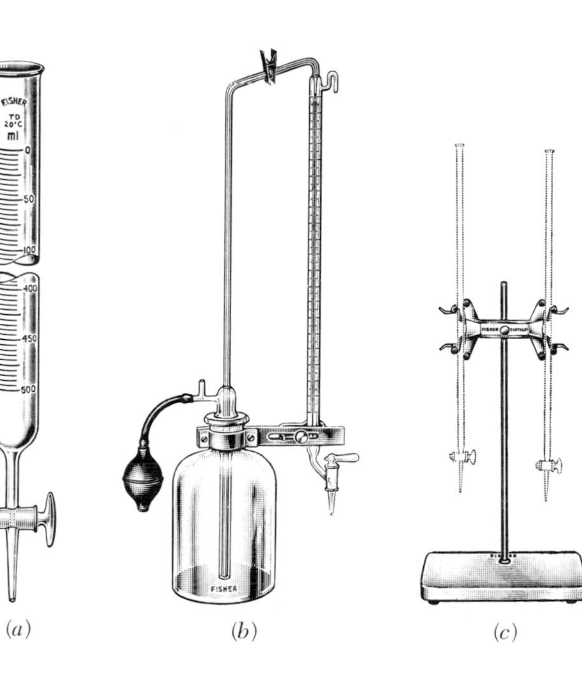

FIGURE 25-13
Burets. (*a*) Single-dispensing buret with graduated etched scale, standard taper, and stopcock. Such burets are available with volumes from 10 to 1000 cm^3. The newer burets are equipped with plastic (e.g., Teflon®) stopcocks. (*b*) Automatic-filling buret. Pumping the rubber bulb fills it to a precise 0.02-cm graduation, and overfill automatically returns to storage. Such burets are used where many titrations with the same solution are to be made. (*c*) Titrating assembly and stand with white base for easy observation of color changes *(Fisher Scientific Company)*.

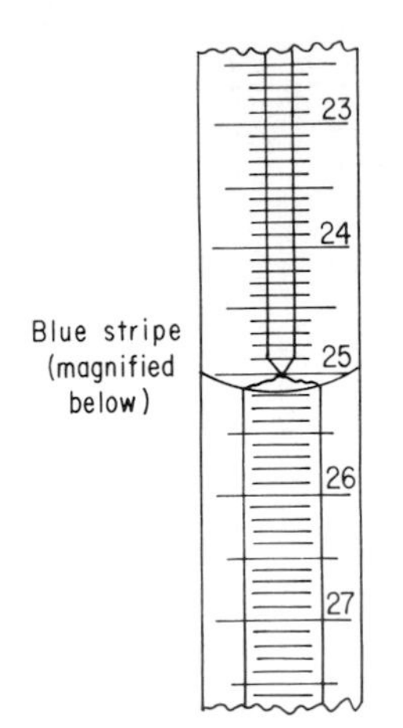

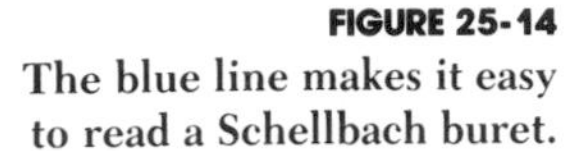

FIGURE 25-14
The blue line makes it easy to read a Schellbach buret.

filled. Others with side arms are filled by gravity. For more accurate work, Schellbach burets (Fig. 25-14) are employed. These have a white background with a blue stripe and can be read at the point of least magnification. When unstable reagents are employed, a buret with a reservoir bottle and pump may be employed (Fig. 25-13*b*).

Directions for the Use of a Buret

Before being placed in service, a buret must be scrupulously clean. In addition, it must be established that the stopcock is liquid-tight.

NOTE Grease films that appear unaffected by cleaning solution may yield to treatment with such organic solvents as acetone or benzene. Thorough washing with detergent should follow such treatment. (Refer to Washing and Cleaning Laboratory Glassware in this chapter.)

Filling

Make certain that the stopcock is closed. Add 5 to 10 mL of solution and carefully rotate the buret to wet the walls completely; allow the liquid to drain through the tip. Repeat this procedure two more times. Then fill the buret above the zero mark. Free the tip of air bubbles by rapidly rotating the stopcock and allowing small quantities of solution to pass. Finally, lower the level of the solution to, or somewhat below, the zero mark; after allowing about a minute for drainage, take an initial volume reading.

Holding the Stopcock

Always push the plug into the barrel while rotating the plug during a titration. A right-handed person points the handle of the stopcock to the right, operates the plug with the left hand, and grasps the stopcock from the left side as shown in Fig. 25-15.

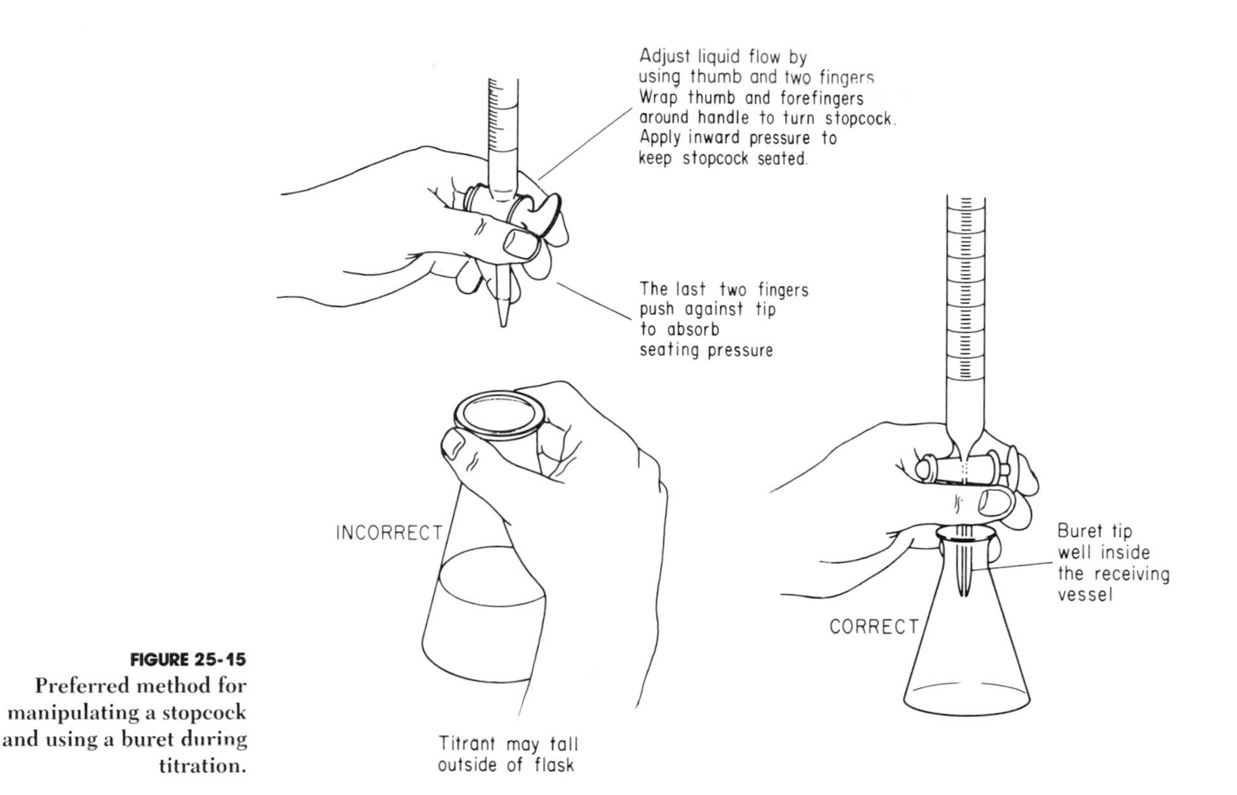

FIGURE 25-15
Preferred method for manipulating a stopcock and using a buret during titration.

Procedure

1. Test the buret for cleanliness by clamping it in an upright position and allow it to drain. *No water drops should adhere to the inner wall.* If they do, *reclean the buret.*

2. Grease the stopcock with clean grease, *after cleaning it.*

   (**a**) Remove the stopcock.

   (**b**) Clean the barrel and stopcock with a swab soaked in benzene.

   (**c**) Regrease as shown in Fig. 25-16. Improperly applied grease will spread and obstruct the holes, making *recleaning* necessary.

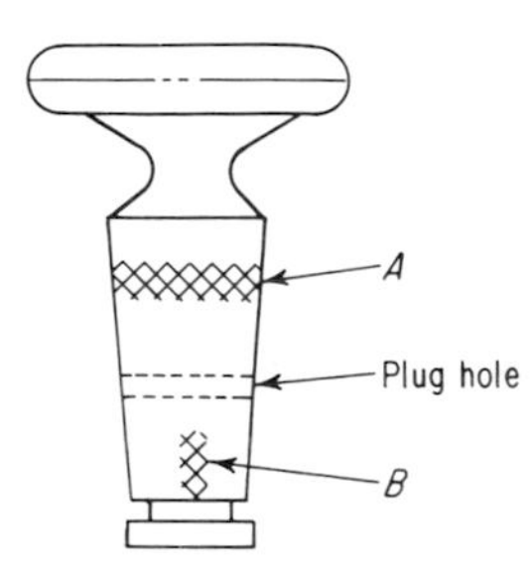

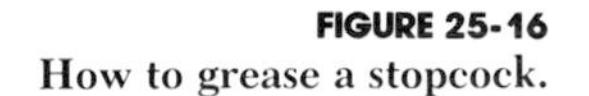
FIGURE 25-16
How to grease a stopcock.

PERFORMING A TITRATION

1. Use a setup such as that shown in Fig. 25-13*c*. Rinse the cleaned buret three times with the solution to be used, draining completely each time.

2. Fill the buret above the zero graduation.

3. Drain slowly until the tip is free of air bubbles and completely filled with liquid and the meniscus of the liquid is at the zero graduation.

4. Add the titrant to the titration flask slowly, swirling the flask with the right hand until the end point is obtained. To avoid error, with the tip well within the titration vessel, introduce solution from the buret in increments of a milliliter or so. Swirl (or stir) the sample constantly to assure efficient mixing. Reduce the volume of the additions as the titration progresses; in the immediate vicinity of the end point, the reagent should be added a drop at a time. When it is judged that only a few more drops are needed, rinse down the walls of the titration. Allow a minute or so to elapse between the last addition of reagent and the reading of the buret. Keep your eye level with the meniscus for all readings.

5. Rinse the walls of the flask frequently (Fig. 25-17).

6. For precision work, volumes of less than one drop can be rinsed off the tip of the buret with wash water (Fig. 25-18). (See Splitting a Drop of Titrant, below.)

7. Near the end point, the trail of color from each drop is quite long (Fig. 25-19). The end point is reached when the color change does not disappear after 30 s.

8. Allow the buret to drain for 30 s; then read the final position of the meniscus.

FIGURE 25-17
Rinse the walls of the flask frequently during a titration.

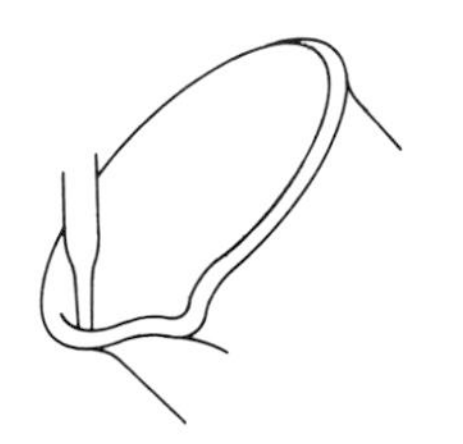

FIGURE 25-18
Removal of an adhering drop.

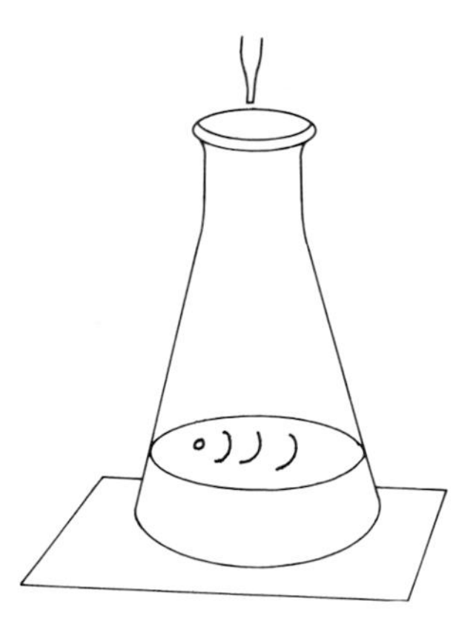

FIGURE 25-19
As the end point is approached, the color trail from each drop gets larger and persists longer.

CAUTION *Never allow reagents to remain in burets overnight. The stopcock may "freeze" because of prolonged contact,* especially with bases such as KOH and NaOH.

9. The difference between the "before" and "after" readings on the buret is the volume of liquid delivered.

Splitting a Drop of Titrant

There are times in precision titration when one drop is more than is needed to reach the equivalence point, and only part of a drop is needed. In these cases you may use method 1 or 2; method 1 is preferable.

1. Carefully open the stopcock so that only part of a drop appears, and then close the stopcock. Carefully direct a small stream of water from the wash bottle on the tip of the buret to wash the part drop into the solution. Mix the solution thoroughly before adding another part drop.

2. Quickly but carefuly spin the closed stopcock 180°. A small stream of liquid will shoot out.

CAUTION *Spin quickly, but carefully.* Hold the stopcock securely so that it does not become loose in the socket. Be sure that the stopcock is closed after the spin.

ACID-BASE TITRATIONS AND CALCULATIONS

A common reaction in chemistry involving an acid and a base is called a *neutralization* and can be represented by this simple equation:

Acid + base → salt + water

The concentration of an unknown solution of acid or base can be determined by reacting a measured quantity of the unknown solution with a measured volume of an appropriate acid or base of known concentration. This process is called an *acid-base titration*. In a typical titration, a measured volume of the unknown solution is placed in a flask, and a solution of known concentration is added (measured) from a buret until an equivalent amount of both the acid and base are present. The point at which the acid and base have been added in equivalent amounts is called the *equivalence point*. Usually the equivalence point of a titration is determined by using an acid-base indicator. These *acid-base indicators* have characteristic colors in acidic and basic solutions (see Chap. 26, "pH Measurement," for details) and are selected to undergo these color changes at the specific equivalence point or pH.

In a typical acid-base titration, the concentration of an unknown sodium hydroxide solution, NaOH, can be determined by titration with a known concentration of hydrochloric acid, HCl, using a phenolphthalein indicator.

Phenolphthalein is colorless in an acidic solution and is pink in a basic solution. The chemical reaction is shown by the equation below:

$NaOH + HCl \rightarrow NaCl + H_2O$

A *simple acid-base equation* for calculating concentrations is as follows:

$$C_{acid} \times V_{acid} = C_{base} \times V_{base}$$

where C_{acid} = concentration of the acid
V_{acid} = volume of the acid
C_{base} = concentration of the base
V_{base} = volume of the base

Acids and bases can also be standardized (concentration accurately determined) by using a primary standard. A primary standard is usually a solid acid or base compound of high purity, stability, and known molecular weight as shown in Table 25-2.

Procedure for Standardization of a Base or Acid

1. Review the section in this chapter entitled Performing a Titration.

2. Using a volumetric pipet, transfer exactly 25.00 mL of standardized acid solution into a clean 25-mL Erlenmeyer flask. If a primary acid standard were used instead of the standardized acid solution, a very accurately

TABLE 25-2
Acid-Base Primary Standards

Compound	*Formula*	*Molecular wt.*
	Primary acids	
Potassium hydrogen phthalate	$KHC_8H_4O_4$	204.23
Sulfamic acid	HNH_2SO_3	97.10
Oxalic acid dihydrate	$HOOC-COOH(H_2O)$	126.07
Benzoic acid	C_6H_5COOH	122.13
	Primary bases	
Sodium carbonate	Na_2CO_3	106.0
Tris(hydroxylmethyl)aminomethane	$(CH_2OH)_3CNH_2$	121.4

weighed quantity would be added to the flask with enough water to dissolve the solid. Add three drops of phenolphthalein indicator solution to the Erlenmeyer flask.

3. Titrate the standardized acid solution slowly by releasing a basic solution from a filled (starting volume reading known) buret through the stopcock. Swirl the Erlenmeyer flask gently as the base solution is added. A white piece of paper under the flask will allow the indicator color change to be more visible. As the base is added to the acid, a faint pink color will develop at the point of contact. As the end point (equivalence point) is approached, this pink coloration will persist for longer and longer periods. When this occurs, add the base drop by drop until the end point is reached. This is indicated by the first drop of base which causes the entire solution to become faintly pink and to remain pink for at least 30 s. Record the final (end point) buret reading to the nearest 0.01 mL.

4. Calculate the concentration of the basic solution using the simple acid-base equation given above.

OXIDATION-REDUCTION TITRATIONS AND CALCULATIONS

Review sections entitled Oxidation-Reduction Reactions and Equivalent Mass in Redox Reactions in Chap. 28, "Electrochemistry." A review of the section entitled Laboratory Solutions in Chap. 24, "Chemicals and Preparation of Solutions," may be helpful in dealing with the concepts of normality and equivalents. The acid-base reactions and calculations in the previous section assumed that all reactions are one-to-one. In reality, molar quantities of the reactants in a chemical equation cannot always be reduced to a one-to-one ratio. For example, the following common acid-base reaction is not a simple one-to-one reaction.

$$H_2SO_4 + 2NaOH \rightarrow Na_2SO_4 + 2H_2O$$

TABLE 25-3
Common Titrimetric Oxidizing and Reducing Agents

Agent	*Formula*	*Product*	*Subscript, s*
	Oxidizing		
Permanganate (in acid)	MnO_4^-	Mn^{2+}	5
Dichromate (in acid)	$Cr_2O_7^{2-}$	Cr^{3+}	6
Chlorite (in acid)	ClO_3^-	Cl^-	6
Peroxide (in acid)	H_2O_2	H_2O	2
Nitrate (in dilute acid)	NO_3^-	NO	3
Ferric (in acid)	Fe^{3+}	Fe^{2+}	1
	Reducing		
Iodide (acid or neutral)	I^-	I_2	1
Oxalate (in acid)	$C_2O_4^{2-}$	CO_2	2
Thiosulfate (in acid)	$S_2O_3^{2-}$	$S_4O_6^{2-}$	1
Sulfur dioxide (in acid)	SO_2	SO_4^{2-}	2
Ferrous (in acid)	Fe^{2+}	Fe^{3+}	1
Zinc (in acid)	Zn	Zn^{2+}	2

Two moles of sodium hydroxide are required to neutralize one mole of sulfuric acid. It should be obvious that the simple acid-base equation which assumes a one-to-one reaction ratio will not yield the correct answer if these two substances are titrated together. Table 25-3 lists some common titrimetric oxidizing and reducing reagents.

A mathematical formula that will always yield a correct answer even with different equation-balancing ratios is the *stoichiometric equation* given below. *Stoichiometry* calculations involve converting the mass of one substance (reactant or product) in a chemical reaction into equivalent terms for all other substances (reactant or product) in the balanced reaction.

$$N_A + V_A = N_B + V_B$$

where N_A = normality of substance A
V_A = volume of substance A
N_B = normality of substance B
V_B = volume of substance B

Normality and molarity have a simple mathematical relationship. The subscript (s) term is detailed in Table 25-4.

$$N = M \times s$$

where N = normality
M = molarity
s = subscript on the H^+ ion, OH^- ion, or the change in the charge on the ion

EXAMPLE A 4 M H_2SO_4 solution would have what normality?

$$N = M \times s$$
$$N = 4 \times 2$$
$$N = 8$$

EXAMPLE A 0.4 M $KMnO_4$ solution would have what normality?

$$N = M \times s$$
$$N = 0.4 \times 5$$
$$N = 2.0$$

TABLE 25-4
Stoichiometric Relationship of Compounds as Acids, Bases, or Redox Reagents

Compound	*Formula*	*Subscript, s*
	H^+ ions	
Hydrochloric acid	HCl	1
Nitric acid	HNO_3	1
Potassium hydrogen phthalate	KHC_8O_4	1
Sulfuric acid	H_2SO_4	2
Phosphoric acid	H_3PO_4	3
	OH^- ions	
Sodium hydroxide	$NaOH$	1
Potassium hydroxide	KOH	1
Calcium hydroxide	$Ca(OH)_2$	2
Aluminum hydroxide	$Al(OH)_3$	3
	Redox agent change	
Ferric to ferrous	Fe^{3+}/Fe^{2+}	1
Stannic to stannous	Sn^{4+}/Sn^{2+}	2
Oxalate to carbon dioxide	$C_2O_4^{2-}/CO_2$	2
Permanganate to manganous	MnO_4^-/Mn^{2+}	5

TABLE 25·5
Commonly Used Chelating Agents

Abbreviation	*Name*	*Formula of the anion*
DCyTA	1,2-diamino-cyclohexane-tetraacetic acid	$\begin{array}{l} \quad\;\; CH_2 \\ H_2C \qquad CH{-}N(CH_2COO^-)_2 \\ \;\; \vert \qquad\quad \vert \\ H_2C \qquad CH{-}N(CH_2COO^-)_2 \\ \quad\;\; CH_2 \end{array}$
DTPA	Diethylene-triamine-pentaacetic acid	$-OOC{-}CH_2{-}N\left(CH_2{-}CH_2{-}N\begin{array}{l} CH_2COO^- \\ CH_2COO^- \end{array}\right)_2$
EDTA	Ethylenediamine-tetraacetic acid	$(^-OOC{-}CH_2)_2N{-}CH_2{-}CH_2{-}N(CH_2{-}COO^-)_2$
EGTA	Ethylene glycol bisaminoethyl ether tetra-acetic acid	$\begin{array}{r} (^-OOCCH_2)_2N{-}CH_2{-}CH_2{-}O{-}CH_2{-}CH_2{-}O \\ \vert \\ (^-OOCCH_2)_2N{-}H_2C{-}CH_2 \end{array}$
EEDTA	Ethyl ether diaminetetra-acetic acid	$\begin{array}{r} (^-OOCCH_2)_2{=}N{-}CH_2{-}CH_2{-}O \\ \vert \\ (^-OOCCH_2)_2{=}N{-}H_2C{-}CH_2 \end{array}$
HEDTA	N′-hydroxyethylethyl-enediamine-triacetic acid	$\begin{array}{l} \quad -OOCCH_2 \\ \qquad\qquad \vert \\ \qquad\qquad N{-}CH_2{-}CH_2{-}N(CH_2COO^-)_2 \\ \qquad\qquad \vert \\ HO{-}CH_2CH_2 \end{array}$
MEDTA	1-methylethylenedi-aminetetra-acetic acid	$\begin{array}{l} \qquad\qquad\qquad\quad CH_3 \\ \qquad\qquad\qquad\quad \vert \\ (^-OOCCH_2)_2N{-}CH{-}CH_2{-}N(CH_2COO^-)_2 \end{array}$
NTA	Nitriloacetic acid	$N(CH_2COO^-)_2$
Penten	Pentaethylene-hexamine	$\begin{array}{r} H_2N{-}CH_2{-}CH_2{-}NH{-}CH_2{-}CH_2{-}N{-}H \\ \vert \\ CH_2 \\ \vert \\ H_2N{-}CH_2{-}CH_2{-}NH{-}CH_2{-}CH_2{-}NH{-}CH_2 \end{array}$
Tetren	Tetraethylene-pentamine	$NH(CH_2{-}CH_2{-}NH{-}CH_2{-}CH_2{-}NH_2)_2$
Trien	Triethylene-tetraamine	$H_2NCH_2{-}CH_2{-}NH{-}CH_2{-}CH_2{-}NH{-}CH_2{-}CH_2{-}NH_2$

CHELATES AND COMPLEXOMETRIC TITRATIONS

Chelates are a special class of coordination compounds which result from the reaction of a metal ion and a ligand that contains two or more donor groups. The result is an ion that has different properties than the parent metal ion. Chelates containing two groups that coordinate are called *bidentate;* those which have a lone pair, *monodentate.* Various chelating agents (chelating ligands) are used to remove troublesome ions from water solutions, as in water softening. In photographic processes, the thiosulfate ion removes silver ions from the film by the formation of the soluble $[Ag(S_2O_3)_2]^{3-}$ ion.

A chelate is a cyclic coordination complex in which the central metal ion is bonded to two or more electron-pair donors from the same molecule or ion, forming five- or six-membered rings in which the metal ion is part of the ring structure. These chelate ring structures are very stable rings, even more stable than complexes involving the same metal ion and same molecules or ion ligands which do not form rings. Aminopolycarboxylic acids and polyamines are excellent chelating agents; of these, EDTA, ethylenediaminetetraacetic acid, is among the best. These chelating agents form complexes with other substances that are only slightly dissociated. Some of the commonest are listed in Table 25-5.

The ethylenediaminetetraacetic acid ion reacts with practically every metal in the periodic table to form stable one-to-one five-membered chelate ring complexes. EDTA can be used to titrate metals in neutral or alkaline solutions, and titration curves can be drawn. When the negative logarithm of the metal-ion concentration is plotted vs. the volume of EDTA titrant added, the curves show a sharp break at the equivalence point (Fig. 25-20).

The equivalence point for complexometric titrations can be shown by colored indicators, the *metallochromic indicators.* These indicators, although

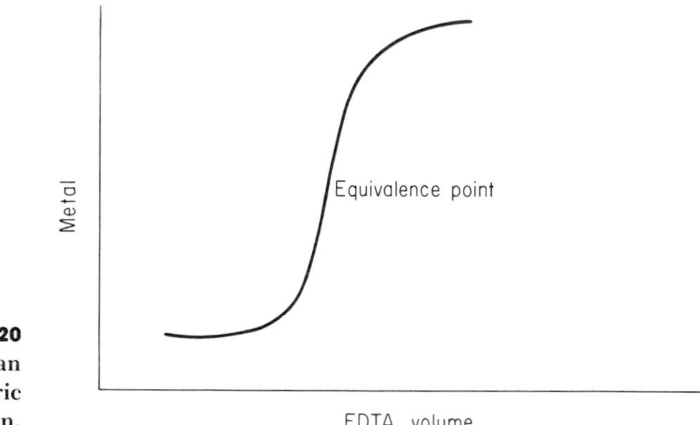

FIGURE 25-20
Titration curve of an EDTA complexometric titration.

they form complexes with metal ions, are also acid-base indicators. One form of the indicator has a specific color, and the complex has another. The point of color change on the addition of the EDTA is the equivalence point. Common complexometric indicators are Eriochrome Black T, Eriochrome Blue Black B, and Calmagite.

Complexometric titrations can also be monitored by potentiometric and coulometric procedures.

26

pH Measurement

INTRODUCTION

An *acid* is a substance that yields *hydrogen ions* when dissolved in water. *Bases* are substances that yield *hydroxyl ions* when dissolved in water. An acid which ionizes in dilute aqueous solution to produce many hydrogen ions (nearly complete ionization) is classified as a *strong* acid. An acid that ionizes slightly in water to produce few hydrogen ions is classified as a *weak* acid. A strong base ionizes in water nearly completely to produce many hydroxyl ions. A weak base ionizes slightly in water to produce few hydroxyl ions. The hydrogen ion is actually hydrated and can be represented as $H^+ \cdot H_2O$ or H_3O^+, commonly known as the hydronium ion. The ionization of HCl in water should be written as

$$HCl + xH_2O + yH_2O \rightarrow H^+ \cdot xH_2O + Cl^- \cdot yH_2O$$

but for simplicity it is represented by

$$HCl \xrightarrow{H_2O} H^+ + Cl^-$$

The relative strength of an acid or base is found by comparing the concentration of H^+ in solution with that of water. Pure water ionizes to a very small extent to produce a few hydrogen and hydroxyl ions in equilibrium with the water molecules.

$$H_2O \rightleftharpoons H^+ + OH^-$$

Each molecule of water ionizes to produce one H^+ and one OH^-. Pure water always contains equal amounts of each of these ions and is *neutral.*

Any solution which contains equal concentrations of H^+ and OH^- is *neutral.* A solution which contains an excess of OH^- over H^+ is *basic.*

At 25°C, the concentration of both the H^+ and OH^- ions in a liter of pure water amounts to only 1×10^{-7} gram-ions. The equilibrium expression for the dissociation of water is

$$H_2O \rightleftharpoons H^+ + OH^-$$

The dissociation constant k is

$$k = \frac{[H^+][OH^-]}{[H_2O]}$$

But, because the concentration of water (55.6 mol/L) is practically constant (55.6 − 0.0000001), it can be incorporated into the expression

$$k[H_2O] = [H^+][OH^-]$$

to give a new constant K_w:

$$K_w = [H^+][OH^-]$$

This new constant is called the ion-product constant for water, and

$$K_w = (1 \times 10^{-7})(1 \times 10^{-7})$$
$$= 1 \times 10^{-14}$$

At 25°C, in any aqueous solution, the product of the concentrations of the H^+ and OH^- ions will always be equal to this constant, 1×10^{-14}. *The concentration must be expressed in moles per liter.*

In a neutral solution the concentration $[H^+] = [OH^-] = 1 \times 10^{-7}$.

In an acid solution the hydrogen-ion concentration is greater than the hydroxyl-ion concentration, and the hydrogen-ion concentration is greater than 1×10^{-7}. The hydroxyl-ion concentration will be less than 1×10^{-7}. The product of the hydrogen-ion and the hydroxyl-ion concentration will always equal 1×10^{-14}.

In a basic solution the hydroxyl-ion concentration is greater than the hydrogen-ion concentration and will exceed 1×10^{-7}, while the hydrogen-ion concentration will be less than 1×10^{-7}. The product of the concentrations $[H^+]$ and $[OH^-]$ will always equal 1×10^{-14}. The hydrogen-ion concentration may vary from 1 mol/L in strongly acid solutions down to exceedingly small numbers in strongly basic solutions, for example, 4.2×10^{-12}. A simpler method of notation to avoid the use of awkward exponential numbers for expressing $[H^+]$ is the use of the pH scale to express small hydrogen-ion concentrations. Keep in mind that pH decreases as a solution becomes more acidic.

The hydrogen-ion concentration expressed as a power of 10 is known as the pH. By definition, the pH is equal to the negative logarithm of the hydrogen-ion concentration, or

$$\mathrm{pH} = -\log [H^+] = \log \frac{1}{[H^+]}$$

and similarly, the pOH scale is often used to express the hydroxyl-ion concentration. pOH is defined as the negative logarithm of the hydroxyl concentration; that is,

$$\mathrm{pOH} = -\log [OH^-] = \log \frac{1}{[OH^-]}$$

The sum of the pH and pOH is 14 because they originated from the hydrogen- and hydroxyl-ion concentrations (expressed in moles per liter). Thus if hydrogen-ion concentration $[H^+] = 1 \times 10^{-10}$ mol/L, then

$$\text{pH} = \log \frac{1}{1 \times 10^{-10}}$$

$$= \log 1 \times 10^{10} = \log 1 + \log 10^{10}$$

$$\text{pH} = 0 + 10 = 10$$

Table 26-1 summarizes the relation between acid solutions, basic solutions, pH, and pOH.

PROBLEM Calculate the pH of a solution which has $[H^+] = 1 \times 10^{-5}$ mol/L.

Solution

$$\text{pH} = \log \frac{1}{[H^+]}$$

$$= \log \frac{1}{1 \times 10^{-5}} = \log 1 \times 10^{5}$$

$$= \log 1 + \log 10^{5}$$

$$= 0 + 5 = 5$$

PROBLEM A solution has a hydrogen-ion concentration of 5×10^{-6}. What is the pH?

Solution

$$\text{pH} = \log \frac{1}{[H^+]}$$

$$= \log \frac{1}{5 \times 10^{-6}}$$

$$= \log \frac{1 \times 10^{6}}{5} = \log 1 + \log 10^{6} - \log 5$$

$$= 0 + 6 - 0.70 = 5.3$$

The acidity or alkalinity of an aqueous solution in terms of its pH in the laboratory can be found by using an indicator which has a characteristic

TABLE 26-1
Relation between Acidic Solutions, Basic Solutions, pH, and pOH

	Acidic solutions							*Pure water*	*Basic solutions*						
H_3O^+ *	10^{-0}	10^{-1}	10^{-2}	10^{-3}	10^{-4}	10^{-5}	10^{-6}	10^{-7}	10^{-8}	10^{-9}	10^{-10}	10^{-11}	10^{-12}	10^{-13}	10^{-14}
OH^- *	10^{-14}	10^{-13}	10^{-12}	10^{-11}	10^{-10}	10^{-9}	10^{-8}	10^{-7}	10^{-6}	10^{-5}	10^{-4}	10^{-3}	10^{-2}	10^{-1}	10^{-0}
pH	0	1	2	3	4	5	6	7	8	9	10	11	12	13	14
pOH	14	13	12	11	10	9	8	7	6	5	4	3	2	1	0

* Concentration in moles per liter.

color at certain pH levels. For example, the common indicator phenolphthalein is colorless from 1 to 8.0 and red in the range from 9.8 to 14.0.

To a solution of 0.1 *M* HCl, pH 1.0, is added the indicator phenolphthalein. The solution remains colorless. However, if phenolphthalein is added to a basic solution, pH = 10, the solution will turn red, indicating that the solution is basic and has a pH greater than 9.8.

Some acid-base indicators frequently show changes of color over a range of two pH units. Within a pH range that is included within its own interval of color change, such a substance provides a visual indication of the pH

TABLE 26-2
Acid-Base Indicators

Name of indicator	*pH range*	*Color change*	*Preparation*
Methyl violet	0.2–3.0	Yellow to blue	0.05% dissolved in water
Cresol red	0.4–1.8	Red to yellow	0.1 g in 26 mL 0.01 *m* NaOH + 200 mL H_2O
Thymol blue	1.2–2.8	Red to yellow	Water + dil. NaOH
Orange IV	1.3–3.0	Red to yellow	H_2O
Benzopurpurin 4B	1.2–4.0	Violet to red	20% EtOH*
Methyl orange	3.1–4.4	Red to orange-yellow	H_2O
Bromphenol blue	3.0–4.6	Yellow to blue-violet	H_2O + dil. NaOH
Congo red	3.0–5.0	Blue to red	70% EtOH
Bromcresol green	3.8–5.4	Yellow to blue	H_2O + dil. NaOH
Methyl red	4.4–6.2	Red to yellow	H_2O + dil. NaOH
Chorphenol red	4.8–6.8	Yellow to red	H_2O + dil. NaOH
Bromcresol purple	5.2–6.8	Yellow to purple	H_2O + dil. NaOH
Litmus	4.5–8.3	Red to blue	H_2O
Alizarin	5.6–7.2	Yellow to red	0.1 g in MeOH†
Bromthymol blue	6.0–7.6	Yellow to blue	H_2O + dil. NaOH
Phenol red	6.6–8.2	Yellow to red	H_2O + dil. NaOH
Thymol blue	8.0–9.6	Yellow to blue	H_2O + dil. NaOH
o-Cresolphthalein	8.2–9.8	Colorless to red	0.04% in EtOH
Phenolphthalein	8.3–9.8	Colorless to red	70% EtOH
Thymolphthalein	9.4–10.5	Yellow to blue	70% EtOH
Alizarin yellow R	10.0–12.0	Yellow to red	95% EtOH
Indigo carmine	11.4–13.0	Blue to yellow	50% EtOH
Trinitrobenzene 135	12.0–14.0	Colorless to orange	70% EtOH

* Ethyl alcohol.
† Methyl alcohol.
Universal indicators are mixtures of several indicators and may be used for estimation of pH values.

TABLE 26-3
pH Values of Acids and Bases (Approximate)

Acids	*pH*	*Bases*	*pH*
Hydrochloric acid 1 *N*	0.1	Sodium bicarbonate 0.1 *N*	8.4
Sulfuric acid 1 *N*	0.3	Borax 0.1 *N*	9.2
Hydrochloric acid 0.1 *N*	1.1	Calcium carbonate, saturated	9.4
Sulfuric acid 0.1 *N*	1.2	Iron (II) hydroxide, saturated	9.5
Orthophosphoric acid 0.1 *N*	1.5	Sodium sesquicarbonate 0.1 *M*	10.1
Sulfurous acid 0.1 *N*	1.5	Magnesium hydroxide, saturated	10.5
Oxalic acid 0.1 *N*	1.6	Ammonium hydroxide, 0.01 *N*	10.6
Hydrochloric acid 0.01 *N*	2.0	Potassium cyanide 0.1 *N*	11.0
Tartaric acid 0.1 *N*	2.2	Ammonium hydroxide 0.1 *N*	11.1
Formic acid 0.1 *N*	2.3	Ammonium hydroxide 1 *N*	11.6
Acetic acid 1 *N*	2.4	Sodium carbonate 0.1 *N*	11.6
Acetic acid 0.1 *N*	2.9	Potassium hydroxide 0.01 *N*	12.0
Benzoic acid 0.01 *N*	3.1	Trisodium phosphate 0.1 *N*	12.0
Alum 0.1 *N*	3.2	Calcium hydroxide, saturated	12.4
Acetic acid 0.01 *N*	3.4	Sodium metasilicate 0.1 *N*	12.6
Carbonic acid, saturated	3.8	Sodium hydroxide 0.1 *N*	13.0
Hydrogen sulfide 0.1 *N*	4.1	Potassium hydroxide 0.1 *N*	13.0
Arsenious acid, saturated	5.0	Sodium hydroxide 1 *N*	14.0
Hydrocyanic acid 0.1 *N*	5.1	Potassium hydroxide 1 *N*	14.0
Boric acid 0.1 *N*	5.2		

value of the solution (Table 26-2). The pH of common laboratory solutions is listed in Table 26-3.

Visual methods work fine in colorless solutions; the change in color of the indicator can be observed. However, in colored soutions indicators cannot be used, nor can we determine accurately in any solution the *exact* pH of the solution, regardless of its color, by purely visual means.

BRÖNSTED-LOWRY THEORY

In some kinds of chemical work, it is useful to know and apply the Brönsted-Lowry theory of acids and bases. According to this system, an *acid* is defined as a *proton donor*, and a *base* as a *proton acceptor.** No reference is made to solvents or to ionic dissociation.

* A proton is a positively charged particle consisting of the nucleus of the ordinary hydrogen atom. It is a fundamental nuclear unit.

TABLE 26-4

Acid		*Proton*	*Conjugate base*
$HClO_4$	=	H^+	+ ClO_4^-
H_3O^+	=	H^+	+ H_2O
NH_4^+	=	H^+	+ NH_3
C_2H_5OH	=	H^+	+ $C_2H_5O^-$
HCO_3^-	=	H^+	+ CO_3^{2-}
$C_5H_5NH^+$	=	H^+	+ C_5H_5N

$$\text{Acid} = H^+ + \text{base}$$

$$HA = H^+ + A^-$$

Examples are shown in Table 26-4.

Thus, an acid is a proton combined with its conjugate base, while a base is a substance which will accept a proton to form its conjugate acid. The intrinsic acidity of acid HA can be expressed as K acidity, (K_A), while the intrinsic basicity of base A^- will be its reciprocal.

This fact is of tremendous importance in understanding acid-base measurements. The only way in which we can determine acid or base strength is by reacting an acid with a reference base and, vice versa, reacting a base with a reference acid.

$$HA = H^+ + A^- \qquad K_A = \frac{[H^+][A^-]}{[HA]}$$

$$B + H^+ = BH^+ \qquad K_B = \frac{[BH^+]}{[B][H^+]}$$

$$HA + B = BH^+ + A^- \quad K = \frac{[BH^+][A^-]}{[HA][B]} = K_A K_B$$

Referring to the above equations, we can determine acid strength only by the extent of an acid-base reaction, or neutralization, in which a proton is transferred from an acid HA to a base B to form the new weaker acid-base pair BH^+ and A^-. Of course, the base B must be stronger, i.e., must have a greater affinity for protons than the conjugate base of the acid HA. Otherwise the reaction will proceed preferentially in the opposite direction. The overall acid-base reaction can be considered as a sum of the two half-reactions, one expressing the intrinsic acidity of acid HA (measured by K_A) and the other the intrinsic basicity of base B (measured by K_B). Neither half-reaction can be evaluated individually, only their sum (measured by $K_A \times K_B = K$).

Solvents can be classifed into four groups depending on their behavior with respect to proton transfer: (1) basic, (2) acidic, (3) both acidic and basic, and (4) inert. Of course, these classifications cannot be rigid and hold only to a degree. That is, one cannot say that a solvent is exclusively acidic, with absolutely no basic properties, or that a solvent is definitely inert. We can classify them only in terms of their most common behavior.

1. *Basic* solvents are those which have a pronounced tendency to accept protons from solutes, forming the protonated solvent (or solvated proton) as the product of the reaction. Examples of common protophilic solvents used in titrations are the amines pyridine, *n*-butylamine, ethylenediamine, and piperidine; and water, being amphiprotic (both acid and base), acts as a base toward the common acids.

2. *Acidic* solvents are those which have a tendency to donate protons to solutes. The most common ones in acid-base titrations are glacial acetic acid and acetic anhydride; trifluoroacetic, formic, and concentrated sulfuric acids are others. None of them are exclusively protogenic, because they all can act as bases under more or less drastic conditions and become protonated themselves. Thus, they can all be classified as amphiprotic—capable of both donating and accepting protons. For the sake of the discussion it is therefore convenient to include water in this class, because it acts as a proton donor to many common bases.

3. *Amphiprotic* solvents are those which can act as both proton donors and proton acceptors; i.e., they have both acidic and basic properties. Water is the prime example. Alcohols are the other common examples. Depending on the intrinsic acidity or basicity of a solute, an amphiprotic solvent will act as an acid or as a base toward it and will form either the protonated solvent or the conjugate base of the solvent.

Relative Strengths of Acids

In Table 26-5 the acids are listed in order of decreasing strengths. Conjugate bases are listed in order of increasing strength.

METHODS FOR DETERMINING pH

Colorimetric Determinations: Indicators

Principle

The hydronium-ion (hydrogen-ion) concentration of a solution can be measured by adding an indicator. An acid-base indicator is a complex organic compound that is a weak acid or a weak base. The molecular form of the indicator is different in color than that of the ionic form. The color changes are rapid and reversible and take place over a pH range of 2 units. The color of the solution indicates the pH of the solution.

TABLE 26-5
Relative Strengths of Acids

Substance	*Acid*	*Conjugate base*	K_A	*pK* *
Perchloric acid	$HClO_4$	ClO_4^-	very large	
Hydrochloric acid	HCl	Cl^-	very large	
Nitric acid	HNO_3	NO_3^-	very large	
Sulfuric acid	H_2SO_4	HSO_4^-	large	
Oxalic acid	$H_2C_2O_4$	$HC_2O_4^-$	3.8×10^{-2}	1.42
Hydrogen sulfate ion	HSO_4^-	SO_4^{2-}	1.2×10^{-2}	1.92
Sulfurous acid ($SO_2 + H_2O$)	H_2SO_3	HSO_3^-	1.2×10^{-2}	1.92
Phosphoric acid	H_3PO_4	$H_2PO_4^-$	7.1×10^{-3}	2.15
Iron(III) ion (ferric	$Fe(H_2O)_6^{3+}$	$Fe(H_2O)_5OH^{2+}$	6.7×10^{-3}	2.17
Hydrofluoric acid	HF	F^-	6.9×10^{-4}	3.16
Nitrous acid	HNO_2	NO_2^-	4.5×10^{-4}	3.50
Chromium(III) ion (chromic)	$Cr(H_2O)_6^{3+}$	$Cr(H_2O)_5OH^{2+}$	1.6×10^{-4}	3.80
Hydrogen oxalate ion	$HC_2O_4^-$	$C_2O_4^{2-}$	5.0×10^{-5}	4.30
Acetic acid	$HC_2H_3O_2$	$C_2H_3O_2^-$	1.8×10^{-5}	4.75
Aluminum ion	$Al(H_2O)_6^{3+}$	$Al(H_2O)_5OH^{2+}$	1.1×10^{-5}	4.96
Carbonic acid ($CO_2 + H_2O$)	H_2CO_3	HCO_3^-	4.4×10^{-7}	6.35
Hydrogen sulfide	H_2S	HS^-	1.0×10^{-7}	7.00
Dihydrogen phosphate ion	$H_2PO_4^-$	HPO_4^{2-}	6.3×10^{-8}	7.20
Hydrogen sulfite ion	HSO_3^-	SO_3^{2-}	5.6×10^{-8}	7.25
Boric acid	H_3BO_3	$H_2BO_3^-$	6.0×10^{-10}	9.22
Ammonium ion	NH_4^+	NH_4OH ($NH_3 + H_2O$)	5.6×10^{-10} ($K_b = 1.8 \times 10^{-5}$)†	9.25
Hydrocyanic acid	HCN	CN	4×10^{-10}	9.4
Zinc ion	$Zn(H_2O)_4^{2+}$	$Zn(H_2O)_3OH^+$	2.5×10^{-10}	9.60
Hydrogen carbonate ion	HCO_3^-	CO_3^{2-}	4.7×10^{-11}	10.33
Hydrogen peroxide	H_2O_2	HO_2	2.4×10^{-12}	11.62
Monohydrogen phosphate ion	HPO_4^{2-}	PO_4^{3-}	4.4×10^{-10}	12.36
Hydrogen sulfide ion	HS^-	S^{2-}	1.3×10^{-13}	12.89
Water	H_2O	OH^-	1.0×10^{-14}	14.00
Hydroxide ion	OH^-	O^{2-}	$<10^{-36}$	>36.00
Ammonia	NH_3	NH_2^-	very small	

* $pK = -\log K_A$.
† If the K_B (the constant for the conjugate base) is wanted, it can be calculated by using the formula: $K_B = K_W/K_A$.

Procedure

1. Choose the indicator which has a range covering the pH interval to be measured. Refer to the section on indicators.

2. Add several drops of indicator to the solution being tested.

3. Compare the color of the solution with the color range of the indicator (see Fig. 26-1).

4. Record the pH observed and determined by the color of the solution.

Preparation of Indicator Solutions

Alizarin Yellow R: 0.01%

Dissolve 1.0 g of the dye, the sodium salt of 5-(*p*-nitrophenylazo) salicylate, in 200 mL 95% ethanol; dilute to 1 L with 95% ethanol.

Bromcresol Purple: 0.1%

Grind 1.0 g of the dye, dibromo-*o*-cresolsulfonephthalein, with 18.5 mL of 0.1 *M* NaOH in a mortar; dilute to 1 L with water.

Bromthymol Blue: 0.1%

Grind 1.0 g of the dye, dibromothymolsulfonephthalein, with 16.0 mL of 0.1 *M* NaOH in a mortar; dilute to 1 L with water.

Indigo Carmine: 0.1%

Dissolve 1.0 g of the dye, sodium indigo disulfonate, in 500 mL 95% ethanol; dilute to 1 L with distilled water. (Although this dye could be made up with water, an aqueous solution does not keep well.)

Methyl Orange: 0.1%

Dissolve 1.0 g of the dye in water; dilute to 1L.

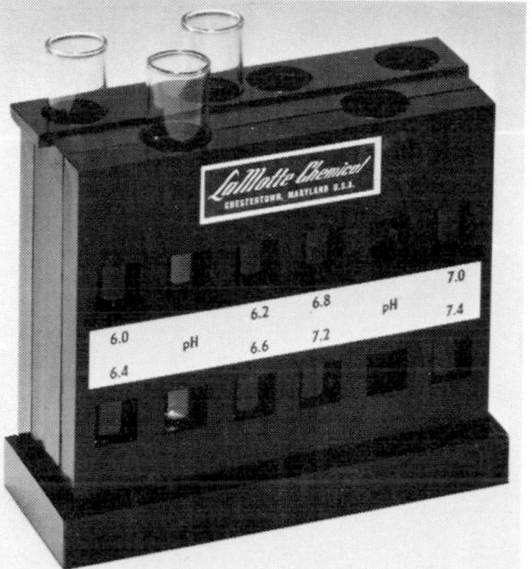

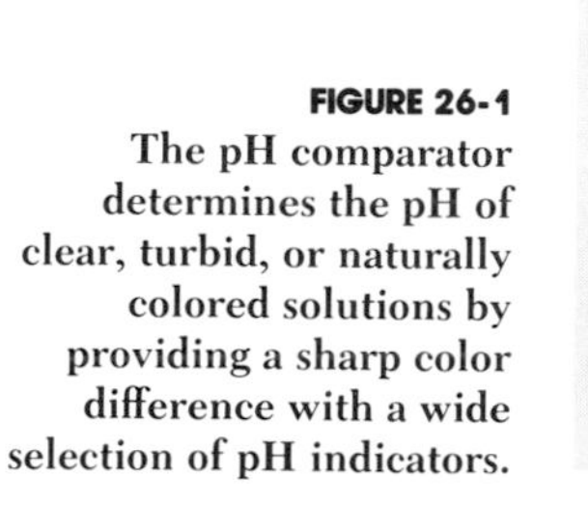

FIGURE 26-1
The pH comparator determines the pH of clear, turbid, or naturally colored solutions by providing a sharp color difference with a wide selection of pH indicators.

Methyl Red: 0.1%

Grind 1.0 g of the dye with 37 mL of 0.1 *M* NaOH in a mortar; dilute to 1 L with water.

Orange IV: 0.1%

Dissolve 1.0 g of the dye, the sodium salt of *p*-(*p*-anilinophenylazo) benzenesulfonic acid, in water; dilute to 1 L.

Phenolphthalein: 0.1%

Dissolve 1.0 g of the dye in 700 mL of 95% ethanol; dilute to 1 L with water.

Phenol Red: 0.1%

Grind 1.0 g of the dye, phenolsulfonephthalein, with 28.2 mL of 0.1 *M* NaOH in a mortar. Dilute to 1 L with water.

The Importance of Selecting the Proper Indicator

When color-changing indicators are used to indicate the equivalence point of acid-base titrations, the selection of the correct indicator for that particular system is most important. This is especially true when strong acid–weak base or weak acid–strong base titrations are carried out.

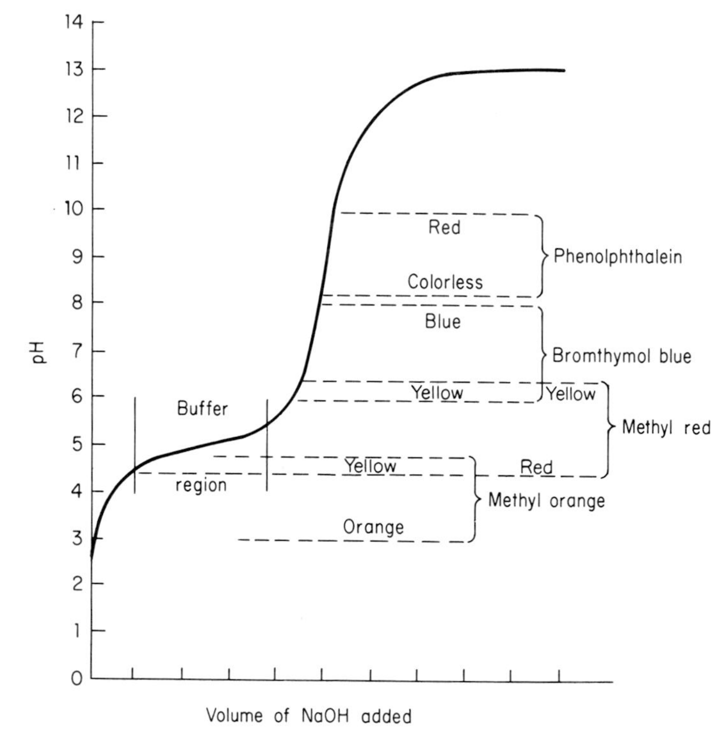

FIGURE 26-2
The pH range over which the indicator color changes determines the correct indicator for the procedure.

EXAMPLE Weak Acid–Strong Base (see Fig. 26-2).

Methyl red (pH interval 4.4–6.2) should not be used for a weak acid–strong base titration, because the equivalence point is between pH 7 and pH 10.

Methyl orange (pH interval 3.1–4.4) should not be used for the weak acid–strong base titration.

Bromthymol blue (pH interval 6.0–7.6) would also give a false indication before the equivalence point was attained.

Phenolphthalein (pH interval 8.3–9.8) would be satisfactory. It would change color at the true equivalence point.

pH Test Paper

The pH range over which the color of the indicator changes determines the correct indicator for the procedure.

pH test papers are available for every value of pH; they cover both broad and narrow pH ranges (Fig. 26-3). The paper is impregnated with the indicator. A strip of test paper is wet with the liquid to be tested and then immediately compared with the standard color chart provided for each paper and range. The pH can be determined visually by comparison of the colors.

pH Meter

Fortunately, in the case of hydrogen-ion measurements a convenient and direct determination of the pH can be made with an instrument known as the pH meter (Fig. 26-4). This instrument measures the concentration of the hydrogen ions in a solution by using a calomel electrode as a reference electrode and a glass electrode as "indicator" electrode (an electrode whose potential will vary with the concentration). The actual measurement of the hydrogen-ion concentration (or pH) is made by dipping the two electrodes into the same solution of unknown pH.

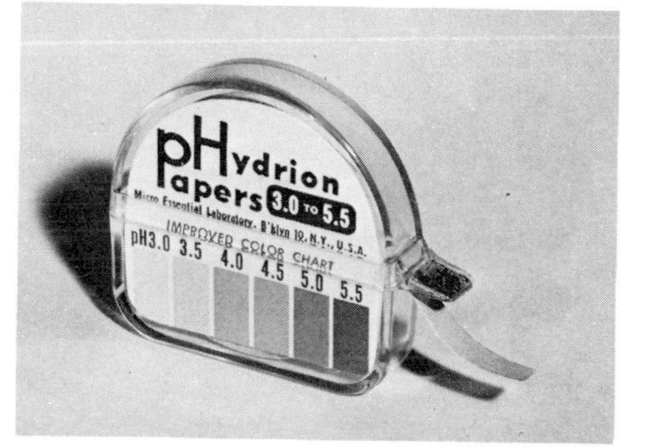

FIGURE 26-3
pHydrion® test paper determines pH directly by immediate comparison of the dipped paper with the color chart provided for each paper and range.

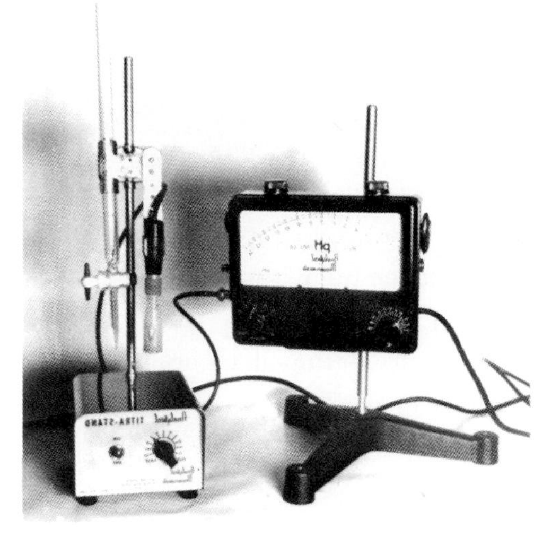

FIGURE 26-4
pH meter in use. This is a single (combination-probe) pH meter with magnetic mixer and buret-holder assembly.

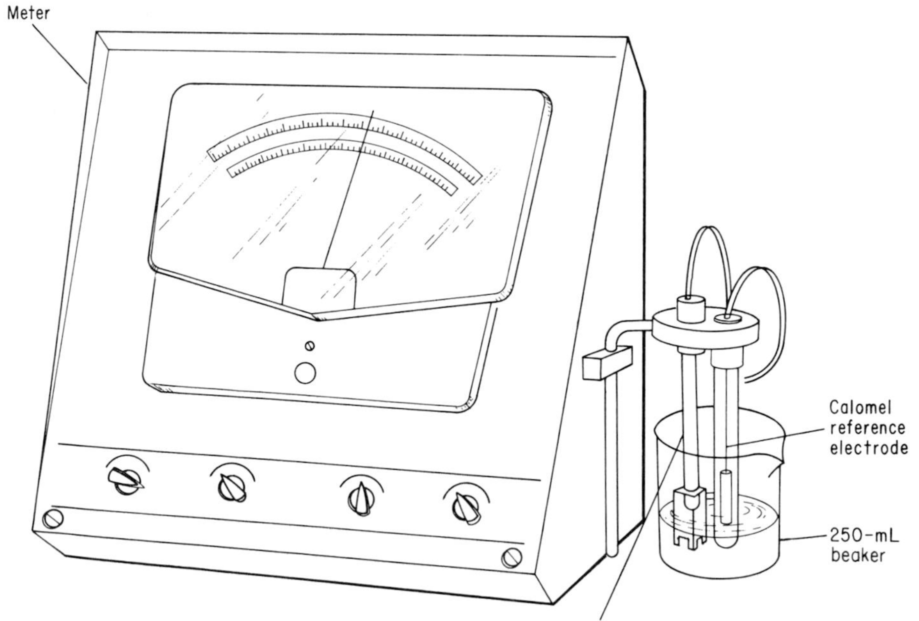

FIGURE 26-5
Parts of a pH meter.

Components

A pH meter (Fig. 26-5) is basically composed of a glass electrode, a calomel electrode, and a solid-state voltmeter. (Figure 26-6 shows a magnified drawing of the electrodes.)

1. The glass electrode has a fixed acid concentration inside. It is immersed in a solution of varying hydrogen-ion content. The electrode responds to the change in H_3O^+-ion concentration, yielding a corresponding voltage.

2. The calomel electrode is the reference electrode. Its voltage is independent of the H_3O^+-ion concentration.

3. The two electrodes constitute a galvanic cell whose electromotive force is measured by the solid-state voltmeter. The meter is calibrated to read in pH units, reflecting the H_3O^+-ion concentration.

Procedure

1. Rotate the switch to STANDBY. Allow to warm for 30 min.

CAUTION When the instrument is not in use, keep on STANDBY.

NOTE Different manufacturers designate the neutral position of the selector switch with various markings, such as BAL for balance.

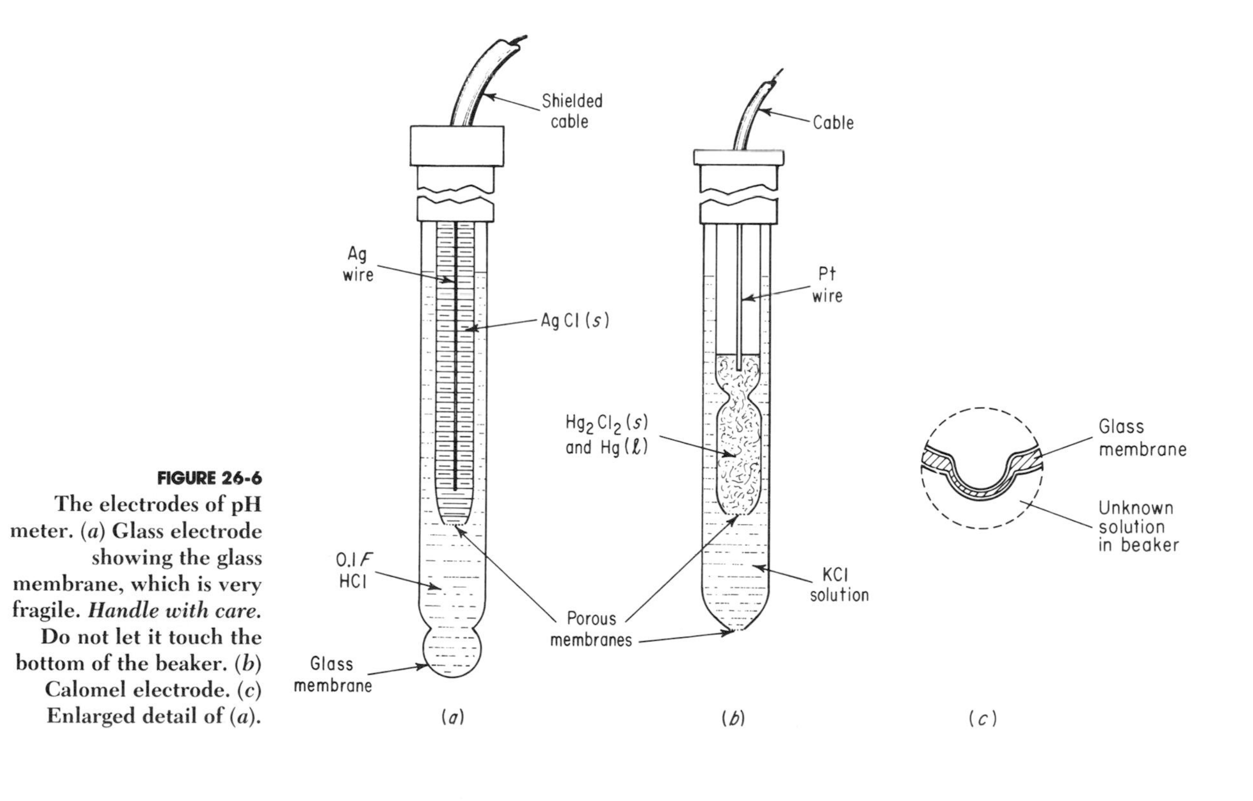

FIGURE 26-6 The electrodes of pH meter. (*a*) Glass electrode showing the glass membrane, which is very fragile. ***Handle with care.*** Do not let it touch the bottom of the beaker. (*b*) Calomel electrode. (*c*) Enlarged detail of (*a*).

2. Raise the electrode from the storage solution in the beaker.

CAUTION Always keep the electrodes in distilled water when not in use.

3. Rinse the electrodes thoroughly with distilled water. Blot with absorbent tissue.

4. Standardize against an appropriate buffer solution. (See Buffer Standardization of the pH Meter, below.)

5. Place the beaker of solution to be tested beneath the electrodes. This is the measurement of the pH of the solution.

6. Lower the electrodes carefully into the solution. Adjust temperature compensator to the temperature of the solution.

7. Rotate the selector knob to pH. Read the pH of the solution directly from the meter. Record the value.

8. When the determination is complete:

   (a) Switch to STANDBY. Raise the electrodes.

   (b) Rinse the electrodes with distilled water.

   (c) Store the electrodes in distilled water.

NOTE Leave the instrument connected to the power line at all times, except when it is not to be used for extended periods. This will ensure stable, drift-free performance, and the slight temperature rise will eliminate humidity troubles. Component life also will be extended by elimination of repeated current surges. Turn selector switch to the balance position when the instrument is not in use.

Buffer Standardization of the pH Meter*

Principle

All pH meters must be standardized daily by means of a buffer solution of known pH value. For maximum accuracy, use a buffer in the range of the sample to be tested. A buffer solution is one which tends to remain at constant pH. For example: use a pH 4 buffer for standardizing when work is to be done in the acid range; a pH 7 buffer when the work is near neutral; and a pH 9 buffer when the work is in the alkaline range.

The buffer temperature should be as close as possible to the sample temperature. Try to keep this temperature difference within 10°C.

---

* This section reprinted by permission of Analytical Measurements, Chatham, N.J.

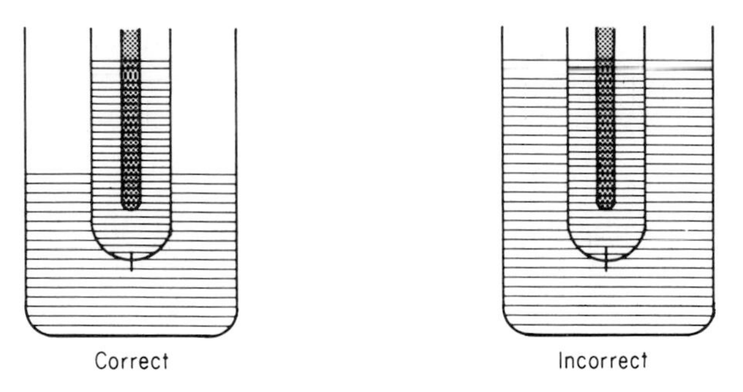

FIGURE 26-7
When using a pH meter, the electrode depth in solutions is important.

Procedure

1. Determine the temperature of the buffer solution with a thermometer. Adjust the TEMPERATURE knob on the unit to that temperature.

2. Calibrate the meter by immersing the electrodes (Fig. 26-7) in a buffer solution of known pH.

   (a) Rotate the selector switch to pH.

   (b) Turn the adjustment knob to CALIBRATION to read the pH of the known buffer solution.

NOTE Some manufacturers designate this position as ASYMMETRY CONTROL.

CAUTION Do not allow the electrodes to touch the sides of the beaker.

3. Rotate the selector knob to STANDBY.

   (a) Raise the electrodes carefully.

   (b) Remove the buffer solution.

   (c) Rinse the electrodes thoroughly with distilled water.

   (d) Blot the electrodes with absorbent, lint-free tissue.

Electrode Maintenance

When not in use, the electrodes should be left soaking in water or buffer solution. Electrodes which have dried out should be soaked in water for several hours prior to use.

Keep the reference electrode reservoir filled, using saturated KCl solution. There should be some KCl crystals in the reservoir.

The fiber junction at the tip of the reference electrode may become plugged when used for long periods of time in samples containing suspended material. A plugged junction can usually be opened by boiling the tip of the electrode in dilute nitric acid.

Glass electrodes used in biological samples may start to drift due to the formation of a coating of protein. The coating may be removed by cleaning with a strong solution of detergent and water.

Hints for Precision Work

For accurate work, fresh buffer solution should be used. Buffer solution bottles should be capped when not in use to prevent evaporation and contamination. Never pour buffer solutions back into the bottle.

Between buffers and samples, rinse the electrodes with distilled water and blot with an absorbent tissue to prevent carryover of solution.

Immerse completely the pH sensitive bulb portion of the glass electrode into the solution to be measured. Be sure that the glass electrode does not touch the wall or bottom of the sample container.

Keep the electrodes and the samples measured at the same constant temperature. Temperature cycling of the electrodes will result in small drifts that may make reading of very small changes in pH difficult.

---

pH TITRATION*

Introduction

In an acid-base titration, the change in $[H^+]$ may be very large, that is, from 10^{-1} to 10^{-10}. This is a 100-million-fold change in concentration, and

* This section reprinted by permission of Analytical Measurements, Chatham, N.J.

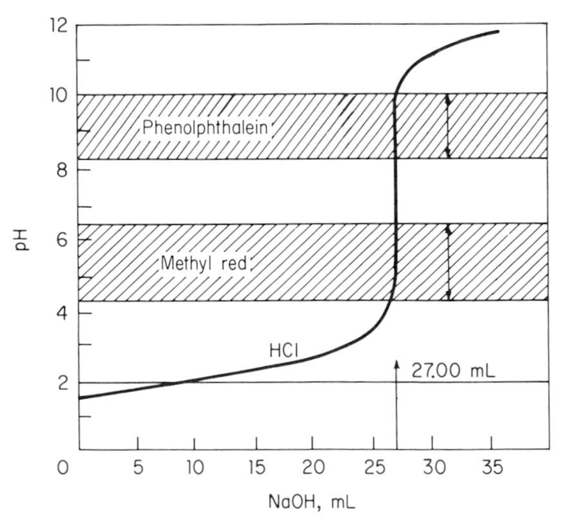

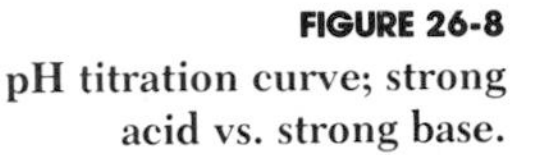
FIGURE 26-8
pH titration curve; strong acid vs. strong base.

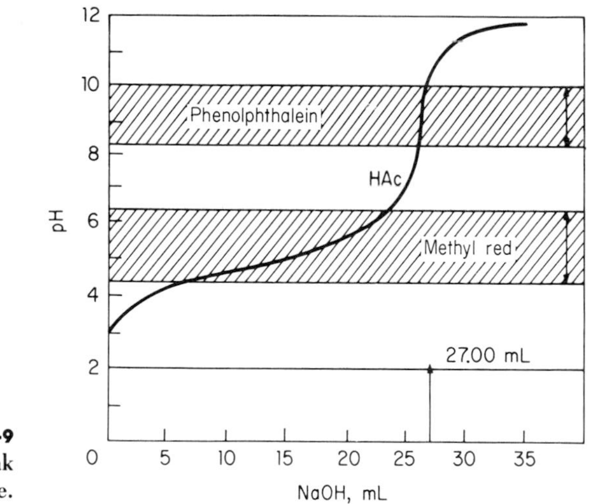

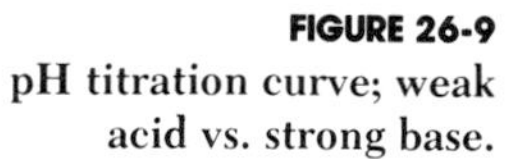
FIGURE 26-9
pH titration curve; weak acid vs. strong base.

it would be rather inconvenient to plot such numbers. However, the change in pH is only from 1.0 to 11.0, and these numbers may conveniently be plotted. When the volume of titrant in milliliters is plotted against the pH, we obtain what is called a *titration curve.* (See Figs. 26-8 to 26-10.)

Acid base titrations can usually be grouped as follows:

1. Strong acid–strong base:

$$H^+ + OH^- \rightarrow H_2O$$

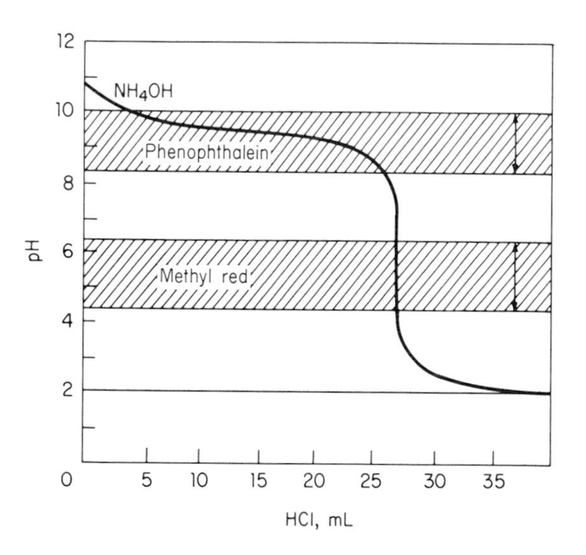

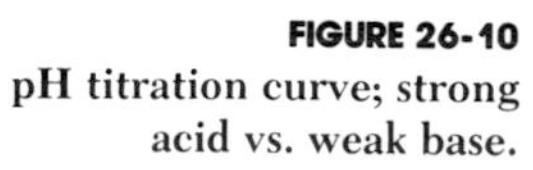
FIGURE 26-10
pH titration curve; strong acid vs. weak base.

2. Weak acid–strong base:

$HAc + OH^- \rightarrow Ac^- + H_2O$

NOTE The symbol Ac^- stands for $C_2H_3O_2^-$.

3. Strong acid–weak base:

$H^+ + NH_4OH \rightarrow NH_4^+ + H_2O$

The hydrogen-ion concentration may be calculated during a titration from the amount of acid or base that has been added. We can therefore calculate the pH of any solution resulting from the reactions of type 1, 2, or 3. Let us now look at some sample calculations to see how the pH is computed at various points during the titrations.

Strong Acid–Strong Base

PROBLEM If 25 mL of a 0.10 *M* HCl solution is diluted to 100 mL with distilled water, what is the pH of the resulting solution?

Solution Since HCl is considered to be 100% ionized, the number of moles of H^+ in solution will be:

$(0.025\ \text{L})(0.10\ \text{mol/L}) = 2.5 \times 10^{-3}\ \text{mol}$

which is now dissolved in a total volume of 100 mL (0.1 L).

$$[H^+] = \frac{2.5 \times 10^{-3}\ \text{mol}}{0.10\ \text{L}} = 25 \times 10^{-3} = 2.5 \times 10^{-2}\ \text{mol/L}$$

The pH is calculated as follows:

$$\begin{aligned} \text{pH} &= \log \frac{1}{[H^+]} = \log \frac{1}{2.5 \times 10^{-2}} \\ &= \log \frac{1 \times 10^2}{2.5} \\ &= \log 1 + \log 10^2 - \log 2.5 \\ &= 0 + 2 - 0.32 = 1.68 \end{aligned}$$

PROBLEM What will be the pH of the solution after 10 mL of 0.10 *M* NaOH has been added?

Solution 1. Original number of moles of HCl = (0.025 L)(0.10 mol/L) = 0.0025 mol

2. Moles of NaOH added = (0.01 L)(0.10 mol/L) = 0.001 mol

3. Moles of HCl remaining in solution = (0.0025 − 0.0010) = 0.0015 mol

4. The 0.0015 mol of HCl is now dissolved in 110 mL of solution. The $[H^+] = 0.0015$ mol/0.11 L = 0.0136 or 1.36×10^{-2}.

5. The pH can now be calculated as

$$\text{pH} = \log \frac{1}{1.36 \times 10^{-2}} = \log \frac{1 \times 10^2}{1.36}$$
$$= \log 1 + \log 10^2 - \log 1.36$$
$$= 0 + 2 - 0.13 = 1.87$$

PROBLEM What will be the pH of the solution when 25 mL of NaOH has been added? This is the equivalence point, and we should expect that neutralization, $H^+ + OH^- \rightarrow H_2O$, has occurred, and the only H^+ in solution would be from the water, that is, 1×10^7 or a pH of 7.

Solution 1. Original number of moles of HCl = (0.025 L)(0.10 mol/L) = 0.0025 mol

2. Moles of NaOH added = (0.025 L)(0.10 mol/L) = 0.0025 mol

3. Moles of HCl remaining in solution = (0.0025 − 0.0025) = 0.00 mol

This means that we have completely reacted the HCl, and only water and NaCl are in solution. Since neither Na^+ or Cl^- reacts with the water, the source of H^+ in solution will be from the dissociation of water. Thus, $[H^+] = 1 \times 10^{-7}$ and the pH = 7.

PROBLEM What will be the pH of the solution after 30.0 mL of NaOH has been added?

Solution 1. Original number of moles of HCl = (0.025 L) (0.10 mol/L) = 0.0025 mol

2. Moles of NaOH added = (0.030 L) (0.10 mol/L) = 0.003 mol

3. Moles of NaOH in excess = (0.003 − 0.0025) = 0.0005 mol

4. The 0.0005 mol of NaOH is now dissolved in 130 mL of solution. The $[OH^-] = 0.0005$ mol/0.13 L = 0.0038 or 3.8×10^{-3}

5. The $[H^+]$ can be found from the expression for the ion product of water:

$$K_w = 1 \times 10^{-14} = [H^+][OH^-]$$
$$1 \times 10^{-14} = [H^+](3.8 \times 10^{-3})$$
$$\frac{1 \times 10^{-14}}{3.8 \times 10^{-3}} = [H^+]$$
$$2.6 \times 10^{-12} = [H^+]$$

6. The pH can now be calculated:

$$\text{pH} = \log \frac{1}{2.6 \times 10^{-12}} = \log \frac{1 \times 10^{12}}{2.6}$$
$$= \log 1 + \log 10^{12} - \log 2.6$$
$$= 0 + 12 - 0.42 = 11.58$$

Weak Acid–Strong Base

PROBLEM If 25 mL of a 0.10 *M* HAc solution is diluted to 100 mL with distilled water, what is the pH of the resulting solution?

Solution Since HAc is ionized only to a slight extent, the $[H^+]$ and thus the pH can be calculated from the ionization constant.

$$HAc \rightarrow H^+ + Ac^-$$

$$K_i = \frac{[H^+][Ac^-]}{[HAc]}$$

Let $X = [H^+]$. Since the HAc ionizes to form equal amounts H^+ and Ac^-, the $[Ac^-]$ will also be equal to X. The concentration of the HAc is (0.025 L)(0.10 mol/L) = 0.0025 mol, which is now dissolved in 100 mL (0.1 L), or 0.0025 mol/0.1 L = 2.5×10^{-2} mol/L.

$$K_i = 1.8 \times 10^{-5} = \frac{(X)(X)}{(2.5 \times 10^{-2} - X)}$$

As X will be very small, we may disregard it when it is to be subtracted from the much larger 2.5×10^{-2}, with very little error. Thus,

$$K_i = 1.8 \times 10^{-5} = \frac{X^2}{2.5 \times 10^{-2}}$$

$$(2.5 \times 10^{-2})(1.8 \times 10^{-5}) = X^2$$

$$4.5 \times 10^{-7} = X^2$$

$$6.7 \times 10^{-4} = X = [H^+]$$

The pH is calculated as follows:

$$\text{pH} = \log \frac{1}{(6.7 \times 10^{-4})} = \log \frac{1 \times 10^{4}}{6.7}$$
$$= \log 1 + \log 10^{4} - \log 6.7$$
$$= 0 + 4 - 0.83 = 3.17$$

PROBLEM What will be the pH of the solution after 10 mL of 0.10 *M* NaOH has been added?

Solution 1. Original number of moles of HAc = (0.025 L)(0.10 mol/L) = 0.0025 mol

2. Moles of NaOH added = (0.01 L)(0.10 mol/L) = 0.001 mol

3. Moles of HAc remaining in solution = (0.0025 − 0.001) = 0.0015 mol

4. The 0.0015 mol of HAc is now dissolved in 110 mL of solution. The [HAc] = 0.0015 mol/0.11 L = 0.0136 = 1.36×10^{-2}.

5. Moles of acetate ion = acetate-ion concentration produced by the ionization of the remaining HAc ($HAc \rightarrow H^+ + Ac^-$) and which we designate X. In addition, there are the acetate ions produced during the titration.

$$HAc + OH^- \rightarrow Ac^- + H_2O$$

Each mole of NaOH forms a mole of Ac^-, and since we added 0.001 mol of NaOH, we form 0.001 mol of Ac^-. Therefore, the total Ac^- concentration is (X + 0.001 mol). We assume that X is small compared to the 0.001 mol of Ac^- produced in the reaction and approximate that the amount of Ac^- = 0.001 mol. Since the volume of the solution is now 110 mL (0.11 L), the $[Ac^-]$ = 0.001 mol/0.11 L = 0.091 mol/L.

6. The $[H^+]$ can be found from the ionization constant:

$$1.8 \times 10^{-5} = \frac{[H^+](9.1 \times 10^{-2})}{(1.36 \times 10^{-2})}$$

$$\frac{(1.8 \times 10^{-5})(1.36 \times 10^{-2})}{(9.1 \times 10^{-2})} = [H^+]$$

$$2.7 \times 10^{-6} = [H^+]$$

7. The pH can now be calculated as follows:

$$\begin{aligned} pH &= \log \frac{1}{(2.7 \times 10^{-6})} = \frac{\log 1 \times 10^6}{2.7} \\ &= \log 1 + \log 10^6 - \log 2.7 \\ &= 0 + 6 - .43 = 5.57 \end{aligned}$$

PROBLEM What will be the pH of the solution after 25 mL of 0.10 M NaOH has been added? This is the equivalence point.

Solution 1. Original number of moles of HAc = (0.025 L)(0.10 mol/L) = 0.0025 mol.

2. Moles of NaOH added = (0.025 L)(0.10 mol/L) = 0.0025 mol. Thus the solution is neutralized and no HAc should remain.

3. Moles of Ac^- produced = moles of NaOH added = 0.0025 mol. Since the volume of solution is now 125 mL (0.125 L), the $[Ac^-]$ = 0.0025 mol/0.125 L = 0.02 mol/L.

4. The Ac^- formed will react with the water (hydrolyze):

$$Ac^- + H_2O \rightarrow HAc + OH^-$$

The $[OH^-]$ may be found from the ion product of water:

$$1 \times 10^{-14} = [H^+][OH^-]$$

If $[H^+] = X$, the $[OH^-] = \dfrac{1 \times 10^{-14}}{X}$ and the [HAc]

$$= \frac{1 \times 10^{-14}}{X} \text{ since the } [HAc] = [OH^-]$$

5. The $[H^+]$ can be found from the ionization constant:

$$1.8 \times 10^{-5} = \frac{X\,0.02}{\dfrac{1 \times 10^{-14}}{X}}$$

$$1.8 \times 10^{-5} = \frac{X^2\,0.02}{1 \times 10^{-14}}$$

$$\frac{(1.8 \times 10^{-5})(1 \times 10^{-14})}{2 \times 10^{-2}} = X^2$$

$$9 \times 10^{-18} = X^2$$

$$3 \times 10^{-9} = X$$

6. The pH can be calculated as:

$$pH = \log \frac{1}{3 \times 10^{-9}} = \log \frac{1 \times 10^{9}}{3}$$

$$= \log 1 + \log 10^9 - \log 3$$

$$= 0 + 9 - 0.48 = 8.52$$

Note that the pH at the equivalence point is not 7 as in the case of a strong acid–strong base titration.

PROBLEM What will be the pH of the solution after 30 mL of NaOH has been added?

Solution **1.** Original moles of HAc = (0.025 L)(0.10 mol/L) = 0.0025 mol

2. Moles of NaOH added = (0.03 L)(0.10 mol/L) = 0.0030 mol

3. Moles of NaOH in excess = (0.0030 − 0.0025) = 0.0005 mol

4. The 5×10^{-4} mol of NaOH is now dissolved in 130 mL of solution. The $[OH^-] = 5 \times 10^{-4}$ mol/0.13 L = 3.8×10^{-3} mol/L.

5. Since the excess OH^- will drive the reaction

$$Ac^- + H_2O \rightleftharpoons HAc + OH^-$$

back to the left, the hydrolysis is repressed and the $[OH^-]$ can be assumed to be 3.8×10^{-3} mol/L with very little OH^- being formed by the hydrolysis reaction.

6. The $[H^+]$ may be found from the ion product of water:

$$1 \times 10^{-14} = [H^+]\,(3.8 \times 10^{-3})$$

$$\frac{1 \times 10^{-14}}{3.8 \times 10^{-3}} = [H^+]$$

$$2.6 \times 10^{-12} = [H^+]$$

7. The pH can be calculated:

$$\begin{aligned} pH &= \log \frac{1}{(2.6 \times 10^{-12})} = \log \frac{1 \times 10^{12}}{2.6} \\ &= \log 1 + \log 10^{12} - \log 2.6 \\ &= 0 + 12 - 0.42 = 11.58 \end{aligned}$$

We have now calculated the pH during several steps of the titration of a strong or weak acid with a strong base. If we plot the pH vs. mL of titrant (NaOH) for these, or similar, reactions we obtain a titration curve. The general form of such curves is given in Figs. 26-8 and 26-9.

Strong Acid–Weak Base

Even though we have not calculated the pH values for the titration of the third group of reactions (strong acid–weak base), the computations are similar and the general form of the titration curve is shown in Fig. 26-10.

BUFFER SOLUTIONS

Mixtures of weak acids and their salts or of weak bases and their salts are called *buffer solutions* because they resist changes in their hydrogen-ion concentration upon addition of small amounts of acid or base. Their pH tends to remain unchanged. Buffer solutions are used in the laboratory to standardize pH meters and whenever solutions having a definite pH are required. Tables 26-6 to 26-10 provide alternative procedures for preparing buffer solutions of any desired pH value.

Premixed, ready-to-use buffer solutions are also available and can be purchased in pint and quart containers. Also available, however, are vials or packets of concentrate (Fig. 26-11), which can be quantitatively transferred to specified volumetric flasks and diluted to the mark with distilled water. Thus a stock of buffer-solution concentrates of any desired pH can be kept in a very small space.

FIGURE 26-11 Buffer solution concentrates; tablets and envelopes (*a*), and capsules (*b*) for preparing standard pH solutions by using premeasured quantities to be dissolved in definite volume of distilled water.

Preparation of Buffer Solutions

Method 1

Procedure

Prepare the stock solutions listed in Table 26-6 according to standard volumetric methods. Select the pH of the buffer solution desired. Mix the indicated volumes of the stock solutions required to prepare the desired buffer solution.

Method 2

Procedure

Prepare the following stock solutions from reagent-grade chemicals and mix according to Table 26-7 to obtain 1 L of a solution having the desired pH.

1. Citric acid, 0.1 *M*. Dissolve 21.0 g of $H_3C_6H_5O_7 \cdot H_2O$ in distilled water and then dilute to 1 L. Add a preservative such as toluene.

2. Disodium phosphate of 0.2 *M*. Dissolve 71.6 g $Na_2HPO_4 \cdot 12H_2O$ in distilled water and then dilute to 1 L.

3. Borax 0.05 *M*. Dissolve 19.07 g $Na_2B_4O_7 \cdot 10H_2O$ in distilled water and dilute to 1 L.

4. Sodium carbonate 0.05 *M*. Dissolve 5.3 g Na_2CO_3 in distilled water and dilute to 1 L.

Method 3

Procedure

Mix according to procedure in Table 26-8.

TABLE 26-6
Preparation of Buffer Solutions, Method 1 (mixture diluted to 100 mL)

Desired pH, (25°C)	0.2 M KCl, mL	0.1 M potassium hydrogen phthalate, mL	0.1 M potassium dihydrogen phosphate, mL	0.2 M HCl, mL	0.1 M HCl, mL	0.1 M NaOH, mL	0.025 M borax, mL	0.05 M disodium hydrogen phosphate, mL	0.2 M NaOH, mL
1.0	25			67					
1.5	25			20.7					
2.0	25			6.5					
2.5		50			38.8				
3.0		50			22.3				
3.5		50			8.2				
4.0		50			0.1				
4.5		50				8.7			
5.0		50				22.6			
5.5		50				36.6			
6.0			50			5.6			
6.5			50			13.9			
7.0			50			29.1			
7.5			50			40.9			
8.0			50			46.1			
8.5					15.2		50		
9.0					4.6		50		
9.5						8.8	50		
10.0						18.3	50		
10.5						22.7	50		
11.0						4.1		50	
11.5						11.1		50	
12.0						26.9		50	
12.5	25								20.4
13.0	25								66.0

TABLE 26-7
Preparation of Buffer Solutions, Method 2

pH	Solution 1, mL	Solution 2, mL	Solution 3, mL	Solution 4, mL
4	615	385		
5	485	515		
6	368	632		
7	176	824		
8	27	973		
9			1000	
10			225	775
11				1000

TABLE 26-8
Preparation of Buffer Solutions, Method 3

Desired pH	*Preparation procedure*
1	0.10 *M* HCl
2	0.01 *M* HCl
3	500 mL of 0.1 *M* potassium hydrogen phthalate and 223 mL of 0.1 *M* HCl
4	500 mL of 0.1 *M* potassium hydrogen phthalate and 10 mL of 0.1 *M* HCl
5	500 mL of 0.1 *M* potassium hydrogen phthalate and 226 mL of 0.1 *M* NaOH
6	500 mL of 0.1 *M* potassium dihydrogen phosphate and 56 mL of 0.1 *M* NaOH
7	500 mL of 0.1 *M* potassium dihydrogen phosphate and 291 mL of 0.1 *M* NaOH
8	500 mL of 0.025 *M* borax (sodium tetraborate) and 205 mL of 0.1 *M* HCl
9	500 mL of 0.25 *M* borax (sodium tetraborate) and 46 mL of 0.1 *M* HCl
10	500 mL of 0.025 *M* borax (sodium tetraborate) and 183 mL of 0.1 *M* NaOH
11	500 mL of 0.05 *M* sodium bicarbonate ($NaHCO_3$) and 227 mL of 0.1 *M* NaOH
12	500 mL of 0.05 *M* Na_2HPO_4 (sodium monohydrogen phosphate) and 269 mL of 0.1 *M* NaOH
13	0.01 *M* NaOH
14	0.1 *M* NaOH

Method 4

Procedure

Mix the two solutions below according to the instructions indicated in Table 26-9 to prepare 1 L of a buffer solution of the desired pH.

1. Dissolve 12.37 g of anhydrous boric acid, H_3BO_3, and 10.51 g of citric acid, $H_3C_6H_5O_7 \cdot H_2O$, in distilled water and dilute to 1 L in a volumetric flask. This makes a 0.20 *M* boric acid and a 0.05 *M* citric acid solution.

2. Dissolve 38.01 g of $Na_3PO_4 \cdot 12H_2O$ in distilled water and dilute to 1 L in a volumetric flask. This makes a 0.10 *M* tertiary sodium phosphate solution.

Method 5

Procedure

This series of buffer solutions requires the preparation of five stock solutions. Large enough volumes of these solutions should be prepared to

TABLE 26-9
Preparation of Buffer Solutions, Method 4

Desired pH	*Solution 1, mL*	*Solution 2, mL*
2.0	975	25
2.5	920	80
3.0	880	120
3.5	830	170
4.0	775	225
4.5	720	280
5.0	670	330
5.5	630	370
6.0	590	410
6.5	545	455
7.0	495	505
7.5	460	540
8.0	425	575
8.5	390	610
9.0	345	655
9.5	300	700
10.0	270	730
10.5	245	755
11.0	220	780
11.5	165	835
12.0	85	915

meet the need. The maximum volume of any one solution needed to prepare 1 L of a buffer solution of desired normality is 500 mL.

1. 0.1 *N* NaOH. Dissolve 4 g NaOH per liter of solution.

2. 0.1 *M* potassium hydrogen phthalate. Dissolve 20.42 g $KHC_8H_4O_4$ in enough distilled water to make 1 L of solution.

TABLE 26-10
Preparation of Buffer Solutions, Method 5

Desired pH	*Volume of each solution, mL*				
	1	*2*	*3*	*4*	*5*
4	4	500		496	
5	226	500		274	
6	56		500	444	
7	291		500	209	
8	39			461	500
9	208			292	500
10	437			63	500

3. 0.1 *M* monopotassium phosphate. Dissolve 13.62 g KH_2PO_4 per liter of solution.

4. 0.1 *M* KCl. Dissolve 7.46 g per liter of solution.

5. 0.1 M boric acid. Dissolve 6.2 g H_3BO_3 in 0.1 *M* KCl (solution 4) to make 1 L of solution.

Mix the stock solutions according to the volumes indicated in Table 26-10 and then dilute the mixture to 1 L with distilled water.

27
BASIC ELECTRICITY

INTRODUCTION

The atomic theory of matter is the basis for understanding electricity. The atoms of all substances consist of electrons, protons, and neutrons. In some solids, called *conductors,* the electrons can move easily from atom to atom. In other substances, called *insulators,* the electrons are bound very tightly, each in its own atom, and do not move freely. Metallic elements conduct electricity, and most nonmetallic elements do not conduct electricity. We call the flow of electrons in the conductor *electricity* or electric current. Electric current is useful when it does work, and is not merely flowing through conductors. It can do work when it transforms the electric energy of the moving electrons into power, heat, or light energy.

Electron Flow

Electrons flow from the more negative to the more positive points. The spring in the upper part of Fig. 27-1 represents the state of electrons in a cell. Since nature tends toward equilibrium states, electrons flow from the point where they are in excess to the point where they are deficient.

ELECTRICITY COMPARED TO FLUIDS

Electricity can be compared to water (Table 27-1). Water flows through pipes; electric current (I) flows through conductors. Water flows through pipes at varying rates, so many gallons per minute. Electrons also flow through conducting wires at a definite rate. A flow of billions of electrons per second is equal to one *ampere,* the unit of electric current ($\sim 6 \times 10^{18}$ $e/s = 1$ A). Electric current (I) is measured in amperes (A).

Water in a pipe is under a definite pressure which can range from fractions of a pound per square inch to hundreds of thousands of pounds per square inch. Water can be in pipes and under pressure, yet stand still and not flow because the faucet is closed. Electricity can be in a conductor, under pressure, and not flow because the conductor (wire) does not make a complete circuit: the switch is open. The higher the electron pressure, the greater the tendency for electrons to flow. Electron pressure (E) (also called electromotive force, emf, or voltage) is measured in volts (V).

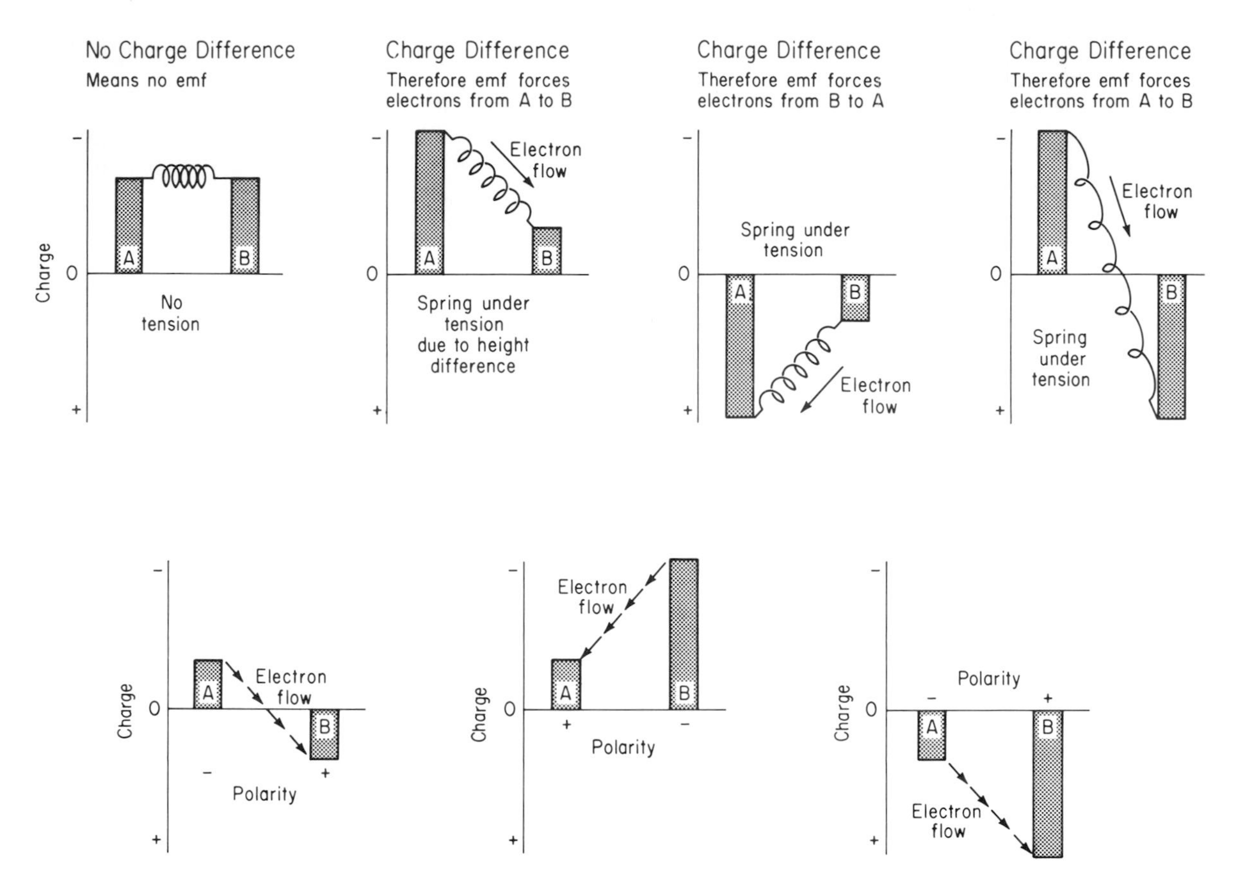

FIGURE 27-1
Direction of electron flow: electrons flow from the more negative to the more positive points.

In water systems there is a certain amount of resistance to the flow of water through the pipes because of the turns, curves, fittings, scale, indentations, constrictions, and rough inner surface. A certain amount of power is required to overcome this resistance. Similarly, energy is required to overcome the energy bond by which electrons are attached to the atom in a conductor. This opposition to the flow of electricity is called resistance (R), and its unit of measurement is the ohm (Ω). Different substances offer a wide range of resistance to the flow of electrons, from extreme ease to complete blockage and no flow.

OHM'S LAW

Just as the energy obtainable from the flow of water is related to the fluid pressure, the amount of flow, and the resistance, the components of elec-

TABLE 27-1
Analogy of Electrical and Water Units of Measurement

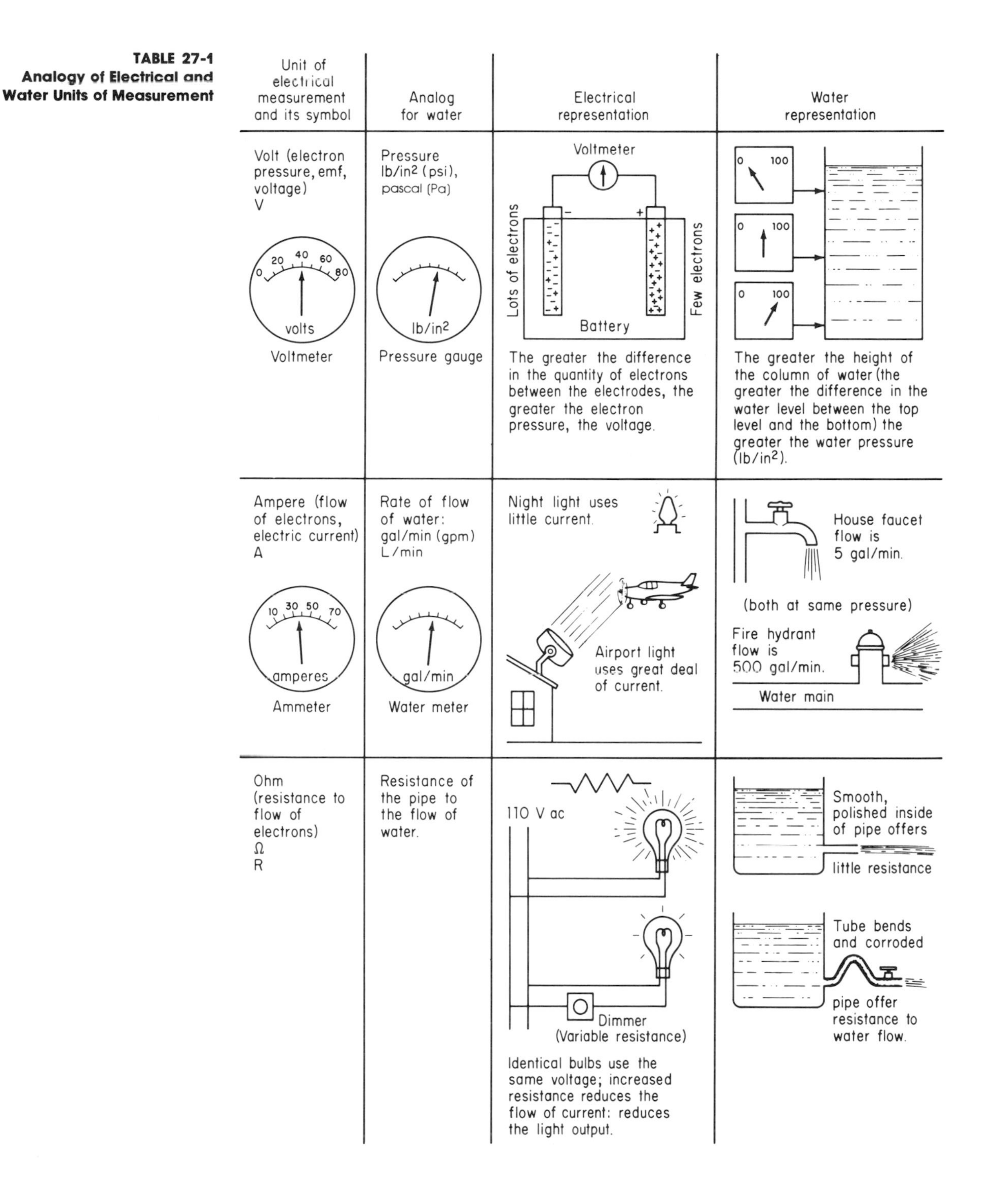

Unit of electrical measurement and its symbol	Analog for water	Electrical representation	Water representation
Volt (electron pressure, emf, voltage) V 0 20 40 60 80 volts Voltmeter	Pressure lb/in² (psi), pascal (Pa) lb/in² Pressure gauge	Voltmeter Lots of electrons – + Few electrons Battery The greater the difference in the quantity of electrons between the electrodes, the greater the electron pressure, the voltage.	0 100 The greater the height of the column of water (the greater the difference in the water level between the top level and the bottom) the greater the water pressure (lb/in²).
Ampere (flow of electrons, electric current) A 10 30 50 70 amperes Ammeter	Rate of flow of water: gal/min (gpm) L/min gal/min Water meter	Night light uses little current. Airport light uses great deal of current.	House faucet flow is 5 gal/min. (both at same pressure) Fire hydrant flow is 500 gal/min. Water main
Ohm (resistance to flow of electrons) Ω R	Resistance of the pipe to the flow of water.	110 V ac Dimmer (Variable resistance) Identical bulbs use the same voltage; increased resistance reduces the flow of current: reduces the light output.	Smooth, polished inside of pipe offers little resistance Tube bends and corroded pipe offer resistance to water flow.

tricty are related by a law known as Ohm's law, which states that:

Voltage (in volts) = current (in amperes) × resistance (in ohms)

or

$E = I \times R$

EXAMPLE If an electric device operates at 110 V and has a resistance of 5 Ω, the current required is 22A.

$110 = \text{amperage (current, } I) \times 5\ \Omega$

$5\ \Omega \times \text{amperage (current, } I) = 110\ \text{V}$

$I = 22\text{A}$

When you know any two of the three factors, the third can be calculated by means of the equation:

$E = IR$

POWER

Electric devices can be characterized by the amount of power they consume in order to operate. The formula for calculating the power (P) in watts (W) is:

Watts = volts × amperes

This can be expressed simply as follows:

$P = IE$

but

$E = IR$

therefore

$$P = I \times IR$$
$$= I^2R$$

This relationship can tell you when resistance is too high and becomes a fire hazard.

If a device uses 5 A on a 110-V line, the power is equal to 110 V × 5 A, or 550 W. Another similar but larger unit which used 10 A on the same 110-V electric line would use 110 V × 10 A, or 1100 W.

Power is also expressed in horsepower (hp): 1 hp = 746 W. Thus a 1-hp motor running on a 110-V line would require

746 W = 110 V × ? A

110 × ? A = 746 W

? A = 6.78 A

You can calculate the current of any device by the above equation if you know the horsepower or the wattage of the unit and the voltage on which it runs.

However, many times, electric devices can operate on either 110 or 220 V. The choice of 110 or 220 V depends on several factors, including safety and the current to be used.

As far as safety is concerned, even 40 V can be dangerous. Accidents involving higher voltages are more likely to be lethal than are those with 40 V, but all electricity should be handled with care and respect.

CONDUCTORS

Wire

Conductors are used to conduct electricity. They are usually termed "wire," and may be a single strand or many strands braided together. Braided wire is usually used where the conductor must be flexible; single-strand wire is used where there will be no movement. The larger the diameter of the wire, the more easily the electrons will flow through it. The smaller its diameter, the more resistance it offers to the flow of electricity. Copper is the most commonly used conductor, and aluminum is next. Wires (in the United States) are measured in thousandths of an inch (mils). A 1-mil wire has a diameter of 0.001 in, therefore a 100-mil wire has a diameter of 0.1 in. In normal usage, wires are designated by the AWG (American Wire Gage), which is standard and specifies the diameter and area of wires. *The larger the diameter and the larger the area of a wire, the more electricity it will conduct.* (See Table 27-2.) But the ease with which electrons flow in conductors, as noted before, depends upon the relative looseness of the atomic structure. The looser the bonds holding the electrons in the atom, the less the resistance to electron flow; and the more tightly the electrons are held, the higher the resistance.

Wire size (Fig. 27-2) is important, because of the current-carrying capacity and the voltage drop in the wire. All conductors have a certain amount of resistance, and there will always be a power loss in the form of heat because of the current flowing through the wire. That power loss is reflected in the voltage available at the end of the wire and is known as the voltage drop (Fig. 27-3). This drop in voltage can be significant if the diameter of the wire is small (resistance is high) and the current flow is large. Therefore, do not use long extension cords; your delivered voltage may be too low to allow the appliance to operate. The voltage drop can be measured with a voltmeter. When in doubt, use heavy extension cords to minimize the voltage drop.

TABLE 27-2
Wire Size and Resistance

AWG No.	*Diameter, in*	*Ohms/1000 ft*	*Ohms/meter*
40	0.00315	1049.0	3.44
38	0.00397	659.6	2.16
36	0.00500	414.8	1.36
34	0.00631	260.9	0.855
32	0.00795	164.1	0.538
30	0.01003	103.2	0.338
28	0.01264	64.9	0.213
26	0.01594	40.8	0.134
24	0.02010	25.7	0.084
22	0.02535	16.1	0.053
20	0.03196	10.15	0.033
18	0.04030	6.385	0.021
16	0.05082	4.016	0.013
14	0.06408	2.525	0.008
12	0.08081	1.588	0.005
10	0.1019	0.999	0.003
8	0.1285	0.628	0.002
6	0.1620	0.395	0.0013
4	0.2043	0.249	0.0008
2	0.2576	0.156	0.0005

Standard annealed copper wire at 20°C.

00 0 2 4 6 8 10 12 14 16 18

FIGURE 27-2
Actual diameters of different AWG sizes of copper wire, without insulation.

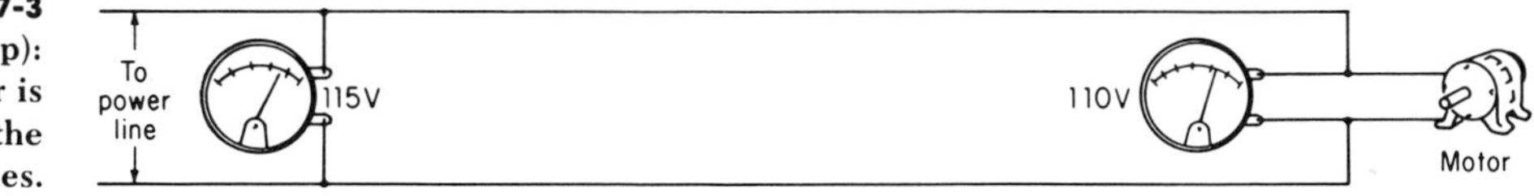

FIGURE 27-3
The voltage drop (*IR* drop): voltage at the motor is lower than that at the power lines.

TABLE 27-3
Electric Resistivity of Metals

Metal	*Resistivity, μΩ-cm*
Aluminum	2.6548
Copper	1.6730
Gold	2.35
Nickel	6.84
Silver	1.59
Tungsten	5.65

Conductivity of Various Metals

Resistivity or specific resistance is the reciprocal of conductivity. By comparing the resistivities of various metals, we can see which ones offer least resistance and therefore conduct electricity the best. Table 27-3 gives some of these values, and you can see that silver is the best conductor of all the metallic elements. Nickel is the worst of the group, but certain alloys of nickel, such as Nichrome, have even higher resistivity and are used as heating elements in hot plates, for example. Copper is used in electric wiring because it is the next best conductor to silver.

ALTERNATING AND DIRECT CURRENT

There are two types of current, alternating and direct. The type normally found in industry and in homes all over the country is of the alternating type. The electrons flow in one direction, then stop and reverse themselves, then stop and continue the cycle many times a second. In the United States, the frequency of the cycling is 60 Hz (hertz); in European countries 50 Hz; one cycle per second equals one hertz.

Modern power stations generate between 11 and 14 kilovolts (kV) ac, which may be boosted by transformers as high as 275 kV for long-distance transmission. The power needed in various areas is brought to substations, where the voltage is reduced to around 2500 V, and then reduced further at secondary points to about 115–230 V ac where it is delivered to industry and residences (Fig. 27-4). The figure "115 V ac" is rather flexible. Actually the voltage may range from 105 V to as high as 122 V, depending upon the conditions. Power companies at times arbitrarily reduce the voltage 5 to 10 percent to conserve energy. When this happens, disconnect as much equipment as possible to avoid damage to it.

Direct current is that current which flows in only one direction. This is the type found in batteries and dry cells (Fig. 27-5). The current is considered to flow from the negative terminal to the positive terminal, because the negative terminal of the battery has an excess of electrons and

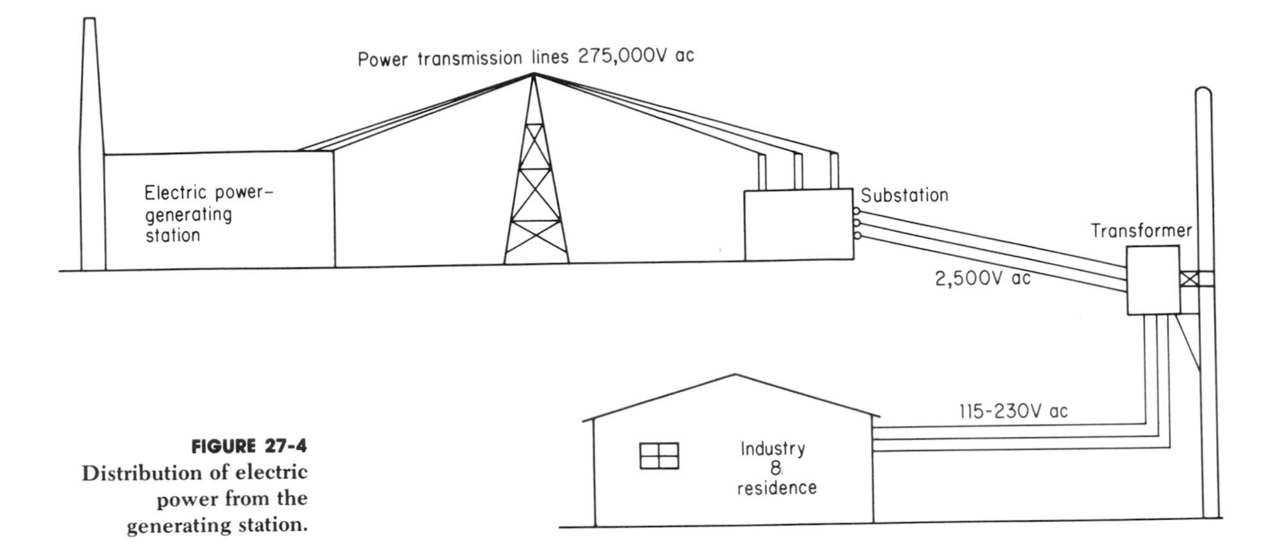

FIGURE 27-4
Distribution of electric power from the generating station.

the positive terminal has a deficiency of electrons. The difference of the electron pressure gives the voltage (emf) (Fig. 27-1).

Since ac equipment can be permanently damaged by connecting it to direct current, and vice versa, it is wise to determine whether a piece of equipment is meant to operate ac or dc. This information is usually marked on the equipment: if the label states "ac only" or "dc only," the equipment must be hooked up *only* to ac or dc as required. A label stating ac/dc means that either type of current can be used. *Transformers* (Fig.

(*a*)

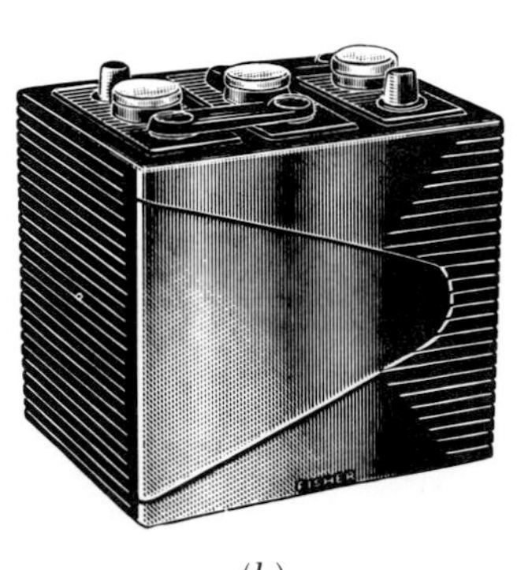

(*b*)

FIGURE 27-5
(*a*) Dry-cell battery is used to power electric and electronic instruments. Observe the expiration date so that you can use fresh batteries for dependable service. (*b*) Storage batteries are used to power instruments or for general laboratory use. They are rechargeable. CAUTION: Such batteries contain sulfuric acid.

FIGURE 27-6
Transformer with variable voltage from 0 to 140 V. There is full power at all voltages; rotation of knob controls voltage. The transformer is protected by a fuse and has an on-off switch. *Do not overload*. When used with a rectifier, it will deliver variable-voltage direct current.

27-6) operate only on alternating current; some electric devices that state "ac only" have transformers in them.

FUSES AND CIRCUIT BREAKERS

Fuses are safety devices used to limit the flow of current (Fig. 27-7). Fuses are rated by the current they safely conduct: 5, 10, 20, 30 A—or more. When the current exceeds that stated amount, sufficient heat is generated in the fuse to cause the metal alloy in the fuse to melt; thus the circuit is broken, and electricity ceases to flow.* The wires in a particular circuit are of a definite size, rated to carry up to a certain amount of current safely. If too much current passes through this wire, heat is generated and fire can result. The amount of current to be carried by a conductor is therefore limited by placing a fuse in the path of the electrons. The fuse

* Ferrule-type fuses (Fig. 22-7*b*) come with replaceable elements. When the fuse "blows," the end is screwed off, the burned element is discarded, and a new element is inserted. The replaced end is then tightened securely to make positive contact.

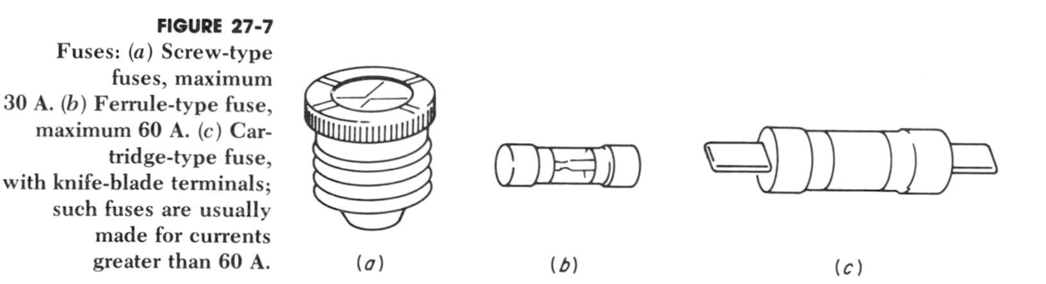

FIGURE 27-7
Fuses: (*a*) Screw-type fuses, maximum 30 A. (*b*) Ferrule-type fuse, maximum 60 A. (*c*) Cartridge-type fuse, with knife-blade terminals; such fuses are usually made for currents greater than 60 A.

will "blow" (burn out) when the current-carrying capacity of the fuse is exceeded. Thus a 15-A fuse will carry 15 A without blowing.

Types of Fuses and How to Replace Them

Time-Lag Fuses

The ordinary fuse will carry its rated capacity indefinitely, but will blow when the current exceeds its rating. Electric monitors are rated by the number of amperes they draw when they are running continuously. A 15-A motor therefore draws 15 A. However, all electric motors draw excessive amounts of current *when they are starting*, as much as two to three times their normal rating; however, they drop down to their normal rating when they have reached full speed. It is during this starting period, the time of excess current flow, that fuses may blow, because their rating is based upon the motor's normal requirements. To prevent fuses from blowing every time a motor starts, time-lag fuses were developed. They will carry a large overload of current safely for a few seconds, but will blow just as quickly for continuous overloads as will an ordinary fuse. They have the appearance of ordinary fuses, but they are constructed differently and are used wherever electric motors are on the circuit.

CAUTION Fuse panels are electrically "hot" (current flows through them). Be extremely careful when replacing fuses.

Screw-Type Fuses

A visual inspection reveals which fuse has blown—the element has melted. Unscrew the fuse and replace it with a fuse having the same rating. *Do not* replace with a higher-rated fuse; larger currents can then be carried by the circuit; this will result in overheating and cause fire.

Ferrule- and Cartridge-Type Fuses

These fuses are securely gripped by spring contacts, and even may be "frozen" to the contacts because of corrosion. For these fuses, *always use a fuse puller* (Fig. 27-8); it is safe to use because your hands are away from "hot" contacts (live wires) and your hand motion is away from the box.

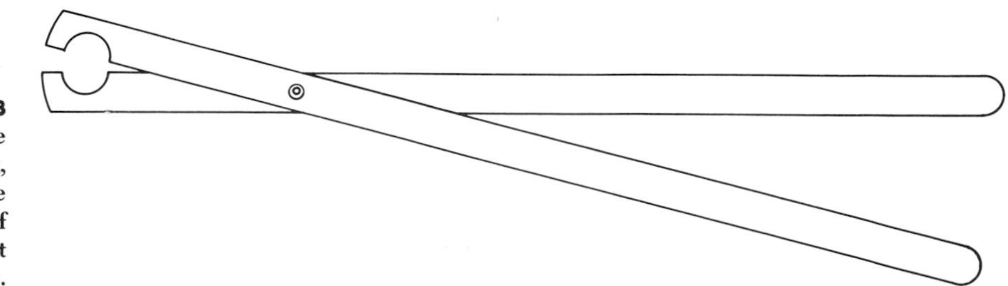

FIGURE 27-8 Fuse puller. This is made of a nonconducting, rigid plastic material. Use it as you would a pair of pliers to remove and insert fuses.

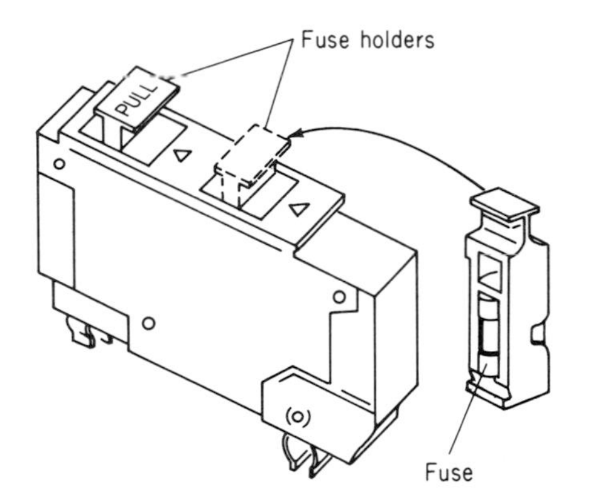

FIGURE 27-9
Plug-in fuse block.

Procedure

1. If practicable, throw the power switch to OFF to disconnect all power from the fuse box. Unfortunately, the power switch may control other important circuits, and total disconnect may not be practical.

2. Use a fuse puller to remove fuse.

3. Replace fuse with the fuse puller.

Fuse-Holding Blocks

Some fuse panels have fuse-holding blocks which are removable (Fig. 27-9). The cabinet is fitted with copper bus bars, and the fuse blocks engage the bars to make contact. A red light goes on as a signal that a fuse block is burned out. To replace a fuse, remove the entire block, which contains two fuses. Once the burned fuse is replaced, the entire block, made of insulating plastic, is pushed back into the cabinet. This piece of equipment is superior from a safety standpoint.

Overview: When a Fuse Blows

1. Replace the fuse. If the fuse does not blow again, there may have been a temporary overload, because the system is probably very near the capacity of the circuit.

2. If the fuse blows again, change some of the electrical units to another circuit.

3. If the fuse still blows, you probably have a short circuit, which you must isolate.

4. Disconnect all units from the circuit and replace the fuse. The fuse should not blow now because no units are connected and nothing is drawing current. If it does blow, call an electrician immediately.

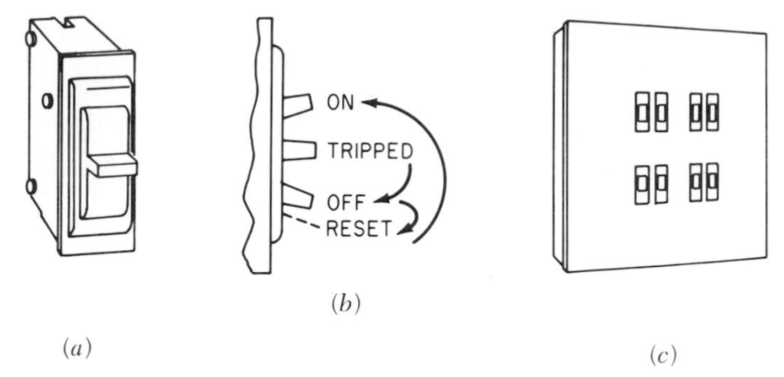

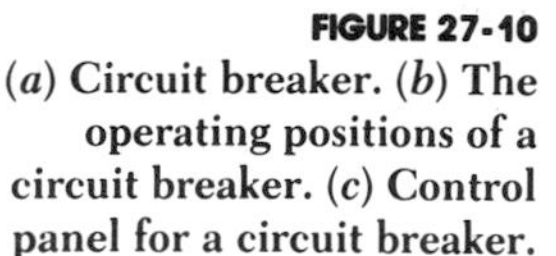

FIGURE 27-10
(*a*) Circuit breaker. (*b*) The operating positions of a circuit breaker. (*c*) Control panel for a circuit breaker.

5. Reconnect one of the units which was originally connected to the circuit; if the fuse does not blow, connect the second, then the third, etc. Eventually, reconnection of one of the units will blow the fuse.

6. Disconnect all the units except the last one which was connected. Replace the fuse. If the fuse blows again, that is the defective unit, and it should be serviced.

Circuit Breakers

Circuit breakers, often used today instead of fuses, operate in precisely the same way. They open the circuit in the event of an overload; but unlike fuses, they may be reset and reused. When a circuit breaker does break the circuit, it is cooled down in a few minutes. If the overload continues to exist, it will break again just as a fuse does. Circuit breakers are more expensive than fuses, but their use avoids the annoyance of replacing fuses and the replacement cost of fuses.

Circuit breakers look much like toggle switches (Fig. 27-10), but they usually have four positions; ON, OFF, TRIPPED, and RESET. When the circuit breaker is carrying current, the position is ON, and just like a fuse, it will carry its rated capacity indefinitely. However, even small overloads will cause the circuit breaker to trip, breaking the circuit. When this happens, push the handle to RESET and, when you feel it engage, push the handle to ON. Circuit breakers are designed to carry small overloads for a short period of time, as for motor starting, but they trip on continuous overload.

SERIES AND PARALLEL CIRCUITS

Electric circuits are designed to be either series or parallel. In series circuits, every electron flows through every electric device. An example is a string of old-fashioned Christmas-tree lights. These are connected in series; the electric circuit is broken when one bulb burns out, and electricity ceases to flow. (See Fig. 27-11*a*.) In parallel circuits, each electric device uses its own current or number of electrons independently of the other (see Fig. 27-11*b*). In the series circuit, when the electrons enter the conductor through one wire of the conductor, they must pass through every piece of equipment before returning to the other wire of the conductor. The current in the wire is therefore the same at all points. In the parallel circuit, the total current that enters the wire of the conductor at the plug is divided between the requirements of each electrical unit, each using its own amperage. Therefore, when several electric devices are connected in parallel, as when such devices are connected to the triple outlet of an extension cord, caution must be exercised that the sum of the amperes used by each unit does not exceed the safe carrying capacity of the extension cord. This concept must be understood in order to explain why fuses blow when additional units are added in parallel.

EXAMPLE A circuit is fused for 20 A. An electric motor drawing 10 A and an electric heater drawing 7 A are connected to the same outlet. Thus the current through the circuit is 17 A. If another heater or unit that draws 5 A is also connected to the same outlet, then the total current will be 22 A, exceeding the safe current-carrying capacity of the wire conductors, and the fuse blows. The answer to this problem is to use another circuit, protected by another fuse. Never remove a fuse and replace it with a solid-wire conductor. This removes the safety valve, and if too much current is drawn through the wires, fire may result.

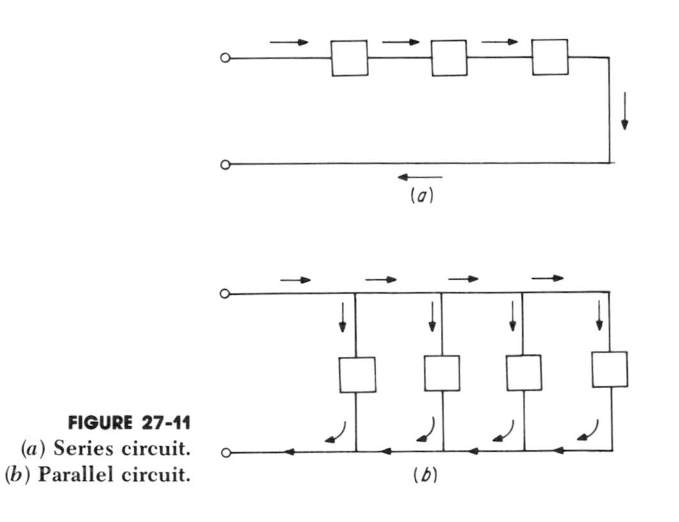

FIGURE 27-11
(*a*) Series circuit.
(*b*) Parallel circuit.

ELECTRIC POWER

In a typical power supply (Fig. 27-12), the electric company brings in service of, say, 110 V to a building with wire conductors capable of carrying 100 A. This passes through the master switch (*A*), which can disconnect all electricity coming into the building. From there it is divided into smaller branch circuits, each fused to protect that circuit from being overloaded. All the current, up to 100 A at 110 V, passes through *A*.

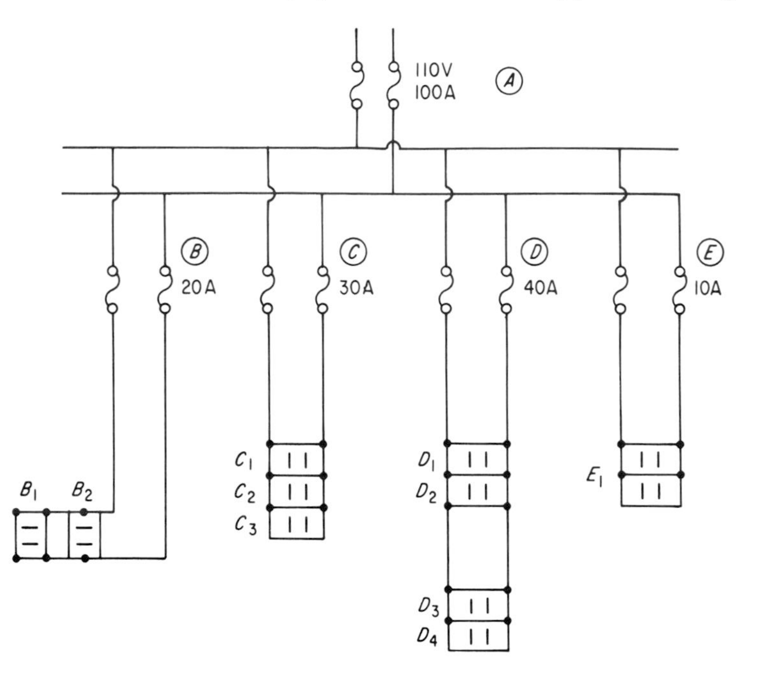

FIGURE 27-12
Current distribution in a typical installation.

Only that current passes through fuse *B* which has been drawn from outlets B_1 and B_2. Fuse *B* (rated at 20 A) will blow when the sum of the currents through these two outlets exceeds 20 A.

A maximum of 30 A can pass through fuse *C*. When the current drawn from outlets C_1, C_2, and C_3 exceeds 30 A, that fuse and only that fuse will blow.

Fuse *D* is capable of carrying 40 A and will pass all the current drawn from outlets D_1, D_2, D_3, and D_4, but it will blow when the current exceeds 40 A.

Only current drawn from outlet E_1 will pass through fuse *E*, and the capacity of circuit *E* is 10 A. To summarize:

Total current available		100 A
Circuit *B*	Capacity 20 A	
Circuit *C*	Capacity 30 A	

Circuit *D*	Capacity 40 A	
Circuit *E*	Capacity 10 A	
Total of circuits *B*, *C*,	*D*, and *E*	100 A

The main fuse will blow when the total current drawn exceeds 100 A. If one of the subfuses, *B*, *C*, *D*, or *E*, is defective, and the current drawn through all circuits exceeds 100 A, then the main fuse, *A*, will blow.

SERVICING INOPERATIVE DEVICES

If a device does not work when connected to an electric outlet:

1. First check the outlet to see if electricity is available (if the outlet is "hot"), by plugging a desk lamp (which is known to work) into the socket.
2. Next check the electric plug to be sure that the wires are connected to the plug.
3. Check the cord connecting the plug to the unit to see if the wire is broken.
4. Examine the fuse in the electrical unit. It may have blown.
5. See that the switch is turned on.
6. If none of these steps results in operation, call the electrician.

ELECTRICAL TESTING INSTRUMENTS: VOLTMETER, AMMETER, AND MULTIMETER

Voltage is tested with a *voltmeter* (Fig. 27-13). The two probes of the voltmeter are touched to the two terminals being checked. The voltage is read off the dial. The probes must touch the base wire, which is not covered with insulation. The voltmeter must be connected to the two conductors when in use. Meters must be selected for the determination they are to make, the range to be covered, and the precision desired. *EXERCISE CAUTION WHEN YOU CHOOSE THE METER. DO NOT OVERLOAD.*

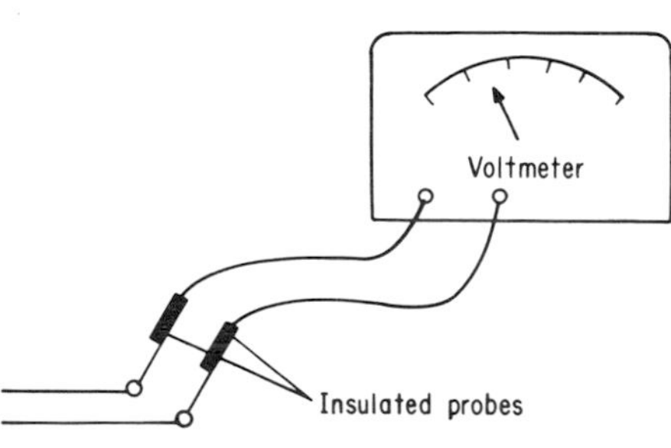

FIGURE 27-13 Measuring voltage with a voltmeter.

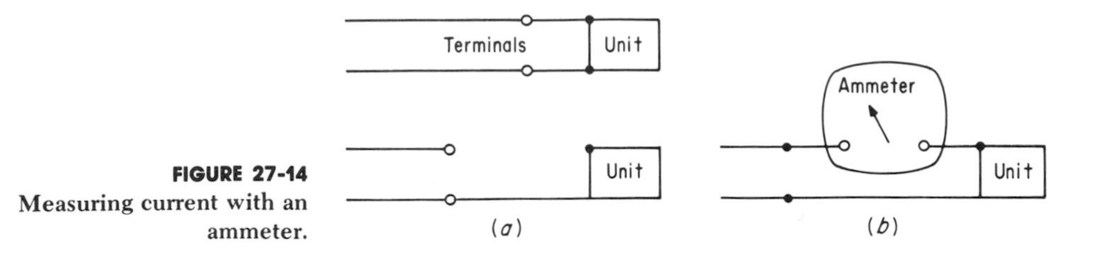

FIGURE 27-14
Measuring current with an ammeter.

Current is tested with an *ammeter* (Fig. 27-14), which must be inserted in the line so that the total amount of current can be read off the meter. The unit must be disconnected (one of the conductors either cut or disconnected from itself at the junction), and the ammeter is inserted in series with the wire conductor.

In work with alternating current, one may use an external type of ammeter, which has a circular portion that can be opened, placed around a *single* wire, then closed. The current passing through the wire creates a magnetic field around the wire conductor which actuates the needles of the ammeter and gives the reading in amperes of the current passing through the conductor.

The *multimeter* (Fig. 27-15) is a compact instrument which is used to measure the electrical characteristics of circuits. Among the characteristics that can be measured, depending upon the instrument, are voltage, amperage, resistance of alternating and direct current, decibels, and output voltages. It can be extremely useful for determining the operating characteristics and for locating abnormalities. Follow the directions given in the instruction manuals because of variations in design.

FIGURE 27-15
The multimeter is a combined voltmeter-ammeter-ohmmeter. It has a selector switch that can be set to measure voltage, amperage, or resistance. Flexible probes allow access to the interior of the instrument to be tested.

CAUTIONS 1. Do not overload.

2. Insert probes in proper holes.

3. Set selector to proper range.

4. When measuring resistances, do not measure the components of units that have electricity flowing through them. This relatively high current will burn out the meter if it is used when the item is electrically "hot."

SAFETY

Electricity can be handled safely, provided the rules by which it works are carefully observed. Normally the electricity available for use in the laboratory and plant has one side connected to the ground; the other side is "hot" at 110 V. If an electric device feels tingly when it is touched, the plug should be removed and the prongs reversed when it is replugged. This keeps the device at ground potential. In order to observe safety rules, the electric device should always be *grounded;* that is, an electric conductor, a wire, should be connected from the metal casing of the device to a water pipe or any metal object which is connected to the ground. This conductor will then carry any dangerous current to the ground instead of through any individual, thus affording positive protection against accidental electrocution by touching a shorted device.

Since electricity is conducted through two conductors, never touch the two at the same time. In fact, never touch *one* base conductor; because there may be water on the ground, you may act as an alternative conductor for the electrons and get a severe or fatal electrical shock.

Know how to handle electricity safely, and you should not fear working with it. (Refer to Chap. 1, "The Chemical Technician.")

ELECTRIC MOTORS

Electric motors convert electric energy to mechanical energy and do work. They must be controlled with safe and positive-acting controls. Control means to govern or regulate, to start, stop, reverse, or alter the rate of revolution of the motor (normally rated in revolutions per minute or rpm's). Any piece of equipment which is used to accomplish this end is a control component or motor control.

Starting Motors

Small motors, fractional horsepower to ½ horsepower, can be started with a simple, knife-type, on-off switch. Many small motors use nothing more than a cord and plug, with a toggle switch (Fig. 27-16).

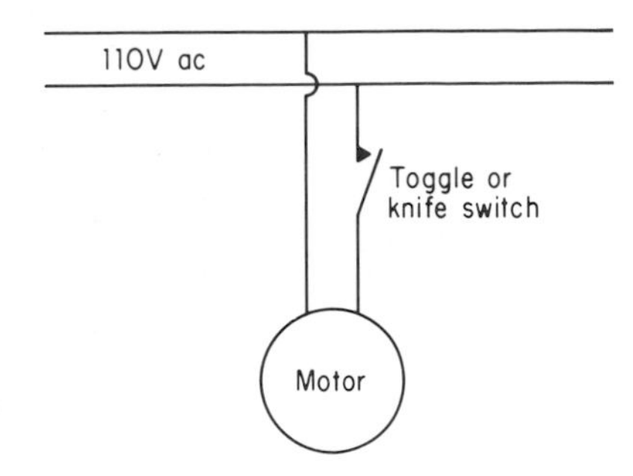

FIGURE 27-16
Toggle-switch starter.

Manual Starters

For larger motors, 1- to 10-hp, manual motor starters are used to give manual control and both overvoltage and undervoltage protection. In the event of a drastic voltage variation, the starter disconnects the motor from the line voltage, thus protecting the motor. Smaller motors may use 110-V or 220-V single-phase ac, but larger motors invariably use 220-V three-phase ac.

The contacts are made by manually pushing the START button which is physically linked to the contacts (Fig. 27-17). Once the contacts are closed, the linkage is latched into position. Should an overvoltage or undervoltage condition develop, a safety component will unlatch the linkage, disconnecting the motor. When the STOP button is pushed, the linkage is released and the contacts disconnect.

Magnetic Starters

In contrast to the manual starter, which is normally located in the immediate vicinity of the motor, a magnetic starter (Fig. 27-18) can control a motor which is not in the immediate vicinity, as long as the magnetic

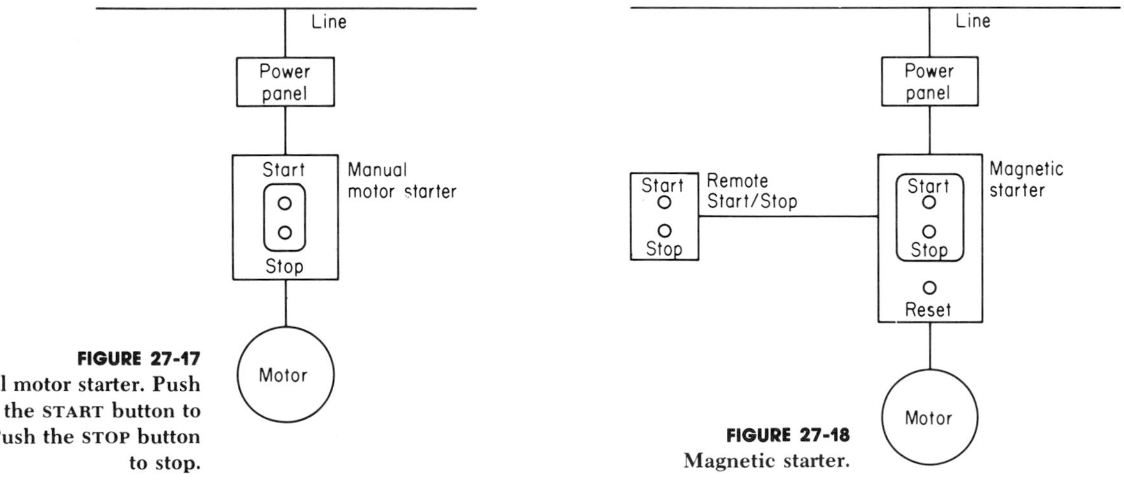

FIGURE 27-17
Manual motor starter. Push the START button to start. Push the STOP button to stop.

FIGURE 27-18
Magnetic starter.

starter is electrically connected to it. When the START button is pressed, it energizes a small coil in the starter which magnetically causes the contact points to close, starting the motor. Only a momentary push to start motion is necessary. The contacts on closing also complete the electric circuit for the small coil, thus keeping the coil energized and the contacts closed. When the motor is to be stopped, the STOP button is pushed. This push breaks the electrical circuit of the energized small coil, causing all the contacts in the starter to disengage, including that of the small coil. Thus the motor is stopped at that point. The cycle can be repeated at will. Magnetic starters also have overvoltage and undervoltage protection which automatically breaks the electrical circuit of the small coil, in effect causing the same result as pushing the STOP button.

Magnetic starters may have several remote START-STOP switches as well as a set on the starter itself. When a magnetic starter becomes deenergized as a result of an overvoltage or undervoltage condition, the starter will not be operative again until the RESET button has been pushed.

Rating Motors

Motors are rated by horsepower (abbreviated hp), ranging from fractional horsepower for small equipment to many horsepower for larger devices.

Electric motors will last for years provided:

1. The proper motor is used for the job.
2. The motor is not abused by overworking.
3. The motor is properly maintained.

Using Motors Correctly

Motors can be used for jobs which normally require twice the rated horsepower, *but only for a short time.* No motor can continually provide more than its rated horsepower.

Most 60-Hz ac motors run at around 1725 to 1750 rpm. Special-purpose motors run at higher specified speeds.

Motors develop heat when they run continuously, usually about 40°C *(above room temperature)*. Therefore, motors running in a hot room may feel too hot to the touch, but if they are not running overloaded, they are running normally and are safe.

Motors will be damaged if excessive current flows through them for a considerable time. They will burn out.

Protect motors by following the simple instructions below.

Three-phase motors: Use a correct fuse on each line.

Single-phase motors: Use only one fuse on 115 V (the black ungrounded line) and two fuses on 230 V (one in each line).

Thermal-overload devices carry a small overload (when starting the motor) for a short time without tripping, but trip quickly if a large overload is imposed. When they do trip, they can be reset by pressing the RESET button after a short time lapse.

CAUTION Always use the proper motor for the job. Abusing motors by overloading them will damage them and burn them out. Chemicals and water can also cause motors to burn out. Keep them *clean* and away from possible spillage of water and chemicals.

Types of Motors

Capacitor motors are fitted with a cylindrical "capacitor" attached to the housing. They are used to start heavier loads, such as a vacuum pump.

Split-phase motors are usually fractional-horsepower motors, less than ⅓ hp. They have no brushes or commutators but are not used for hard-to-start items such as compressors.

Repulsion-induction motors operate only on single-phase ac. They have the ability to start heavy-load items.

Universal motors operate on ac or dc, but they do not run at constant speed. The greater the load, the lower the speed. Rheostats can be used to control and vary their speeds, as in mixers.

Three-phase motors require three-phase ac. They cannot be run on single-phase ac. Their direction of rotation can be reversed by changing any two of the three leads to the motor.

Some larger motors, from about ½ hp up, have four leads; this means that the motor will run on 115- or 230-V ac, depending upon the connection. These are called *dual-voltage* motors. At the lower voltage, the coils are connected in parallel, which means that two leads will be joined at *each* terminal as in Fig. 27-19. At the higher voltage, the two coils of the motor are connected in series. Two of the coil terminals (one from each coil) are joined together and are not connected to the input voltage, as in Fig. 27-20.

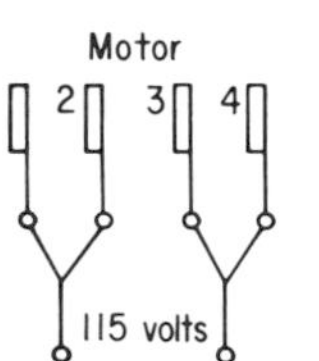

FIGURE 27-19
Low-voltage connection.

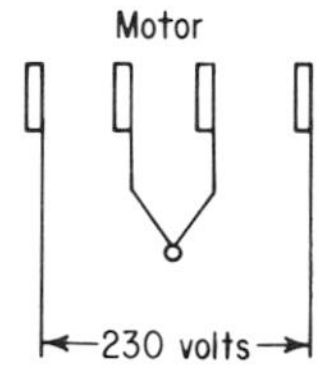

FIGURE 27-20
High-voltage connection.

BELTS, PULLEYS, AND POWER TRANSMISSION

Motors may be coupled directly to the shaft of a piece of equipment with a rigid coupling for stirrers or with a flexible coupling to a machine. Rigid couplings should have the correct internal-diameter opening to fit the shaft of the motor and the machine. Force is not required to assemble the unit, *provided the shaft is not scarred or burred.* If the shafts are burred, smooth them with a fine file (on the motor while running) just enough to remove the burrs. Always tighten the locking screws firmly so that the shafts will not slip and become burred. *Do not* apply excessive pressure to tighten the locking screws.

How to Change the Speed of Machines Using Pulleys

V belts should not be excessively tight. Too tight a belt will cause the belt to become worn too quickly, and the *excessive pressure on the motor bearings* will cause the motor to wear out the bearings. V belts can be operated with considerable slack and fairly loosely. If possible, use the larger pulleys. Pulleys of too small a diameter cause sharp bends in the belt and shorten belt life.

When you wish to change the speed of any machine which is using a pulley-belt method of power transmission, you may change the speed by varying the diameter of the pulleys. Of course, you may have to lengthen or shorten the belt. *You may not have to do this provided that* (1) the motor can be moved on the base or (2) you selectively change both motor and machine pulleys.

Use Table 27-4 to determine the pulley combination:

1. Measure the diameter of the motor pulley with a ruler as in Fig. 27-21.
2. Determine the speed at which you wish the machine to run.
3. Read across the table to find the diameter of the machine pulley.

NOTE If you wish the machine to run at 1725 rpm, use the *same diameter machine pulley as your motor pulley.* The size of the pulley makes no difference except for excessive belt wear.

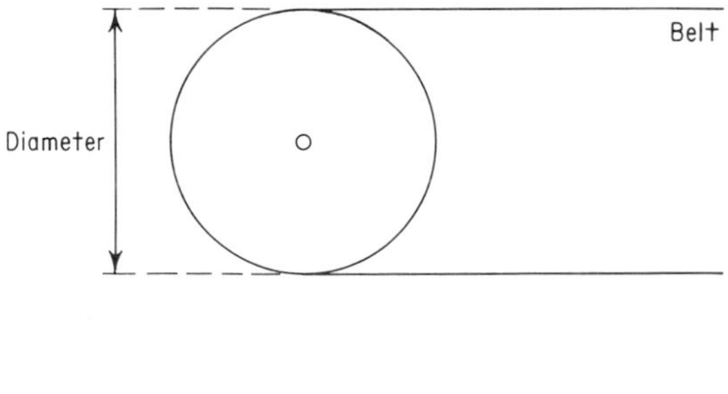

Belt

Diameter

FIGURE 27-21
Measuring the diameter of a pulley.

TABLE 27-4

Diam. motor pulley	*Diam. of pulley on machine, in*														
	1¼	*1½*	*1¾*	*2*	*2¼*	*2½*	*3*	*4*	*5*	*6½*	*8*	*10*	*12*	*15*	*18*
1¼	1725	1435	1230	1075	950	850	715	540	430	330	265	215	175	140	115
1½	2075	1725	1475	1290	1140	1030	850	645	515	395	320	265	215	170	140
1¾	2400	2000	1725	1500	1340	1200	1000	750	600	460	375	315	250	200	165
2	2775	2290	1970	1725	1530	1375	1145	850	685	530	430	345	285	230	190
2¼	3100	2580	2200	1930	1725	1550	1290	965	775	595	485	385	325	255	215
2½	3450	2870	2460	2150	1900	1725	1435	1075	850	660	540	430	355	285	240
3	4140	3450	2950	2580	2290	2070	1725	1290	1070	800	615	515	430	345	285
4	5500	4575	3950	3450	3060	2775	2295	1725	1375	1060	860	700	575	460	375
5	6850	5750	4920	4300	3825	3450	2865	2150	1725	1325	1075	860	715	575	475
6½	8950	7475	6400	5600	4975	4480	3730	2790	2240	1725	1400	1120	930	745	620
8		9200	7870	6900	6125	5520	4600	3450	2750	2120	1725	1375	1140	915	765
10			9850	8620	7670	6900	5750	4300	3450	2650	2150	1725	1430	1140	950
12					9200	8280	6900	5160	4130	3180	2580	2075	1725	1375	1140
15							8635	6470	5170	3970	3230	2580	2150	1725	1425
								7750	6200	4770	3880	3100	2580	2070	1725

Data based on standard 1750-rpm motors.

GENERAL RULE For the machine to run at *higher speeds*, the machine pulley is *smaller* than the motor pulley. For the machine to run at *lower speeds*, the machine pulley is *larger* than the motor pulley.

MAKING ELECTRIC CONNECTIONS

You must strip the insulation from the end of electric wires before you can make a connection. *Use care* when removing the insulation.

1. Do not nick the copper wire. Bending of nicked wire leads to breaks (Fig. 27-22).

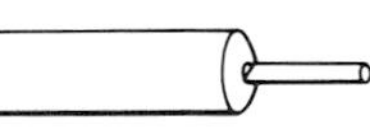

FIGURE 27-22
Incorrectly stripped wire.

2. Apply the knife at an angle when cutting away insulation to attach the wire to a terminal. This decreases the chance of nicking the wire and gives the correct insulation angle cut (Fig. 27-23).

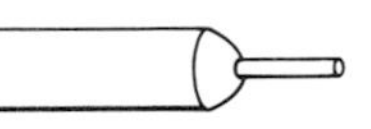

FIGURE 27-23
Wire correctly stripped for terminal connection.

3. Apply the knife at a greater angle when preparing the wire to be spliced to another wire (Fig. 27-24).

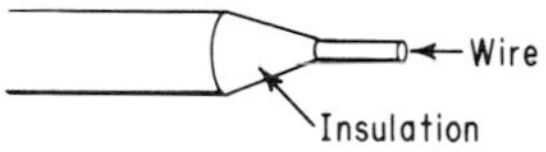

FIGURE 27-24
Wire correctly stripped for splicing.

Electric-Terminal Connections

Faulty connections of electric wire to terminals of equipment are a major source of trouble and improper performance. *Use care* when making connections.

1. Strip away only sufficient insulation.
2. Bend the wire into a ¾ circle with needlenose pliers (Fig. 27-25).

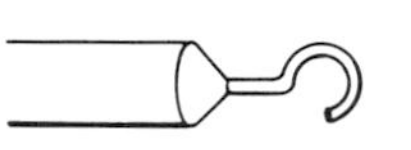

FIGURE 27-25
Loop for terminal connection.

3. Tighten the screw firmly, after you insert the loop in the correct direction (Fig. 27-26*a*). Loop will open if it is inserted in the opposite direction and the screw is tightened (Fig. 27-26*b*).

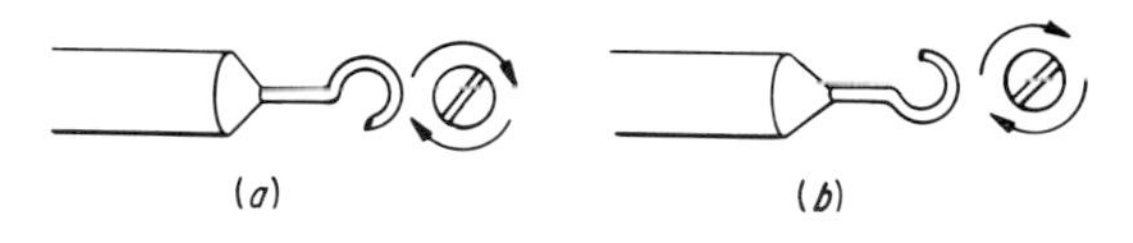

FIGURE 27-26
(*a*) Loop *closes* when screw is tightened (correct).
(*b*) Loop *opens* when screw tightens (incorrect).

4. Solderless lug connections should be used on larger-diameter wires (Fig. 27-27).

FIGURE 27-27
(*a*) Screw tightens to secure wire. (*b*) Soft-metal lug is squeezed with tool to secure wire.

Braided wires: After stripping insulation, twist the braided wires with the fingers to make essentially one strand.

CAUTION *Avoid short circuits by making clean connections. Prevent wires from touching other terminals.*

Splicing Wires Temporarily

Electric wires can be connected together easily with solderless connectors, provided there is no strain on the wires.

1. Strip the insulation from the end of the wires as shown in Fig. 27-24.

2. Twist the two (or more) wires together as shown in Fig. 27-28. If one wire has a smaller diameter than the others, let it extend beyond (project some beyond) the others.

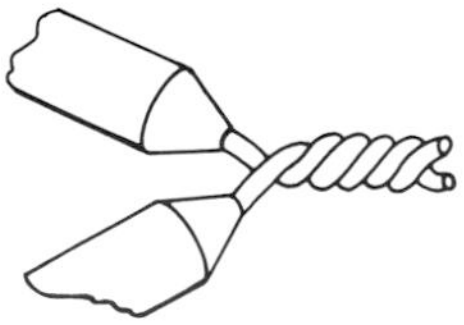

FIGURE 27-28
Splicing wires together before securing with solderless connector.

3. Slip the connector over the bare twisted wires and screw the insulating shell over the wires, tightening the connector securely by twisting. The insulating shell should *extend* over the insulation of the wires; then you will not have to tape the connection (see Fig. 27-29).

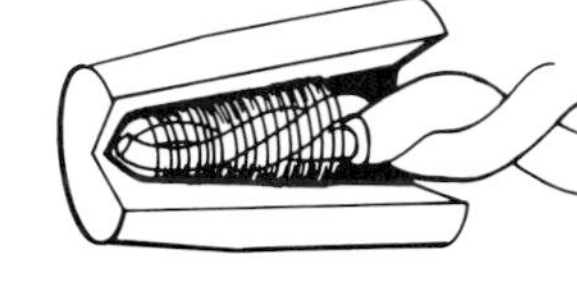

FIGURE 27-29
The completed splice.

Continuous Wire

To connect one wire to another wire which is not to be cut, use the following procedure:

For solid-strand wires:

1. Strip the continuous wire for about ½ in at the point desired (see Fig. 27-24).

2. Wrap the end of the stripped wire to be connected to the continuous wire as in Fig. 27-30.

FIGURE 27-30
Splicing wire (one uncut, one cut).

3. If it is a permanent connection, solder and then tape. If it is to be a temporary connection, merely tape with insulating tape.

For braided wires (stranded wires):

1. Strip the insulation from both wires as needed.
2. Separate the continuous wire into two groups of wires (Fig. 27-31).

FIGURE 27-31
Separating braided wire.

3. Separate the terminal wire into two sections (Fig. 27-32).

FIGURE 27-32
Separating terminal wire.

4. Insert the wires as shown in Fig. 27-32 into separated wires (Fig. 27-31).
5. Twist the ends of the terminal wires in opposite directions as shown in Fig. 27-33.

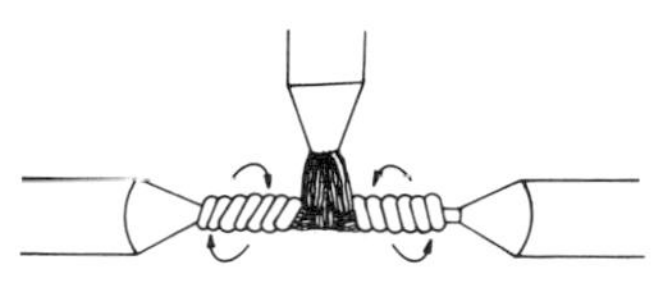

FIGURE 27-33
Completing the splice.

6. Solder and then tape if the connection is permanent. Tape if the connection is temporary.

Splicing Wires Permanently

1. Strip the ends of the wires as shown in Fig. 27-24.
2. Hook the two wires together as shown in Fig. 27-34.

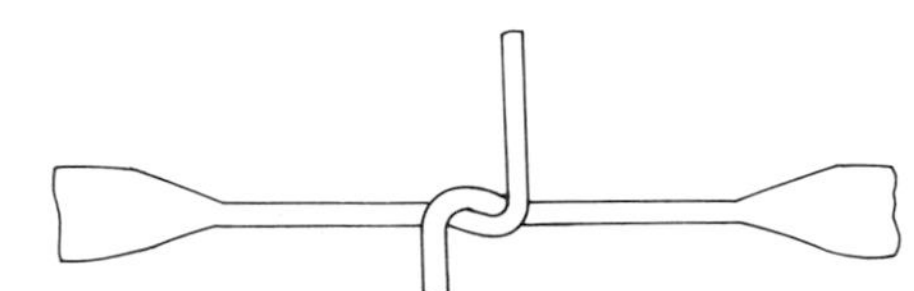

FIGURE 27-34
Hooking the wires.

3. Twist the wires in opposite directions as in Fig. 27-35.

FIGURE 27-35
Twisting the wires together.

4. Solder and then tape the twisted wires.

Soldering

To solder wires or tinned or copper objects:

1. Use an electric soldering iron and flux-core solder.

2. Clean, by scraping, the wire or object to be soldered.

3. Clean the soldering-iron tip by allowing it to heat until the residual solder on the tip melts and then quickly wiping the tip firmly with a heavy piece of cloth (*quickly* so that the cloth does not burn). The tip will now be clean and shiny with a film of melted solder.

4. Apply the *hot* soldering iron to the surface to be soldered (Fig. 27-36).

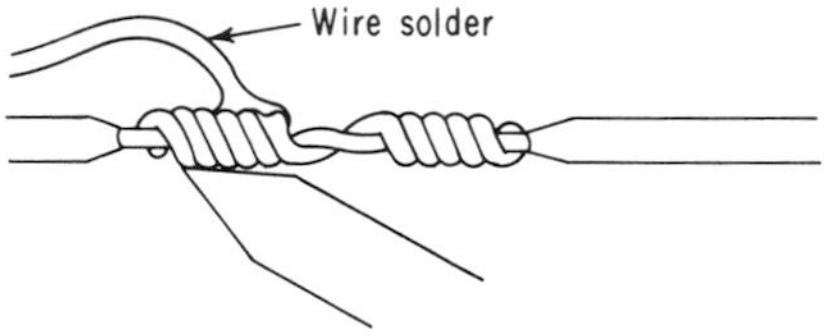

Wire solder

FIGURE 27-36 Correct method of soldering: iron heats wire.

5. Touch the solder to the area to be soldered. *Do not touch the solder to the hot iron.*

6. When the wire is hot enough to melt the solder, the solder will melt and flow evenly and smoothly, making good contact (Fig. 27-37*b*).

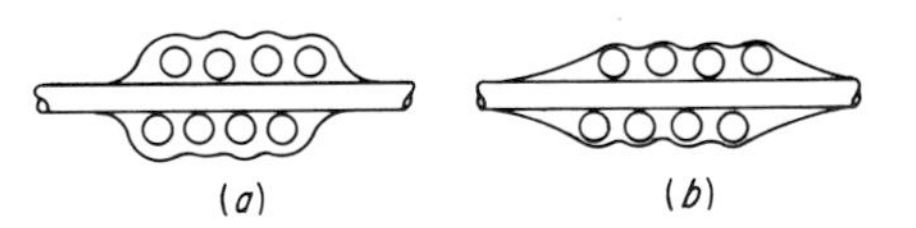

FIGURE 27-37 Soldering: (*a*) Incorrect: hot solder is applied on cold wire. (*b*) Correct: solder is melted by hot wire providing a good flow of solder.

7. Remove the iron when the solder has flowed.

NOTE You will get good results if:

1. The surface of the wires is clean.

2. The wires are heated to melt the solder; never use the soldering iron to melt the solder.

3. The joint *is never jiggled or moved until the solder has cooled and solidified.* The joint will be weakened unless this is done. The surface of the solder will turn from *shiny* to *dull* when the solder has cooled.

Taping

Tape electric connections to prevent short circuits and electric shocks.

1. Use plastic tape which is specially made for electrical work. Although it is thin, it possesses high insulating qualities.

2. Start the tape at one end of the insulation. Wrap spirally to the other insulation; then reverse and return to the start. Continue the process of wrapping until the taped section is as thick as original wire.

CAUTION *Keep the tape stretched at all times.*

GROUNDING ELECTRIC EQUIPMENT

Electricity is dangerous and can be fatal. It is not necessarily the high voltage that is dangerous. It is the number of amperes that can cause injury and death. Electricity is transported through electric wires that are covered by nonconducting materials (insulation). Most electric equipment is constructed of metal. When that metal comes in accidental contact with the wires carrying the electricity, a dangerous situation arises. When that happens, and you touch the normally neutral metal which is now "hot" (carrying current) because of the accidental contact, your body becomes the path for the electric current. It is the flow of the electric current through your body that is dangerous.

Approximate limits the body can stand:

1 mA is detectable.

8 mA causes shock and surpise.

15 mA causes muscular freeze.

75 mA can cause death.

NOTES The higher the voltage, the greater the number of milliamperes that can pass through your body under identical conditions. It is therefore more dangerous to touch a 230-V line than a 115-V line.

The effectiveness of your contact with the hot wire and with the ground determines the number of milliamperes that pass through your body.

When you stand on *dry* surfaces, the contact is poor and few milliamperes pass through you.

When you stand on wet surfaces, the contact is much better and many milliamperes pass through you.

CAUTION *Avoid working with hot electric wires while standing on wet surfaces.*

Summary of Safety Precautions

Wherever there are electric outlets, plugs, wiring, or connections, there is danger of electric shock.

1. Do not use worn wires, replace them.
2. Use connections that are encased in heavy rubber.
3. Ground all apparatus: use three-prong plugs or pigtail adapters.
4. Do not handle any electric connections with damp hands or while standing on a wet or damp surface.
5. Do not continue to run motors after there has been a spill on the motor.

Importance of Grounding

Electric equipment is grounded for safety. Improperly grounded items can cause injury, fires, damage to other units, and death. When you refer to *ground*, you are referring to the earth, which is neutral and therefore safe. A "grounded piece of equipment" means that electrically the unit is at the same electric potential as the earth, which is what you want.

Grounding wires, called "grounds," are electric-wire connections to a point such as a water pipe buried in the ground or a ground rod which has been driven into the ground. Ground wires normally carry no electric current. They are connected to a metal component of the device, which itself is electrically connected to the motor, switch, case, cabinet, motor shell, base, or outlet box.

Properly Grounded Items

Should a motor become defective, the insulation breaks, and the hot electric wire touches the casing. The casing becomes hot. Because the item is grounded properly, the current flows through the grounding wire to ground and not through you (Fig. 27-38).

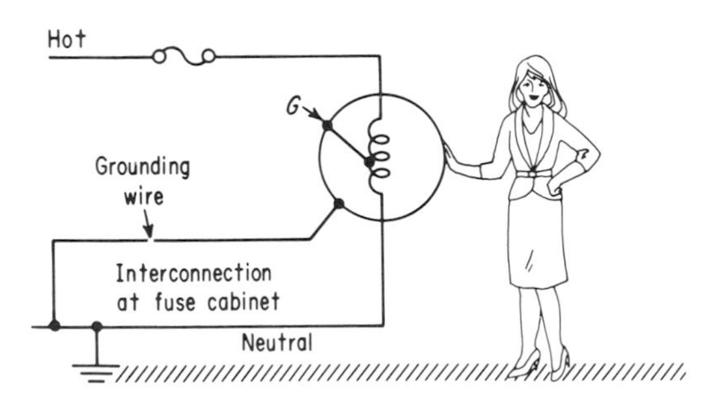

FIGURE 27-38
Safe installation, grounded machine.

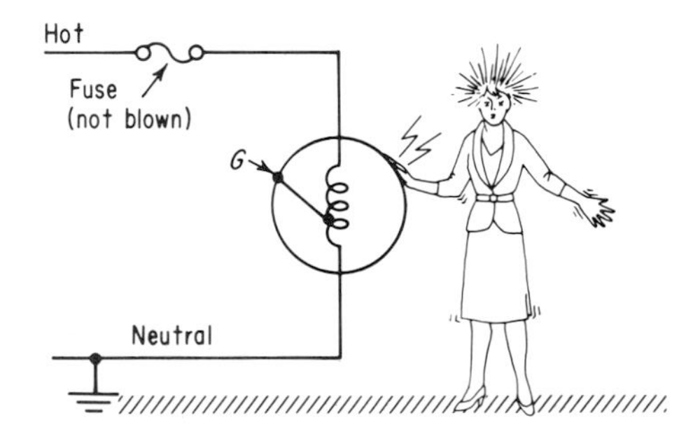

FIGURE 27-39
Dangerous installation, no electric ground.

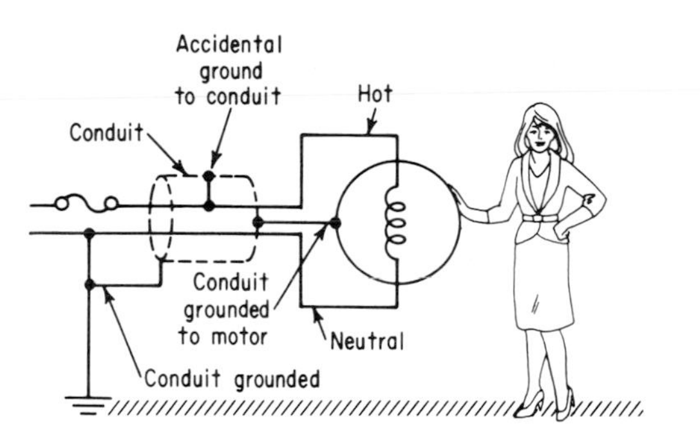

FIGURE 27-40
Safe installation, properly grounded.

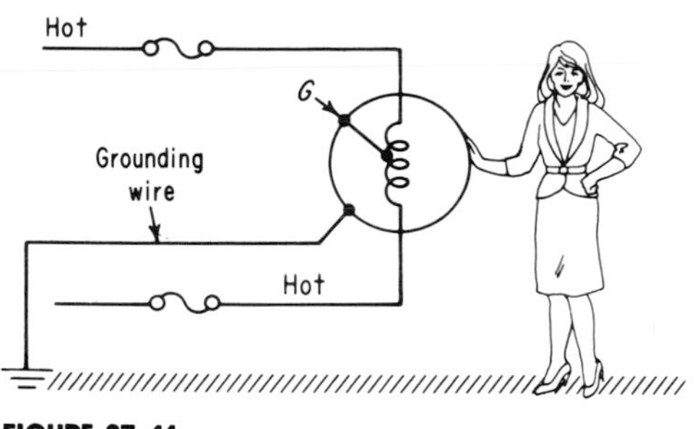

FIGURE 27-41
Safe installation, properly grounded.

Improperly Grounded Items

As stated above, the casing became hot, but because there is no ground, your body becomes the conductor when you touch the casing (Fig. 27-39), and electric shock results, causing injury or death.

Proper Grounding

1. Merely to attach an electric ground to a piece of apparatus is not sufficient. Be *certain* that *every* metal component in the apparatus is electrically grounded. Many times, the mechanical connections and fittings on equipment may themselves be insulators and thus will insulate the other parts from the electrical ground.

2. Be *certain* that the electric ground connection is *actually* an electric ground and not itself insulated from ground. *Consult an electrician* to determine what is and what is not a "true ground." Then *use that ground.*

EXAMPLE An electric motor may be grounded at the switch box but not be grounded itself. Should the motor become defective, with insulation breaking on the wires or coils, it will become hot, because it is insulated from the ground. This is a dangerous situation. Safe installations are shown in Figs. 27-40 and 27-41.

How to Determine If Your Equipment is Grounded

1. Complete your assembly.

2. Electrically disconnect your equipment from the power source.

3. Obtain a multimeter and set the selector to resistance, at the zero range.

4. Firmly press one probe onto your previously determined safe electric ground.

5. Then touch all pieces of metal of your assembly with the other probe. *All readings should be zero, showing that all are electrically connected to each other.*

6. Remove your multimeter, and connect your assembly to electric power.

7. Adjust the selector switch to the voltage used, for example 115 V ac, on your multimeter.

8. Firmly press one probe of the multimeter to the electric ground, and then press the other probe to all metal surfaces. *All readings should be zero volts.*

9. Your equipment is electrically safe under normal conditions for use.

Receptacle Grounding

Most modern electric outlets are fitted with three openings, two for the electric wires and one for the grounding wire. These outlets look like Fig. 27-42.

CAUTION Just because the outlet looks as if it is a grounded outlet, do not assume it has been electrically grounded. *Always test.*

Older installations have only the two-opening receptacle (Fig. 27-43), and you must affix a grounding wire.

If your electrical device is fitted with a three-prong plug but your outlet receptacles are only the two-opening type, you first must use an adapter

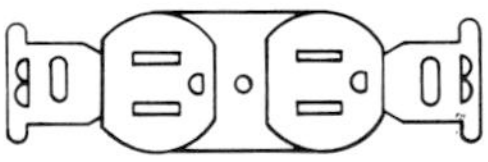

FIGURE 27-42
This new type of electric outlet has a ground.

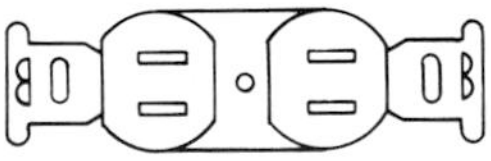

FIGURE 27-43
The old type of electric outlet has no ground.

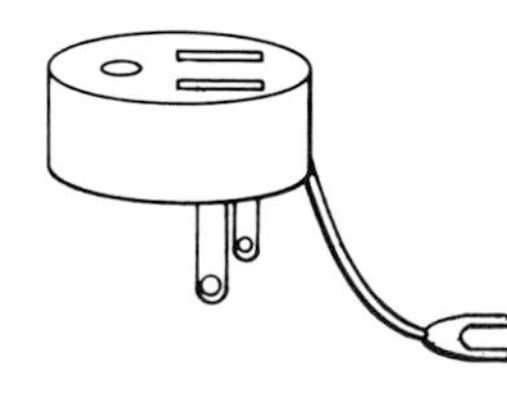

FIGURE 27-44
Adapter for using a three-prong electric plug in a two-opening outlet.

FIGURE 27-45
Adapter plug (bottom view).

TABLE 27-5
Standard Electric-Circuit Symbols

Element	*Symbol*
Connecting wire (negligible resistance)	
Resistance (fixed)	R
Variable resistance (two terminals) such as a dial box	
Rheostat or potentiometer (three terminals)	
Capacitance (fixed)	
Capacitance (variable)	
Inductance	
Inductance (iron core) or choke	
Transformer (or mutual inductance)	
Transformer (iron core)	
Autotransformer	
Battery (two or more cells in series)	+
Voltaic cell	+
Fuse	
Switch [single-pole single-throw (SPST)]	
Switch [double-pole double-throw (DPDT)]	
Tap key (SPST momentary contact)	
Galvanometer	G
Ammeter	A
Milliammeter	MA
Voltmeter	V
Wattmeter	W

which converts from three-plug to two-plug. This adapter has a "grounding wire" hanging from it. *Connect the grounding wire to ground. Under no circumstances* should you ever break off the grounding-plug prong of the plug, merely to avoid locating an adapter. (See Figs. 27-44 and 27-45.)

ELECTRIC-CIRCUIT ELEMENTS

Standard symbols for the various elements encountered in electric circuits are given in Table 27-5.

28 ELECTROCHEMISTRY

OXIDATION-REDUCTION REACTIONS

Many chemical reactions can be classified as oxidation-reduction reactions (redox reactions) and can be considered as the resultant of two reactions, one oxidation and the other reduction. An element is said to have undergone oxidation if it loses electrons or if its oxidation state has increased—that is, it has attained a more positive charge. An element is said to have undergone reduction if it gains electrons or if its oxidation state has been reduced—that is, it has attained a more negative charge. Atoms of elements in their elemental state have a zero charge.

A typical reaction occurs if a piece of metallic zinc is dropped into a solution of lead nitrate: The zinc atoms (zero charge) lose electrons to become zinc ions with a +2 charge; they are oxidized. Lead ions (with a +2 charge) become metallic lead with a zero charge; they are reduced.

$$Zn(s) + Pb^{2+}(aq) \rightarrow Zn^{2+}(aq) + Pb(s)$$

The overall reaction can be considered the result of two half-reactions:

$Zn(s) \rightarrow Zn^{2+}(aq) + 2$ electrons   oxidation reaction

$Pb^{2+}(aq) + 2$ electrons $\rightarrow Pb(s)$ reduction reaction

Lead is a *stronger oxidizing agent* than zinc because, in the reaction that took place, the lead has a stronger attraction for the electrons and takes them away from the zinc. By the same reasoning, the zinc is a stronger reducing agent than the lead because it releases its electrons more easily. Oxidation-reduction reactions involve competition of substances for electrons. Each substance has it own characteristic affinity for electrons, and that affinity is the basis for constructing the electrochemical-reaction or electromotive series table, which reflects that affinity in terms of volts. (See Table 28-1.)

The relative tendencies of substances to gain or lose electrons has made it possible to arrange the various substances in a series of "oxidation-reduction couples," a table which reflects the relative affinity for electrons between an element and its ion, such as Li-Li^+, K-K^+, Cs-Cs^+, and so on. A substance may act as an oxidizing agent toward one substance, and a reducing agent toward another, depending upon its relative position in the series. Any oxidation-reduction reaction has the possibility of taking

TABLE 28-1
Oxidation-Reduction Potentials for Substances, Unit Acitivty at 25°C

Couple	*Volts*	*Couple*	*Volts*
$Li \longrightarrow Li^+$	+3.04	$Sn^{2+} \longrightarrow Sn^{4+}$	−0.15
$K \longrightarrow K^+$	+2.92	$Cu^+ \longrightarrow Cu^{2+}$	−0.15
$Cs \longrightarrow Cs^+$	+2.92	$Cu \longrightarrow Cu^{2+}$	−0.34
$Ba \longrightarrow Ba^{2+}$	+2.90	$Cu \longrightarrow Cu^+$	−0.52
$Sr \longrightarrow Sr^{2+}$	+2.89	$I^- \longrightarrow I_2$	−0.54
$Ca \longrightarrow Ca^{2+}$	+2.87	$U^{4+} \longrightarrow UO_2^+$	−0.62
$Na \longrightarrow Na^+$	+2.71	$H_2O_2 \longrightarrow O_2(H^+)$	−0.67
$Mg \longrightarrow Mg^{2+}$	+2.37	$Fe^{2+} \longrightarrow Fe^{3+}$	−0.77
$Al \longrightarrow Al^{3+}$	+1.66	$Hg \longrightarrow Hg_2^{2+}$	−0.79
$Mn \longrightarrow Mn^{2+}$	+1.18	$Ag \longrightarrow Ag^+$	−0.80
$Zn \longrightarrow Zn^{2+}$	+0.76	$Hg_2^{2+} \longrightarrow Hg^{2+}$	−0.92
$Cr \longrightarrow Cr^{3+}$	+0.74	$NO \longrightarrow NO_3^-(H^+)$	−0.96
$U^{3+} \longrightarrow U^{4+}$	+0.61	$Br^- \longrightarrow Br_2$	−1.07
$Ga \longrightarrow Ga^{3+}$	+0.53	$H_2O \longrightarrow O_2(H^+)$	−1.23
$Fe \longrightarrow Fe^{2+}$	+0.44	$Mn^{2+} \longrightarrow MnO_2(H^+)$	−1.23
$Eu^{2+} \longrightarrow Eu^{3+}$	+0.43	$Cr^{3+} \longrightarrow Cr_2O_7^{2-}(H^+)$	−1.33
$Cr^{2+} \longrightarrow Cr^{3+}$	+0.41	$Cl^- \longrightarrow Cl_2$	−1.36
$Cd \longrightarrow Cd^{2+}$	+0.40	$Au \longrightarrow Au^{3+}$	−1.50
$In \longrightarrow In^{3+}$	+0.34	$Mn^{2+} \longrightarrow Mn^{3+}$	−1.51
$Co \longrightarrow Co^{2+}$	+0.28	$Mn^{2+} \longrightarrow MnO_4^-(H^+)$	−1.51
$Ni \longrightarrow Ni^{2+}$	+0.25	$IO_3 \longrightarrow H_5IO_6$	−1.60
$Sn \longrightarrow Sn^{2+}$	+0.14	$MnO_2 \longrightarrow MnO_4^-(H^+)$	−1.70
$Pb \longrightarrow Pb^{2+}$	+0.13	$H_2O \longrightarrow H_2O_2(H^+)$	−1.77
$H_2 \longrightarrow H^+$	0.00	$F^- \longrightarrow F_2$	−2.87
$H_2S \longrightarrow S(H^+)$	−0.14	$HF \longrightarrow F_2(H^+)$	−3.06

place, but the electrochemical reaction table makes no predictions or guarantees that the rate of the reaction will be practical. Some will take place at too slow a rate.

We cannot determine absolute electrode potentials; therefore all electrode potentials are based on an arbitrary standard, the standard hydrogen electrode (Fig. 28-1). The arbitrary potential of the standard hydrogen electrode or the normal hydrogen electrode is assigned a value of 0 V. Three conditions govern the acceptance of this value: a temperature of 25°C, a hydrogen ion concentration of 1 mol/L (unit molarity), and a partial pressure of 1 atm of hydrogen gas at the surface of the platinum electrode. These conditions must be maintained if the normal hydrogen electrode is used for reference purposes. To achieve the largest possible

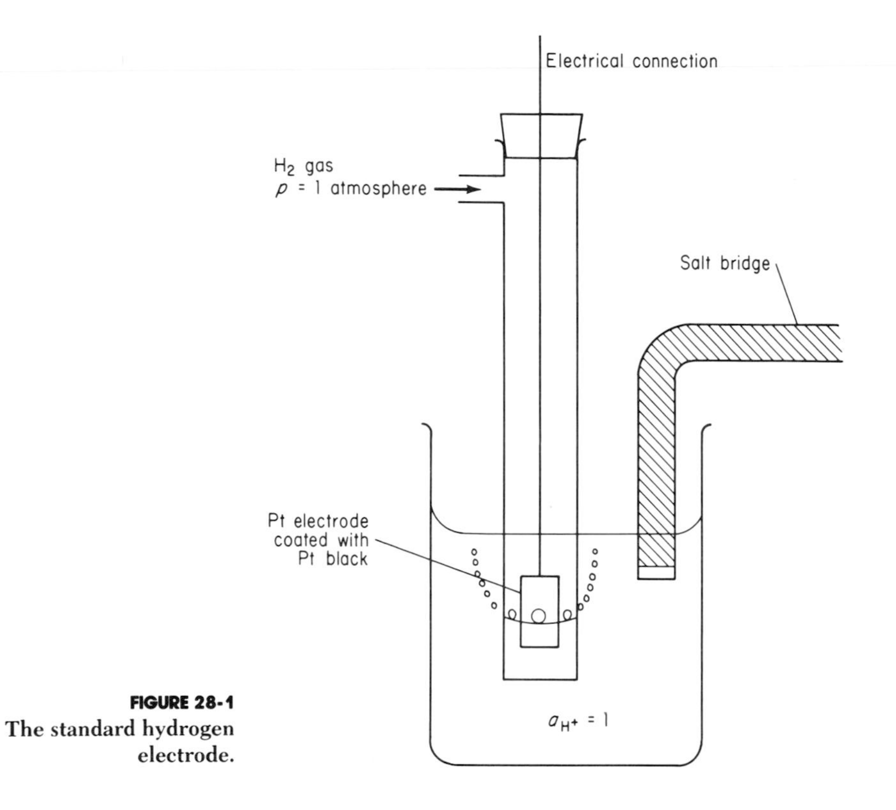

FIGURE 28-1
The standard hydrogen electrode.

surface area, the platinum electrode is coated with a layer of finely divided platinum, known as *platinum black*.

The electromotive force, emf (in volts), of the total reaction can be considered to be the sum of the emf of the oxidation reaction and the emf of the reduction reaction.

$$E_{\text{oxidation–reduction reaction}} = E_{\text{oxidation reaction}} + E_{\text{reduction reaction}}$$

ACTIVITY OF METALS

Metals vary in their activity, or in the ease with which they give up electrons to form ions. Some metals are extremely difficult to change into their corresponding ions, while others are extremely easy to convert. Lithium is the most active metal. The activity decreases with each succeeding metal. Metals above hydrogen yield hydrogen when treated with an acid. Those below hydrogen are unreactive with acids (unless the acids are oxidizing agents as well). Potassium, sodium, and calcium are so active that they react with cold water. Zinc, iron, tin, and lead react with water if it is in the form of steam. Tin and lead do not react with water. Metals in

TABLE 28-2
Relative Activities of Metals

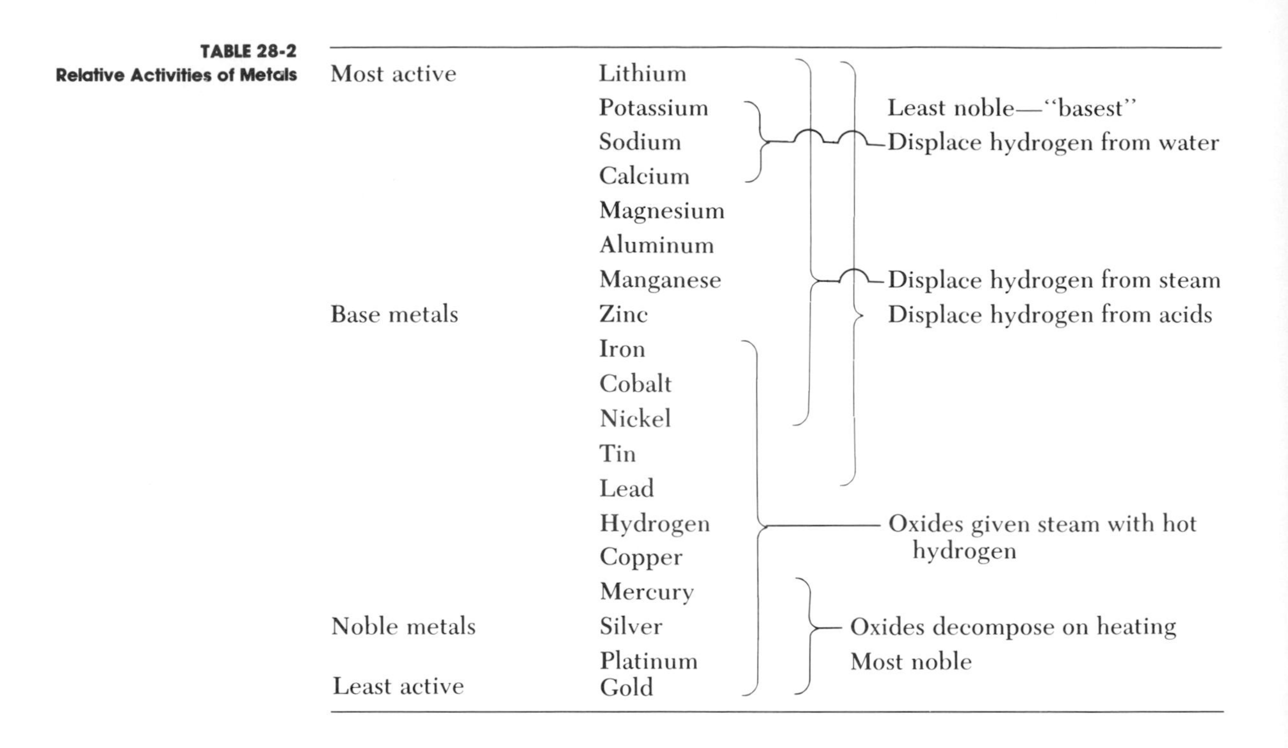

	Metal	
Most active	Lithium	
	Potassium	Least noble—"basest"
	Sodium	Displace hydrogen from water
	Calcium	
	Magnesium	
	Aluminum	
	Manganese	Displace hydrogen from steam
Base metals	Zinc	Displace hydrogen from acids
	Iron	
	Cobalt	
	Nickel	
	Tin	
	Lead	
	Hydrogen	Oxides given steam with hot hydrogen
	Copper	
	Mercury	
Noble metals	Silver	Oxides decompose on heating
	Platinum	Most noble
Least active	Gold	

the series will displace a less active metal from its salt. When copper sulfate is treated with metallic zinc, the zinc displaces the copper, forming zinc sulfate and copper, thus:

$$CuSO_4 + Zn \rightarrow Cu + ZnSO_4$$

See Table 28-2. Any metal above hydrogen in the table will react with hydrogen ions to liberate hydrogen gas and form the metal ion. Those below hydrogen in the table will not react (under normal conditions) with the metal; they are inert to the hydrogen ion. Similarly, any metal which is above another metal will displace that metal in solution. In other words, any metal which is higher in the table will oxidize any metal which is below it.

Prevention of Corrosion

The relative activities of metals is the basis for using *cathodic protection* to prevent corrosion of metals in contact with moisture. This method involves the use of a more active metal electrically coupled to the apparatus. The more active metal will preferentially give up electrons more easily, reacting instead of the apparatus, thus protecting the metal of the apparatus (Fig. 28-2). The active metal, when consumed, can be easily replaced.

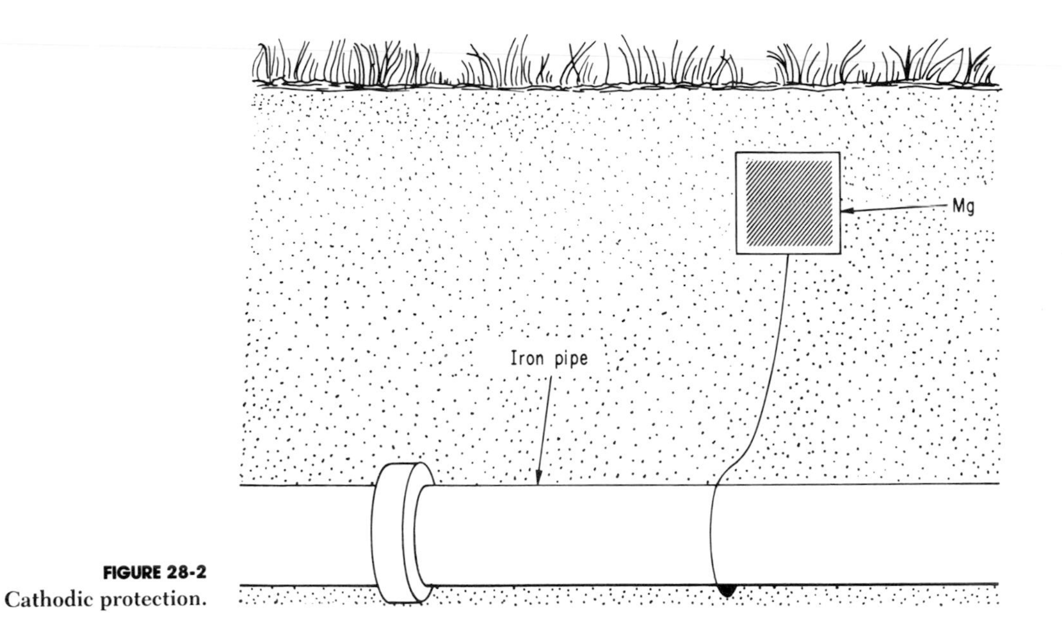

FIGURE 28-2
Cathodic protection.

NOTE Refer to the activity table of the elements (Table 28-2) to choose a more active element than that element from which the apparatus is made.

Corrosion of metal apparatus and equipment can also be prevented by:

1. Painting the surface to exclude air and moisture
2. Coating surfaces with a thin film of oil
3. Electroplating
4. Galvanizing (zinc coating—hot dip)
5. Tinning (tin coating—hot dip)
6. Coating with transparent self-adhesive tape or lacquer

OXIDATION NUMBERS

The oxidation number is sometimes called the *oxidation state.* The basic rules to remember are:

1. Elements in the free state have an oxidation number of zero.

2. In simple ions, those which contain one atom, the oxidation number is equal to the charge on the ion.

3. In the majority of compounds which contain oxygen (except perox-

ides), the oxidation number of each oxygen atom is −2. In peroxides, the oxidation number of oxygen equals −1.

4. In the majority of compounds which contain hydrogen, the oxidation number of hydrogen is +1, except in hydrides, where it is −1.

5. In molecules, which are neutral, the sum of the positive charges equals the sum of the negative charges.

6. In radicals, complex ions, the sum of all the oxidation numbers making up the radical must equal the sum of the charge on the ion.

Redox reactions always involve a change in the valence or the oxidation number of *two* elements. That element which is oxidized attains a more positive (a less negative) charge, and the element which is reduced always attains a more negative (a less positive) charge. This change takes place with the transfer of one or more electrons from one element to another. The element accepting the electrons is reduced, the element donating the electrons is oxidized. The tendency for oxidation-reduction reactions to occur can be measured with a voltmeter, provided the reactions occur in separate regions connected by a barrier porous to ions. Table 28-3 illustrates some of these relations.

EQUIVALENT MASS IN REDOX REACTIONS

In acid-base reactions, the equivalent mass of the substance is equal to the molecular mass divided by the number of available hydrogen or hydroxyl groups; it designates the relative amounts of substances which are chemically equivalent, that is, the amount of substance which will just react with or replace another in a chemical reaction. In redox reactions, the equivalent mass involves the transfer of one mole of electrons. Thus, the redox equivalent weight of a substance can be calculated by dividing its molecular mass (formula mass) by the number of electrons which it loses or gains in the half-reaction which takes place. See Table 28-4.

A 1 *N* oxidizing solution or a 1 *N* reducing solution has one equivalent mass (redox equivalent mass) of the reactant in one liter of solution. Solutions which have different redox normalities have proportional amounts of the equivalent mass in one liter of solution.

INDICATORS

In acid-base reactions, the color of the indicator was changed by the addition or the removal of a hydrogen ion. In redox reactions, redox indicators change color as they are oxidized or reduced, and they are selected so that their oxidation-reduction potential is within the range of the particular reaction being conducted.

TABLE 28-3 Change in Oxidation States

Direction	Oxidation number	Acid condition	Basic condition
		Chromium	
OXIDATION ↑ REDUCTION ↓	+6	$Cr_2O_7^{2-}$	CrO_4^{2-}
	+3	Cr^{3+}	$Cr(OH)_4^-$
	+2	Cr^{2+}	$Cr(OH)_2$
	0		Cr (metal)
		Manganese	
OXIDATION ↑ REDUCTION ↓	+7	MnO_4^-	MnO_4^-
	+6	MnO_4^{2-}	MnO_4^{2-}
	+4	MnO_2	MnO_2
	+3	Mn^{3+}	$Mn(OH)_3$
	+2	Mn^{2+}	$Mn(OH)_2$
	0		Mn (metal)
		Chlorine	
OXIDATION ↑ REDUCTION ↓	+7	Cl_2O_7	ClO_4^-, $HClO_4$
	+5	—	ClO_3^-, $HClO_3$
	+4	ClO_2	— —
	+3	—	ClO_2^-, $HClO_2$
	+1	Cl_3O_7	ClO^-, $HClO$
	0	Cl_2 (gas)	— —
	−1	Cl^-	— —
		Oxygen	
OXIDATION ↑ REDUCTION ↓	0	O_2	—
	−1	H_2O_2	HO_2^-
	−2	H_2O	OH^-

Direction	Oxidation number	Products
		Nitrogen
OXIDATION ↑ REDUCTION ↓	+5	HNO_3, NO_3^-, N_2O_5 (gas)
	+4	NO_2 (gas), N_2O_4
	+3	HNO_2, NO_2^-, N_2O_3 (gas)
	+2	NO (gas)
	+1	N_2O (gas)
	0	N_2 (gas)
	−1	NH_3OH^+, NH_2OH
	−2	N_2H_4 (gas) $N_2H_5^+$
	−3	NH_3 (gas), NH_4OH, NH_4^+
		Sulfur
OXIDATION ↑ REDUCTION ↓	+6	HSO_4, SO_4^{2-}, H_2SO_4, SO_3
	+4	HSO_3^-, SO_3^{2-}, H_2SO_3, SO_3^{2-}
	+2	$S_2O_3^{2-}$ (the average)
	0	S_8
	−2	H_2S, HS^-, S^{2-}

TABLE 28-4
Equivalent Masses in Redox Reactions

Reactant	*Product*	*Formula mass*	÷ *electron change*	= *equivalent mass*
MnO_2	Mn^{2+}	86.94	2	43.47
HNO_3	NO_2	63.02	1	63.02
$KMnO_4$	Mn^{2+}	158.04	5	31.61
HNO_3	NO	63.02	3	21.01
$FeSO_4$	$Fe_2(SO_4)_3$	151.91	1	151.91
$K_2Cr_2O_7$	$2Cr^{3+}$	294.19	6	49.03
H_2S	S^{2-}	34.08	2	17.04

THE VOLTAIC CELL

In redox reactions, the electrons are transferred from the oxidizing agent to the reducing agent by actual contact in the solution. In the voltaic cell (Fig. 28-3), the electrons are transferred through wires so that electrical work can be done. The strength of an electric cell, the voltaic cell, depends upon the relative strengths of the redox agents and their tendencies to attract or release electrons, and that driving force is measured in volts (V).

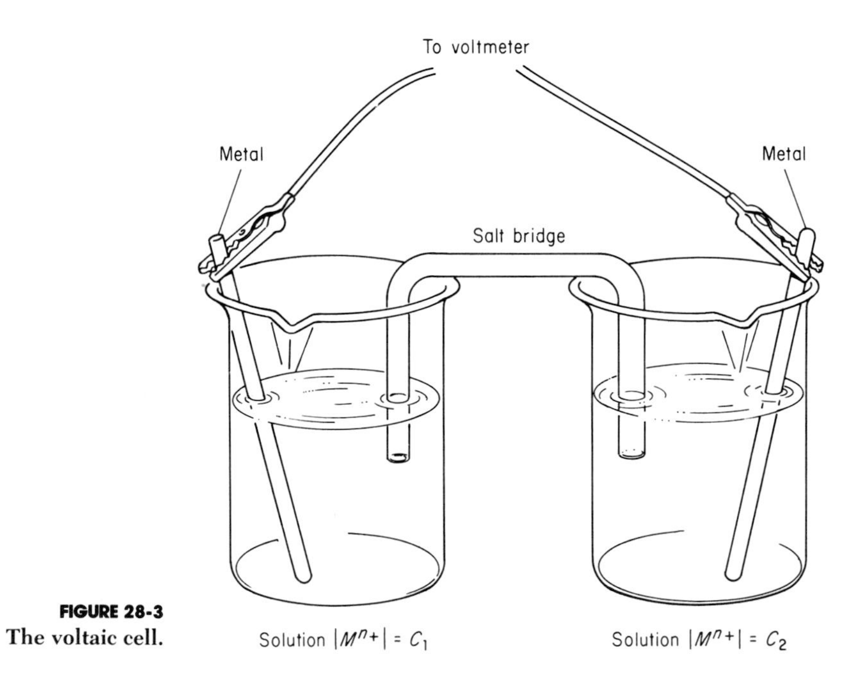

FIGURE 28-3
The voltaic cell.

Effect of the Concentration of the Ions on the emf of the Cell

Cell potentials are related to unity concentrations, where the molarities are 1. However, the concentrations of the reactants, the solute concentrations, are important and their effect on the emf of the oxidation-reduction reaction at 25°C can be expressed by a relation called the *Nernst equation:*

$$E_{cell} = E^\circ_{cell} \pm \frac{0.06}{n} \log \frac{[\text{product ion}]}{[\text{reactant ion}]}$$

where E° is the standard cell potential (at unit concentration)
n is the number of electrons in either reaction
concentrations of the product and reactant ions are expessed in moles per liter

---

ELECTROLYSIS

When direct current is applied to electrodes and the ions migrate toward the oppositely charged electrodes, a chemical reaction takes place, called electrolysis. Metal ions are reduced to the elemental metal when they accept electrons:

$$Cu^{2+} + 2 \text{ electrons at the cathode} \rightarrow \underset{\text{metal}}{Cu^0}$$

$$Ag^+ + 1 \text{ electron} \rightarrow \underset{\text{metal}}{Ag^0}$$

This reduction of the cation to the elemental metal is the basis for electroplating procedures. Any metal ion can be deposited on a cathode in a thin layer. Concentrations of ions, temperature, voltage, and current density (the relationship of the current to the area of the object being plated) must be carefully adjusted and controlled for optimum electroplated finishes.

At the anode, which is positively charged, the anions lose their electrons to form molecules:

$$2Br^- \rightarrow \underset{\text{molecules}}{Br_2} + 2 \text{ electrons at the anode}$$

Electrolytic Separation of Substances

In the process of electrolysis, an externally applied emf causes oxidation to take place at the anode, because of its deficiency of electrons, and reduction to take place at the cathode, because of the excess of electrons. The cathode (negative electrode) will react most completely with the strongest oxidizing agent in the solution with which it is in contact. In a solution of ionic salts of various metals—silver, gold, and copper—the gold ion will be reduced first. It is the strongest oxidizing agent, and all the gold will be converted to the metal form. Silver is next in line (refer to Table 28-1), and the silver ions will then be converted into metallic silver. Finally, the copper, which is the least powerful of these three, will be reduced. However, if, in the experimental setup, a sufficiently high voltage to plate out copper is initially applied (with an extremely high current density), then even the silver and copper will be reduced along with the gold.

CALOMEL ELECTRODE

The hydrogen electrode is a cumbersome and awkward electrode to use in the laboratory. Another, more convenient electrode that can be used as a standard to make measurements is the calomel electrode (refer to Chap. 26, "pH Measurement") which has a standard potential of 0.2415 at 25° C. The calomel electrode consists of metallic mercury which is in contact with mercury(I) ions. The emf of the metallic mercury depends upon the concentration of the mercury(I) ions, Hg_2^{2+}, which are produced by saturating a KCl solution with Hg_2Cl_2 (calomel).

$$\tfrac{1}{2}\, Hg_2Cl_2 + e \rightarrow Hg^0 + Cl^-$$

CONDUCTIVITY

Pure water is a nonconductor of electricity; however, when certain substances are added to water, it does conduct electric current. These substances are called electrolytes. They dissociate in water into positive and negative ions which will carry the electric current. The ability of solutions of electrolytes to carry electric current is called *electrolytic conduction*.

When a voltage is applied to electrodes immersed in the solution (Fig. 28-4), the positive ions (cations) are attracted to the negative electrode (cath-

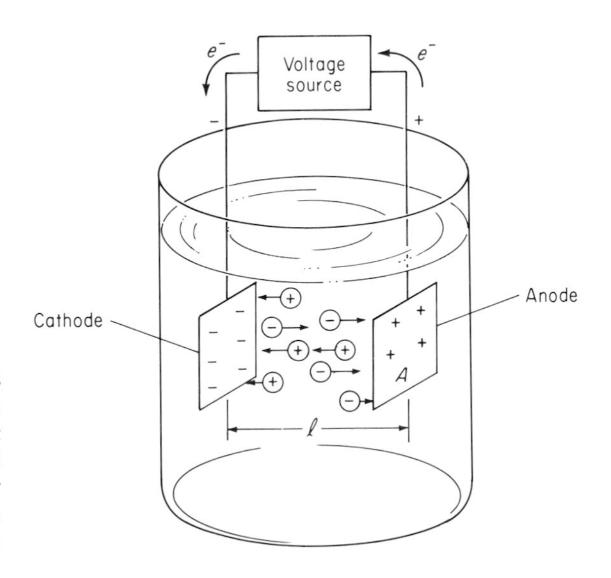

FIGURE 28-4
Migration of ions in solution: A and l represent the cross-sectional area and distance of separation for the electrodes, respectively.

ode) while the negative ions (anions) are attracted to the positive electrode (anode).

Substances can be divided into two classes, those which conduct electricity (electrolytes) and those which do not (nonelectrolytes). Some electrolytes are composed totally of ions, even in the pure state (not in solution), and when the ions dissolve, they merely separate in the solution. Other compounds are nonionic in the pure state; they are covalent and therefore do not conduct electricity when they are in the pure state. However, some of these covalent compounds are attacked by water and then "ionize" in water, becoming conductors of electricity. Hydrogen chloride is such an example:

$$H_2O + HCl \rightleftharpoons H_3O^+ + Cl^-$$

$$\underset{\displaystyle H}{H:\ddot{\underset{\cdot\cdot}{O}}:} + H:\ddot{\underset{\cdot\cdot}{Cl}}: \rightleftharpoons \underset{\displaystyle H}{H:\ddot{\underset{\cdot\cdot}{O}}:H^+} + :\ddot{\underset{\cdot\cdot}{Cl}}:^-$$

Hydronium ions and chloride ions carry the electric current.

Electrolytes may be divided into two classes, those which ionize completely (strong electrolytes) and those which ionize only partially (weak electrolytes). The magnitude of the current carried through a solution depends upon the number of ions present, which therefore depends upon the concentration of the electrolyte and the degree of ionization (Fig. 28-5).

The flow of current depends upon the magnitude of the applied voltage and the resistance of the solution between the electrodes, as expressed by Ohm's law:

$$\text{Current (in amperes)} = \frac{\text{voltage (in volts)}}{\text{resistance (in ohms)}}$$

Instrumentation

Conductance values are obtained from resistance measurements made with a conductivity bridge (a Wheatsone bridge) by actually measuring the electrical resistance of solutions (Fig. 28-6).

Samples are usually placed in a conductivity cell, which consists of two platinum electrodes covered with a colloidal deposit of platinum (platinum black) to provide a large surface area, the electrodes being insulated from each other (Fig. 28-7). The platinum black facilitates the absorption of any reaction products, preventing them from accumulating in the solution and affecting the cell resistance.

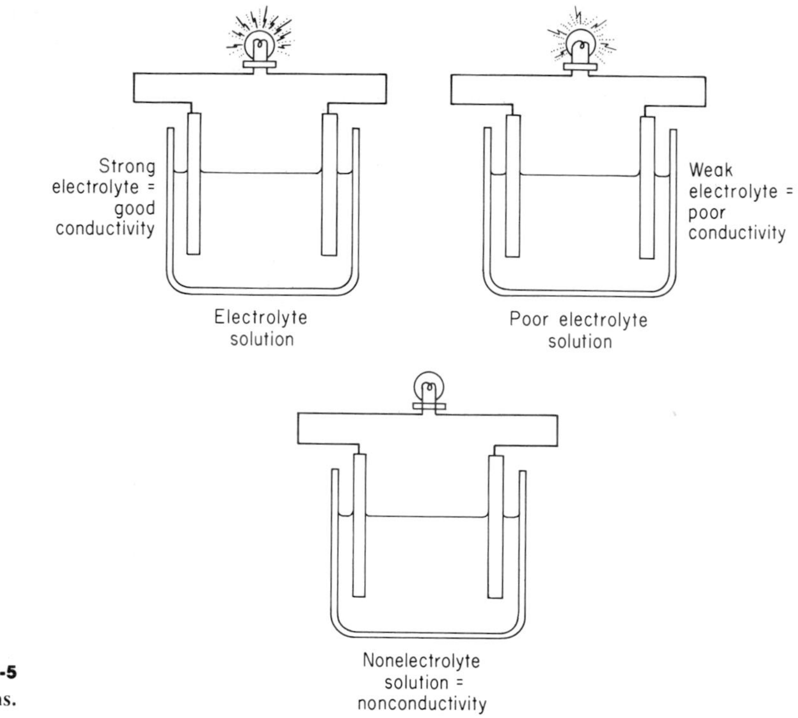

FIGURE 28-5
Conductivity of solutions.

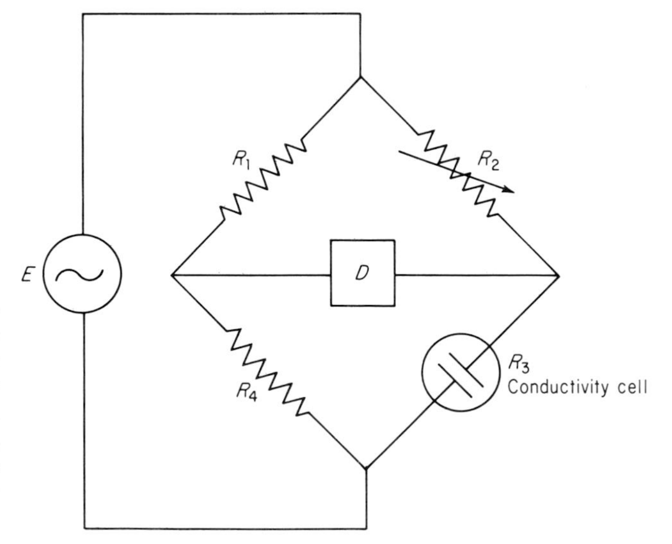

FIGURE 28-6
Schematic diagram of an ac Wheatstone bridge circuit for conductivity measurements. R_1, R_2, and R_4 are known resistances; E is the source of ac voltage; D is the detector (meter).

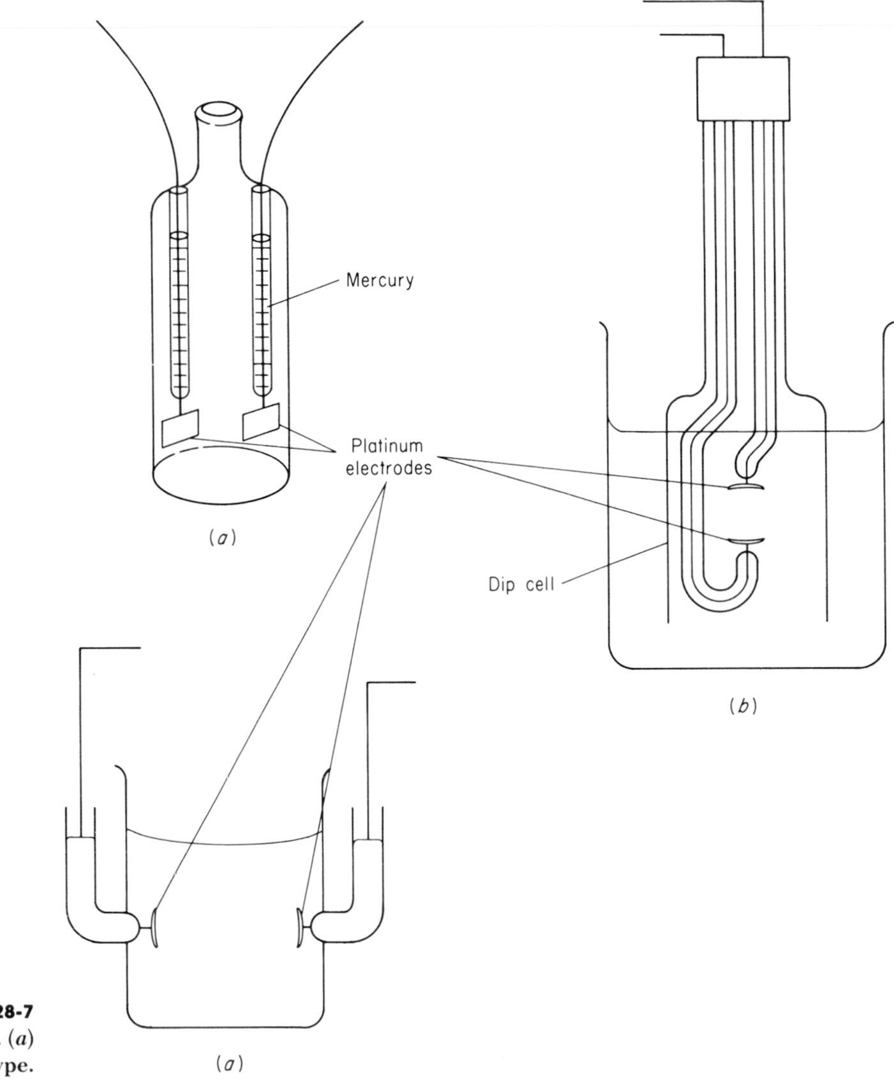

FIGURE 28-7
Conductivity cells. (*a*) Freas type. (*b*) Dip type.

NOTE Extremely pure water is required for accurate conductivity studies (refer to Chap. 24).

The reciprocal of the resistance R is the conductance L, which is directly proportional to the cross-sectional area A of the electrodes and inversely proportional to the distance l between them (Fig. 28-2):

$$\frac{1}{R} = L = K\frac{A}{l}$$

where R is resistance, in ohms

L is conductance, in reciprocal ohms (ohm^{-1}, or mho) or siemens (S)

K is specific resistance in $ohm^{-1}\ cm^{-1}$ ($\Omega^{-1}\ cm^{-1}$ or S/cm)

The values of L and K increase as electrolyte concentration increases; the value of R decreases.

Specific Conductance

The area-to-length ratio is determined by measuring the resistance of a standard calibrating solution in the conductivity cell. Standard solutions, such as those of KCl, have a known specific conductance, and the area-to-length ratio needed for a cell can be found by the equation

$$L = kK$$

where k is the cell constant, in cm.

A conductivity cell with large electrodes very close together has a low cell constant, for example, 0.1. If the distance is increased and the area of the electrodes is decreased, the constant increases. The choice of a cell having the right cell constant is important for accurate measurement. To measure distilled water (conductance 2 $\mu\Omega$/cm), use a cell having a constant of 0.1. A reading of 50,000 Ω will result, and this value falls within the limits tolerable for industrial measurements of 50 to 100,000 Ω measured resistance.

$$\frac{1{,}000{,}000}{2} \times 0.1 = 50{,}000\ \Omega$$

CONDUCTOMETRIC TITRATIONS

Conductometric titrations are extremely useful and easily performed and are a way to determine the stoichiometry of the titrations. As nondissociated nonconducting molecules are formed in the reaction process, changes in conductance of the solution take place. By monitoring the volume of titrant and plotting that volume against the conductance of the solution, the equivalence or stoichiometric point, that is, the point of minimum conductance, can be easily obtained (Fig. 28-8).

Actually, only a few measurements are necessary, some as the conductance decreases, and some as it increases. The equivalence point can be obtained by extrapolation (Fig. 28-9).

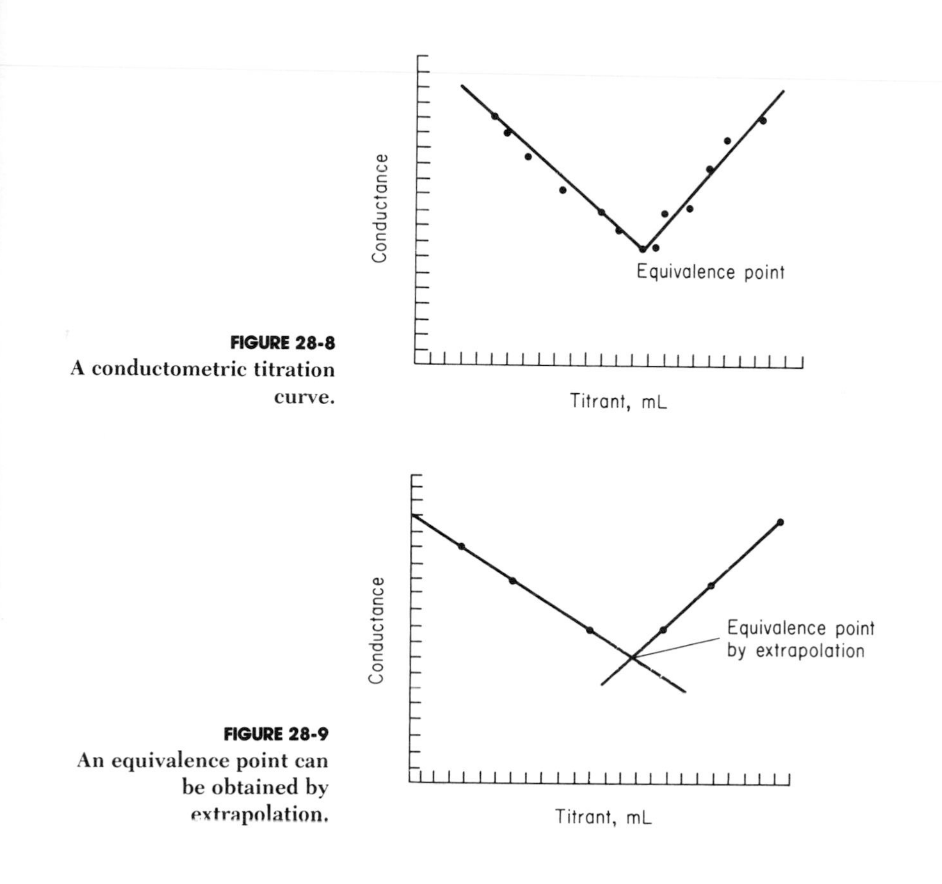

FIGURE 28-8
A conductometric titration curve.

FIGURE 28-9
An equivalence point can be obtained by extrapolation.

COULOMETRY

Principle

There is a direct proportion between the amount of oxidation-reduction occurring at electrodes and the quantity of electricity used. Coulometric analysis or titration relates the quantity of electricity and the amount of chemical substance in solution to result in a determination of that substance (Fig. 28-10).

Basic Techniques

1. The emf of the working electrode is maintained at a fixed value. When current flow approaches zero, the reaction is approaching completion. The chemical coulometric or amperage-time curve measures the quantity of electricity.

2. The amperage is kept constant. Indicators signal the completion of the reaction. The quantity of current is obtained from the amperage and the time.

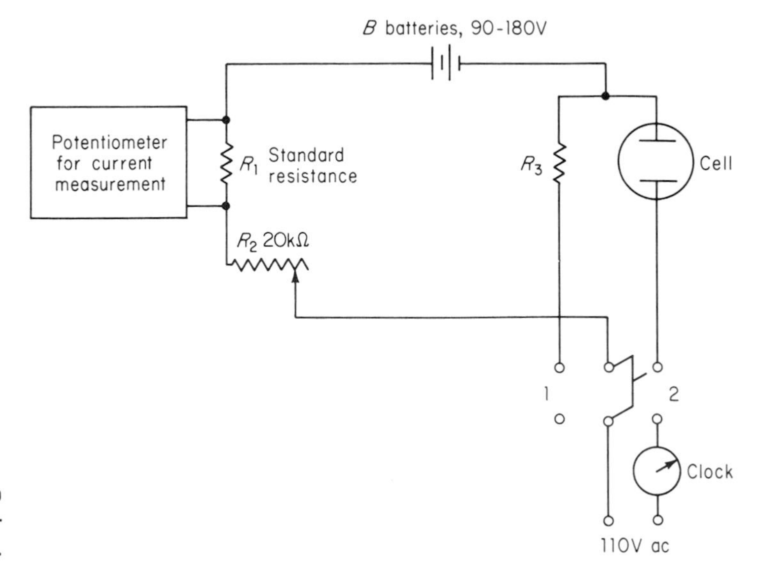

FIGURE 28-10
A simple apparatus for coulometric titrations.

The Chemical Coulometer

All coulometric procedures require operation at 100 percent efficiency: each faraday of electricity must result in the conversion of one equivalent of the substance being tested. The hydrogen-oxygen coulometer (Fig. 28-11) collects hydrogen and oxygen, and their volumes are measured, providing the data needed to calculate the amount of current.

POLAROGRAPHY

Principle

Polarography measures the electric current as a function of the emf applied to a particular type of electrolytic cell consisting of an easily polarized microelectrode (a few square millimeters in area) and a large nonpolarizable electrode. The microelectrode is usually a dropping mercury electrode (Fig. 28-12) which provides a constant stream of identical droplets of mercury. Analyses can be performed on samples with volumes of 1 drop. When the voltage-current curves are plotted, graphs called polarograms (Fig. 28-13) are obtained. These provide the desired qualitative and quantitative information.

Definitions

Some useful definitions follow. (See Fig. 28-12.)

Residual current is the amount of current that passes through a cell in the complete absence of a reducible substance.

Decomposition potential is the voltage at which the electrode-reactive substance leaves the residual-current baseline.

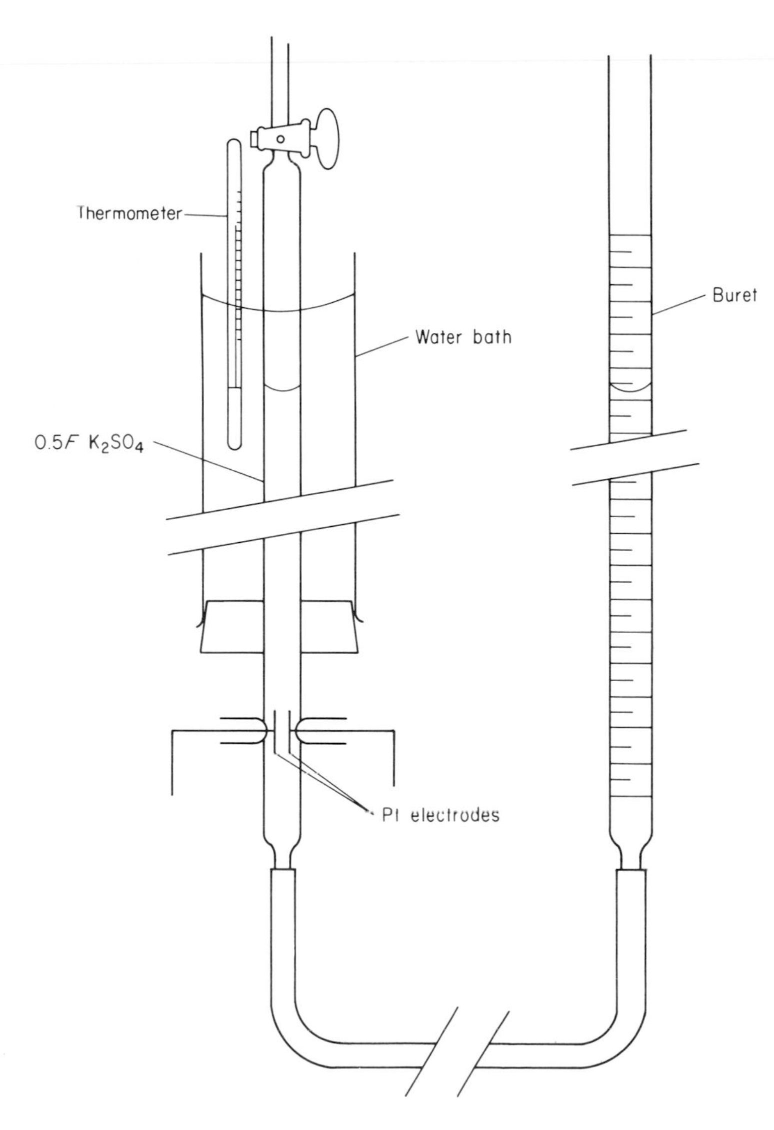

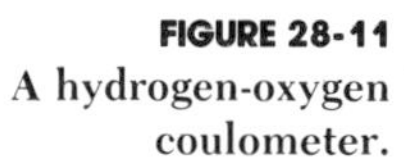
FIGURE 28-11
A hydrogen-oxygen coulometer.

Limiting current is the region where the current becomes independent of the applied emf.

Diffusion current is proportional to the concentration of the reactive component.

Half-wave potential is the emf which corresponds to current equal to one-half of the diffusion current.

Use

About 30 metals may be analyzed at levels of 50 ppb (parts per billion) by a modified polarography technique, differential-pulse polarography

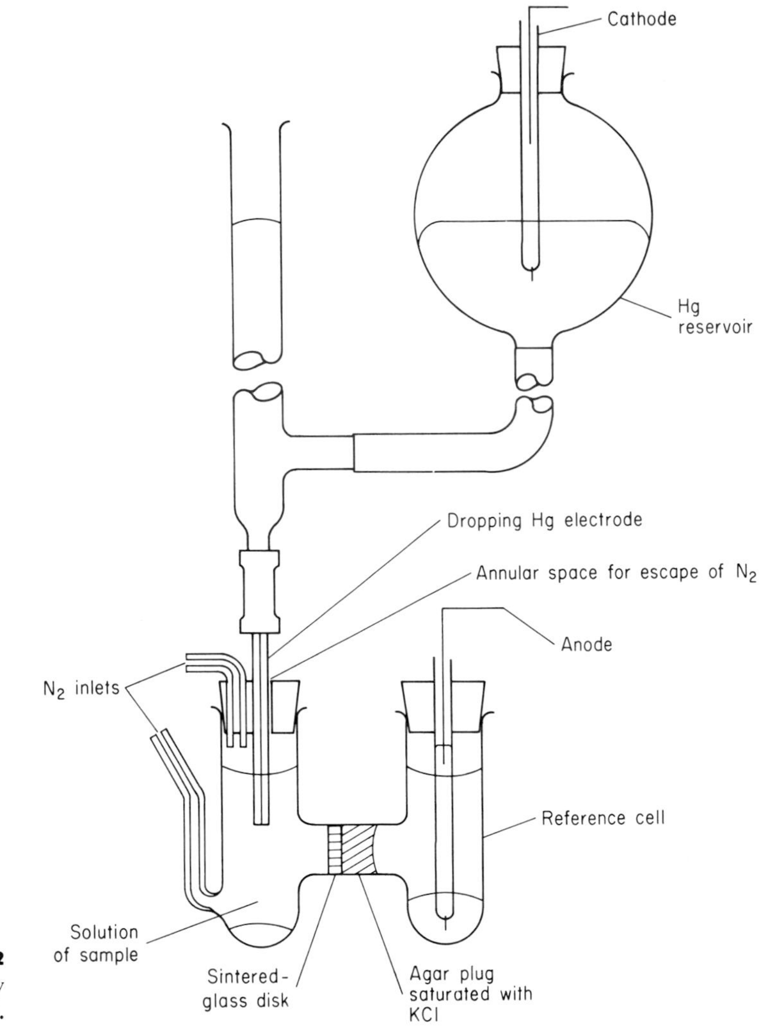

FIGURE 28-12
A dropping mercury electrode and cell.

(DPP). Another variant of polarography, differential-pulse stripping anodic voltametry (DPSAV) can analyze for about 15 of these same 30 metals at levels below 1 ppb.

AMPEROMETRIC PROCEDURES

Amperometric procedures can be used for the determination of the equivalence point, provided at least one of the reactants can be either oxidized or reduced at the dropping mercury electrode (Fig. 28-12). The volume of the titrant is plotted against the current, and, since the data plots on either side of the equivalence point are straight lines, the equivalence point can be found by extrapolation (Fig. 28-14).

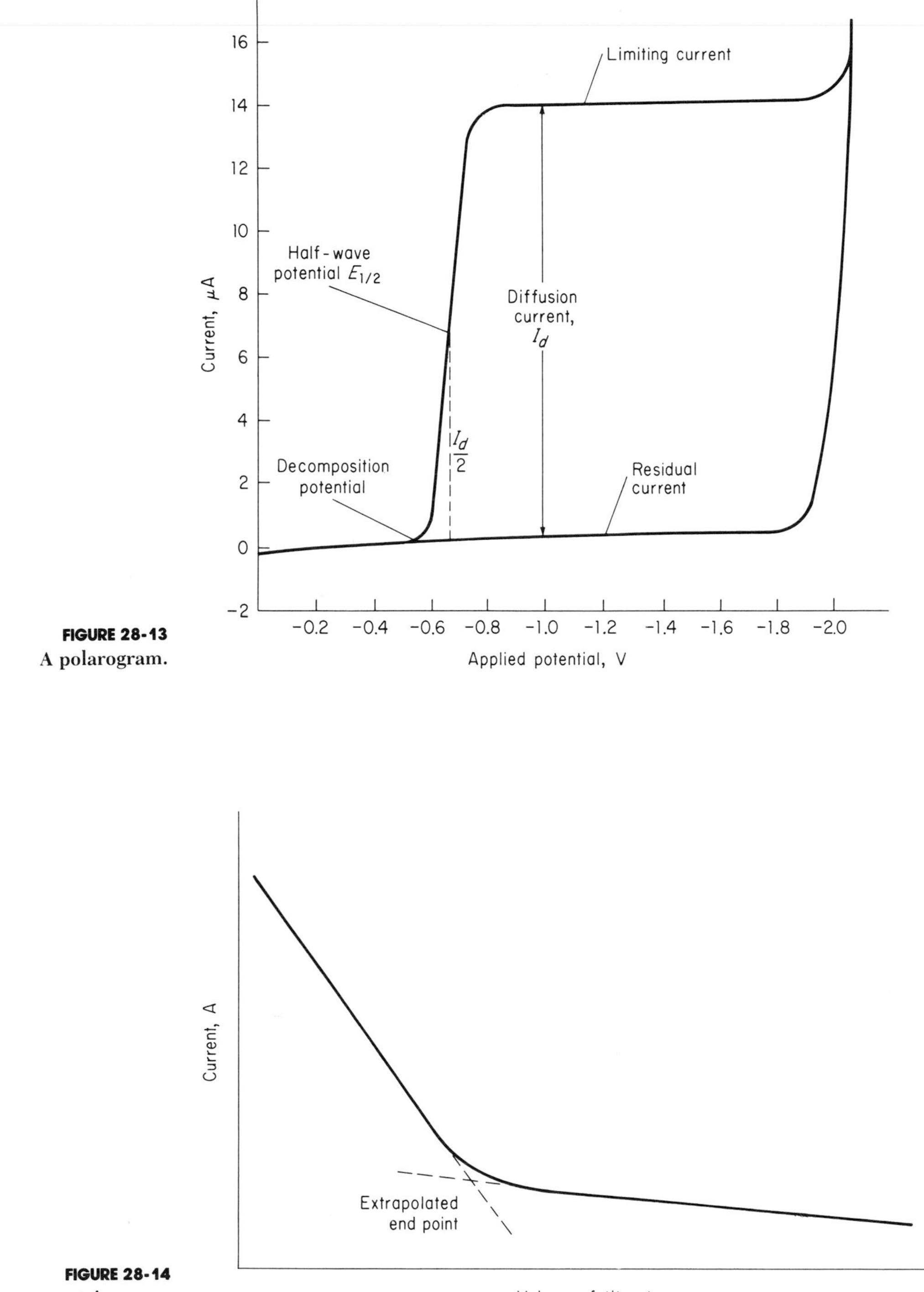

FIGURE 28-13
A polarogram.

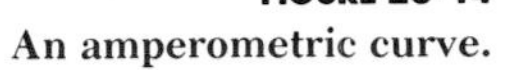

FIGURE 28-14
An amperometric curve.

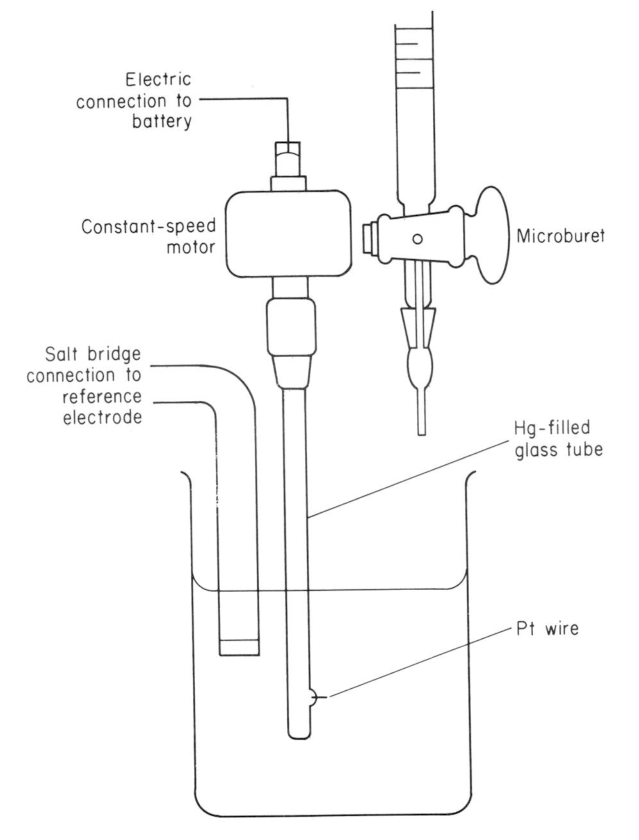

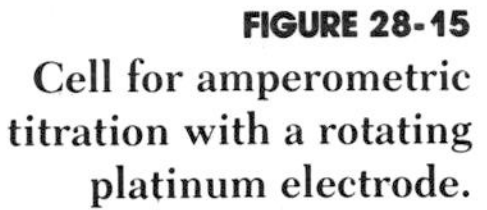

FIGURE 28-15
Cell for amperometric titration with a rotating platinum electrode.

In plotting the amperometric curve (Fig. 28-14) the volume of a titrant is plotted against the current. The reactants can be dissolved in highly concentrated solutions of electrolytes, which do not affect the data, provided they are nonreactive at the electrode.

The cell may be a micro dropping mercury electrode or a micro platinum (wire) electrode (Fig. 28-15) with a reference electrode, such as a calomel half-cell, and have a volume of about 100 mL.

The buret should be mounted so that the tip can be touched to the side of the receiving vessel, enabling any fraction of a drop remaining on the tip to be washed down into the solution by means of a fine spray of distilled water from the washbottle.

29
THE ELECTROMAGNETIC SPECTRUM

INTRODUCTION

Energy can be transmitted by electromagnetic (em) waves, and there are many types of these. They are characterized by their frequency f, the number of waves passing a fixed point per second, and their wavelength λ, the distance between the peaks of any two consecutive waves (Fig. 29-1).

Wavelengths of waves in the electromagnetic family or spectrum vary greatly, from fractions of angstroms to kilometers (Fig. 29-2). An examination of this spectrum reveals that as the frequency increases, the wavelength decreases, and as the frequency decreases, the wavelength increases.

The types of radiations are distinguished by the different characteristics exhibited by radiations of different frequencies. For example, visible light waves cannot pass through opaque substances, whereas radio waves, x-rays, and gamma rays can. The human eye cannot detect x-rays or ultraviolet rays, yet they are there. The lines of demarcation between the types of radiations are not sharp and distinct, and ranges may overlap.

In vacuum, the whole family of electromagnetic (em) waves has a constant velocity of 2.998×10^{10} cm/s (usually written as 3×10^{10} cm/s). In any

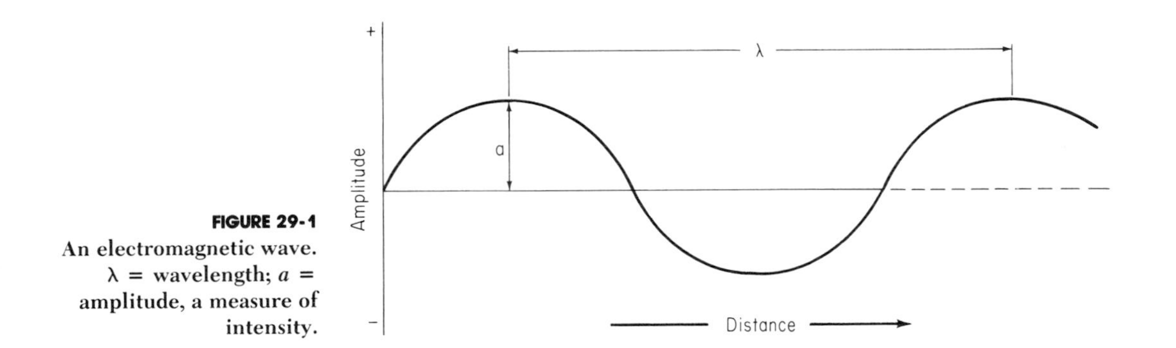

FIGURE 29-1
An electromagnetic wave. λ = wavelength; a = amplitude, a measure of intensity.

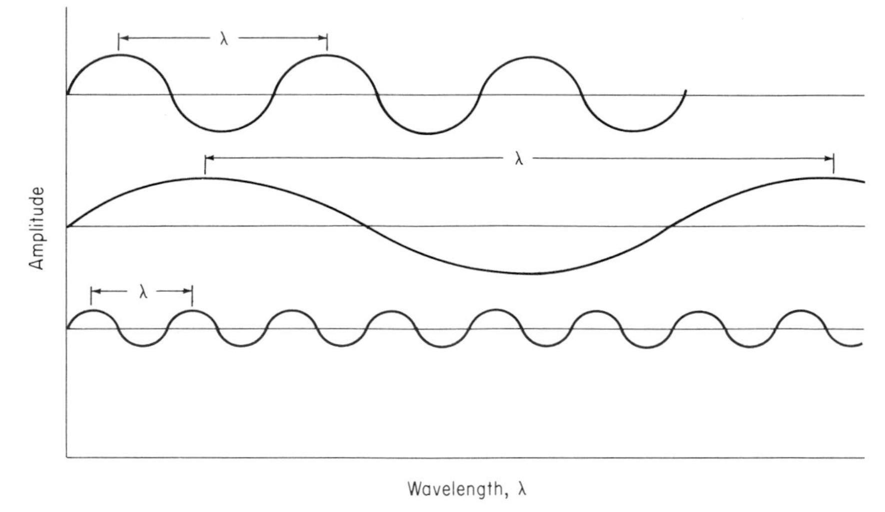

FIGURE 29-2
Variation of wavelength; as wavelength increases, frequency decreases.

em wave, the product of the wavelength λ and the frequency f equals the velocity of the wave:

Wavelength × frequency = velocity

or

$$\lambda \times f = v$$

Types of Electromagnetic Radiation

The values of the wavelength (λ) and frequency (f) are what differentiate one kind of radiation from another within the em radiation family (Fig. 29-3).

Conversion of Units

Wavelengths are often expressed in different units, and this may lead to confusion and error when one attempts to determine equivalent values. Table 29-1 enables you to interconvert wavelengths, regardless of how they are expressed.

Frequency and Wavelength Relationship

Figure 29-4 shows values for wave frequencies in hertz and wavelengths in centimeters for the band-spectrum spread of the various types of em radiations.

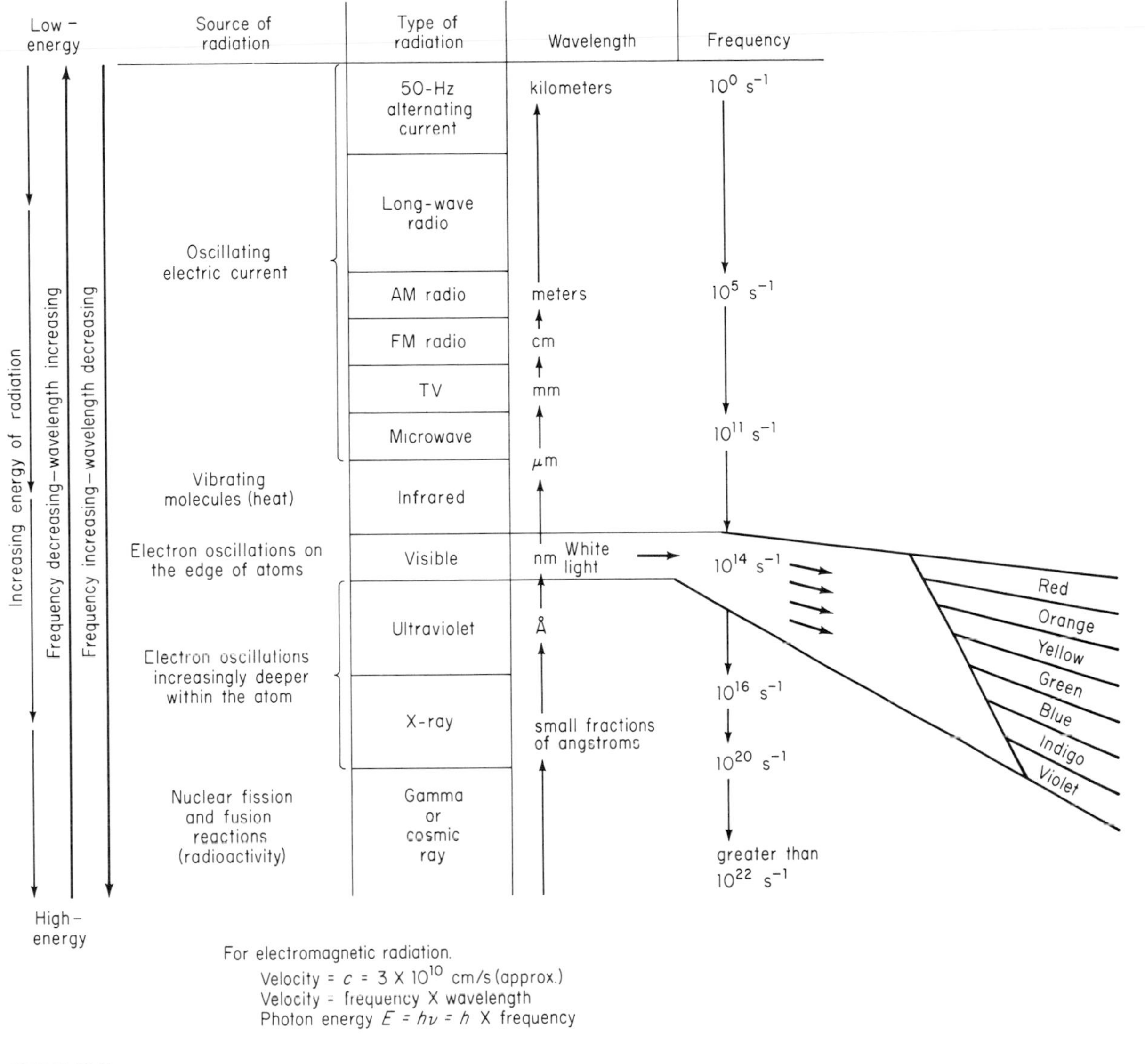

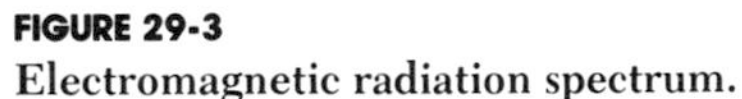

FIGURE 29-3
Electromagnetic radiation spectrum.

THE VISIBLE SPECTRUM AND REFRACTION

White light, emitted from the sun or from an incandescent tungsten-filament bulb, produces a continuous spectrum consisting of all the wavelengths visible to the human eye (Fig. 29-5).

TABLE 29-1
Conversion Table

Multiply number of → ↓ *To obtain number of* *By* ↘	*Angstroms (Å)*	*Millimicrons* (nm)*	*Microns† (μm)*	*Millimeters (mm)*	*Centimeters (cm)*	*Meters (m)*
Angstroms (Å)	1	10	10^4	10^7	10^8	10^{10}
Millimicrons (nm)*	10^{-1}	1	10^3	10^4	10^7	10^9
Microns (μm)†	10^{-4}	10^{-3}	1	10^3	10^4	10^6
Millimeters (mm)	10^{-7}	10^{-4}	10^{-3}	1	10	10^3
Centimeters (cm)	10^{-8}	10^{-7}	10^{-4}	10^{-1}	1	10^2
Meters (m)	10^{-10}	10^{-9}	10^{-6}	10^{-3}	10^{-2}	1

*More properly called nanometers (therefore nm).
† More properly called micrometers (therefore μm).

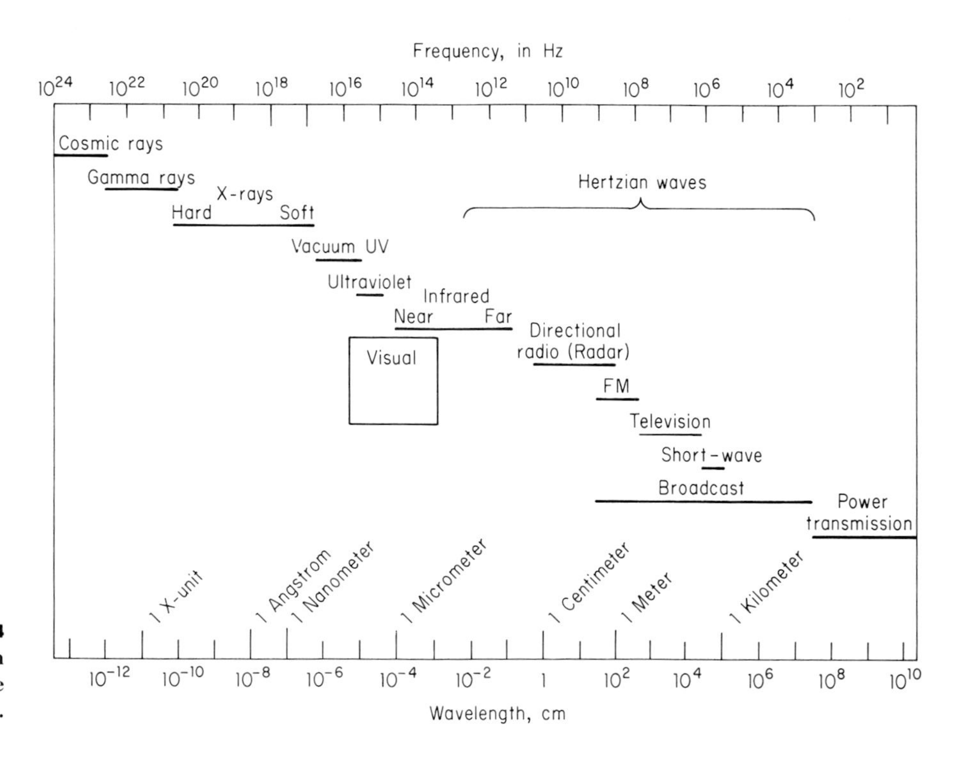

FIGURE 29-4
Frequency and wavelength relationships of the electromagnetic spectrum.

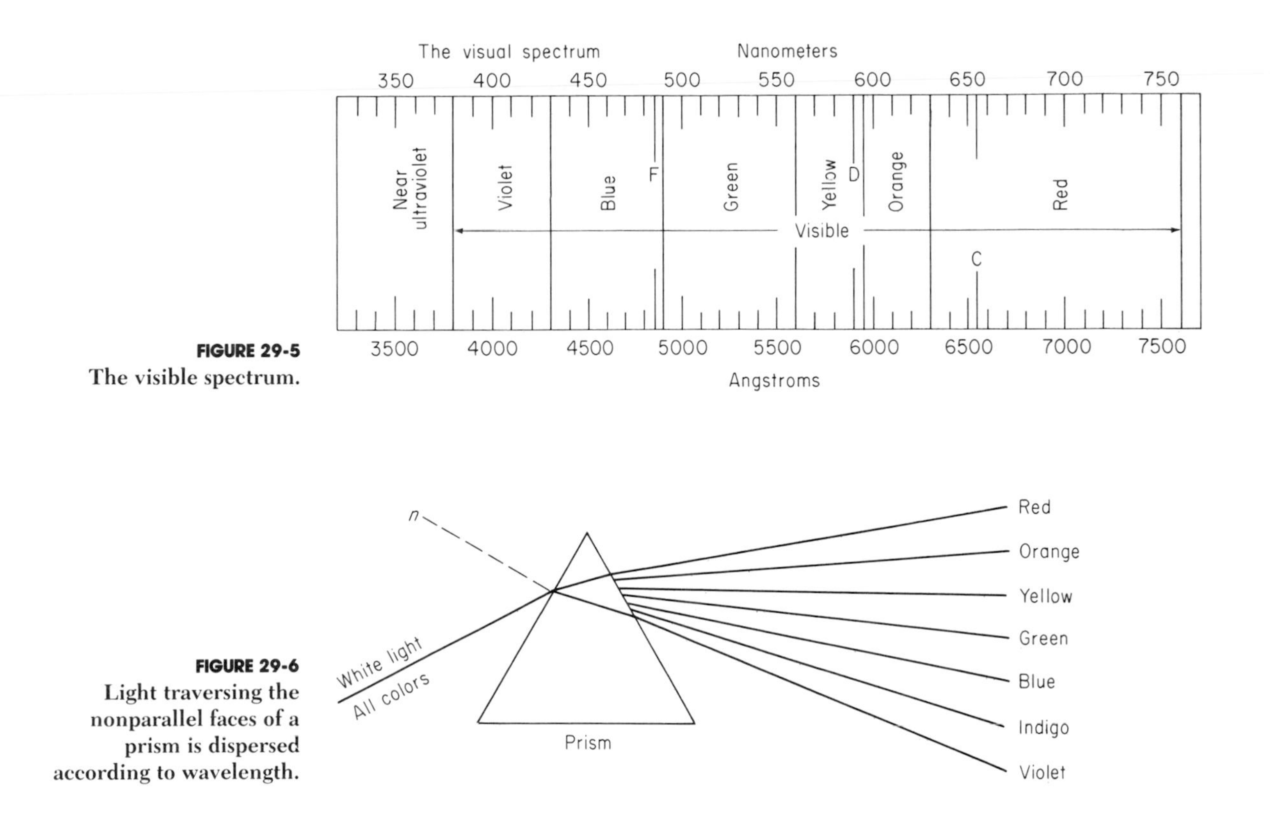

FIGURE 29-5
The visible spectrum.

FIGURE 29-6
Light traversing the nonparallel faces of a prism is dispersed according to wavelength.

When white light is passed through nonparallel surfaces such as those of a prism, there is a permanent bending of the beam, because different wavelengths (each color has its own wavelength) passing through a material are slowed down to different extents; each one is bent or refracted differently. Generally red is not slowed down as much as blue, and therefore the blue is bent or refracted through a larger angle (Fig. 29-6).

DIFFRACTION

When waves of em radiation strike an obstacle which does not reflect or refract them, a change occurs in their amplitude or phase; this change is called *diffraction*.

A *diffraction grating* is a flat piece of metal, glass, or plastic which has a great many parallel grooves ruled on it. Nonvisible electromagnetic radiations which are directed onto its surfaces are separated into their individual wavelength components (Fig. 29-7). In effect, such gratings separate nonvisible radiations just as the prism separates the components of

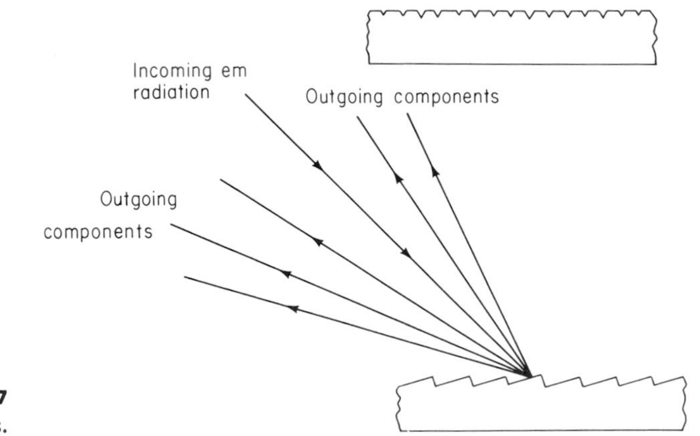

FIGURE 29-7
Diffraction gratings.

the visible spectrum. Crystals of salt have been used as diffraction gratings for x-rays.

SODIUM VAPOR AND ULTRAVIOLET LAMPS

Although white light is a combination of all the components of the visible spectrum, not all incandescent bulbs emit "white" light. Some of them emit higher intensities of one component than of the others.

Some lamps are designed to emit light of a specific wavelength. The sodium vapor lamp emits yellow light (Fig. 29-8), and the ultraviolet (UV) lamp provides a concentrated beam of ultraviolet radiation on demand.

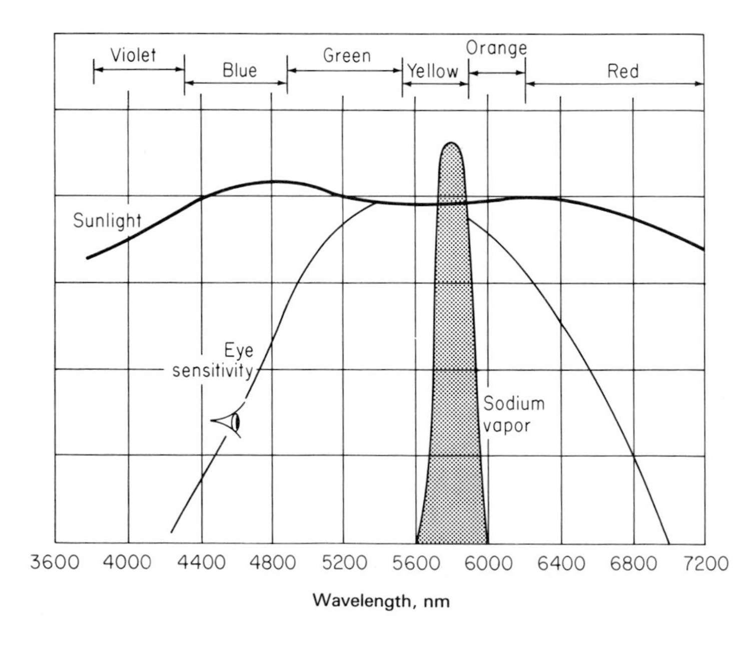

FIGURE 29-8
The sodium vapor lamp spectrum compared to sunlight and the sensitivity of the human eye.

USE OF THE ELECTROMAGNETIC SPECTRUM IN THE STUDY OF SUBSTANCES

The use of the electromagnetic spectrum in the study of substances is divided according to the region or range of radiations used in the procedure. These regions are shown in Table 29-2.

TABLE 29-2
Regions of Study in the Electromagnetic Spectrum

Region	*Range*
Vacuum ultraviolet	100–180 nm
Ultraviolet	180–400 nm
Visible	400–750 nm
Near-infrared	0.75–2.5 nm
Infrared	2.5–15 μm
Far-infrared	15–300 μm

COLORIMETRY

The solutions of many compounds have characteristic colors. The intensity of such a color is proportional to the concentration of the compound. By using a colorimeter to match samples, one can perform many analyses quickly and easily. One of these instruments is the Duboscq colorimeter.

Duboscq Colorimeter

Materials

All that is required is a solution of known concentration and one of unknown, a pipet, a graduated cylinder, and the colorimeter (Fig. 29-9).

Procedure

1. Be sure that the cups and plungers are clean both before and after use. Use a soft cloth or lens tissue to wipe optical glass.

2. Test the zero point of the scale by carefully raising the cups until they touch the plungers. The zero point of the scale is adjusted by screws at the bottom of the cup holders or at the side of the holders. Cups should not be changed from one side to the other.

3. Adjust the instrument for equal light intensity on both sides by filling cups with standard solution (three-fourths full). Set both sides at the same value, and adjust the position of the light until it is equal on both sides. This may be accomplished by shifting the position of the instrument or mirror if daylight or an external light source is employed or by changing the position of the bulb or reflector. If a light source is attached to the instrument, both rods should appear equally bright.

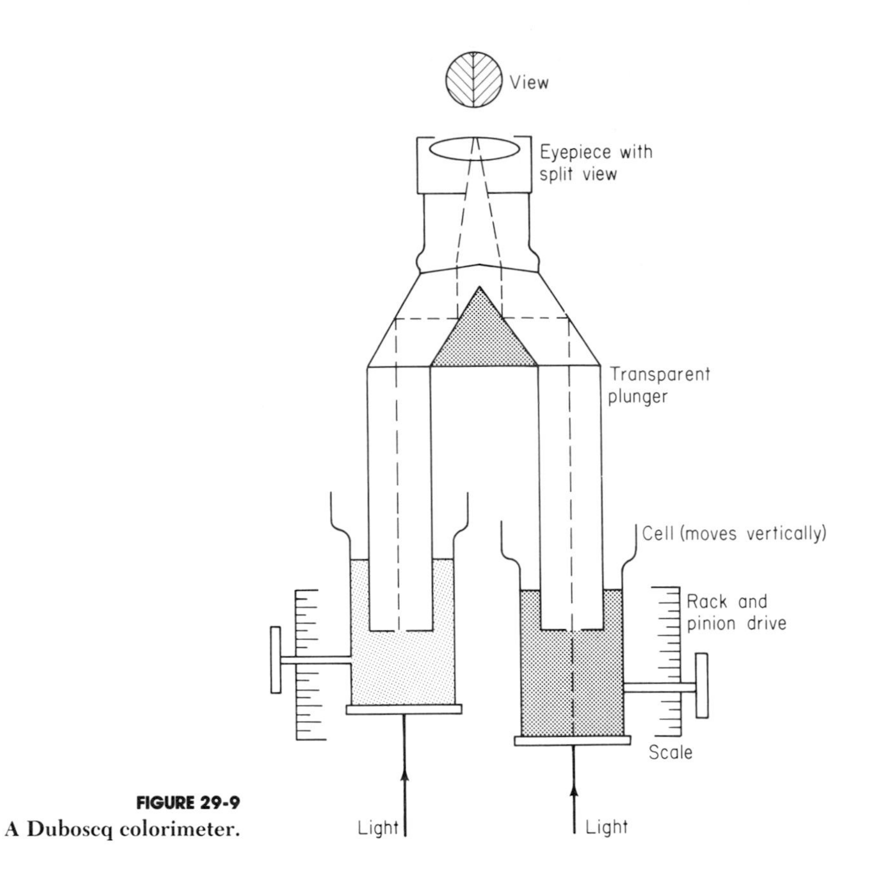

FIGURE 29-9
A Duboscq colorimeter.

4. Test the light adjustment by filling both cups with the standard solution and setting one cup at a convenient depth; then move the other cup until a balance is obtained. Half of the balance point should be obtained by approaching the point of balance from one direction, and the other half from the opposite direction. Repeat this balancing until 6 to 10 readings have been obtained. The average should agree within 1 or 2 percent with the reading of the stationary cup. If this is not the case, then use one side as a fixed reference as follows: Set the cup just moved (cup 2) at the average reading obtained for it. Leave the standard solution in this cup. Replace the standard solution in the other cup (cup 1) with an unknown. Adjust the unknown cup (cup 1) until a balance is obtained. Then

$$C_{\text{unknown}} = C_{\text{standard}} \frac{R_{\text{standard}}}{R_{\text{unknown}}}$$

where R_{standard} = reading of standard solution in cup 1
R_{unknown} = reading of unknown in cup 1
C_{standard} = concentration of standard
C_{unknown} = concentration of unknown

5. Be sure there are no air bubbles beneath the plungers when the plungers are inserted beneath the liquid.

NEPHELOMETRY

Nephelometry is the measurement of the light which is reflected from a finely divided suspension or dispersion of small particles, the Tyndall effect. Optically controlled beams of light directed into a suspension of finely divided particles in a fluid will measure the concentration of the particles in a linear relationship. Nephelometric analysis measures the intensity of the Tyndall light which results from the incidence of a controlled light beam of constant intensity upon the suspension. The Tyndall light is detected and measured by photocells mounted at right angles to the incident beam (Fig. 29-10), and their output is measured on the detector, usually a galvanometer or a precisely calibrated potentiometer.

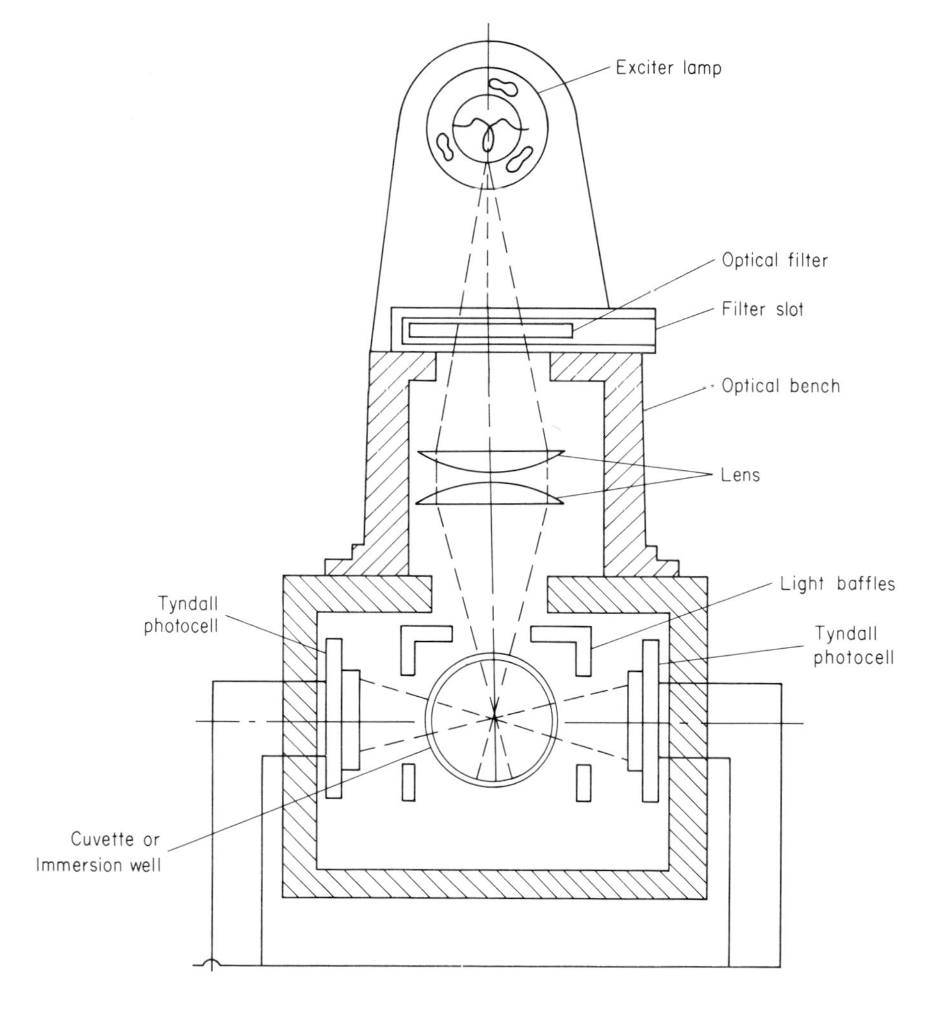

FIGURE 29-10
In the nephelometer, the Tyndall photocells are placed at right angles to the incident beam.

SPECTROSCOPY AND SPECTRO-PHOTOMETRY

Spectroscopy is the study of the interaction of electromagnetic radiations with matter. Instruments which measure em *emission* are called spectroscopes or spectrographs. Those which measure em *absorption* are called spectrophotometers. All spectroscopic instruments separate electromagnetic radiation into its component wavelengths to enable one to measure the intensity or strength of the radiation at each wavelength.

There are three kinds of emission spectra:

1. Continuous spectra which are emitted by incandescent solids (Fig. 29-11)

FIGURE 29-11
A continuous spectrum.

2. Line spectra, which are characteristic of atoms that have been excited and are emitting their excess energy (Fig. 29-12)

FIGURE 29-12
A line spectrum.

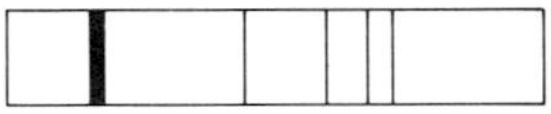

3. Band spectra, which are emitted by excited molecules

Electrons in an atom are normally in their lowest energy states, known as the *ground state*. When sufficient energy is added, electrically or thermally, one or more electrons may be raised to higher-energy states. When these electrons lose their energy and return to their ground state, they emit electromagnetic radiations. In returning to their ground state, they may do this in several discrete jumps or energy changes, emitting light of different wavelengths for each jump. When high-energy excitation is used, more lines appear in the spectrum.

Definitions and Symbols

Spectroscopic terminology frequently uses different names and symbols to identify the same property, and there are both modern terms and terms that have been used in the past.

Radiant power (P) is the rate at which energy in a beam of radiation arrives at some fixed point. ***Intensity*** (***I***) is the same term.

Transmittance (T) is the ratio of the radiant power (P) in a beam of radiation after it has passed through a sample to the power of the incident beam (P_0). See Fig. 29-13. It is also referred to as ***percent T*** or $T \times 100$.

$$T = \frac{P}{P_0}$$

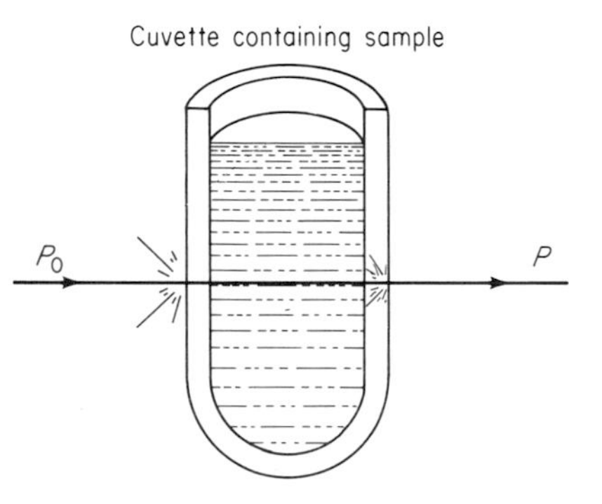

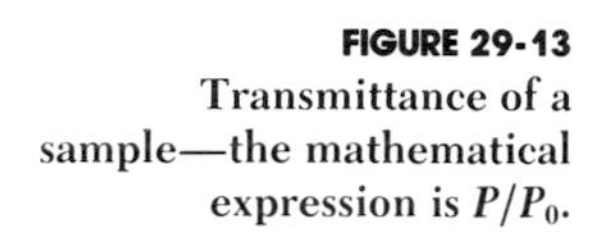
FIGURE 29-13
Transmittance of a sample—the mathematical expression is P/P_0.

Absorbance (A) is also called the ***optical density,*** and it is the logarithm (base 10) of the reciprocal of the transmittance (T).

$$A = \log \frac{1}{T} = \log \frac{P_0}{P}$$

where P = radiation transmitted by the solution
P_0 = radiation transmitted by the pure solvent

Molar absorptivity (also known as the ***molar extinction coefficient***) ϵ is the absorbance of a solution divided by the product of the optical path b in centimeters and the molar concentration c of the absorbing molecules or ions:

$$\epsilon = \frac{A}{bc}$$

Spectrophotometric Concentration Analysis

According to Beer's law, the absorbance of a solute in solution is a function of its concentration at a particular wavelength, and therefore, absorbance measurements can be used to determine the concentration of solutions. When the optimum wavelength is selected for an analysis, the concentration of the solute can be established. (See Using the Spectrophotometer in Chap. 30, "Visible and Ultraviolet Spectroscopy.")

Absorption of Radiant Energy

When radiation of a specified wavelength is passed through a solution containing only one solute which absorbs that wavelength, the absorbance (which has no units) can be calculated by the formula:

$$A = \log \frac{P_0}{P} = \log \frac{1}{T} = \epsilon bc$$

where P = radiation transmitted by the solution
P_0 = radiation transmitted by the pure solvent
b = optical length of the solution cell
c = molar concentration of the absorbing solute
ϵ = the molar extinction coefficient (in liters per mole-centimeter)

Practical application of Beer's law is shown in Figs. 29-14 and 29-15.

When a beam of polychromatic radiation is directed into a sample of some substance, the wavelengths of certain components may be absorbed while others pass through essentially undisturbed. See Fig. 29-16. Components of the radiation are absorbed only if its energy exactly matches that energy which is required to raise molecular or ionic components of the sample from one energy level to another. Those energy transitions may involve vibrational, rotational, or electronic states. After it has been absorbed, that energy may be emitted as fluorescence, utilized to initiate chemical reactions, or actually dissipated as heat energy.

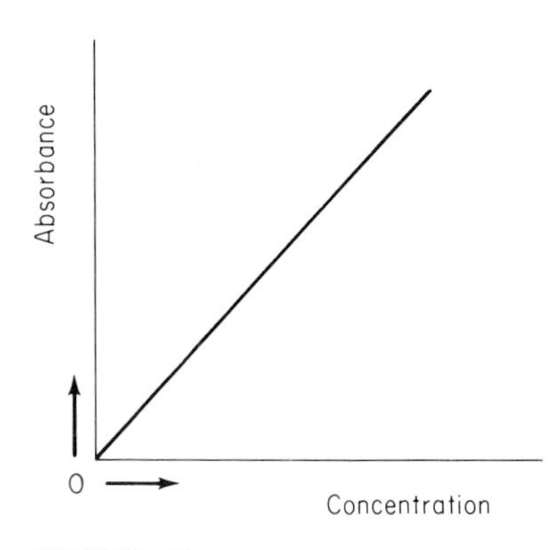

FIGURE 29-14
Plot of absorbance vs. concentration.

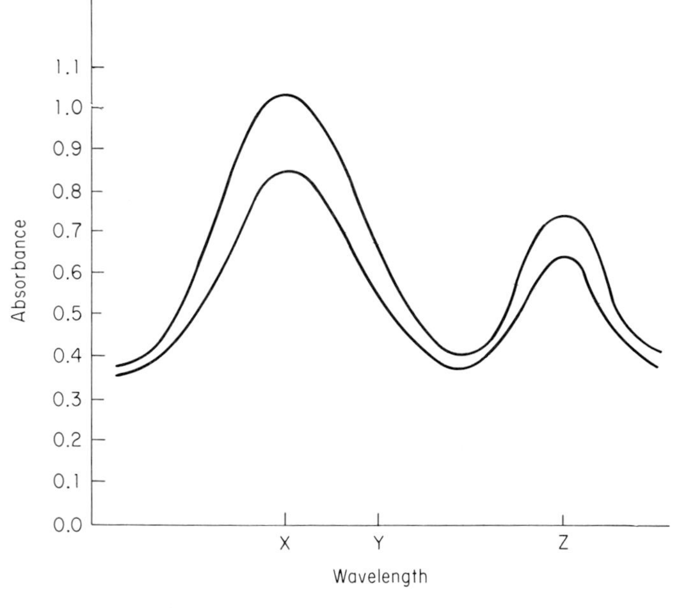

FIGURE 29-15
Absorption spectra for solutions of the same substance at two different concentrations.

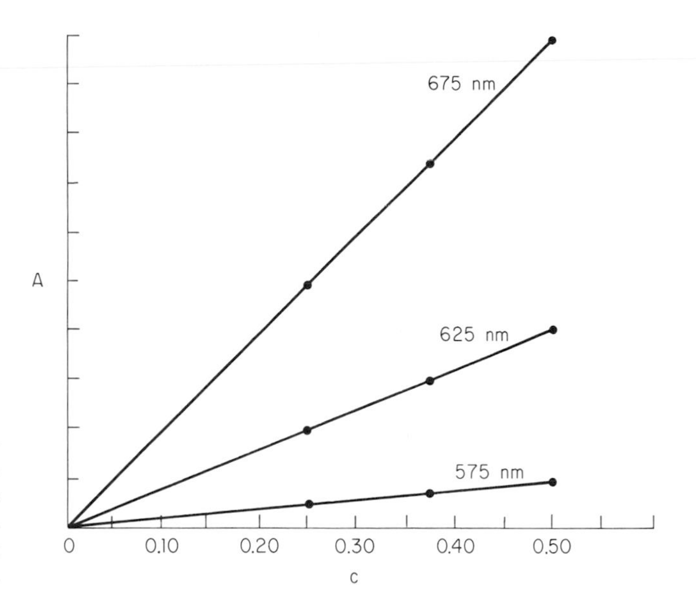

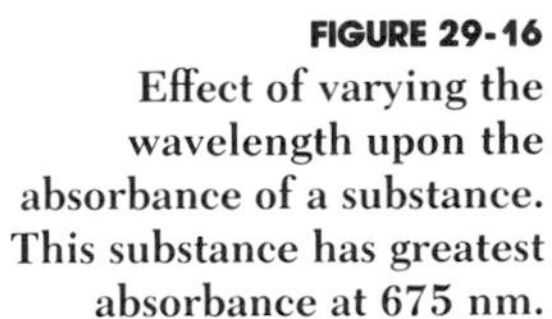

FIGURE 29-16
Effect of varying the wavelength upon the absorbance of a substance. This substance has greatest absorbance at 675 nm.

Measurement of Absorption Spectra

The spectrophotometer (Fig. 29-17) is used to measure radiation absorption. Its basic components are:

1. A monochromator, a device which attempts to provide a beam of radiation at a single frequency which is to be passed through a sample.
2. A sample holder to contain the solution.
3. A detector to measure the intensity of the transmitted radiation after it has passed through the sample.

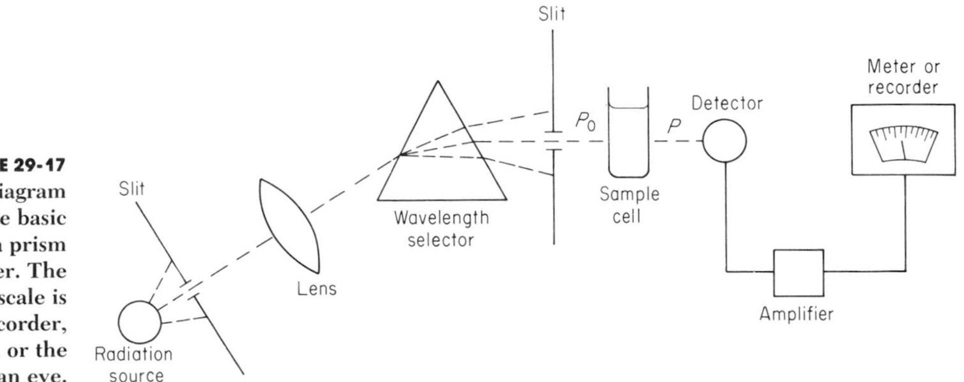

FIGURE 29-17
Schematic diagram showing the basic components of a prism spectrophotometer. The detector screen-scale is viewed by meter, recorder, photographic film, or the human eye.

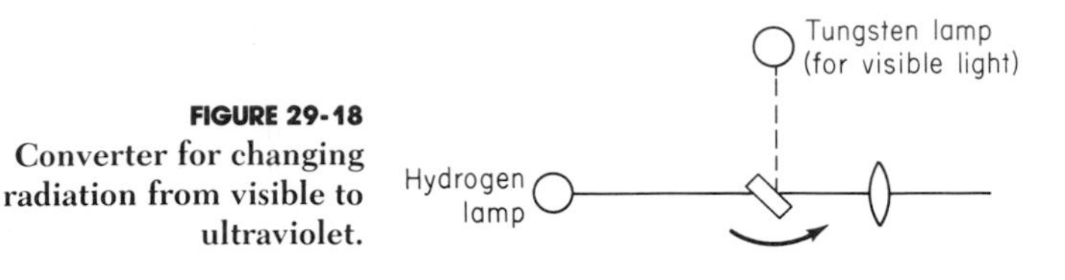

FIGURE 29-18
Converter for changing radiation from visible to ultraviolet.

Visible spectrophotometers can be converted into ultraviolet spectrophotometers by changing the source of radiation, from a tungsten filament—for visible light—to a hydrogen or deuterium discharge lamp (Fig. 29-18)—for ultraviolet.

The prisms, lens, and cells must also be changed from glass to quartz or fused silica, because glass absorbs UV radiations. At wavelengths less than 180 nm the system must be operated in a vacuum (vacuum UV), because air will absorb UV radiations which are shorter than 180 nm.

Monochromator

Radiation sources emit a broad band of wavelengths which pass through a narrow slit and are focused by lenses and mirrors onto a wavelength selector. The selector is a prism or diffraction grating which disperses the continuous radiation; however, only a narrow band directed toward a slit in the screen is able to pass through the sample. By rotating the prism or grating, it is possible to direct components of different wavelengths through the sample and thus to select desired wavelengths. Those instruments which pass a broad band through the sample are simply called *photometers.*

Sample Cell or Cuvette

A sample solution is contained in an optically transparent cell or *cuvette* (Fig. 29-19) with a known width and optical length. Cells are made of optical glass. Protect them, therefore, from scratches. Avoid the use of abrasive cleaning agents; clean them well with soft cloths, avoid finger marks and lint or dirt, and handle them only by the top edge when inserting them into the instrument. Because of additional reflection from air-to-glass surfaces, empty cuvettes transmit less radiation than do cells filled with reference standards, such as distilled water.

Detector

Transmitted radiation through the solution is picked up by a photosensitive detector. Photocells are used for visible and ultraviolet radiations; infrared (IR) instruments use a temperature-sensitive detector, a bolometer or a thermopile, because infrared radiation does not have sufficient energy to activate photocells. Radiations which strike the detector generate electric current, which is amplified with an amplifier and is then

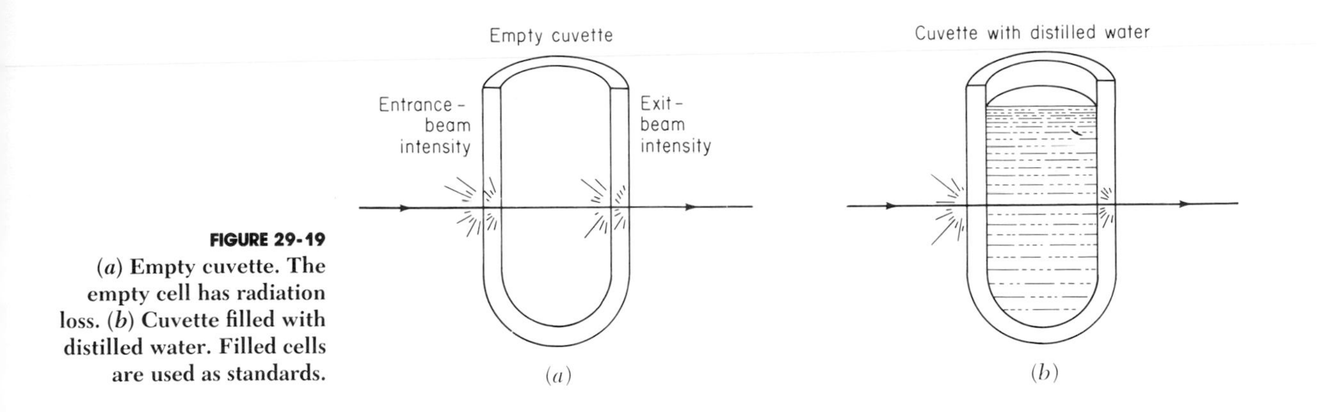

FIGURE 29-19
(*a*) Empty cuvette. The empty cell has radiation loss. (*b*) Cuvette filled with distilled water. Filled cells are used as standards.

transmitted to a recorder or a meter. Anything that is sensitive to infrared radiations (heat) can be used; some instruments, however, are more sensitive and useful at certain wavelengths.

Absorption Spectra

When the absorbance of a sample is measured as a function of wavelength, the result is an absorption spectrum. Some instruments vary wavelength manually, while others vary it automatically. The absorption spectrum can be used to identify unknown substances, because particular molecules and ions have specific absorptions at characteristic wavelengths. The comparison of an unknown spectrum with spectra of known substances enables one to identify unknown substances. In the infrared spectrum, particularly, specific wavelengths are characteristic of certain structural components of molecules.

30

VISIBLE AND ULTRAVIOLET SPECTROSCOPY

INTRODUCTION

A review of Chap. 29, "The Electromagnetic Spectrum," is recommended as a preview of spectroscopy. The visible and ultraviolet (UV) spectra of ions and molecules were the first to be used to obtain both qualitative and quantitative chemical information. The typical visible-UV spectrum consists of a plot of the *molar absorptivity* (ϵ) as a function of the wavelength expressed in nanometers as shown in Fig. 30-1. The symbol lambda with subscript "max" (λ_{max}) stands for the *absorption maximum,* the wavelength at which a maximum absorbance occurs. Molar absorptivity can be defined from Beer's law by rearranging the Beer's law equation (see Chap. 29, pages 721 and 722, under Spectrophotometric Concentration Analysis) to:

$$\epsilon = \frac{A}{bc}$$

where ϵ = molar absorptivity
A = absorbance = $-\log T$ (where T represents transmittance)
b = cell path length, cm
c = concentration, mol/L

The absorption of visible and/or UV radiation is associated with the transition of electrons between different energy levels within the ion group or

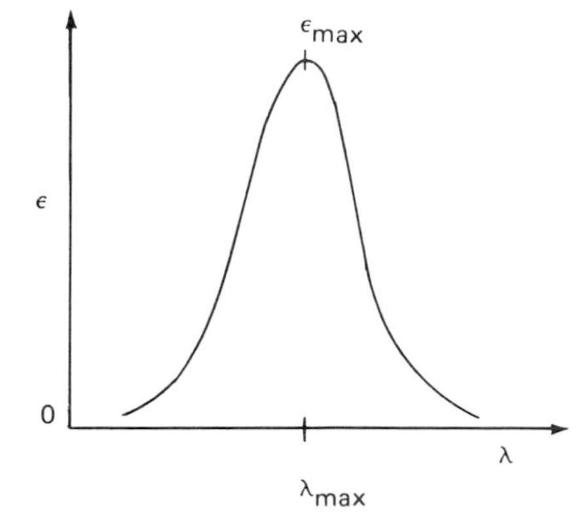

FIGURE 30-1 Typical visible-UV spectrum with molar absorptivity vs. wavelength.

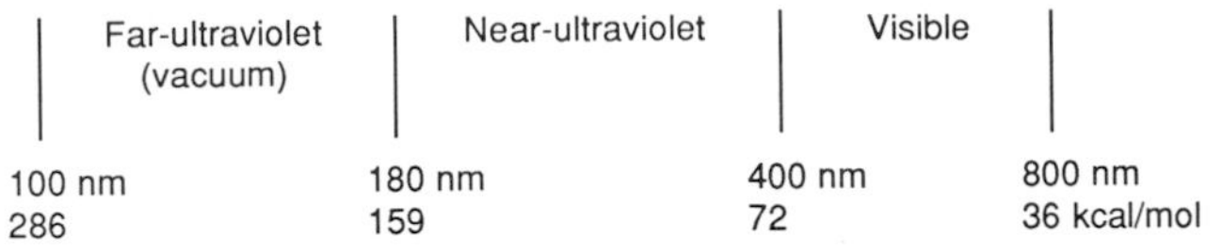

Far-ultraviolet (vacuum)	Near-ultraviolet	Visible	
100 nm	180 nm	400 nm	800 nm
286	159	72	36 kcal/mol

FIGURE 30-2 Visible and UV spectroscopic regions and associated energy levels.

molecule. Visible and UV radiation is energetic enough to cause not only various electronic transitions, but vibrational and rotational changes (see Chap. 31, "Infrared Spectroscopy"); thus these absorption bands tend to be very broad. As described in Chap. 29, "The Electromagnetic Spectrum," the visible region is considered to be from 800 to 400 nm and the UV region is subdivided into two distinct regions: near-UV (400 to 180 nm) and the far-UV (180 to 100 nm). Figure 30-2 shows the visible and UV spectroscopic regions and the associated energy levels. Some visible-UV spectrophotometers are capable of reaching into the near-infrared (approximately 2600 nm). The newer "UV-Vis" instruments have Fourier transform for very rapid analysis and computer capabilities for data manipulation, storage, and full-spectrum searches.

ULTRAVIOLET SPECTROPHOTOMETER COMPONENTS

The radiation source for a UV spectrophotometer is either a mercury or deuterium (hydrogen) lamp. The optical components (prisms, lens, and cells) are usually composed of quartz or fused silica because glass absorbs ultraviolet radiation. A special UV-sensitive photomultiplier tube is used as a detector. Most routine UV spectroscopy work is limited to the near-UV because the silica optics and atmospheric oxygen absorb an excessive amount of the radiation. Far-ultraviolet is sometimes referred to as *vacuum ultraviolet* (vacuum UV) because the instrument is evacuated to remove the atmospheric oxygen in the monochromator.

CHROMOPHORES

Functional groups that absorb visible and/or UV are called *chromophores*. Chromophores tend to have unsaturated bonds or contain functional groups with multiple bonds as shown in Table 30-1.

When a beam of light is passed through an absorbing solution, the intensity of the incident radiation will be greater than the intensity of the emerging radiation. In general, the excited electrons, atoms, and/or molecules resulting from the absorption of radiation return very rapidly to the ground state by losing electromagnetic radiation. Table 30-2 shows the

TABLE 30-1
Characteristic Ultraviolet Absorption Bands for Some Chromophores

Type	*Example*	*Absorption band, nm*
Alkenes	$CH_2{=}CH_2$	165–193
Alkynes	$HC{\equiv}CH$	195–225
Aldehydes	CH_3CHO	180–290
Ketones	CH_3COCH_3	188–279
Carboxylic acids	CH_3COOH	208–210
Aromatics	C_6H_6	204–254

TABLE 30-2
Absorption of Visible Light and the Corresponding Colors

Wavelength, nm	*Color (absorbed)*	*Color (observed)*
400–435	Violet	Yellowish green
435–480	Blue	Yellow
480–490	Greenish blue	Orange
490–500	Bluish green	Red
500–560	Green	Purple
560–580	Yellowish green	Violet
580–595	Yellow	Blue
595–650	Orange	Greenish blue
650–800	Red	Bluish green

wavelength, color absorbed, and corresponding color observed in various colorimetric applications.

SOLVENTS USED IN VISIBLE AND ULTRAVIOLET SPECTROSCOPY

Most compounds with only single bonds tend to absorb radiation in the far-UV (less than 180 nm), but make reasonable solvents for working in the near-UV region. Water, which has an absorption-band maximum at 167 nm, should provide a solvent medium for most polar chromophores that absorb in the visible and/or near-UV regions. Other commonly used polar solvents are 95% ethanol and methanol, while nonpolar aliphatic hydrocarbons (like hexane and cyclohexane) can be used, if the alkenic and aromatic trace impurities have been removed. Table 30-3 provides a listing of solvents and their wavelength cutoff limits. *Cutoff limits* for the solvents listed below are defined as the wavelength at which the absorbance approaches one in a 10-mm cell.

TABLE 30-3
Ultraviolet Cutoff Limits for Various Solvents

Solvent	*Cutoff point, nm*	*Boiling point, °C*
Acetonitrile	190	81.6
Hexane	195	68.8
Cyclohexane	205	80.8
Ethanol (95%)	204	78.1
Water	205	100.0
Methanol	205	64.7
Diethyl ether	215	34.6
1,4-Dioxane	215	101.4
Carbon tetrachloride	265	76.9
Benzene	280	80.1
Toluene	285	110.8
Acetone	330	56.0

SAMPLE PREPARATION

All volumetric glassware and UV cells must be clean and dry. Solid samples should be dried to constant weight and stored in a desiccator to minimize moisture and solvent interference. A typical 1-cm cell holds approximately 3 mL of solution; however by preparing samples in greater volume the inherent error that results from weighing small solute quantities can be reduced. Most solutions are analyzed at very low concentrations (10^{-2} to 10^{-6} M) to minimize solvent interactions and shifts in the absorbance maxima. Since both visible and UV spectroscopy are nondestructive techniques, the sample can be readily recovered if a volatile solvent is selected. In general, a sample should transmit in the range from 20% to 80% (0.7 to 0.1 A) for greatest accuracy in reading the meter. Most UV-Vis meters are normally calibrated in both percent transmittance (%T) and absorbance (A), but %T readings are used because the scale is linear and easier to interpolate. The %T can be converted to its equivalent in absorbance value by the equation:

$$A = 2 - \log \%T$$

Matched Cuvettes

If more than one cuvette is used in an analysis, they should be determined to be "matched." Matched cuvettes have identical path lengths plus similar reflective and refractive properties in the spectrophotometer beam. Sets of matched cuvettes can be purchased, and some are marked with a reference line on the cuvette wall to assure proper and reproducible placement in the beam. Obviously, these cells can become mismatched with use, cleaning procedures, and scratches. A guideline for selecting a set of matched cuvettes is to analyze a 50% transmittance solution and use only those cuvettes in a single analysis with a 1% deviation or less.

USING THE SPECTROPHOTOMETER

Bausch and Lomb Spectronic 20

Principle

The spectrophotometer measures the intensity of visible light after passage through a sample. (See Figs. 30-3 and 30-4.)

Components

1. A monochromatic source of light vibrating at a single frequency (between 375 and 650 nm) (1 nm = 1×10^{-7} cm = 10 Å).

2. A sample holder for supporting the solution in the beam of light.

3. A detector to measure the intensity of light passing through the sample.

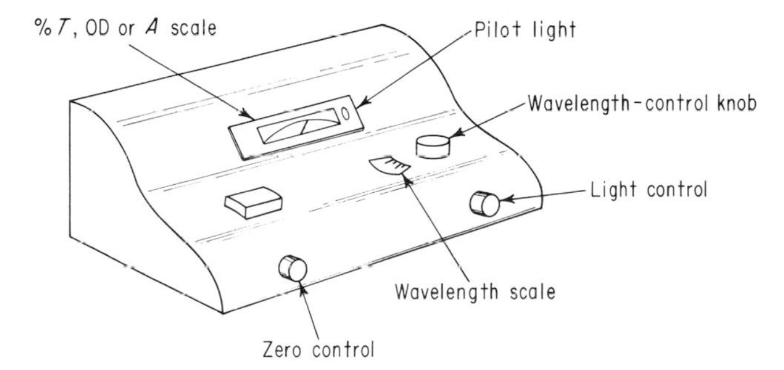

FIGURE 30-3 Bausch and Lomb Spectronic 20.

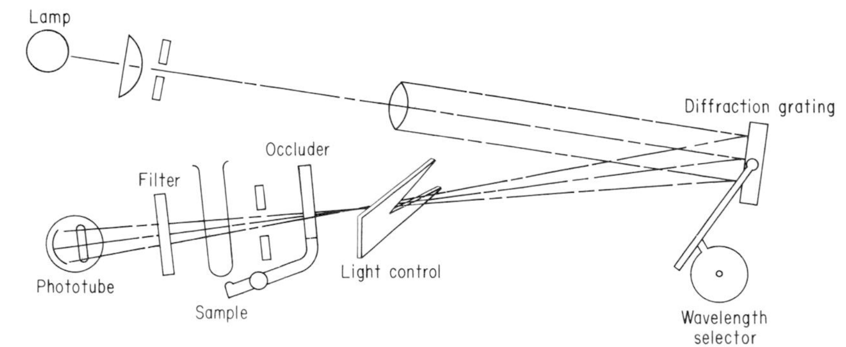

FIGURE 30-4 Schematic optical diagram of the Spectronic 20.

Procedure

1. Switch to ON by rotating the ZERO CONTROL knob clockwise until a click is heard. This turns the power on. Allow the unit to warm up for 20 min.

2. Adjust the ZERO CONTROL knob until the meter reads 0 percent *T* (no tube in the sample holder).

3. Obtain two Spectronic 20 colorimeter tubes.

   (a) Use one for a solvent blank.

   (b) Use the other as a sample holder.

   (c) *Handle the tubes with extreme care:* Wipe them clean with soft, absorbent, lint-free paper. Never touch the lower part of the tubes with your fingers. Fingerprints absorb and scatter light.

4. Rotate the right-hand LIGHT CONTROL knob counterclockwise as far as it will turn. *Do not force.*

5. Insert a solvent-blank cuvette.

   (a) Zero the meter.

   (b) Set the wavelength.

   (c) Close the cover lid on the sample holder.

   (d) Rotate the LIGHT CONTROL knob clockwise until the meter reads 100 percent *T*.

6. Remove the blank cuvette holder.

7. Insert the sample cuvette. *Close the cover.*

8. Read the absorbance on the dial.

9. Repeat step 5 each time the wavelength is changed.

CAUTIONS

1. It may be necessary to rezero the meter because of fluctuations in the electronic components.

2. The meter needle should drop to zero whenever the colorimeter tube is removed from the sample holder and the occluder drops into position.

3. *Always close the sample cover* to exclude stray light when making measurements.

REFERENCES

Association of Official Analytical Chemists, *Official Methods of Analysis of the Association of Official Analytical Chemists* 14th edition, AOAC, 1985.

Bauman, R. P., *Absorption Spectroscopy*, John Wiley and Sons, New York, 1962.

Ewing, G. W., *Instrumental Methods of Chemical Analysis,* 4th ed., McGraw-Hill Book Co., New York, 1975.

Jaffe, H. H., and Orchin, M., *Theory and Applications of Ultraviolet Spectroscopy,* John Wiley and Sons, New York, 1962.

Lambert, J. B., Shurvell, H. F., Lightner, D., and Cooks, R. G., *Introduction to Organic Spectroscopy,* Macmillan Publishing Co., New York, 1987.

National Institute of Occupational Safety and Health, *NIOSH Manual of Analytical Methods*, 2nd ed., Vols. 1–7, 1977–81.

Olsen, E. D., *Modern Optical Methods of Analysis,* McGraw-Hill Book Co., New York, 1975.

Sadtler Research Laboratories, *A Comprehensive Catalog of Ultraviolet Spectra of Organic Compounds*, Vols. 1–62, Philadelphia, 1980.

Willard, H. H., Merritt, L. L., Dean, J. A., and Settle, F. A., *Instrumental Methods of Analysis,* 6th ed., Wadsworth Publishing Co., New York, 1981.

31

INFRARED SPECTROSCOPY

INTRODUCTION

Infrared spectroscopy is primarily used as a nondestructive technique to determine the structure, identity, and quantity of organic compounds. As explained in Chap. 29, "The Electromagnetic Spectrum," the infrared region extends from approximately 0.75 to 400 micrometers (μm) (1 μm = 1×10^{-6} m); however, the region from 2.5 to 16 μm is most often used by organic chemists for structural and quantitative determinations (Fig. 31-1). There are two less useful infrared regions: 0.75 to 2.5 μm, which is referred to as the *near-infrared* (near the visible region), and the region continuing from 16 to 400 μm, referred to as the *far-infrared*. The prefix "infra," which means "inferior to," was added to "red" because this particular radiation is adjacent to but slightly lower in energy than red light in the visible spectrum.

The interpretation of an infrared spectrum is not a simple exercise, and typically an organic compound is ultimately identified by comparing its

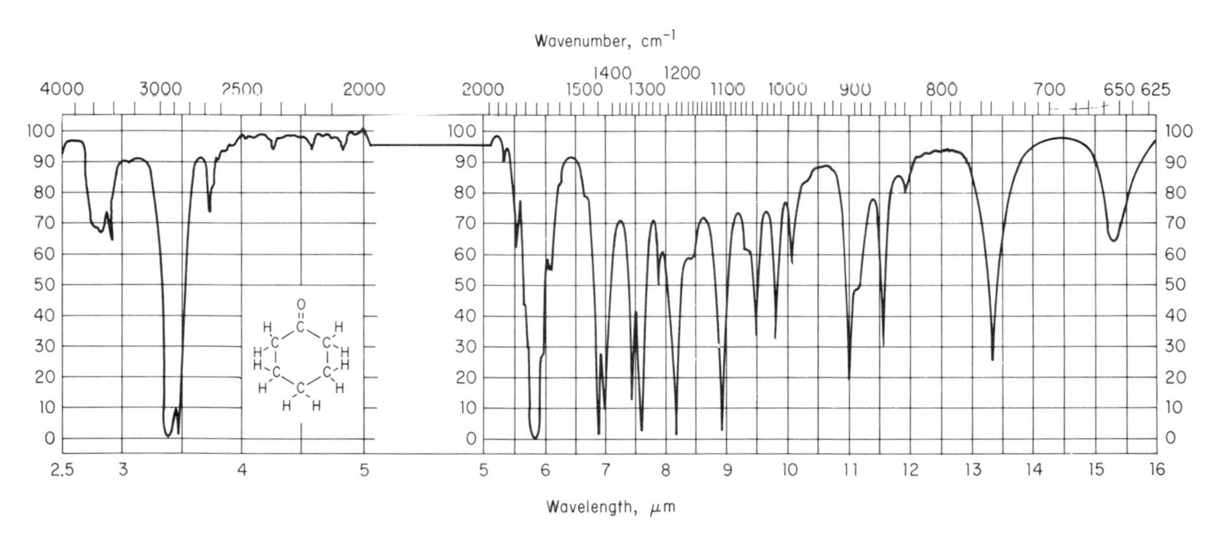

FIGURE 31-1
Typical infrared absorption spectrum of an organic compound (cyclohexanone) with grating change at 2000 cm^{-1}. The ordinate reads in percent transmittance and the abscissa indicates the wavelength expressed in micrometers (μm) or wave numbers (cm^{-1}).

infrared absorption spectrum to that of a known compound. Most molecules are in constant modes of rotation and/or vibration at temperatures above absolute zero. This rotation, stretching, and bending of molecular bonds happens to occur at frequencies (typically on the order of 100 trillion per second) found in the infrared spectral region. Of course, these vibrational frequencies will vary as the bonded atoms, functional groups, and bond strengths are changed; as a result, each molecule has a distinctive infrared absorption spectrum or "fingerprint." To be more specific, in the *functional-group* region (2.50 to 7.14 μm), it is relatively easy to assign structures to the absorption bands, while in the *fingerprint* region (7.14 to 16.0 μm) it is more difficult to assign structures; nonetheless, this fingerprint region is very useful because it contains many very characteristic and unique absorption bands of the compounds.

WAVELENGTH VS. WAVE NUMBER

The frequency or energy level at which these characteristic infrared absorptions occur can be expressed in terms of wavelengths (μm) or wave numbers (cm^{-1}). Research has shown that a typical organic molecule requires an average energy of 1.6×10^{-20} joules (J) for vibrational excitation. Given *Planck's equation* which relates energy and frequency:

$$E = h\nu$$

where E = energy, in J
h = Planck's constant = 6.6×10^{-34} J·s
ν = frequency, in s^{-1} or the equivalent measure hertz (Hz)

Substituting the average energy found necessary to induce vibrational excitation in a typical organic molecule:

$$1.6 \times 10^{-20}\ \text{J} = (6.6 \times 10^{-34}\ \text{J·s})\ \nu$$

$$\nu = \frac{1.6 \times 10^{-20}\ \text{J}}{6.6 \times 10^{-34}\ \text{J·s}}$$

$$\nu = 2.4 \times 10^{13}\ \text{s}^{-1}$$

Finally, substituting our calculated frequency into the following *frequency-to-wavelength equation:*

$$\lambda = \frac{c}{\nu}$$

where λ = wavelength, in cm
c = velocity of light = 3.0×10^{10} cm/s
ν = frequency, in s^{-1}

we find that

$$\lambda = \frac{3.0 \times 10^{10} \frac{\text{cm}}{\text{s}}}{2.4 \times 10^{13}\ \text{s}^{-1}}$$

$$\lambda = 1.2 \times 10^{-3}\ \text{cm}$$

$$\lambda = (1.2 \times 10^{-3}\ \text{cm})\left(1 \times 10^{4} \frac{\mu\text{m}}{\text{cm}}\right) = 12\ \mu\text{m}$$

Note that an average energy of 1.6×10^{-20} J found necessary to induce vibrational adsorption corresponds to a wavelength of 12 μm (infrared radiation). (In the infrared region, the term "micrometer" (μm) is the preferred one, not the older term "micron" (μ) used to describe the same unit.)

The reciprocal of the wavelength is called the *wave number* and represents the number of cycles passing a fixed point per unit of time. The use of a scale expressed in wave numbers (cm^{-1}) in infrared spectroscopy is preferred to a scale of wavelengths (μm) because wave numbers are linear and proportional to the energy and frequency being absorbed. Wave numbers are sometimes incorrectly referred to as frequencies, an error because the wave number is expressed in reciprocal centimeters (cm^{-1}) and frequency is expressed in units of reciprocal time (s^{-1}). Figure 31-2

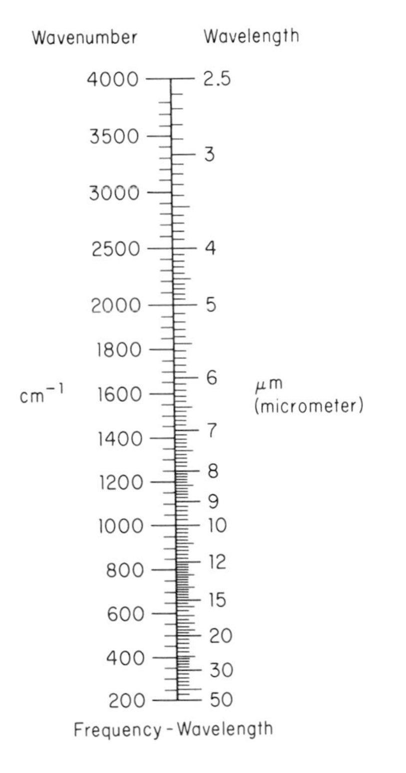

FIGURE 31-2
Frequency-to-wavelength conversion scale for infrared data.

gives a frequency-to-wavelength conversion scale for the infrared spectral region. Wave numbers can be calculated from wavelengths using the following equation:

$$\text{Wave number (expressed in cm}^{-1}) = \frac{1 \times 10^4}{\text{wavelength (expressed in } \mu\text{m)}}$$

For example, if the wavelength is 12 μm, the wave number in cm^{-1} is calculated as follows:

$$\text{Wave number} = \frac{1 \times 10^4}{12}$$
$$= 830 \text{ cm}^{-1}$$

Vibrational and rotational absorptions in the infrared region tend to occur over a rather broad range of wavelengths called *bands*. These infrared absorption bands do not usually occur as sharp lines as might be the case with visible or UV spectral absorptions involving electron excitations.

INFRARED INSTRUMENT COMPONENTS AND DESIGN

A schematic diagram of a simple dispersive-type infrared spectrophotometer is shown in Fig. 31-3. An infrared source of energy is usually provided by a *Nernst glower* which is composed of rare earth oxides (zirconium, cerium, etc.) formed into a cylinder of 1 to 2 mm diameter and 20 mm long. This filament material is electrically heated to approximately 1500°C, and the resulting infrared radiation split into a reference and a sample beam by mirrors. The reference beam is passed through air or a beam attenuator, and the sample beam is passed through a sample where selective radiation absorption occurs.

The reference and sample beams are then passed alternately by a *chopper* (a rotating half mirror) to a dispersive device which separates the radiation into specific frequencies (wavelengths). A *prism* is one type of dispersive device which separates radiation of different wavelengths based on differences in refractive indexes at different wavelengths (see Chap. 29, "The Electromagnetic Spectrum," for details on monochromator components). A prism requires the radiation to pass through the material, and since glass will absorb most infrared radiation, sodium chloride or other special materials are required. A *grating* can be used as a dispersive device and consists of a series of closely spaced parallel grooves scribed into a flat surface. A grating simply reflects the diffracted light from its surface and absorbs very little radiation. Many infrared spectrophotometers require special temperature and/or desiccants to prevent moisture damage to the optics in the monochromator.

The radiation reflecting from the grating is directed through a series of narrow *slits* which eliminate most undesirable frequencies and onto a

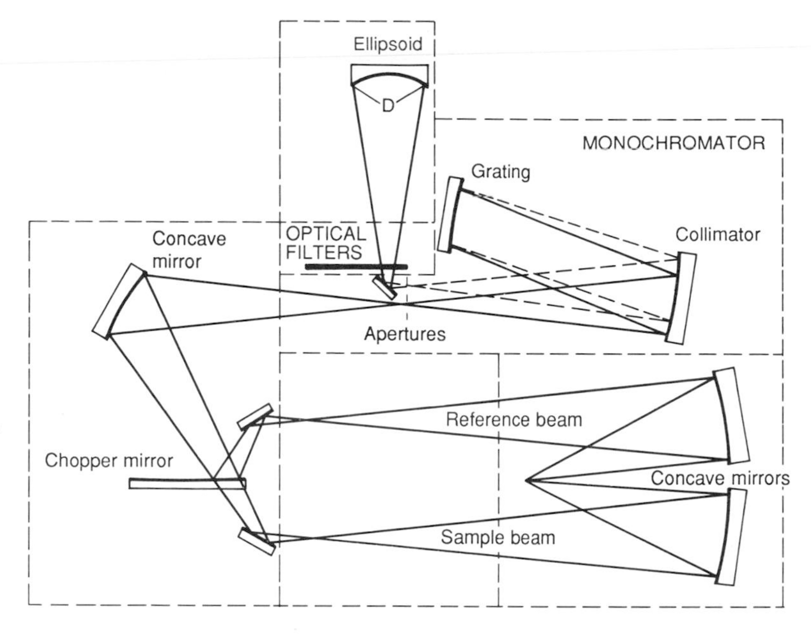

FIGURE 31-3
Schematic diagram of Perkin-Elmer Model 700. A typical double-beam, optical null infrared dispersive-type monochromator. S = source; D = detector.

detector. An infrared detector must sense very small energy changes (typically in the 1×10^{-9} W range). *Thermocouples* are routinely used in infrared detectors and consist of two dissimilar metals fused together at a junction. A potential difference develops as the temperature varies between the thermocouple junctions monitoring the sample beam and the reference beam. These thermocouples can respond to temperature differences in the 1×10^{-6}°C range. *Golay* or gas thermometers filled with xenon are used as infrared detectors above 50 μm.

Most dispersive-type infrared spectrophotometers are of the double-beam *optical null* type (Fig. 31-3). The radiation passing through the reference beam is reduced or *attenuated* to match the intensity of the sample beam. The attenuator is normally a fine-toothed comb which moves in and out of the reference beam's path. This movement is synchronized with a recorder so that its position gives a measure of the relative intensity of the two beams and thus the transmittance of the sample.

INFRARED CELL MATERIALS

Many different inorganic salts are used in infrared spectroscopy to provide optical components and cells. Table 31-1 lists some of the properties of various window materials.

TABLE 31-1
Properties of Some Infrared Window Materials

Material	Transmission range, cm^{-1}	Water solubility @ 20°C, g/100 g H_2O	Comments
KBr	40,000–400	65	Hygroscopic, inexpensive.
NaCl	40,000–625	36	Hygroscopic, easy to polish.
CaF_2	67,000–1110	0.0015	Water insoluble; do not use with ammonium salts.
AgCl	20,000–435	0.00015	Water insoluble; darkens with UV.
CsBr	20,000–250	110	Hygroscopic, fogs easily, easily deformed.
ZnS	10,000–715	0.0008	Water insoluble; called Irtran-2® (trademark of Kodak); attacked by oxidizing agent.

INFRARED LIQUID CELLS

Infrared cells for liquid samples are available in three basic types:

1. *Demountable cells* (Fig. 31-4) are used primarily for "mulls" and highly viscous liquids. These cells have no inlet ports and must be disassembled to clean and fill.

2. *Semipermanent cells* (Fig. 31-5) allow the sample to be introduced through injection ports and can be disassembled for cleaning and varying the cell's pathlength.

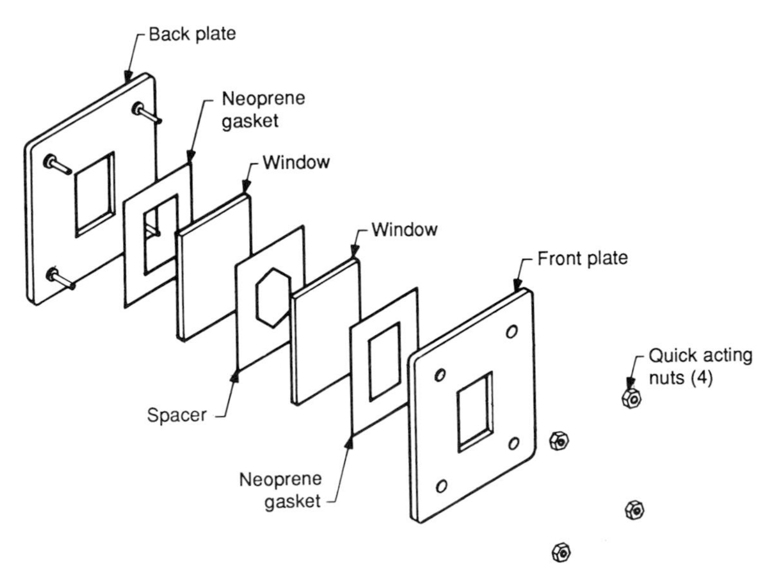

FIGURE 31-4
Demountable liquid cell.

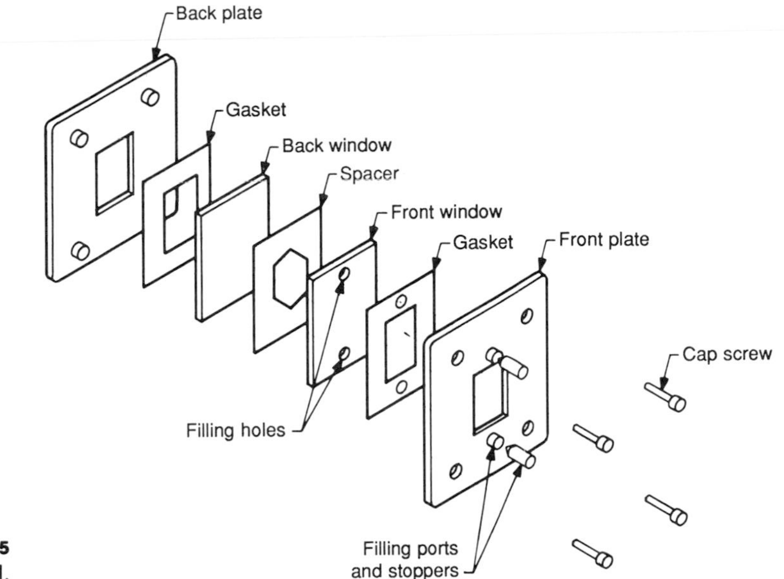

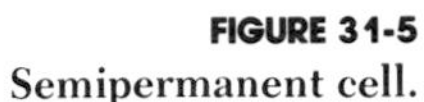
FIGURE 31-5
Semipermanent cell.

3. *Permanent cells* (Fig. 31-6) are similar to the semipermanent cells except that the windows and spacers are sealed together forming a leak-proof cell. This type of cell is recommended for highly volatile samples and precise quantitative work requiring reproducible pathlengths. These liquid cells are permanently sealed by the manufacturer and are not designed to exchange spacers. This type of cell is sometimes referred to as *amalgamated,* meaning literally "merged into a single body," and mercury is sometimes used for creating this seal.

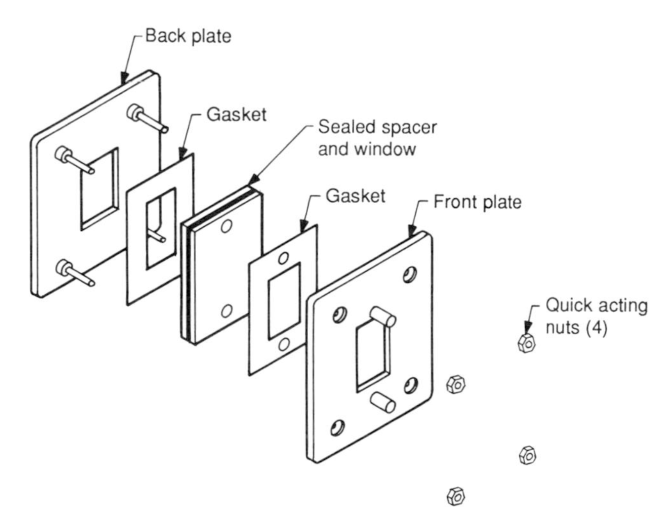

FIGURE 31-6
Permanent cell.

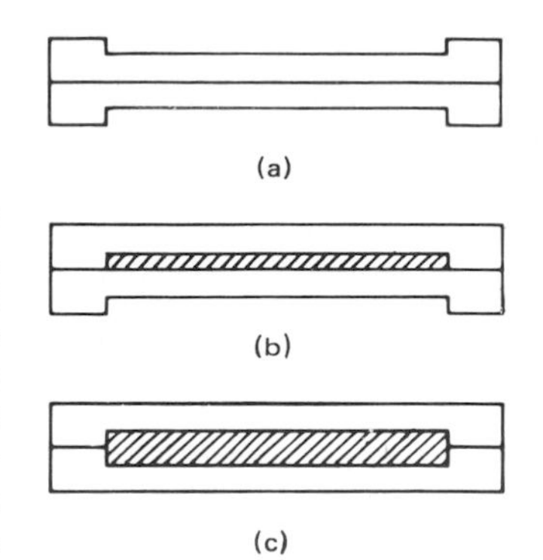

FIGURE 31-7 Pathlength variations with minicell windows. (*a*) Window positioned for smears. (*b*) Window positioned for 0.025-mm pathlength. (*c*) Window positioned for 0.050-mm pathlength.

An economical variation of a liquid cell is the liquid *minicell*. These cells are useful only for qualitative analysis and consist of a threaded, two-piece plastic body and two silver chloride windows. The AgCl window material withstands much abuse and is not as affected by water as most other infrared window materials. However, some darkening will occur upon exposure to sunlight. The AgCl windows each contain a 0.025-mm circular depression. The 0.025-mm circular depression in each window allows the cell's pathlength to be varied as shown in Fig. 31-7.

Cell pathlengths in infrared spectroscopy normally range from 0.01 to 1.0 mm and can be produced by using *spacers* made of lead, copper, Teflon®, etc. Some manufacturers provide a *variable-pathlength cell* equipped with a vernier scale in which the pathlength can be continuously adjusted from 0.005 to 5 mm reproducible. This type of cell can be used for solvent compensation and differential analysis. The cell is filled with solvent and placed in the reference beam, and the pathlength is adjusted to compensate for the unwanted solvent absorption in the sample beam. Pure liquids are referred to as *neat* liquids in infrared analysis and usually require shorter pathlengths (0.01 to 0.025 mm) because of the high concentration.

CALCULATING CELL PATHLENGTH

The pathlength for a cell (or thickness of a film) can be calculated using the cyclic interference pattern produced by the reflected radiation from the walls of an empty cell (or surface of the film). The following formula can be used to determine the cell pathlength in centimeters. An infrared spectrum showing the interference fringe pattern from a 0.150-mm empty cell is given in Fig. 31-8.

$$b = \frac{n}{2(W_1 - W_2)}$$

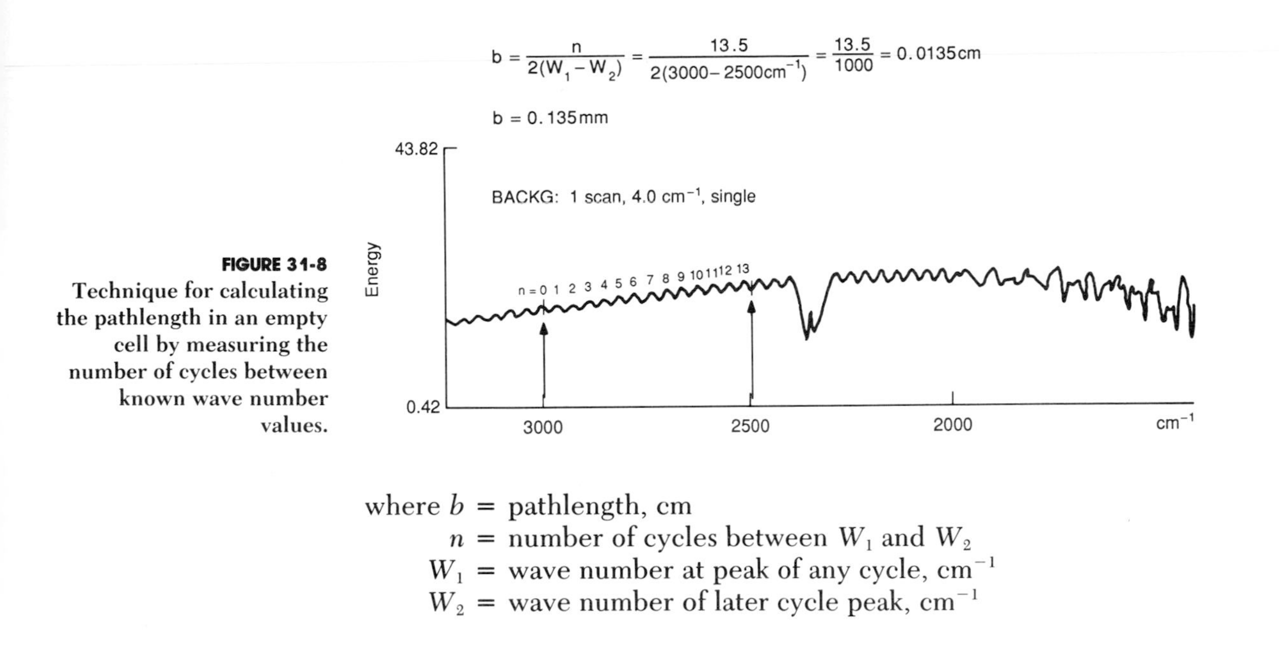

FIGURE 31-8 Technique for calculating the pathlength in an empty cell by measuring the number of cycles between known wave number values.

where b = pathlength, cm
n = number of cycles between W_1 and W_2
W_1 = wave number at peak of any cycle, cm^{-1}
W_2 = wave number of later cycle peak, cm^{-1}

SOLID SAMPLES

There are two primary methods of analyzing solids by infrared spectroscopy. The *mull method* requires that the solid sample be pulverized (mulled) and mixed in a mortar and pestle to an ointment consistency with an oil medium. Only mortars and pestles made of stainless steel or agate are recommended for preparing infrared samples because of possible contamination problems in grinding. The two most common suspending media are *Nujol®* (trademark of Plough, Inc.—a high-molecular-weight hydrocarbon recommended for use from 1370 cm^{-1} to the far-infrared) and *Fluorolube®* (trademark of Hooker Chemical Co.—a high-molecular-weight fluorinated hydrocarbon recommended for use from 4000 to 1370 cm^{-1}). The infrared absorption spectra of pure Nujol® and Fluorolube® are given in Figs. 31-9 and 31-10.

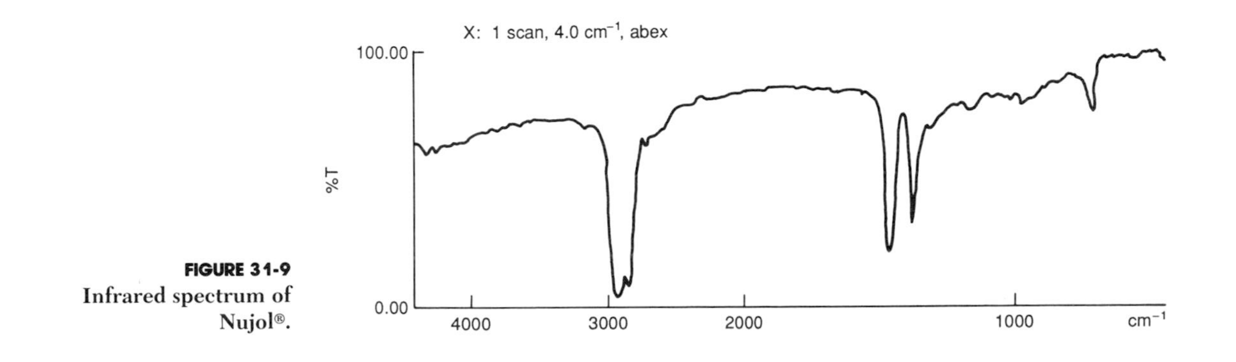

FIGURE 31-9 Infrared spectrum of Nujol®.

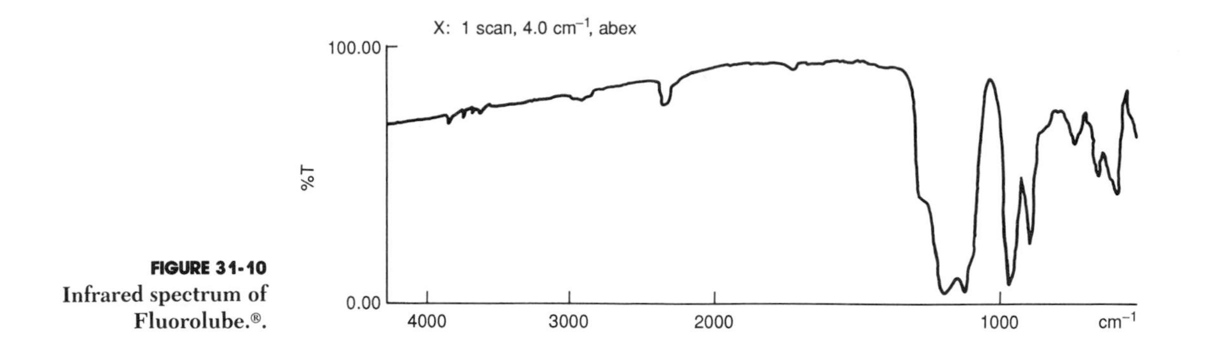

FIGURE 31-10
Infrared spectrum of Fluorolube.®.

It should be noted that both Nujol® and Fluorolube® have undesirable absorption bands, but they complement each other in that these bands are on opposite ends of the spectrum. A common technique is to run the sample in both mulling agents separately to cover the entire 4000- to 650-cm^{-1} range.

The mull (suspended solids) is then spread between two salt plates and placed in a demountable cell holder. A good mull will have a brown-blue color when held to the light and be spread uniformily between the salt plates. The mull technique is quick, but the resulting spectra are more complicated than those resulting from the KBr pellet technique described below because the Nujol® and Fluorolube® have infrared absorption bands. If there is a large slope to the baseline (greater than 0.5 absorbance units from 4000 to 1500 cm^{-1}), then the mull should be prepared again. Always disassemble the salt plates in a mull by sliding the plates (shearing them) apart. Do not attempt to pull the plates apart. The salt plates should be wiped clean with paper tissue and finally cleaned with chloroform and stored dry.

A second method of preparing solid samples for infrared analysis is called the *potassium bromide pellet technique.* The solid sample is ground and incorporated into a thin, transparent disc (pellet) of pure KBr. The KBr becomes transparent (Fig. 31-11) when subjected to pressures of 10,000 to 15,000 psi. The KBr must be pure and moisture-free. A major advantage to pressing pellets by the die technique is that the sample pellet can be saved as a reference. This KBr die technique produces a high-quality (usually 13-mm diameter) pellet, but does require a hydraulic press. Some KBr dies are equipped with a vacuum port to facilitate evacuating the cell and reducing moisture contamination. Prior to pressing, 1 mg or less of finely ground sample is mixed intimately with approximately 100 mg of KBr. The KBr bottle should always be returned to a desiccator when not in use. A grinding mill or *WIG-L-BUG*® is an ideal sample-mixing and

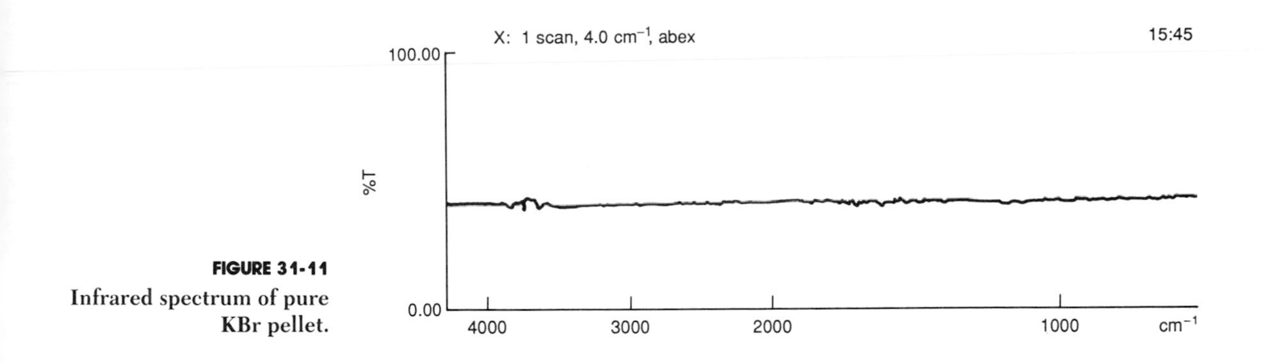

FIGURE 31-11
Infrared spectrum of pure KBr pellet.

-grinding accessory. Most samples can be prepared in less than 10 s using Plexiglas®, agate, or stainless steel ball pestles. A major disadvantage to the KBr pellet technique is its reactivity toward some samples as compared to the common mulling materials.

KBr pellets can be formed in a simpler mechanical device called a minipress which does not require hydraulic pressure. The *minipress* consists of two highly polished bolts which are manually tightened against each other in a rugged steel cylinder as shown in Fig. 31-12. Some minipress cells come equipped with a vacuum outlet for removing moisture. The KBr pellet formed inside the cylinder can then be placed directly into the infrared spectrophotometer. Some manufacturers provide special wrenches and sockets to facilitate pressing the pellet.

CAUTION Never apply pressure to a die or tighten bolts without sample and KBr present as scoring might occur. One major disadvantage to the minipress is that the pellet must be destroyed in removing it from the cylinder. A newer KBr die technique called a *quick press* is now available which does not require a hydraulic press or wrenches. The quick press has three die sets (1 mm, 3 mm, and 7 mm) and is manually operated (Fig. 31-13).

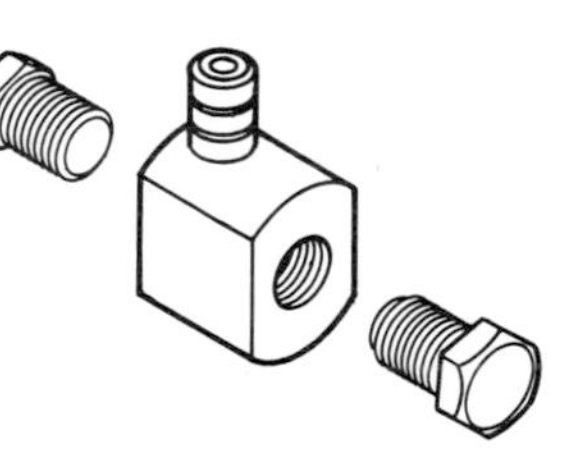

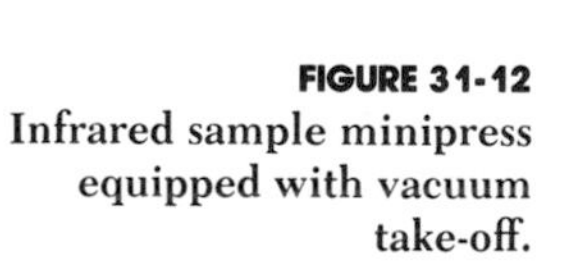

FIGURE 31-12
Infrared sample minipress equipped with vacuum take-off.

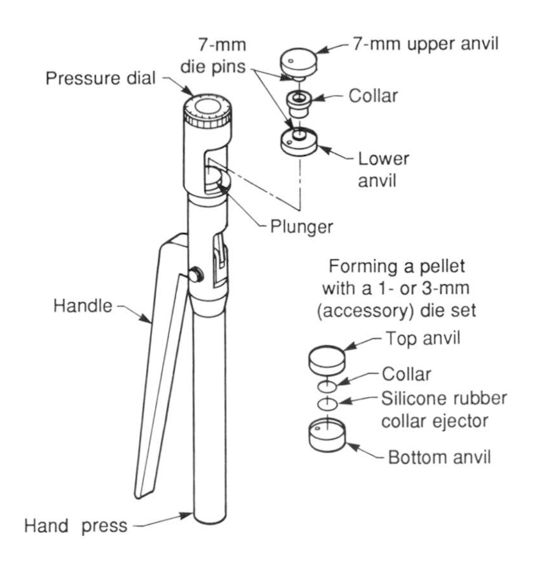

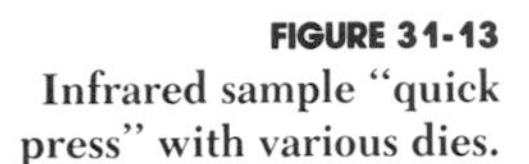

FIGURE 31-13
Infrared sample "quick press" with various dies.

FOURIER TRANSFORM INFRARED SPECTROSCOPY

The Fourier transform spectroscope provides speed and sensitivity in making infrared measurements. The Michelson interferometer is a basic component of the Fourier transform instrument. A fundamental advantage is realized as the interferometer "scans" the infrared spectrum—in fractions of a second at moderate resolution, a resolution which is constant throughout its optical range. These "scans" can be co-added tens, hundreds, or thousands of times. Co-addition of interferometer signals (called interferograms) reduces the background noise of the infrared spectrum dramatically. An interferometer has no slits or grating; its energy throughput from the infrared source is high. High efficiency of energy passage through the spectral discriminating device means more energy at the detector, where it is most needed.

The Michelson Interferometer

The Michelson interferometer (Fig. 31-14) consists of two mirrors and a beam splitter. The beam splitter transmits half of all incident radiation from a source to a moving mirror and reflects half to a stationary mirror. Each component reflected by the two mirror returns to the beam splitter, where the amplitudes of the waves are combined to form an interferogram as seen by the detector. It is the interferogram which is then Fourier-transformed into the frequency spectrum (Fig. 31-15).

The use of the interferometer allows almost all the source energy to be passed through the sample, since no radiation dispersion is required. This total energy availability is known as *Jacquinot's advantage* and increases

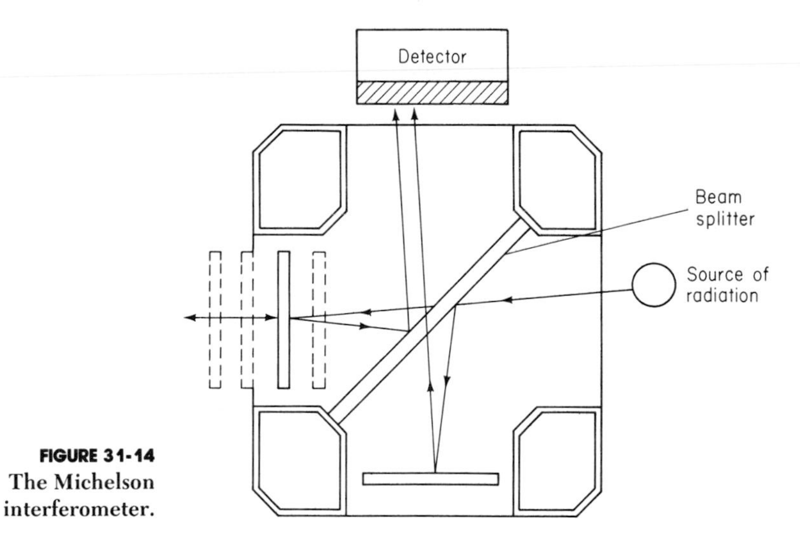

FIGURE 31-14
The Michelson interferometer.

both the speed of analysis and the sensitivity. The use of computers in Fourier transform spectroscopy has improved the signal-to-noise ratio in interferograms by allowing the co-addition of several scans, known as the *Fellgett's advantage.* The interferometer is coupled to the dedicated microprocessor which quickly yields on-line Fourier transformations. This

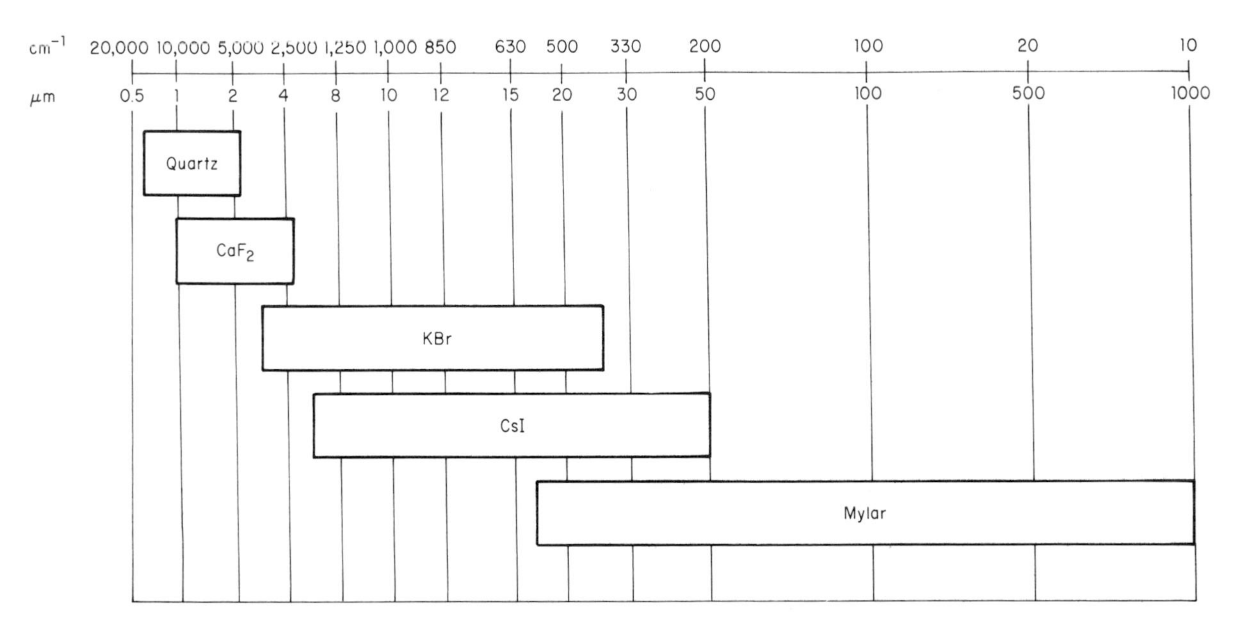

FIGURE 31-15
The wavelength ranges covered by various beam splitters.

immediate data processing and display allows the spectrometer parameters and results to be used almost instantaneously.

INFRARED ABSORPTION SPECTRA

Pure liquids can be analyzed directly in a liquid cell provided a suitable thickness is available. The liquid can be analyzed as a solution if a suitable solvent is available. A major requirement of the solvent is a lack of absorption bands in the spectral range of the solute. There are no nonabsorbing solvents in the infrared region, and the most commonly used ones are the nonpolar, non-hydrogen-containing CCl_4 (see Fig. 31-16) and CS_2 (see Fig. 31-17).

CAUTION CCl_4 and CS_2 are excellent solvents for infrared analyses but are potentially hazardous and should only be used with proper ventilation and precautions. A rough guide for analyzing solutions is 2% (wt/vol) solution for a cell with a 0.5-mm pathlength.

It would be impossible to include comprehensive coverage of infrared spectra on all common organic compounds in this Handbook, but the characteristic absorption bands for the major organic families (Table 31-2) and

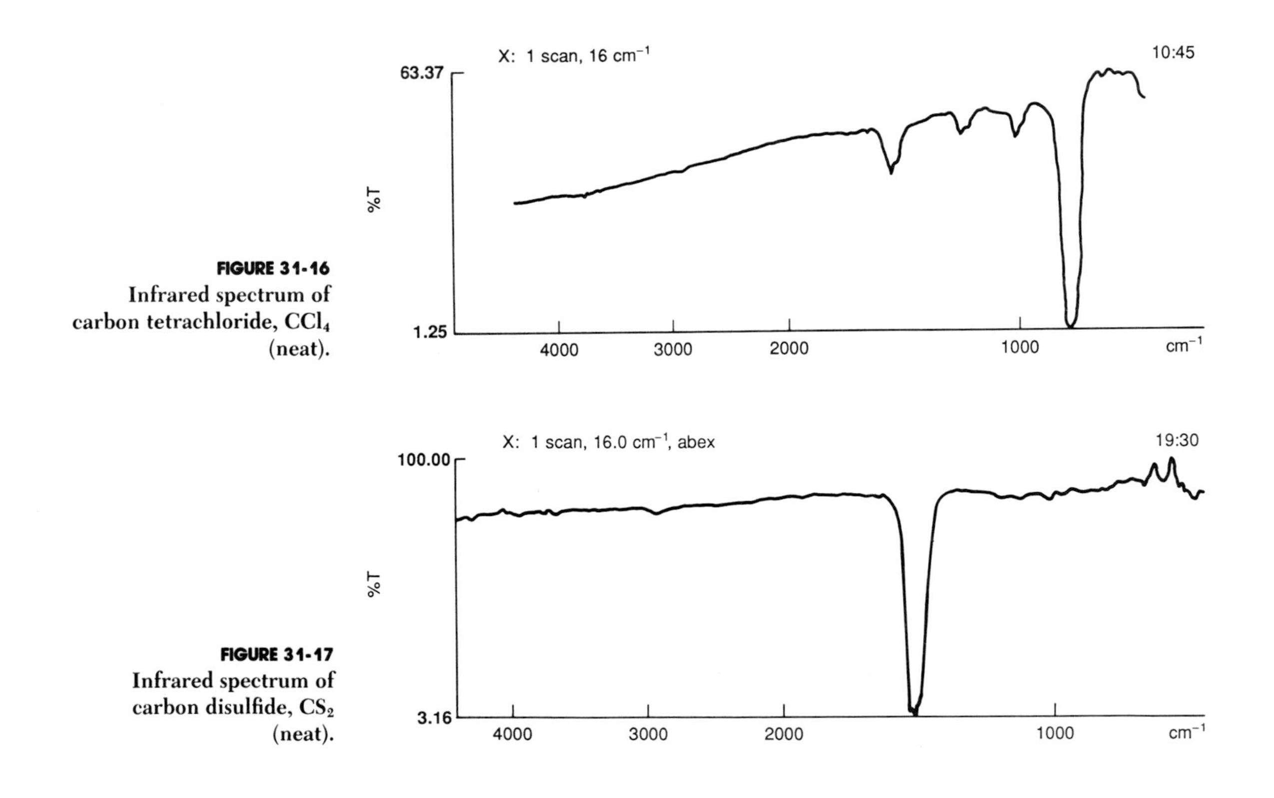

FIGURE 31-16 Infrared spectrum of carbon tetrachloride, CCl_4 (neat).

FIGURE 31-17 Infrared spectrum of carbon disulfide, CS_2 (neat).

TABLE 31-2
Characteristic Infrared Absorption Bands for Organic Families

Family	*Bond*	*Types of vibration*	*Frequency range, cm^{-1}*
Alkanes	C—H	Streching	2800–3000
	C—H	Bending	1370 and 1385
Alkenes	C—H	Stretching	3000–3100
	C=C	Stretching	1620–1680
	cis	Bending	675–730
	trans	Bending	950–975
Alkynes	C—H	Stretching	2100–2260
	C≡C	Stretching	3000
Aromatics	C—H	Stretching	3000–3100
		Monosubstituted	690–710 and 730–770
		o-Disubstituted	735–770
		m-Disubstituted	680–725 and 750–810
		p-Disubstituted	790–840
Alkyl halides	C—F	Stretching	1000–1350
	C—Cl	Stretching	750–850
	C—B	Stretching	500–680
	C—I	Stretching	200–500
Alcohols	O—H	Hydrogen-bonded	3200–3600
	O—H	Not hydrogen-bonded	3610–3640
	C—O	Primary	1050
	C—O	Secondary	1100
	C—O	Tertiary	1150
	C—O	Phenol	1230
Aldehydes	C=O	Stretching	1690–1740
Ketones	C=O	Stretching	1650–1730
Esters	C—O	Stretching	1080–1300
	C=O	Stretching	1735–1750
Ethers	C—O	Stretching	1080–1300
Amines	N—H	Stretching	3200–3500
	N—H	Bending	650–900 and 1560–1650
	C—H	Stretching	1030–1230
Amides	N—H	Stretching	3050–3550
	N—H	Bending	1600–1640 and 1530–1570
	C=O	Stretching	1630–1690
Carboxylic acids	C—O	Stretching	1080–1300
	C=O	Stretching	1690–1760

a listing of references on interpretation of infrared spectra (see References below) are given.

REFERENCES

Bellamy, L. J., *The Infrared Spectra of Complex Organic Molecules*, 2nd ed., John Wiley and Sons, New York, 1958.

Cook, B. W., and Jones, K., *A Programmed Introduction to Infrared Spectroscopy*, Heyden & Son, London, 1975.

Dyer, J. R., *Applications of Absorption Spectrosocpy of Organic Compounds*, Prentice-Hall, Englewood Cliffs, N.J., 1965.

Meloan, C. E., *Elementary Infrared Spectroscopy*, Macmillan, New York, 1963.

Sadtler Research Laboratories, *Infrared Prism Spectra*, Vols. 1–36; and *Infrared Grating Spectra*, Vols. 1–16; in the *Sadtler Standard Spectra*, Philadelphia, a continually updated subscription service.

Silverstein, R. M., Bassler, G. C., and Morrill, T. C., *Spectrometric Identification of Organic Compounds*, 3rd ed, John Wiley and Sons, New York, 1974.

32
ATOMIC ABSORPTION SPECTROSCOPY

INTRODUCTION

Atomic absorption spectroscopy (AA) is a spectrophotometric technique based on the absorption of radiant energy by atoms. Flame photometry (see Chap. 33, "Optical Emission Spectroscopy") is based on a principle that requires free atoms to be excited by a source of energy (flame). These excited atoms emit a characteristic radiation as they return to their ground (unexcited) state. However, the majority of atoms suspended in a flame remain in the ground state and never reach an excited state. If a beam of light is passed through the flame, these ground-state atoms will absorb this energy. In order for absorption to occur, the wavelength of this radiation must be characteristic for the atoms (element) present. Table 32-1 shows the characteristic wavelengths for some specific elements.

Approximately 70 elements can be determined by AA in concentrations ranging from perhaps 10 ppm for some of the difficult rare earths to less than 1 ppb for mercury by the graphite-furnace method (described below). Table 32-2 gives some typical detection limits for several elements by AA spectroscopy.

TABLE 32-1
Characteristic Atomic Absorption Wavelengths for Some Elements

Element	*Wavelength, nm*	*Element*	*Wavelength, nm*
Aluminum	309.3	Lead	217.0
Antimony	217.6	Manganese	279.5
Arsenic	193.7	Nickel	232.0
Barium	553.6	Potassium	766.5
Beryllium	234.9	Silicon	251.6
Cadmium	228.8	Silver	328.1
Chromium	357.9	Sodium	589.0
Copper	324.7	Tin	286.3
Gold	242.8	Vanadium	318.5
Iron	248.3	Zinc	213.9

TABLE 32-2
Typical Detection Limits by Atomic Absorption

Element	Detection limit, ng/mL	Element	Detection limit, ng/mL
Aluminum	0.1	Lead	0.1
Antimony	0.2	Manganese	0.01
Arsenic*	0.002	Nickel	0.04
Barium	0.1	Potassium	0.01
Beryllium	0.005	Silicon	0.02
Cadmium	0.005	Silver	0.01
Chromium	0.05	Sodium	0.002
Copper	0.02	Tin	0.8
Gold	0.1	Vanadium	0.2
Iron	0.03	Zinc	0.005

* By gaseous hydride method.
SOURCE: Analytical Division, Corporate Research Center, Allied Chemical Corporation, Morristown, N.J.

The absorption of this specific radiation follows Beer's law and is directly proportional to the concentration of atoms in the flame (Fig. 32-1). These AA spectral lines (described in Chap. 29, "The Electromagnetic Spectrum") are approximately 0.02 Å wide; therefore a continuous radiation source, like a tungsten lamp, cannot provide sufficient energy. A source called a hollow cathode lamp (see description on the next page) is usually required for each element to be investigated. Since each element usually requires a different source, AA is a poor qualitative tool but is extremely useful for quantitative determinations.

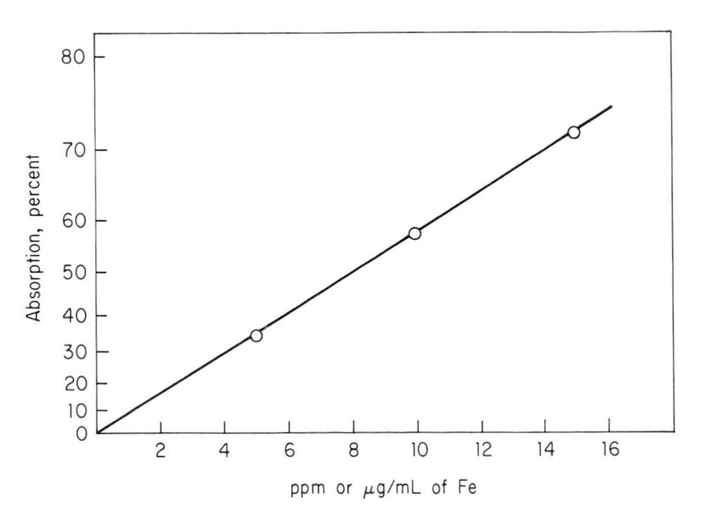

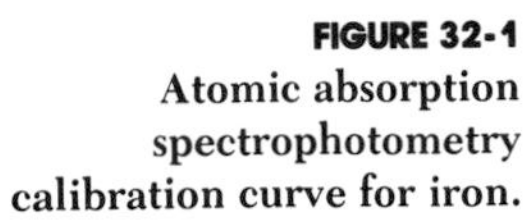
FIGURE 32-1
Atomic absorption spectrophotometry calibration curve for iron.

ATOMIC ABSORPTION SPECTROPHOTOMETER COMPONENTS

A block diagram of a double-beam AA spectrophotometer is shown in Fig. 32-2. The source of radiant energy is provided by a *hollow cathode lamp* in most conventional AA applications. These hollow cathode lamps have a cathode made of the particular element being investigated. These lamps are sealed and usually contain a noble gas: helium, argon, or neon gas at 1 to 2 torr pressure. When energized, the metal atoms contained in the lamp's cathode are bombarded by the ionized noble gas, elevated to an excited state, and ultimately emit light energy as they return to their ground state. The window material in the lamp is normally made of silica for wavelengths shorter than 250 nm; glass is used for wavelengths longer than 250 nm. These lamps routinely have an operating life of approximately 1000 h (at 5- to 25-milliamp range), but the lamp can be abused with higher operating currents. Most hollow cathode lamps require between 5 to 30 min for "warmup"; if a lamp requires more than a 30-min warmup, it probably should be replaced. Multi-element lamps are also available in which a metallic powder from several highly purified elements is combined into a sintered cathode. These multi-element lamps are normally limited to two or three elements because of their more complex spectral emissions.

Each double-beam AA monochromator has a *chopper* located in the optical beam between the cathode-lamp source and the flame. This chopper is actually a rotating half mirror; the beam is allowed to pass directly through the flame half the time and is reflected to bypass the flame for half the time. This allows the detector to compare the radiation from the source and the flame itself. A *photodetector* measures the light passing through the flame both before and after the sample is introduced into the flame.

The characteristic line spectrum emitted from a lamp is directed through a flame that contains unexcited atoms of a metal sample. Only if the metal atoms suspended in the flame and the hollow cathode filament atoms are

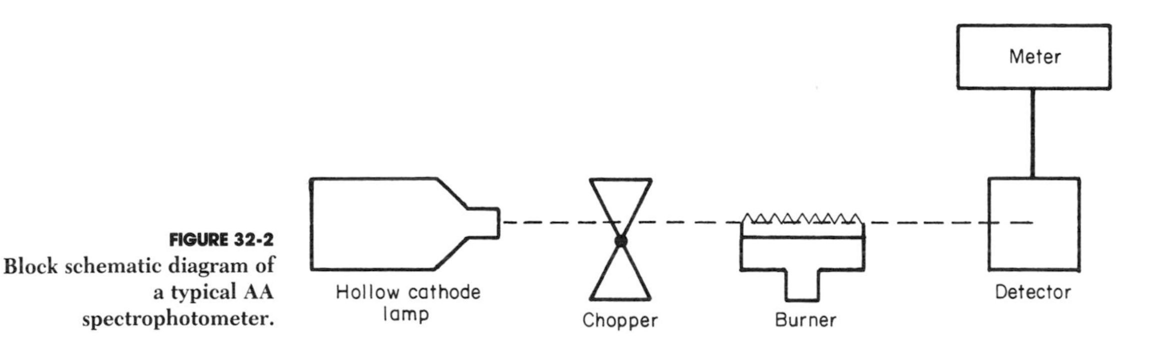

FIGURE 32-2
Block schematic diagram of a typical AA spectrophotometer.

TABLE 32-3
Elements Analyzed by Atomic Absorption Spectroscopy and Flame Mixtures Commonly Used

Air–acetylene mixture	*Nitrous oxide–acetylene mixture*
Antimony	Aluminum
Chromium	Beryllium
Cobalt	Germanium
Copper	Silicon
Lead	Tantalum
Iron	Titanium
Magnesium	Tungsten
Zinc	Zirconium

the same will light-energy absorption occur. This absorbance is proportional to the concentration of metal atoms in the flame.

All conventional AA instruments require a burner and a flame. The *burner head* provides a means for suspending and exposing the free metal atoms to the hollow cathode light beam. There are two commonly used gas mixtures: (1) air–acetylene, which burns at approximately 2300°C and (2) a nitrous oxide–acetylene combination that produces flame temperatures in the 3000°C range. Usually a burner with a 10-cm slot is used with air–acetylene mixtures and a smaller, 5-cm-slot burner with the hotter nitrous oxide–acetylene mixtures. Table 32-3 provides a listing of some common elements which may be determined with different AA flames.

Unlike in flame photometry, most AA burners are not of the *total-consumption* type. Atomic absorption spectroscopy burners tend to be of the *premix* (laminar-flow) type which aspirate the sample solution through a capillary tube into a mixing chamber where the sample is atomized and mixed with the fuel (acetylene). Some AA instruments use a *nebulizer* which simply reduces the sample solution by a pneumatic process to a spray of droplets of various sizes. The larger droplets of the sample settle to the bottom of the mixing chamber because of a series of baffles. A majority (approximately 90%) of the aspirated sample drains from the bottom of the mixing chamber and is discarded; only a small (but reproducible) portion of the sample solution passes through the burner head.

SAFETY

Acetylene is the primary fuel source in AA applications. This particular gas is very flammable. A phenomenon called *flashback*—a minor explosion inside the burner assembly—can occur with an improper mixture of fuel and air. Most instruments come equipped with safety cables and shields to protect the operator; these safety devices should never be removed. The potential of flashback can be reduced by periodically clean-

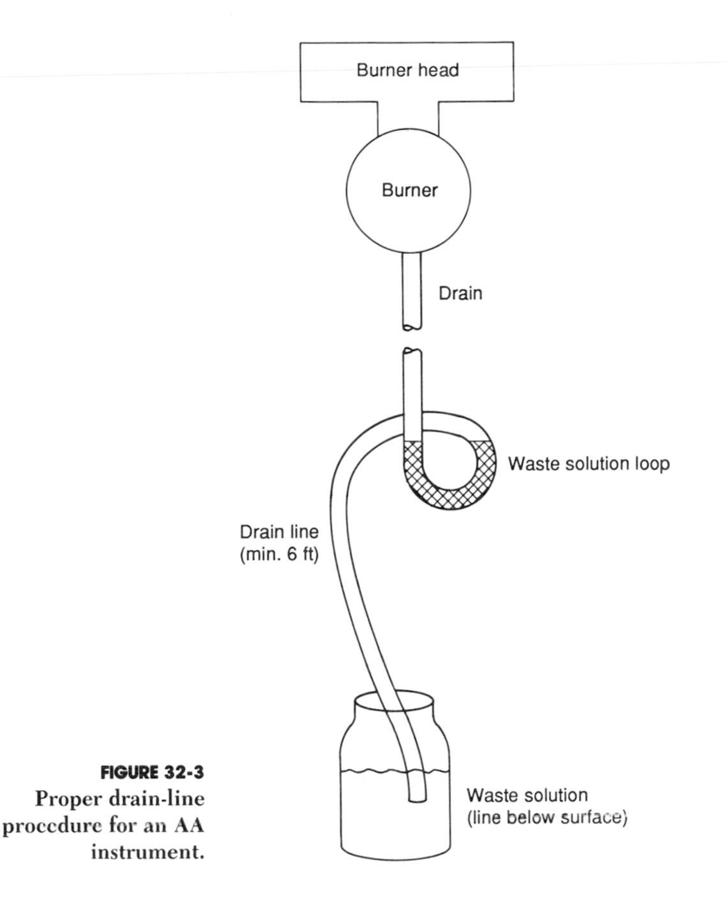

FIGURE 32-3 Proper drain-line procedure for an AA instrument.

ing the burner nebulizer and head following the manufacturer's suggested procedures. Flashbacks can be caused by air being drawn through the drain line on the premix burner. This drain line should always contain a loop filled with solution, and the end of the drain tube should be below the surface of the waste solution as shown in Fig. 32-3. As with any potential source of hazard, follow the directions provided by the manufacturer for your specific AA instrument.

CAUTION *A fire extinguisher should always be available and safety glasses worn while operating an AA instrument.*

CAUTION *Always turn the air on first and off last when igniting the AA burner.*

Each AA instrument should have a dedicated ventilation system above the burner. The actual operation of the instrument can create a potential hazard with free metal atoms being exhausted into the laboratory atmosphere. Elements like beryllium and mercury are extremely toxic in the free atomic state. Remember that the instrument could be exhausting potentially toxic metal atoms, even if not analyzing for them specifically.

CAUTION *Always turn the AA ventilation system on before starting any work.*

NONBURNER TECHNIQUES (GRAPHITE FURNACE)

A recently developed flameless AA technique uses a device called a *graphite furnace.* This technique is more efficient than conventional AA because most of the sample is atomized and stays in the optical beam. The typical graphite furnace consists of a small, cylindrical graphite tube equipped with an injection opening on the top. A measured amount of sample (usually just a few microliters or micrograms) is directly placed into the interior of the furnace. The furnace is normally heated electrically in three stages: (1) a relatively low temperature to drive off the solvent, (2) a higher temperature to ash the sample, and (3) heating to incandescence (approximately 2500°C) to atomize the sample. The graphite tube must be surrounded by a nitrogen or argon atmosphere to prevent air oxidation. The detection limits of this technique, of course, depend upon the element and operating conditions, but results in the 1×10^{-12} g range can be obtained. Mercury is routinely analyzed by this technique where a detection limit of 0.03×10^{-12} g/mL is not uncommon.

ATOMIC ABSORPTION SPECTROPHOTOMETER CALIBRATION AND INTERFERENCES

The performance of an AA spectrophotometer is generally expressed in terms of *sensitivity* and *detection limit. Sensitivity* is defined as the concentration of an element necessary to absorb 1% of the incident light energy. *Detection limit* is defined as the concentration of an element necessary to cause a reading equal to twice the standard deviation of the background signal.

Atomic absorption spectroscopy is not subject to spectral interferences because of the narrow band width created by the hollow cathode lamp. A potential source of interference can be certain anions with specific cations. For example, sulfates, phosphates, and silicates are known to interfere with some alkaline earth cations (namely, calcium, strontium, and barium). This particular type of chemical interference can be reduced by the addition of lanthanum ions to the sample solution which compete for the interfering anions.

Another potential source of error in AA has to do with sample viscosity. Unknown solutions can vary widely in their viscosity, and viscosity affects the rate of sample aspiration into the flame. The technique of "standard additions" can reduce this particular matrix effect. Known quantities of the element of interest are spiked directly into the unknown solution. The calibration technique of standard additions consists of adding small increments of known concentrations directly to the unknown solution and determining the absorbance after each addition. This technique virtually eliminates errors caused by the sample's complexed matrix (background) and viscosity, since these variables are constant in both the sample and standard addition samples. The most accurate technique for using standard additions would be to plot the standard additions concentrations against the absorbance data and extrapolate the graph to zero absorbance. The length of the x-axis (concentration) from the origin to the point of intersection represents the original concentration of the unknown. Figure 32-4 shows a typical set of data and standard additions graph.

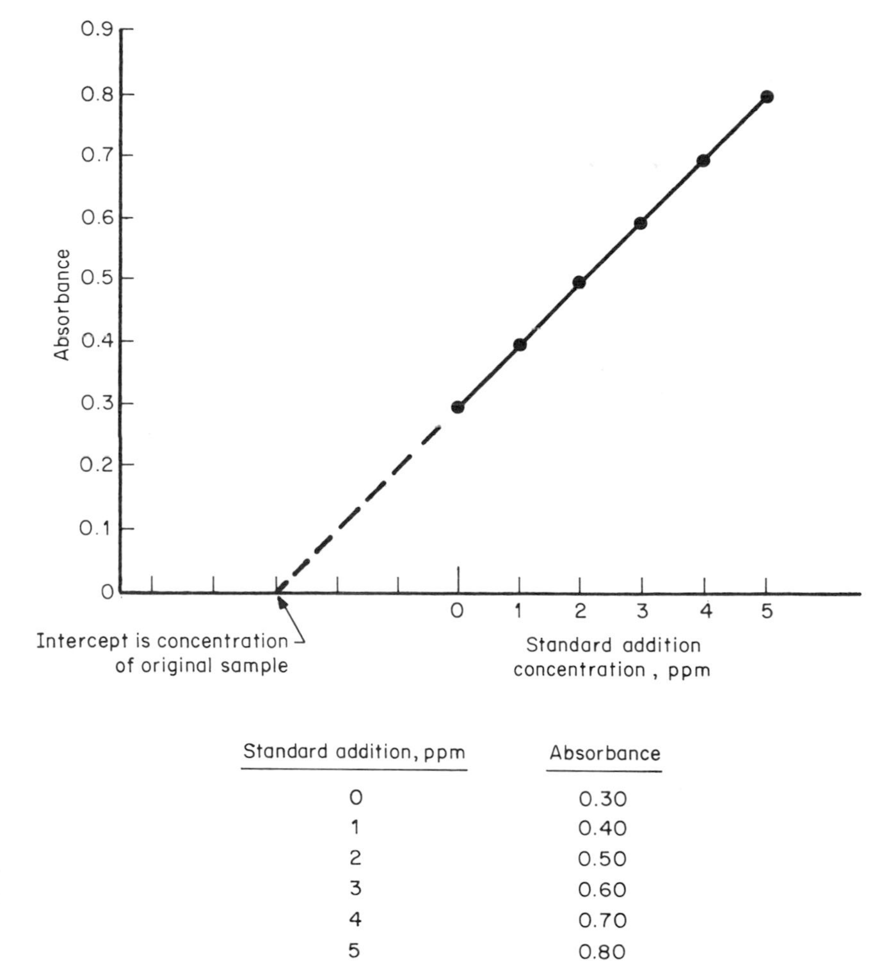

Standard addition, ppm	Absorbance
0	0.30
1	0.40
2	0.50
3	0.60
4	0.70
5	0.80

FIGURE 32-4
Standard additions calibration curve for AA calibration.

REFERENCES

Grove, E. L. (ed.), *Applied Atomic Spectroscopy,* Vols. 1 and 2, Plenum Press, New York, 1978.

Slavin, M., *Atomic Absorption Spectroscopy,* John Wiley and Sons, New York, 1978.

Van Loon, J. C., *Analytical Atomic Absorption Spectroscopy—Selected Methods,* Academic Press, New York, 1980.

33

OPTICAL EMISSION SPECTROSCOPY

FLAME PHOTOMETRY

In the section entitled Flame Tests for Identification of Elements in Chap. 15, "Gravimetric Analysis," the classical technique of identifying metal ions in solution by exposing them to a flame was discussed. Each metal atom yields a characteristic color upon being energized by a flame. This phenomenon is called *atomic emission* because the flame actually reduces the metal cations to their free atomic state and the process of light emission involves atoms, not ions. If this simple flame emissions test is made more sophisticated with a reproducible sampler-burner system, monochromator, and photomultiplier detector, it is known as *flame photometry* (see Fig. 33-1). The number of metals with which flame photometric methods can be used depends primarily on the flame temperature. The alkali and alkaline earth metals tend to be the easiest elemental families to undergo excitation and emission. Flame photometry has found extensive use in clinical analysis, especially for sodium, potassium, and calcium.

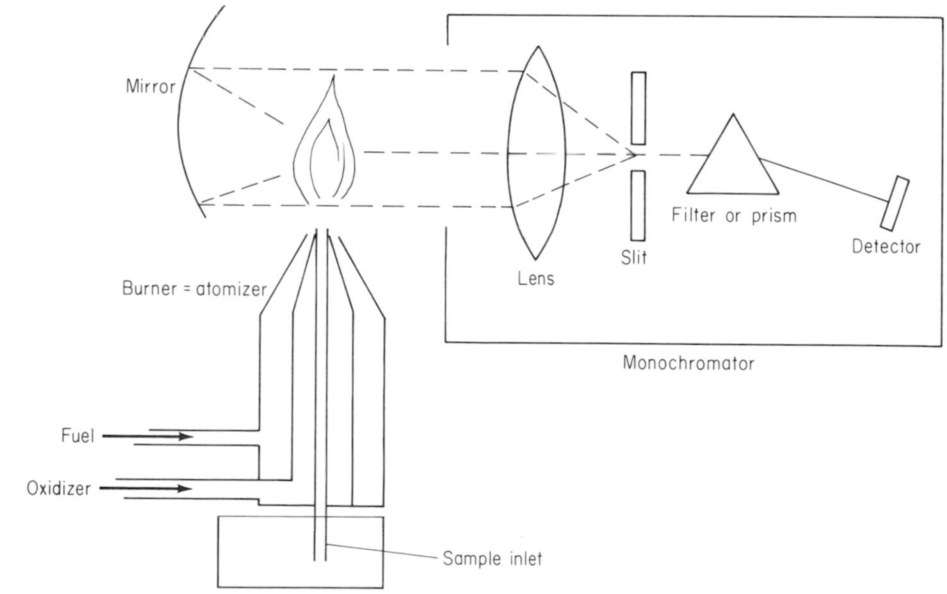

FIGURE 33-1
Flame photometer.

TABLE 33-1
Detectability of Metals by Flame Photometry

Element	Detectability, ppm
Barium	0.03
Calcium	0.005
Cesium	0.005
Gallium	0.01
Indium	0.03
Lanthanum	0.1
Lithium	0.00003
Potassium	0.0001
Rubidium	0.002
Sodium	0.0001
Strontium	0.004

SOURCE: Analytical Division, Corporate Research Center, Allied Chemical Corporation, Morristown, N.J.

Flame Photometric Quantitative Analysis

The intensity I of light emitted from the flame is directly proportional to the concentration c of the metal atoms. The mathematical relationship is:

$$I = kc$$

The proportionality constant k is simply an empirically determined constant for the specified metal. In most laboratory applications, the intensity of a series of standard solutions is plotted against their known concentrations. According to the equation above, a linear plot should be obtained, and the concentration of an unknown solution can be determined directly from the graph. The proportionality constant is normally not determined in flame photometric work. Advanced techniques that employ lithium as an internal standard allow routine analyses in the parts per billion range. Table 33-1 gives some typical detectabilities of various metals by flame photometry.

FLUOROMETRY

Most of the spectrophotometric applications discussed thus far have been based on the absorption of radiant energy (visible, UV, infrared, etc.). Some substances will not only absorb radiation, but will simultaneously re-emit radiation of a different wavelength. A common example of this phenomenon can be found with the exposure of novelty posters treated with fluorescing chemicals to black light radiation (UV). This phenomenon is called *fluorescence* or sometimes *phosphorescence* and can be used to determine the quantity of substances with similar absorption but different emission spectral characteristics. Since only a few substances are found to fluoresce in nature, this technique can be very selective and sensitive. Fluorescence spectroscopy is primarily used in biomedical (fused benzene ring systems) applications, some high-pressure liquid chromatography (HPLC) applications, and with metals that complex with fluorescent ligands.

An instrumental technique called *x-ray fluorescence spectroscopy* uses x-ray as the incident beam and analyzes the emitted (fluorescence spectrum) radiation. This technique can provide a very rapid, nondestructive, elemental analysis for both metals and nonmetals. Most routine x-ray fluorescence applications are generally limited to elements with greater atomic numbers than 21 (scandium) because of air absorption.

Since fluorescing substances have two characteristic spectra, the excitation spectrum and the emission spectrum, two monochromators or two different filters are required for the *fluorometer* (sometimes spelled "fluorimeter"). Fluorometers normally generate the excitation energy from a source similar to conventional spectrophotometers (xenon, mercury vapor lamp, etc.). A second monochromator is normally placed at a 90° angle to the sample to isolate and measure the fluorescent radiation energy. Many compounds fluoresce at reduced temperatures, therefore some spectrofluorometers are equipped with cryogenic (liquid nitrogen at 77 K) accessories. A block diagram of a typical fluorometer is shown in Fig. 33-2.

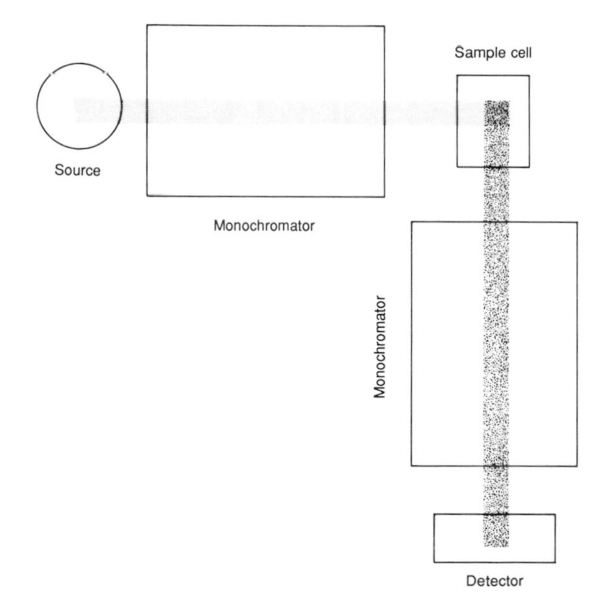

FIGURE 33-2
Block diagram of a typical fluorometer.

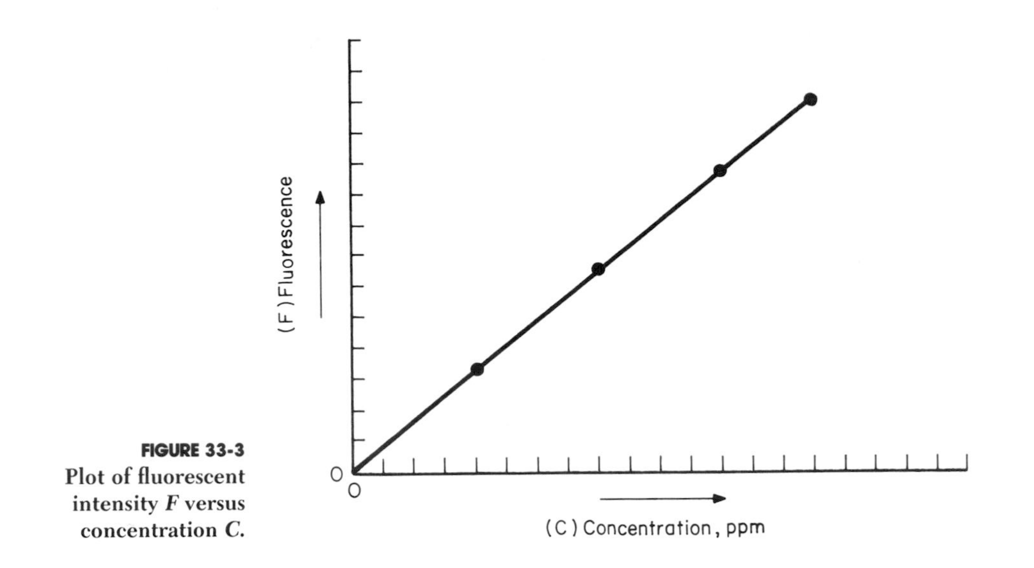

FIGURE 33-3
Plot of fluorescent intensity F versus concentration C.

Fluorometer Quantitative Analysis

The total fluorescent intensity F of a fluorescing species is directly proportional to its concentration c. The mathematical relationship is very similar to the one derived for flame photometry. The proportionality constant k depends upon the intensity and wavelength of the incident radiation, temperature, pH, sample cross-sectional area, etc.

$$F = kc$$

As was the case with flame photometric analysis, the proportionality constant is rarely determined. The fluorescent intensities of various known concentrations are determined and ratioed to the unknown's fluorescent intensity under similar operating conditions. A plot of these fluorescent intensities versus concentration should yield a linear plot (see Fig. 33-3), and the unknown's concentration can be determined directly from the graph by interpolation.

INDUCTIVELY COUPLED ARGON PLASMA SPECTROSCOPY

Several techniques for analyzing metal samples without a flame have been introduced in the past decade. One device called an *inductively coupled argon plasma torch* (ICAP torch) provides an excitation source by creating a *plasma* (stream of charged particles) with argon. An induction coil is wrapped around a quartz tube in which a stream of charged argon particles and sample solute is flowing. The sample must be in solution and is normally introduced through a nebulizer as in atomic absorption (AA) spectroscopy applications. The interaction between the induced magnetic field from the coil and the argon plasma create an extremely high temperature (approximately 10,000 K). The ICAP is usually more sensitive, can do multi-elemental analyses, and has fewer matrix problems than conventional AA. However, ICAP is very expensive both to purchase and to operate; it is economically competitive with AA when large numbers of samples and multi-elemental analyses are required.

REFERENCES

Barnes, R. M. (ed.), *Emission Spectroscopy*, Halsted, New York, 1976.

Christian, G. D., *Analytical Chemistry*, 4th ed., John Wiley and Sons, New York, 1986.

Ewing, G. W., *Instrumental Methods of Analysis*, 4th ed., McGraw-Hill, New York, 1975.

34

NUCLEAR MAGNETIC RESONANCE SPECTROSCOPY

INTRODUCTION

Nuclear magnetic resonance (NMR) spectroscopy provides a rapid, accurate, and nondestructive method for determining the structure of simple inorganic to complexed biochemical compounds. Unlike ultraviolet, visible, and/or infrared spectroscopy, NMR does not involve electrons or bonds but the nuclei of compounds.

The nuclei of all atoms possess a charge and spin about an axis; this creates a magnetic field which is analogous to the field produced when an electric current (electrons) passes through a coil of wire. The spin number for both the electron and the proton is ½; however, not all nuclei possess a magnetic moment (see Table 34-1). Only nuclei with an odd sum of protons and neutrons have magnetic moments; ^{1}H and ^{13}C are the most often studied. An instrument designed to study the magnetic resonance of ^{1}H nuclei, or protons, is referred to as a *proton magnetic resonance* (PMR) spectrometer.

When a substance containing hydrogen nuclei (almost any organic compound) is placed in a powerful magnetic field, the tiny magnetic moment of each proton will orient itself either in alignment with this external magnetic field or against the external field. Thus the applied external magnetic field causes all the sample protons to be in either a low energy state (aligned with the magnetic field) or a high energy state (aligned against

TABLE 34-1
Spectral and Magnetic Properties of Some Common Nuclei

Nucleus	*Spin number*	*Absorption frequency,* MHz*	*Isotopic abundance, %*
^{1}H	½	60.0	99.98
^{2}H (deuterium)	1	9.2	0.016
^{12}C	0	—	98.9
^{13}C	½	15.1	1.11
^{19}F	½	56.6	100.0
^{31}P	½	24.3	100.0

* At 14,092 G external magnetic field.

the magnetic field). The nuclei in the lower energy state can be caused to absorb electromagnetic radiation (in the radio-frequency range) to make the transition from the lower to the higher energy state. When a radio-frequency field is superimposed over a stationary magnetic field containing specific nuclei, the energy transition (ΔE) from the lower energy state to the higher state can be calculated from *Planck's equation:*

$$E = \frac{\gamma H_0}{2\pi}$$

where ΔE = frequency
γ = gyromagnetic ratio = $2.674 \times 10^4\ G^{-1*} \cdot s^{-1}$
H_0 = magnetic-field strength = 14,092 G for ^1H
π = 3.142

Calculations with Planck's equation yield a frequency of $60.00 \times 10^6\ s^{-1}$ (or 60.00 MHz) for ^1H. The quantity 1 MHz represents a million (M, the SI prefix for "mega" or "1 million") hertz (Hz) or cycles per second.

CONTINUOUS-WAVE NMR INSTRUMENT DESIGN

The absorption of energy by the specific nuclei can be obtained by changing the radio frequency while holding the magnetic field constant or varying the magnetic field systematically while holding the radio-frequency oscillator constant. It is electronically simpler to hold the radio frequency constant and vary the magnetic field continuously. For a given nucleus, both the radio frequency and magnetic-field strength are directly proportional and can be plotted against the NMR absorption spectrum. A block diagram of a permanent-magnet-type, continuous-wave NMR spectrometer is shown in Fig. 34-1.

Permanent magnets which produce fields in the range of 10,000 G and electromagnets of approximately 25,000 G are most often used. These permanent magnets (14,092 G) tend to be found in cheaper instruments and are very temperature-sensitive. Electromagnets and superconductive magnets are found in the more expensive NMR instruments. A *field-sweep generator* consisting of a pair of coils located parallel to the permanent magnet's faces is used to alter the applied magnetic field over a very small range. For example, a PMR instrument fixed at 60 MHz radio frequency might have a magnetic-sweep range maximum of 16.7 ppm (1000 Hz) or its equivalent of 235 mG [(1000 Hz/60,000,000 Hz) × 14,092G]. These sweep ranges are normally expressed in parts per million (ppm), which is

* Magnetic-field strength is measured in gauss (G).

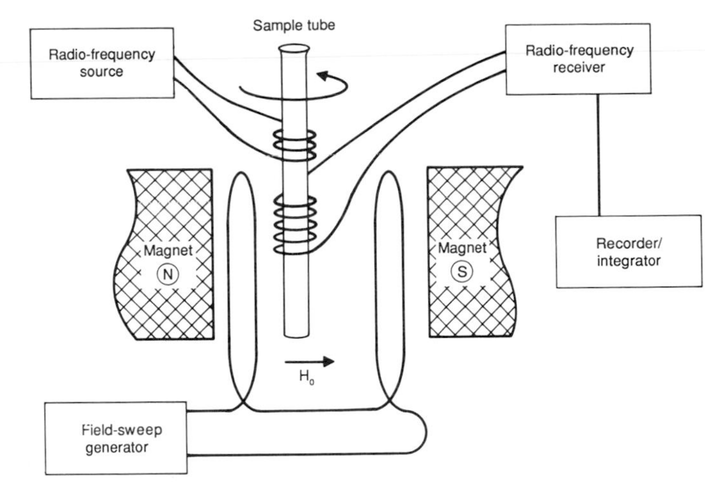

FIGURE 34-1
A block diagram of a permanent-magnet-type, continuous-wave NMR spectrometer.

not a concentration expression. It simply indicates the proportion change in the fixed-magnetic-field strength as shown in the example below:

$$\text{ppm} = \frac{235 \times 10^{-3}\ \text{G}}{14{,}092\ \text{G}} \times 10^{6} = 16.7$$

A *radio-frequency source* generates a signal from a transmitter fed into a pair of coils mounted perpendicular to the magnetic-field path. In a PMR instrument this frequency is fixed at 60 MHz and must be constant to 1 ppb for high-resolution work. A *radio-frequency receiver* detects the transmitted signal after it passes through the sample.

A *signal recorder* is used to trace the absorption of energy (ordinate) versus the abscissa drive which is connected to a radio-frequency receiver coil and is synchronized to the magnetic sweep expressed in ppm as discussed above.

An *integrator* is also available on the NMR recorder which serves the same purpose as an integrator in gas chromatography. (See Chap. 37, "Gas Chromatography," under Recorder-Integrator.) It provides a relative area under each absorption peak or grouping. This integration is normally displayed in a stepwise fashion and is superimposed over each spectrum. The integrator ratio permits an estimation of the relative number of absorbing nuclei in each proton environment. Figure 34-2 is an integrated PMR spectrum of ethylbenzene. Notice the ratio of 5:2:3 on the integrator pattern superimposed over the spectrum. The 5 represents the aromatic protons, the 2 represents the methylene group protons, and the 3 indicates the number of methyl protons.

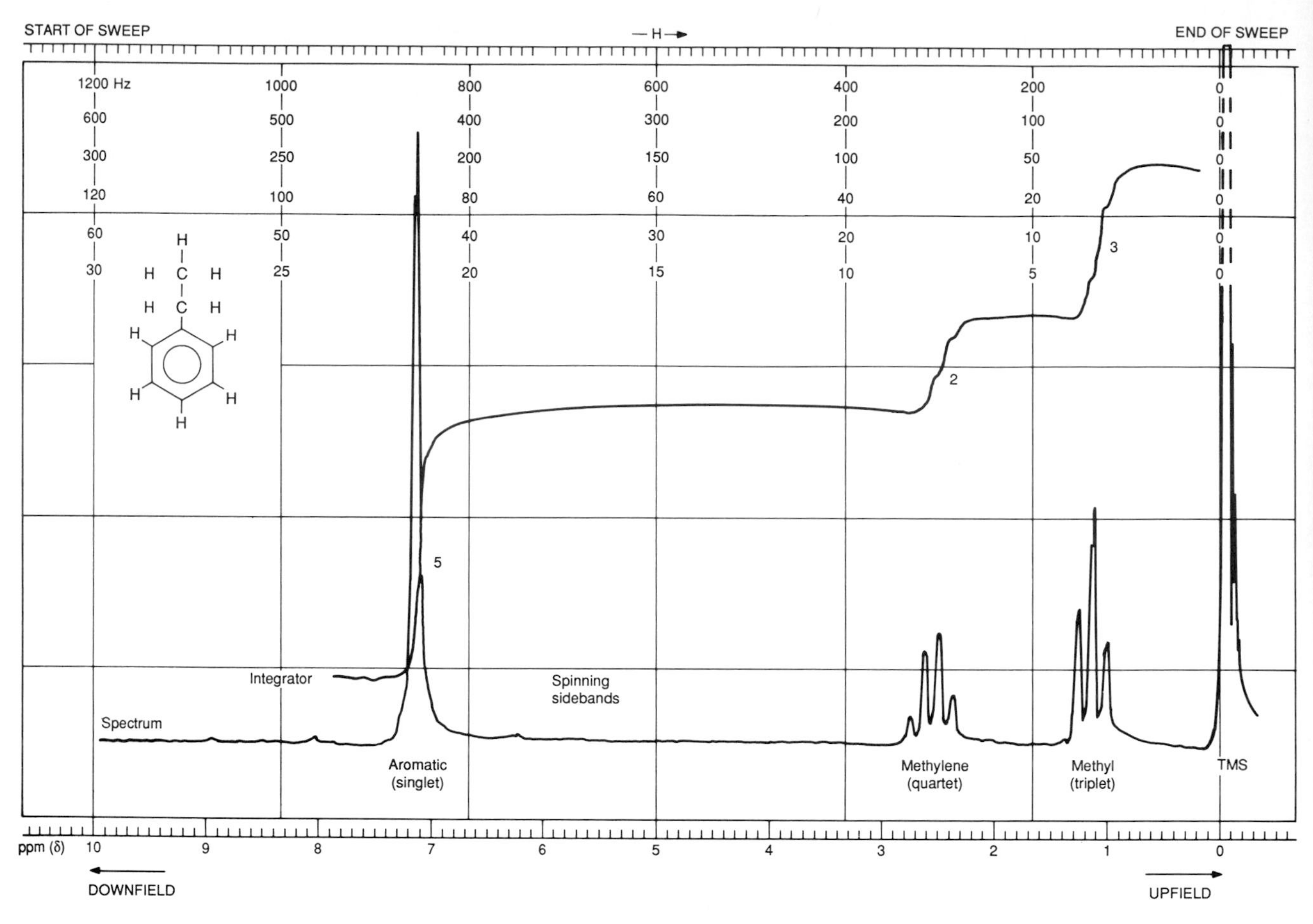

FIGURE 34-2
The PMR spectrum of ethylbenzene.

NMR Cells

NMR cells are usually 5-mm (OD) glass tubes designed to contain 0.5 mL of sample (a column approximately 2.5 cm high). These tubes are usually fitted with a spinner turbine which is set at the proper depth by a depth gauge supplied with the instrument.

CAUTION Always wipe the tube with a lint-free cloth or material before inserting into the instrument.

Microtubes are also available for smaller volumes of samples. These NMR tubes should never be cleaned with dichromate cleaning solution because possible paramagnetic residues may be introduced. Dirty tubes are best cleaned with a moderately hot soap solution introduced by a suitable pipet. Rinse the soap solution away with water, rinse with acetone, and blow dry.

Cell Spinning Rate

The spinning of the NMR tube is very important and is normally accomplished by an air-driven turbine. Rotation of the sample is necessary to average out the effects of inhomogeneities in cell walls, sample mixture, and especially the magnetic field. If the sample is spinning too slowly, broad, ill-defined peaks will result. If the spin rate is too high, *vortexing* can occur which draws a whirlpool of air and sample down into the sensing area of the tube resulting in an inhomogenous sample and poorly defined spectrum. The recommended spinning rate is usually between 30 to 60 revolutions per second (rps). A pair of undesirable peaks called *spinning sidebands* or *satellite peaks* can appear on each side of a signal peak if the spin rate is improperly set. The peaks result from improperly shaped NMR tubes or spinning at too slow a rate. The spin rate can be easily determined by measuring the distance from the center of the peak to the first sideband. Since the ppm scale can be expressed in milligauss or hertz units, the separation between the peak and the first satellite peak can be used to estimate the rate of spin. The positions of the sidebands change directly as the spin rate changes.

THE CHEMICAL SHIFT

The applied magnetic-field strength at which a given nucleus (for example, a proton) in a molecule absorbs energy relative to an arbitrary reference is called its *chemical shift*. Surprisingly not all protons in a given molecule absorb radiation at the same magnetic-field strength. This phenomenon is caused by the influence of surrounding magnetic fields on each proton. Electrons in covalent bonds tend to "shield" the nuclei from the effects of an applied magnetic field. The extent of this shielding is directly dependent on the electron density around the proton, which also depends on the structure (bonding, electronegative elements, etc.) of the compound. The most common reference material in PMR is tetramethylsilane, $(CH_3)_4Si$, abbreviated TMS. All 12 protons in this compound are identical and have a very high degree of shielding from the electropositive central silicon atom. Tetramethylsilane exhibits a single sharp resonance line at a high magnetic applied field, well beyond where most organic protons will resonate. Tetramethylsilane is chemically inert and is soluble in most nonpolar solvents. However, TMS is fairly volatile (bp = 26.5°C) and is not soluble in water. Chemical-shift values are usually expressed as "delta" values in parts per million (ppm) as shown by the following equation:

$$\text{Chemical shift, ppm} = \frac{(RF_S - RF_{TMS})}{RF_{TMS}} \times 10^6$$

where RF_S = resonance frequency of sample, in Hz
RF_{TMS} = resonance frequency of TMS reference, in Hz

TABLE 34-2
Some Typical Chemical-Shift Values

Type of proton	*Chemical shift, ppm*
Primary	0.9
Secondary	1.3
Tertiary	1.5
Vinylic	4.6 to 6.0
Acetylenic	2.0 to 3.0
Aromatic	6.0 to 8.5
Iodo (primary)	2.0 to 4.0
Bromo (primary)	2.5 to 4.0
Chloro (primary)	3.0 to 4.0
Fluoro (primary)	4.0 to 4.5
Alcohol	3.4 to 4.0
Hydroxyl	1.0 to 5.6
Alkoxyl	3.3 to 4.0
Carbonyl	2.0 to 2.7
Aldehydic	9.0 to 10.0
Carboxylic	10.5 to 12.0

A high "ppm" value would indicate that the proton is experiencing *deshielding* (lower electron density), while a lower "ppm" value would indicate that the electrons are providing shielding to the proton similar to TMS. Table 34-2 lists some typical chemical-shift values with respect to TMS.

SPIN-SPIN COUPLING

Spin-spin coupling is the effective field around one nucleus being further enhanced or reduced by local fields generated by the hydrogen nuclei bonded to an adjacent atom. Spin-spin coupling is normally only important for atoms that are three or fewer bonds apart. The splitting pattern tells how many hydrogen atoms are on the carbon atom adjacent to the proton causing the signal. The multiplicity of bands is determined by the number n of magnetically equivalent protons on neighboring atoms and is equal to $n + 1$. Consult Table 34-3 for details.

SAMPLE PREPARATION

The sample must be as pure as possible and free of particulate matter and certainly any magnetic contaminates. Most samples are analyzed as solutions in the 2 to 15% (wt/wt) concentration range. Neat samples are possible if not too viscous. The best solvents for PMR do not contain hydro-

TABLE 34-3
Splitting Patterns from Spin-Spin Coupling

Number of hydrogen neighbors (n)	*Total number of peaks (n + 1)*	*Name*	*Splitting pattern*
0	1	Singlet	1
1	2	Doublet	1:1
2	3	Triplet	1:2:1
3	4	Quartet	1:3:3:1
4	5	Quintet	1:4:6:4:1
5	6	Sextet	1:5:10:10:5:1

gen. Carbon tetrachloride is an excellent solvent for nonpolar solutes. Some laboratories prepare CCl_4 solvent in advance for NMR use by spiking with 1 to 2% TMS. Deuterated solvents like deuterated chloroform, deuterated acetone, deuterated methanol, and deuterium oxide (heavy water) are used for dissolving polar solutes. It should be noted that most deuterated solvents are never completely free of protons and some small peaks can be observed.

NMR OPERATIONAL PROCEDURE

The procedure below is by no means comprehensive and will vary for different NMR instruments.

1. Fill a clean NMR tube with enough sample and TMS to produce a liquid depth of approximately 2.5 cm.

2. Set the turbine spinner on the NMR tube using the manufacturer's depth gauge and wipe the tube clean with a lint-free cloth.

3. Carefully insert the tube into the instrument; check for proper spinning.

4. Align recorder paper in position and set all controls to manufacturer's suggested settings.

5. Sweep the field manually, locating the highest peak.

6. Adjust the spectrum amplitude control to cause the highest peak to remain on scale.

7. Select the lower and upper spectrum-range-control limits (usually 0 and 10 ppm). Set the chart zero by lining up the TMS reference peak.

8. Set the recorder baseline to the desired height, usually approximately 10% scale.

9. Install the pen/holder.

10. Scan the spectrum from either direction and record all instrument parameters.

11. Remove sample cell from instrument.

NOTE Some manufacturers recommend leaving a spinning cell in the turbine at all times.

INTERPRETATION OF NMR SPECTRA

The NMR spectrum provides four different kinds of information.

1. The number of signal groupings in a spectrum provides information about the number of different proton environments in a compound.

2. The chemical shift indicates the electron density around the various hydrogen nuclei and therefore helps to identify the proton type.

3. The integrator allows the relative number of equivalent protons of each type to be determined.

4. The splitting pattern for a given signal indicates the number of protons on the adjacent carbons.

REFERENCES

Dyer, J. W., *Applications of Absorption Spectroscopy of Organic Compounds,* Prentice-Hall, Englewood Cliffs, N.J., 1965.

Grasselli, J. G. (ed.), *Atlas of Spectral Data and Physical Constants for Organic Compounds,* Chemical Rubber Company, Cleveland, Ohio, 1973.

Pople, J. A., Schneider, W. G., and Bernstein, H. J., *High-Resolution Nuclear Magnetic Resonance,* McGraw-Hill, New York, 1959.

Roberts, J. D., *Nuclear Magnetic Resonance,* McGraw-Hill, New York, 1959.

Silverstein, R. M., Bassler, G. C., and Morrill, T. C., *Spectrometric Identification of Organic Compounds,* 3rd ed., John Wiley and Sons, New York, 1974, Chap. 4.

35

RADIOACTIVITY

INTRODUCTION

Radioactivity is the spontaneous disintegration of an atom with the emission of radiation. There are many radioactive elements which are isotopes (have the same atomic number but different atomic mass) of nonradioactives. Among these are hydrogen, carbon, iodine, and cobalt. An atom of a radioactive isotope has the same number of orbital electrons as an atom of its nonradioactive counterpart and will, in general, behave chemically and biologically like the nonradioactive species. Therefore experimental and diagnostic as well as analytical procedures can utilize atoms of radioactive isotopes as tracers. The difference between the radioactive and the nonradioactive atoms of identical elements is the number of neutrons in the nucleus, the number of protons and electrons being the same for all. (Some elements have more than two isotopes.)

Atomic radiation is generally of three types:

Alpha radiation, in which the emitted particles (heavy, positively charged particles: helium nuclei or alpha particles) travel about an inch in air and have very little penetrating effect—a piece of paper or the dead outer layer of the human skin stops them.

Beta radiation, in which the emitted particles (lighter, negatively charged particles: electrons or beta particles) have a range of up to 30 ft in the air. The most energetic can penetrate about ½ in of skin, but they are almost all stopped by ⅛ in of aluminum, ½ in of Lucite, or a like thickness of similar materials.

Gamma and x-radiation, which consist of true electromagnetic waves of higher frequency and much greater penetrating ability than those in the visible spectrum. They cannot be entirely stopped, but they can be attenuated (i.e., their intensity can be reduced) as they pass through lead or concrete barriers. Substances emitting such radiation are dangerous and must be handled with due regard to all safety precautions (Fig. 35-1).

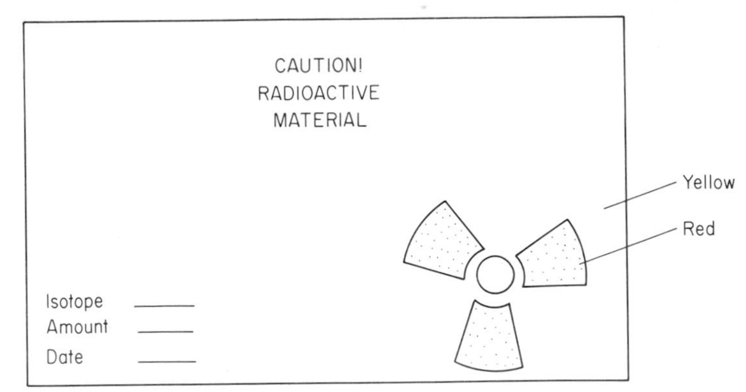

FIGURE 35-1 Standard radiation warning label.

UNITS AND CHARACTERISTICS OF RADIOACTIVITY

Specific activity is the term used to describe the rate of radioactive decay of a substance whose energies are measured in millions of electronvolts, or megaelectronvolts, MeV. The specific activity is expressed as the number of curies per gram of substance; the curie (Ci) is defined as 3.7×10^{10} disintegrations per second, which is equivalent (approximately) to 1 g of radium. However, this unit is too large and has been replaced by the rutherford (rd), which is defined as 10^6 disintegrations per second. However, some submultiples of the curie are still sometimes used (see Table 35-1).

TABLE 35-1 Conversion Factors for Units Smaller Than the Curie

Unit	*Conversion factor*	*dps (disintegrations per second)*
Ci curie	basic unit	3.7×10^{10}
mCi millicurie	10^3 mCi Ci^{-1}	3.7×10^7
μCi microcurie	10^4 μCi Ci^{-1}	3.7×10^4
nCi nanocurie	10^9 nCi Ci^{-1}	37
pCi picocurie	10^{12} pCi Ci^{-1}	3.7×10^{-2}

RADIATION HAZARDS AND SAFEGUARDS

Nuclear radiations produce biological injury by damaging cells as a direct or secondary result of their ionizing properties. Gamma and beta rays have relatively the same biological effect. Alpha particles produce about 10 times more ionizing effects than beta or gamma rays, but, since alpha particles are unable to penetrate the skin, they are not hazardous *if kept outside the body*. On the other hand, gamma and beta radiations are hazardous *both inside and outside the body.*

Specific information on the properties and hazards of various isotopes and radioactively tagged compounds is available from the supplier. (National Bureau of Standards Handbooks No. 42 and 52 also give information.) When you are planning shielding or other control measures, be sure that you consider the complete range of radiations of the original material and all its decay products. For example, ruthenium 106 has a very-low-energy beta emission, but its "daughter," rhodium 106, emits both high-energy beta rays and gamma rays.

In radiochemistry laboratories, contamination may be encountered on tools, glassware, working surfaces, clothing, hands, biological or other specimen materials, wastes, and in the air. From these sources it can be transferred unwittingly to contaminate the entire laboratory area. Unless adequate precautions are taken, radioactive material can get into the body by inhalation, ingestion, or through cuts or abrasions on exposed skin surfaces. Some materials can even be absorbed through the intact skin. Once materials are inside the body, they can be removed only by natural elimination—an uncertain process at best—or by radioactive decay, which may be extremely slow. The difficulty of removal makes prevention of absorption extremely important.

Safeguards against radioactive dust are of prime importance, since materials inhaled as dusts or taken in through wounds are more readily dissolved in the blood stream and deposited in the body than materials ingested. The laboratory area should be kept clean, with special air monitors and protective respiratory equipment wherever they are needed. Any equipment in which radioactive materials are used should be maintained as a closed system. Protective gloves and shoe covers—even complete clothing changes—may be necessary. Areas in which radioactive materials are being used should be clearly marked and restricted. Special areas should be set up for smoking, eating, applying cosmetics, etc., and for decontamination upon leaving restricted areas.

Special markings or tags should be used to warn all persons who come in contact with radioactive materials. All apparatus, containers, etc., used at each stage of the laboratory procedure from storage to disposal should be plainly marked with the date, isotope, decay products, and the type of radiation emitted.

RADIATION-DETECTION INSTRUMENTS

Both fixed and portable instruments are available for determining radiation levels in the laboratory. They detect and measure the radiation by means of an ionizing chamber or a Geiger-Müller tube and display the measurement on an appropriate indicating device. Among the larger fixed instruments are penetrating-radiation monitors, continuous air samplers, and counters for personal monitoring. Usually pocket dosimeters or film badges are used for personnel monitoring.

Since the cumulative effect of many small exposures may cause trouble, it is desirable to maintain complete records not only of individual exposure (as shown by pocket monitoring devices), but also of environmental conditions (as shown by frequent air monitoring). If there is a possibility of slight air contamination in the laboratory, samples should be taken during the entire length of the work shift. If radiation is confined to a closed system, spot checks for leaks, taken at frequent intervals, will suffice. At least once a week a thorough inspection should be made for leaks in the radiation system and for contamination of the area.

Wash your hands before eating or smoking and at the end of the shift. Before leaving the radiation area, check your hands, feet, and clothing for contamination. Schedule a medical examination at least once a year.

Geiger-Müller Counter

The Geiger-Müller counter (Fig. 35-2) is used to detect and count radioactive particles; it is also called a Geiger counter. It consists of a probe, an amplifier, and a meter to indicate the intensity. The probe is a tube

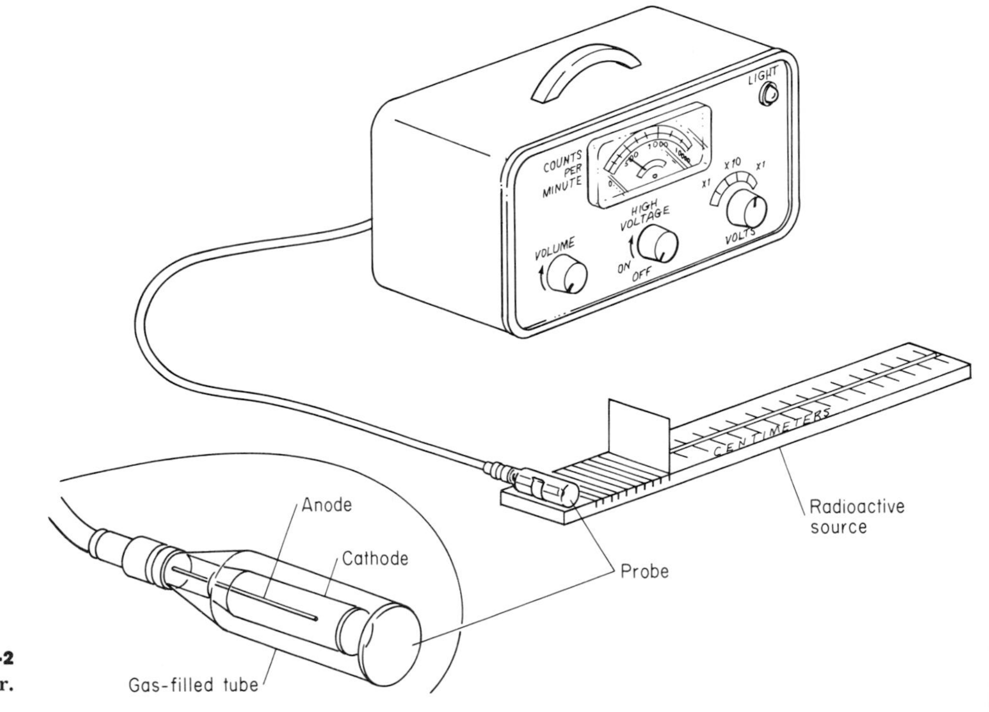

FIGURE 35-2
A Geiger counter.

filled with an inert gas, such as argon, and an organic substance such as acetone. A thin window permits radiation to enter the tube and cause ionization of some of the gas molecules inside it. The ions produced conduct electric current, which is amplified and is then displayed on the meter. The ionization that normally takes place without the presence of radioactive particles is called background radiation; it shows up on the counter as the "background count," which is subtracted from the meter reading when a radioactive substance is being tested.

Scintillation Counters

Certain substances (called *phosphors or fluors*) will emit light when they are subjected to radiations: NaI is activated by gamma rays, and anthracene and other substances are activated by beta radiations. The more intense the radiation, the greater the intensity of the light emitted. Scintillation counters (Fig. 35-3) are instruments that have light-amplification stages (photomultiplier tubes), which can amplify this light emission thousands of times. Their output signal is applied to readout instruments, such as scalers, which provide a visual record of the intensity. Just as with the Geiger counter, the background effect must be subtracted from the experimental result.

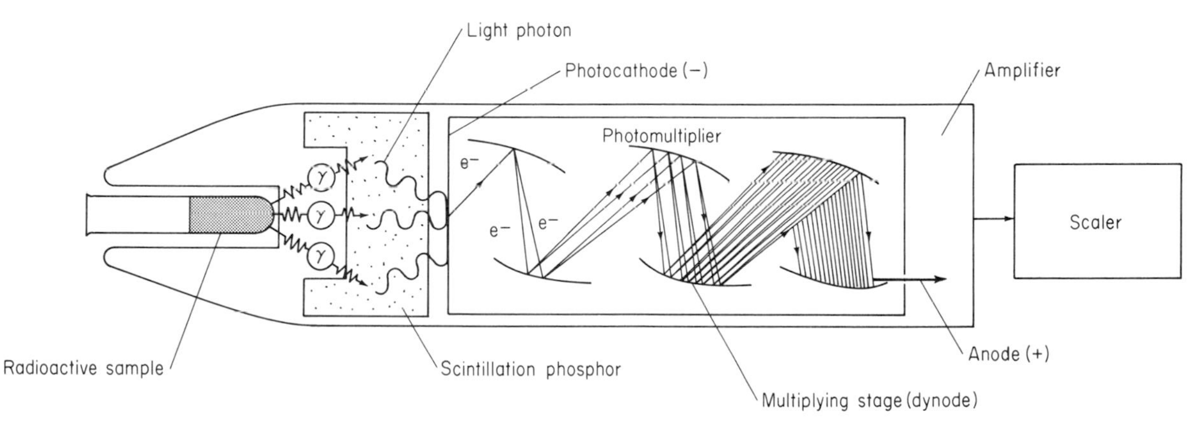

FIGURE 35-3
A scintillation counter.

DECONTAMINATION OF AREAS IN WHICH RADIOACTIVES ARE USED

Remove surface contamination as a solid whenever possible, since a solvent will embed the material in porous surfaces and make decontamination more difficult. To facilitate decontamination procedures, all working surfaces and equipment in the area should be nonporous and resistant to chemicals. Stainless steel is easily decontaminated and not too costly if a reaonably long-term operation is contemplated. Various acid-resistant varnishes and plastics can be applied to porous surfaces to simplify decontamination.

When working with liquids, place absorbent paper on the nonporous surface to absorb most of any material accidentally spilled and to prevent its distribution by air currents. Paper used for this purpose should be replaced at frequent intervals; the used paper should be discarded as contaminated waste.

Waste disposal is important to the safety program of a laboratory using radioactives. One disposal method is to dilute the waste with a convenient material such as water, or with materials containing stable isotopes of the radioactive substance.

HALF-LIFE

Radioactive substances decay (disintegrate) at a constant rate which *positively cannot be altered* by any treatment to which such a substance may be subjected.

This means that the rate is unvarying. In order to measure this rate, scientists determine how long it takes for one-half of the substance to decay, and this is called the half-life period. One-half of the amount of the substance will decay in that time; then one-half of what is left will decay in that same length of time; and so on. Some substances have a half-life measured in fractions of seconds, other in thousands of years. As a radioactive substance decays, it forms another element or an isotope of the same element.

Half-Life Calculations

Procedure

1. Make a time chart, using the half-life period as the time increment.
2. Record the amount of the substance at the start.
3. Divide that amount in half for each half-life period.
4. Determine the amount remaining at the end of the specified number of time periods.

Alternative Procedure

Use the formula

$$Q = A(\tfrac{1}{2})^n$$

where Q = amount remaining
A = starting amount
n = number of half-life periods

EXAMPLE Iodine 131 has a half-life period of 8 days. If you start with 48 mg, how much is left after 48 days?

Solution 1 The half-life period = 8 days. Therefore, using procedure 1:

Time:	Start	8 days	16 days	24 days	32 days	40 days	48 days
Sample size:	48 mg	24 mg	12 mg	6 mg	3 mg	1.5 mg	0.75 mg

Solution 2 Using the alternative procedure, the formula $Q = A(\frac{1}{2})^n$:

where $n = 48/8 = 6$

$$A = 48 \text{ mg}$$
$$Q = 48(\tfrac{1}{2})^6$$
$$= 48(\tfrac{1}{64}) = \tfrac{48}{64} = \tfrac{6}{8} = \tfrac{3}{4} = 0.75 \text{ mg}$$

Biologic Half-Life

Radioactive elements and their compounds are utilized by the body exactly as their nonradioactive counterparts. Since the body constantly eliminates waste substances through normal biological processes, some of the radioactive substances taken into the body will be eliminated along with the nonradioactive ones. A substance such as strontium 90 poses serious problems because a long period of time is required to eliminate one-half of it from the body through natural processes. It is said to have a long biologic half-life: the time that is required for half of the amount of radioactive substance taken into the body to be eliminated through natural processes. Substances that are eliminated quickly are said to have a short biologic half-life.

Radioactive Decay

The decay of radioactivity can be calculated by using the following equation:

$$\frac{A_t}{A_0} = e^{-0.693t/T}$$

where A_0 = initial activity
A_t = activity which remains after time t
t = time elapsed, in half-life units
T = half-life, expressed in some convenient units
e = base of natural logarithms (2.71828)

INVERSE-SQUARE LAW

The intensity of radiation upon a person or object varies inversely as the square of the distance. The further away from a radiation source, the less the effect by the square of the distance (Fig. 35-4).

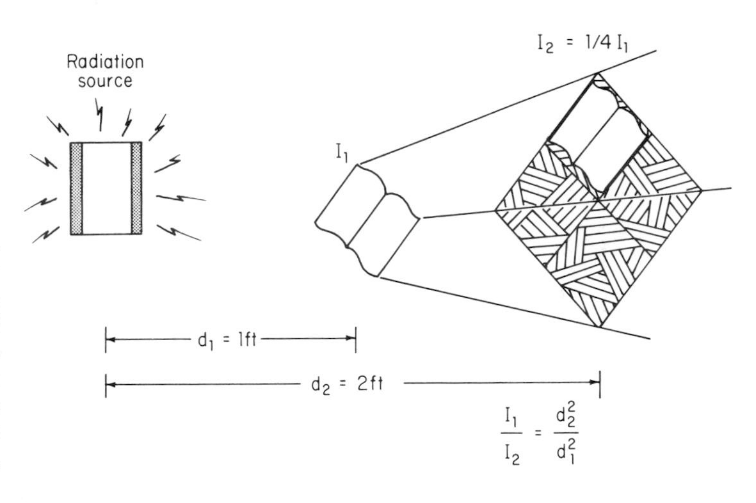

FIGURE 35-4
The inverse-square law describes the relationship between the intensity (*I*) of a radiation source and the distance (*d*) from that source. When the distance is doubled, the intensity is reduced to one-fourth.

Procedure

1. Measure the final distance.
2. Measure the original distance.
3. Divide the final distance by the original distance:

$$\frac{\text{Final distance}}{\text{Original distance}}$$

4. Invert the fraction (reduce to lowest terms first if possible).
5. Square the inverted fraction. The value obtained is the intensity of the radiations at the final distance relative to the intensity of the radiations at the original distance.

RADIOACTIVE TRACERS

Radioactive tracers are used to follow the behavior of atoms or groups of atoms in a chemical reaction or physical transformation. A radioactive tracer unequivocally labels the particular atoms to be traced, regardless of what may happen to them in a complicated system. The specificity and sensitivity of detection of radioactive isotopes is extremely good. Less than 10^{-18} g can often be detected.

Sample

Radioactive materials used for this purpose can be solid, liquid, or gaseous. Sample preparation is normally kept to a minimum. For liquid scintillation counting, the radioactive substance is put into intimate contact with the scintillation medium by dissolving it, suspending it, or immersing it in the liquid solution of fluors.

Liquid scintillation counting is commonly used for tracing a wide range of alpha- and beta-emitting radioisotopes in many chemical forms. The radioactivity is detected by means of a solution of fluors and a multiplier phototube. The scintillation solution converts the energy of the primary particle emitted by the radioactive sample to visible light, and the phototube responds to this light energy by producing a charge pulse which can be amplified and counted by a scaling circuit.

Many radioactive isotopes are available from specialty suppliers for use by the chemical, industrial, electronic, and specialty industries. The most commonly used isotopes are carbon 14, hydrogen 3 (tritium), and krypton 85. These mixtures are provided in precisely controlled concentrations which range from 0.01 μCi/L to 100 mCi/L. Each of the isotopes emits its own particular radiation, and the choice of the isotope depends upon the half-life and needs and requirements of the particular job.

REFERENCES

Brune, D., Forkman, B., and Persson, B., *Nuclear Analytical Chemistry*, Verlag Chemie, Deerfield Beach, Florida, 1984.

Choppin, G. R., and Rydberg, J., *Nuclear Chemistry Theory and Application*, Pergamon Press, Oxford, 1980.

36

CHROMATOGRAPHY

INTRODUCTION

The term "chromatography" was coined in 1906 by a Russian botanist, Mikhail Tswett, after the words *chromatus* and *graphein*, meaning "color" and "to write." Tswett's coinage was appropriate for his technique for separating colored plant pigments. Tswett discovered that by washing the compounds through a column packed with an adsorbent medium (calcium carbonate) the least-adsorbed pigments were washed through the column quickly, while the strongly adsorbed pigments were immobilized by their attraction to the column packing. (In all chromatographic processes, one medium is fixed in the system and is called the *stationary phase* while a second medium flows through the fixed medium and is called the *mobile phase.*) Adsorption in this column process is directly related to the affinity of the solute for either the stationary adsorbent or the flowing solvent, and the process is generally referred to as *column chromatography.* Later applications of this adsorption-desorption process showed that colorless substances could also be separated by this technique, but the term "chromatography" remained in use. Today, all types of chromatographic processes are used primarily for the nondestructive separation of complexed mixtures.

ADSORPTION CHROMATOGRAPHY

Adsorption chromatography typically uses silica gel or alumina as the stationary (solid) phase and organic solvents for the mobile (liquid) phase. *Adsorption chromatography* occurs when the sample components transfer from the mobile phase to the stationary phase where they are selectively adsorbed on the surface. This chromatographic process has the advantages of being simple and applicable over a wide range of temperatures, but is best for resolving heat-labile substances at low temperatures. The main disadvantages are that many substances undergo chemical changes with these very active adsorbents, separations tend to be concentration-dependent, and the adsorbents have low capacities for adsorbing solutes.

PARTITION CHROMATOGRAPHY

A second and more versatile type of chromatography called *partitioning chromatography* is based on the differential distribution of the sample components between the two phases. Partition chromatography differs from adsorption chromatography in that two liquid phases are used in which the sample components vary in their degrees of solubility. The term "partition" was first encountered in Chap. 20, "Extraction," when describing the distribution of components between two immiscible liquids. The solubility and polarity of the individual components dictate the ultimate distribution of these components in the mobile and stationary liquids. The stationary phase consists of a thin layer of liquid coated on a porous inert solid and the mobile phase can be a pure liquid or a combination of liquids. One of the first applications of this technique used water adsorbed into silica gel as the stationary phase and a mixture of butanol and chloroform as the mobile phase to separate complexed mixtures of amino acids. In addition to silica gel, other common adsorbents are cellulose, starch, diatomaceous earth, and even powdered rubber. The versatility of this system is much greater than that of adsorption chromatography, and it tends to be less concentration-dependent.

The different types of adsorption and partition chromatography based on the types of mobile and stationary phases used and the mechanism of the phase equilibria are outlined in this chapter, and in Chap. 37, "Gas Chromatography," and Chap. 38, "Liquid Chromatography."

ION-EXCHANGE CHROMATOGRAPHY

Ion-exchange chromatography (IC) is a modern day example of adsorption chromatography. As the name implies, this technique deals with the separation of ion mixtures. Many naturally occurring inorganic substances (clays, zeolites, etc.) have a strong attraction for certain ions in solution. In addition, stationary phases composed of solid polymeric material of styrene cross-linked with divinylbenzene in "bead" form having anion or cation sites on the surface have been developed (Fig. 36-1).

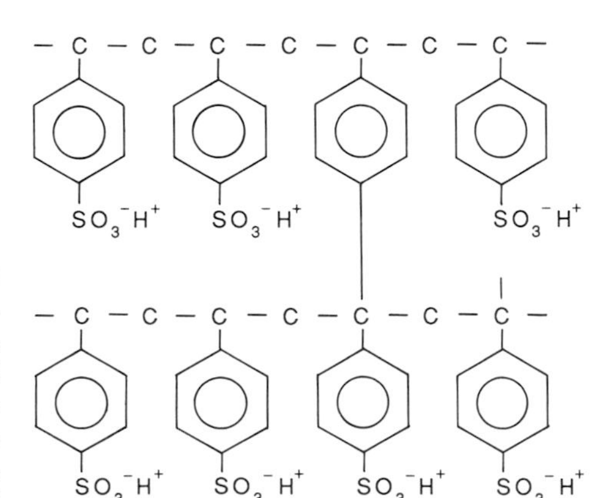

FIGURE 36-1
Typical ion-exchange resin. (*Hajian and Pecsok,* Modern Chemical Technology, *rev. ed., vol. 6, Prentice-Hall, Englewood Cliffs, N.J., 1973.*)

TABLE 36-1
Common Ion-Exchange Functional Groups

Cation exchangers		*Anion exchangers*
SO_3H	strong	$N^+(CH_3)_3Cl^-$
$COOH$	↓	$(CH_3Cl)_2$
CH_2SO_3H	↓	N
OH	↓	CH_2OH
SH	↓	NR_2
	weak	NHR
		NH_2

SOURCE: Hajian and Pecsok, *Modern Chemical Technology*, rev. ed., vol. 6, Prentice-Hall, Englewood Cliffs, N.J., 1973.

An acidic functional group, like sulfonic acid (SO_3H) or carboxylic acid (COOH), is bonded to the polymeric material, and the acidic hydrogen group tends to dissociate leaving a negative site on the bead's surface to attract other cations. Similarly, basic functional groups such as a tertiary amine (R_3N) tend to convert to quaternary ammonium groups with a positive charge on the polymer bead and attract anions. These cation and anion functional groups are classified as strong or weak depending on their tendency to dissociate (Table 36-1). Water is normally used with these cationic and anionic exchange resins as the mobile phase, but changes in pH may be required to flush various ions from the beads selectively.

Many laboratories use a combination of anionic and cationic resins to purify (deionize) tap water, instead of using the more expensive and slower distillation process. Deionized water is not as pure (see Chap. 24, "Chemicals and Preparation of Solutions") as distilled water, since this process removes only cation and anion contaminants. These deionizing columns release an equivalent number of H^+ ions for each cation adsorbed and an equivalent number of OH^- for each anion adsorbed. Thus, a molecule of water is released or substituted for each equivalent of cation and anion contaminants retained on the column. Since ion-exchange resins attract only changed particles, few organic contaminants are removed by this process.

Preparing an Ion-Exchange Column

The following list gives information that might be found on a typical label on an ion-exchange-resin bottle and an interpretation of each item.

- *Strong-acid resin:* Contains acid functional group such as SO_3H or COOH that dissociates readily in water.
- *Sodium form:* The exchangeable ion is Na^+.
- *20–60 mesh:* The bead spheres are 20 to 60 particles per inch.
- *10X:* The polymer is cross-linked with 10% divinylbenzene.

- *Capacity 2.0 meq/mL:* The resin has the capacity to exchange ions up to an equivalent of 2.0 milliequivalents per milliliter of wet resin.

Fresh ion-exchange resins should be washed respectively with 2 *M* HCl, rinsed with water, washed with 2 *M* NaOH, and again rinsed with water until the wash solution is neutral and salt free. Ion-exchange resins labeled "Analytical Grade" have already been treated.

Procedure

1. Soak a calculated quantity (based on equivalents of ions to be exchanged) of the freshly washed resin in water for at least 2 h in a large beaker. Resins with greater X values require less time for soaking. More highly cross-linked (for example, 10X) resins swell less than those with lower X values (for example, 2X).

2. Using a conventional liquid chromatographic column, place a glass-wool plug in the bottom, and fill it half full with water.

3. With the aid of a powder funnel, transfer the soaked resin into the column and drain the excess water without allowing any of the resin to "go dry."

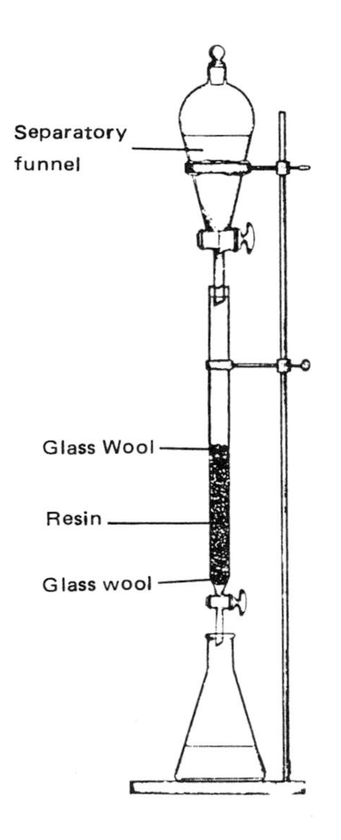

FIGURE 36-2
Ion-exchange column with separatory funnel. (*Hajian and Pecsok,* Modern Chemical Technology, *rev. ed., vol. 6, Prentice-Hall, Englewood Cliffs, N.J., 1973.*)

4. Backflush the packed resin column with a stream of water to remove any air bubbles and then allow the resin to settle.

5. Open the stopcock at the bottom of the column and using a graduated cylinder determine the flow rate. Never at any time allow the resin to go dry as "channeling" can occur and thus reduce the efficiency of the packing drastically.

6. The column is now ready for use. A separatory funnel is a convenient device for adding sample to the ion-exchange column. (Fig. 36-2).

GEL-PERMEATION CHROMATOGRAPHY

Gel-permeation chromatography (GPC) is also called *size-exclusion chromatography* or *gel filtration* and is another example of adsorption chromatography. It is a separation process which employs a gel composed of an insoluble, cross-linked polymer as the stationary phase. These gels swell in aqueous solution and create cavities that can trap molecules of various sizes. The degree of polymeric cross-linking determines the size of the holes in the gel matrix. Because of the sievelike molecular structure of these materials, compounds ranging from molecular weights of one hundred to several millions can be concentrated and separated.

When the molecules are too large to enter the pore, they will be excluded from the pore (this is size-exclusion chromatography); therefore they must travel with the solvent front. The selection of the gel with its corresponding exclusion range enables the technician to achieve separation; in order to achieve that separation, gels must be selective and must possess exclusion ranges greater than those of the molecules which are to be separated. High-pressure liquid chromatography (HPLC) is also applicable here.

Gel-permeation procedures are well suited for the separation of polymers, copolymers, proteins, natural resins and polymers, cellular components, viruses, steroids, and dispersed high-molecular-weight compounds. The gels prepared should have a narrow or broad range of pore sizes, depending upon the specific need of the technician and the composition of the substance being separated (Fig. 36-3). Gels are available in pore sizes ranging from 25 to 25,000 Å, and most have excellent temperature stability; the silica gels remain stable up to 500°C. The eluting solvents can be either polar or nonpolar. The volume of solvent which is required for a procedure elution equals the volume of solvent in the column, because there is no affinity of the substances in the sample for the packing. All sample components elute completely with one column volume of solvent.

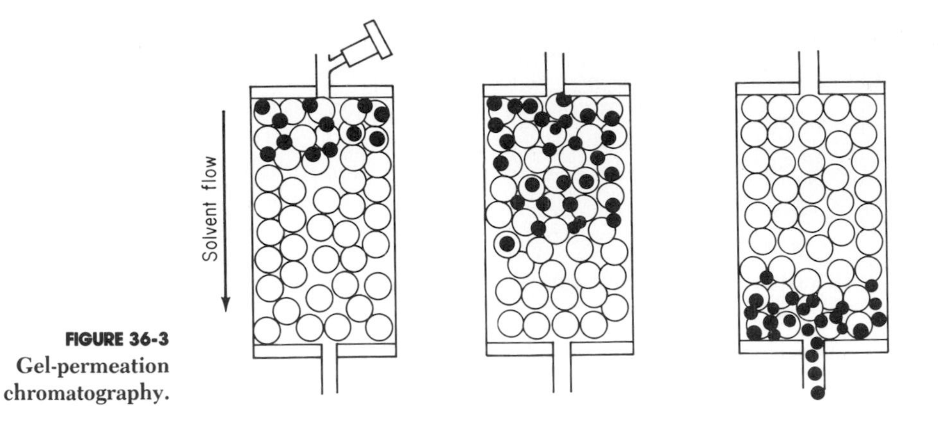

FIGURE 36-3
Gel-permeation chromatography.

Preparation of Gels

Gel-permeation sorbents may be polymers or copolymers of vinyl acetate, polyethylene glycol dimethylacrylate, polystyrene, phenol-formaldehyde, and others. See Table 36-2. Gels should be prepared at least 6 h in advance. They must be allowed to swell in the eluting solvent. After the swelled gel has been prepared, it is stirred well, and the resultant slurry is fed into the column in small portions. Be certain that each portion added settles continuously while the cylinder is being rotated. This procedure provides a more uniform bed and results in better separations.

Only plastic discs, not fritted-glass discs are recommended for use with gel columns. Fungal and bacterial growth in gel media can be controlled with a 0.2% sodium azide solution.

THIN-LAYER CHROMATOGRAPHY

Thin-layer chromatography (TLC) is another application of adsorption chromatography. Unlike conventional liquid chromatography, the mobile phase ascends a thin layer of adsorbent (stationary phase) as opposed to descending a column of adsorbent. The adsorbent is coated in a thin film on a flat plate of glass, plastic, or metal. The two most common adsorbent materials are alumina G (aluminum oxide) and silica gel G (silicic acid). The G stands for addition of approximately 10% by weight gypsum (calcium sulfate) which serves as a binder to the plate and adsorbent.

In TLC, the adsorbent is deposited on an appropriate supporting plate and the edge of the plate is in contact with the solution. The liquid rises through capillary attraction. The components are separated by the differences in the distance they rise. See Fig. 36-4.

TABLE 36-2
Sorbent Selection Guide for Gel-Permeation Chromatography*

Sample	Sample solubility	Column packing		Exclusion limit (molecular mass)	Maximum pressure (approximate)	Additional comments
SAMPLE	Aqueous solvents	LiChrospher(TM)	SI 100 SI 500 SI1000 SI4000	1 × 10^5 5 × 10^5 5 × 10^6 approx. Undetermined	3000 psi (200 atm)	Values determined using polystyrenes in tetrahydrofuran (THF). Must compromise between resolution and separation time. High linear velocities can adversely affect diffusional equilibrium in the column.
		EM GEL type	OR-PGM	2000	hydrostatic	Values determined using polyethylene glycols in water and in THF.
	Organic solvents	LiChrospher(TM)	SI 100 SI 500 SI1000 SI4000	1 × 10^5 5 × 10^5 5 × 10^6 approx. Undetermined	1100 psi (75 atm)	Note remarks above for LiChrospher.
		EM GEL type OR-PVA	500 2000 6000 20,000 80,000 300,000 1,000,000	500–300 2,000–1,000 6,000–4,000 20,000–14,000 80,000–50,000 300,000–200,000 up to 1,000,000	600 psi (40 atm)	Values determined using oligophenylenes and polystyrenes in THF.
		EM GEL type SI	200Å 500Å 1000Å	50,000 400,000 1,000,000	Incompressible at relatively high pressure. Upper limit undetermined.	Values determined using linear polystyrenes in chloroform.
		EM GEL type SI	2500Å 5000Å 10,000Å 25,000Å	2,500Å 5,000Å 10,000Å 25,000Å	Requires low flow rates to insure equilibrium. For column ID 10 mm and length 120–360 cm, flow rate = 2–60 mL/h	Used for particulate sample (rather than molecular). Gels are calibrated in angström (Å) units. Ratio of excluded particle radius to mean pore radius is approximately 2:3.

* This guide is based upon the most frequently used chromatographic systems. Exceptions are possible.

SOURCE: ALLTECH Associates. Reproduced by permission.

Preparation of the Plate for Thin-Layer Chromatography

1. Before preparing the slurry, treat the mixing flask and glass rod with a hydrophobic substance, such as dimethyldichlorosilane, thus rendering them hydrophobic. Dissolve 2 mL of the dimethyldichlorosilane in 100 mL of toluene and thoroughly wash the mixing vessel and glass rod with it. Use this solution as a rinse to waterproof both the mixing vessel and rod (to be used for preparing the slurry) and all other glassware to be used. Finally, rinse all glassware with methanol and distilled water prior to use.

2. Prepare a slurry of the adsorbent. In the case of aluminum oxide or silica gel, use Table 36-3 or 36-4 to determine the proportions of the slurry.

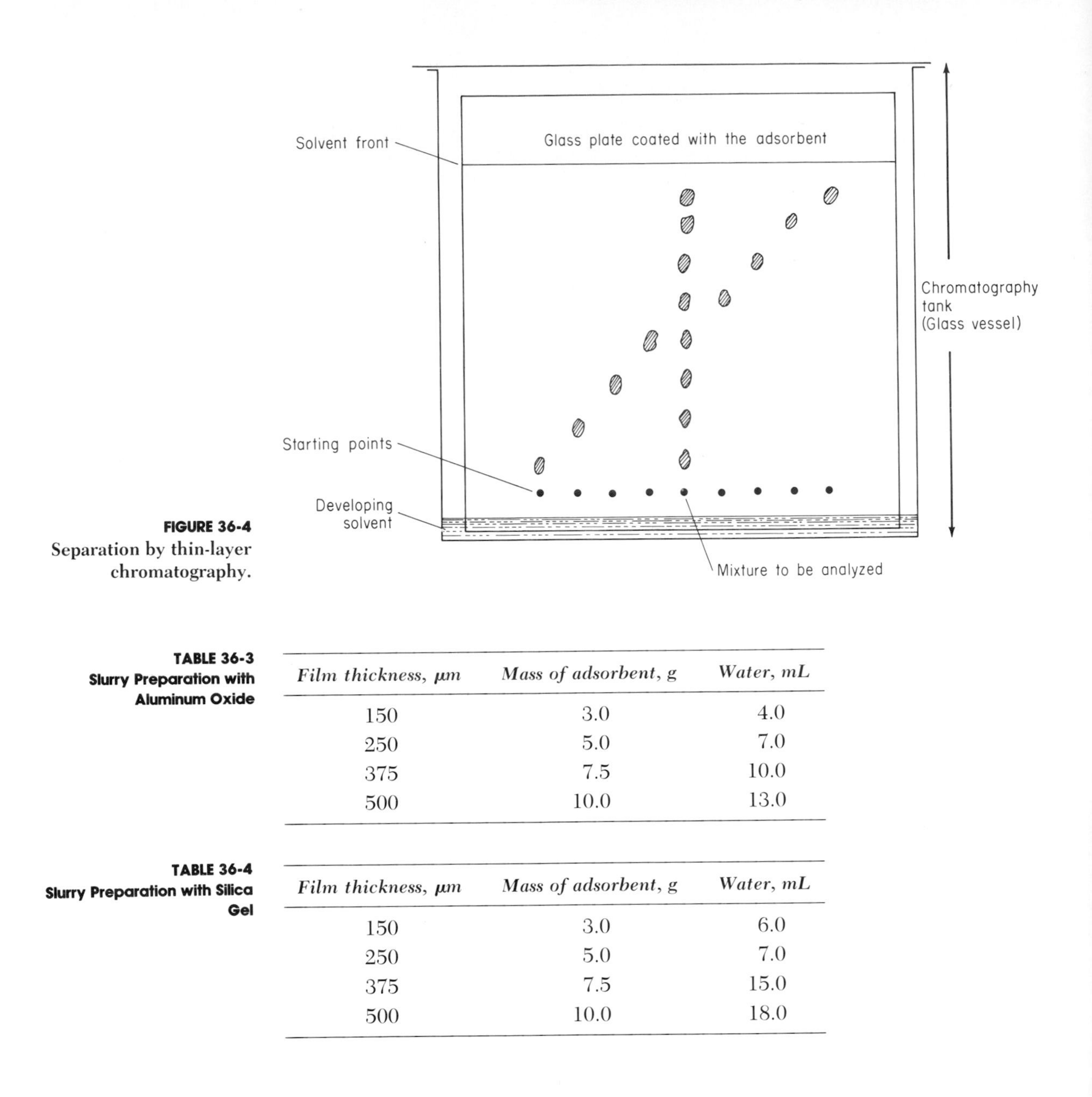

FIGURE 36-4
Separation by thin-layer chromatography.

TABLE 36-3
Slurry Preparation with Aluminum Oxide

Film thickness, μm	*Mass of adsorbent, g*	*Water, mL*
150	3.0	4.0
250	5.0	7.0
375	7.5	10.0
500	10.0	13.0

TABLE 36-4
Slurry Preparation with Silica Gel

Film thickness, μm	*Mass of adsorbent, g*	*Water, mL*
150	3.0	6.0
250	5.0	7.0
375	7.5	15.0
500	10.0	18.0

Other adsorbents used for TLC are cellulose and polyamides; some of them contain binders that make them stick to the glass.

3. Put the required mass of adsorbent into the mixing flask and then add the specified volume of water. Shake thoroughly for about 5 s.

4. Place a clean, dry, glass plate (Fig. 36-5) on a paper towel and pour the slurry evenly across the carrier plate near the bottom edge. Use the glass rod to spread the slurry with a smooth steady motion to the top edge of the plate (but not over it); slide the rod, don't roll it. A commercial spreader (Fig. 36-6) may also be used.

CAUTION This operation may be repeated to obtain a smooth coating, but cannot be repeated once the slurry hardens.

5. Without allowing further flow of the coating, dry the coated plate in an oven at 89 to 90°C for about 1 h. This permits the plate to be handled and at the same time activates the coating. Store dried plates in a suitable desiccator.

FIGURE 36-5
Glass plate used in thin-layer chromatography.

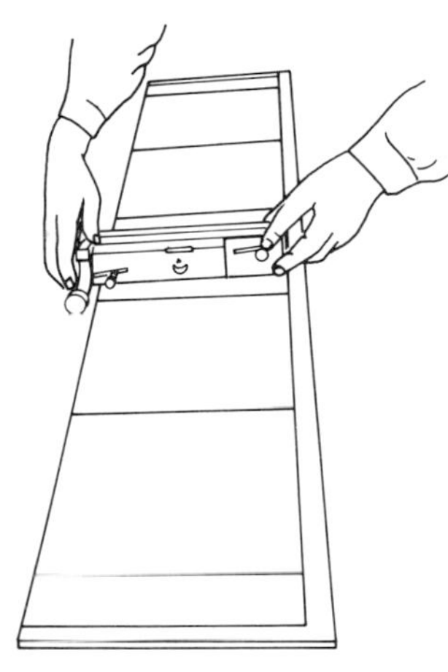

FIGURE 36-6
Preparing a plate using a commercial spreader.

Thin-Layer Chromatography Procedure

Standard TLC procedure involves three steps.

1. The substance to be separated into fractions is spotted on the edge of the plate with a micropipet in such a manner as to yield a minimum area. Better separation and development are obtained with small sample spots.

2. The prepared solvent mixture is placed in the bottom of a developing tank (Fig. 36-7) and the plate (or plates) is positioned in the tank with the upper part of the carrier plate leaning against the side of the tank. The tank is securely covered with a glass plate and the developing begins as the solvent rises up the plate by capillary attraction.

An alternate method uses a sandwich technique: A blank plate and the spotted plate are placed channel-to-channel and clamped together. They

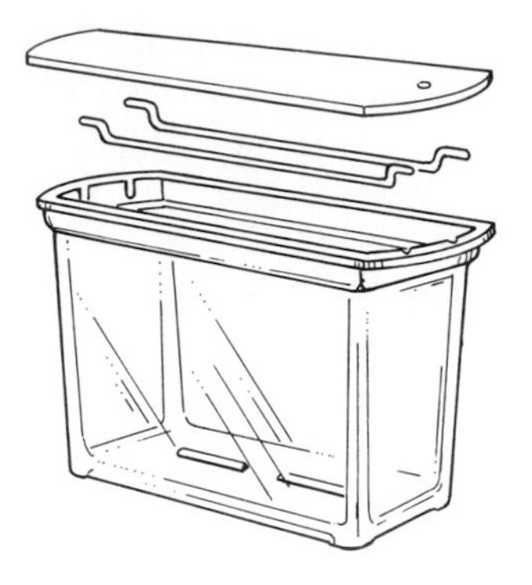

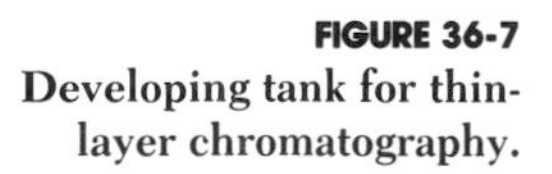

FIGURE 36-7
Developing tank for thin-layer chromatography.

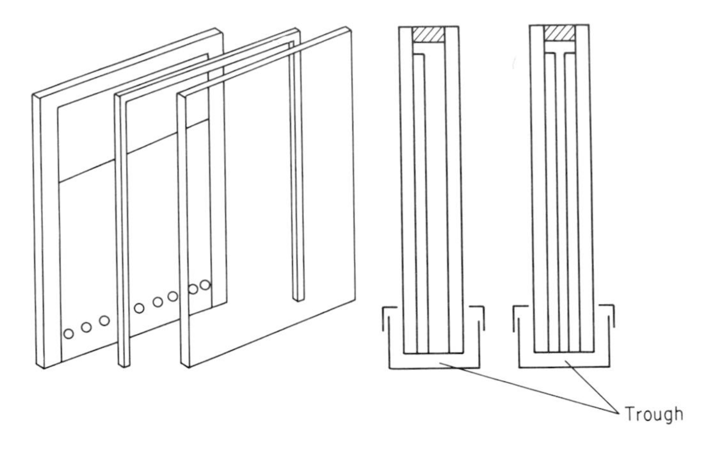

FIGURE 36-8
Assembly showing normal and saturated sandwich chamber.

are placed in the trough (Fig. 36-8) which contains the solvent in a vertical position, with the level of the solvent lower than the level of the spotting.

The TLC plate for this variation is prepared in the normal manner, but a strip about ½ in wide is scraped off the sides and upper edge. An inverted-U cardboard frame is placed over the cleared strip, and the cover plate placed over the U frame. Clamps securely hold the sandwich together. Only about 15 mL of solvent are needed to fill the trough for developing the chromatogram.

3. The spots must be located. Once the solvent front has reached the desired level, the plate is removed and allowed to dry. Now the location of the spot must be determined. This location can be used as a criterion for the identification of the substance, particularly if controls have been set up; the intensity of the spot is a quantitative measure of the concen-

TABLE 36-5
Spraying Reagents

Type	*Application*
Strong acids [H_2SO_4; H_3PO_4 (heated to 120°C)]	Natural substances
Strong acids [H_2SO_4; H_3PO_4 with 0.5–1.0% aldehyde (such as anisaldehyde)]	Natural substances
Antimony(V) chloride (20%) and carbon tetrachloride* (80%) heated to 20°C	Resins, terpenes, and oils
Iodosulfuric acid solution [1.0 *N* iodine, 16% H_2SO_4 (1:1)]	Organic nitrogenous compounds
Iodine 0.5% solution in chloroform or alcohol	Organic nitrogenous compounds
Antimony(III) chloride (25%) in chloroform; may yield fluorescence when heated	Carotenoids, steroid glycosides

* Tetrachloromethane.

tration of the substance. Chromophoric substances can be located visually; colorless substances require other means.

Some substances fluoresce under UV light, and irradiation of the plate will indicate the position of the spot. Other plates are coated with fluorescent materials; the spot will obscure this fluorescence when the plate is irradiated. Spray reagents, selected to react with the spot (Table 36-5), reveal its location. The spray must be applied uniformly to the dried plate. Finally, exposure of the plate to chemical vapors can also reveal the location of the spot.

Tips on Technique for Thin-Layer Chromatography

Precoated, Commercially Available Plates

Precoated aluminum sheets are commercially available, and they offer certain advantages over glass plates. Aluminum (or rigid plastic) sheets can be cut easily with scissors, and they are stiff enough to stand without supports. The coatings are abrasion-resistant and uniform.

Activation of Plates

All prepared or precoated plates should be stored and protected from environmental contamination. They must be "activated" by heating them in an oven before use and then cooled and stored in a suitable desiccator prior to being spotted and developed.

Spotting

Capillary tubes or micropipets are used to spot the sample solutions. The capillary is positioned in the desired location over the TLC plate and momentarily touched to the plate, with due care to avoid disturbing the coating. The solvent is allowed to evaporate. The procedure is repeated until the whole sample has been spotted, keeping the size of the spot as small as possible for better separation.

Placing the Plate in the Tank

Do not touch the plate on the sides. Hold it by its edges and place it squarely in the solvent surface. Cover the tank securely.

Edge Effects

The best results are obtained when at least 1 in of the outer edge of the carrier plate is not coated with the slurry. Samples should not be spotted closer than 1 in from the edge in the direction of development, and no closer than 1 in to the edge which is parallel to the development. Edge effects such as distortion of the solvent boundary or distortion of the spot shape can result from disregarding these tips.

PAPER CHROMATOGRAPHY

Paper chromatography is physically similar in techniques to TLC, but really is a special type of liquid-liquid chromatography in which the stationary phase is simply water adsorbed on paper. The technique is very simple; using a moisture-containing (about 22%) cellulose sheet, small spots of a mixture are placed approximately 3 cm from one edge. *Spotting* of sample is usually done by the repeated application of sample solution with a capillary tube onto the paper. Capillary action causes the solution to be drawn from the tube into the paper forming a uniform spot. The spotted sheet is then placed in a developing chamber, not unlike what was described in the section on TLC, which prevents the loss of paper moisture and maintains a saturated atmosphere of mobile phase. The cellulose can be kept saturated with water by adding water to the developing solvent (mobile phase). The mobile phase is approximately 1 cm deep in the bottom of the chamber. The solvent level should never be allowed to be above the spotted samples.

In *ascending chromatography*, the paper is suspended vertically in the development chamber and the sample spot is transported in an upward direction as the solvent wicks up the paper. The disadvantage to ascending chromatography is that the length of chromatograms is limited to 20 cm because of gravity. In *descending chromatography* the paper is suspended vertically but the solvent enters the paper at the top. An antisiphon rod is required to prevent the solvent from being siphoned from the solvent reservoir. This technique permits the development of longer chromatograms and increases resolution. In *reversed-phase chromatography* (RPC) the paper is thoroughly dried and impregnated with a nonpolar solvent. Water or other polar solvents can now be used as the mobile phase to resolve difficult mixtures like steroids and high-molecular-weight fatty acids.

Procedure

Assemble necessary materials: capillary tubes, filter paper, solution to be separated into fractions. Then:

1. Prepare micropipets.

2. Cut filter paper to size desired.

3. Draw a *light* pencil line about 1 in from edge and make marks 1 in apart, leaving 1 in from each vertical edge, and place a spot of the solution at each mark (Fig. 36-9).

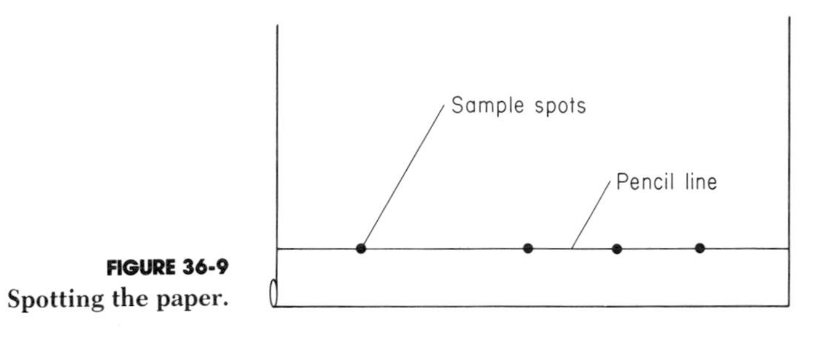

FIGURE 36-9
Spotting the paper.

4. Roll paper into a cylinder, stapling ends together (Fig. 36-10). Alternatively, use a strip of paper having approximately the size shown in Fig. 36-11. In either case, let the spots dry.

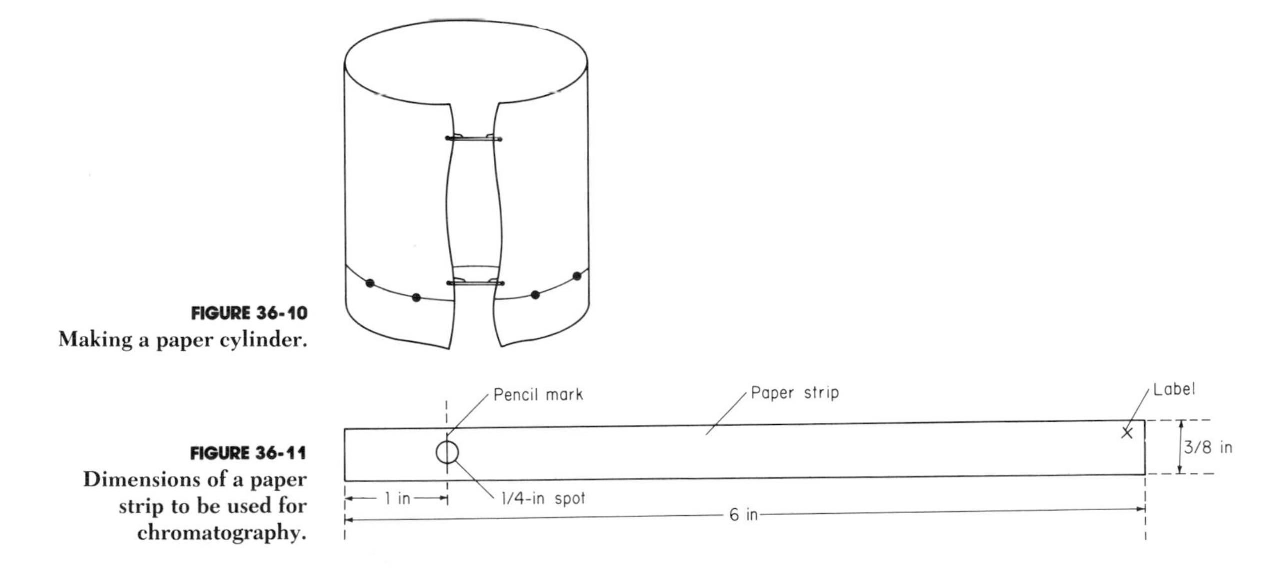

FIGURE 36-10
Making a paper cylinder.

FIGURE 36-11
Dimensions of a paper strip to be used for chromatography.

5. Prepare the solvent mixture; the composition of this mixture depends upon the substances which are to be fractionated by the chromatography process.

6. Transfer the solution to the bottom of the developing tank or flask, being careful none touches the sides of the tank.

7. Place the dried, spotted paper cylinder in the developing tank (or beaker of suitable size) so that the spots are above the level of solvent in the container, cover the container securely with a glass plate, aluminum foil, plastic wrap, or Parafilm. See Fig. 36-12.

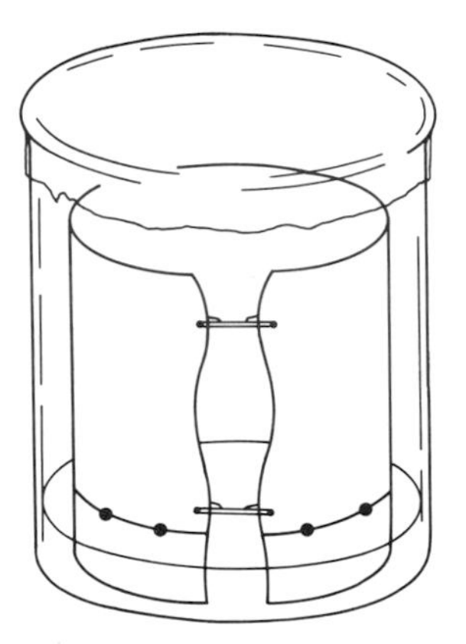

FIGURE 36-12 Developing a chromatogram on a paper cylinder.

If you are using strips of paper, develop the chromatogram as follows: Hang the dried, spotted strips by placing two of them opposite each other on the side of a cork, and place the cork in the Erlenmeyer flask containing the solvent. The sample spots should not be immersed in the solvent; only the bottom edge of the strips (Fig. 36-13).

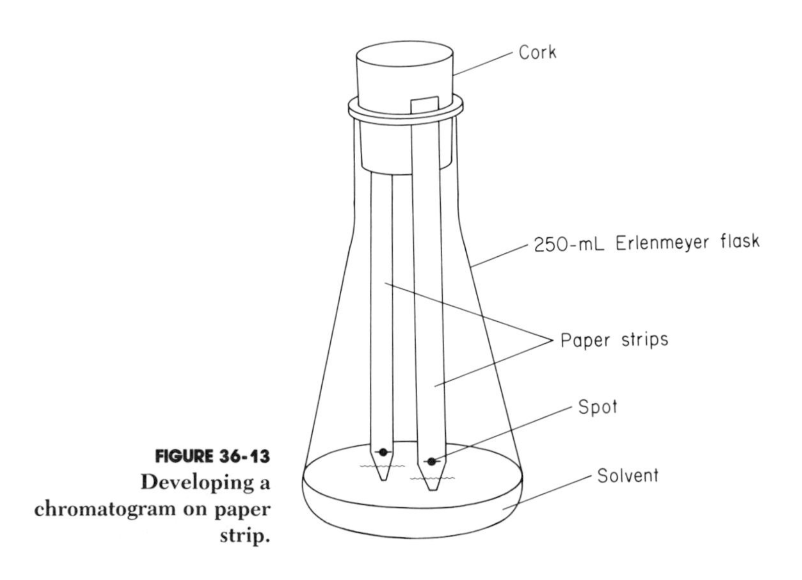

FIGURE 36-13 Developing a chromatogram on paper strip.

8. Remove the paper cylinder (or paper strips), marking initial and final solvent levels with a pencil. Allow the paper to dry, and measure and mark the distance traveled by any chromophoric substances you can see. Depending upon the substances tested, the procedure must be adaptable at this point, to determine the distance moved by the substance if it carries no chromophores and cannot be followed visually. The spots can be rendered visible by the use of UV light, a chromogenic spray, or a chemical reagent vapor, as with TLC.

R_f Value

The R_f is the relative rate of movement of a dissolved compound and is found by the formula:

$$R_f = \frac{\text{distance (in cm) a compound moves}}{\text{distance (in cm) the solvent moves}}$$

ELECTROPHORESIS

Electrophoresis (Fig. 36-14) is basically a modified paper-chromatographic separation technique which includes the influence of an applied electrical potential to assist in separating charged species. The migration of charged particles in this electrochromatographic process will depend on the magnitude and charge on the solute, the applied voltage, and the solute's adsorptivity for the paper. Routinely, voltages up to 1000 V are applied to parallel electrodes which develop a potential in one direction on the chromatographic paper, while simultaneously a flow of sample moves in a perpendicular direction. The combined influences of the chromatographic process and electromigration cause the charged solute particles to separate. Electrophoresis has primarily been applied in labile, biological mixture separations of colloids, proteins, and enzymes.

The R_f value can be used to identify an unknown compound by calculating the migration rates of both knowns and unknowns on the same sheet or plate under identical conditions. R_f values are calculated by measuring the distance that the compound traveled from the original spot and dividing this by the solvent's distance traveled as shown under R_f Value above. If the migrated spots are small, the center of the spot is measured. For large spots with "tails" the center of gravity of the spot is used or the chromatogram repeated using smaller quantities of sample.

If the mixture components are colored, a visual examination will reveal their relative migration with respect to the solvent front. Various visualization methods are available if the spots are colorless—for example, the paper can be exposed to heated iodine vapors which tend to react and form colored spots with various organic compounds. Another visualization method consists of simply exposing the developed paper to UV light, since

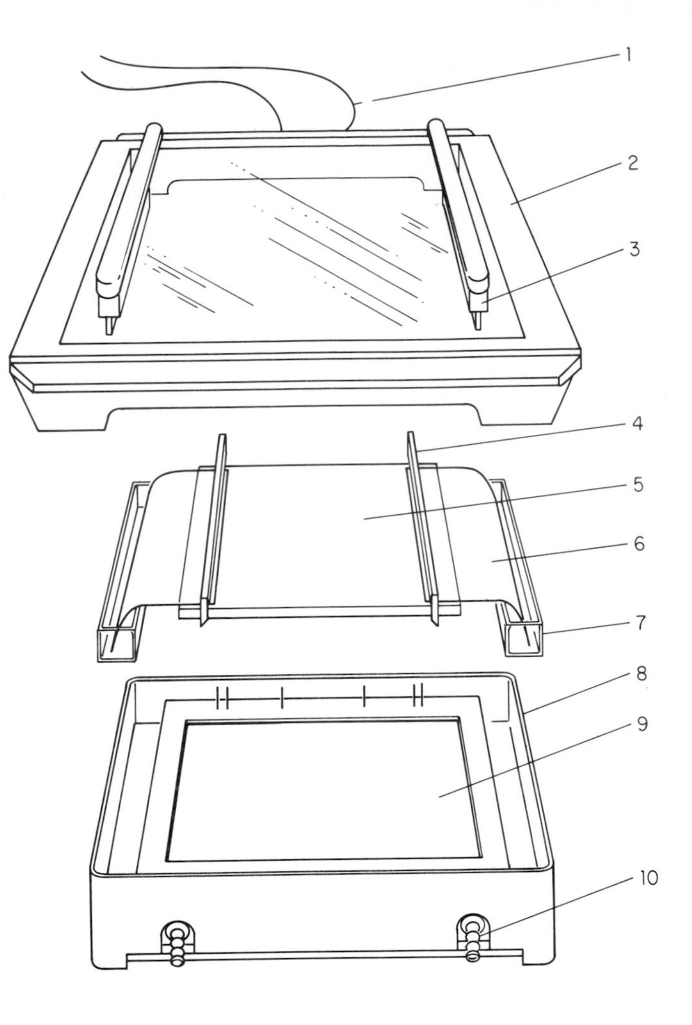

FIGURE 36-14 Electrophoresis equipment, showing component parts: (1) electric power leads; (2) cell lid with safety microswitch; (3) electrode unit; (4) clamping rod; (5) thin-layer plate: (6) conductor wicks: (7) buffer trough: (8) polypropylene housing: (9) insulated cooling block: (10) cooling-water connections.

some substances fluoresce. A solution spray called *Ninhydrin* is used to develop colored spots with amino acids in paper chromatography. The same spraying reagents and visualization methods (Table 36-5) described under Thin-Layer Chromatography are sometimes useful. The most common utilization of paper chromatography is for the separation of polar and/or polyfunctional natural compounds such as amino acids, carbohydrates, and plant pigments.

REFERENCES

Chromatography: A Laboratory Handbook of Chromatography and Electrophoretic Methods, 3rd ed., Van Nostrand Reinhold, New York, 1975.

37

GAS CHROMATOGRAPHY

INTRODUCTION

Gas chromatography (GC) is one of the fastest and most useful separation techniques available in the laboratory (Fig. 37-1). Gas chromatographic analysis is basically limited to organic compounds that are volatile and not thermally labile. There are two types of GC: gas-solid (adsorption) chromatography and gas-liquid (partition) chromatography. Gas-Liquid chromatography (GLC) is used more extensively than gas-solid chromatography (GSC). Both types of GC require that the sample be converted to or exist in the vapor state and be transported by an inert carrier gas through a column packed with some type of a liquid phase coated on a solid support (GLC) or simply a solid adsorbent with no liquid-phase coating (GSC).

A sample is injected into a heated block where it is immediately vaporized and swept as a concentrated vapor into a column. Separation occurs as the various compound vapors are selectively adsorbed by the stationary phase and then desorbed by fresh carrier gas. This sorption-desorption process occurs repeatedly as the compounds move through the column toward a detector. The compounds will be eluted from the column with those having a high affinity for the column packing being slower than those with little affinity.

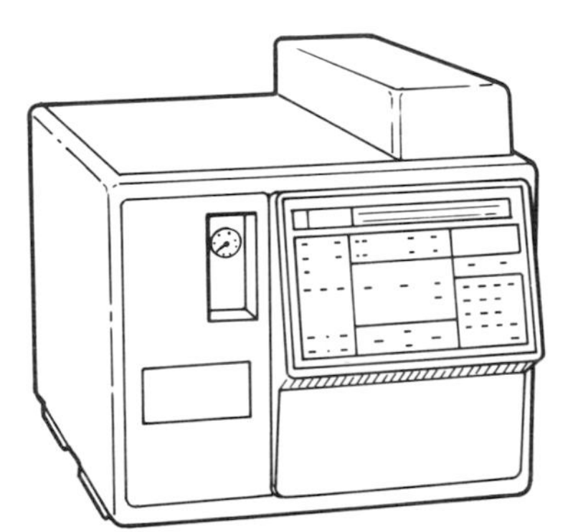

FIGURE 37-1
Basic gas chromatograph *(Photo courtesy of Varian Associates, Inc.)*.

INSTRUMENT DESIGN AND COMPONENTS

A typical gas chromatograph consists of a carrier-gas supply, sample-injection port, column, column oven, detector, and a recorder-integrator system. A block diagram showing these main components is shown in Fig. 37-2.

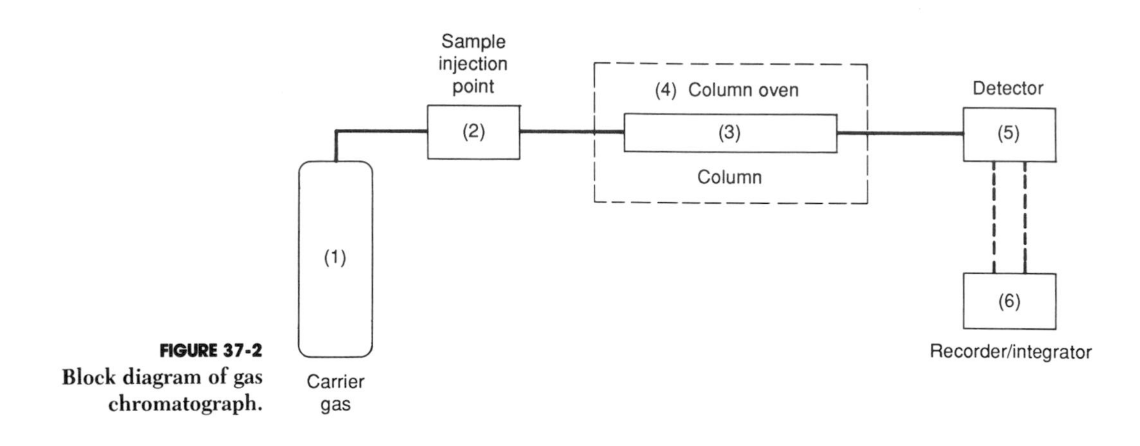

FIGURE 37-2 Block diagram of gas chromatograph.

Carrier Gas

The carrier gas is used to transport the sample molecules from the injection port to the detector and provide the means for partitioning the sample molecules from the stationary phase. The most common carrier gases are helium and nitrogen. These gases are supplied in high-pressure tanks which require a two-stage pressure regulator for reducing the inlet gas pressure and controlling the gas velocity through the column. This gas must be of high purity with minimal moisture or other contaminants present to reduce erroneous detector signals.

Sample-Injection Port

A sample-inlet system must be provided which allows liquid samples in the range of 1 to 10 μL to be injected with a microsyringe (see the section entitled Microliter Handling Technique in this chapter) through a self-sealing silicone rubber septum into a block that is heated to a temperature in excess of the boiling points of the compound(s). The liquid sample is immediately vaporized as a "plug" and swept through the column by the carrier gas. Gas and liquid samples can also be introduced using a gastight valve and calibrated volume loop system. These valves have multiport arrangements and can have sample loops for injecting liquid or gas sam-

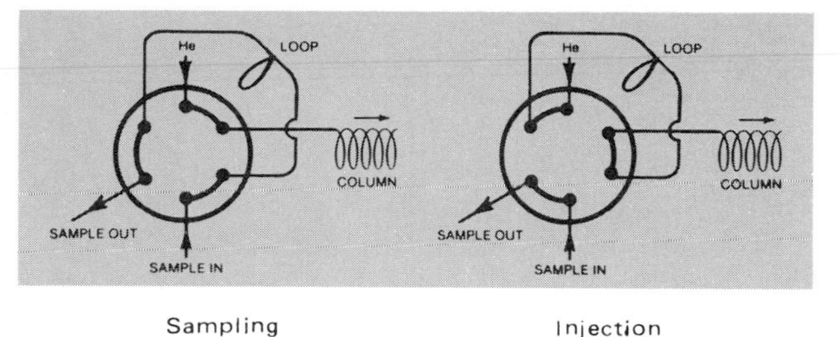

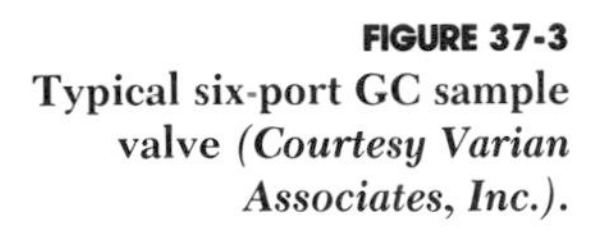

FIGURE 37-3
Typical six-port GC sample valve *(Courtesy Varian Associates, Inc.)*.

ples. A typical six-port valve with sample purge, sample volume loop, and column connection is shown in Fig. 37-3.

Samples can be introduced using a programmable injector for the continuous operation of chromatographic systems. These automatic injectors use the same type of microsyringes but are capable of a higher degree of reproducibility than a manual technique. This unattended operation releases the operator for other duties and allows 24-h operation of a GC system equipped with an automated integration-recorder system. Figure 37-4*a* shows a typical gas chromatograph equipped with an automatic sampler, and Fig. 37-4*b* provides a close-up view of the microsyringe entering the injection port.

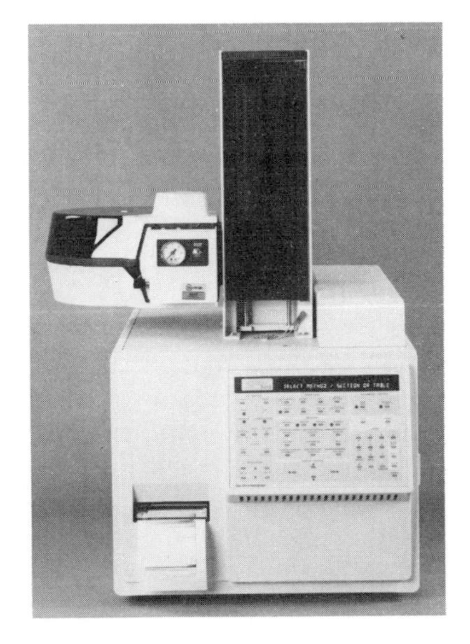

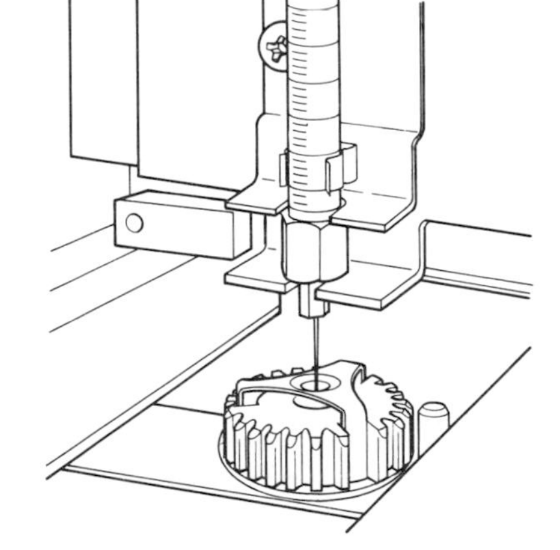

FIGURE 37-4
(*a*) Gas chromatograph equipped with an automatic sampler. (*b*) Automatic sampler with microliter syringe *(Photos courtesy of Varian Associates, Inc.)*.

Special injector splitters are used with capillary columns (see under Column below) which usually require samples of less than 1 μL. These injector splitters mix the vaporized sample and split (ratio between 1:10 and 1:1000) the originally injected volume by venting the excess.

Column

Two basic types of columns are currently being used: packed and capillary. Packed columns will usually have 1000 to 3000 plates per meter, while capillary columns can exceed 4000 plates per meter. Packed columns are normally made of copper, stainless steel, or glass with common bores of 1.6, 3.2, 6.4, or 9.5 mm and lengths of 1 to 3 m. Glass columns and glass injection-port liners are necessary when dealing with labile compounds that might react or decompose on contact with metal surfaces especially at elevated temperatures. These columns have been "packed" with a coated, sieved (ranging 60 mesh to 120 mesh) inert solid support (see under Solid Supports in this chapter). The solid support is coated (usually 1 to 10% by weight) with a liquid phase (see under Liquid Phase). Capillary columns do not contain solid support coatings and simply have the liquid phase (less than 1 μm thick) coated directly onto the interior walls of the column. This wall-coated open tubular (WCOT) technique provides an open, unrestricted carrier-gas path through the typical 0.25-mm diameter column. Since capillary columns present very little restriction to gas flow, they can be made extremely long (50 to 150 m) for greater compound resolution. A newer form of capillary column technology called support-coated open tubular (SCOT) has recently been developed. A layer of solid support is adsorbed onto the interior of the walls of the capillary tubing and the liquid phase applied. The primary advantage of this technique is that it increases the sample capacity of the column and, in some cases, sample splitting is not necessary as was the case with conventional capillary columns.

Column Oven

Operating the column at a constant temperature (isothermal) during an analysis is critical for reproducible results. However, it may be necessary to change the column temperature in a reproducible way (temperature programming) during an analysis to separate components having greatly different boiling points and/or polarities in a reasonable period of time. Good resolution is obtained for low-boiling components at moderate oven temperatures; however, the analysis will require an excessive amount of time to elute any high-boiling compounds, and the chromatographic peaks will be too broad for proper quantitative interpretation. Temperature programming can reduce this difficulty by allowing the low-boiling compounds to elute at initially low temperatures; the column-oven tempera-

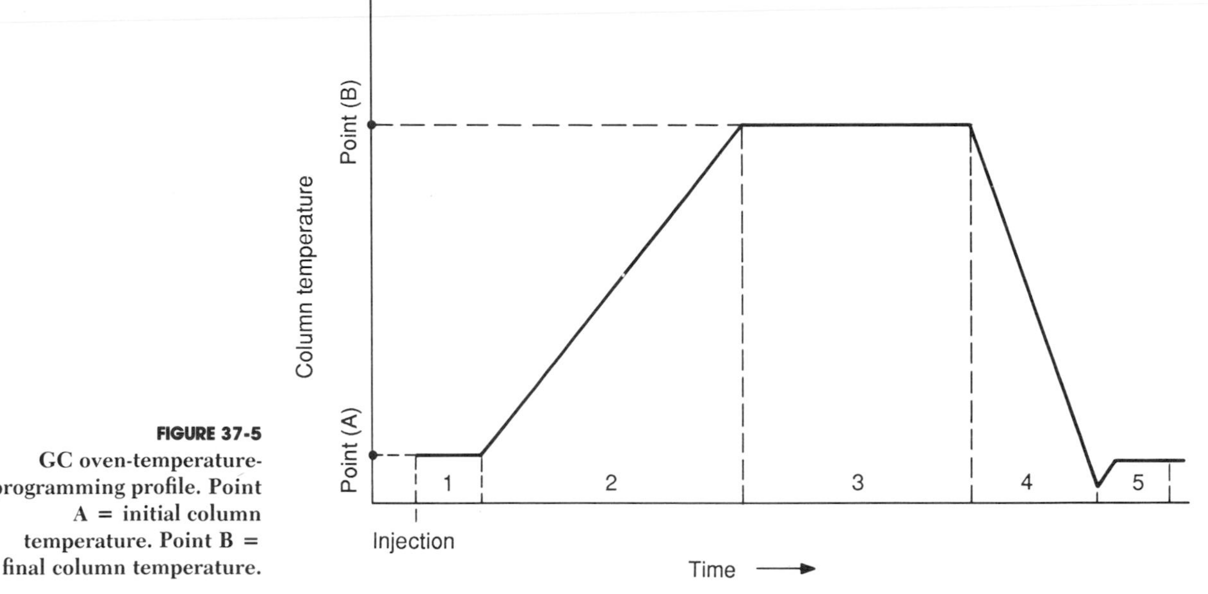

FIGURE 37-5 GC oven-temperature-programming profile. Point A = initial column temperature. Point B = final column temperature.

ture is then increased at a reproducible rate to elute the high-boiling compounds in a reasonable time. A typical temperature program would consist of the following steps as shown in Fig. 37-5.

1. Post-injection period
2. Temperature-programming rate (°C/min)
3. Upper temperature period
4. Automatic cool-down
5. Isothermal recovery to point 1

Detectors

Each chromatograph has a detection device at the exit of each column to monitor the gas composition. There are many types of detectors: thermal-conductivity, flame-ionization, electron-capture, photoionization, Hall-electrolytic-conductivity, flame-photometric, thermionic-specific, and coulometric detectors are the most common. Some chromatographic systems have dual-channel detectors to analyze a single sample by two dissimilar detectors simultaneously. This technique will provide the operator with additional information on the identification of compounds. This dual-detector capability is also necessary for stabilizing the detector baseline when temperature programming is being used. As the column temperature is deliberately increased in the programming cycle, some of the liquid phase bleeds from the column and increases the detector background signal. A matched column with the same bleed rate is installed to produce an opposite, but identical, signal to the detector; the two opposing signals cancel each other, and the baseline remains relatively straight.

Thermal-Conductivity Detectors (TCDs)

Some thermal-conductivity (hot wire) detectors use a thin filament of metal, while other TCD types use thermistors. In TCDs, a tungsten filament is heated by a continuous current flow and cooled by the carrier gas as it exits from the column.

CAUTION *Thermal-conductivity detectors can be damaged by turning on the filament current without proper carrier-gas flow.*

Hydrogen and helium are the best TC carrier gases because of their very high thermal conductivities; however, hydrogen is flammable and helium is expensive. Hydrogen has the highest thermal conductivity (53 cal/°C·mol) and produces the greatest sensitivity of all carrier gases. Helium is also an excellent TC carrier gas (42 cal/°C·mol) and is much safer than hydrogen. Most other gases have thermal conductivities of less than 10 cal/°C·mol. The elution of a sample component causes the detector filament or thermistor to heat up because of the diluted cooling effect of the carrier gas. This increase in temperature (resistance) is measured by a Wheatstone bridge circuit and the imbalance signal is sent to a recorder-integrator. A typical TCD is shown in Fig. 37-6.

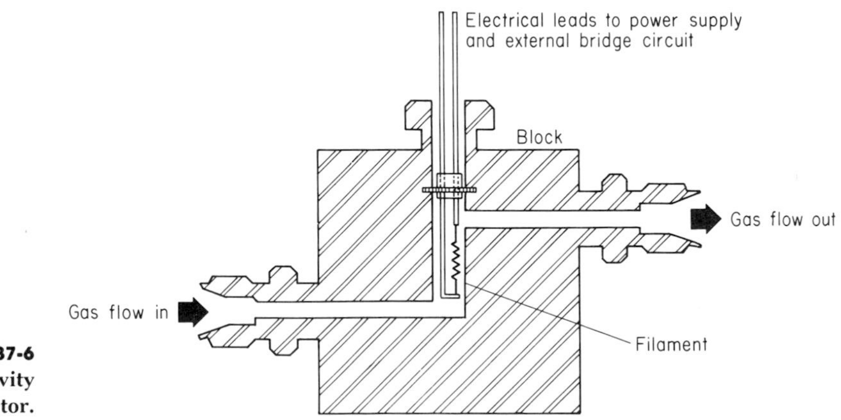

FIGURE 37-6 Thermal-conductivity detector.

Flame-Ionization Detector (FID)

Hydrogen and air are mixed in a burner to produce a very hot (approximately 2100°C) flame which can ionize carbon-containing compounds. A collector electrode with a dc potential is placed above the flame to measure its conductivity. As the column effluent passes through the burner jet, certain compounds are ionized and create a current flow which is proportional to the concentration of the molecules in the flame. Flame-ionization detectors are much more sensitive (perhaps 1000 times) than TCDs. Flame-ionization detectors are primarily limited to use with organic compounds and do not respond very well to air, water, or most inorganic compounds (Table 37-1). Thermal-conductivity detectors are more universal because they can detect any gaseous compound with a different thermal conductivity than that of the carrier gas. A cross-sectional view of a typical FID is shown in Fig. 37-7.

TABLE 37-1
Gaseous Substances Giving Little or No Response in the Flame-ionization Detector

He	CS_2	NH_3
Ar	COS	CO
Kr	H_2S	CO_2
Ne	SO_2	H_2O
Xe	NO	$SiCl_4$
O_2	N_2O	$SiHCl_3$
N_2	NO_2	SiF_4

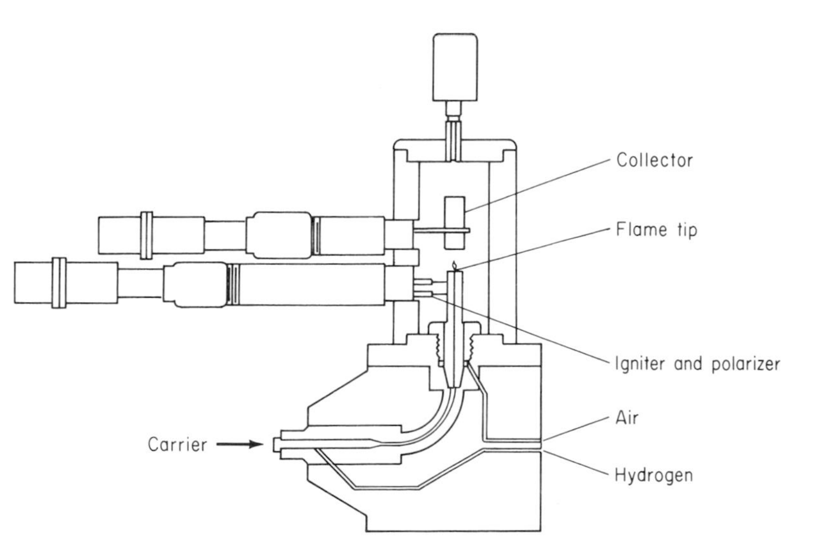

FIGURE 37-7
Flame-ionization detector.

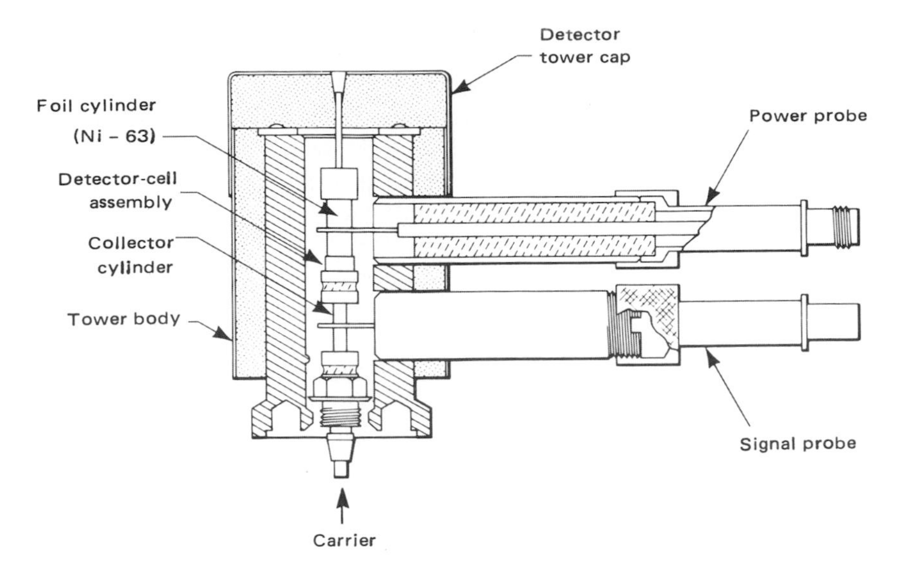

FIGURE 37-8
Electron-capture detector. *(Courtesy of Varian Associates, Inc.)*.

Electron-Capture Detector (ECD)

This type of detector uses a radioactive element (usually tritium or ^{63}Ni) as a source of electrons (beta particles) to provide an ionization current. The ^{63}Ni radioisotope has proved to be very reliable and can withstand temperatures of up to 400°C. A special carrier-gas mixture of argon with 5 to 10% methane is usually required with this type of detector. These beta emissions provide a steady current between electrodes in the detector as shown in Fig. 37-8. A portion of this electron flow can be "captured" by certain column effluent molecules and produce a measurable signal. The actual loss of the steady-state current is dependent upon the electron affinity of the eluting compounds. For example, halogenated compounds are very sensitive to this electron-capture process; thus pesticides, which are usually highly halogenated, are routinely analyzed using electron-capture-type detectors. Electron-capture detectors are insensitive to most organic compounds except those containing halogens, sulfur, nitrogen, and other electronegative groups.

Gas Chromatograph–Mass Spectrometry (GC–MS)

The most sophisticated detector system for both qualitative and quantitative analysis in a gas chromatograph is a mass spectrometer. The effluent from a packed GC column with flow rates normally in excess of 20 mL/min must be directed through a *molecular separator* which effectively separates the carrier gas from the sample components. Most newer GC–MS systems use capillary columns which have carrier-gas flow rates of approximately 2 mL/min and present no major problem for the mass spectrometer's pumping system. Once these effluent components enter the evacuated mass spectrometer chamber, they are fragmented by a powerful electron beam. These fragmented, charged particles are then accelerated into the center of four parallel rods called a *quadrupole mass filter.* These component ions can then be separated according to mass-to-charge ratio and ultimately identified by their characteristic fragmentation patterns. The quantity of fragmentation information and very short duration of analysis time necessitates that all GC–MS instruments have an integrated computer. Like the other GC detectors, this mass spectrometer system provides retention-time data and peak-area measurements, but structural information obtained from the fragmentation patterns can be used to positively identify the mixture components. The identification of molecules by their complicated fragmentation patterns has been simplified by computerization of mass spectral data into libraries stored in the GC–MS computer memory.

Recorder-Integrator

The signal from the detector can be sent to a servo (potentiometric) strip-chart recorder where the magnitude of the voltage is measured and recorded. The recorder (x-axis) draws a straight line when only carrier gas is passing through the detector and any eluting compounds cause the recorder to respond in proportion to the quantity of that compound. These strip-chart recorders generally have selectable voltage ranges of 1 mV to 5 V. Gas chromatographs normally use signals of 10 mV or less; however extensive-voltage-range recorders permit other laboratory uses (liquid chromatography, atomic absorption spectroscopy, etc.). The y-axis of the recorder measures the retention time for each eluting compound from its time of injection. Most GC recorder systems allow for different chart speeds (usually 1 to 30 cm/min range) to improve manual integration measurements and/or to conserve paper. Some newer gas chromatographs have computerized integration capabilities which automatically record the retention time to the thousandth of a minute, if desired, and the proportional detector response of each compound. Figure 37-9 shows

FIGURE 37-9
GC recorder-integrator *(Photo courtesy of Varian Associates, Inc.).*

an electronic integrator displaying retention time (RT), area responses for each peak, and the chromatogram. This type of integrator can be used in both gas and liquid chromatography.

MICROSYRINGE-HANDLING TECHNIQUES

In chromatography, your results are often only as good as the reproducibility of your sample. And the reproducibility of your sample quantity depends on many factors, among which are the operator and an accurate syringe. The following lists contain some syringe-handling techniques known to experienced operators in the field.

Filling the Syringe

1. First be certain the syringe and plunger are clean. Following is one recommended method of cleaning a syringe: Pump cold chromic acid solution through the syringe with the plunger, rinse both with distilled water, blow the syringe dry, and carefully wipe the plunger with lint-free tissue, being careful not to touch the plunger shaft with your fingers.

The chromic acid solution does a good job of destroying organic detergent residues and fingerprints on the plunger. The more stubborn oxide stains in the syringe-body neck may be removed with aqua regia, but it is recommended only as a last resort and should never come in contact with the plunger or needle.

2. Pressurize your sample bottle by using a gastight syringe filled with inert gas. Repeat as needed to build pressure in the bottle. This is particularly recommended for syringes with a detachable needle.

3. To assure an accurate measurement, wet the interior surfaces (barrel and plunger) of the syringe with the sample by pumping the plunger before filling the syringe. When properly done, this technique neutralizes the liquid movement by capillary forces.

4. Overfill the syringe in the sample bottle, withdraw from the bottle, and move the plunger to the desired calibration line, discharging the excess sample. This is your best assurance of a "full" sample. Discharging the excess sample while the needle is still in the sample bottle may cause a loss of sample upon withdrawing.

5. Read the syringe graduation from the same angle each time you fill the syringe. Recommended practice is to read the sample at the "top" (flange end) of the calibration line. This provides an accurate visual check and reduces the problems of "line thickness." Strive to develop a smooth uniform loading operation to minimize slight involuntary errors.

6. Check the syringe visually for bubbles or foreign matter in the sample.

7. Syringes with a detachable needle require care when filling because of the dead volume in the needle. (Refer to point 2.)

8. Before injecting, wipe the needle clean with a lint-free tissue, using a quick motion and taking care neither to wipe sample out of the needle nor to transfer body heat from your fingers to the needle.

Injecting the Sample

1. Develop a rhythm in your motion that is used each time you inject; that is, do the same things in the same manner at the same time.

2. Hold the syringe (Fig. 37-10) as close to the flange (in the unmarked area) as possible. This will prevent the heat transfer that occurs when you hold the needle or barrel with your fingers. Another way to prevent heat transfer is to use a Kel-F syringe guide. The guide also makes septum penetration easier.

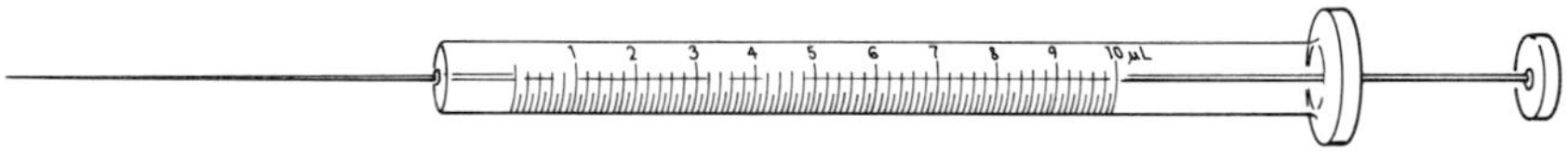

FIGURE 37-10
A microliter syringe.

3. Handle the plunger by the button, not the plunger shaft. This reduces the possibility of damage or contamination.

4. Develop a smooth rhythm that allows you to inject the sample as quickly as you can with accuracy.

5. For greatest accuracy, it is recommended that the syringe be used at less than maximum capacity.

COLUMN FITTINGS

Column fittings and various plumbing techniques are discussed in Chap. 18, "Laboratory Hardware."

SOLID SUPPORTS

Solid supports provide the surface area for a liquid phase to be exposed to a mobile gas phase. The ideal solid support should be inert, not pulverize readily, and have a high surface area (greater than 1 m^2/g). Solid supports are normally graded into *mesh sizes*—for example, an 80 mesh means a screen with 80 holes per linear inch. Mesh ranges (60/80) are used to indicate the largest to smallest solid support particles. Naturally occurring silicates and *diatomaceous earth,* which is composed of the skeletons of thousands of single-celled plants, are used extensively as GC solid-support materials. This type of support derived from diatomaceous earth is known by the trade name Chromosorb, and several varieties are available with various chemical and physical properties. These solid supports can be *silanized* by treatment with dimethyldichlorosilane or other silanes to reduce the surface activity of the silicates in the diatomaceous earth. Acid-washing treatments are also beneficial in reducing the tailing effects caused by solid-support adsorption.

Chromosorb P is a pink material with a rather high surface area (4 to 6 m^2/g), has the highest liquid-phase holding capacity or adsorptivity, but is the least inert of the Chromosorbs.

Chromosorb W is a white material, prepared by mixing with sodium carbonate flux at about 900°C. This material tends to be more rugged than Chromosorb P, but has a very low surface area (approximately 1 m^2/g).

TABLE 37-2
Typical and Maximum Loadings for Solid Supports

		Maximum loading, w/w%	
Support material	*Typical loading, w/w%*	*Typical liquid phase*	*Sticky or gum phases**
Anakrom	3–10	25	20
Chromosorb G	2–6	12	7
Chromosorb P	10–30	35	20
Chromosorb T	1–2	6	3
Chromosorb W	3–10	25	20
Chromosorb 101, 102 through 108	3–10	25	10
Gas Chrom Q	3–10	25	20
Porapak N, Q, Q–S, R, S	1–5	8	4
Porapak T	1–2	6	4
Teflon®	1–2	6	3
Tenax®	1–4	5	4

* OV-1, SE-30, SE-52, OV-275, DEGS, Apiezon L are typical sticky or gum phases.
SOURCE: Varian Associates, Inc., Sunnyvale, California.

Chromosorb G combines the high surface area of P and ruggedness and inertness of W.

Chromosorb T is not made from diatomaceous earth, but is actually a fluorocarbon polymer (Teflon®). This solid support is much more inert than the other Chromosorbs and is used, when coated, to separate very polar compounds, like water, without the typical adverse tailing effects.

Porapak® is a trade name for stryrene–vinyl benzene cross-linked polymers. This polymer is formed into porous beads which serve as both the liquid phase and the solid support. The Porapaks are excellent for high-temperature (usually 250°C maximum) separations of polar mixtures. These polymeric beads can also be coated with conventional liquid phases, usually less than 5% by weight. Table 37-2 shows the typical and maximum liquid loadings for specific support materials.

LIQUID PHASES

Selecting the proper liquid phase for a particular GC separation can be the most difficult task in operating such a system. Generally, the liquid phase must

1. Be nonvolatile or at least have boiling point of 100°C greater than the maximum column operating temperature
2. Be thermally stable, not decompose with heat
3. Have good solubility for the sample components

Historically, hundreds of substances have been used as GC liquid phases, but the most important ones number fewer than fifty. These liquid phases can be classified as *polar, nonpolar,* or of *intermediate polarity.* The old laboratory expression, "like dissolves like," can be readily applied in selecting GC liquid phases. *Polar* substances are usually better separated on *polar* columns and *nonpolar* substances on *nonpolar* columns. Table 37-3 gives a classification of common compounds or functional groups listed in decreasing order of polarity.

Some forty GC liquid phases are listed in Table 37-4 classified as to their degree of polarity. This table should be used in conjunction and as a guide with Table 37-3's classification of compound polarities.

TABLE 37-3
Solute Classification

Class	*Solute*
Class I (most polar)	Water Glycol, glycerol, etc. Amino alcohols Hydroxy acids Polyphenols Dibasic acids
Class II (polar)	Alcohols Fatty acids Phenols Primary and secondary amines Oximes Nitro compounds with α-H atoms Nitriles with α-H atoms NH_3, HF, N_2H_4, HCN
Class III (intermediate)	Ethers Ketones Aldehydes Esters Tertiary amines Nitro compounds with no α-H atoms Nitriles with no α-H atoms
Class IV (low polarity)	$CHCl_3$ CH_2Cl_2 CH_3CHCl_2 CH_2ClCH_2Cl $CH_2ClCHCl_2$ etc. Aromatic hydrocarbons Olefinic hydrocarbons
Class V (nonpolar)	Saturated hydrocarbons CS_2 Mercaptans Sulfides Halocarbons not in Class IV, such as CCl

SOURCE: Gow-Mac Instrument Co., Bound Brook, N.J. Reproduced by permission.

Generally, as the components increase in molecular weight, their respective boiling points and retention times increase. Table 37-5 lists the most common liquid phases, their maximum operating temperatures, the solvent recommended for column preparation, and the McReynold's constants used to determine the polarity. *McReynolds constants* are used to compare the potential separating ability of GC liquid phases for compounds containing specific functional groups. McReynolds actually used 10 reference compounds in his study at 120°C. The larger the McReynolds constant for a particular functional group, the greater the retention time. For example, if a liquid phase is required that will retain ketones

TABLE 37-4
Liquid-Phase Classification

Class	*Liquid phase*
Class A (I)	FFAP
	20M-TPA
	Carbowaxes
	UCONs
	Versamid 900
	Hallcomid
	Quadrol
	THEED (tetrahydroxyethylenediamine)
	Mannitol
	Diglycerol
	Castorwax
Class B (II)	Tetracyanoethyl pentaerythritol
	Zonyl E-7
	Ethofat
	β,β-Oxydipropionitrile
	XE-60 (nitrile gum)
	XF-1150
	Amine 220
	Epon 1001
	Cyanoethyl sucrose
Class C (III)	All polyesters
	Dibutyl tetrachlorophthalate
	SAIB (sucrose acetate isobutyrate)
	Tricresyl phosphate
	STAP
	Benzyl cyanide
	Lexan
	Propylene carbonate
	QF-1 (silicone, fluoro-)
	Polyphenylether
	Dimethylsulfolane
	OV-17 (50% phenyl silicone)
Class D (IV & V)	SE-30
	SF-96
	DC-200
	Dow 11
	Squalane
	Hexadecane
	Apiezons
	OV-1 (methyl silicone)

SOURCE: Gow-Mac Instrument Co., Bound Brook, N.J.

more strongly than alcohols, a phase with a large Z′ value and a low Y′ value should be selected. According to Table 37-5, an OV-210 would be a better liquid-phase selection than an OV-101. For additional information on McReynolds constants consult the three references given with Table 37-5.

TABLE 37-5
Liquid Phases

Liquid phase	*Maximum temp., °C*	*Solvent**	*McReynolds constants†* X′	Y′	Z′	U′	S′	H	J
Apiezon	300	4, 7	32	22	15	32	42	13	35
Bentone 34	200	2, 4, 7							
BMEA [bis(2-methoxyethyl) adipate]	120	4							
Carbowax 20M	250	4, 6	322	536	368	572	510	387	282
Carbowax 1540	175	4	371	639	453	666	641	479	325
D.C. series (see silicone)									
DDP (didecyl phthalate)	125	1, 4, 6	84	173	137	218	155	133	83
DEGA (diethylene glycol adipate)	190	1, 4	377	601	458	663	655	477	328
DEGS (diethylene glycol succinate)	190	1, 4	502	755	597	849	852	599	427
Dexsil 300	500	2, 3, 7	47	80	103	148	96		
DNP (dinonyl phthalate)	150	1, 4	83	183	147	231	159	141	82
EGA (ethylene glycol adipate)	200	1h	372	576	453	655	617	462	325
FFAP	250	4	340	580	397	602	627	423	298
IGEPAL CO 880 (nonyl phenoxypoly-oxyethylene ethanol)	200	4h	259	461	311	482	426	334	227
OV-1 (methyl silicone)	350	3	16	55	44	65	42	32	4
OV-11 (phenylmethyl dimethyl silicone)	375		102	142	145	219	178	103	92
OV-17 (50% phenyl silicone)	325	1, 4	119	158	162	243	202	112	119
OV-25 (75% phenyl silicone)	300	3	178	204	208	305	280	144	160
OV-101 (liquid methyl silicone)	350	4, 7	17	57	45	67	43	33	4
OV-202 (trifluoropropyl methyl silicone)	275		146	238	358	468	310	206	56
OV-210 (trifluoropropyl methyl silicone)	350		146	238	358	468	310	206	56
OV-225 (cyanopropylmethyl phenyl-methyl silicone)	275	1	228	369	338	492	386	282	226
OV-275 (cyano silicone)	250	1	629	872	763	1106	849		
Polyethylene glycol (see Carbowaxes)									
Polyglylcol (see Carbowaxes)									
SE-30 (methyl silicone), GC grade	325	3h, 4h	15	53	44	64	41	31	3
Silicone, D.C. 200, 200 CS	200	4, 7	16	57	45	66	43	33	3
Silicone, D.C. 200, 500 CS	200	4, 7	16	57	45	66	43	33	3
Silicone, D.C. 200, 12,500 CS	250	4, 7	16	57	45	66	43	33	3
Silicone, D.C. 550	275	4, 7	81	124	124	189	145	87	81

COLUMN EFFICIENCY

Fractional distillation and chromatographic processes can be compared in their respective abilities to separate mixtures. Review Chap. 21, "Distillation and Evaporation," for a discussion of fractional distillation, theoretical plates, and column efficiency calculations. *Resolution* is defined as the degree of separation between adjacent peaks on a chromatogram. For symmetrical peaks (gaussian-shaped), the resolution R can be calculated with the following equation. For symmetrical peaks, a resolution R cal-

TABLE 37-5
(Continued)

Liquid phase	*Maximum temp., °C*	*Solvent**	McReynolds constants† *X′*	*Y′*	*Z′*	*U′*	*S′*	*H*	*J*
Silicone (fluoro) QF-1 (FS 1265)	250	1, 4	144	233	355	463	305	203	136
Silicone GE SF-96	250	1, 4	12	53	42	61	37	31	0
Silicone GE XE-60 (nitrile gum)	250	1, 4	204	381	340	493	367	289	203
Silicone gum rubber SE-52 (phenyl)	300	4h, 7	32	72	65	98	67	44	23
Squalane	100	4, 7	0	0	0	0	0	0	0
Squalene	100	4, 7	152	341	238	329	344	248	140
Tributyl phosphate	50	4							
UCON 50 HB 2000 polar	200	4, 6	202	394	253	392	341	277	173
Versamid 900	250	8							

* **Solvent definitions for coating:**

1 = Acetone
2 = Hexane
3 = Chloroform
4 = Methylene chloride
5 = Ethyl acetate
6 = Methanol
7 = Toluene
8 = 1:1 butanol and phenol or 87% chloroform and 13% methanol. **Note that Versamid is unstable at high temperatures in the presence of oxygen.**
h = Hot

† A limited number of McReynolds constants are shown.

X′ = V Benzene
Y′ = V Butanol
Z′ = V 2-Pentanone
U′ = V Nitropropane
S′ = V Pyridine
H = V 2-Methyl-2 pentanol
J = V 1-Iodobutane

Further information about McReynolds constants is contained in the following references:

L. Rohrschneider, *J. Chromatogr.*, **22**, 6, 196.
W. O. McReynolds, *J. Chromatogr. Sci.*, **8**, 685, 1970.
R. A. Keller, *J. Chromatogr. Sci.*, **11**, 49, 1973.
L. S. Ettre, *Chromatographia*, **6**, 489, 1973 and **7**, 39, 1974.
SOURCE: Varian Associates, Inc., Sunnyvale, California.

culation of 1.0 corresponds to approximately a 2% peak overlap. Figure 37-11 shows a chromatogram with two unsymmetrical peaks and a technique for estimating the resolution.

$$R = \frac{2(t_2 - t_1)}{w_1 + w_2}$$

where t_2 and t_1 are retention times for peak 2 and 1, respectively
w_1 and w_2 represent the estimated peak base widths for peaks 1 and 2

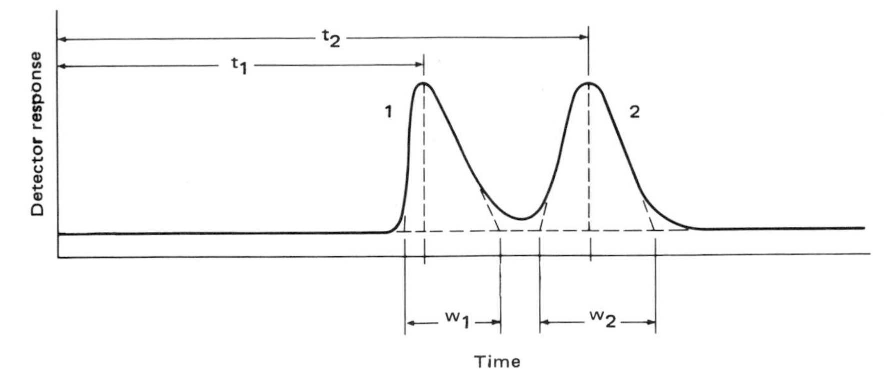

FIGURE 37-11
Chromatogram with two unsymmetrical peaks.

A simple mathematical relationship can be drawn between the number of theoretical plates (number of transfer equilibria) in a fractional distillation and the resolution of components in a chromatographic column. In a chromatographic column, the number of theoretical plates n can be calculated by the equation given below. A better term for this value is "column efficiency" since it depends on column flow rate, temperature, column length, and even the compound itself. The n term is unitless: therefore R, and W must be in the same units of length (usually millimeters). The greater the number of theoretical plates n, the more efficient and better the resolution in the column. Figure 37-12 shows a typical chromatogram and the appropriate measurements.

$$n = 16\left(\frac{t_R}{W}\right)^2$$

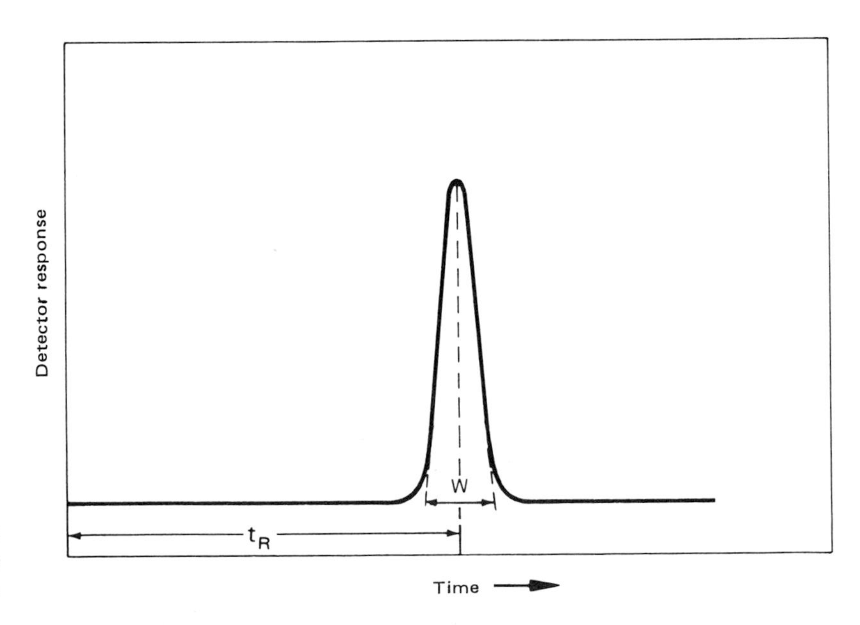

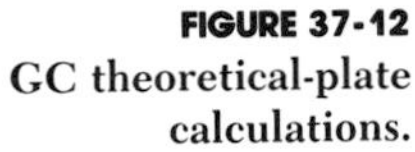

FIGURE 37-12
GC theoretical-plate calculations.

where t_R = retention time
W = width of peak

Another term used to describe column efficiency is the *height equivalent to a theoretical plate* (HETP). This value can be calculated with the equation given below with L representing length of column expressed in centimeters. Obviously, the smaller the HETP (length of column required to produce the equivalent of a distillation plate), the better the resolution of mixture components.

$$\text{HETP} = \frac{L}{n}$$

In general, resolution can be improved by changing one or more of the following parameters in the following ways:

1. Increasing the retention time t_R
2. Increasing the column length
3. Using smaller-diameter stationary phase
4. Decreasing the diameter of the column
5. Optimizing the carrier-gas flow rate
6. Reducing sample size

FLOWMETERS

There are two basic ways to determine the carrier-gas flow rate in a gas chromatograph. The simplest and most versatile technique is to use a *bubble flowmeter* which consists of a reservoir of soap solution, a calibrated glass tube, and a squeeze bulb as shown in Fig. 37-13. A rubber tube connects the GC exit port to a side arm on the calibrated flowmeter. The col-

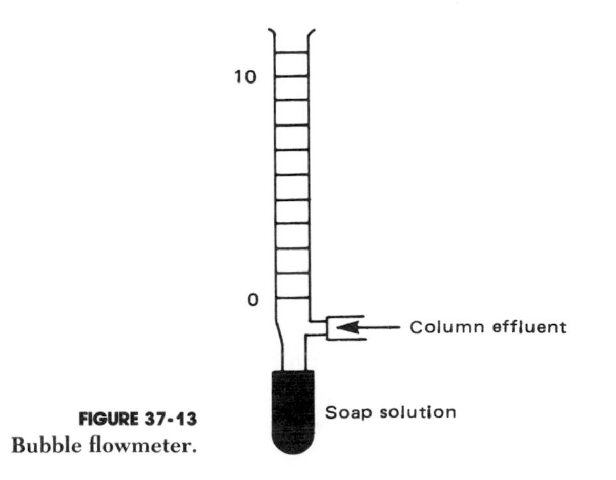

FIGURE 37-13
Bubble flowmeter.

umn effluent causes the soap film to form bubbles that traverse the calibrated column as the rate is timed, usually in minutes. Assume that a stopwatch is started as a bubble passes the "0 mL" calibration mark, stopped at the "10 mL" mark, and found to take exactly 15 s (0.25 min). The flow rate would be 10 mL/0.25 min—a better expression would be 40 mL/min. This type of flowmeter normally is calibrated for 0–2 mL, 0–10 mL, and 0–60 mL volume ranges for capillary and standard GC applications. This type of flowmeter is also available in a digital model which provides more convenience and accuracy.

A second device used to determine carrier-gas flow rate is called a *rotameter.* Again a calibrated glass tube is used, but fitted with a ball that rises in the tube as the column effluent flows passes it. This type of flowmeter is normally built into the instrument directly in the carrier-gas stream between the gas source and the injection port. This type of flowmeter usually contains a valve for adjusting the carrier-gas flow rate through the column. Rotameters must be calibrated for the different carrier gases; otherwise the numerical scale is only in arbitrary units.

QUALITATIVE ANALYSIS

The time required for each compound to pass through the column is called the *retention time* and is characteristic of that compound and can be used to identify each substance in a mixture (Fig. 37-14*a*). Peak A in Fig. 37-14*a* represents air and indicates that even air has a column retention time or at least residence time. Caution should be exercised in assuming that a retention time definitely identifies a compound; several compounds can have identical retention times. The identification of compounds causing GC peaks is performed by comparing the retention time of each eluted compound with the retention time of a pure standard under identical conditions. In other words, retention time for a known sample is determined immediately before or after the unknown's analysis on the same column using identical column temperature, flow rate, sample size, etc.

A second technique of peak identification is called *spiking.* A known compound is deliberately added to the unknown mixture, and the size of the original peak is observed (Fig. 37-14*b*). If the unknown peak increases in size after the spiking, it would indicate that the two compounds have the same retention time. However, since several compounds can have identical retention times, one should be cautious about assuming the identity of a compound using GC retention data only.

INTEGRATION TECHNIQUES

The concentration of each eluting compound is directly proportional to the area under the recorded peak. Several methods are currently used to determine this area response: peak height, height times one-half base

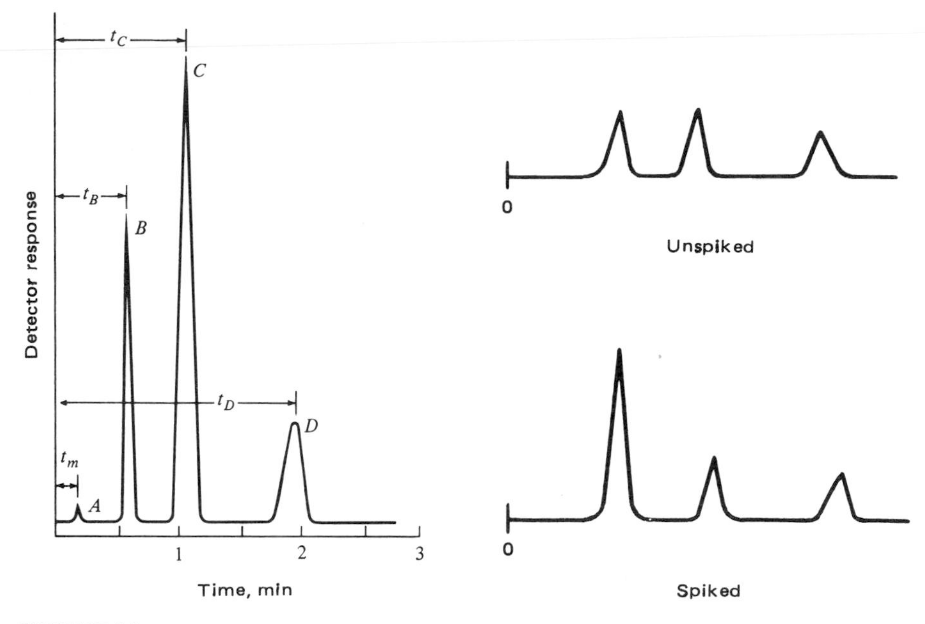

FIGURE 37-14
(*a*) Retention times on a chromatogram. (*b*) Chromatograms of unspiked and spiked sample.

width, height times half-width, planimeter, Disc Integrator®, and electronic integration. At one time, chromatographers had to cut out each peak on the chromatogram, weigh the paper, and assume the mass to be proportional to the compound's concentration. A summary of the precision of various integration techniques is given in Table 37-6.

Peak Height

Peak-height measurement is the quickest and least accurate of these quantifying techniques. The assumption is made that if a compound's concen-

TABLE 37-6
Precision of Various Integration Techniques for GC Peaks

Type	*Precision, %*
Computer	0.5 and better
Electronic integrator	0.5
Disc Integrator®	1
Height times half-width	3
Triangulation	4
Planimeter	4
Peak height	5

SOURCE: Hajian and Pecsok, *Modern Chemical Technology*, rev. ed., vol. 6, Prentice-Hall, Englewood Cliffs, N.J., 1973.

tration is proportional to its area response then its peak height should be directly proportional to the compound's concentration. Peak-height measurements are more prone to errors from temperature and flow variations and nonuniform injection techniques than are the area measuring techniques. Capillary chromatography lends itself to peak-height integration techniques since the peaks are very narrow and closely spaced; however, area determinations are generally the preferred method.

Triangulation

The *triangulation integration technique* assumes that the GC peak is a triangle and the area can be calculated using the formula:

Area = ½base × height

This triangulation technique is shown in Fig. 37-15. The "half-width method" is a slight modification of this technique in that the peak height is measured and one-half base width is assumed to be the width at half this measured peak height. This technique is demonstrated in Fig. 37-16. The advantage with this modification comes from eliminating the necessity to measure the rounded corners of the triangle at the baseline.

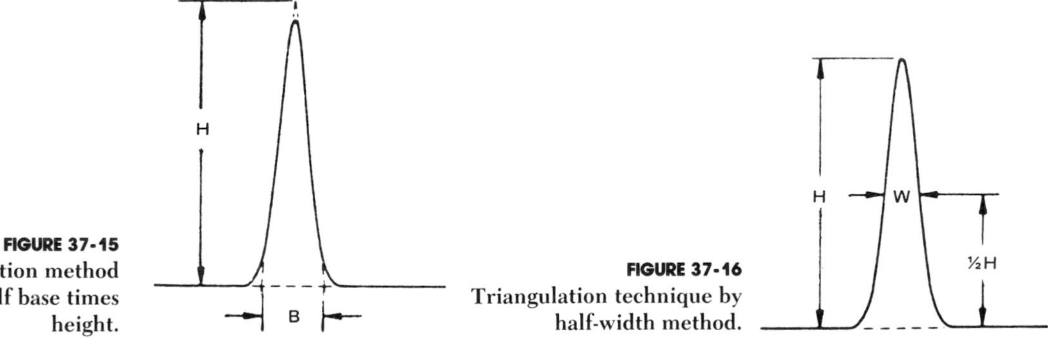

FIGURE 37-15 Triangulation method using one-half base times height.

FIGURE 37-16 Triangulation technique by half-width method.

Planimeter

The *planimeter* is a mechanical device used to trace out the perimeter of geometric shapes and, using conversion factors, converting this distance to area. Figure 37-17 explains the procedure for integrating an area using a planimeter. A weighted reference point is located at point A so that stylus C can trace the entire perimeter of the chromatographic peak after the base has been enclosed by a straight line. Start stylus C at any convenient point on the peak, but preferably at a corner like point D. A wheel mechanism located at point B follows the motion at the apex (also point B). A reading is taken from the wheel mechanism before moving the stylus from point D; then the GC peak and straight-line base are traced completely around until point D is again reached. The change in the wheel-

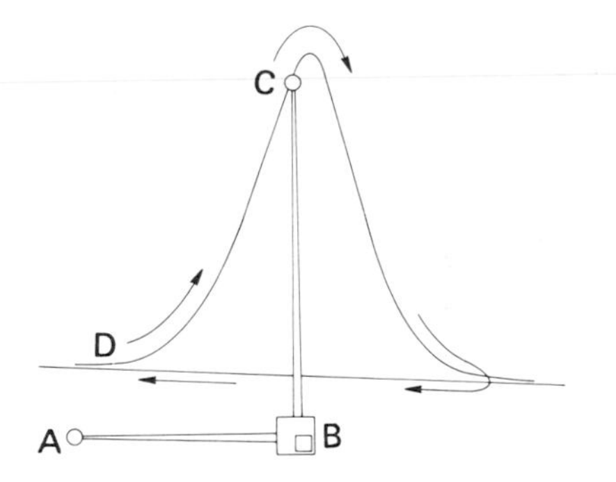

FIGURE 37-17
Planimeter integration calculation *(Courtesy of American Chemical Society).*

mechanism readings is calculated, which gives the perimeter distance (area enclosed) in arbitrary units. A conversion factor can be calculated by tracing a square of known area and recording the equivalent arbitrary units from the planimeter. The actual area can then be calculated by multiplying the arbitrary number determined from a GC peak by the conversion factor. In most GC integration work, it is not necessary to convert to actual area since relative ratios between arbitrary numbers comparing standards with unknowns work just as well.

Disc Integrator

The Disc Integrator® is a mechanical device built into the recorder and is used to integrate the area under a chromatographic peak. Electronic integrators have pretty much replaced this mechanical device; however, Fig. 37-18 shows a typical pattern of this type of integrator. The integrator

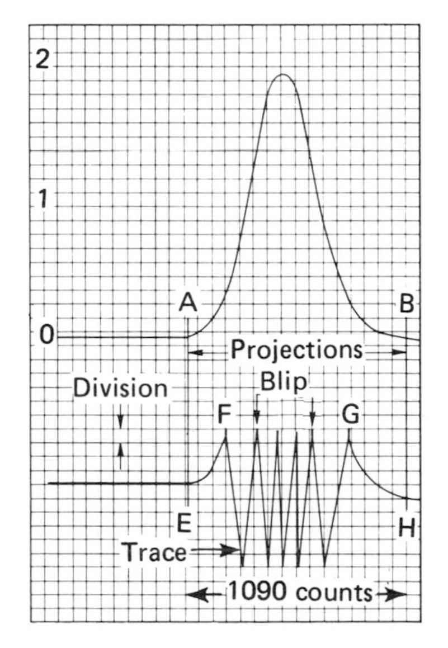

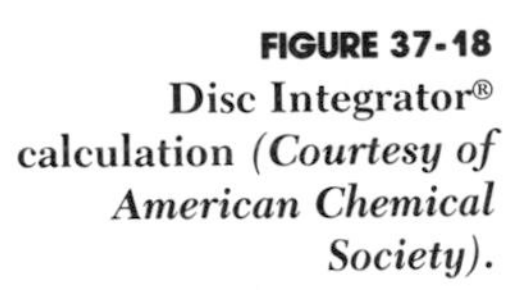

FIGURE 37-18
Disc Integrator® calculation *(Courtesy of American Chemical Society).*

trace is interpreted by first determining the start and end points of the chromatographic peak (points A and B) and projecting a line directly down to the trace pattern (points E and H). The relative area is obtained by counting the chart horizontal lines crossed by the integrator trace. Each chart division is given an arbitrary value of 10 with a complete traverse having a value of 100. Notice that every sixth full traverse produces a counting blip. The 600 counts between successive blips makes the counting easier. Figure 37-18 is interpreted as having 1090 counts because there are 10 full traverses (F to G) or 1000 counts. Partial traverse from E to F yields 40 counts and partial G to H an additional 50 counts.

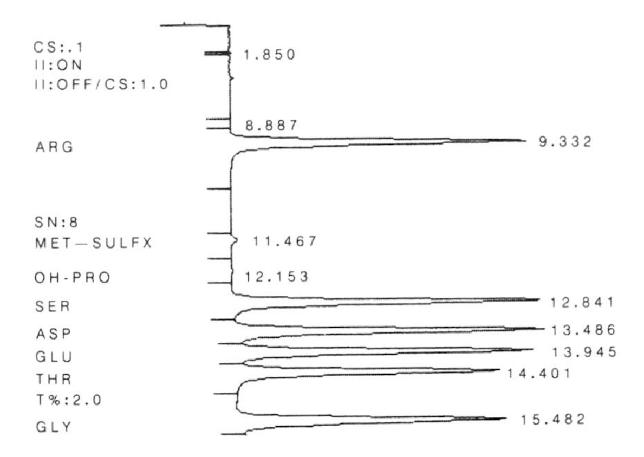

RECALCULATE ON FILE: AMSTD
CHANNEL: 3A - 1 TITLE: BORATE 0.1M; 9090 FMOC-CL 15:37 22 JAN 87
SAMPLE: TYPEHPH8.5 METHOD: AMINOTAG1 CALCULATION: ES - ANALYS

PEAK NO	PEAK NAME	RESULT PICOMO	TIME (MIN)	TIME OFFSET	AREA COUNTS	RRT	SEP CODE	W1/2 (SEC)
1	ARG	134.6110	9.332	0.002	1084938	0.60	VB	7.56
2	MET-SULFX	2.7967	11.467	-0.033	24319	0.74	BV	8.06
3	OH-PRO	0.6929	12.153	-0.047	5363	0.78	VV	? 12.94
4	SER	134.8622	12.841	0.001	1138428	0.83	VV	7.69
5	ASP	134.7084	13.486	0.006	1162682	0.87	VV	7.81
6	GLU	133.9743	13.945	0.005	1151814	0.90	VV	8.06
7	THR	133.5645	14.401	0.001	1109164	0.93	VV	8.25
8	GLY	133.8057	15.482	0.002	1201640	1.00	VV	9.19
TOTALS:		809.016		-0.063	6879348			

TOTAL UNIDENT AREA/HT: 402883
DETECTED PKS: 16 REJECTED PKS: 1
DIVISOR: 1.00000 MULTIPLIER: 1.00000
RESOLUTION: 4.38
NOISE: 6.9 OFFSET: -19131
RACK: 1 VIAL: 9 INJ: 1

RUN LOG:	TIME:	RUN LOG:	TIME:
CMPA: 73	0.00	CMPB: 0	0.00
CMPC: 27	0.00	FLOW: 1.40	0.00
AOUT: %B	0.00	PMAX: 210	0.00
PMIN: 20	0.00	PRES: 132	0.00
NM: 269	0.00	ZERO: N	0.00
ATTN: .1	0.00	TC: .5	0.00
CMPA: 58	11.50	CMPB: 0	11.50

NOTES:
FMOC-CL DERIVATIZED AMINO ACIDS ON THE 9090 A/S.
FLUORICHROM DETECTOR, 2541 EXC., 304 AND 345 EMISSION FILTERS.

FIGURE 37-19
Typical electronic-integrator display *(Courtesy of Varian Associates, Inc.)*.

Electronic Integrator

The *electronic integrator* measures the area under each peak automatically, calculates the relative areas, and prints the calculated results with the retention time for each peak to the nearest thousandth of a minute, if desired. This technique is by far the most accurate integration technique, but these integrators cost several thousand dollars. These integrators can be calibrated by internal and external standardization techniques as described below and are used in both GC and liquid chromatography (LC); a typical LC electronic-integrator display is shown in Fig. 37-19.

QUANTITATIVE ANALYSIS

The various integration techniques listed above simply provide areas or numbers proportional to the area under each GC peak. There are four mathematical procedures for converting area responses into actual concentrations: normalization, external standard, internal standard, and the standard-addition method.

Normalization

In Fig. 37-20, assume that the total area under all peaks (A, B, and C) represents 100% of the sample.

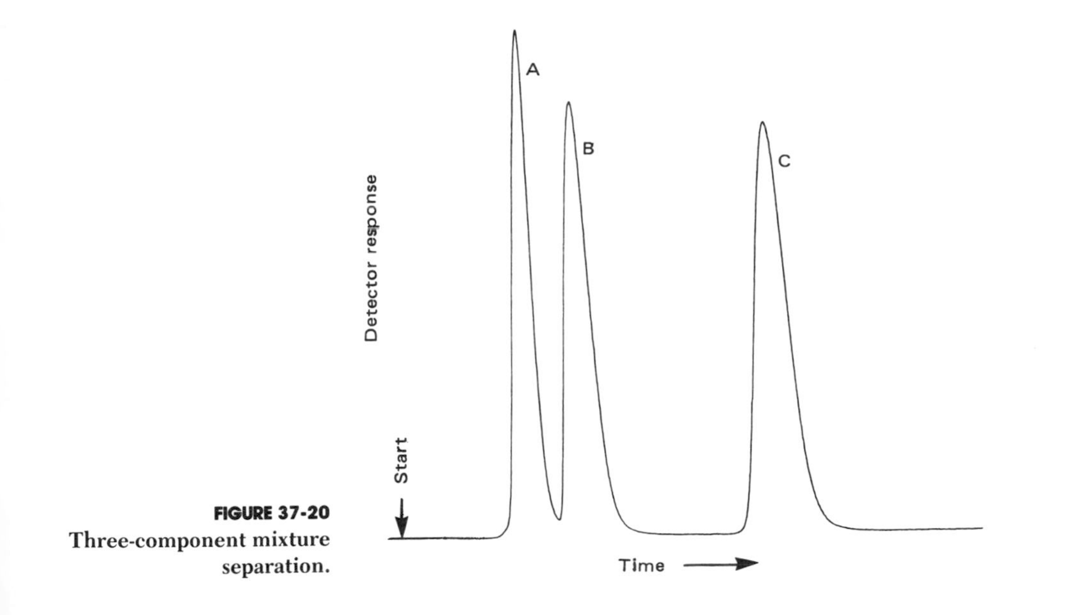

FIGURE 37-20 Three-component mixture separation.

Each peak's area (A, B, and C) represents that compound's fraction of the total composition, and the relative percentage can be calculated using the following equation:

$$\%A = \frac{\text{area of peak A}}{\text{total peak areas}} \times 100$$

The denominator (total peak areas) in this equation is the sum of peak areas of compounds A, B, and C. There are two assumptions with this "simple" normalization method: (1) all compounds were eluted, and (2) all compounds have the same detector response. By design, the percent figure for all components should add up ("normalize") to 100%.

If *response factors* are known for each compound, then each peak area can be corrected before using the normalization formula. Absolute response factors can be determined by plotting a calibration curve with a minimum of three concentrations for each of the compounds expressed in mass and plotted against the corresponding peak areas as shown in Fig. 37-21.

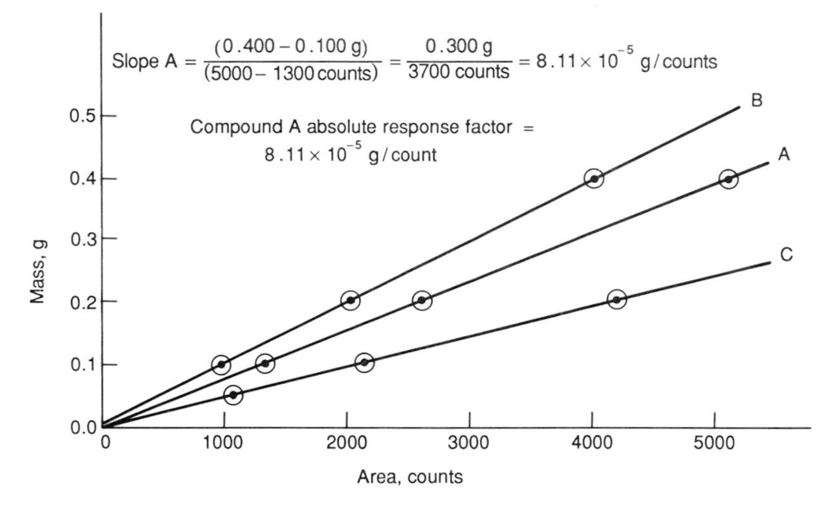

FIGURE 37-21 Absolute-response-factor calibration curve.

The best straight line through the data points is then selected. The slope of this curve (mass/area) is the absolute response factor for that particular compound. The corrected area for each compound must now be used in the normalization formula by multiplying the peak area of each compound by its absolute response factor.

$$\%A = \frac{\text{area of peak A} \times \text{correction factor}}{\text{total corrected peak areas}} \times 100$$

External Standard

The *external-standard* technique uses the exact procedure described above for determining the absolute response factor for each compound of interest. The advantage to this technique is that calibration plots are necessary only for compounds of interest and it is not necessary to resolve or to determine response factors for the remaining peaks. The main disadvantage of this external-standard calibration technique is that the sample injection size must be very reproducible.

Internal Standard

The *internal-standard* technique requires that all samples and standards be spiked with a fixed amount of a substance called the *internal standard.* The substance selected as the internal standard must meet the following criteria:

1. It must not be present in the sample to be analyzed.
2. It must be completely resolved from all peaks in the sample.
3. It must be available in high purity.
4. It is added at a concentration level similar to that of the unknown compound(s).
5. It does not react with the unknown mixture.

This internal standard serves as a reference point for all subsequent peak-area measurements. This procedure minimizes any error from variations in injection size or GC parameters. The calibration procedure is to measure the peak areas for both the compound of interest and the internal-standard peak. The internal-standard technique does not require that all peaks be measured, only those of the compounds of interest and the internal standard itself. As was described for external standards above, a response factor for each compound of interest must be determined and an additional determination for the fixed amount of internal standard must also be included. Chromatograms for each concentration level of the known standards are run with the peak areas for both the standard and internal standard being determined. A typical chromatogram with internal standard is shown in Fig. 37-22.

The ratio of peak area for each compound divided by the peak area for the internal standard is calculated. This ratio is plotted against the known concentrations of each compound as shown in Fig. 37-23.

Standard Addition

Standard addition is a method that actually combines both external and internal standardization techniques and is only used in GC work when the sample matrix has many interferences. A standard amount of one of the compounds of interest is spiked into the sample. This standard-addition

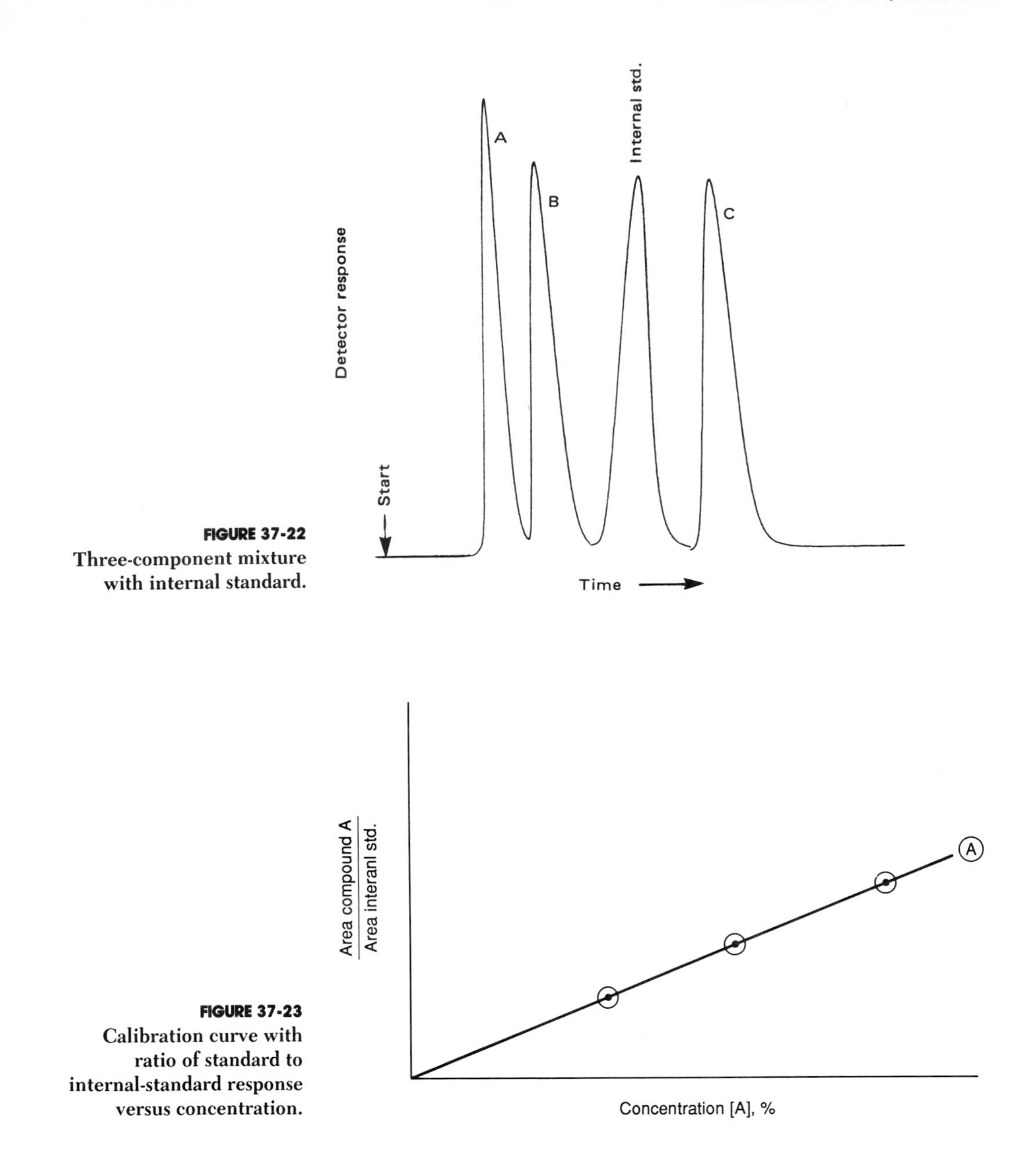

FIGURE 37-22
Three-component mixture with internal standard.

FIGURE 37-23
Calibration curve with ratio of standard to internal-standard response versus concentration.

technique is performed ideally by adding one-half the original concentration and twice the concentration. The resulting peak areas are measured and plotted versus the concentration added. The unknown's original concentration is then determined graphically by extrapolating the curve back to the x-axis (zero peak size) as shown in Fig. 37-24.

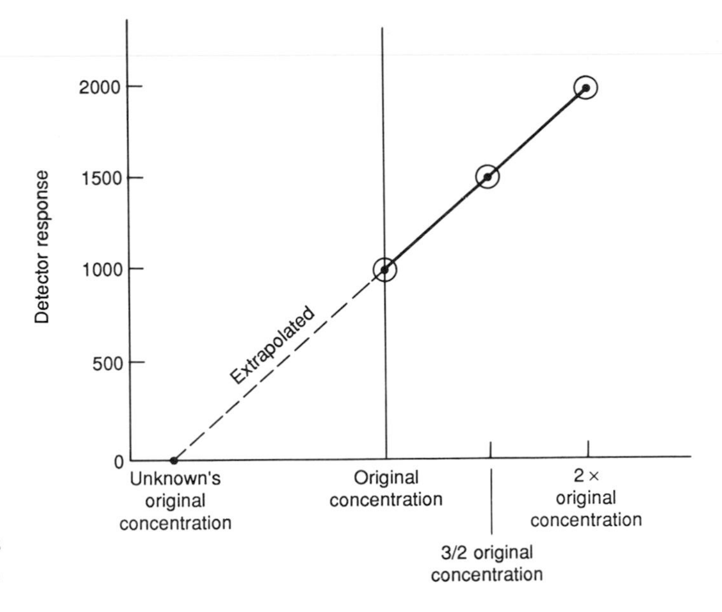

FIGURE 37-24
Standard additions curve.

A more mathematical calculation simply uses the following ratio formula where X represents the unknown concentration:

$$\frac{X}{X + \text{addition}} = \frac{\text{area original}}{\text{area with addition}}$$

This procedure as described does not compensate for sample-size variations as did the internal-standard procedure. In addition, if the standard addition amounts are kept small with respect to the total volume of sample, then dilution effects should not be a major source of error; otherwise it will be necessary to dilute all samples to the same volume before analysis.

REFERENCES

Grob, R. L., *Modern Practice in Gas Chromatography*, John Wiley and Sons, New York, 1977.

Jennings, W., *Gas Chromatography with Glass Capillary Columns*, 2nd ed., Academic Press, New York, 1980.

McNair, H. M., and Bonelli, E. J., *Basic Gas Chromatography*, 5th ed., Varian Associates, 1969. Available from Varian Associates, 2700 Mitchell Drive, Walnut Creek, CA 94598.

Perry, T. A., *Introduction to Analytical Gas Chromatography*, Vol. 14 of Chromatographic Science Series, Marcel Dekker, New York, 1981.

Yancy, J. A., ed., *Guide to Stationary Phases for Gas Chromatography*, Analabs, Inc., North Haven, Connecticut, updated on a regular basis.

38

LIQUID CHROMATOGRAPHY*

INTRODUCTION

Liquid chromatography (LC) is a separation technique which uses two phases in contact with each other; the stationary phase can be an immiscible liquid or solid, but the mobile phase must be a liquid. Liquid chromatography has many similarities to other chromatographic systems discussed in Chap. 36, "Chromatography," and Chap. 37, "Gas Chromatography" (for example, in gas chromatography (GC), liquid and solid stationary phases are used, but the mobile phase is a gas). The sample is introduced as a liquid into the mobile phase; it does not have to be volatile as is necessary in GC. Therefore, LC is very useful for analyzing mixtures of nonvolatile and thermally labile compounds. Liquid chromatography can be conveniently divided into three classifications: liquid-solid chromatography (LSC), liquid-liquid chromatography (LLC), and high-pressure liquid chromatography (HPLC). Ordinarily, LC processes use a polar stationary phase like silica gel or alumina and a nonpolar mobile phase. *Reverse-phase chromatography* (RPC) uses a nonpolar stationary phase and a polar mobile phase. See the section in this chapter titled Bonded Stationary Phases for more details on reverse-phase chromatography.

A basic LC system consists of a solvent reservoir, pump, injection port, column, detector, recorder, and fraction collector as shown in Fig. 38-1 and a typical liquid chromatogram is shown in Fig. 38-2. These LC components will be described in detail later in this chapter.

Technically, ion-exchange chromatography, gel-permeation chromatography, thin-layer chromatography, and even paper chromatography could be classed as LC systems because they all use liquid mobile phases. These chromatographic systems are discussed in detail in Chap. 36, "Chromatography."

COLUMN CHROMATOGRAPHY

The simplest example of LSC is represented by column chromatography as shown by Fig. 38-3.

* The authors wish to thank ALLTECH Associates; Gow-Mac Instrument Co.; Varian Associates, Inc.; and Waters Associates, Inc. for their assistance and contributions to this chapter.

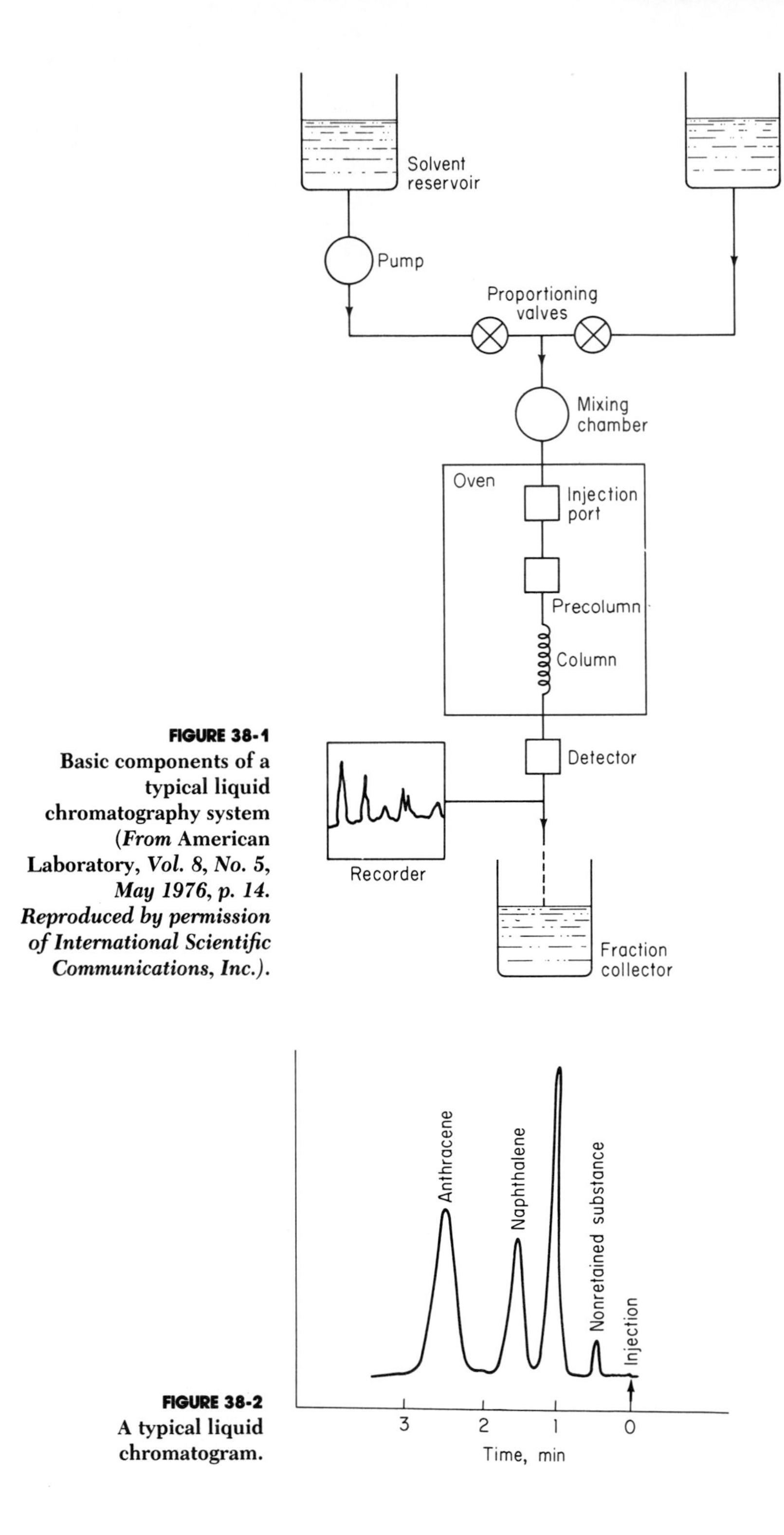

FIGURE 38-1
Basic components of a typical liquid chromatography system (*From* American Laboratory, *Vol. 8, No. 5, May 1976, p. 14. Reproduced by permission of International Scientific Communications, Inc.).*

FIGURE 38-2
A typical liquid chromatogram.

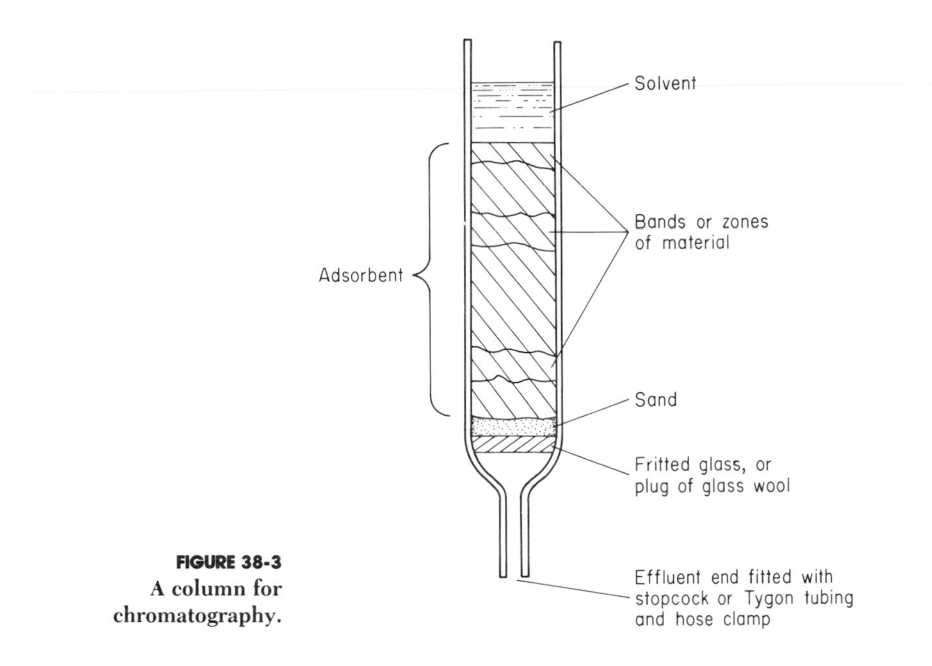

FIGURE 38-3
A column for chromatography.

Techniques

Column chromatography is one of the most useful and versatile of the chromatography classifications. Some techniques and tips for using column chromatography follow.

1. Use a column with a minimum length-to-diameter ratio of 20:1.

2. Check any plugs (cotton, for example) under a UV lamp to avoid contamination from an optical brightener.

3. Determine the optimum quantity of sorbent required to effect the separation. Generally adsorption requires between 50 and 100 g of sorbent per gram of sample, and partition requires between 500 and 1000 g of sorbent per gram of sample.

4. When you use a dry-packed column, cover the sorbent with solvent immediately after carefully packing the column; do not allow the solvent to evaporate so that the sorbent dries out. Allow any heat which has developed during the addition of the solvent to the sorbent to dissipate, and let the column come back to normal (or room) temperature. If you desire to shorten the cooling time, the use of cooling jackets will accelerate this heat dissipation.

5. Slurry pack instead of dry pack when you desire the highest resolution in a procedure. Isooctane is well suited for use in nonpolar (nonaqueous) systems. Gently tap or bump the column continuously while you fill it with a slurry of the sorbent in isooctane by pouring in one small portion at a time. This disturbance will free any trapped air bubbles and pack the column more tightly and more uniformly, so that it will yield better results.

6. The way in which the sample is introduced into the system bears directly upon the results. The sample should be introduced uniformly and symmetrically, without disturbing the column sorbent. You can (1) slurry the sample with some sorbent and pour this slurry on top of the column bed, or (2) seat a filter disk or pad of filter paper on top of the column bed and then gently pipet the sample onto the filter disk.

GENERAL OPERATING PROCEDURE

There are almost limitless combinations of modes, solvents, sorbents, and procedures to select from, and yet there are no ironclad rules to guide you in your selection. However, Table 38-1 offers some general guidelines which provide you with a starting point, and the following list* gives hints on LC operations.

TABLE 38-1
Suggested Chromatographic Techniques

Categories of samples	*Liquid chromatography modes*
Positional isomers, moderate-polarity molecules	Liquid-solid
Insect molting steroids	Liquid-solid
Compounds with similar functionality	Liquid-solid or liquid-liquid
Polar and polynuclear aromatics	Liquid-solid
Barbiturates	Liquid-solid
Ionizable species	Ion exchange
Polysulfonated hydroxynaphthalenes	Ion exchange
High-polarity compounds	Liquid-liquid
Metallic chelates	Liquid-liquid
Compounds with differing solubilities	Liquid-liquid
Mixtures of varied sizes of molecules	Gel-permeation
Lubricating oils	Gel-permeation

1. The septum should be checked daily for leaks and must be changed often.

2. Check the flow rate regularly at a specified pressure to detect buildup of pressure (or decrease of flow).

* Used by permission of Gow-Mac Instrument Co., Bound Brook, N.J.

3. Pressure buildup can be caused by small pieces of septum which become deposited at the head of the column after many injections. To correct this situation, remove a few millimeters of sorbent from the top of the column and repack with new material.

4. Allow sufficient time for the LC system to stabilize after being turned on. Plan ahead.

5. The activity of a solid stationary phase can vary with the purity of the solvents being used and the polarities of the samples. It may be necessary to regenerate the column if it appears to have lost its separating capability.

6. If possible, samples should be dissolved in the liquid mobile phase.

7. Exercise care with flammable and/or toxic solvents.

8. Only high-purity solvents should be used as mobile phases. Some may require distillation prior to use.

9. Try to dissolve samples in the mobile phase or in a less polar solvent than the mobile phase. This technique tends to concentrate the injection on the tip of the column and yields better resolution.

10. When filling the pump, hold the funnel slightly above the opening in the pump; this maneuver allows air to escape from the reservoir.

11. Never remove or loosen the lower ¼-in column fitting; this disturbs the column bed and destroys column efficiency.

12. If the syringe is pushed too far into the column packing, the needle becomes plugged. To clear the needle, hold the syringe with the needle pointed down, allow some solvent to collect around the plunger, and then rapidly remove the plunger, causing a vacuum to form inside the syringe barrel. The vacuum sucks in some of the liquid. If you now replace the plunger, pushing the liquid through the needle, you will force out the plug of packing material.

13. After the standard column has been used for a period of time, its chromatographic properties may change. The column may be restored to its previous activity by pumping through it 50 mL each of ethyl alcohol, acetone, ethyl acetate, chloroform, and hexane. This treatment should leave the column as active as it was when you received it.

14. If you want to change from a hexane mobile phase to water, pump a solvent miscible in both liquids through the system before making the change. This removes all traces of hexane remaining in the system.

15. Stop-flow injections can be made easily by opening the three-way valve, releasing the pressure, and then making the injection and repressurizing the system by turning the three-way valve back to the OPERATE position.

16. Many times it is possible to inject very large samples (100 to 200 mL) in LC when more sensitivity is needed. If the sample is dissolved in a solvent less polar than the mobile phase, even a 1- or 2-L sample is possible with no deleterious effects apparent in the separation.

General Precautions

General precautions for solvent compatibility, flow limitations, and general care should be thoroughly read prior to any analysis. Failure to do this can result in misleading analytical information and could terminate the usefulness of the column(s). Careful attention will increase column life and allow you to return to a stored column with a knowledge of the purging steps required prior to analytical use. Always tag a column to indicate the last solvent passed through it. For lengthy storage, refer to the column-maintenance booklet provided with each column.

It is important to filter all solvents which are used in the liquid chromatograph. The liquid chromatograph is a high-precision instrument, and particulate matter (>0.5 μm) could be the source of problems. It is also important to filter all samples before they are injected into the instrument.

Sorbents for Column Chromatography

Since the sorbent used in column chromatography is packed in a vertical column, there is no need for any binders such as those used in thin-layer chromatography (TLC). However, the critical parameters for sorbents are particle size and size distribution. These factors directly affect the flow of the solvent, of which the driving force may be either hydrostatic or low pump pressure. A narrow particle-size distribution usually will provide better separation, all other factors remaining the same (Fig. 38-4).

In TLC, binders are needed for the sorbent, and generally a sorbent with smaller particle size is used. (See Thin-Layer Chromatography in Chap. 36, "Chromatography.") Table 38-2 offers some suggestions on choosing sorbents for TLC and column chromatography.

Solvent Mixture for Use in Column Chromatography

The possible combinations of solvent mixtures are almost unlimited because of the number of solvents used and the possible proportions of the concentrations of the components of the mixtures. Representative solvent mixtures used with different sorbents are shown in Table 38-3. These may be used as a basis for creating special mixtures as they are needed. Solvent mixtures are formulated by experiment. For example:

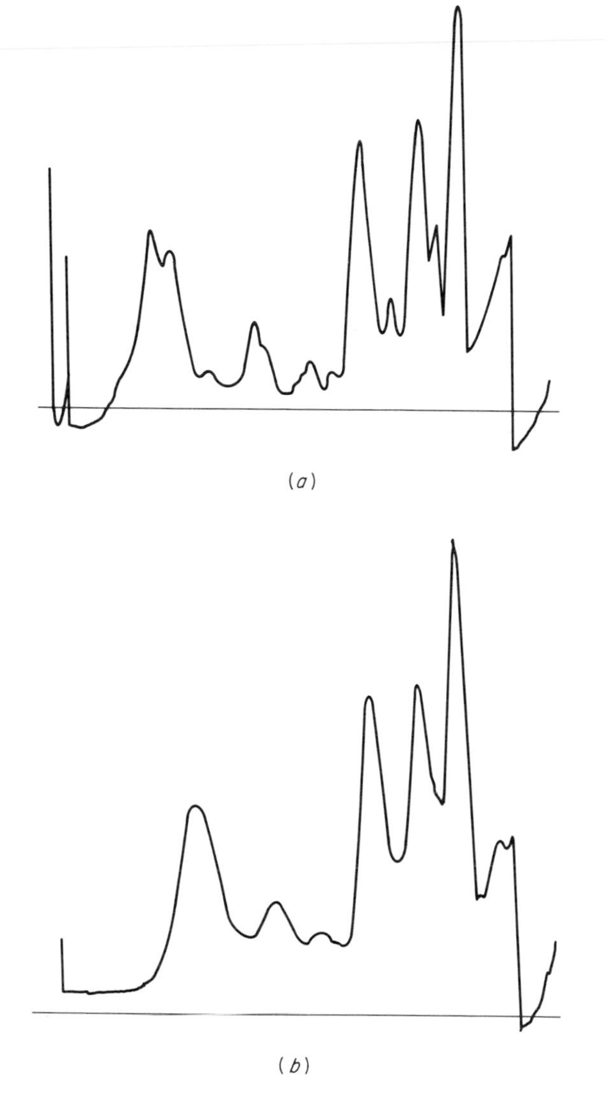

FIGURE 38-4
Effect of particle-size destribution on band resolution. (*a*) Narrow particle-size distribution. (*b*) Wide particle-size distribution.

1. Choose a single solvent on the basis of its polarity (Fig. 38-5) and the nature of the sorbent, starting with the least polar such as the alkanes, hexane, or petroleum ether.

2. Pass the solvent through the column as long as the bands appear to move. When they stop moving, modify the solvent by adding a small quantity of a more polar solvent, such as chloroform.

TABLE 38-2
Sorbent Selection Guide for Thin-Layer and Column Chromatography*

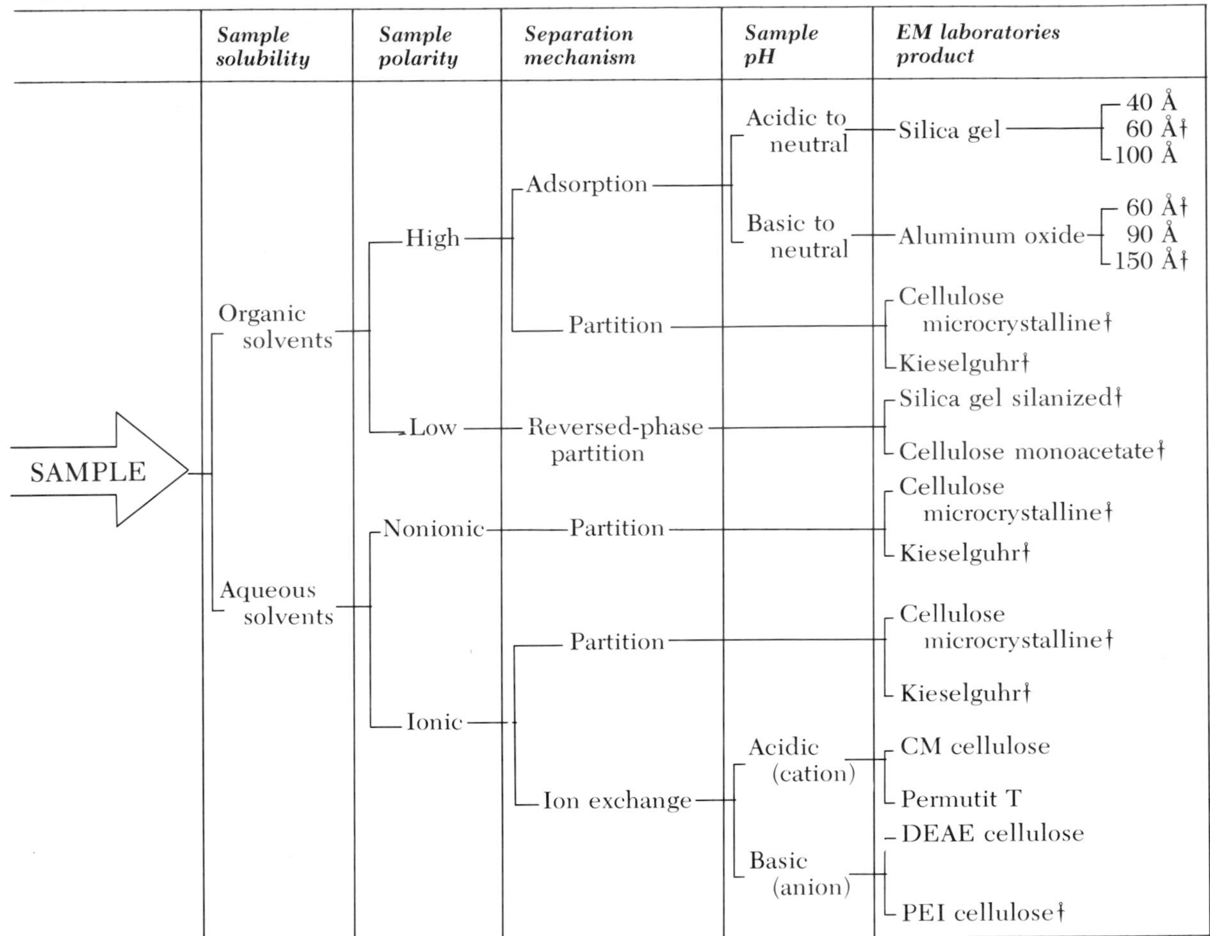

* This guide is based upon the most frequently used chromatographic systems. Exceptions are possible.
† Also available as a precoated product.
SOURCE: ALLTECH Associates. Reproduced by permission.

3. Continue increasing the proportion of the more polar solvent, retaining each mixture as long as the bands move. When they stop moving, increase the proportion of the more polar solvent again. Continue repeating the process as often as necessary.

4. You can formulate empirical mixtures by arbitrarily selecting components and determining their activity; one component must be much more polar or nonpolar than the other.

TABLE 38-3
Solvent Mixes for Column Chromatography

Solvent mixture	*Proportion*	*Sorbent*
Ethyl alcohol	60	Silica gel
Acetic acid	30	
Water	10	
Hexane	90	Silica gel
Chloroform	10	
Benzene	45	Silica gel
Methyl alcohol	10	
Acetic acid	5	
n-Butanol	10	Aluminum oxide
Methanol	1	
Dioxane	50	Silica gel
Acetic acid	5	
Benzene	1	

LIQUID CHROMATOGRAPHY STATIONARY PHASES

The simplest LC system is one which contains a stationary phase of a solid adsorbent and a mobile (pure solvent) phase. The most common adsorbents are alumina and silica gel with magnesium silicate, charcoal, calcium carbonate, sucrose, starch, powdered rubber, and powdered cellulose being used less frequently.

Alumina

Alumina (Aluminum oxide, Al_2O_3) is a very polar adsorbent with a surface area of approximately 150 m^2/g; it normally contains approximately 3% water, and the degree of activity can be controlled by its water content. It can be reactivated by dehydration at 360°C for 5 h and then allowing the desired moisture content to be readsorbed.

Silica Gel

Silica gel (silicic acid, H_2SiO_3) is a very polar adsorbent with a surface area of approximately 500 m^2/g; it is less chemically active than alumina and is preferred when dealing with chemically active organic compounds. Silica gel is an acidic compound and very stable in acidic or neutral solvents, but will dissolve in solvents of pH greater than 7.5.

Bonded Stationary Phase

The first commercial LC nonpolar phases were called C18 or ODS because of their octyldecylsilane (18-carbon) groups. This stationary

Eluotropic Series: Relative Solvent Strengths

Increasing activity (polarity) ↓

Aluminun Oxide
Fluoroalkanes
Pentane
Isooctane
Cyclohexane
Tetrachloromethane
Xylene
Toluene
Benzene
Diethyl ether
Chloroform
Methylene chloride
Tetrahydrofuran
1, 2-Dichloroethane
Ethylmethyl ketone
1, 4-Dioxane
Ethyl acetate
1-Pentanol
Dimethyl sulfoxide
Aniline
Nitromethane
Acetonitrile
Pyridine
1-Propanol
Ethanol
Methanol
Ethylene glycol
Acetic acid

Silica Gel
Cyclohexane
Heptane
Tetrachloromethane
Carbon disulfide
Ethylbenzene
Toluene
Benzene
2-Chloropropane
Chloroform
Nitrobenzene
Di-isopropyl ether
Diethyl ether
Ethyl acetate
2-Butanol
Ethanol
Water
Acetone
Methanol

Magnesium Silicate
Pentane
Tetrachloromethane
Benzene
Chloroform
Methylene chloride
Diethyl ether

FIGURE 38-5
The eluotropic activity (eluting strength) of various solvents in different sorbents.

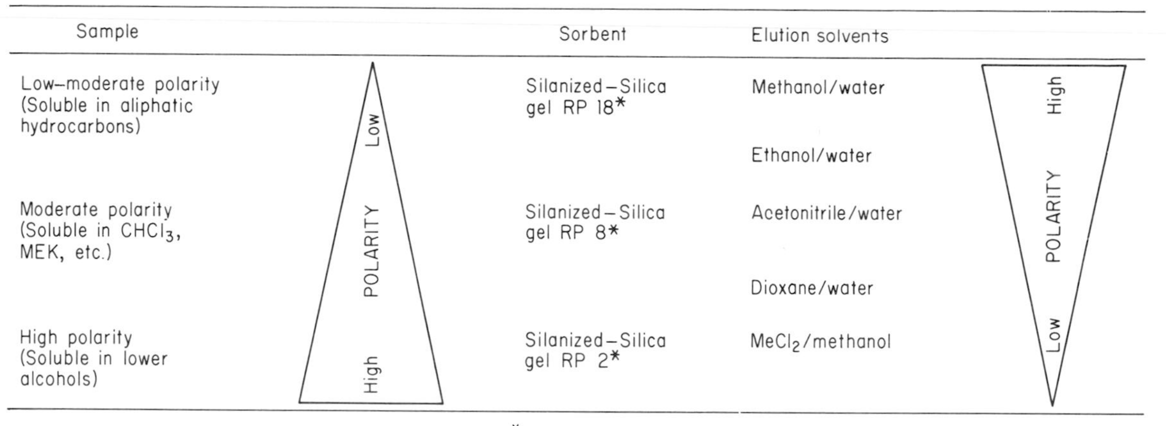

FIGURE 38-6
Sample sorbent and elution solvents for reverse-phase chromatography.

phase was prepared by *silylation* of silica gel to produce a less polar column by reacting the silicic acid stationary phase with chlorodimethylalkylsilane and/or chlorotrimethylalkylsilane. The very polar sililol (SiOH) groups on the silicic acid surface are blocked or made sterically inaccessible by these large, nonpolar trimethyl groups (see Chap. 37, "Gas Chromatography," under Column). This modified (polar to nonpolar) column requires a polar mobile phase to elute the solute components. Since this is the opposite polarity to the phases found with silica gel columns, this technique is referred to as reverse-phase chromatography (RPC). Thus, nonpolar compounds are retained by the nonpolar stationary phase. The sorbent chosen (Fig. 38-6) is selected because it closely matches the chemical properties of the sample.

Other RPC stationary-phase materials contain phenyl groups which are more polar than the C18 and have an affinity for double-bonded compounds. The C8 (octyl) and the cyano (—RCN) bonded phases have polarities that are intermediate between C18 and silica gel.

LIQUID CHROMATOGRAPHY MOBILE PHASES

The mobile-phase liquid not only transports the solute through the stationary phase, but its solvent power is critical for the proper distribution of the solute between the mobile phase and the adsorption sites on the stationary phase. A solvent that readily elutes the solute from the column will not resolve the components of the mixture, and a solvent that elutes the solute too slowly will be too time-consuming. A large number of LC solvents is available, and the use of mixed solvents expands the versatility even more. However, solvents should have a viscosity of 0.5 cP or less and be "Spectrograde" or "HPLC-grade" for best results. The use of only one solvent throughout the entire LC analysis is called *isocratic elution*. If necessary, a series of solvent mixtures can be used in LC applications where the polarity of the solvent mixture is gradually changed to effectively resolve components in a mixture; this technique is called *gradient elution*. Gradient systems can be created in an LC system by using two or more pumps to introduce different solvents from separate reservoirs and mixing before introduction of the sample. A newer gradient-elution system uses a single pump fitted with multiple solvent lines, a controller, solvent-proportioning valves, a series of solenoids, and a mixing chamber.

CAUTION Certain mobile phases (halogenated hydrocarbons, buffer solutions, etc.) can corrode stainless steel components and should not be allowed to stand in the HPLC for extensive periods (over 3 h) of time without pumping. Most manufacturers recommend that the system be flushed with isopropanol for organic mobile phases and distilled water followed by isopropanol for aqueous mobile phases.

HIGH-PERFORMANCE LIQUID CHROMATOGRAPHY

High-performance liquid chromatography, sometimes called *high-pressure liquid chromatography* (HPLC), is at maximum efficiency at low flow rate because of typically slow diffusion rates between liquid phases. The time required to elute solute through a column can be decreased by using a shorter column, but short columns tend to have less separating ability. Thus, LC systems have been developed which use short columns, but with very small particles of stationary phase to increase the resolution of solute compounds. Particles of 10 μm diameter and smaller are found to be the most efficient, but tend to be almost impermeable in long columns. Particles of 3 μm diameter are theoretically the optimum size for these columns. *Pellicular beads* are solid glass beads of 20 to 50 μm diameter that are manufactured with a thickness of approximately 1 μm of porous material on the surface. This porous surface is then used to coat a thin layer of liquid phase and to provide a large surface area. Most thin films are simply washed away by the mobile phase; thus the chemical bonding of the liquid phase to the solid support has been developed. Porous resin beads have been developed as a column packing where the stationary support and the surface are composed of the same insoluble material.

Liquid chromatography columns packed with these very small beads or bonded particles require extremely high pressure to force liquid to flow. Pressures as high as 10,000 psi can be handled with small-diamater (2 to 3 mm) columns, but most analytical LC applications occur at pressures of 4000 psi or less. High-pressure pumps have been perfected within the past few years that can produce an almost pulse-free flow rate of several milliliters per minute.

HPLC INSTRUMENT DESIGN AND COMPONENTS

Most basic HPLC systems have five major components: (1) a solvent pump, (2) a sample injector, (3) a column, (4) a detector, and (5) a recorder-integrator as shown in Figure 38-7.

More sophisticated HPLC systems might include additional sources of solvent, a gradient pump, proportioning valves, mixing chambers, oven(s), automatic integrator, programmable autosampler, and a computer-controlled work station (Fig 38-8).

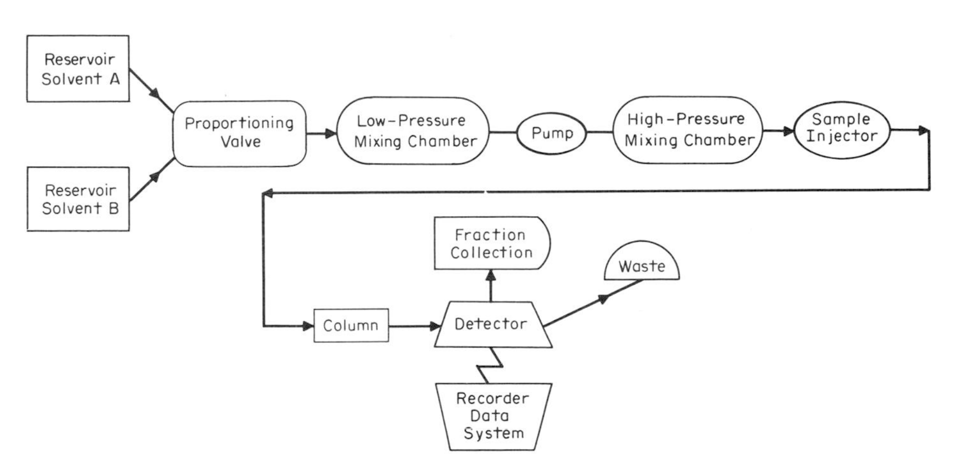

FIGURE 38-7
Schematic showing the components of a one-pump HPLC system.

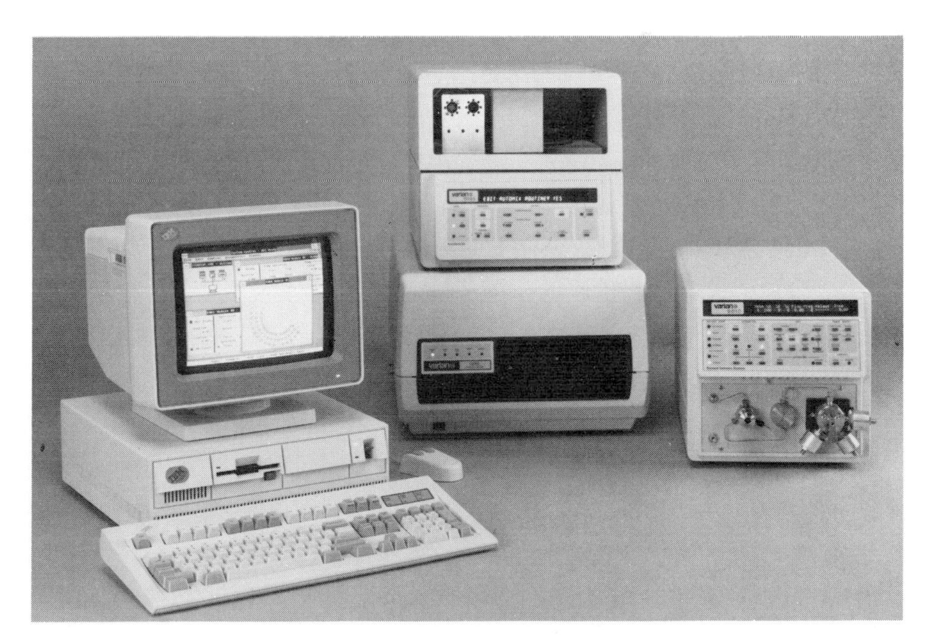

FIGURE 38-8
Modern HPLC system and work station *(Photo courtesy of Varian Associates, Inc.)*.

Solvent Pumps

There are two basic types of high-pressure pumps: (1) The *continuous-displacement* (syringe-type) model forces solvent from a single reservoir and tends to be very smooth and pulse-free; however, the chromatographic run is limited by the volume of solvent housed in the reservoir or requires frequent shutdown of the system for refilling. (2) A more complicated delivery system used the *intermittent-displacement* approach. This type of pump refills itself intermittently which disrupts the column flow and introduces a fluctuation in the detector baseline. Manufacturers have reduced this undesired pulsation by installing an air-compression baffle or long, narrow-bore coil of tubing between the pump and injector. A second pulse-damping technique requires greater expense but involves a pump equipped with two reciprocal pistons. One piston is pumping solvent into the system while the other is filling for the next stroke.

As a precautionary step, these pumps should not be allowed to stand for long periods of time containing potentially corrosive solvents (acidic solvents, aqueous buffer of low pH, etc. [see Caution on p. 840]). In addition, the entire LC system should be flushed before storage with a nonaqueous solvent like methanol to minimize bacterial growth.

It is strongly recommended that all solvent systems be filtered and degassed before starting an HPLC analysis. In-line solvent filters (0.2 to 10 μm) are highly recommended with HPLC systems because of the clogging and contamination potential of the pump, column packing, column frits, etc. These filters can be placed on the inlet of the solvent reservoir and/or near the pump inlet. Guard columns are also used to reduce column contamination and degradation (see the section titled Column in this chapter).

Degassing can be accomplished by several techniques:

1. Purging the solvent with a steady stream of pure nitrogen or helium
2. Applying a vacuum to the solvent and drawing the liquid through a filter membrane
3. Heating the solvent under reduced pressure to minimize the quantity of dissolved gases
4. Exposing the untreated solvent to ultrasonification in a vacuum environment.

Degassing is necessary because dissolved gases tend to form tiny bubbles as the solvent leaves the HPLC system and enters the lower pressure of the detector.

Sample Injector

In GC the sample was introduced using a syringe and a self-sealing septum inlet. This technique is not very practical with HPLC because of the high

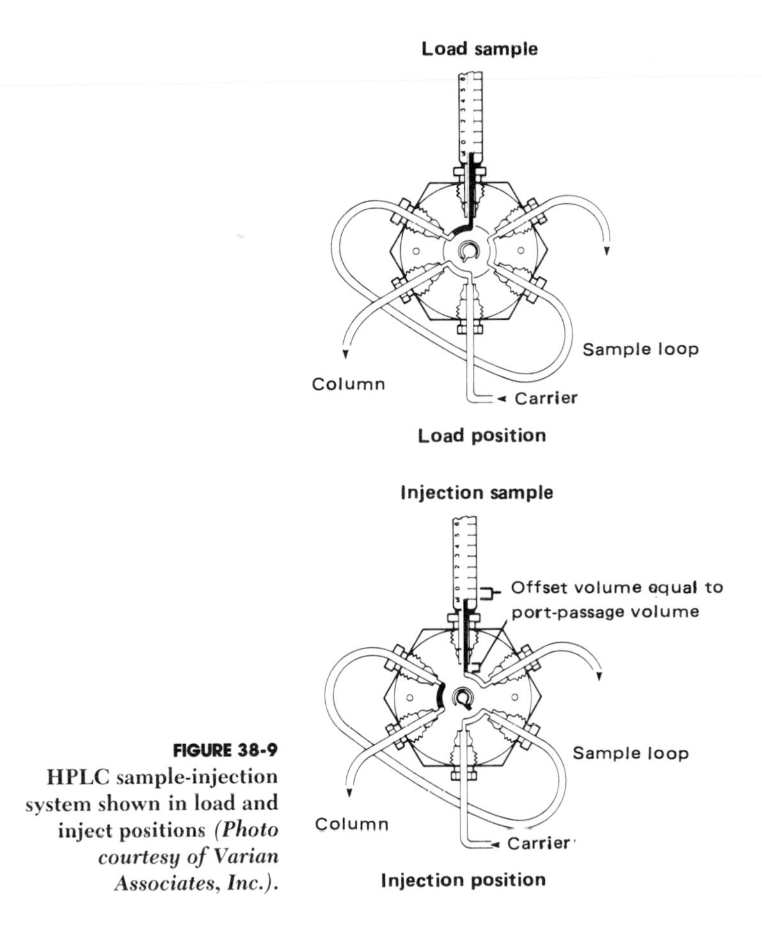

FIGURE 38-9
HPLC sample-injection system shown in load and inject positions *(Photo courtesy of Varian Associates, Inc.)*.

column pressures and the rapid degradation of septum material by the various liquid phases. Most HPLC systems use a loop injector system which consists of a rotary valve fitted with several ports and loops. Samples are simply purged through a loop (normally 10-μL to 2-mL capacity) with a liquid syringe, and at injection time the valve is rotated to cause the filled loop to be an integral part of the mobile-phase flow as shown by the "load" and "inject" positions in Fig. 38-9. The entire loop volume of liquid sample is swept away by the liquid (carrier) phase and introduced onto the column.

All solute components must be soluble in the mobile phase, and it may become necessary to dissolve the sample in a solvent different from the mobile phase. If a different solvent is used, it must be tested (before injection) to demonstrate that the solute does not precipitate upon mixing with the mobile phase. This precipitation or immiscibility problem can block the chromatographic system very quickly.

Column

HPLC columns are made of type-316 stainless steel for higher pressures (less than 10,000 psi) and glass for lower pressures (less than 1500 psi). Three types of particles are found in HPLC columns (solid, pellicular, and porous resin beads) and are discussed in the section titled Liquid Chromatography Adsorbents in this chapter.

In general, the slower the solvent flow rate, the better the resolution of sample components. Thus, analytical LC columns are normally 4.6 mm diameter by 5 cm to 25 cm long and packed with 5-μm particles, have maximum sample loads of 150 μg, and flow rates of approximately 1 mL/min. Preparative LC columns are 4.6 mm to 21 mm diameter by 25 cm long and packed with 40- to 50-μm particles, have maximum sample loads of 5 g, and flow rates of approximately 30 mL/min.

Most manufacturers recommend the use of a *precolumn* or a *guard column* to protect the chromatographic column from strongly retained compounds and/or between the injector and the pump to protect against contaminants in a sample. Most new columns are shipped from the manufacturer with a test chromatogram and test mixture. Column performance should be evaluated upon installation and periodically by analyzing this test sample under the listed instrumental conditions. Manufacturer and actual retention times and operating pressures should be comparable; however, differences of as much as 20% can be attributed to instrument parameter differences (injector design, detector geometry, connecting tubing, etc.) other than column efficiency.

Detectors

Flow-through detectors with low-dead-volume designs are required for these high-pressure–low-flow-rate systems. There are currently two types of detectors in gerneral use: (1) UV spectrophotometers and (2) differential refractometers. The UV detectors rely on a mercury lamp and a characteristic wavelength of 254 nm where most organic compounds with double bonds or aromatic groups tend to absorb. Additional UV and visible wavelengths of 214, 229, 280, 308, 360, 410, 440, and 550 nm are available. A *chromophore* is defined as a group of atoms that absorb energy in the near-UV region (400–190 nm). Certain solvents are strong UV absorbers and cannot be used with a UV detector because they would "cut off" any detector response. Only solvents with cutoffs below the detector absorption wavelength can be used. Table 38-4 lists some useful HPLC solvents with their boiling points, refractive indexes, and their respective UV-transmittance cutoffs.

The differential refractometer monitors the refractive index difference between pure mobile phase entering the column and the column effluent. For additional discussion of this principle, review the section on refractive index in Chap. 19, "Determination of Physical Properties." This type of detector is almost universal since all substances have their own unique

TABLE 38-4 Common HPLC Solvents with Boiling Points, Refractive Indexes, and UV-Transmittance Cutoffs

Solvent	*UV cutoff (max), nm*	*Refractive index at 20°C*	*Boiling point, °C*
Acetone	330	1.358	56–57
Acetonitrile-UV	190	1.344	81–82
Chloroform*	245	1.445	60–61
Ethyl acetate	256	1.372	77–78
Heptane	200	1.387	98–99
Hexane-UV	195	1.375	68–69
Methanol	205	1.328	64–65
Methylene chloride	233	1.424	40–41
Propanol-2	205	1.376	82–83
Tetrahydrofuran-UV	212	1.407	66–67
Toluene	284	1.496	110–111
2,2,4 Trimethylpentane	215	1.391	99–100

* Contains hydrocarbon preservative.
SOURCE: Varian Associates, Inc., Sunnyvale, California.

refractive index but is not well suited to gradient LC because the baseline (reference) will always be shifting. Refractometer detectors tend not to be very sensitive and are subject to drift with even minor temperature fluctuations.

Some commercially available HPLC systems are equipped with detectors based on effluent fluorescence which are very sensitive, but selective. As the fluorescing solutes elute from the column, a photocell measures the light intensity and generates a proportional electrical signal. High-performance liquid chromotography systems utilizing ion-exchange columns (review Chap. 36, "Chromatography") and conductivity detectors for measuring the column effluent have recently been perfected. These HPLC ion-exchange systems have found limited use in cation analysis, but are excellent for multicomponent anion analysis.

Recorder-Integrator

The same recording-integrating systems described for GC (Chap. 37) are applicable with HPLC. In addition, computer-controlled automatic-injection systems, multipump gradient controllers, sample fraction collectors, and other devices can be added to increase instrument efficiency.

QUALITATIVE AND QUANTITATIVE TECHNIQUES

The same separation evaluations, spiking techniques, integrating procedures, and calculation methods described in Chap. 37, "Gas Chromatography," are used in HPLC.

REFERENCES

Braun, R. D., *Introduction to Chemical Analysis*, McGraw-Hill, New York, 1982.

Johnson, E. L., and Stevenson, R. L., *Basic Liquid Chromatography*, Varian Associates, Palo Alto, CA, 1978.

Snyder, L. R., and Kirkland, J. J., *Introduction to Modern Liquid Chromatography*, 2nd ed., John Wiley and Sons, New York, 1979.

Appendix A
LABORATORY FIRST AID

Any person who attempts to render first aid before professional treatment by a physician is available should remember that such assistance is *only* a stopgap: an emergency procedure to be followed until the physician arrives. Stop bleeding, start breathing, treat for poisoning, treat for shock, and then care for the wound(s)—in that order.

GENERAL RULES

The prime rule—a difficult one to obey—is *KEEP CALM.* First aid is largely a matter of using common sense. If an accident occurs in the laboratory, make certain that the right person administers first aid (ideally, a trained first-aider should be designated for each shift). Have someone call for professional assistance immediately. Most companies have an emergency number posted; if yours does not, make sure that you have the number of the nearest hospital and poison-control center taped to your telephone.

No more than two persons should attend to the patient at any one time; a patient needs room to breathe: Unless fire, fumes, or some other hazardous condition makes it necessary, *do not move* an injured person. The moving may do more damage than the injury has caused.

If you need an assistant to help stop the bleeding, remember that some persons become ill at the sight of blood. Be sure, then, to choose a helper who can tolerate the conditions. Never ask for help from a person who responds slowly or is squeamish, excitable, or easily panicked.

Checklist: What to Do

1. Call an ambulance; state type of accident, its location, type of injuries, and the number of persons injured.

2. *Keep calm* and keep crowds away; give the patient fresh air. *Do nothing else* unless you are certain of the proper procedure.

---

3. Stop any bleeding, as directed under Wounds and Fractures below.

4. Restore breathing by administering artificial respiration.

5. Treat for poisoning if necessary.

6. Treat for physical shock, as directed under Shock below, in all injuries.

7. Remove the patient from a hazardous environment involving spillage of chemicals or high concentrations of noxious gases, vapors, or fumes.

8. Wear proper protective clothing and a respirator so as not to expose yourself to the environment.

9. Never give liquids to an unconscious person or to persons with abdominal or lower chest injuries.

10. Do not move a person with possible broken bones or possible head or internal injuries unless fumes or fire necessitate it. Then, improvise a splint to support and prevent aggravating the fracture.

THERMAL BURNS

Smother flames by rolling the victim in a coat or blanket. If clothing adheres to the burned skin surface, do not attempt to remove it, but carefully cut away the clothing around the burned area.

A slight burn should be immersed in cold water, with ice added, to relieve pain. For first- or second-degree burns with no open wounds, apply cold-water compresses; for open wounds or third-degree burns, apply a dry sterile dressing and transport the patient as soon as possible. Fasten dressings securely but not tightly. *Do not* use tannic acid on any burn. *Do not* use ointments on severe burns. A physician will apply the proper medication after the degree and extent of the burn have been determined. Treat for shock.

CAUTION Do not open blisters.

CHEMICAL BURNS

1. Remove contaminated articles of clothing as well as the source of contamination. If necessary, wear protective clothing and respiratory equipment during this process so as not to contaminate yourself.

2. Flush the contaminated skin area with large quantities of water for at least 15 min.

3. *Do not* use oils, fats, or sodium bicarbonate on the burned area unless specifically advised to do so by a physician. *Do not* apply salves or ointments, since these may increase skin absorption of the noxious chemical.

Soap may be used, especially where phenol (carbolic acid) and its derivatives are in the contaminant. The patient should be taken to a hospital immediately for further treatment.

WOUNDS AND FRACTURES

1. If bleeding is copious, it must be controlled before other aid can be given. Apply a large compress on the wound with direct pressure on the wound (or even in it, if it is large enough). If the wound is on an extremity, pressure can be applied to one of the two pressure points shown in Fig. A-1. When there are chest and abdominal injuries, cover the wound(s) with a sterile dressing moistened with physiological saline solution. *Do not* attempt to replace protruding viscera.

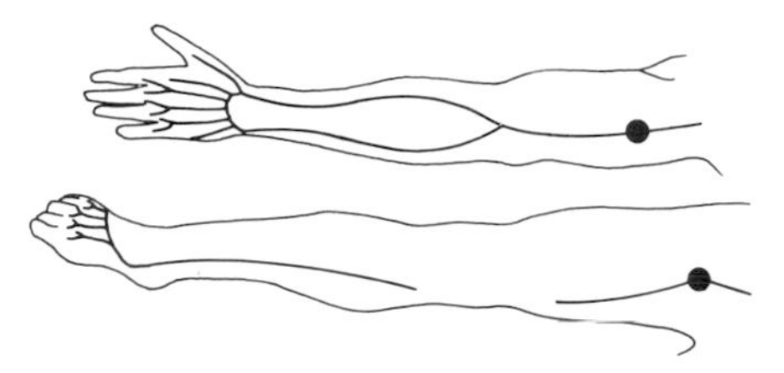

FIGURE A-1
Pressure points on arm and leg. (*Fisher Safety Manual*)

2. If the cut is slight and bleeding is not profuse, remove all foreign material (glass, dirt, etc.) projecting from the wound (but do not probe for embedded material). Removal is best accomplished by careful washing with soap and water. Apply an antiseptic to all parts of the cut and to approximately one-half inch of skin around the cut.

3. All wounds should be securely, but not tightly, bandaged.

4. In cases of puncture wounds (from broken thermometers, glass tubing, etc.) the patient should be sent to a hospital or physician's office. Danger of foreign material in the wounds and inability to reach the bottom of the wound with antiseptic make this action mandatory.

5. In cases of possible fracture, *do not* move the patient unless fumes or fire necessitates it. Treat for bleeding and shock and leave splinting to a physician. When necessary to transport the victim to treatment, improvise a splint support to prevent aggravating the fracture in transit.

SHOCK

Shock occurs to some extent in all injuries, varies with the individual, and can cause death. Some easily recognized symptoms are pallor, cold and moist skin with perspiration on the forehead and palms of the hands, nausea, shallow breathing, and trembling.

Place the patient in a reclining position, with the head lower than the body. (In cases of severe hemorrhage of the head, fractured skull, or stroke, the head should be elevated.) Control any bleeding. Wrap a cold patient in blankets, and elevate the patient's legs if there are no broken bones. If the patient is overheated, try to reduce body temperature by sponging with cold water. The objective is to attain and maintain normal body temperature.

Keep the patient's airway open. If vomiting occurs, turn the head to one side so that the neck is arched. If there is no bleeding, rub the extremities briskly toward the heart to restore circulation, and give stimulants either by mouth or inhalation (inhalation *only*, if the patient is unconscious). A stimulant cannot be given until bleeding is controlled and should not be given at all in cases of fractured skull, abdominal injuries, or stroke.

The preferred liquid stimulant is a formula consisting of 1 teaspoon salt and ½ teaspoon of sodium bicarbonate in a quart of warm water; even plain warm water will do, however. For inhalation, use aromatic spirits of ammonia, dilute ammonia, dilute acetic acid, or amyl nitrite on a cloth. *Do not* give alcoholic liquids, and *never* administer any liquids while a person is unconscious. Reassure the patient and *remain calm* yourself.

ELECTRIC SHOCK

1. *Shut off the current* and then cautiously remove the wire or other contact while protecting yourself with an insulator. Use heavy rubber or asbestos gloves or a hand inside a glass beaker to push the victim or current source aside—*a dry stick or dry towel will not suffice.*

2. Start artificial respiration immediately (see page 851).

3. Do not regard early rigidity or stiffening as a sign for ceasing artificial respiration. Efforts for revival should be continued for at least 4 h, or until a physician certifies death, even though there is no sign that the patient is regaining consciousness.

4. Keep the patient warm, using blankets.

POISONS: INGESTED

1. If the patient is conscious, give two to four glasses of water (or milk if water is unavailable) immediately:

2. Call the local poison-control center immediately, and then call an ambulance.

3. Induce vomiting *except* when poisoning is due to phenothiazine, strong acids, strong alkalies, cyanide, strychnine, gasoline, kerosene, or other hydrocarbons, or when the patient is already having convulsions. Have

the patient place an index finger far back on the tongue and stroke it from side to side (do this for the patient if necessary). Emetics such as ipecac, a teaspoonful of powdered mustard in sufficient warm water to make a paste, clear warm water, or even *soap suds* can be used. Repeat dosage of the emetic until the vomited fluid is clear. (Ipecac is the best emetic to use.) When poisoning is due to strong acids or strong bases, give the patient a glass of milk. For cyanide, use artificial respiration if the patient is not breathing. *Maintain consciousness* by allowing inhalation of amyl nitrite for 20 to 30 s out of each minute. For strychnine, give 1 oz of powdered charcoal and keep the patient quiet. If the patient is having convulsions, treat for shock while waiting for the ambulance.

4. For methyl alcohol poisoning, give the patient beer to drink.

CAUTION *Do not* give mineral oils, fats, or alcohol except at the advice of a physician.

5. Prevent shock, keep the patient warm.

POISONS: INHALED

1. Call an ambulance.

2. Wear protective clothing and respiratory gear. Remove the patient to fresh air immediately. Give oxygen if necessary, using an inhalator to avoid overdosage.

3. Allow the patient to rest. At any sign of cessation of breathing, begin artificial respiration.

4. Treat for shock; keep the patient warm.

5. Note that carbon monoxide poisoning often mimics food poisoning when it is not severe enough to produce unconsciousness.

ARTIFICIAL RESPIRATION

If breathing stops because of electrocution, sedative poisoning, gas poisoning, or suffocation, start artificial respiration immediately. Don't delay—seconds count. As soon as possible, send for a physician.

Mouth-to-mouth resuscitation (Fig. A-2) is the only currently accepted method.

1. *Position of patient.* Place the patient face up and loosen tight clothing. To make sure that the tongue does not obstruct the air passage, tilt the head back so the chin is pointing upward. Insert your left thumb in the mouth, grasp the lower jaw, and lift it forcibly but gently upward and forward.

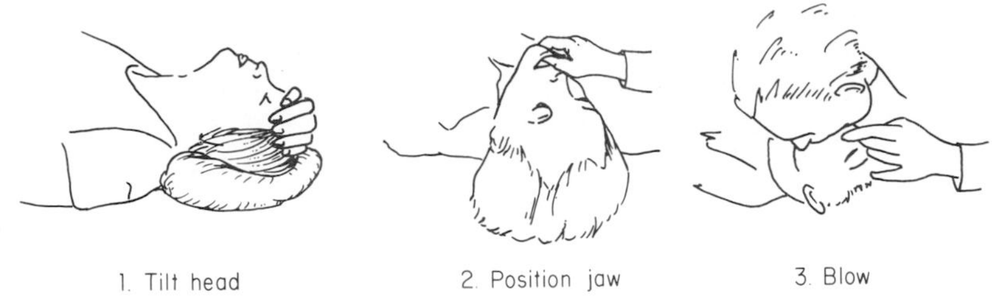

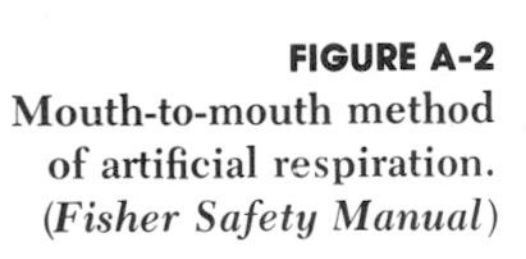

FIGURE A-2
Mouth-to-mouth method of artificial respiration. (*Fisher Safety Manual*)

2. *Position of operator.* Hold the lower jaw up and pinch the patient's nostrils together to close them. Open your mouth wide and place it lightly over the subject's mouth. Alternatively, you may close the mouth and place your mouth over the nose.

3. *Exchange of air.* While breathing into the patient, watch the chest rise to make sure the air passage is clear. Remove your mouth, turn your head to the side, and listen for the return rush of air that indicates air exchange. Repeat the blowing effort. Blow vigorously at the rate of about 12 breaths per minute for an adult. Children require a rate of 20 breaths per minute.

4. *Mouth-to-mouth resuscitator.* Plastic mouth-to-mouth resuscitators are available; these remove the danger of the tongue blocking the air passage. However, don't use one unless you are thoroughly familiar with it.

General Precautions

It is all-important that artificial respiration, when needed, be started quickly. A patient can die within 3 min! Irreversible brain damage can also occur. There should be a slight inclination of the body in such a way that fluid drains well from the respiratory passage. The patient's head should be extended, not flexed forward, and the chin should not sag, lest obstruction of the respiratory passages occur.

The operator should be alert to the presence of any foreign matter in the mouth. Wipe it out quickly with your fingers or with a cloth wrapped around your fingers. If vomiting occurs, quickly turn the patient on the side, wipe out the mouth, and then reposition the patient. Do not give any liquids by mouth until the patient is fully conscious.

A revived victim should be kept as quiet as possible until regular breathing is restored. Keep the patient from becoming chilled and otherwise treat for shock. Continue artificial respiration until the patient begins to breathe unaided, or until a physician states that death has occurred. Because respiratory and other disturbances may develop as an aftermath, a doctor's care is necessary during the recovery period.

OBJECTS IN THE EYE

If a piece of glass or any other foreign body flies into an eye, do not attempt to remove it. Cover the eye with a sterile gauze pad and *rush* the patient to a hospital or doctor's office *immediately*. Have someone alert the hospital or doctor that you are on the way, so that treatment can be given as soon as possible. It is most important that the patient keep hands away from the eye.

If any kind of chemical splashes in the eye, the recommended treatment is a 10-min (minimum) washing of the eye with water. An eye-washing fountain is best for this purpose; if you have none, a stream of water from a slowly running faucet (or even water from a washbottle), directed from the inner corner of the eye outward, will serve. Cover the eye with a sterile gauze pad and then get the patient to a hospital or doctor's office. It is *most important* to remove or dilute any chemical in the eye immediately.

Appendix **B**

Table of Four-place Logarithms

Natural numbers	*0*	*1*	*2*	*3*	*4*	*5*	*6*	*7*	*8*	*9*	*Proportional parts* 1	2	3	4	5	6	7	8	9
10	0000	0043	0086	0128	0170	0212	0253	0294	0334	0374	4	8	12	17	21	25	29	33	37
11	0414	0453	0492	0531	0569	0607	0645	0682	0719	0755	4	8	11	15	19	23	26	30	34
12	0792	0828	0864	0899	0934	0969	1004	1038	1072	1106	3	7	10	14	17	21	24	28	31
13	1139	1173	1206	1239	1271	1303	1335	1367	1399	1430	3	6	10	13	16	19	23	26	29
14	1461	1492	1523	1553	1584	1614	1644	1673	1703	1732	3	6	9	12	15	18	21	24	27
15	1761	1790	1818	1847	1875	1903	1931	1959	1987	2014	3	6	8	11	14	17	20	22	25
16	2041	2068	2095	2122	2148	2175	2201	2227	2253	2279	3	5	8	11	13	16	18	21	24
17	2304	2330	2355	2380	2405	2430	2455	2480	2504	2529	2	5	7	10	12	15	17	20	22
18	2553	2577	2601	2625	2648	2672	2695	2718	2742	2765	2	5	7	9	12	14	16	19	21
19	2788	2810	2833	2856	2878	2900	2923	2945	2967	2989	2	4	7	9	11	13	16	18	20
20	3010	3032	3054	3075	3096	3118	3139	3160	3181	3201	2	4	6	8	11	13	15	17	19
21	3222	3243	3263	3284	3304	3324	3345	3365	3385	3404	2	4	6	8	10	12	14	16	18
22	3424	3444	3464	3483	3502	3522	3541	3560	3579	3598	2	4	6	8	10	12	14	15	17
23	3617	3636	3655	3674	3692	3711	3729	3747	3766	3784	2	4	6	7	9	11	13	15	17
24	3802	3820	3838	3856	3874	3892	3909	3927	3945	3962	2	4	5	7	9	11	12	14	16
25	3979	3997	4014	4031	4048	4065	4082	4099	4116	4133	2	3	5	7	9	10	12	14	15
26	4150	4166	4183	4200	4216	4232	4249	4265	4281	4298	2	3	5	7	8	10	11	13	15
27	4314	4330	4346	4362	4378	4393	4409	4425	4440	4456	2	3	5	6	8	9	11	13	14
28	4472	4487	4502	4518	4533	4548	4564	4579	4594	4609	2	3	5	6	8	9	11	12	14
29	4624	4639	4654	4669	4683	4698	4713	4728	4742	4757	1	3	4	6	7	9	10	12	13
30	4771	4786	4800	4814	4829	4843	4857	4871	4886	4900	1	3	4	6	7	9	10	11	13
31	4914	4928	4942	4955	4969	4983	4997	5011	5024	5038	1	3	4	6	7	8	10	11	12
32	5051	5065	5079	5092	5105	5119	5132	5145	5159	5172	1	3	4	5	7	8	9	11	12
33	5185	5198	5211	5224	5237	5250	5263	5276	5289	5302	1	3	4	5	6	8	9	10	12
34	5315	5328	5340	5353	5366	5378	5391	5403	5416	5428	1	3	4	5	6	8	9	10	11
35	5441	5453	5465	5478	5490	5502	5514	5527	5539	5551	1	2	4	5	6	7	9	10	11
36	5563	5575	5587	5599	5611	5623	5635	5647	5658	5670	1	2	4	5	6	7	8	10	11
37	5682	5694	5705	5717	5729	5740	5752	5763	5775	5786	1	2	3	5	6	7	8	9	10
38	5798	5809	5821	5832	5843	5855	5866	5877	5888	5899	1	2	3	5	6	7	8	9	10
39	5911	5922	5933	5944	5955	5966	5977	5988	5999	6010	1	2	3	4	5	7	8	9	10
40	6021	6031	6042	6053	6064	6075	6085	6096	6107	6117	1	2	3	4	5	6	8	9	10
41	6128	6138	6149	6160	6170	6180	6191	6201	6212	6222	1	2	3	4	5	6	7	8	9
42	6232	6243	6253	6263	6274	6284	6294	6304	6314	6325	1	2	3	4	5	6	7	8	9
43	6335	6345	6355	6365	6375	6385	6395	6405	6415	6425	1	2	3	4	5	6	7	8	9
44	6435	6444	6454	6464	6474	6484	6493	6503	6513	6522	1	2	3	4	5	6	7	8	9
45	6532	6542	6551	6561	6571	6580	6590	6599	6609	6618	1	2	3	4	5	6	7	8	9
46	6628	6637	6646	6656	6665	6675	6684	6693	6702	6712	1	2	3	4	5	6	7	7	8
47	6721	6730	6739	6749	6758	6767	6776	6785	6794	6803	1	2	3	4	5	5	6	7	8
48	6812	6821	6830	6839	6848	6857	6866	6875	6884	6893	1	2	3	4	4	5	6	7	8
49	6902	6911	6920	6928	6937	6946	6955	6964	6972	6981	1	2	3	4	4	5	6	7	8

Table of Four-place Logarithms Continued

Natural numbers	*0*	*1*	*2*	*3*	*4*	*5*	*6*	*7*	*8*	*9*	*Proportional parts* 1	2	3	4	5	6	7	8	9
50	6990	6998	7007	7016	7024	7033	7042	7050	7059	7067	1	2	3	3	4	5	6	7	8
51	7076	7084	7093	7101	7110	7118	7126	7135	7143	7152	1	2	3	3	4	5	6	7	8
52	7160	7168	7177	7185	7193	7202	7210	7218	7226	7235	1	2	2	3	4	5	6	7	7
53	7243	7251	7259	7267	7275	7284	7292	7300	7308	7316	1	2	2	3	4	5	6	6	7
54	7324	7332	7340	7348	7356	7364	7372	7380	7388	7396	1	2	2	3	4	5	6	6	7
55	7404	7412	7419	7427	7435	7443	7451	7459	7466	7474	1	2	2	3	4	5	5	6	7
56	7482	7490	7497	7505	7513	7520	7528	7536	7543	7551	1	2	2	3	4	5	5	6	7
57	7559	7566	7574	7582	7589	7597	7604	7612	7619	7627	1	2	2	3	4	5	5	6	7
58	7634	7642	7649	7657	7664	7672	7679	7686	7694	7701	1	1	2	3	4	4	5	6	7
59	7709	7716	7723	7731	7738	7745	7752	7760	7767	7774	1	1	2	3	4	4	5	6	7
60	7782	7789	7796	7803	7810	7818	7825	7832	7839	7846	1	1	2	3	4	4	5	6	6
61	7853	7860	7868	7875	7882	7889	7896	7903	7910	7917	1	1	2	3	4	4	5	6	6
62	7924	7931	7938	7945	7952	7959	7966	7973	7980	7987	1	1	2	3	3	4	5	6	6
63	7993	8000	8007	8014	8021	8028	8035	8041	8048	8055	1	1	2	3	3	4	5	5	6
64	8062	8069	8075	8082	8089	8096	8102	8109	8116	8122	1	1	2	3	3	4	5	5	6
65	8129	8136	8142	8149	8156	8162	8169	8176	8182	8189	1	1	2	3	3	4	5	5	6
66	8195	8202	8209	8215	8222	8228	8235	8241	8248	8254	1	1	2	3	3	4	5	5	6
67	8261	8267	8274	8280	8287	8293	8299	8306	8312	8319	1	1	2	3	3	4	5	5	6
68	8325	8331	8338	8344	8351	8357	8363	8370	8376	8382	1	1	2	3	3	4	4	5	6
69	8388	8395	8401	8407	8414	8420	8426	8432	8439	8445	1	1	2	2	3	4	4	5	6
70	8451	8457	8463	8470	8476	8482	8488	8494	8500	8506	1	1	2	2	3	4	4	5	6
71	8513	8519	8525	8531	8537	8543	8549	8555	8561	8567	1	1	2	2	3	4	4	5	5
72	8573	8579	8585	8591	8597	8603	8609	8615	8621	8627	1	1	2	2	3	4	4	5	5
73	8633	8639	8645	8651	8657	8663	8669	8675	8681	8686	1	1	2	2	3	4	4	5	5
74	8692	8698	8704	8710	8716	8722	8727	8733	8739	8745	1	1	2	2	3	4	4	5	5
75	8751	8756	8762	8768	8774	8779	8785	8791	8797	8802	1	1	2	2	3	3	4	5	5
76	8808	8814	8820	8825	8831	8837	8842	8848	8854	8859	1	1	2	2	3	3	4	5	5
77	8865	8871	8876	8882	8887	8893	8899	8904	8910	8915	1	1	2	2	3	3	4	4	5
78	8921	8927	8932	8938	8943	8949	8954	8960	8965	8971	1	1	2	2	3	3	4	4	5
79	8976	8982	8987	8993	8998	9004	9009	9015	9020	9025	1	1	2	2	3	3	4	4	5
80	9031	9036	9042	9047	9053	9058	9063	9069	9074	9079	1	1	2	2	3	3	4	4	5
81	9085	9090	9096	9101	9106	9112	9117	9122	9128	9133	1	1	2	2	3	3	4	4	5
82	9138	9143	9149	9154	9159	9165	9170	9175	9180	9186	1	1	2	2	3	3	4	4	5
83	9191	9196	9201	9206	9212	9217	9222	9227	9232	9238	1	1	2	2	3	3	4	4	5
84	9243	9248	9253	9258	9263	9269	9274	9279	9284	9289	1	1	2	2	3	3	4	4	5
85	9294	9299	9304	9309	9315	9320	9235	9330	9335	9340	1	1	2	2	3	3	4	4	5
86	9345	9350	9355	9360	9365	9370	9375	9380	9385	9390	1	1	2	2	3	3	4	4	5
87	9395	9400	9405	9410	9415	9420	9425	9430	9435	9440	0	1	1	2	2	3	3	4	4
88	9445	9450	9455	9460	9465	9469	9474	9479	9484	9489	0	1	1	2	2	3	3	4	4
89	9494	9499	9504	9509	9513	9518	9523	9528	9533	9538	0	1	1	2	2	3	3	4	4
90	9542	9547	9552	9557	9562	9566	9571	9576	9581	9586	0	1	1	2	2	3	3	4	4
91	9590	9595	9600	9605	9609	9614	9619	9624	9628	9633	0	1	1	2	2	3	3	4	4
92	9638	9643	9647	9652	9657	9661	9666	9671	9675	9680	0	1	1	2	2	3	3	4	4
93	9685	9689	9694	9699	9703	9708	9713	9717	9722	9727	0	1	1	2	2	3	3	4	4
94	9731	9736	9741	9745	9750	9754	9759	9763	9768	9773	0	1	1	2	2	3	3	4	4
95	9777	9782	9786	9791	9795	9800	9805	9808	9814	9818	0	1	1	2	2	3	3	4	4
96	9823	9827	9832	9836	9841	9845	9850	9854	9859	9863	0	1	1	2	2	3	3	4	4
97	9868	9872	9877	9881	9886	9890	9894	9899	9903	9908	0	1	1	2	2	3	3	4	4
98	9912	9917	9921	9926	9930	9934	9939	9943	9948	9952	0	1	1	2	2	3	3	4	4
99	9956	9961	9965	9969	9974	9978	9983	9987	9991	9996	0	1	1	2	2	3	3	3	4

Appendix 

ABBREVIATIONS, SYMBOLS, AND UNIT CONVERSIONS

COMMONLY USED ABBREVIATIONS

Å	angstrom (unit of wavelength measure)
abs	absolute
ac	alternating current
amor or amorph	amorphous
A	ampere
anhyd	anhydrous
aq	aqueous
atm	atmosphere
at no	atomic number
at wt	atomic weight
Bé	Baumé degree
bp	boiling point
Btu	British thermal unit
°C	degree Celsius
ca.	approximately
Cal	large calorie (kilogram calorie or kilocalorie)
cal	small calorie (gram calorie)
cg	centigram
cgs	centimeter-gram-second (system of units)
Ci	Curie
cm	centimeter
cm^2	square centimeter
cm^3	cubic centimeter
conc'd.	concentrated
cos	cosine
cp	candlepower
cP	centipoise
cu	cubic
D	density diopter
dB	decibel
dc	direct current
deg or °	degree
dg	decigram
dil.	dilute
dr	dram
dyn	dyne
E	electric tension; electromotive force
e.g.	for example
emf	electromotive force
esu	electrostatic unit
etc.	and so forth
et seq.	and the following
eV	electronvolt
°F	degree Fahrenheit;
F	frictional loss
f	frequency
ft	foot
ft^2	square foot
ft^3	cubic foot
ft·c	foot-candle
ft·lb	foot-pound
g	acceleration due to gravity
g	gram
G	gauss
g·cal	gram-calorie
gal	gallon
gr	grain
h	Planck's constant
h	hour
hp	horsepower
hp·h	horsepower-hour

hyg	hygroscopic
Hz	hertz (formerly cycles per second, cps)
I	electric current
${}^{a}_{z}I$	symbol for isotope with atomic number *z* and atomic mass *a*
ibid.	in the same place
i.e.	that is
in	inch
in^2	square inch
in^3	cubic inch
insol.	insoluble
iso	isotropic
J	joule; mechanical equivalent of heat
k	kilo (1000)
K	Kelvin
kc	kilocycle
kcal	kilogram-calorie
kg	kilogram
kW	kilowatt
kWh	kilowatt-hour
L	liter
l	lumen
l	length
λ	lambda; wavelength; coefficient of linear expansion
lb	pound
lb/ft^3	pound per cubic foot
ln	natural logarithm
log	logarithm
MW	molecular weight; mass
M	molar, as 1 *M*
m	meter
m^2	square meter
m^3	cubic meter
μ	micro- (10^{-6})
μm	micrometer (micron)
meq	milliequivalent
MHz	million (or mega-) hertz
MeV	million (or mega-) electronvolt
mg	milligram
min	minute
mks	meter-kilogram-second (system of units)
mL	milliliter
mm	millimeter
mm^2	square millimeter
mm^3	cubic millimeter
mm Hg	millimeters of mercury
mol	mole
mp	melting point
mph	miles per hour
N	newtons
N	normality, as 1 *N*
n	index of refraction; neutron (component of atomic nucleus)
nm	nanometer
oz	ounce
P	poise
Pa	pascal
pH	measure of hydrogen-ion concentration of a solution
ppb	parts per billion
ppm	parts per million
ppt	precipitate
psi	pounds per square inch
p. sol.	partly soluble
Q	energy of nuclear reaction
qt	quart
q.v.	which see
R	roentgen (international unit for x-rays)
rpm	revolutions per minute
rps	revolutions per second
sat'd.	saturated
s	second
S	Siemens
sin	sine
sol'n.	solution
sp	specific
sp gr	specific gravity

sp ht	specific heat
sq	square
T	temperature
t	time
tan	tangent
V	volt
W	watt
Wh	watt-hour
yr	year

COMMONLY USED SIGNS AND SYMBOLS

$+$	plus; add; positive
$-$	minus; subtract; negative
$\pm$	plus or minus; positive or negative
$\times$ or $\cdot$	times; multiplied by
$\div$ or $/$	is divided by
$=$ or $::$	is equals; as
$\equiv$	is identical to; congruent with
$\not\equiv$	is not identical
$>$	is greater than
$<$	is less than
$\geqq$, $\geq$	is equal to or greater than
$\leqq$, $\leq$	is equal to or less than
$:$	is ratio of
$\propto$	varies as (is proportional to)
$\sqrt{a}$ or $a^{1/2}$	is square root of a
a^2	a squared, the second power of a; $a \times a$
a^3	a cubed, the third power of a; $a \times a \times a$
a^{-1}	is $\frac{1}{a}$ (reciprocal of a)
a^{-2}	is $\frac{1}{a^2}$ (reciprocal of a^2)
Σ	summation of
Δ	difference in
k	a mathematical constant
π	the ratio of the circumference of a circle to its diameter; roughly equal to 3.1416

UNIT-CONVERSION TABLES*

TABLE C-1
Units of Area

Units	*Square inches*	*Square feet*	*Square yards*	*Square miles*	*Square centimeters*	*Square meters*
1 square inch	**1**	0.006 944 44	0.000 771 605	0.000 000 000 249 1	6.451 626	0.000 645 162 6
1 square foot	**144**	**1**	0.111 111 1	0.000 000 035 870 1	929.0341	0.092 903 41
1 square yard	**1 296**	**9**	**1**	0.000 000 322 831	8361.307	0.836 130 7
1 square mile	**4 014 489 600**	**27 878 400**	**3 097 600**	**1**	25 899.984 703	2.589 998
1 square centimeter	0.154 996 9	0.001 076 387	0.000 119 598 5	0.000 000 000 038 610 06	**1**	**0.000 1**
1 square meter	**1549.9969**	10.763 87	1.195 985	0.000 000 386 100 6	**10 000**	**1**

TABLE C-2
Units of Length

Units	*Inches*	*Feet*	*Yards*	*Miles*	*Centimeters*	*Meters*
1 inch	**1**	0.083 333 3	0.027 777 8	0.000 015 782 8	2.540 005	0.025 400 05
1 foot	**12**	**1**	0.333 333	0.000 189 393 9	30.480 06	0.304 800 6
1 yard	**36**	**3**	**1**	0.000 568 182	91.440 18	0.914 401 8
1 mile	**63 360**	**5 280**	**1 760**	**1**	160 934.72	1609.3472
1 centimeter	**0.3937**	0.032 808 33	0.010 936 111	0.000 006 213 699	**1**	**0.01**
1 meter	**39.37**	3.280 833	1.093 611 1	0.000 621 369 9	**100**	**1**

TABLE C-3
Units of Volume

Units	*Cubic inches*	*Cubic feet*	*Cubic yards*	*Cubic centimeters*	*Cubic decimeters*	*Cubic meters*
1 cubic inch	**1**	0.000 578 704	0.000 021 433 47	16.387 162	0.016 387 16	0.000 016 387 16
1 cubic foot	**1 728**	**1**	0.037 037 0	28 317.016	28.317 016	0.028 317 016
1 cubic yard	**46 656**	**27**	**1**	764 559.4	764.5594	0.764 559 4
1 cubic centimeter	0.016 023 38	0.000 035 314 45	0.000 001 307 94	**1**	**0.001**	**0.000 001**
1 cubic decimeter	61.023 38	0.035 314 45	0.001 307 943	**1 000**	**1**	**0.001**
1 cubic meter	61 023.38	35.314 45	1.307 942 8	**1 000 000**	**1000**	**1**

* The figures in boldface type signify exact values.

TABLE C-4
Units of Liquid Measure

Units	*Fluid ounces*	*Liquid pints*	*Liquid quarts*	*Gallons*	*Milliliters*	*Liters*	*Cubic inches*
1 fluid ounce	**1**	**0.0625**	**0.031 25**	**0.007 812 5**	29.5729	0.029 572 9	1.804 69
1 liquid pint	**16**	**1**	**0.5**	**0.125**	473.167	0.473 167	**28.875**
1 liquid quart	**32**	**2**	**1**	**0.25**	946.333	0.946 333	**57.75**
1 gallon	**128**	**8**	**4**	**1**	3785.332	3.785 332	**231**
1 milliliter	0.033 814 7	0.002 113 42	0.001 056 71	0.000 264 178	**1**	**0.001**	0.061 025 0
1 liter	33.8147	2.113 42	1.056 71	0.264 178	**1000**	**1**	61.0250
1 cubic inch	0.554 113	0.034 632 0	0.017 316 0	0.004 329 00	16.3867	0.016 386 7	**1**

TABLE C-5
Units of Mass

Units	*Grains*	*Apothecaries' scruples*	*Pennyweights*	*Avoirdupois drams*	*Apothecaries' drams*	*Avoirdupois ounces*
1 grain	**1**	**0.05**	0.014 666 67	0.036 571 43	0.016 666 7	0.002 285 71
1 apoth. scruple	**20**	**1**	0.833 333 3	0.731 428 6	0.333 333	0.045 714 3
1 pennyweight	**24**	**1.2**	**1**	0.877 714 3	**0.4**	0.054 857 1
1 avdp. dram	**27.343 75**	**1.367 187 5**	1.139 323	**1**	0.455 729 2	**0.0625**
1 apoth. dram	**60**	**3**	**2.5**	2.194 286	**1**	0.137 142 9
1 avdp. ounce	**437.5**	**21.875**	18.229 17	**16**	7.291 67	**1**
1 apoth. or troy ounce	**480**	**24**	**20**	17.554 28	**8**	1.097 142 9
1 apoth. or troy pound	**5760**	**288**	**240**	210.6514	**96**	13.165 714
1 avdp. pound	**7000**	**350**	291.6667	**256**	116.6667	**16**
1 milligram	0.015 432 356	0.000 771 618	0.000 643 014 8	0.000 564 383 3	0.000 257 205 9	0.000 035 273 96
1 gram	15.432 356	0.771 618	0.643 014 85	0.564 383 3	0.257 205 9	0.035 273 96
1 kilogram	15 432.356	771.6178	643.014 85	564.383 32	257.205 94	35.273 96

Units	*Apothecaries' or troy ounces*	*Apothecaries' or troy pounds*	*Avoirdupois pounds*	*Milligrams*	*Grams*	*Kilograms*
1 grain	0.002 083 33	0.001 173 611 1	0.000 142 857 1	64.798 918	0.064 798 918	0.000 064 798 9
1 apoth. scruple	0.041 666 7	0.003 472 222	0.002 856 143	1295.9784	1.295 978 4	0.001 295 978
1 pennyweight	**0.05**	0.004 366 667	0.003 428 571	1555.1740	1.555 174 0	0.001 555 174
1 avdp. dram	0.056 966 146	0.004 747 178 8	**0.003 906 25**	1771.8454	1.771 845 4	0.001 771 845
1 apoth. dram	**0.125**	0.010 416 667	0.008 571 429	3887.9351	3.887 935 1	0.003 887 935
1 avdp. ounce	0.911 458 3	0.075 954 861	**0.0625**	28 349.527	28.349 527	0.028 349 53
1 apoth. or troy ounce	**1**	0.083 333 33	0.068 571 43	31 103.481	31.103 481	0.031 103 48
1 apoth. or troy pound	**12**	**1**	0.822 857 1	373 241.77	373.241 77	0.373 241 77
1 avdp. pound	14.583 333	1.215 277 8	**1**	**453 592.427 7**	**453.591 427 7**	**0.453 592 427 7**
1 milligram	0.000 032 150 74	0.000 002 679 23	0.000 002 204 62	**1**	**0.001**	**0.000 001**
1 gram	0.034 150 74	0.002 679 23	0.002 204 62	**1000**	**1**	**0.001**
1 kilogram	32.150 742	2.679 228 5	2.204 622 341	**1 000 000**	**1000**	**1**

Appendix D GLOSSARY OF CHEMICAL TECHNOLOGY TERMS

Abscissa The horizontal axis in a graph, usually symbolized by x.

Absolute A chemical substance that is not mixed with any other substances. An absolute ethanol solution would be without any other substances (for example, water).

Absolute zero The lowest temperature theoretically possible, called zero degrees Kelvin (−273.15°C).

Accuracy The degree to which a given answer agrees with the true value. (See Chap. 7.)

Activity The effective concentration of ions in solution.

Activity coefficient The ratio of activity to actual concentration.

Aliquot A measured volume of a liquid which is a known fractional part of a larger volume. (See Chap. 25.)

Allotropes Two or more forms of an element that exists in the same physical state (example: coal and diamond). (See Chap. 22.)

Alpha (α) particle A helium nucleus (two neutrons and two protons) emitted from a nuclear decay. (See Chap. 35.)

Amphoteric Compound with the ability to act as an acid or base. (See Chap. 22.)

Anion A negatively charged atom or group of atoms. (See Chap. 22.)

Anode The electrode at which oxidation occurs in an electrochemical cell. (See Chap. 28.)

Anodizing The coating of a metal surface with a metal oxide by anodic oxidation. (See Chap. 28.)

Atmosphere A unit of pressure equivalent to the normal pressure of our atmosphere experienced at sea level, 760 torr, 14.7 pounds per square inch. (See Chap. 5.)

Atom The smallest particle of an element retaining the properties of that element. (See Chap. 22.)

Atomic mass or atomic weight The relative weight of an atom with the lightest isotope of carbon being arbitarily set at 12.0000 atomic mass units. (See Chap. 22.)

Avogadro's number The number (6.023×10^{23}) of atoms, molecules, particles, etc. found in exactly one mole of that substance. (See Chap. 22.)

Background radiation Radiation extraneous to an experiment. Usually the low-level natural radiation from cosmic rays and trace radioactive substances present in our environment. (See Chap. 35.)

Beta (β) particle [beta (β) radiation, beta (β) ray] An electron which has been emitted by an atomic nucleus.

Capillary tube A tube having a very small inside diameter.

Carbanion An organic ion carrying a negative charge on a carbon atom.

Carbonium ion An organic ion carrying a positive charge on a carbon atom.

Carcinogen *A substance capable of causing or producing cancer in mammals.* (See Chap. 2.)

Catalyst A substance which changes the rate of a reaction, but is not itself consumed.

Cathode The electrode at which reduction occurs in electrochemical cells. (See Chap. 28.)

Cation A positively charge ion.

cis- Prefix used to indicate that groups are located on the same side of a bond about which rotation is restricted. (See Chap. 23.)

Coefficient of expansion The ratio of the change in length or volume of a body to the original length or volume for a unit change in temperature. (See Chap. 19.)

Combustible *Classification of liquid substances that will burn on the basis of flash points.* (See Chap. 2.)

Conjugated Two double bonds separated by one single bond ($-C=C-C=C-$).

Corrosion The slow conversion of a metal to an oxidized form. Example: iron to rust. Most acids are corrosive to active metals.

Coulometry The quantitative application of Faraday's law to the analysis of materials. The current and the time are the usual variables measured. (See Chap. 28.)

Curie (Ci) The basic unit used to describe the intensity of radioactivity in a sample of material. One curie equals 37 billion disintegrations per second or approximately the amount of radioactivity given off by one gram of radium. (See Chap. 35.)

Daughter A nuclide formed by the radioactive decay of a different (parent) nuclide. (See Chap. 35.)

Decay (radioactive) The change of one radioactive nuclide into a different nuclide by the spontaneous emission of alpha (α), beta (β), or gamma (γ) rays. (See Chap. 35.)

Denaturation A process pertaining to a change in structure of a protein from a regular to an irregular arrangement of the polypeptide chains.

Denatured Commercial term used to describe ethanol which has been rendered unfit for human consumption by the addition of harmful ingredients in order to make its sale tax-exempt.

Derivative A compound which can be imagined to arise from a parent compound by replacement of one atom with another atom or group of atoms. Used extensively in organic chemistry to assist in identifying compounds.

Dermal toxicity Adverse health effects resulting from skin exposure to a substance. (See Chap. 2.)

Deuterium An isotope of hydrogen whose atoms are twice as massive as ordinary hydrogen; deuterium atoms contain both a proton and a neutron in the nucleus.

Differential scanning calorimetry (DSC) A technique for measuring the temperature, direction, and magnitude of thermal transitions in sample materials by heating and/or cooling the material and comparing the amount of energy required to maintain its rate of temperature increase or decrease with an inert reference material under similar conditions. (See Chap. 19.)

Differential thermal analysis (DTA) A technique for observing the temperature, direction, and magnitude of thermally induced transitions in a material by heating and/or cooling a sample and comparing its temperature with that of an inert reference material under similar conditions.

Distilland The material in a distillation apparatus that is to be distilled. (See Chap. 21.)

Distillate The material in a distillation apparatus that is collected in the receiver. (See Chap. 21.)

Dosimeter A small, calibrated electroscope worn by laboratory personnel designed to detect and measure incident ionizing radiation or chemical exposure.

DOT U.S. Department of Transportation. (See Chap. 2.)

Doublet Two peaks or bands of about equal intensity appearing close together on a spectrogram.

DP number The "degree" of polymerization; more specifically, the average number of monomer units per polymer unit.

Electrolytic cell An electrochemical cell in which chemical reactions are forced to occur by the application of an outside source of electrical energy. (See Chap. 28.)

Electrophile Positively charge or electron-deficient particle.

Electrophoresis A technique for separation of ions by rate and direction of migration in an electric field.

Electroplating The deposition of a metal onto the surface of a material by an electrical current.

Eluant or Eluent The solvent used in the process of elution, for instance, in liquid chromatography. (See Chap. 36.)

Eluate The liquid obtained from a chromatographic column; same as "effluent."

Enantiomer One of the two mirror-image forms of an optically acive molecule.

Equilibrium A state in which two opposing processes are occurring at the same rate in a closed system.

Essential oil An extract of a plant which has a pleasant odor or flavor.

Evaporation rate The rate at which a particular substance will vaporize (evaporate) when compared to the rate of a known substance like ethyl ether. This term is especially useful for health and fire-hazard considerations. (See Chap. 21.)

Explosive limits The range of concentrations over which a flammable vapor mixed with proper ratios of air will ignite or explode, if a source of ignition is provided. (See Chap. 2.)

Extrapolate To estimate the value of a result outside the range of a series of known values. Technique used in standard additions calibration procedure.

Film badge A small patch of photographic film worn on clothing to detect and measure accumulated incident ionizing radiation.

Flammable A liquid defined by NFPD and DOT as having a flash point below 100°F (37.9°C). (See Chap. 2.)

Flash point The temperature at which a liquid will yield enough flammable vapor to ignite. There are various recognized industrial testing methods; therefore the method used should also be stated.

Free radical A highly reactive chemical species carrying no charge and having a single unpaired electron in an orbital.

Fuel cell A voltaic cell that converts the chemical energy of a fuel and an oxidizing agent directly into electrical energy on a continuous basis.

Functional groups Groups of atoms having the same properties when appearing in various molecules, especially organic compounds. (See Chap. 23.)

Gamma (γ) ray A highly penetrating type of nuclear radiation similar to x-ray radiation, except that it comes from within the nucleus of an atom and has a higher energy. Energywise, very similar to cosmic ray except that they originate from outer space. (See Chap. 35.)

Galvanizing Placing a thin layer of zinc on a ferrous material to protect the underlying surface from corrosion.

Geiger counter A gas-filled tube which discharges electrically when ionizing radiation passes through it. (See Chap. 35.)

Gem-dimethyl group Two methyl groups of the same carbon atom.

Half-life The time in which half the atoms of a particular radioactive nuclide disintegrate. (See Chap. 35.)

Heat of combustion The quantity of heat released when one mole of a substance is oxidized.

Heat of fusion The quantity of heat required to melt one gram of a solid substance. For example, heat of fusion of ice is 80 calories per gram at 0°C. (See Chap. 19.)

Heat of vaporization The quantity of heat required to vaporize one gram of a liquid substance. For example, heat of vaporization is 540 calories per gram for water at 100°C.

Heavy metals Those metals which have ions that form an isoluble precipitate with sulfide ion.

Heavy water Water formed of oxygen and deuterium; deuterium oxide. Similar to water except approximately 1.1 times heavier than ordinary water.

Heterocyclic molecule A molecule containing a ring of atoms of which at least one atom is not carbon.

HETP Height equivalent to a theoretical plate. The column length divided by the number of equilibrium steps in the column. (See Chap. 36.)

Hybridization The mixing of atomic orbitals in the bonding process to produce an equal number of molecular orbitals of identical character.

Hydrogen bond A loose bond between two molecules, one of which contains an active hydrogen; the other contains primarily a very electronegative atom of O, N, or F.

Hydrolysis A reaction in which water molecules react with another species with the release of H^+ or OH^- ions from the water molecules.

Incompatible Materials which could cause dangerous reactions from direct contact with one or more other substances.

Interpolate To find or insert an unknown point between two known points in a series, like a graph.

In situ At the original location.

In vitro In an artificial environment, such as a test tube.

Ionizing radiation Radiation that is capable of producing ions either directly or indirectly.

Irradiate To expose to some form of radiation.

Irritant A substance that with sufficient concentration and period of exposure can cause an inflammatory response or reaction of the eye, skin, or respiratory system.

Isoelectric point The pH at which there is no migration in an electric field of dipolar ions.

Isotopes Atoms which have the same number of protons, but a different number of neutrons in their nuclei. (See Chap. 22.)

Kinetic molecular theory The theory which states that all matter is composed of particles that are in constant motion. (See Chap. 5.)

LEL or LFL Lower explosive limit or lower flammable limit; the lowest concentration of vapor that will produce a fire or flash when an ignition source is available. (See Chap. 2.)

Ligands (coordination groups) The molecules or anions attached to the central metal ion.

Lipid A nonpolar, solvent-soluble product occurring in nature, usually limited to oils, fats, waxes, and steroids.

Lyophilization A process whereby the material is frozen, a vacuum applied, and the water and low-boiling compounds removed by sublimation. (See Chap. 15.)

MAC Maximum allowable concentration of a toxic substance under prescribed conditions in an atmosphere to be breathed by humans. (See Chap. 2.)

MSDS Material Safety Data Sheets; these documents are required by current OSHA regulations on all chemicals as to their possible health, fire, and other hazards. (See Chap. 1.)

Manostat A device for maintaining a constant pressure. (See Chap. 5.)

Manometer An instrument for measuring the pressure of gases in a system. (See Chap. 5.)

Mantissa The decimal part of a logarithm. In the logarithm 3.6222, the characteristic is 3 and the .6222 is called the mantissa. (See Chap. 6.)

Mass number Approximately the sum of the numbers of protons and neutrons found in the nucleus of an atom. (See Chap. 22.)

Melting point The temperature at which a pure substance changes from the solid state to the liquid state. An impure substance will have a melting range and tend to be lower than the pure substance. (See Chap. 19.)

Meniscus The curved upper surface of a liquid column. (See Chap. 25.)

Metallic bond The type of bond found in metals as the result of attraction between positively charged ions and mobile electrons.

Metalloid An element that exhibits properties intermediate between metals and nonmetals. (See Chap. 22.)

Metric system The preferred reference is Système International (SI), a system of measurement in which the units are related by powers of ten. (See Chap. 6.)

Metric ton Unit of mass in the metric system equal to 1000 kilograms. Also called a megagram.

Molecular distillation The distillation of viscous materials under high vacuum so that the mean free path is longer than the distance from the material to the condenser. (See Chap. 21.)

Molecular orbital Overlapping of atomic orbitals or hybrid orbitals in covalent bonding.

Molecular weight The sum of the atomic weights of all atoms in a molecule. The relative weight of a molecule based on the standard carbon-12 = 12.0000.

Molecule The smallest particle of a compound which retains the properties of that substance.

Monosaccharide A small carbohydrate molecule which is the monomeric unit from which the polymeric carbohydrates are composed.

Mutarotation The interconversion of two forms of a sugar molecule.

Neat liquid Pure liquids as opposed to solutions.

Nonpolar molecule A molecule in which the electrical charges are symmetrically distributed around the center.

Nucleon A constituent of the nucleus; a proton or a neutron. (See Chap. 35.)

Nuclide Any species of atom that exists for a measurable length of time, distinguished by its atomic weight, atomic number, and energy state. (See Chap. 35.)

OSHA The U.S. Depatment of Labor's Occupational Safety and Health Administration. (See Chap. 1.)

Optical activity The property of a molecule involving rotation of plane-polarized light. (See Chap. 29.)

Ordinate The vertical axis on a graph, normally symbolized by y.

Overtone band A band which occurs at twice the frequency of the fundamental band, as in infrared analysis. (See Chap. 31.)

Oxidation A reaction with oxygen or oxygen compounds; a loss of electrons; a positive change in valence. (See Chap. 28.)

Oxidation number An arbitrary number assigned to represent the number of elecrons lost or gained by an atom in a compound or ion. (See Chap. 28.)

Oxidizing agent A substance that oxidizes another substance while it is reduced. (See Chap. 28.)

Parent A radionuclide that decays to another nuclide which may be either radioactive or stable. (See Chap. 35.)

Parts per million An expression for concentration; one part of the substance per million total parts. (See Chap. 24.)

PEL Permissible exposure level; an exposure limit established by OSHA which can be a maximum concentration exposure limit or a time-weighted average (TWA). (See Chap. 2.)

Peptide Two or more amino acids joined by peptide linkages (—CONHC—) between the terminal carboxylic acid group of one amino acid and an amine group located on another molecule.

Peroxide A molecule containing —O—O— bonds.

Pi (π) bond Covalent bond formed by overlapping of *p* atomic orbitals on adjacent atoms in a molecule.

Plasticizer A liquid of low volatility (high boiling point) which is added to soften polymers.

Polar molecule A molecule in which the electrical charges are not symmetrically distributed; the center of positive charge is separated from the center of negative charge.

Potentiostat An instrument designed to maintain a constant potential between two electrodes in a solution.

ppb Parts per billion parts, a unit of concentration expression primarily used for trace impurities. Parts of gas, liquid, or solid per billion parts of sample. (See Chap. 24.)

ppm Parts per million parts, a unit of concentration expression primarily used for trace impurities. (See Chap. 24.)

Precision A measure of the reproducibility of results. (See Chap. 7.)

Proton The nucleus of a hydrogen atom having a mass of approximately one atomic mass unit and carrying one unit of positive charge.

Pyrolysis The breakdown of a material by heating, usually in the absence of oxygen.

Pyrophoric Any substance that will spontaneously ignite in air at temperatures of less than 130°F (54.4°C).

Pyrophoric material A substance which can heat up substantially or ignite spontaneously upon exposure to air.

Racemate A mixture having no optical activity which consists of equal amounts of enantiomers.

Radioactive date A technique for estimating the age of an object by measuring the amounts of various radioisotopes compared to stable isotopes. Carbon-13 dating is an example of this technique. (See Chap. 35.)

Reactivity The tendency of a substance to undergo a chemical reaction, especially with the rapid release of energy. Other properties like rapid pressure changes and formation of toxic or corrosive products are also included in reactivity descriptions.

Reflux ratio In distillation, the ratio of the amount of material returned to a column compared to the amount collected per unit time. (See Chap. 21.)

Resonance Characterized by stability shown by a molecule, ion, or radical to which two or more structures differing only in the distribution of electrons can be assigned. (See Chap. 23.)

Resonance wavelength The wavelength which corresponds to the energy required to shift a ground-state electron to a higher energy level. (See Chap. 29.)

Reducing agent A substance that reduces another while it is being oxidized. (See Chap. 28.)

Reduction The removal of oxygen or an addition of hydrogen; a gain of electrons; a negative change in valence. (See Chap. 28.)

R_f value Ratio of the distance a compound moved to the distance the solvent moved on a paper or thin-layer chromatogram. (See Chap. 36.)

Scaler An electronic instrument for counting radiation-induced pulses from radiation detectors such as a Geiger-Müller tube. (See Chap. 35.)

Scintillation counter An instrument that detects and measures gamma (γ) radiation by counting the light flashes (scintillations) induced by the radiation. (See Chap. 35.)

Sigma (σ) bond Covalent bond formed by overlapping of s-s, s-sp^3, s-sp^2, or s-sp orbitals on adjacent atoms in a molecule.

Specific gravity The ratio of the density of a substance to the density of a standard substance. For solids and liquids, water is the standard substance; for gases, air is the standard substance. Specific gravity is unitless because of this definition. (See Chap. 19.)

Specific heat The number of calories required to raise the temperature of one gram of a substance by one degree Celsius. The specific heat of water is 1 calorie per gram degree Celsius.

Stability The tendency of a substance to remain unchanged with time, storage, or exposure to other chemicals.

Standard conditions In dealing with gases, the set of conditions adopted for standard pressure and temperature (STP) which is one atmosphere and 0°C. (See Chap. 5.)

Standard electrode potential The potential compared to a hydrogen electrode which exists when the electrode is immersed in a solution of its ions at unit activity. (See Chap. 28.)

Stereoisomers Isomers having different arrangement of the atoms in space but the same number of atoms, kinds of atoms, and sequence of bonding of the atoms.

Steric hindrance A condition in which a reaction is slowed or stopped because the size of the groups near the reaction site causes a blocking of approach to the site.

Thermogravimetric analysis (TGA) An analytical technique in which the weight of a sample is measured continuously as its temperature is increased or decreased.

TLV Threshold limit value. Concentration expression for airborne substances which have no adverse effects to most people on a daily basis. (See Chap. 2.)

TLV-TWA The short-term exposure limit or maximum concentration for continuous 15-min exposure period. The exposure episodes must not exceed four per day and must allow at least 60 min between exposure periods (See Chap. 2.)

TLV-C The ceiling exposure limit is a concentration that should never be exceeded. (See Chap. 2.)

Tracer A small amount of radioactive isotope introduced into a system in order to follow the behavior of some component of that system.

trans- Prefix used to indicate that groups are located on the opposite sides of a bond about which rotation is restricted. (See Chap. 23.)

UEL Upper explosive limit. Concentration of flammable substance that will not support combustion because the vapor is too rich in fuel for the amount of oxygen present. (See Chap. 2.)

Vapor density The mass of gas vapor compared to the mass of an equal volume of air at the same temperature. A gas with a vapor density less than 1.0 will tend to rise and dissipate, even though some mixing will occur. A gas with a vapor density of greater than 1.0 will tend to sink in air and concentrate in low places. (See Chap. 19.)

Vapor pressure The pressure exerted by a saturated vapor in equilibrium with the pure liquid phase in a closed container. The vapor pressure of a substance is directly related to the temperature; as the temperature increases, the vapor pressure increases. By convention, most substances have their vapor pressures reported at 100°F. (See Chap. 21.)

Voltaic cell An electrochemical cell in which an electrical current is generated by a chemical reaction. (See Chap. 28.)

Water-reactive Any substance that reacts with water to release a flammable gas (alkali metals) or toxic substance.

Zwitterion An ion which contains both a negative and a positive charge, same as a dipolar ion. Some neutral amino acids are examples of zwitterions.

INDEX

ABOUT THE AUTHORS

Gershon J. Shugar is Professor of Engineering Technologies, Essex County College, Newark, N.J. In 1947, he founded a chemical manufacturing business which became the largest exclusive pearlescent pigment manufacturing company in the United States. In 1968, Dr. Shugar was appointed Assistant Professor of Chemistry at Rutgers University, where he taught until his appointment at Essex County College.

Jack T. Ballinger is Professor of Chemistry at St. Louis Community College at Florissant Valley. In 1970, he implemented for the College a chemical technology training program, which has graduated over 300 chemical technicians and has been selected as a model program by the American Chemical Society. He has had extensive experience in the chemical, petrochemical, and environmental industries.